PHYSICAL CONSTANTS IN SI UNITS

Absolute zero temperature	−273.3 °C
Acceleration due to gravity, g	9.807 m s^{-2}
Avogadro's number, N_A	6.022×10^{23}
Base of natural logarithms, e	2.718
Boltsmann's constant, k	$1.381 \times 10^{-23} \text{ J K}^{-1}$
Faraday's constant, k	$9.648 \times 10^{4} \text{ C mol}^{-1}$
Gas constant, R	$8.314 \text{ J mol}^{-1} \text{ K}^{-1}$
Permeability of vacuum, μ_0	$1.257 \times 10^{-6} \text{ H m}^{-1}$
Permittivity of vacuum, ε_0	$8.854 \times 10^{-12} \text{ F m}^{-1}$
Planck's constant, h	$6.626 \times 10^{-34} \text{ J s}^{-1}$
Velocity of light in vacuum, c	$2.998 \times 10^{8} \text{ m s}^{-1}$
Volume of perfect gas at STP	$22.41 \times 10^{-3} \text{ m}^3 \text{ mol}^{-1}$

CONVERSION OF UNITS

Angle, θ	1 rad	57.30°
Density, ρ	1 lb ft^{-3}	16.03 kg m^{-3}
Diffusion Coefficient, D	$1 \text{ cm}^3 \text{ s}^{-1}$	$1.0 \times 10^{-4} \text{ m}^2 \text{ s}^{-1}$
Force, F	1 kgf	9.807 N
	1 lbf	4.448 N
	1 dyne	1.0×10^{-5} N
Length, L	1 ft	304.8 mm
	1 inch	25.40 mm
	1 Å	0.1 nm
Mass, M	1 tonne	1000 kg
	1 short ton	908 kg
	1 long ton	1107 kg
	1 lb mass	0.454 kg
Specific Heat, Cp	1 cal gal^{-1} °C	4.188 kJ kg^{-1} °C
	Btu lb^{-1} °F	4.187 kJ kg^{-1} °C
Stress Intensity, K_{IC}	$1 \text{ ksi} \sqrt{\text{in}}$	$1.10 \text{ MN m}^{-3/2}$
Surface Energy, γ	1 erg cm^{-2}	1 mJ m^{-2}
Temperature, T	1 °F	0.556 K
Thermal Conductivity, λ	$1 \text{ cal s}^{-1} \text{cm}$ °C	418.8 W m^{-1} °C
	1 Btu h^{-1} ft °F	1.731 W m^{-1} °C
Volume, V	1 Imperial gall	$4.546 \times 10^{-3} \text{ m}^3$
	1 US gall	$3.785 \times 10^{-3} \text{ m}^3$
Viscosity, η	1 poise	0.1 N s m^{-2}
	1 lb ft s	$0.1517 \text{ N s m}^{-2}$

Renewable Energy Focus Handbook

Note from the Publisher

This book has been compiled using extracts from the following books within the range of Renewable Engineering books in the Elsevier collection:

Yang (2007) *Bioprocessing for Value-Added Products from Renewable Resources*, 9780444521149
Da Rosa (2005) *Fundamentals of Renewable Energy Processes*, 9780120885107
Gupta and Roy (2007) *Geothermal Energy*, 9780444528759
Doble and Kumar (2007) *Green Chemistry and Engineering*, 9780123725325
Breeze (2005) *Power Generation Technologies*, 9780750663137
Sorensen (2004) *Renewable Energy* 9780126561531
Suppes and Storvick (2007) *Sustainable Nuclear Power* 9780123706027
Encyclopedia of Energy (2004) 9780121764807
Kalogirou (2004) Solar thermal collectors and applications, *Progress in Energy and Combustion Science* 30, 0360-1285

The extracts have been taken directly from the above source books, with some small editorial changes. These changes have entailed the re-numbering of Sections and Figures. In view of the breadth of content and style of the source books, there is some overlap and repetition of material between chapters and significant differences in style, but these features have been left in order to retain the flavour and readability of the individual chapters.

End of chapter questions
Within the book, several chapters end with a set of questions; please note that these questions are for reference only. Solutions are not always provided for these questions.

Units of measure
Units are provided in either SI or IP units. A conversion table for these units is provided at the front of the book.

Upgrade to an Electronic Version
An electronic version of the Desk reference, the *Renewable Energy Focus Handbook e-Mega Reference*, 9780123747068

- A fully searchable Mega Reference eBook, providing all the essential material needed by Renewable Energy Engineers on a day-to-day basis.
- Fundamentals, key techniques, engineering best practice and rules-of-thumb at one quick click of a button
- Over 1,500 pages of reference material, including over 1,000 pages not included in the print edition

Go to http://www.elsevierdirect.com/9780123747051 and click on **Ebook Available**

Renewable Energy Focus Handbook

ELSEVIER

Amsterdam · Boston · Heidelberg · London · New York · Oxford
Paris · San Diego · San Francisco · Sydney · Tokyo
Academic Press is an imprint of Elsevier

Academic press is an imprint of Elsevier
Linacre House, Jordan Hill, Oxford OX2 8DP, UK
525 B Street, Suite 1900, San Diego, CA 92101-4495, USA

First edition 2009

British Library Cataloguing in Publication Data
A catalogue record for this book is available from the British Library

Library of Congress Cataloging-in-Publication Data
A catalog record for this book is available from the Library of Congress

ISBN: 978-0-12-374705-1

For information on all Academic press publications visit our web site at elsevierdirect.com

Printed and bound in the United States of America

09 10 11 11 10 9 8 7 6 5 4 3 2 1

Contents

Author Biographies

Dr. Paul Breeze is a journalist and energy consultant who has specialised in the power generation industry for twenty five years. As well as writing for many UK newspapers, including *The Financial Times* and *The Guardian*, he has, over the past ten years, produced a series of detailed reports into individual renewable and traditional power generation technologies.

Professor Aldo Vieira da Rosa is currently Emeritus Professor of Electrical Engineering at Stanford University. He has taught Introductory Electronics, Space Physics and, at present teaches a course in Non-Traditional Energy Processes. Professor da Rosa, a retired Brigadier General in the Brazilian Air Force, was Chairman of the Brazilian National Research Council, Director of the Aeronautical Technical Center, and founder and first Chairman of the Brazilian NASA (Instituto de Pesquisas Espaciais). In his more adventurous days, he was also a test pilot of helicopters under development in the Research and Development Institute of the Brazilian Ministry of Aeronautics.

Dr Mukesh Doble is Professor at the Department of Biotechnology, IIT Madras. Prior to teaching, he served in the technology centres of ICI India and GE India. He has co-authored three books, presented 25 conferences, published about a hundred technical papers and files 3 Indian patents. Amongst his awards is the Indian Institute of Chemical Engineers 'Herdilla award' for Excellence in Basic Research in Chemical Engineering. He is a member of both the American and Indian Institutes of Chemical Engineers.

Dr. Harsh Gupta is an eminent geophysicist, currently at Raja Ramanna. He is Fellow at the NGRI, Hyderabad, India, and President of the Geological Society of India. He is also Vice President of the IUGG and a member of the CSPR of ICSU. Earlier he was Secretary to Government of India, looking after the Department of Ocean Development; Director, NGRI; Vice Chancellor, Cochin University of Science and Technology; and Adjunct Professor at the University of Texas at Dallas, USA. Dr. Gupta was invited to deliver the Brunn Memorial Lecture by the Intergovernmental Oceanographic Commission, Paris on Gas Hydrates. He received the Waldo E. Smith Medal of the American Geophysical Union for 2008.

Dr. Soteris Kalogirou is an Instructor of Mechanical Engineering at the Department of Mechanical Engineering and Materials Sciences and Engineering of the Cyprus University of Technology, Limassol, Cyprus. For more than 25 years, he has been actively involved in research in the area of solar energy. He has 17 book contributions and published 173 papers; 77 in international scientific journals and 96 in refereed conference proceedings. He is Associate Editor of *Renewable Energy* and *Energy* journals and Editorial Board Member of another nine journals.

Preben Maegaard is a Danish renewable energy pioneer, author and expert. He is Executive Director of the Nordic Folkecenter for Renewable Energy. Since the 1973 oil crisis he has worked locally, nationally and internationally for the transition from fossil fuels to renewable energy. He has served on several Danish governmental committees and councils for the deployment of renewable energy. Since 2001 he has been an Associated Member of the Chairmen Committee of the World Council for Renewable Energy and the President of the World Wind Energy Association, WWEA.

Gianfranco Pistoia was formerly Research Director for the National Research Council, Rome, Italy.

Sukanta Roy is leading the Geothermal Studies program at the National Geophysical Research Institute (NGRI), Hyderabad, India. He has generated extensive datasets on heat flow, thermal properties of rocks and radiogenic heat production characteristics of continental crust. He has also published his work in peer-reviewed national and international journals and has co-authored a book on Geothermal Energy. He has been a Visiting Scholar at the University of Utah and is currently serving as an executive member of the International Heat Flow Commission of the IASPEI (IUGG).

Dr. Bent Sørensen is professor at Roskilde University, Department of Environmental, Social and Spatial Change President of NOVATOR Advanced Technology Consulting, and has formerly held academic positions at Berkeley, Yale, Golden, Kyoto, Grenoble and Sydney. He has been an advisor to the OECD, the Japanese and Australian governments, various UN agencies, and has served as technical director and board member of Cowiconsult Inc., and as lead author in the IPCC working group on climate change mitigation, where he is recognised for his contribution to the 2007 Nobel Peace Prize. He served as chairman of the Danish Energy

Agency Solar Energy Committee and the Hydrogen Energy Committee, and received the Australian-European Award for Eminent Scholars. He was knighted by HRH Queen Margrethe of Denmark.

Dr. Truman Storvick is an emeritus professor at the University of Missouri, Columbia, and has been an active professor of chemical engineering since 1959. His distinguished scientific career includes research in: Thermodynamic and transport properties of dilute and moderately dense gases, Kinetic theory of transition flow phenomena, and Separation of domestic spent nuclear fuel. Dr. Storvick's work includes 46 technical journal publications and he has co-edited and co-authored two books.

Shang-Tian Yang is Professor of Chemical and Biomolecular Engineering at the Ohio State University, where he has been teaching and researching since 1985. He is also Director of Ohio Bioprocessing Research Consortium, which works with industry in developing novel bioprocesses for economical production of value-added products from food processing wastes and agricultural commodities. Dr. Yang has over one hundred scientific publications and a dozen US patents in the bioengineering field. He is an elected fellow of the American Institute of Medical and Biological Engineering and an active member of the American Institute of Chemical Engineers (AIChE) and American Chemical Society (ACS).

Dr. Anil Kumar Kruthiventi is Senior Lecturer in the Chemistry Department of Sri Sathya Sai University, India.

Section One

Introduction

Chapter 1.0

1.0

Introduction

1.0.1 Units and constants

Although many different units are employed in energy work, we shall adopt, whenever possible, the "Système International," SI. This means **joules** and **watts**. If we are talking about large energies, we'll speak of MJ, GJ, TJ, and EJ—that is, 10^6, 10^9, 10^{12}, and 10^{18} joules, respectively.

We cannot entirely resist tradition. Most of the time we will express pressures in **pascals,** but we will occasionally use **atmospheres** because most of the existing data are based on the latter. Sometimes **electron-volts** are more convenient than joules. Also, expressing energy in **barrels of oil** or **kWh** may convey better the idea of cost. On the whole, however, we shall avoid "quads," "BTUs," "calories," and other non-SI units. The reason for this choice is threefold: SI units are easier to use, they have been adopted by most countries, and are frequently better defined.

Consider, for instance, the "calorie," a unit preferred by chemists. Does one mean the "international steam table calorie" (4.18674 J)? Or the "mean calorie" (4.19002 J)? Or the "thermochemical calorie" (4.18400 J)? Or the calorie measured at 15 C (4.18580 J)? Or at 20 C (4.18190 J)?

Americans like to use the BTU, but, again, there are numerous BTUs: "steam table," "mean," "thermochemical," at 39 F, at 60 F. The ratio of the BTU to the calorie of the same species is about 251.956 with some variations in the sixth significant figure. Remember that 1 BTU is roughly equal to 1 kJ, while 1 quad equals roughly 1 EJ. The conversion factors between the different energy and power units are listed in Table 1.0-2. Some of the fundamental constants used in this book are listed below.

1.0.2 Energy and utility

In northern California, in a region where forests are abundant, one cord of wood sold in 1990 for about $110. Although one cord is a stack of 4 by 4 by 8 ft (128 cubic feet), the actual volume of wood is only 90 cubic feet—the rest is empty space between the logs. Thus, one cord contains 2.5 m^3 of wood or about 2200 kg. The heat of combustion of wood varies between 14 and 19 MJ/kg. If one assumes a mean of 16 MJ per kilogram of wood burned, one cord delivers 35 GJ. Therefore, the cost of energy from wood was $3.2/GJ in northern California.

In 1990, the price of gasoline was still approximately $1.20 per gallon, the equivalent of $0.49 per kg. Since the heat of combustion of gasoline is 49 MJ/kg, gasoline energy costs $10/GJ, or three times the cost from burning wood.

Notwithstanding electricity being inexpensive in California, the domestic consumer paid $0.04 per kWh or $11.1/GJ.

From the above, it is clear that when we buy energy, we are willing to pay a premium for energy that is, in a more convenient form—that is, for energy that has a higher **utility**.

Utility is, of course, relative. To stoke a fireplace in a living room, wood has higher utility than gasoline and, to drive a car, gasoline has higher utility than electricity, at least for the time being. For small vehicles, liquid fuels have higher utility than gaseous ones. For fixed installations, the opposite is true.

The relative cost of energy is not determined by utility alone. One barrel contains 159 liters or 127 kg of oil. With a heat of combustion of 47 MJ/kg, this corresponds to 6 GJ of energy. In mid-1990, the price was $12/barrel or $2/GJ, somewhat less than the price of wood at that

Renewable Energy Focus Handbook 2009; ISBN: 9780123747051

Table 1.0-1 Fundamental constants

Quantity	Symbol	Value	Units
Avogadro's number	N_0	6.0221367×10^{26}	per kmole
Boltzmann constant	k	1.380658×10^{-23}	$J\ K^{-1}$
Charge of the electron	q	$1.60217733 \times 10^{-19}$	C
Gas constant	R	8314.510	$J\ kmole^{-1}K^{-1}$
Gravitational constant	G	6.67259×10^{-11}	$m^3s^{-2}kg^{-1}$
Planck's constant	h	$6.6260755 \times 10^{-34}$	J s
Permeability of free space	μ_0	$4\pi \times 10^{-7}$	H/m
Permittivity of free space	ε_0	$8.854187817 \times 10^{-12}$	F/m
Speed of light	c	2.99792458×10^{8}	$m\ s^{-1}$
Stefan-Boltzmann constant	σ	5.67051×10^{-8}	$W\ K^{-4}m^{-2}$

time notwithstanding oil being, in general, more useful. However, oil prices are highly unstable depending on the political circumstances of the world.

Government regulations tend to depress prices below their free market value. During the Carter era, natural gas was sold in interstate commerce at the regulated price of $1.75 per 1000 cubic feet. This amount of gas corresponds to 1 GJ of energy. Thus, natural gas was cheaper than oil or wood.

1.0.3 Conservation of energy

Energy can be utilized but not consumed.[†] It is a law of nature that energy is conserved. Instead of consuming it, we degrade or randomize energy, just as we randomize mineral resources when we process concentrated ores into metal and then discard the final product as we do, for example, with used aluminum cans. All energy we use is degraded into heat and eventually radiated out into space.

The consumable is not energy; the consumable is the fact that energy has not yet been randomized. The degree of randomization of energy is measured by the entropy of the energy. This is discussed in some detail in Chapter 2.1.

1.0.4 Planetary energy balance

The relative stability of Earth's temperature suggests a near balance between planetary input and output of energy. The input is almost entirely that of the solar radiation incident on Earth. This amounts to 173,000 TW ($173{,}000 \times 10^{12}$ W).

Besides solar energy, there is a contribution from tides (3 TW) and from heat sources inside the planet, mostly radioactivity (32 TW).

Some 52,000 TW (30% of the incoming radiation) is reflected back to the interplanetary space: it is the **albedo** of Earth. All the remaining energy is degraded to heat and re-emitted as long-wave infrared radiation. Figure 1.0-1 shows the different processes that take place in the planetary energy balance mechanism.

The recurrence of ice ages shows that the equilibrium between incoming and outgoing energy is oscillatory in nature. Some fear that the observed secular increase in atmospheric CO_2 might lead to a general heating of the planet resulting in a partial melting of the Antarctic glaciers and consequent flooding of sea level cities. The growth in CO_2 concentration is the result of the combustion of vast amounts of *fossil*[††] fuels and the destruction of forests in which carbon had been locked.

1.0.5 The energy utilization rate

The energy utilization rate throughout the ages can only be estimated in a rough manner. In early times, man was totally nontechnological, not even using fire. He used energy only as food, probably at a rate somewhat below the modern average of 2000 kilocalories per day,

[†] It is convenient to distinguish *consumption* from *utilization*. The former implies destruction—when oil is consumed, it disappears being transformed mainly into carbon dioxide and water, yielding heat. On the other hand, energy is never consumed—it is utilized but entirely conserved (only the entropy is increased).

[††] Fuels derived from recent biomass, such as ethanol from sugar cane, do not increase the amount of carbon dioxide in the atmosphere—such fuels only recycle this gas.

Table 1.0-2 Conversion coefficients

To convert from	to	multiply by
Energy		
BARREL OF OIL	GJ	≈6
BRITISH THERMAL UNIT (Int. Steam Table)	joule	1055.04
BRITISH THERMAL UNIT (mean)	joule	1055.87
BRITISH THERMAL UNIT (thermochemical)	joule	1054.35
BRITISH THERMAL UNIT (39 F)	joule	1059.67
BRITISH THERMAL UNIT (60 F)	joule	1054.68
CALORIE (International Steam Table)	joule	4.18674
CALORIE (mean)	joule	4.19002
CALORIE (thermochemical)	joule	4.1840
CALORIE (15 C)	joule	4.1858
CALORIE (20 C)	joule	4.1819
CUBIC FOOT (Methane, STP)	MJ	≈1
ELECTRON VOLT	joule	1.60206×10^{-19}
ERG	joule	1.0×10^{-7}
FOOT LBF	joule	1.3558
FOOT POUNDAL	joule	4.2140×10^{-2}
kWh	joule	3.6×10^{6}
QUAD	BTU	1.0×10^{15}
TON of TNT	joule	4.2×10^{9}
Power		
FOOT LBF/SECOND	watt	1.3558
FOOT LBF/MINUTE	watt	2.2597×10^{-2}
FOOT LBF/HOUR	watt	3.7662×10^{-4}
HORSEPOWER (550 Foot LBF/sec)	watt	745.70
HORSEPOWER (electric)	watt	746
HORSEPOWER (metric)	watt	735
Other		
ATMOSPHERE	pascal	1.0133×10^{5}
DALTON	kg	1.660531×10^{-27}

LBF stands for pounds (force).

equivalent to 100 W. Later, with the discovery of fire and an improved diet involving cooked foods, the energy utilization rate may have risen to some 300 W/capita.

In the primitive agricultural Mesopotamia, around 4000 B.C., energy derived from animals was used for several purposes, especially for transportation and for pumping water in irrigation projects. Solar energy was employed for drying cereals and building materials such as bricks. Per capita energy utilization may have been as high as 800 W.

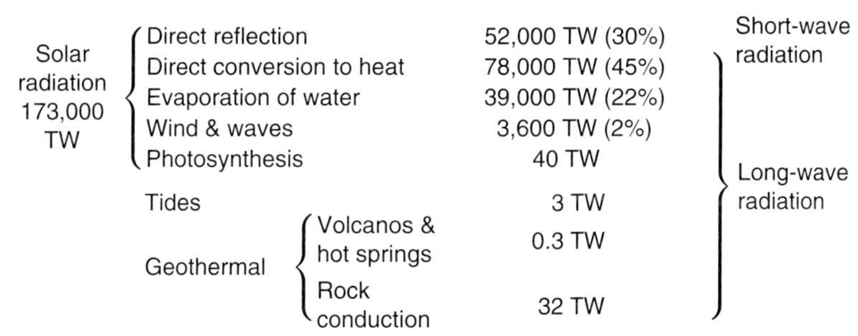

Figure 1.0-1 Planetary energy balance.

The idea of harnessing wind, water and fire to produce useful work is ancient. Wind energy has been in use to drive sailboats since at least 3000 B.C. and windmills were described by Hero of Alexandria around 100 A.D. Extensive use of windmills started in Persia around 300 A.D. and, only much later, spread to China and Europe.

Hero described toy steam engines that apparently were built and operated. Vitruvius, the famous Roman architect and author whose book, first published at the time of Hero, is still on sale today, describes waterwheels used to pump water and grind cereals.

In spite of the availability of the technology, the ancients limited themselves to the use of human or animal muscle power. Lionel Casson (1981), a professor of ancient history at New York University, argues that this was due to cultural rather than economic constraints and that only at the beginning of the Middle Ages did the use of other energy sources become "fashionable." Indeed, the second millennium saw an explosion of mechanical devices starting with windmills and waterwheels.

The energy utilization rate in Europe was likely 2000 calories per capita around 1200 A.D. when there was widespread adoption of advanced agriculture, the use of fireplaces to heat homes, the burning of ceramics and bricks, and the use of wind and water. Since the popular acceptance of such activities, energy utilization has increased rapidly.

Figure 1.0-2 illustrates (a wild estimate) the number of kilowatts utilized per capita as a function of the date. If we believe these data, we may conclude that the annual rate of increase of the per capita energy utilization rate behaved as indicated in Figure 1.0-3. Although the precision of these results is doubtful, it is almost certain that the general trend is correct— for most of our history the growth of the per capita energy utilization rate was steady and quite modest. However, with the start of the industrial revolution at the beginning of the 19th century, this growth accelerated dramatically and has now reached a worrisome level.

One driving force behind the increasing worldwide per capita energy utilization was the low cost of oil before 1973 when the price of oil was substantially lower than what it is currently.[†] Perez Alfonso, the Venezuelan Minister of Oil in 1946, was among those who recognized that this would lead to future difficulties. He was instrumental in creating OPEC in 1954, not as a cartel to squeeze out higher profits but to "reduce the predatory oil consumption to guarantee humanity enough time to develop an economy based on renewable energy sources."

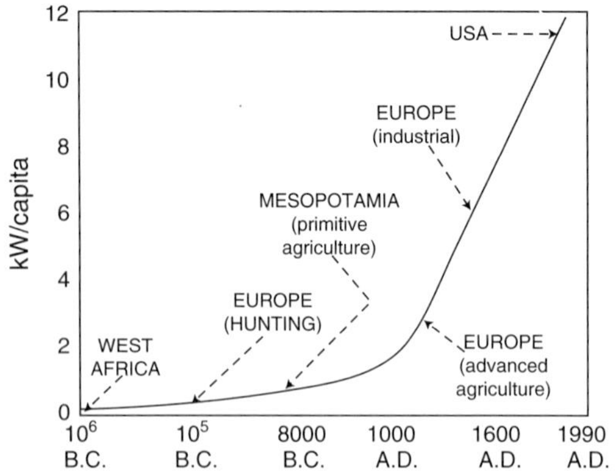

Figure 1.0-2 A very rough plot of the historical increase in the per capita energy utilization rate.

Alfonso also foresaw the ecological benefits stemming from a more rational use of oil.

OPEC drove the oil prices high enough to profoundly alter the world economy. The result was that the overall energy utilization rate slowed its increase. Owing to the time delay between the price increase and the subsequent response from the system, several years elapsed before a new equilibrium was established in the oil markets. The result was a major overshooting of the oil producing capacity of OPEC and the softening of prices that we witnessed up to the 1991 Iraqi crisis.

The recent effort of less developed countries (LDCs) to catch up with developed ones has been an important factor in the increase in energy demand. Figure 1.0-4 shows the uneven distribution of energy utilization rate throughout the world. 72% percent of the world population uses less than 2 kW/capita whereas 6% of the population uses more than 7 kW/capita.

There is a reasonable correlation between the total energy utilization rate of a country and its corresponding annual gross national product. About 2.2 W are used per dollar of yearly GNP. Thus, to generate each dollar, 69 MJ are needed. These figures, which are based on 1980 dollars, vary with time, in part owing to the devaluation of the currency, but also due to changing economic circumstances. It fact, it has been demonstrated that during an energy crisis, the number of megajoules per dollar decreases, while the opposite trend occurs during financial crises.

Further industrialization of developed countries may not necessarily translate into an increase of the per capita energy utilization rate—the trend toward higher efficiency in energy use may have a compensating effect. However, in the USA, the present decline in energy utilization[††] is due mainly to a change in the nature of

[†] In 1973, before the OPEC crisis, petroleum was sold at between \$2 and \$3 per barrel. The price increased abruptly traumatizing the economy. In 2000 dollars, the pre-1973 petroleum cost about \$10/bbl (owing to a 3.8-fold currency devaluation), a price that prevailed again in 1999. However, in 2004, the cost had risen to over \$50/bbl.

[††] The use of energy by the American industry was less in 1982 than in 1973.

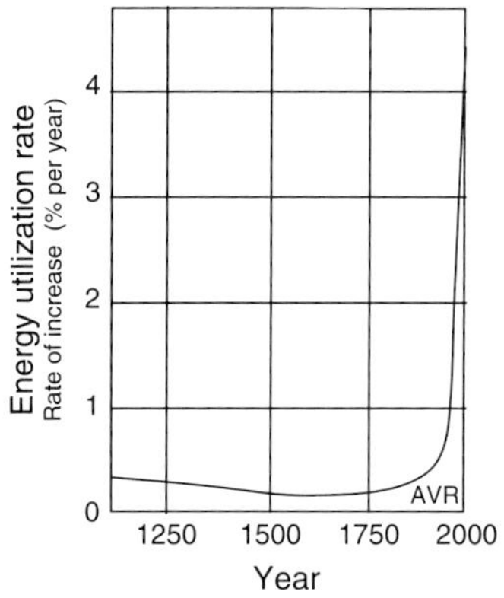

Figure 1.0-3 The annual rate of increase of per capita energy utilization was small up to the 19th century.

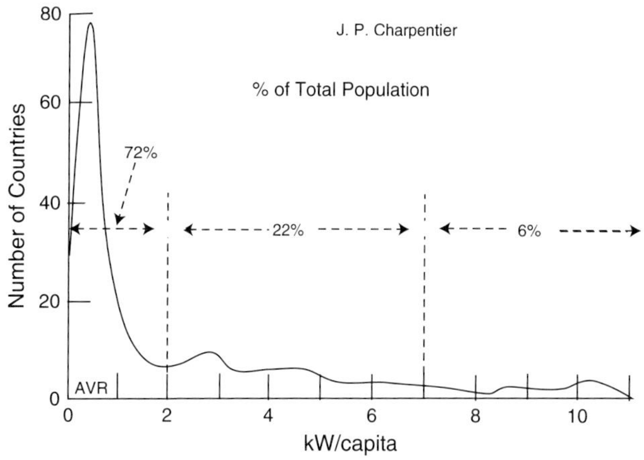

Figure 1.0-4 Most countries use little energy per capita while a few developed ones use a lot.

industrial production. Energy intensive primary industries (such as steel production) are phasing out owing to foreign competition, while sophisticated secondary industries (such as electronics and genetic engineering) are growing.

Technological innovation has resulted in more efficient use of energy. Examples of this include better insulation in houses and better mileage in cars. Alternate energy sources have, in a small measure, alleviated the demand on fossil fuels. Such is the case of using ethanol from sugar cane for the propulsion of automobiles. It is possible that the development of fusion reactors will, one day, bring back the times of abundant energy.

Introduction of a more efficient device does not immediately result in energy economy because it takes a considerable time for a new device to be widely accepted. The reaction time of the economy tends to be long. Consider the privately owned fleet of cars. A sudden rise in gasoline price has little effect on travel, but it increases the demand for fuel efficiency. However, car owners don't rush to buy new vehicles while their old ones are still usable. Thus, the overall fuel consumption will only drop many years later, after a significant fraction of the fleet has been updated.

Large investments in obsolete technologies substantially delay the introduction of more desirable and efficient systems. A feeling for the time constants involved can be obtained from the study of the "market penetration function," discussed in Section 1.7.

1.0.6 The population explosion

In the previous section we discussed the *per capita* energy utilization rate. Clearly the total rate of energy utilization is proportional to the planetary population which has been growing at an accelerated rate.†

The most serious problem that confronts mankind is the rapid growth in population. The planet has a little more than 6 billion inhabitants, and the growth rate these last few decades has been around 1.4% per year. Almost all projections predict a population of about 7 billion by the year 2010. This will be the case even if, right now, everyone were to agree on a limit of two children per family. Under present-day actuarial conditions, the population would eventually stabilize at around 11 billion by the year 2050. Thus, population growth alone could account for 1.4% a year increase in energy demand, in the next few decades.

If, in 2050, all the estimated 11 billion inhabitants of Earth were to use energy at the present day USA level (11 kW/capita), the world energy utilization rate would reach 122 TW—a 16-fold increase over the present 7.6 TW. Such a rate is probably one order of magnitude higher than can be supplied unless fusion energy becomes practical and inexpensive.

A more modest scenario views the worldwide energy utilization rate stabilizing at the present level of Eastern Europe: 5 kW per capita. This would lead to an overall rate of 65 TW in 2050, which is still too high. Finally, if the world average kept its present 2 kW per capita, the rate would grow to 26 TW by the middle of next century. Clearly, it is difficult to provide adequate energy for 11 billion people. This is one more reason for attempting to limit the planetary population growth.

The constant population increase has its Malthusian side. About 10% of the world's land area is used to raise crops—that is, it is **arable land,** (See "Farming and Agricultural Technology: Agricultural Economics: Land, output,

† On 10/12/99, a 3.2 kg baby was born in Bosnia. Kofi Annan, General Secretary of the United Nations was on hand and displayed the new Bosnian citizen to the TV cameras because, somewhat arbitrarily, the baby was designated as the 6,000,000,000th inhabitant of this planet.

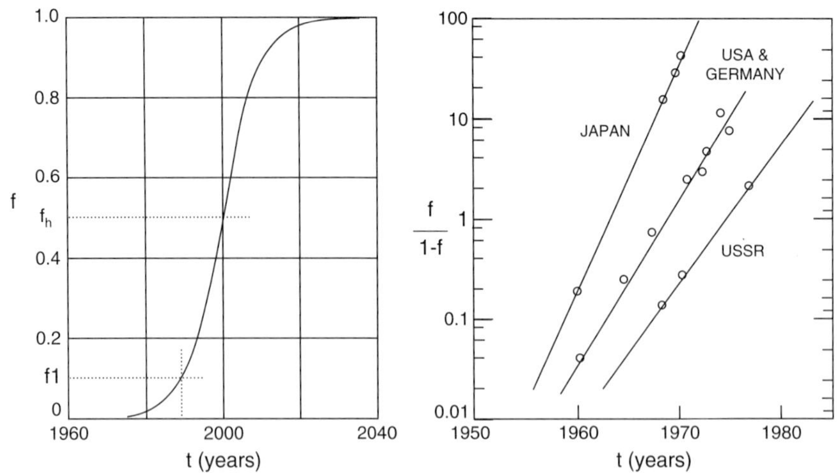

Figure 1.0-5 Left: Typical penetration function vs time. Right: The penetration function of oxygen steel technology fits accurately the Fisher-Pry rule.

and yields." Britannica Online.) This means that roughly 15 million km^2 or 1.5×10^9 hectares are dedicated to agriculture. Up to the beginning of the 20th century, on average, each hectare was able to support 5 people (*Smil*), thus limiting the population to 7.4 billion people. More arable land can be found, but probably not enough to sustain 11 billion people. What limits agricultural productivity is nitrogen, one kilogram of which is (roughly) needed to produce one kilogram of protein. Although it is the major constituent of air, it is, in its elemental form, unavailable to plants and must either be "fixed" by appropriate micro-organisms or must be added as fertilizer.

Nitrogen fertilizers are produced almost exclusively from ammonia, and when used in adequate amounts can increase land productivity by nearly an order of magnitude. The present day and the future excess population of the planet can only exist if sufficient ammonia is produced. Although there is no dearth of raw materials for this fertilizer (it is made from air and water), its intensive use has a serious adverse environmental effect as discussed in the article by Smil.

1.0.7 The market penetration function

A new technology, introduced in competition with an established one, may take over a progressively larger fraction of the market. Is it possible to forecast the rate at which such penetration occurs?

Let f be the fraction of the total market captured by the new technology. As time progresses, f grows from 0 to some value equal or less than 1. The latter corresponds to the new technology having totally replaced all competition. In due time, f may decrease again when a even newer technologies is introduced.

An empirical plot of the ascending phase of f vs time, t, has an "S" shape as exemplified by Figure 1.0-5 (left). A **market penetration time** is defined as $\Delta T \equiv (t_h - t_1)$, where t_h is the time at which $f = 0.5 \equiv f_h$, and t_1 is the time at which $f = 0.1 \equiv f_1$. ΔT may be negative if the technology in question is being replaced. It is then called the **abandonment time**. Fisher and Pry (1971) and Pry (1973) showed that when $\ln \frac{f}{1-f}$ is plotted versus time, a straight line results. Figure 1.0-5 (right) illustrates an example of how the Fisher-Pry equation provides an excellent fit to the empirical data. The data show how, in four different countries, the use of oxygen in steel converters is gradually substituted for the older open-hearth and Bessemer technologies. The straight lines in the plots correspond to a regression of the type:

$$\ln \frac{f}{1-f} = at + b. \tag{1.0.1}$$

Constants a and b characterize the market and the particular technology considered. One would expect that the fractional rate of technology penetration of the market, $\frac{1}{f}\frac{df}{dt}$, is proportional to the fraction, $(1 - f)$, of the market that has not yet been penetrated:

$$\frac{1}{f}\frac{df}{dt} = a(1 - f). \tag{1.0.2}$$

The empirical evidence of Figure 1.0-5 (right) and of Equation 1.0.1 supports the model of Equation 2, because the former is the integral of the latter.

The quantities, a and b depend on the nature of the technology and on the specific location where the technology is being introduced. It is possible to generalize the Fisher-Pry equation by making it independent of these parameters.

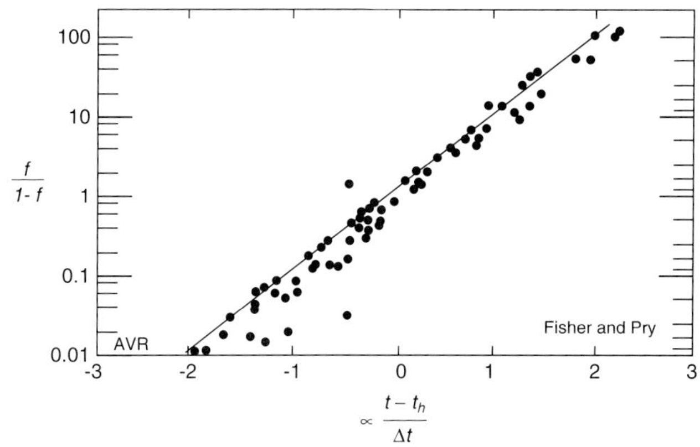

Figure 1.0-6 Fisher-Pry plot for 17 different substitutions.

For $t = t_h$, $f = 0.5$ and

$$\ln \frac{f}{1-f} = at_h + b = 0 \quad \therefore \quad b = -at_h. \tag{1.0.3}$$

For $t = t_1$, $f = 0.1$ and

$$\ln \frac{f}{1-f} = at_1 + b = -2.2. \tag{1.0.4}$$

Subtracting one equation from the other,

$$2.2 = a(t_h - t_1) = a\Delta t \quad \therefore \quad a = \frac{2.2}{\Delta t}. \tag{1.0.5}$$

Thus, the market penetration formula can be written as:

$$\ln \frac{f}{1-f} = 2.2 \frac{(t - t_h)}{\Delta t}. \tag{1.0.6}$$

Equation 1.0.6 is a function of only the normalized independent variable, $(t - t_h)/\Delta t$. This permits presenting data with different a's and b's in a single graph. An example of such a plot is shown in Figure 1.0-6, prepared by Fisher and Pry. Data for 17 different cases of technology penetration are shown, with a surprisingly small scatter of points.

The Fisher-Pry model is insensitive to the overall market volume. Many factors that affect the market as a whole don't appear to influence its distribution among different technologies.

Figure 1.0-5 shows that the take over time for oxygen steel differed among countries: in Japan it was 5 years, in West Germany and in the USA 6, and in the Soviet Union 8 years. The rapid penetration of the technology was partially due to the fast depreciation of plants allowed by law.

Marchetti (1978) showed that the market penetration law is also applicable to energy. Figure 1.0-7 illustrates the fraction of the market supplied by a particular energy source as a function of time. The data are for the USA. The graph shows how energy from wood started abandoning the market in the 19th century owing to the introduction of coal as a source of fuel.

Coal, after penetrating the market for half a century, was forced out by oil and natural gas. Owing to the dispersed nature of the market, the time constants of both penetration and abandonment of energy products is much longer than that of most other technologies. Table 1.0-3 lists the different takeover times (abandonment times have a "minus" sign).

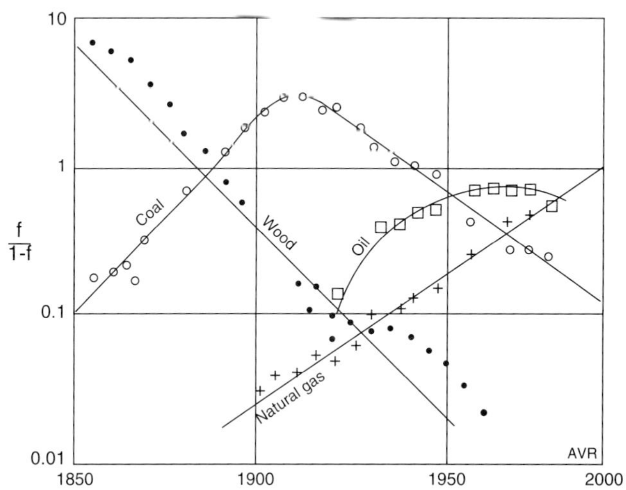

Figure 1.0-7 Long before the OPEC intervention, the Fisher-Pry model would have predicted the current decline in oil consumption.

Table 1.0-3 Take-over and abandonment times

Wood	−38 years
Coal	38 years
Coal	−50 years
Oil	30 years

Examine the period beginning in 1920. Wood, coal, and natural gas seem to have behaved according to the Fisher-Pry model. During this period, hydroelectric energy made a constant contribution of about 3.6% of the total. The regression coefficients for wood, coal and gas are shown in Table 1.0-4.

Since $\Sigma f = 1$, the fraction of the energy market supplied by oil can be calculated by subtracting from 1 the fractional contributions of the remaining fuels. When this is done, one arrives at the curve for oil penetration shown in Figure 1.0-7. It can be seen that it matches reasonably well the actual data (open squares).

Table 1.0-4 Fisher-Pry coefficients for wood, coal, and gas

	a	b
Wood	−0.0585	110.20
Coal	−0.0439	85.18
Gas	0.0426	−84.64

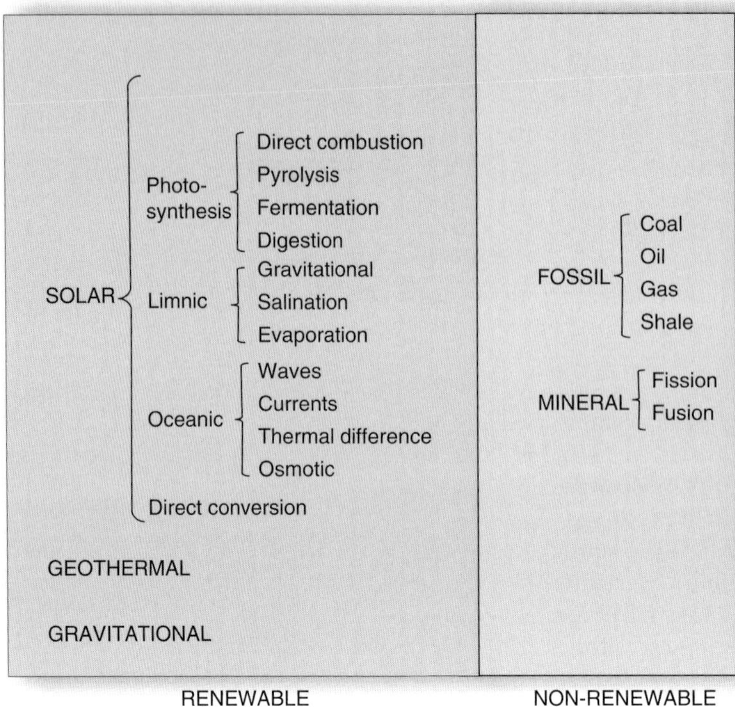

Figure 1.0-8 The energy resources of Earth.

The regression coefficients were obtained from data for 1920 through 1950 only; the rest of the information for these items resulted from extending the straight lines in the graph. Yet, the derived oil penetration curve shows a decline starting around 1970, which, in fact, did occur. The recent decline in relative oil consumption could have been predicted back in 1950, years before the creation of OPEC! One can therefore conclude that the reduction in relative oil usage would have occurred regardless of the actions of OPEC. All OPEC did was to affect the overall price of energy.

1.0.8 Planetary energy resources

In Section 1.0.5, we pointed out that the rate of per capita energy utilization rose rapidly in the last century. This, combined with the fast increase in population mentioned in Section 1.0.6, leads one to the inescapable conclusion that we are facing a serious challenge if we hope to maintain these trends in the foreseeable future. To investigate what can be done to resolve this difficulty we must first inquire what energy resources are available (Section 1.0.8) and next (Section 1.0.9) how we are using the resources at present.

Figure 1.0-8 shows the planetary energy resources. These can be renewable or nonrenewable.

Geothermal energy has been used for a very long time in Iceland and more recently in Italy, New Zealand, and the United States. Great expansion of its contribution to the total energy supply does not seem probable.

Gravitational energy—that is, energy from tides (see Chapter 7.2) has been used in France. Tides can only be harnessed in certain specific localities of which there is a limited number in the world.

Of the renewable resources, solar energy is by far the most abundant. A small part of it has been absorbed by plants and, over the eons, has been stored as coal, oil, and gas. Estimates of fossil reserves (as well as of nuclear fuel reserves) are extremely uncertain and are sure to be greatly underestimated because of incomplete prospecting. Table 1.0-5 gives us an idea of our fossil fuel reserves and Table 1.0-6 shows roughly estimated reserves of fissionable materials. These estimates do not include the old Soviet Union and China.

Table 1.0-5 Known fossil fuel reserves

Methane clathrate	>100,000 EJ (1998)
Coal	39,000 EJ (2002)
Oil	18,900 EJ (2002)
Gas	15,700 EJ (2002)
Liquefied gas	2,300 EJ (2002)
Shale	16,000 EJ (?)

Table 1.0-6 Known reserves of fissionable materials[†]

^{235}U	2,600 EJ
^{238}U	320,000 EJ
^{232}Th	11,000 EJ

[†] Does not include the former USSR and China.

The values given in the table are very far from precise. They may, however, represent a *lower* limit. People who estimate these numbers tend to be conservative as testified by the fact that there is actually a secular *increase* in proved reserves. As an example, the proved reserves of dry natural gas, 2200 EJ in 1976, rose to 5500 EJ in 2002 not withstanding the substantial consumption of gas in the intervening years.

For oil and gas, the table lists the sum of proved reserves, reserve growth and undiscovered reserves.

Proved reserves are fuels that have been discovered but not yet produced. Proved reserves for oil and gas are reported periodically in the *Oil and Gas Journal*.

Reserve growth represents the increase in the reserves of existing fields owing to further development of these fields and to the introduction of better technology for their extraction.

Undiscovered reserves represent the best possible guess of possible new discoveries.

Reserve growths and undiscovered reserves are estimated by the US Geological Survey (<http://greenwood.cr.usgs.gov/energy/WorldEnergy/DDS-60/>). For example, in 2002 the *Oil and Gas Journal* reported proved reserves of oil of 7280 EJ and the USGS estimated a growth of 4380 EJ and undiscovered oil reserves amounting to 5630 EJ adding up to the total of 18,900 EJ listed in the table.

The indicated reserves also include 3000 EJ of proved dry natural gas that is currently too far from pipe lines to be economically transported to consumers.

In addition to the dry natural gas (mostly methane), a well will also produce other gases (propane, for example) that can be liquefied and shipped. The table lists a worldwide reserve of 2300 EJ in 2002.

For coal, the table shows only proved reserves. The total reserves for this fuel are, thus, substantially larger than listed.

One number in the table that is particularly uncertain is that referring to hydrated methane. William P. Dillon, a geologist of the USGS, testified in the U.S. House of Representatives in 1998, that "the amount of methane contained in the world's gas hydrate accumulations is enormous, but estimates of the amounts are speculative and range over three orders-of-magnitude from about 100,000 to 270,000,000 trillion cubic feet [100,000 to 270,000,000 EJ] of gas." We, being ultraconservative, listed the lower figure.

Methane clathrate

Clathra is the Latin word for "bar" or "cage".

Atoms in a number of molecules group themselves in such a fashion that a cavity (or cage) is left in the center. The most famous of these arrangement is the "buckyball," a molecule consisting of 60 carbon atoms arranged as a hollow sphere capable of engulfing a number of substances. Buckyballs, discovered in the early 1980s, are not alone among "hollow" molecules. Under appropriate circumstances, water will freeze forming a cage consisting, sometimes, of 20 water molecules, but more commonly, of 46 water molecules. The configuration is unstable (it decays into a common ice crystal) unless certain gases become trapped in the central cage of the large molecule. Gases commonly trapped are methane, ethane, propane, iso-butane, n-butane, nitrogen, carbon dioxide, and hydrogen sulfide.

The ice crystal consisting of 46 water molecules is able to trap up to 8 "guest" gas molecules (a water-to-gas ratio of 5.75:1). In natural deposits, methane is by far the most abundant and the one of greatest interest to the energy field. Usually, up to 96% of the cages are fully occupied. These solid hydrates are called **clathrates**.

The density of the clathrate is about 900 kg/m^3. This means that the methane is highly compressed. (See Problem 1.0.28.) Notwithstanding its low density, water ice clathrate does not float up from the bottom of the ocean because it is trapped beneath the ocean sediment.

Clathrates form at high pressure and low temperature under sea and are stable at sufficient depth. The methane is the result of anaerobic digestion of organic matter that continuously rains down on the ocean floor.

There is no mature technology for the recovery of methane from clathrates. Proposed processes all involve destabilizing the clathrate and include:

1. Raising the temperature of the deposits.
2. Depressurization of the deposits.
3. Injecting methanol or other clathrate inhibitor.

The latter process may be environmentally undesirable.

There are dangers associated with methane clathrate extraction. The most obvious ones are the triggering of seafloor landslides and the accidental release of large volumes of methane into the Earth's atmosphere where it has a powerful greenhouse effect.

Read more about clathrates in *Clathrates: little known components of the global carbon cycle* <http://ethomas.web.wesleyan.edu/ees123/clathrate.htm>

1.0.9 Energy utilization

Most of the energy currently used in the world comes from non-renewable sources as shown in Figures 1.0-9 and 1.0-10, which display energy sources in 2001 for the whole world and for the United States, respectively. The great similarity between these two charts should not come as a surprise in view of the US using such a large fraction of the total world consumption.

What may be unexpected is that most of the renewable resources (geothermal, biomass, solar and wind) make such a small contribution to the overall energy picture. Figure 1.0-11 shows that as late as 1997 only 12% of the energy used to generate electricity in the USA came from renewable sources. Of these, 83% came from hydroelectrics. Thus, only 2% of the total came from the remaining renewables.

Disappointingly, so far, the contribution of solar and wind energy has been very small, much less than that of geothermal. Most of the renewable energy comes from hydro electric plants and some, from biomass.

For all sources of energy, the cost of the plant is proportional to the installed capacity, while the revenue is proportional to the energy generated. The **plant utilization factor** is the ratio of the energy produced to that which would be produced if the plant operated uninterruptedly at full capacity (Table 1.0-7).

Observe the extremely high utilization factor of nuclear plants and the rather small factor of wind generators, the latter resulting from the great variability of wind velocity. Although specific data for solar plants are not available, they also suffer from a low utilization factor owing to the day/night cycle and the vagaries of meteorological conditions.

It is of interest to know which are the main users of energy in the United States.

American residences account for nearly 20% of all energy used. Most of it is used for ambient heating, an area in which considerable economy can be realized, especially through better home design.

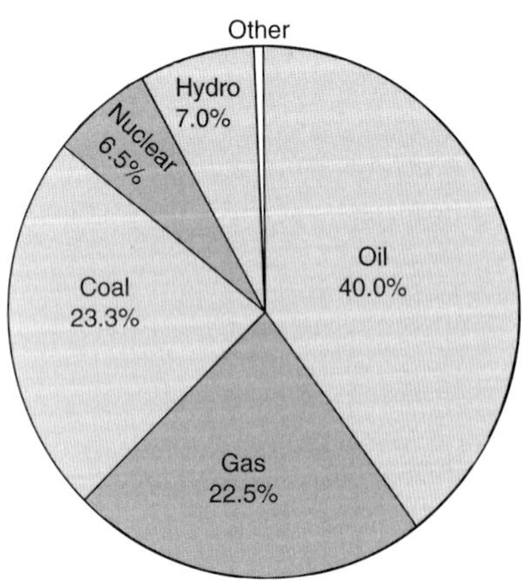

Figure 1.0-9 Energy sources in the world.

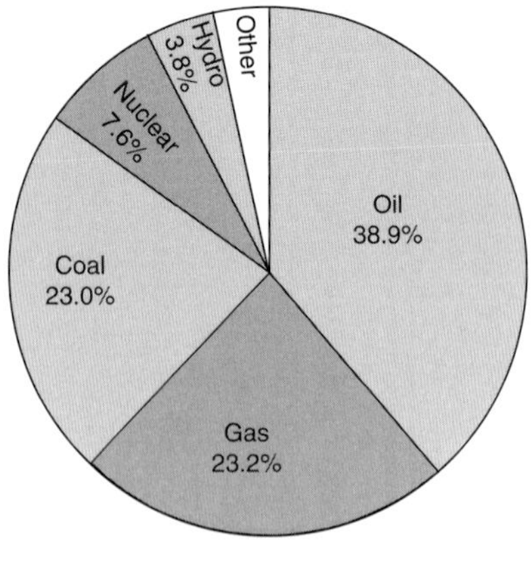

Figure 1.0-10 Energy sources in the USA.

Waste heat from electric power plants can be used for ambient heating in homes and offices. "District heating" is common In Sweden. Thermal power plants in that country, operate with an average 29% efficiency but 24% of the total fuel energy (from the heat rejected by the stem plant) is piped, as hot water, to buildings in the neighborhood. Thus, only 47% of the available combustion energy is wasted. In contrast, in the United States, a total of 68% of the combustion energy is wasted in spite of the larger average steam plant efficiency (32%). District heating requires the location of power plants in densely populated areas. This is, of course, inadvisable in the case of nuclear plants and large fossil fueled installations. However, fuel cell plants (see Chapter 5.1), being noiseless and pollution free, can be placed in a downtown area.

It is probably in the transportation sector (25% of the total energy use) where modern technology can have the most crucial impact. Fuel cell cars promise to increase automobile efficiency while reducing pollution.

1.0.10 The ecology question

We have shown that there is an almost unavoidable trend toward increasing energy utilization. We have also pointed out that at present the energy used is at least 85% of fossil origin. Finally, we have shown that the fossil fuel reserves seem ample to satisfy our needs for a good fraction of the next millennium. So, what is the problem?

Most of the easily accessible sources of oil and gas have already been tapped. What is left is getting progressively more expensive to extract. Thus, one part of the problem is economical. Another is political—most of the fuel used by developed nations is imported (using the large American reserves is unpopular and politicians hesitate

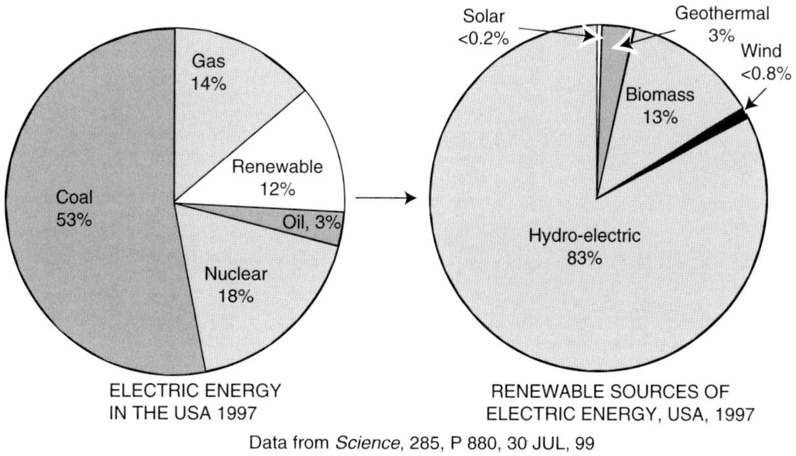

Figure 1.0-11 Sources of electric energy in the United States.

to approve such exploration). This creates an undesirable vulnerability. There are also technological difficulties associated with the identification of new reserves and the extraction of fuels from more remote locations. The major obstacle, however, is ecological. Fossil fuels are still the most inexpensive and most convenient of all energy resources, but their use pollutes the environment, and we are quickly approaching a situation in which we can no longer dismiss the problem or postpone the solution.

By far, the most undesirable gas emitted is carbon dioxide whose progressively increasing concentration in the atmosphere (from 270 ppm in the late 1800 to some 365 ppm at present) constitutes a worrisome problem. It is sad to hear influential people (among them, some scientists) dismiss this problem as inconsequential, especially in view of the growing signs of a possible runaway ecological catastrophe. For instance, in the last few decades, the thickness of the north polar ice has decreased by 40% and on the first year of the current millennium, a summertime hole appeared in the polar ice.

Table 1.0-7 Energy use, USA 2001

Source	Used (EJ)	Capacity (GW)	Utilization factor
Thermal	9.69	600	51.1%
Nuclear	2.71	98	87.6%
Hydro	0.97	99	31.1%
Geothermal	0.26		
Wind	0.0208	4.28†	15.4%
Other	0.015		

† This datum is from AWEA (American Wind Energy Association), all other are from EIA (Energy Informatiion Administration.)

Since increased concentrations of CO_2 can lead to global warming, some people have proposed increasing the emission of SO_2 to stabilize the temperature because of the cooling effect of this gas. Even ignoring the vegetation-killing acid rain that would result, this proposal is equivalent to balancing a listing boat by piling stones on the other side.

The lack of public concern with the CO_2 problem may be due to the focus on planetary temperature rise. Although the growth in CO_2 concentration is very easily demonstrated, the conclusion that the temperature will rise, although plausible, is not easy to prove. There are mechanisms by which an increase of greenhouse gases would actually result in a *cooling* of Earth. For instance, increasing greenhouse gases would result in enhanced evaporation of the tropical oceans. The resulting moisture, after migrating toward the poles, would fall as snow thereby augmenting the albedo of the planet and, thus, reducing the amount of heat absorbed from the sun.

Some scientist and engineers who are less concerned with political correctness, are investigating techniques to reduce (or at least, to stabilize) the concentration of atmospheric carbon dioxide. This can, in principle, be accomplished by reducing emissions or by disposing carbon

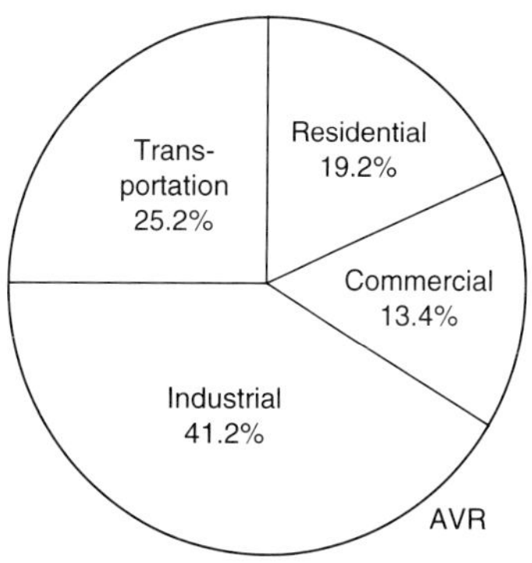

Figure 1.0-12 The different users of energy in the USA.

Table 1.0-8 Stored carbon on Earth

Oceans	45×10^{15} kg
Fossil fuels	10×10^{15} kg
Organic matter	2.4×10^{15} kg
Atmosphere	0.825×10^{15} kg

dioxide in such a way as to avoid its release into the air. Emissions can be reduced by diminishing overall energy consumption (a utopian solution), by employing alternative energy sources, by increasing efficiency of energy use, and by switching to fuels that yield more energy per unit amount of carbon emitted. 1 kilomole of methane, CH_4, when burned yielding liquid water and carbon dioxide, releases 889.6 MJ and emits 1 kilomole of carbon—it generates heat at a rate of 889.6 MJ per kilomole of carbon. n-Heptane, C_7H_{16}, which can represent gasoline, releases 4820 MJ of heat per kilomole burned and emits 7 kilomoles of CO_2—a rate of 688.6 MJ per kilomole of carbon. Clearly, the larger the number of carbon atoms in the hydrocarbon molecule, the lower the ratio of the heat of combustion to the amount of carbon dioxide emitted because the ratio of hydrogen to carbon decreases. This is one reason for preferring methane to oil and oil to coal.

Alternative forms of energy are attractive but, at least for the present, are too expensive to seriously compete with fossil fuels.

In order to select a carbon dioxide disposal technique, it is important to inquire where nature stores the existing carbon.

Table 1.0-8 shows the estimated amount of carbon stored in different places.

Methods to dispose of CO_2 could include:

1.0.10.1 Biological

Photosynthesis removes carbon dioxide from the air. The biomass produced must be preserved if it is to permanently affect the CO_2 concentration. This means it cannot be burned or allowed to rot. There seems to be limited capacity for this method of CO_2 disposal. It should be noted that the biological uptake rate of carbon is, at present, only 0.002×10^{15} kg year.

1.0.10.2 Mineral

CO_2 is removed naturally from the air by forming carbonates (principally of magnesium and calcium). The gas is removed by reacting with abundant silicates. However, this process is too slow to cope with man-made emissions.

Carbon in the atmosphere

How much carbon is there in the atmosphere?

The surface area of earth is 510×10^{12} m^2, while the scale height of the atmosphere is around 8800 m. Consequently the volume of air (all of it compressed to 1 atmosphere pressure) is $510 \times 10^{12} \times 8800 = 4.5 \times 10^{18}$ m^3.

Present day atmospheric CO_2 concentration is 13.5×10^{-6} kmol/m^3. Thus, the atmosphere contains $13.5 \times 10^{-6} \times 4.5 \times 10^{18} = 61 \times 10^{12}$ kmol of CO_2 and, therefore, 61×10^{12} kmol of carbon. Since the atomic mass of carbon is 12 daltons, the mass of carbon in the atmosphere is 0.73×10^{15} kg. Compare with the 0.825×10^{15} kg in Table 1.0-8.

A simpler way to achieve about the same result is to consider that the atmospheric pressure at sea level is 1 kg/cm^2 or 10^4 kg/m^2. Consequently, the total mass of the atmosphere is $510 \times 10^{12} \times 10^4 = 510 \times 10^{16}$. Of this 360×10^{-6} is carbon dioxide and 12/44 of this is carbon. Carbon content of the atmosphere is $510 \times 10^{16} \times 365 \times 10^{-6} \times 12/44 = 0.51 \times 10^{15}$ kg, a result comparable with the previous one.

Ziock et al. propose the use of magnesium silicates to sequester carbon dioxide at the point where fossil fuels are burned. Enormous deposits of magnesium oxide-rich silicates exist in the form of olivines and serpentines.

For serpentine, the net reaction involved is

$$Mg_3Si_2O_5(OH)_4 + 3CO_2 \rightarrow 3MgCO_3 + 2SiO_2 + 2H_2O$$

Notice that the end products are materials that already exist naturally in great abundance.

Substantial additional research is needed to improve the proposed disposal system and to make it economical.

1.0.10.3 Subterranean

CO_2 can be sequestered underground as the oil industry has been doing (for secondary oil recovery) for more than 50 years. The volume of the exhaust gases of a combustion engine is too large to be economically stored away. It is necessary to separate out the carbon dioxide, a task that is not easy to accomplish. One solution is proposed by Clean Energy Systems, Inc. of Sacramento, CA. The suggested equipment extracts oxygen from air (a well developed process) and mixes this gas with the fuel. Combustion produces steam and CO_2 at high temperature and pressure and drives several turbines at progressively lower temperatures. The water in the final exhaust is condensed and recycled leaving the carbon dioxide to be pumped, at 200 atmospheres, into an injection well. At present, no turbines exist capable of operating at the high temperature (over 3000 C) of the combustion products. See Anderson et al., 1998.

1.0.10.4 Undersea

The Norwegian government imposes a stiff carbon dioxide emission tax that has made it economical to install disposal systems. They pump the gas deep into the ocean. It appears that liquid carbon dioxide can be injected into the seas at great depth and that it will stay there for a long time. More work is required to see if such scheme is indeed feasible and economical.

1.0.11 Nuclear energy

Chemical fuels, such as oil or methane, release energy when the atoms in their molecules are rearranged into lower energy configurations. The energies involved are those of molecular binding and are of the order of some tens of MJ/kmol. When the *components* of an atom are arranged into lower energy configurations, then the energy released is orders of magnitude larger (GJ/kmole) because of the much larger intra-atomic binding energies.

The internal structure of atoms can be changed in different ways:

1. An atomic nucleus can be bombarded with a neutron, absorbing it. A different atom emerges.
2. An atom can spontaneously change by emitting either electrons (beta-rays) or helium nuclei (alpha-rays). Such radioactive decay releases energy which can be harvested as, for instance, it is done in **Radioisotope Thermal Generators** (RTGs).
3. Atoms with large atomic number can be made to break up into smaller atoms with the release of energy. This is called **nuclear fission** and requires that the atomic number, Z, be larger than 26.
4. Atoms with low atomic numbers can be assembled into a heavier one, releasing energy. This is called **nuclear fusion** and requires that the final product have an atomic number smaller than 26.[†]

Nuclear energy has developed a bad reputation especially after the Chernobyl accident. Nevertheless it is still a source of substantial amounts of energy in many countries. In 2001, the USA led the world in installed capacity—98 GW, followed by France (60 GW) and Japan (42 GW).[††]

The utilization factor of nuclear plants was excellent. In the USA, the plants generated 87.6% of the energy they would have delivered if they had operated uninterruptedly at full power. In France, this figure was 69.5% and in Japan, 75.4%.[††]

Of the total electricity generated, nuclear plants in the USA contributed a relatively modest 18%, while in France, heavily reliant on this form of energy, the contribution was 76.1%. In Japan, it was 33.4%. In 2000, Germany decided to phase out its 19 nuclear power plants. Each one was assigned a 32-year life after which they would be deactivated. Many plants have already operated more than half of their allotted life time.

The cost of nuclear electricity is high, about double of that from fossil fuel. In the USA (1996) it was 7 cents/kWh, while that of a state of the art natural gas plant was 3 cents/kWh (Sweet, 1997a). Advanced reactor designs may bring these costs down considerably while insuring a greater safety in the operation of the plants (Sweet, 1997b). This promised reduced cost combined with the ecological advantage of no greenhouse gas emission—a growing concern—may lead to renewed popularity of nuclear generators.

The major objection to fission-type reactors is not so much the danger of the operation of the power plants, but rather the problem of disposing of large amounts of long-lived radioactive by-products. If the need for such disposal can be avoided, then there is good reason to reconsider fission generators as an important contributor to the energy supply system.

Specifications of new generation nuclear fission reactors might include (not necessarily in order of priority), the following items:

1. Safety of operation (including resistance to terrorist attacks)
2. Affordability
3. Reliability
4. Absence of weaponizable sub-products
5. Absence of long-lived waste products
6. Ability to transmute long-lived radioactive waste products from old reactors into short-lived radioactive products.

The U.S. Department of Energy was funding research (2004) in several technologies that might realize most of the specifications above. One of these is the **heavy metal nuclear reactor** technology. Although the technology is complicated, it appears that this type of reactor may be able to not only produce wastes with relatively short

[†] All are transmutations, the age-old dream of medieval alchemists.
[††] The French and the Japanese data are for 1996.

half-lives (100 years contrasted with 100,000 years of the current waste), but in addition may be able to use current type waste as fuel thus greatly alleviating the waste disposal problem. Furthermore, because heavy-metal reactors operate at high temperatures (yet at low pressures), the thermolytic production of hydrogen (see Chapter 5.1) for use in fuel cell-driven automobiles looms as a good possibility. For further reading on this topic see Loewen (2004).

The waste disposal problem is absent in fusion devices. Unfortunately, it has been impossible to demonstrate a working prototype of a fusion machine, even after several decades of concerted research.

To do even a superficial analysis of the technical aspects of nuclear reactions, we need to know the masses of some of the atoms involved. (See Table 1.0-9.) Most of the mass values are from Richard B. Firestone. Those marked with a ♣ are from Audi and Wapstra (1993), and the one marked with a • is from a different source. It can be seen that the precision of the numbers is very large. This is necessary because, in calculating the energy released in a nuclear reaction, one uses the small difference between large numbers which is, of course, extremely sensitive to uncertainties in the latter.

Table 1.0-9 Masses of some particles important to nuclear energy

Particle	Symbol	Mass (daltons†)	Mass (kg)
electron ♣	e	0.00054579903	$9.1093897 \times 10^{-31}$
muon •	μ	0.1134381	1.883566×10^{-28}
proton ♣	p	1.007276467	1.672648×10^{-27}
neutron ♣	n	1.008664909	$1.6749286 \times 10^{-27}$
$^{1}_{1}H$		1.007825032	$1.673533967 \times 10^{-27}$
$^{2}_{1}D$		2.014101778	$3.344496942 \times 10^{-27}$
$^{3}_{1}T$		3.016049278	$5.008271031 \times 10^{-27}$
$^{3}_{2}He$		3.016029319	$5.008237888 \times 10^{-27}$
$^{4}_{2}He$		4.002603254	$6.646483555 \times 10^{-27}$
alpha ♣	α	4.001506175	$6.644661810 \times 10^{-27}$
$^{5}_{3}Li$		5.01254	$8.323524107 \times 10^{-27}$
$^{6}_{3}Li$		6.015122794	$9.988353127 \times 10^{-27}$
$^{7}_{3}Li$		7.01600455	$1.165035751 \times 10^{-26}$
$^{10}_{5}B$		10.012937	$1.662688428 \times 10^{-26}$
$^{11}_{5}B$		11.009305	1.82814×10^{-26}

† The dalton is not yet the official name for the atomic mass unit.

The listed values for the masses of the nucleons (the proton and the alpha, in Table 1.0-9) are nearly the values of the masses of the corresponding atoms minus the mass the electron(s). On the other hand, there is a large difference between the the mass of a nucleon and the sum of the masses of the component protons and neutrons. Indeed, for the case of the alpha, the sum of the two protons and the two neutrons (4.03188278 daltons) exceeds the mass of the alpha (4.001506175 daltons) by 0.030376606 daltons—about 28 MeV of mass. This is, of course, the large **nuclear binding energy** necessary to overcome the great electrostatic repulsion between the protons.

1.0.11.1 Fission

There are at least three fissionable elements of practical importance: ^{235}U, ^{239}Pu and ^{233}U. Of these, only ^{235}U is found in nature in usable quantities; ^{239}Pu and ^{233}U must be created by transmutation from "fertile" materials, respectively ^{238}U and ^{232}Th.

A nuclear fission reaction (with a corresponding release of energy) occurs when a fissionable material interacts with thermal, i.e., low energy, neutrons. The collision of high energy neutrons with ^{235}U, for example, is elastic, whereas low energy neutrons are captured:

$$^{235}_{92}U + ^{1}_{0}n \rightarrow ^{236}_{92}U. \tag{1.0.7}$$

The resulting ^{236}U decays with the emission of alpha-particles (lifetime 7.5 seconds). More importantly, the uranium also suffers spontaneous fission:

$$^{236}_{92}U \rightarrow 3^{1}_{0}n + \text{fission products} + \ldots \quad 3 \times 10^{-11} \text{ joules}. \tag{1.0.8}$$

Thus, under the proper circumstances, $^{235}_{92}U$ absorbs a neutron and the resulting atom splits into smaller nuclei simultaneously releasing 3 neutrons and about 3×10^{-11} joules of energy:

$$^{235}_{92}U + ^{1}_{0}n \rightarrow 3^{1}_{0}n + \text{fission products} + 3 \times 10^{-11} \text{ J}. \tag{1.0.9}$$

Per kilogram of $^{235}_{92}U$, the energy released is

$$\frac{3 \times 10^{-11} \frac{\text{J}}{\text{atom}} \times 6 \times 10^{26} \frac{\text{atoms}}{\text{kmol}}}{235 \frac{\text{kg}}{\text{kmol}}} = 77 \text{ TJ/kg}.$$

Compare this with the energy from chemical reactions which is frequently of the order of a few tens of MJ/kg.

When Otto Hahn, in 1939, demonstrated uranium fission, it became immediately obvious that a sustained "chain" reaction would be achievable. To such an end, all that was needed was to use one of the emitted neutrons to split a new uranium atom. In trying to build such a fission reactor, a number of problems had to be overcome.

1. The $^{235}_{92}U + ^{1}_{0}n$ reaction requires slow (thermal) neutrons. The high energy neutrons emitted will not do. Thus, these neutrons must be made to transit through some material that has the property of slowing the particle down without absorbing it. Examples of such "moderating" substances are heavy water and graphite.
2. Fast neutrons may be absorbed by impurities in the fuel or in the moderator. The fuel is a mixture of $^{235}_{92}U$ and $^{238}_{92}U$. The latter is an abundant "impurity" that absorbs fast neutrons but not slow ones. To reduce neutron losses, it may be necessary to "enrich" the fuel, i.e, increase the $^{235}_{92}U/^{238}_{92}U$ ratio.† It is also necessary to place the fuel into a number of long rods embedded in a mass of moderator. This configuration allows most of the fast neutrons to escape the fuel region and reach the moderator where they are slowed and may eventually reenter one of the fuel rods. They now have insufficient energy to interact with the $^{238}_{92}U$ but will do so with $^{235}_{92}U$, perpetuating the reaction.

Clearly, it is essential that exactly one of the released neutrons is, on average, used to trigger a new fission. If more than one, the reaction will grow exponentially, if less, it will die out. Control systems are used to adjust this number to precisely one. Fortunately, the process is, to a degree, self adjusting—if the reaction rate rises, so will the temperature, and this reduces the probability of neutron capture.

Uranium isotopes cover the range from 227 to 240 in atomic mass, but natural uranium consists chiefly of:

Table 1.0-10 Uranium isotopes

Isotope	Abundance (%)	Lifetime (years)
^{238}U	99.283	4.5×10^9
^{235}U	0.711	7.1×10^8
^{234}U	0.005	2.5×10^5

It is estimated that in the Western World there are reserves of uranium oxide (U_3O_8) amounting to some 6×10^9 kg, but only 34×10^6 kg are fissionable, corresponding to an available energy of 2600 EJ. Compare this with the 40,000 EJ of available coal energy.

The relatively modest resources in fissionable uranium led to "breeder reactors" in which fertile materials are transformed into fissionable ones.

Take ^{238}U, which suffers inelastic collisions with high energy neutrons (neutrons from fission):

$$^{238}_{92}U + ^{1}_{0}n \rightarrow ^{239}_{92}U \rightarrow ^{0}_{-1}e + ^{239}_{93}Np \rightarrow ^{0}_{-1}e + ^{239}_{94}Pu \quad (1.0.10)$$

or take ^{232}Th:

$$^{232}_{90}Th + ^{1}_{0}n \rightarrow ^{233}_{90}Th \rightarrow ^{0}_{-1}e + ^{233}_{91}Pa \rightarrow ^{0}_{-1}e + ^{233}_{92}U. \quad (1.0.11)$$

By creating plutonium in a breeder reactor, all uranium can be made to yield energy: 320,000 EJ become available. Even larger amounts of energy could be derived from thorium.

1.0.11.2 Fusion

The main objections to fission reactors are related to

1. lack of safety,
2. dangerously radioactive "ashes," and
3. scarcity of fuel.

Fusion reactors may overcome all of the above objections. The reaction that is, by far, the easiest to ignite is††

$$^{3}_{1}T + ^{2}_{1}D \rightarrow ^{4}_{2}He + ^{1}_{0}n \quad (1.0.12)$$

To estimate the reaction energy released, one calculates the amount of mass lost. The mass of $^{3}_{1}T$ is 5.00827×10^{-27} kg as given by Table 1.0-9. The mass of the deuterium is 3.3434×10^{-27} kg so that the mass of the left side of the equation is 8.3526×10^{-27} kg. On the right-hand side of the equation the sum of the masses of the alpha particle (the helium ion) and the neutron is 8.3214×10^{-27} kg, a deficit of 3.12×10^{-29} kg. When multiplied by c^2, this yields an energy of 2.80×10^{-12} joules per deuterium/tritium pair. The correct value is very slightly larger (it is nearer 2.81×10^{-12} J). The small discrepancy is mainly due to the fact that we used the mass of the atoms instead of that of the corresponding ions. The reaction yields 337 TJ per kg of tritium/deuterium alloy or 562 TJ per kg of tritium.

The energy released by the reaction is carried by both the alphas and the neutrons. The conversion of the neutron energy to usable forms has an efficiency of only some 40% because the particles are uncharged and heat management and mechanical heat engines are involved. On the other hand, the alphas can be directly converted to electricity at a much higher efficiency ($\approx$ 90%). (See Rostoker

† No enrichment is needed if the moderator is heavy water (D_2O) as used in the CANDU reactor. This is the CANadian Deuterium Uranium, pressurized heavy water reactor that uses natural (unenriched) uranium and heavy water as both moderator and coolant.
†† The larger the atomic number, *Z*, the greater the difficulty of reaction owing to the large electrostatic repulsion between nuclei.

et al. (1997); Moir and Barr (1973); Momota *et al.* (1992); Yoshikawa *et al.* (1991); Bloch and Jeffries (1950).) In addition, the heavy neutron flux creates serious radioactivity and material destruction problems. Consequently, it is important to know how the released energy is divided between the alphas and the neutrons. This can be done by assuming that the momenta are equally divided between the two types of particle:

$$m_\alpha v_\alpha = m_n v_n, \tag{1.0.13}$$

and combining this with with the energy equation,

$$\frac{1}{2}m_\alpha v_\alpha^2 + \frac{1}{2}m_n v_n^2 = W = W_\alpha = W_n. \tag{1.0.14}$$

Here, m_α is the mass of the alpha, m_n is the mass of the neutron, v_α is the velocity of the alpha, v_n is the velocity of the neutron, and W is the energy released by one pair of reacting atoms. Solving these simultaneous equations leads to

$$W_\alpha = \frac{W}{\frac{m_\alpha}{m_\alpha} + 1} \tag{1.0.15}$$

and

$$W_\alpha = \frac{W}{\frac{m_n}{m_\alpha} + 1}. \tag{1.0.16}$$

For the reaction under consideration, it is found that neutrons carry about 14 MeV, while the more massive alphas carry only some 3.5 MeV.

The T+D reaction is popular because of its high reactivity, which should facilitate ignition, and because the atomic number of the fuel is $Z = 1$, thus minimizing radiation losses. This is because radiation is a function of Z^2. However, it has drawbacks:

1. One neutron is emitted for each 2.8×10^{-12} J generated, whereas, in fission, the rate is one neutron per 10^{-11} J. Thus, the neutron bombardment is serious: it radioactivates substances and weakens structures by causing dislocations in the crystal lattice and by generating hydrogen bubbles inside materials.
2. As pointed out before, most of the energy is in the neutron stream reducing the recovery efficiency.
3. Although deuterium is not radioactive, tritium is radioactive with a lifetime of 12 years. It has the tendency to "stick around" by replacing normal hydrogen in water molecules.
4. There is no natural source of tritium; it must be obtained from lithium:

$$^6_3\text{Li}^6 + ^1_0\text{n} \rightarrow ^3_1\text{T} + ^4_2\text{He} + \ldots \quad 7.7 \times 10^{-13} \text{ joules} \tag{1.0.17}$$

Thus, each lithium atom yields $2.8 \times 10^{-12} + 7.7 \times 10^{-13} = 3.57 \times 10^{-12}$ J. One kg of lithium yields 350 TJ.

The world reserves of lithium are not known accurately. Conservative estimates are of 10^{10} kg. However, most of this is ^{7}Li. The desired isotope, ^{6}Li, has a relative abundance of 7.4%. Consequently, one can count on only 740×10^6 kg of this material or 260,000 EJ of energy.

In order of ease of ignition, the next two reactions are

$$^2_1\text{D} + ^2_1\text{D} \rightarrow ^3_2\text{He} + ^1_0\text{n} + \ldots \quad 5.23 \times 10^{-13} \text{ joules}, \tag{1.0.18}$$

$$^2_1\text{D} + ^2_1\text{D} \rightarrow ^3_1\text{T} + ^1_1\text{H} + \ldots \quad 6.45 \times 10^{-13} \text{ joules}. \tag{1.0.19}$$

The above reactions have equal probability of occurring.

The tritium produced will react with the deuterium according to Reaction 1.0.12. The average energy of the D+D reaction is

$$\frac{(5.1 + 6.40 + 28.0) \times 10^{-13}}{5} = 7.9 \times 10^{-13} \text{ J per D atom.} \tag{1.0.20}$$

The D+D reaction is still dirty (neutronwise) and still involves a radioactive gas (tritium). However, it does not use a fuel of limited abundance, such as lithium. It uses only deuterium, which is available in almost unlimited amounts. In common water, there is one D_2O molecule for every 6700 H_2O molecules. One can estimate roughly how much deuterium is available:

The oceans cover about 2/3 of the Earth's surface, which is 5.1×10^{14} m^2. Assuming an average depth of 3000 m, the ocean has a volume of 10^{18} m^3 and a mass of 10^{21} kg. Of this, 1/9 is the mass of hydrogen, and 2/6700 of the latter is the mass of deuterium, amounting to some 3.3×10^{16} kg or about 10^{31} J—an amount of energy that, for practical purposes, can be considered unlimited.

Next, in order of ignition difficulty is the ^{2}D + ^{3}He reaction that burns cleanly: no radioactive substances are involved and no neutrons are generated. Also clean is the ^{3}H + ^{3}He reaction.

The catch in these reactions is that there is no natural ^{3}He on earth; it must be made from the (dirty) fusion of Li and H. However, it is estimated that over a billion tons of the material exists on the moon. This may, one day, justify a mining operation on our satellite.

The ^{3}H on the moon comes from the solar wind that has, for billions of years, deposited it there. The ^{3}H on Earth is trapped by the atmosphere and is eventually evaporated away.

Table 1.0-11 Neutron yields

Reaction	% of energy carried by neutrons
D + T	65...75
D + D	20...45
B + H	<0.1

An interesting reaction involves ^{11}B, the common isotope of boron:

$$^{11}_{5}B + ^{1}_{1}H \rightarrow ^{12}_{6}C^{*} \rightarrow ^{4}_{2}He + ^{8}_{4}Be. \quad (1.0.21)$$

$^{12}_{6}C^{*}$ is nuclearly excited carbon which spontaneously decays into an alpha and $^{8}_{4}$Be, a very unstable atom with a lifetime of 2×10^{-16} seconds. Fortunately, it is an alpha-emitter:

$$^{8}_{4}Be \rightarrow 2\,^{4}_{2}He. \quad (1.0.22)$$

The overall reaction is:

$$^{11}_{5}B + ^{1}_{1}H \rightarrow 3\,^{4}_{2}He. \quad (1.0.23)$$

or, using a different notation

$$^{1}_{1}H + ^{11}_{5}B = 3\alpha. \quad (1.0.24)$$

It appears that this **triple alpha** reaction can be made to sustain itself in a **colliding beam fusion reactor** (See Rostoker *et al.*, 1997) but this has not yet been demonstrated. If it does work, we would have a clean fusion reactor using abundantly available fuel and capable of operating in units of moderate size, in contrast with the T+D reaction in a Tokamak which must be 10 GW or more if it can be made to work at all.

It should be noticed that ^{10}B will also yield a triple-alpha reaction when combining with a deuteron:

$$^{10}_{5}B + ^{2}_{1}D \rightarrow 3\,^{4}_{2}He. \quad (1.0.25)$$

Both isotopes of boron considered above are abundant, stable, and nonradioactive. Natural boron consists essentially of 20% ^{10}B and 80% ^{11}B.

The triple alpha reaction may also be an important player in the cold fusion process (if such process exists at all). See the next subsection.

Table 1.0-11 lists the percentage of the energy of a reaction that is carried away by neutrons.

Although fusion reactors have not yet been demonstrated†, there is a possibility that they will become the main source of energy some 50 years from now. If so, they may provide the bulk of the energy needed by humanity and the energy crunch will be over.

1.0.11.3 Cold fusion

At the beginning of the millennium, when this subsection was being rewritten, the cold fusion question remained unresolved. So far, no one has been able to reproduce the claims of Pons and Fleishmann, but, on the other hand, no one has been able to disprove the existence of cold fusion. As a matter of fact, cold fusion can and has been demonstrated. Let us review what we know for sure of this topic.

As indicated in Subsection 1.0.11.2, deuteron will react spontaneously with deuteron in one of these two reactions:

$$^{2}_{1}D + ^{2}_{1}D \rightarrow ^{3}_{1}T + ^{1}_{1}H \quad (1.0.26)$$

$$^{2}_{1}D + ^{2}_{1}D \rightarrow ^{3}_{2}He + ^{1}_{0}n \quad (1.0.27)$$

These reactions have about the same probability of occurrence and they produce a substantial amount of energy. The problem is that the probability of occurrence (under normal conditions) is extremely small, of the order of one fusion per galaxy per century according to a good humored scientist.

It is easy to understand the reluctance of the ^{2}D atoms to get together: they carry positive charges and therefore repel one another. This can be overcome by imparting sufficient kinetic energy to the atoms, as, for instance, by heating them to extreme temperatures as in thermonuclear fusion.

There is a neat trick suggested by Alvarez (late professor of the University of California at Berkeley and Nobel Prize winner) that increases by 85 orders of magnitude the reaction cross-section (read probability). Replacing the orbital electron of the deuterium by a muon, which is 207 times heavier, collapses the orbital by a large factor.†† Muon mediated fusion can be observed in the laboratory as Jones (Brigham Young) demonstrated. The catch is that it takes more energy to create the muon than what one gets from the fusion.

Thus, cold fusion certainly does occur. More than that, cold fusion occurs (almost certainly) even when not mediated by muons.

Jones (1989) described an experiment that appears to prove just that. He used an electrolytic cell consisting of a platinum positive electrode and a palladium (sometimes, titanium) negative electrode. The electrolyte was D_2O (heavy water). Since water is a poor conductor of electricity, salts had to be added to the solution. Here is Jones's extraordinary recipe:

† Fusion research dates back to at least 1938 when Jacobs and Kantrowitz built the first magnetic confinement fusion reactor at NACA's (now NASA) Langley Memorial Aeronautical Laboratory.

†† A deuteron/electron molecule is about 74,000 fm in size, while the deuteron/muon molecule is only 250 fm in size. The nuclei are, therefore, 300 times closer together and this raises enormously the probability of fusion.

"The electrolyte is a mixture of about 160 g of deuterium oxide (D_2O) plus various metal salts in about 0.2 g amounts each: $FeSO_4$, $NiCl_2$, $PdCl_2$, $CaCO_3$, $LiSO_4$, $NaSO_4$, $CaH_4(PO_4)2$, $TiOSO_4$, and a very small amount of AuCN."

A chemist might be horrified by the cocktail above—it would be hard to tell what is going on.†

When a current was forced through the cell, a small flux of neutrons with a characteristic energy of 2.5 MeV was observed. Jones, a physicist, did a good job of neutron detection. Since 2.5 MeV is the energy of the neutrons in Reaction 27, this experiment tends to show that indeed fusion is going on.

Jones observed that some 8 hours after start of operation, the neutron "signal" turned off by itself. This effect was attributed to the poisoning of the palladium electrode by deposition of metals from the solution. In fact, etching the electrode revived the cell.

The reaction rate observed by Jones was small, perhaps 10^{-20} fusions per deuterium pair per second. This could be explained if the deuterium molecules were somehow squeezed from 74,000 fm to half this distance by their residence in the palladium lattice.†† Jones dubs this **piezonuclear fusion.**

Pons and Fleishmann ran similar experiments but, being chemists not physicists, adopted a simpler electrolyte: an LiOH solution in D_2O (heavy water). They also failed to make careful neutron measurements. What they reported is that, after a prolonged pre-cooking, some cells suddenly developed a great deal of heat, billions of time greater than in the Jones experiment. Unfortunately, these results were never reproduced by other experimenters and this casts severe doubts on their validity. Here is where I will don my devil's advocate mantle and, just for the fun of it, will defend the P&F results.

In a lecture delivered at the Utah University on March 31, 1989, Stanley Pons relates the most spectacular of his results. "A cube of palladium with a volume of 1 cm^3 was used as cathode of an electrolyzer with lithium hydroxide dissolved in D_2O as an electrolyte. A current of 250 mA/cm^2 was applied for several weeks/months [sic] with nothing remarkable happening. A Geiger counter detected no radiation. The current was cut to 125 mA/cm^2 late one day, and next morning the cube of palladium and the electrolysis cell were gone. A nearby Geiger counter was also ruined." †††

There was a long delay (several days, at least) before heat evolved. Since the Jones cell poisons itself in 8 hours, this cell will never reach the primed state and no heat can be observed.

Why such a delay? Hydride hydrogen storage systems are well known and are commercially available. One popular system uses a TiFe alloy to absorb H_2. Many other metals and alloys will do the same. Palladium, in particular, is a notorious H_2 absorber. It is not used commercially owing to its high price.

When TiFe powder (after being duly activated) is exposed to hydrogen, it will form a (reversible) hydride, TiFeH. If the amount of hydrogen is small, there will be a mixture of TiFe and TiFeH in the powder. This mixture, called β-phase, has the empirical formula $TiFeH_x$, where x becomes 1 when all the material has been hydrided.

After full hydridization, addition of more hydrogen will cause the formation of a di-hydride, $TiFeH_2$, (γ-phase). Clearly, the hydrogen is more densely packed in the (di-hydride) γ-phase than in the β-phase. It is, therefore, plausible that the fusion will proceed faster once the γ-phase is reached. How long does it take to reach this γ-phase?

In the described experiment, Pons used a current density of 250 mA/cm^{-2}, a total current of 0.042 A. This corresponds to a production of 2.6×10^{17} deuterons/second. Each cubic centimeter of palladium contains 68×10^{21} atoms. Thus, it takes 260,000 seconds or some 72 hours (3 days) for the palladium, in this particular experiment, to start becoming di-hydrided. This assumes that all the deuterons produced are absorbed by the palladium and, thus, the time calculated is a rough lower limit.

Could the heat have resulted from a chemical reaction? The highest enthalpy of formation of any palladium salt seems to be 706 MJ/kmole, for palladium hydroxide. Atomic mass of palladium is 106 daltons and density is 12 g cm^{-3}. This means that one gets 80 kJ cm^{-3} chemically. Pons and Fleishmann have (they say) gotten 5 MJ cm^{-3}, two orders of magnitude more than chemistry allows.

Conclusion:

1. The heat produced cannot be due to classical fusion reaction (insufficient neutrons, tritium, and γ-rays).
2. The heat produced cannot be due to chemical reaction.
3. Then, simplistically, the heat was not produced.

There is at least one more possible reaction which occurs very rarely:

$$^2_1D + ^2_1D \rightarrow ^4_2He \tag{1.0.28}$$

As written above, this reaction cannot take place because two particles are converted into a single particle and

† Jones was trying to create a chemical environment somewhat like the one in the soil because he was trying to show that some of the internal heat of our planet is generated by deuterium fusion.

†† A possible cause of the squeezing would be the increase of the electron mass to a few times its free mass.

††† As related by Patrick Nolan, 1989 (paraphrased).

it is impossible to conserve simultaneously energy and momentum under such conditions. For the reaction to proceed, it is necessary to shed energy and, in classical physics, this is done by emitting a 16 MeV γ-ray. Pons did not report γ-rays. There is still an outside possibility that the energy can be shed by some other mechanism such as a phonon, although physicists tell me that this is nonsense. Observe that Reaction 1.0.28 produces one order of magnitude more energy per fusion than do Reactions 1.0.26 and 1.0.27.

So far, we have attempted to explain the hypothetical cold fusion as the result of deuteron-deuteron reaction. It has been difficult to account for the absence of the expected large fluxes of neutrons or gamma rays. It is even more difficult to imagine such reaction proceeding when common water is used in place of heavy water. Nevertheless, some experimentalists make exactly such a claim.

There have been suggestions that cold fusion actually involves nuclear reactions other than those considered so far. Let us recapitulate what has been said about cold fusion.

1. The results, if any, are not easily reproduced.
2. No substantial neutron flux has been detected. This seems to eliminate the deuteron-deuteron reactions of Equations 1.0.26 and 1.0.27.
3. No substantial gamma ray flux has been detected. This eliminates the classical form of the deuteron-deuteron reaction of Equation 1.0.28.
4. Reactions are reported to be highly dependent on the exact nature of the palladium electrode.
5. Reactions have been reported with an H_2O instead a D_2O electrolyte.

The following cold fusion mechanism fitting the above observations has been recently proposed.

Boron is a common impurity in palladium. Natural boron exist in the form of two isotopes with the relative abundance of 20% for ^{10}B and 80% for ^{11}B. Thus, under some special circumstances, the two triple-alpha reactions of Equations 1.0.24 and 1.0.25 might occur. They emit neither neutrons nor gamma rays and can occur with either normal water or heavy water.

The boron impurity may be interstitial or it may collect in grain boundaries. The reaction may only occur if the boron is in one or the other of these distributions. It may also only occur when the amount of impurity falls within some narrow range. Thus, a palladium rod may become "exhausted" after some time of operation if the boron concentration falls below some given limiting concentration.

Perhaps the worst indictment of the P&F experiment is its irreproducibility. No one has claimed to have seen the large heat production reported from Utah. Pons himself states that his experiment will only work occasionally—he claims that there is *live* palladium and *dead* palladium. This could be interesting. Hydrogen absorbed in metals is known to accumulate in imperfections in the crystal lattice. It is possible that such defects promote the high concentrations of deuterium necessary to trigger the reaction.

I still have an old issue of the CRC handbook that lists the thermoelectric power of silicon as both +170 mV/K and −230 mV/K. How can it be both positive and negative? Notice that the determination of the sign of the Seebeck effect is trivial; this cannot be the result of an experimental error. In both cases "chemically pure" silicon was used. So, how come? We have a good and classical example of irreproducibility. That was back in the 1930s. Now any EE junior knows that one sample must have been *p*-silicon, while the other, *n*-silicon. Both could be "chemically pure"—to change the Seebeck sign, all it takes is an impurity concentration of 1 part in 10 million. Is there an equally subtle property in the palladium that will allow fusion in some cases?

In April, 1992, Akito Takahashi of Osaka University revealed that his cold fusion cell produced an average excess heat of 100 W over periods of months. The electric power fed to the cell was only 2.5 W. The main difference between the Takahashi cell and that of other experimenters is the use of palladium sheets (instead of rods) and of varying current to cause the cell to operate mostly under transient conditions. The excess heat measured is far too large to be attributed to errors in calorimetry. Disturbing to theoreticians is the absence of detectable neutrons. See D. H. Freedman's (1992) report.

In spite being saddled with the stigma of "pseudoscience", cold fusion does no seem to go away. The September 2004 issue of *IEEE Spectrum*, published a report titled "Cold Fusion Back from the Dead," in which recent work on the cold fusion by reputable laboratories is mentioned. It quotes the US Navy as revealing that the Space and Naval Systems Center (San Diego) was working on this subject.[†] It also mentioned the Tenth International Conference on Cold Fusion that took place in Cambridge, MA in August 2003.

It appears that by 2004, "a number of groups around the world have reproduced the original Pons-Fleishmann excess heat effect ..." *Mike McK-ubre* of SRI International maintains that the effect requires that the palladium electrode be 100% packed with deuterium (One deuterium-to-one-palladium atom). This coincides with our wild guess at the beginning of this sub-section.

[†] It is reported that Stanislaw Szpak, of the SNSC, has taken infrared pictures of miniexplosions on the surface of the palladium, when cold fusion appears to be taking place.

At the moment, cold fusion research has gone partially underground, at least as far as the media are concerned. Yet, the consensus is that it merits further study. This is also the opinion of independent scientists such as Paul Chu and Edward Teller who have been brought in as observers. It may be that cold fusion will one day prove practical. That is almost too good to be true and, for the classical fusion researchers, almost too bad to be true.

1.0.12 Financing

Some of the proposed alternative energy sources, such as the fusion reactor, require, for their implementation, a scientific break-through. Others need only technological development, as is the case of wind turbines or of ocean thermal energy converters. Still others have reached a fairly advanced stage of development, but their massive implementation awaits more favorable economic conditions, such as further increase in the price of oil. The production of synthetic fuel from coal falls in this category, as does the utilization of shale.

Finding new sources of energy is not difficult. What is difficult is finding new sources of *economically attractive* energy. It is, therefore, important to estimate the cost of the energy produced by different methods. One of the main ingredients of the cost formula is the cost of financing, examined below.

Frequently, the financing of the development is borne by the government, especially during the early high-risk stages of the work. It is an important political decision for the nation to finance or not to finance the development of a new energy source. For instance, the Solar Power Satellite scheme is one that has possibilities of being economical. However, its development costs, estimated as nearly 80 billion dollars, are too high to be funded by private corporations. Thus, the SPS system will be implemented only if the government feels justified in paying the bill.

Financing the implementation is simpler. Engineers can estimate roughly how the investment cost will affect the cost of the product by using a simple rule of thumb:

"The yearly cost of the investment can be taken as 20%[†] of the overall amount invested."

Thus, if a 1 million dollar power plant is to be built, one must include in the cost of the generated energy, a sum of $200,000 per year.

To allow a comparison of the costs of energy produced by different alternative sources, the Department of Energy has recommended a standard method of calculating the cost of the capital investment.

We will here derive an expression for the cost of a direct reduction loan.

Assume that the payment of the loan is to be made in N equal installments. We will consider a $1.00 loan. Let x be the interest rate of one payment period (say, one month) and let p be the value of the monthly payment. At the end of the first month, the amount owed is

$$1 + x - p \tag{1.0.29}$$

and, at the end of the second month, it is

$$(1 + x - p)(1 + x) - p = (1 + x)^2 - p(1 + 1 + x) \tag{1.0.30}$$

and, at the end of the third month, it is

$$\begin{aligned}&[(1 + x)^2 - p(1 + 1 + x)](1 + x) - p \\ &\quad = (1 + x)^3 - p[1 + (1 + x) + (1 + x)^2].\end{aligned} \tag{1.0.31}$$

At the end of N months, the amount owed is zero because the loan has been repaid. Thus,

$$\begin{aligned}&(1 + x)^N - p[1 + (1 + x) + (1 + x)^2 + \ldots \\ &\quad + (1 + x)^{N-1}] = 0,\end{aligned} \tag{1.0.32}$$

whence

$$\begin{aligned}p &= \frac{1}{(1 + x)^1 + (1 + x)^{-2} + \ldots + (1 + x)^{-N}} \\ &= \left[\sum_{\gamma=1}^{N} z^\gamma\right]^{-1},\end{aligned} \tag{1.0.33}$$

where

$$z \equiv (1 + x)^{-1}. \tag{1.0.34}$$

But

$$\sum_{\gamma=1}^{N} z^\gamma = \frac{1 - z^{N+1}}{1 - z} - 1, \tag{1.0.35}$$

hence

$$p = \frac{1 - z}{z - z^{N+1}} = \frac{x}{1 - (1 + x) - N}. \tag{1.0.36}$$

The formula above yields the magnitude of the monthly payment as a function of the interest rate (per month) and the number of payments.

As an example, consider a small entrepreneur who owns a Diesel-electric generating plant in which he has invested $1000 per kW. The utilization factor is 50%—that is, 4380 kWh of electricity are produced

[†] This percentage is, of course, a function of the current interest rate. In the low interest rate regimen of the early years of this millennium, the percentage is lower than 20%.

yearly for each kW of installed capacity. Taxes and insurance amount to \$50 kW^{-1} $year^{-1}$. Fuel, maintenance and personnel costs are \$436 kW^{-1} $year^{-1}$. In order to build the plant, the entrepreneur borrowed money at 12% per year and is committed to monthly payments for 10 years. What is the cost of the generated electricity?

The monthly rate of interest is

$$(1+x)^{12} = 1.12 \quad \therefore \quad x = 0.009489. \tag{1.0.37}$$

The number of payments is

$$N = 10 \text{ years} \times 12 \text{ months/year} = 120 \tag{1.0.38}$$

The monthly payment is

$$p = \frac{0.009489}{1-(1+0.009489)^{-120}} = \$0.013995 \text{ month}^{-1}. \tag{1.0.39}$$

The yearly payment is

$$P = 12p = \$0.167937 \text{ year}^{-1}. \tag{1.0.40}$$

If there were no interest, the yearly payment would be \$0.1. Thus, the yearly cost of interest is \$0.067937.

All of the above is on a loan of \$1.00. Since the plant cost \$1000 kW^{-1}, the cost of the investment is \$167.94 kW^{-1} $year^{-1}$. But, on a per kW basis, there is an additional expense of \$50 for taxes and insurance, raising the yearly total to \$217.94. Thus, in this example, the yearly investment cost is 21.79% of the total amount.

A total of 4380 kWh per kW installed are generated (and sold) per year. The fixed cost per kWh is, therefore

$$\frac{217.94}{4380} = \$0.0497 \text{ kWh}^{-1} \tag{1.0.41}$$

whereas, the fuel, maintenance and personnel cost is

$$\frac{436}{4380} = \$0.0995 \text{ kWh}^{-1} \tag{1.0.42}$$

Total cost is \$0.1492 kWh^{-1}. This is commonly expressed as 149.2 mils/kWh, an awkward unit. It is better to use \$149.2/MWh or, to stick to the conventional SI units of measure, \$41.4 GJ^{-1}.

When the loan is paid after 10 years, does the entrepreneur own the plant? Maybe. The diesel-generator may have only a 10-year life and a new one may have to be acquired.

References

Anderson, R., H. Brandt, H. Mueggenburg, J. Taylor, and F. Viteri, A power plant concept which minimizes the cost of carbon dioxide sequestration and eliminates the emission of atmospheric pollutants. *Clean Energy Systems, Inc., 1812 Silica Avenue, Sacramento, CA* 95815 **1998**.

Audi, G. and A. H. Wapstra, The 1993 atomic mass evaluation. *Nuclear Physics A 565* **1993**.

Bloch, F. and C. D. Jeffries, *Phys. Rev.* 77, 305 **1950**.

Casson, Lionel, Godliness & Work, *Science* **81**, 2, 36, **1981**.

Firestone, Richard B., <http://ie.lbl.gov/toi2003/MassSearch.asp>

Fisher, J.C., and R. H. Pry, A simple substitution model of technological change, *Report 70-C-215, General Electric, R. & D. Center, June,* **1970**.

Fleishmann, M., and S. Pons, Electrochemically induced nuclear fusion of deuterium, *J. Electroanal. Chem., 261*, 301–308, **1989**.

Freedman, D. H., A Japanese claim generates new heat, *News and Comments, Science, 256,* 24 April **1992**.

Jones, S. E., et al., Observation of cold nuclear fusion in condensed matter, *reprint from Brigham Young University,* March 23, **1989**.

Hafele, W., and W. Sassin, Resources and endowments. An outline on future energy systems, *IIASA, NATO Science Comm. Conf.*, Brussels, April, **1978**.

Loewen, Eric P., Heavy-metal nuclear power, *American Scientist*, November–December **2004**.

Marchetti, C., Primary energy substitution models, *Int. Inst. Appl. Syst. An. (IIASA)* internal paper WP-75-88, June, **1975**.

Moir, R. W., and W. L. Barr, *Nucl. Fusion* **13**, 35 **1973**.

Momota, H. et al. *Fusion Technol.* **21**, 2307 **1992**.

Nolan, Patrick, *e-mail circular*, 31 March **1989**.

Peterka, V., Macrodynamics of technological change: Market penetration by new technologies, *Int. Inst. Appl. Syst. An. (IIASA)*, RR-77-22, November, **1997**.

Rafelski, et al., Theoretical limits on cold fusion in condensed matter, *AZPH-TH/89-19*, March 27, **1989**.

Rostoker, N., H. Monkhorst, and M. Binderbauer, *Office of Naval Research reports*, February, May, and August **1997** (available upon request).

Rostoker, Norman, Michl W, Binderbauer, and Hendrik J. Monkhorst, *Science* **278** 1419, 21 Nov. **1997**.

Smil, Vaclav, Global population and the nitrogen cycle, *Scientific American*, p. 76, July **1997**.

Sweet, William #1, A nuclear reconnaissance, *IEEE Spectrum* 23, (Nov. **1997a**).

Sweet, William #2, Advanced reactor development rebounding, *IEEE Spectrum* 23, (Nov. **1997b**).

Yoshikawa, K., T. Noma, and Y. Yamamoto, *Fusion Technol.* **19**, 870 (**1991**).

Ziock, Hans-J., Darryl P. Butt, Klaus S. Lackner, and Christopher H. Wendt *Reaction Engineering for Pollution Prevention, Elsevier Science.*

Abundant statistical information on energy: http://www.eia.doe.gov/

For more detailed information on some topics in this chapter, read: Sørensen, Bent, Renewable energy, *Academic Press* **2003**.

Problems

1.0.1 Assume that from 1985 on the only significant sources of fuel are:

1. coal (direct combustion)

2. oil,

3. synthetic liquid fuel (from coal), and

4. natural gas.

Sources 1, 2 and 3 are assume 4 to follow the market penetration rule:

$$\ln \frac{f}{1-f} = at + b$$

where f is the fraction of the market supplied by the fuel in question and t is the year (expressed as 1988, for instance, not as simply 88). The coefficients are:

	a	b
for coal:	−0.0475,	92.14;
for oil:	−0.0436,	86.22.

The above coefficients are derived from historical data up to 1975.

The objective of this exercise is to predict what impact the (defunct) federal coal liquefaction program would have had on the fuel utilization pattern.

According to the **first in, first out** rule, the "free" variable, i.e., the one that does not follow the market penetration rule, is the natural gas consumption fraction, f_{ng}. The questions are:

— in what year will f_{ng} peak?

— what is the maximum value of f_{ng}?

Assume that f_{syn} (the fraction of the market supplied by synthetic fuel) is 0.01 in 1990 and 0.0625 in 2000. Please comment.

1.0.2 The annual growth rate of energy utilization in the world was 3.5% per year in the period between 1950 and 1973. How long would it take to consume all available resources if the consumption growth rate of 3.5% per year is maintained?

Assume that the global energy resources at the moment are sufficient to sustain, at the current utilization rate

a. 1000 years,

b. 10000 years.

1.0.3 A car moves on a flat horizontal road with a steady velocity of 80 km/h. It consumes gasoline at a rate of 0.1 liters per km. Friction of the tires on the road and bearing losses are proportional to the velocity and, at 80 km/h, introduce a drag of 222 N. Aerodynamic drag is proportional to the square of the velocity with a coefficient of proportionality of 0.99 when the force is measured in N and the velocity in m/s.

What is the efficiency of fuel utilization? Assuming that the efficiency is constant, what is the "kilometrage" (i.e., the number of kilometers per liter of fuel) if the car is driven at 50 km/h?

The density of gasoline is 800 kg per cubic meter and its heat of combustion is 49 MJ per kg.

1.0.4 Venus is too hot, in part because it is at only 0.7 AU from the sun. Consider moving it to about 0.95 AU. One AU is the distance between Earth and Sun and is equal to 150 million km.

To accomplish this feat, you have access to a rocket system that converts mass into energy with 100% efficiency. Assume that all of the energy of the rocket goes into pushing Venus. What fraction of the mass of the planet would be used up in the project? Remember that you are changing both kinetic and potential energy of the planet.

1.0.5 Consider the following arrangement:

A bay with a narrow inlet is dammed up so as to separate it from the sea, forming a lake. Solar energy evaporates the water causing the level inside the bay to be h meters lower than that of the sea.

A pipeline admits sea water in just the right amount to compensate for the evaporation, thus keeping h constant (on the average). The inflow water drives a turbine coupled to an electric generator. Turbine plus generator have an efficiency of 95%.

Assume that there is heat loss neither by conduction nor by radiation. The albedo of the lake is 20% (20% of the incident radiation is reflected, the rest is absorbed). The heat of vaporization of water (at STP) is 40.6 MJ per kilomole. Average solar radiation is 250 W/square meter.

If the area of the lake is 100 km^2, what is the mean electric power generated? What is the efficiency? Express these results in terms of h.

Is there a limit to the efficiency? Explain.

1.0.6 The thermonuclear (fusion) reaction

$${}^{11}_{5}\text{B} + {}^{1}_{1}\text{H} \rightarrow 3\,{}^{4}_{2}\text{He},$$

is attractive because it produces essentially no radiation and uses only common isotopes.

How much energy does 1 kg of boron produce? Use the data of Problem 1.0.11.

1.0.7 The efficiency of the photosynthesis process is said to be below 1% (assume 1%). Assume also that, in terms of energy, 10% of the biomass produced is usable as food. Considering a population of 6 billion people, what percentage of the **land** area of this planet must be planted to feed these people.

1.0.8 Each fission of ^{235}U yields, on average, 165 MeV and 2.5 neutrons. What is the mass of the fission products?

1.0.9 There are good reasons to believe that in early times, the Earth's atmosphere contained no free oxygen.

Assume that all the oxygen in the Earth's atmosphere is of photo-synthetic origin and that all oxygen produced by photosynthesis is in the atmosphere. How much fossil carbon must there be in the ground (provided no methane has evaporated)? Compare with the amount contained in the estimated reserves of fossil fuels. Discuss the results.

1.0.10 What is the total mass of carbon in the atmosphere?

CO_2 concentration is currently 330 ppm but is growing rapidly!

If all the fossil fuel in the estimated reserves (see Section 1.0.8) is burned, what will be the concentration of CO_2.

1.0.11 Here are some pertinent data:

Particle	Mass (daltons)	Particle	Mass (daltons)
electron	0.00054579903	alpha	4.001506175
muon	0.1134381	^5_3Li	5.01254
proton	1.007276467	^6_3Li	6.015122794
neutron	1.008664909	^7_3Li	7.01600455
^1_1H	1.007825032	$^{10}_5B$	10.012937
^2_1D	2.014101778	$^{11}_5B$	11.009305
^3_1T	3.016049278		
^3_2He	3.016029319		
^4_2He	4.002603254		
Constants			
c			2.998×10^8 m/s
h			6.625×10^{-34} joule-sec

To convert daltons to kg, divide by $6.02213670 \times 10^{26}$.

Deuterium is a very abundant fusion fuel. It exists in immense quantities in Earth's oceans. It is also relatively easy to ignite. It can undergo three different reactions with itself:

$$^2_1D + ^2_1D \rightarrow ^3_1T + ^1_1H \quad (1)$$

$$^2_1D + ^2_1D \rightarrow ^3_2He + {}_0n^1 \quad (2)$$

$$^2_1D + ^2_1D \rightarrow ^4_2He + h\nu \quad (3)$$

For each reaction, calculate the energy released and, assuming equipartition of momenta of the reaction products, the energy of each product.

What is the energy of the photon released in Reaction 3?

1.0.12 Random access memories (RAMs) using the "Zing Effect" were first introduced in 1988 but only became popular in 1990 when they accounted for 6.3% of total RAM sales. In 1994 they represented \$712 million of a total of \$4.75 billion. Sales of all types of RAMs reached \$6 billion in 1997.

A company considering the expansion of Z-RAM production needs to have an estimate of the overall (all manufacturers) sales volume of this type of memory in the year 2000. Assume that the growth rate of the overall dollar volume of RAM sales between 1900 and 2000 is constant (same *percentage* increase every year).

1.0.13 A 1500-kg Porsche 912 was driven on a level highway on a windless day. After it attained a speed of 128.7 km/h it was put in neutral and allowed to coast until it slowed down to almost standstill. The coasting speed was recorded every 10 seconds and resulted in the table below.

From the given data, derive an expression relating the decelerating force to the velocity.

Calculate how much horse power the motor has to deliver to the wheel to keep the car at a constant 80 mph.

Coasting time (s)	Speed (km/h)	Coasting time (s)	Speed (km/h)
0	128.7	100	30.6
10	110.8	110	25.9
20	96.2	120	20.4
30	84.0	130	16.2
40	73.0	140	12.2
50	64.2	150	9.2
60	56.4	160	5.1
70	48.0	170	2.0
80	41.8	180	0
90	35.8		

1.0.14 The California Air Resources Board (CARB) mandated, for 1995, an upper limit of 200 g/km for the emission of CO_2 from a minivan.

This could be achieved by bubbling the exhaust through a $Ca(OH)_2$ bath or through a similar CO_2 sequestering substance. However, this solution does not seem economical. Assume that all the produced CO_2 is released into the atmosphere.

What is the minimum mileage (miles/gallon) that a minivan had to have by 1995? Assume gasoline is pentane (C_5H_{12}) which has a density of 626 kg m^{-3}. A gallon is 3.75 liters and a mile is 1609 meters. The atomic mass of H is 1, of C is 12, and of O is 16.

1.0.15 A geological survey revealed that the rocks in a region of Northern California reach a temperature of 600 C at a certain depth. To exploit this geothermal source, a shaft was drilled to the necessary depth and a spherical cave with 10 m diameter was excavated. Water at 30 C is injected into the cave where it reaches the temperature of 200 C (still in liquid form, owing to the pressure) before being withdrawn to run a steam turbine.

Assume that the flow of water keeps the cave walls at a uniform 200 C. Assume, furthermore that, at 100 m from the cave wall, the rocks are at their 600 C temperature. Knowing that the heat conductivity, λ, of the rocks is 2 W $m^{-1}K^{-1}$, what is the flow rate of the water?

The heat capacity of water is 4.2 MJ m^{-3} K^{-1} and the heat power flux (W m^{-2}) is equal to the product of the heat conductivity times the temperature gradient.

1.0.16 The following data are generally known to most people:

a. The solar constant, C (the solar power density), at earth's orbit is 1360 Wm^{-2};
c. the astronomical unit (AU, the average sun-earth distance) is about 150 million km;
c. the angular diameter of the moon is 0.5°.

Assume that the sun radiates as a black body. From these data, estimate the sun's temperature.

1.0.17 Using results from Problem 1.0.16, compare the sun's volumetric power density (the number of watts generated per m^3) with that of a typical homo sapiens.

1.0.18 Pollutant emission is becoming progressively the limiting consideration in the use of automobiles. When assessing the amount of pollution, it is important to take into account not only the emissions from the vehicle but also those resulting from the fuel production processes. Gasoline is a particularly worrisome example. Hydrocarbon emission at the refinery is some 4.5 times larger than that from the car itself. Fuel cell cars (see Chapter 5.1) when fueled by pure hydrogen are strictly zero emission vehicles. However, one must inquire how much pollution results from the production of the hydrogen. This depends on what production method is used. The cheapest hydrogen comes from reforming fossil fuels and that generates a fair amount of pollution. A clean way of producing hydrogen is through the electrolysis of water; but, then, one must check how much pollution was created by the generation of the electricity. Again, this depends on how the electricity was obtained: if from a fossil fuel steam plant, the pollution is substantial, if from hydroelectric plants, the pollution is zero.

The technical means to build and operate a true zero emission vehicle are on hand. This could be done immediately but would, at the present stage of the technology, result in unacceptably high costs.

Let us forget the economics and sketch out roughly one possible ZEV combination. Consider a fuel-cell car using pure hydrogen (stored, for instance, in the form of a hydride—Chapter 13.1). The hydrogen is produced by the electrolysis of water and the energy required for this is obtained from solar cells (Chapter 6.2). Absolutely no pollution is produced. The system is to be dimensioned so that each individual household is independent. In other words, the solar cells are to be installed on the roof of each home.

Assume that the car is to be driven an average of 1000 miles per month and that its gasoline driven equivalent can drive 30 miles/gallon. The fuel cell version, being much more efficient, will drive 3 times farther using the same energy as the gasoline car.

How many kilograms of hydrogen have to be produced per day?

How large an area must the solar cell collector have?

You must make reasonable assumptions about the solar cell efficiency, the efficiency of the electrolyzer and the amount of insolation.

1.0.19 From a fictitious newspaper story:

A solar power plant in the Mojave Desert uses 1000 photovoltaic panels, each "40 meters square." During the summer, when days are invariably clear, the monthly sale of electricity amounts to $22,000. The average price charged is 3 cents per kWh. The plant is able to sell all the electricity produced.

There is an unfortunate ambiguity in the story: "40 meters square" can be interpreted as a square with 40 meters to its side or as an area of 40 m^2.

From the data in the story, you must decide which is the correct area.

1.0.20 Sport physiologists have a simple rule of thumb: Any healthy person uses about 1 kilocalorie per kilometer per kilogram of body weight when running.

It is interesting to note that this is true independently of how well trained the runner is. A trained athlete will cover 1 km in much less time than an occasional runner but will use about the same amount of energy. Of course, the trained athlete uses much more power.

The overall efficiency of the human body in transforming food intake into mechanical energy is (a surprisingly high) 25%!

A good athlete can run 1 (statute) mile in something like 4 minutes and run the Marathon (42.8 km) in a little over 2 hours.

1. Calculate the power developed in these races. Repeat for a poor performer who runs a mile in 8 minutes and the Marathon in 5 hours. Assume a body weight of 70 kg.
2. Evaporation of sweat is the dominant heat removal mechanism in the human body. Is this also true for a dog? For a horse?
3. Assuming that all of the sweat evaporates, i.e. none of it drips off the body, how much water is lost by the

runners in the four cases above? The latent heat of vaporization of water is 44.1 MJ/kmole.

1.0.21 One major ecological concern is the emission of hot-house gases, the main one being CO_2.

A number of measures can be taken to alleviate the situation. For instance, the use of biomass derived fuels does not increase the carbon dioxide content of the atmosphere.

Fossil fuels, on the other hand are a major culprit. Suppose you have the option of using natural gas or coal to fire a steam turbine to generate electricity. Natural gas is, essentially, methane, CH_4, while coal can be taken (for the purposes of this problem only) as eicosane, $C_{20}H_{42}$. The higher heat of combustion of methane is 55.6 MJ/kg and that of eicosane is 47.2 MJ/kg.

For equal amounts of generated heat, which of the two fuels is preferable from the CO_2 emission point of view? What is the ratio of the two emission rates?

1.0.22 A planet has a density of 2500 kg/m^3 and a radius of 4000 km. Its "air" consists of 30% ammonia, 50% carbon dioxide and 20% nitrogen.

Note that the density, δ_{earth}, of Earth is 5519 kg/m^3.

What is the acceleration of gravity on the surface of the planet?

1.0.23 At 100 million km from a star, the light power density is 2 kW/m^2. How much is the total insolation on the planet of Problem 1.22 if it is 200 million km from the star. The total insolation on earth is 173,000 TW.

1.0.24 $^3_2$He can be used as fuel in "dream" fusion reactions—that is, in reactions that involve neither radioactive materials nor neutrons. Two possible reactions are

$$^2_1D + ^3_2He \rightarrow ^1_1H + ^4_2He \tag{1}$$

and

$$^3_2He + ^3_2He \rightarrow ^1_1H + ^1_1H + ^4_2He \tag{2}$$

1. For each of the above reactions, calculate the energy (in kWh) released by 1 kg of $^3_2$He.

On earth, $^3_2$He represents 0.00013% of the naturally occurring helium. The US helium production amounts, at present, to 12,000 tons per year.

2. If all this helium were processed to separate the helium-three, what would be the yearly production of this fuel?

There are reasons to believe that there is a substantial amount of $^3_2$He on the moon. Let us do a preliminary analysis of the economics of setting up a mining operation on our satellite.

One of the advantages of using "dream" reactions is that only charged particles (protons and alphas) are produced. The energy associated with charged particles can more efficiently be transformed into electricity than when the energy is carried by neutrons, which must first produce heat that is then upgraded to mechanical and electric energy by inefficient heat engines. Thus, it is not necessarily optimistic to assign a 30% efficiency for the conversion of fusion energy into electricity.

3. How many kWh of electricity does 1 kg of $^3_2$He produce? Use the most economical of the two reactions mentioned.

Assume that the plant factor is 70% (the reactor delivers, on average, 70% of the energy it would deliver if running constantly at full power). Assume further that the cost of the fusion reactor is \$2000/kW and that the cost of borrowing money is 10% per year. Finally, the cost of running the whole operation is \$30 kW^{-1}year^{-1}.

4. How much would the electricity cost (per kWh) if the fuel were free?

5. How much can we afford to pay for 1 kg of $^3_2$He and still break even when electricity is sold at 5 cents per kWh?

1.0.25 Between 1955 and 1995, the ocean temperature (Atlantic, Pacific, and Indian) increased by 0.06 C.

Estimate how much energy was added to the water. What percentage of the solar energy incident on earth during these 40 years, was actually retained by the ocean?

1.0.26 There seems to be a possibility that climate changes will cause the polar ice caps to melt. The amount of ice in Antarctica is so large that if it were to melt, it would submerge all port cities such as New York and Los Angeles.

Estimate by how much the sea level would rise if only the North Pole ice is melted, leaving Greenland and Antarctica untouched.

1.0.27 Refueling a modern ICV with 50 liters of gasoline may take, say, 5 minutes. A certain amount of energy was transferred from the pump to the car in a given time. What is the power represented by this transfer? Assume that the overall efficiency of a gasoline car is 15% and that of an electric car is 60%. How much power is necessary to charge the batteries of the electric car in 5 minutes (as in the ICV case)? Assume that the final drive train energy is the same in both the ICV and the EV. Is it practical to *recharge* a car as fast as *refueling* one?

1.0.28 Some of the more attractive fuels happen to be gases. This is particularly true of hydrogen. Thus, storage of gases (Chapter 13.1) becomes an important topic in energy engineering. Lawrence Livermore Labs, for instance, has proposed glass micro-balloons, originally developed for housing minute amounts of tritium-deuterium alloy for laser fusion experiments. When heated, the glass becomes porous and hydrogen under pressure can fill the balloons. Cooled, the gas is trapped.

Clathrate is one of nature's way of storing methane, even though no one is proposing it as a practical method for transporting the gas.

Methane clathrate frequently consists of cages of 46 H_2O molecules trapping 8 CH_4 molecules.

1. What is the gravimetric metric concentration, GC, of methane in the clathrate? Gravimetric concentration is the ratio of the mass of the stored gas to the total mass of gas plus container.

Consider a hermetic container with 1 m^3 internal volume, filled completely with the clathrate described which has a density of 900 kg/m^3. Assume that by raising the temperature to 298 K, the material will melt and methane will evolve. Assume also (although this is not true) that methane is insoluble in water.

2. What is the pressure of the methane in the container?

1.0.29 A Radioisotope Thermal Generator (RTG) is to deliver 500 W of dc power to a load at 30 V. The generator efficiency (the ratio of the dc power out to the heat power in) is 12.6%. The thermoelectric generator takes heat in at 1200 K and rejects it at 450 K. The heat source is plutonium-241. This radioactive isotope has a half-life of 13.2 years and decays emitting α and β^- particles. These particles have an aggregate energy of 5.165 MeV.

Only 85% of the power generated by the plutonium finds its way to the thermoelectric generator. The rest is lost.

How many kilograms of plutonium are required? Note that radioactive substances decay at a rate proportional to the amount of undecayed substance and to a constant decay rate, λ:

$$\frac{dN}{dt} = -\lambda N.$$

1.0.30 In the USA we burn (very roughly) an average of 150 GW of coal, 40 GW of oil and 70 GW of natural gas.

Assume that

Coal is (say) $C_{20}H_{44}$ and that it yield 40 MJ per kg,
Oil is (say) $C_{10}H_{22}$ and yields 45 MJ per kg.
Natural gas is CH_4 and yields 55 MJ per kg.

How many kg of carbon are released daily by the combustion of coal alone? (Clearly, after you have handled coal, the other two fuels can be handled the same way. But, for the sake of time, don't do it.)

Section Two

Energy perspectives

Chapter 2.1

2.1

Energy perspectives

2.1.1 Current penetration of renewable energy technologies in the marketplace

The penetration of renewable energy into the energy system of human settlements on Earth is from one point of view nearly 100%. The energy system seen by the inhabitants of the Earth is dominated by the environmental heat associated with the greenhouse effect, which captures solar energy and stores it within a surface-near sheet of topsoil and atmosphere around the Earth. Only 0.02% of this energy system is currently managed by human society, as illustrated in Fig. 2.1-1. Within this economically managed part of the energy sector, renewable energy sources currently provide about 25% of the energy supplied. As the figure indicates, a large part of this renewable energy is in the form of biomass energy, either in food crops or in managed forestry providing wood for industrial purposes or for incineration (firewood used for heat and cooking in poor countries or for mood-setting fireplaces in affluent countries, or residue and waste burning in combined power and heat plants or incinerators). The additionally exploited sources of renewable energy include hydro, wind and solar. Hydropower is a substantial source, but its use is no longer growing due to environmental limits identified in many locations with potential hydro resources. Passive solar heating is a key feature of building design throughout the world, but active solar heat or power panels are still at a very minute level of penetration. Also, wind has both a passive and an active role. Passive use of wind energy for ventilation of buildings plays a significant role, and active power production by wind turbines is today a rapidly growing energy technology in many parts of the world. The highest penetration reaching nearly 20% of total electricity provided is found in Denmark, the country pioneering modern wind technology. Further renewable energy technologies, so far with small global penetration, include biofuels such as biogas and geothermal power and heat. As indicated in Fig. 2.1-1, the dominant energy sources are still fossil fuels, despite the fact that they are depletable and a cause of frequent national conflicts, due to the mismatch between their particular geographical availability and demand patterns.

From a business point of view, the total renewable energy flows, including free environmental heat, are, of course, not quite as interesting as the energy that can be traded in a market. Current renewable energy markets comprise both consumer markets and markets driven by government demonstration programmes and market-stimulating subsidy schemes. The reason for the initial support is partly industrial policy, aimed at getting new industry areas started, and partly a question of compensation for market distortions created by the fact that conventional energy industries are not fully paying for the negative environmental impacts caused by their products. This is a complex issue, partly because of the difficulty in exact determination of external costs and partly because most countries already levy taxation on energy products that may in part be contributing towards paying for the environmental damage, but often is just a government revenue not specifically used to offset the negative effects associated with using fossil or nuclear fuels .

The current penetration of active uses of renewable energy in national energy systems is growing, and Figures 2.1-2–2.1-14 show the values for the year 2000, which may serve as a reference year for assessing newer data. In cases where the growth rate is particular high, its annual value is mentioned in the caption to the figure showing the national distribution of markets.

Renewable Energy Focus Handbook 2009; ISBN: 9780123747051

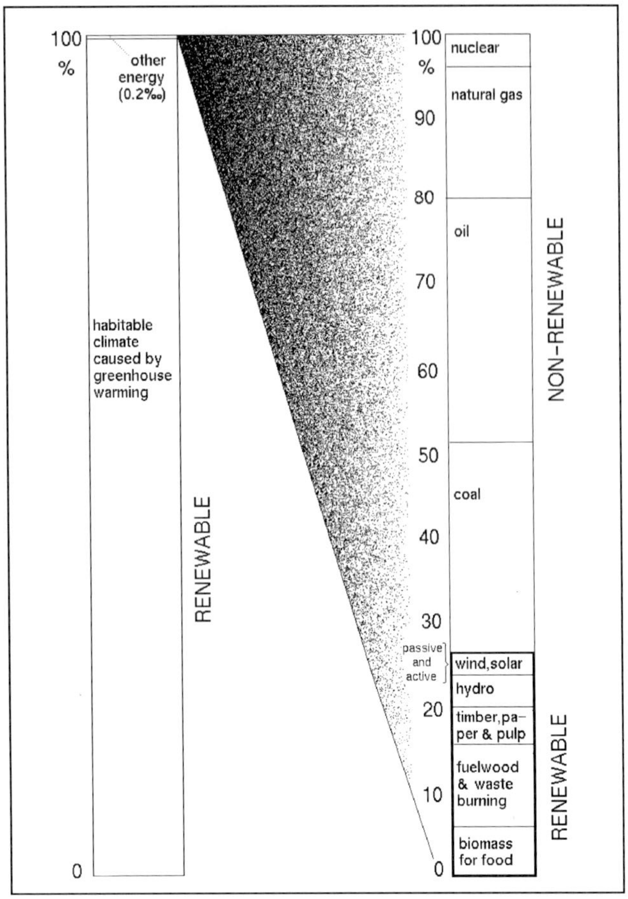

Figure 2.1-1 Renewable energy in the global energy system (Sørensen, 1992c).

As Table 2.1-1 shows, at a global average of 222 W/cap., the traditional use of biomass for combustion is still the dominating use of renewable energy, although it takes more efficient forms in many industrialised countries. Only slightly less (146 W/cap.) is the use of food energy in biomass of animal or vegetable origin (the nutrient value of food being, in any case, more than that of the energy it provides). Next comes hydropower (50 W/cap.) and then geothermal power, which only in part can be classified as renewable (as many steam reservoirs are exploited at a rate that will exhaust the reservoir over periods of decades). At the level of 1 W/cap., i.e. two orders of magnitude under the energy in food intake, one finds biomass waste (used for power or heat), biogas, liquid biofuels (used in the transportation sector), wind power and geothermal heat (used for district heating). At the bottom comes solar heat, tidal power and solar power, with the latter below 0.01 W/cap. However, the fastest growing markets are those of wind and solar power, with both currently adding 35% of installed power each year.

The market characteristics of the various renewable energy forms exhibit differences linked to the nature of each source. For food energy, the price is influenced by variations in production due to climatic variations, the choices made in regard to area use, livestock holdings, fish quotas and the competitive behaviour of the food processing and marketing industry. Yet the bulk prices of different commodities seem remarkably consistent with their energy content, varying only between some 70 US or euro-cents per kWh (heat value) and 200 c/kWh. Translating OECD data (OECD, 2002) to energy units, the current wholesale price of cereals such as rice or wheat is around 70 c/kWh, while the wholesale price of typical meat and diary products is about 100 c/kWh. Only specialised gourmet products obtain higher prices in the marketplace. Consumer retail prices are typically five times higher than the bulk prices just quoted. This is more than 30 times the current consumer price of a kWh of electricity produced from fossil fuels.

Wholesale market prices for biomass waste and fuelwood range from about 1 c/kWh (of "burning value", i.e. energy of combustion) in India (FAO-Asia, 2003) to 2 c/kWh in industrialised countries (e.g. straw, wood chips 1.6 c/kWh and wood pellets 1.9 c/kWh; Danish Energy Agency, 1996; Alakangas *et al.*, 2002). For comparison, the cost of coal before considering externalities is 0.5 c/kWh (Danish Energy Agency, 2002). The production cost of biogas is 3.6–7 c/kWh (Danish Energy Agency, 1992), while that of wind power is 3–7 c/kWh (depending on wind conditions) and that of photovoltaic solar power is 40–130 c/kWh (IEA-PVPS, 2002). The photovoltaic market enjoys substantial public start-up subsidies (often in the form of subsidising customer investments or offering attractive buy-back rates for excess solar power). This is the case in countries such as Germany and Japan, while in Switzerland, the market has largely been created by industries buying photovoltaic panels for reasons of aesthetics or image greening.

Hydropower costs 1–5 c/kWh, while coal- and gas-based power costs about 5 c/kWh to produce (Danish Energy Agency, 2002). To this comes distribution costs from centralised production units to the customers and, in many countries, taxes and environmental externality payments, leading to customer prices in excess of 14 c/kWh. As a result, wind power, being exempt from pollution and CO_2 fees, and biomass-based power are in many countries sold at prices very similar to that of fossil-based power. Also geothermal power is usually competitive with other forms of electricity, while the viability of geothermal heat depends on local costs of district heating distribution.

Current oil production costs vary from well under 1 c/kWh at some Middle East wells to near 2 c/kWh from off-shore facilities in the North Sea. The bulk sales price (in February 2003 – around 2 c/kWh) is not strongly coupled to production prices, but is determined by market and political considerations. Some countries are

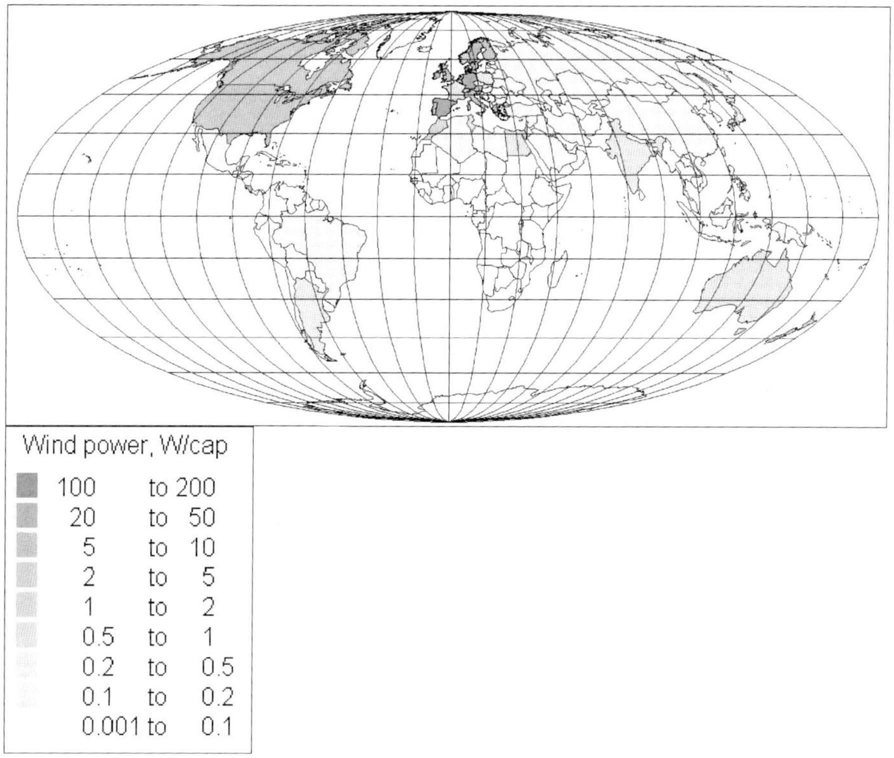

Figure 2.1-2 Wind power production. National average values for the year 2000 are shown, based upon BTM (2001) and an average capacity factor of 0.3. The world average for the year 2000 is 0.92 W/cap. The growth in cumulated installed capacity from 2000 to 2001 was 35% (BTM, 2002). Some observers expect the growth to slow during the following years, for economic and political reasons, but then to resume growth (Windpower Monthly, 2003).

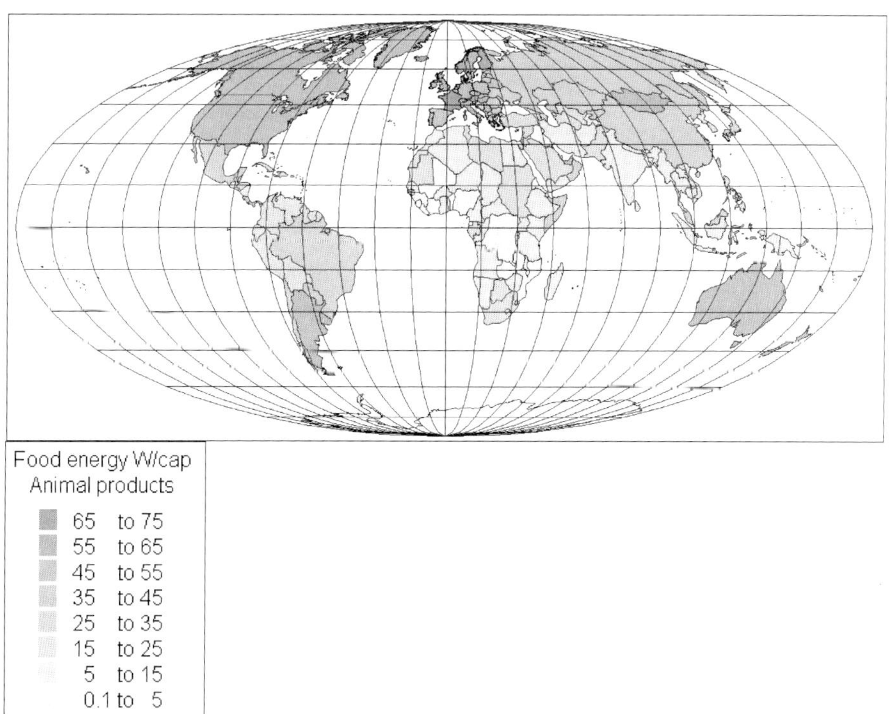

Figure 2.1-3 Biomass energy contained in human time-averaged food intake of animal products. National average values for the year 2000 are shown (FAO, 2003). The world average animal-based food intake for the year 2000 is 22.2 W/cap.

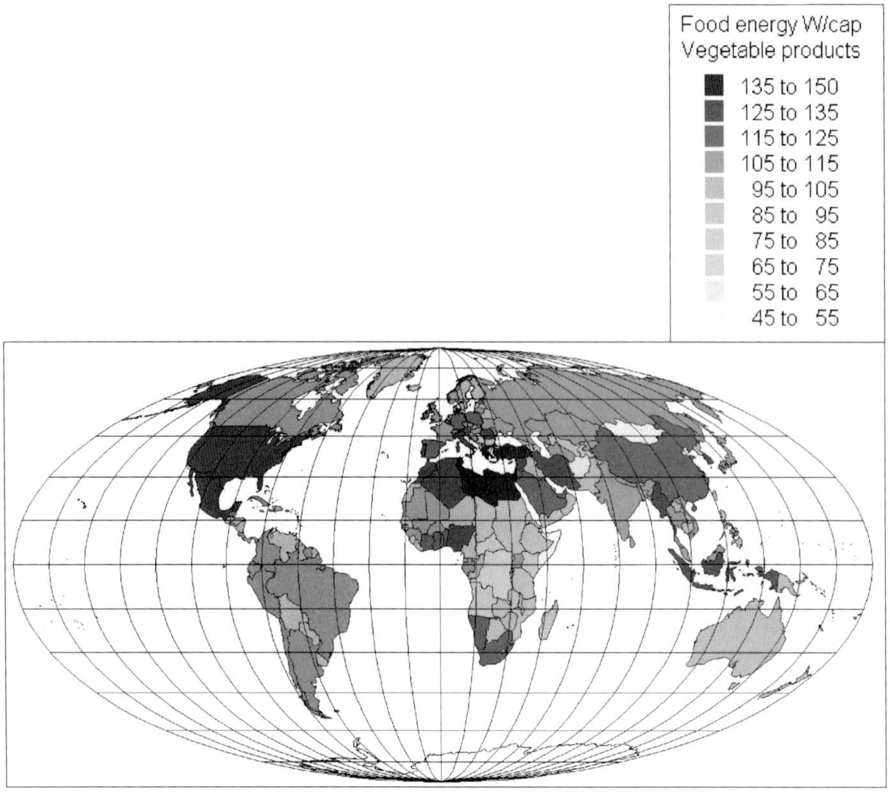

Figure 2.1-4 Biomass energy contained in human time-averaged food intake of vegetable products. National average values for the year 2000 are shown (FAO, 2003). The world average vegetable food intake for the year 2000 is 113.7 W/cap.

willing to wage war against oil-producing countries in order to control prices. Refined products such as gasoline are currently sold at prices around 4 c/kWh, with diesel fuel slightly lower, plus taxes and environmental fees where they apply (Danish Energy Agency, 2002; IEA 2002). Liquid biofuels have production costs of 3–7 c/kWh (ethanol from sugar cane the lowest, ethanol from sugar beet the highest, methanol from woody biomass at 4–5 c/kWh). Hydrogen from woody biomass is about 3 c/kWh (Turkenburg *et al.*, 2000). Natural gas market prices are currently 10% higher than those of oil (IEA, 2002).

Because of the cost of often advanced equipment, it is clear that prices of renewable energy only in particular cases can match those of fossil fuels, even with expected progress in technology and production facilities. The case for increasing the role of renewable energy sources is therefore linked to uncertainty of future fossil fuel prices (for political and resource depletion reasons) and increased awareness of the indirect costs of pollution caused by fossil and nuclear fuels, including in the fossil case emissions of substances contributing to excess greenhouse warming.

The source data for the Figs. 2.1-2–2.1-14 are given in tabular form in Table 2.1-1. It should be kept in mind that several of the numbers involve estimates and modelling, as direct energy production is not always monitored.

2.1.2 The energy scene – its history and present state

Taking now a scientific point of view, an issue more essential than the placement of renewable energy in the marketplace is its place within the physical universe. This view will be developed in the following, as a prerequisite for estimating the amounts of energy that can be extracted for use by human society, at a rate that qualifies the process as renewable.

The speed of the Earth in its orbit around the Sun is about 3×10^4 m s^{-1}, corresponding to a kinetic energy of some 2.7×10^{33} J. The Earth further rotates around its axis with an angular velocity of about 7.3×10^{-5} rad s^{-1}, furnishing an additional kinetic energy of some 2.2×10^{29} J. The work required in order to pull the Earth infinitely far away from the Sun, against the gravitational attraction, is about 5.3×10^{33} J, and the corresponding work required to separate the Earth from its Moon is of the order of 8×10^{28} J.

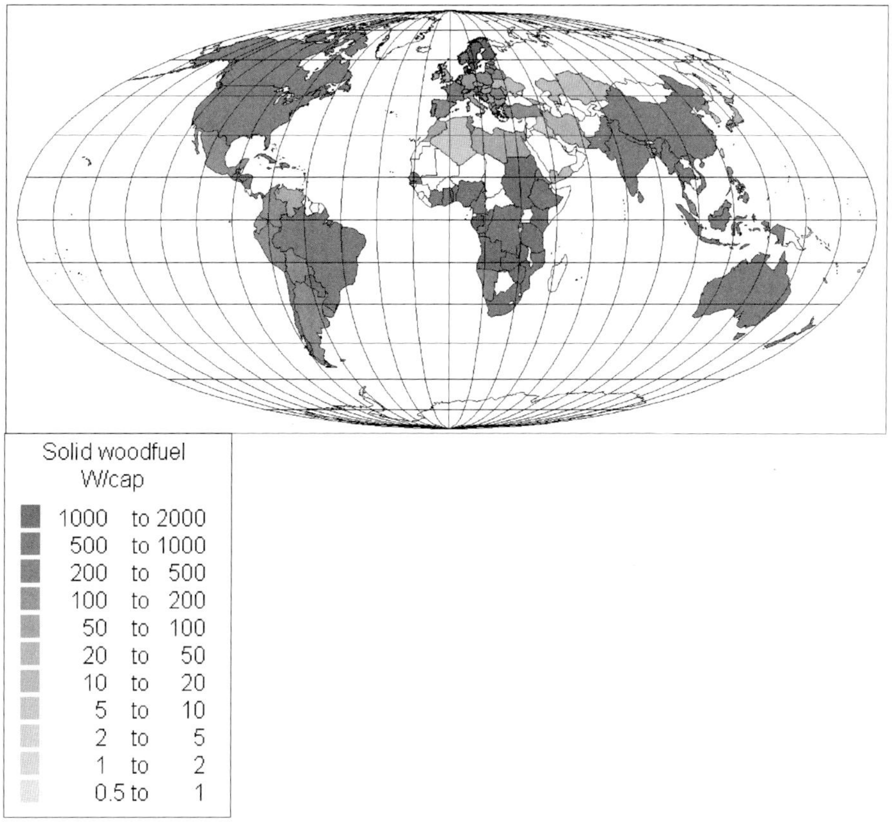

Figure 2.1-5 Biomass energy contained in woodfuel. National average values for the year 2000 are shown (OECD/IEA, 2002a, 2002b). The world average woodfuel use in the year 2000, implied by the data, is 221.9 W/cap. No woodfuel use is reported for Russia. Some other countries are not covered by the data source.

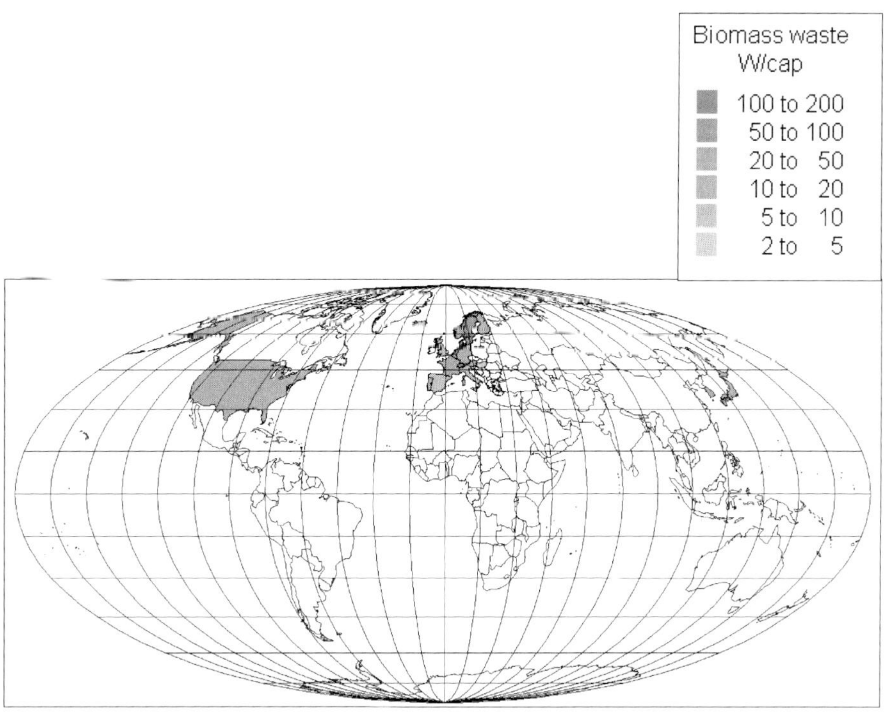

Figure 2.1-6 Energy in biomass waste (refuse) utilised for power or heat production. National average values for the year 2000 are shown, based upon OECD/IEA (2002a, 2002b). The world average for the year 2000 is 3.7 W/cap.

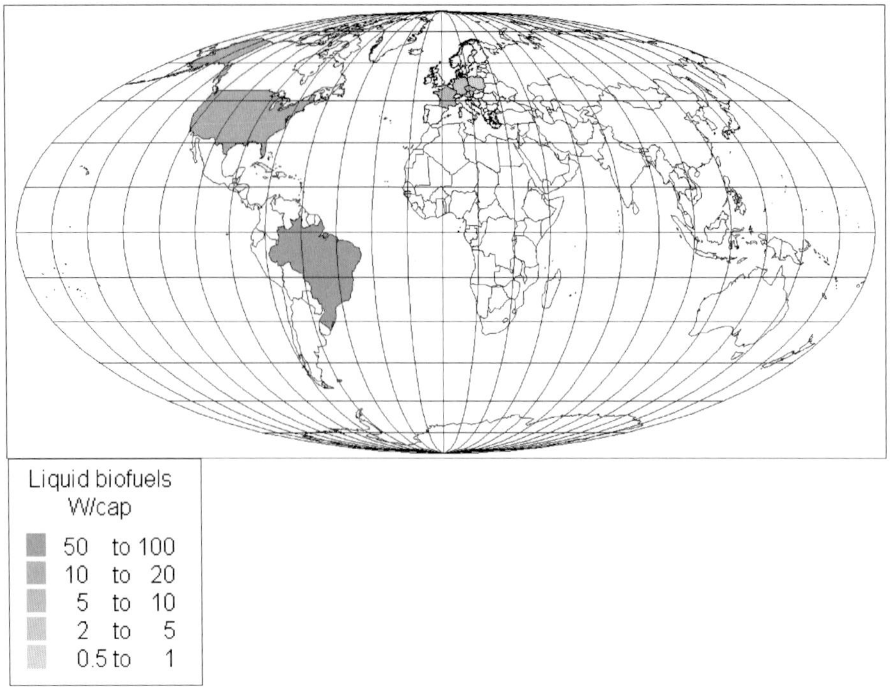

Figure 2.1-7 Energy in liquid biofuels (presently ethanol and biodiesel). National average values for the year 2000 are shown, based upon EC-ATLAS (2003) and OECD/IEA (2002a, 2002b). The world average for the year 2000 is 2.3 W/cap.

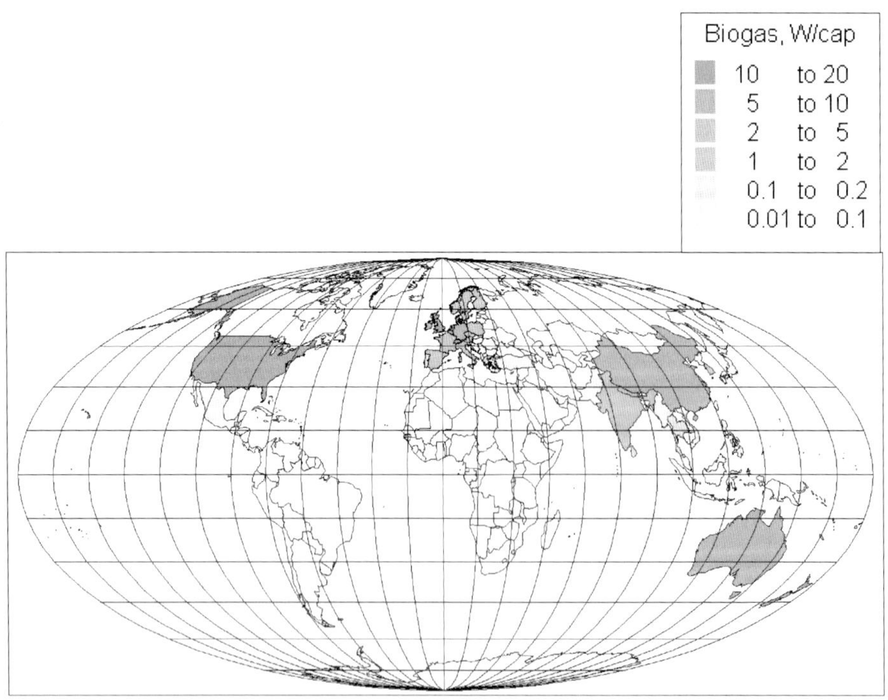

Figure 2.1-8 Energy in biogas. National average values for the year 2000 are shown, based upon converting the number of units (FAO-Asia, 2003), assuming on average for each family unit a biogas production of half that of an optimally working unit fed manure from the equivalent of 2.5 cows and producing 1736 W of biogas. Additional data are from OECD/IEA (2002a, 2002b). The world average for the year 2000 is 2.8 W/cap.

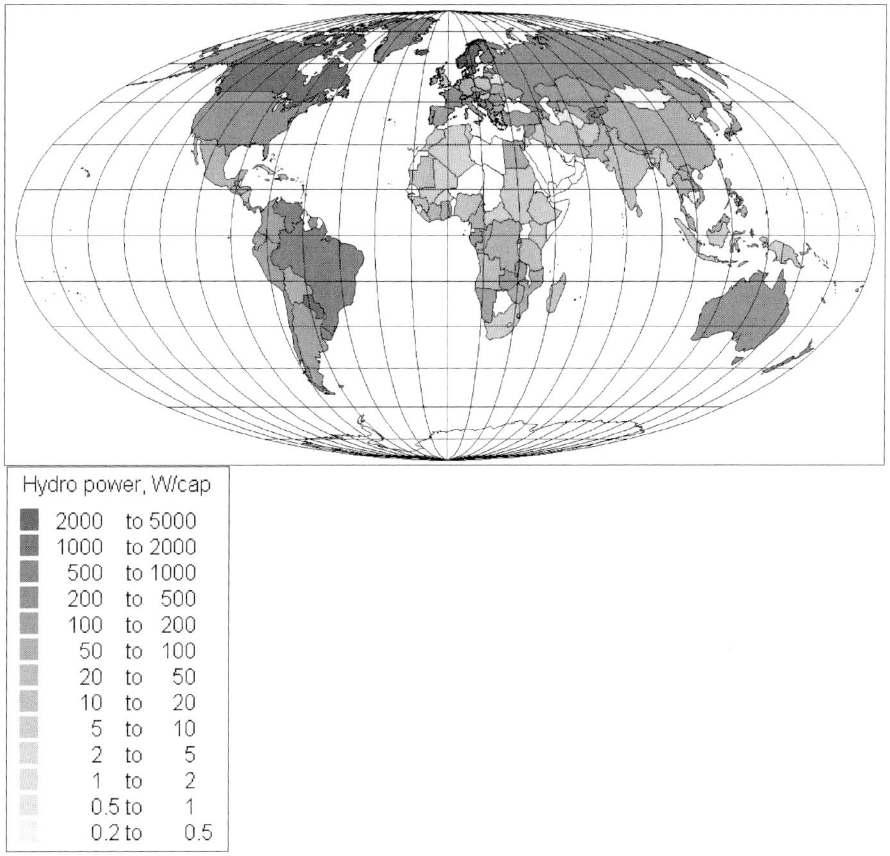

Figure 2.1-9 Hydropower. National average values for the year 2000 are shown, based upon OECD/IEA (2002a, 2002b) and installed power data from Aqua-Media (1997) for countries not covered by IEA, using estimated capacity factors of 0.2 to 0.4. The world average for the year 2000 is 50 W/cap.

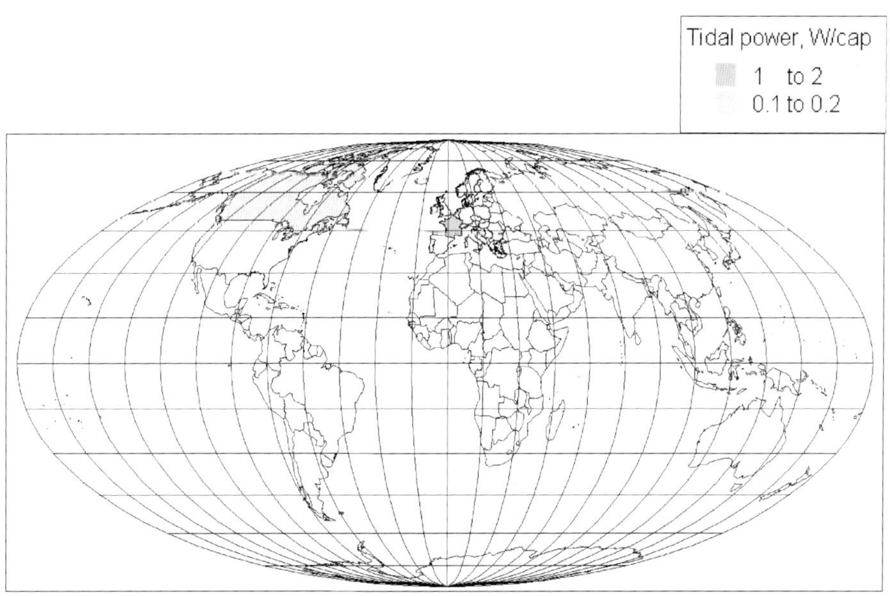

Figure 2.1-10 Tidal power. National average values for the year 2000 are shown, based upon OECD/IEA (2002a, 2002b). The world average for the year 2000 is 0.01 W/cap.

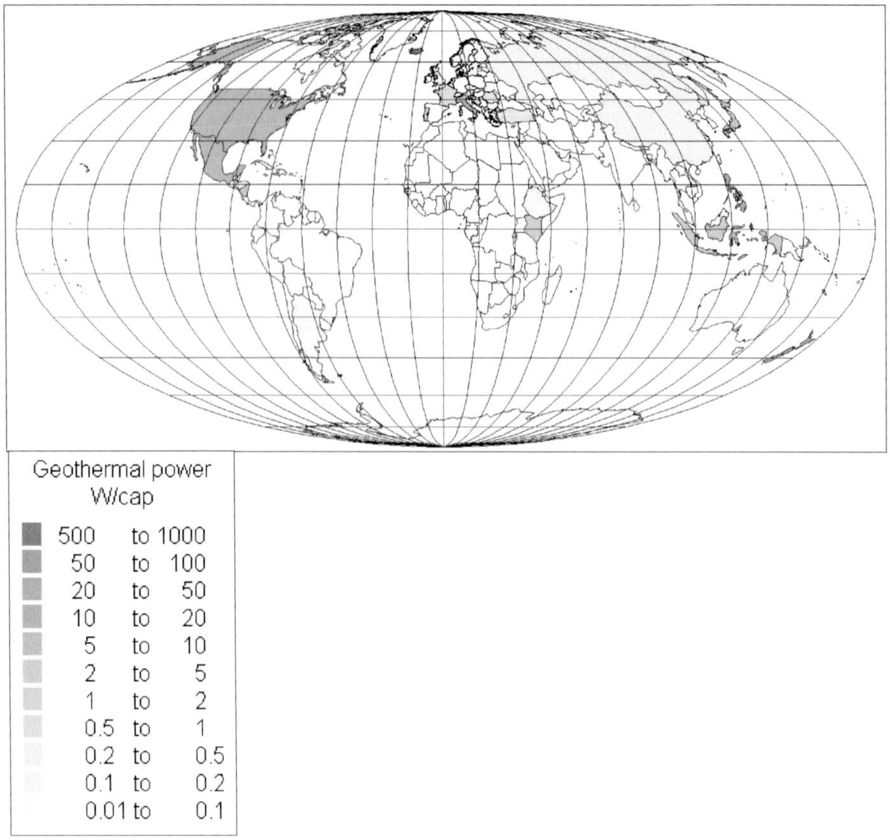

Figure 2.1-11 Geothermal power. National average values for the year 2000 are shown, based upon either 10% of heat input given in OECD/IEA (2002a, 2002b) or 60% of installed power from Barbier (1999). The world average for the year 2000 is 9.3 W/cap.

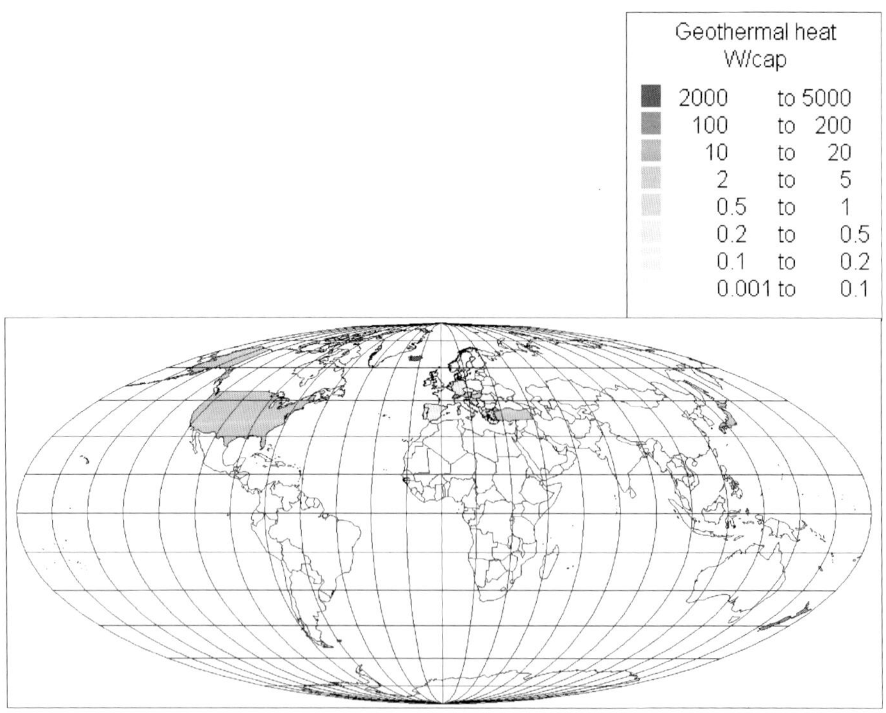

Figure 2.1-12 Geothermal heat (mainly district heating). National average values for the year 2000 are shown, based upon OECD/IEA (2002a, 2002b). The world average for the year 2000 is 0.5 W/cap.

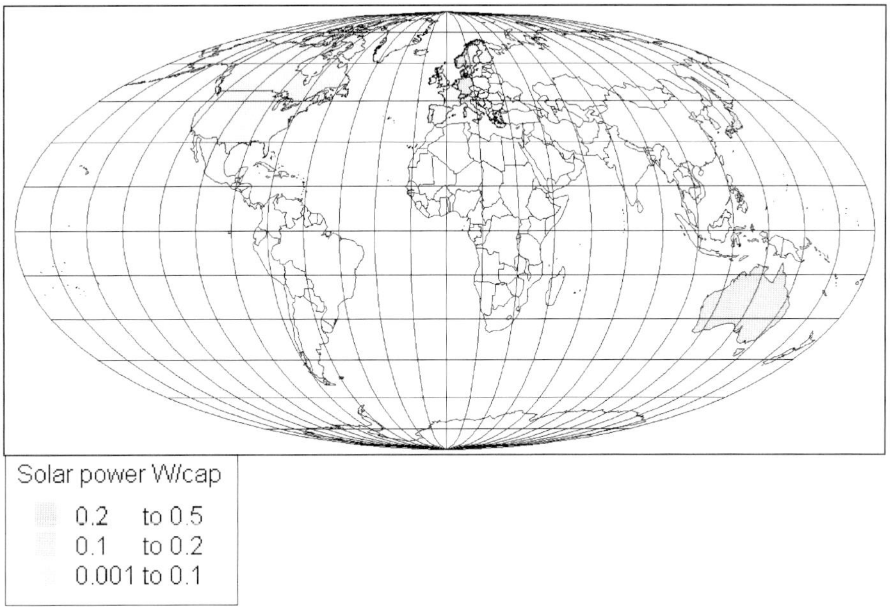

Figure 2.1-13 Solar power. National average values for the year 2000 are shown, based upon OECD/IEA (2002b) or an average energy production equal to 10% of the installed capacity given in IEA-PVPS (2001). The world average for the year 2000 is 0.007 W/cap. The growth rate from 2000 to 2001 was 35% (IEA-PVPS, 2002).

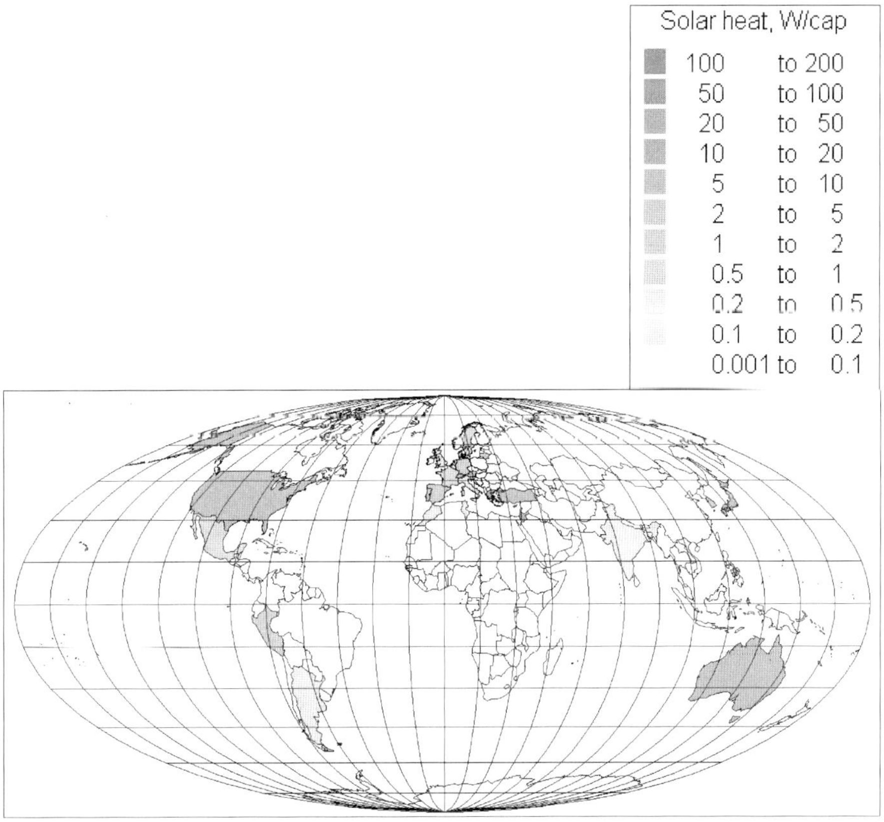

Figure 2.1-14 Solar heat. National average values for the year 2000 are shown, based upon IEA (2002). Both building-integrated and central district heating thermal systems are included. The world average for the year 2000 is 0.11 W/cap.

Table 2.1-1 Year 2000 renewable energy use (W/cap.), cf. Figs. 2.1-2–2.1-14

COUNTRY W/cap. or number	2000 Pop./1000	HYDRO	GEOTH POWR	GEOTH HEAT	PV	SOLAR HEAT	TIDAL	WIND	BIO-RES	BIO-SOLID	BIO-LIQ	BIO-GAS	ANIM. FOOD	VEG. FOOD
Afghanistan	17270	3.6	0	0	0	0	0	0	0	0	0	0	11.67	62.86
Albania	3400	164.13	0	0	0	0	0	0	0	23.45	0	0	37.68	101.02
Algeria	30400	1.8	0	0	0	0	0	0.04	0	3.5	0	0	14.77	127.85
Andorra	70	0	0	0	0	0	0	0	0	0	0	0	44.41	121.07
Angola	13100	8.11	0	0	0	0	0	0	0	572.03	0	0	7.65	84.5
Anguilla	10	0	0	0	0	0	0	0	0	0	0	0	12	65
Antarctica	0	0	0	0	0	0	0	0	0	0	0	0	0	0
Antigua Barb.	70	0	0	0	0	0	0	0	0	0	0	0	38.69	77.34
Argentina	37000	89.05	0.01	0	0	0.22	0	0.13	0	105.93	0	0	48.43	105.62
Armenia	3800	38.46	0	0	0	0	0	0	0	0	0	0	15.01	79.08
Aruba	70	0	0	0	0	0	0	0	0	0	0	0	12	96.85
Australia	19160	100.06	0	0	0.15	6.8	0	0.47	0	349.36	0	9.57	50.8	102.95
Austria	8110	591.74	0.15	0.82	0.05	7.7	0	2.6	9.34	453.31	2.13	4.75	59.23	122.71
Azerbaijan	8000	21.59	0	0	0	0	0	0.05	0	0	0	0	17.53	101.99
Azores Port.	260	0	0	0	0	0	0	0	0	0	0	0	48.43	121.07
Bahamas	320	0	0	0	0	0	0	0	0	0	0	0	34.43	83.87
Bahrain	700	0	0	0	0	0	0	0	0	0	0	0	38.74	121.07
Bangladesh	131100	0.81	0	0	0	0	0	0	0	77.02	0	0	3.24	98.55
Barbados	260	0	0	0	0	0	0	0	0	0	0	0	33.95	112.4
Belarus	10000	0.2	0	0	0	0	0	0.04	0	131.54	0	0	38.79	101.74
Belgium	10250	5.06	0.01	0.13	0	0.26	0	0.6	15.94	33.96	0	3.89	54.29	124.94
Belize	250	0	0	0	0	0	0	0	0	0	0	0	29.06	110.8
Benin	6300	0	0	0	0	0	0	0	0	375.39	0	0	4.84	119.03
Bermuda	70	0	0	0	0	0	0	0	0	0	0	0	37	110.07

Bhutan	1840	64	0	0	0	0	0	0	0	0	0	0	4.84	96.85
Bolivia	8300	27.21	0	0	0	0	0	0	0	115.26	0	0	17.19	90.27
Bosnia Herzeg	4000	146.15	0.33	0	0	0	0	0	0	59.79	0	0	17.53	111.33
Botswana	1480	0	0	0	0	0	0	0	0	0	0	0	18.11	91.09
Brazil	170400	204.29	0	0	0	0	0	0.04	0	324.52	52	0	29.78	114.77
Br. Virgin Isl.	30	0	0	0	0	0	0	0	0	0	0	0	24.21	96.85
Brunei Daruss.	300	0	0	0	0	0	0	0	0	88.58	0	0	24.65	112.49
Bulgaria	8200	37.27	0	0	0	0	0	0	0	93.98	0	0	33.41	86
Burkina Faso	11100	0.7	0	0	0	0	0	0	0	0	0	0	5.62	105.42
Burundi	6030	0	0	0	0	0	0	0	0	0	0	0	1.74	75.98
Cambodia	6820	0.03	0	0	0	0	0	0	0	0	0	0	8.81	91.43
Cameroon	14900	26.75	0	0	0	0	0	0	0	444.07	0	0	6.44	102.76
Canada	30750	1331.4	0	0	0.02	0	0.13	1.36	0	484.27	0	0	45.42	108.28
Cape Verde	440	0	0	0	0	0	0	0.1	0	0	0	0	22.37	136.37
Cayman Isl.	30	0	0	0	0	0	0	0	0	0	0	0	24.21	96.85
Cen. African R	2790	1.7	0	0	0	0	0	0	0	0	0	0	9.49	84.7
Chad	6670	0	0	0	0	0	0	0	0	0	0	0	6.92	92.15
Chile	15200	143.35	0	0	0	0	0	0	0	369.75	0	0	30.22	109.35
China	1262500	20.13	0.02	0	0	0	0	0.08	0	225.66	0	5	28.23	118.45
Colombia	42300	86.69	0	0	0	0	0	0	0	165.22	0	0	20.63	105.18
Comoros	3800	171.33	29.37	0	0	5.59	0	0	0	87.41	0	0	4.46	80.44
Congo	3000	13.29	0	0	0	0	0	0	0	261.3	0	0	6.39	101.26
Cook Islands	20	0	0	0	0	0	0	0	0	0	0	0	24.21	96.85
Costa Rica	2500	74	0	0	0	0	0	6.1	0	0	0	0	25.18	109.64
Croatia	4400	154	0.3	0	0	0	0	0	0	111.73	0	0	24.94	95.3
Cuba	11200	1.19	0	0	0	0	0	0	0	347.58	0	0	17	107.17

Continued

Table 2.1-1 Year 2000 renewable energy use (W/cap.), cf. Figs. 2.1-2–2.1-14—cont'd

COUNTRY W/cap. or number	2000 Pop./1000	HYDRO	GEOTH POWR	GEOTH HEAT	PV	SOLAR HEAT	TIDAL	WIND	BIO-RES	BIO-SOLID	BIO-LIQ	BIO-GAS	ANIM. FOOD	VEG. FOOD
Cyprus	800	0.17	0	0	0	59.79	0	0	0	16.61	0	0	46.15	111.62
Czech R	10270	19.54	0	0	0	0	0	0.05	0	41.14	5.95	4.66	40.53	109.78
Denmark	5340	0.5	0.02	0	0.03	1.99	0	131.5	145.55	220.94	8.46	17.17	63.24	101.21
Djibouti	70	0	0	0	0	0	0	0	0	0	0	0	12.83	86.44
Dominica	80	21	0	0	0	0	0	0	0	0	0	0	33.41	111.53
Dominican R	8400	11.07	0	0	0	0	0	0	0	215.11	0	0	16.46	96.17
Congo/Zaire	50900	12.27	0	0	0	0	0	0	0	355.26	0	0	2.28	71.04
Ecuador	12600	68.54	0	0	0	0	0	0	0	73.81	0	0	20.97	109.44
Egypt	64000	25.33	0	0	0	0	0	0.32	0	27.61	0	0	12.4	149.64
El Salvador	6300	21.09	14.34	0	0	0	0	0	0	291.04	0	0	14.67	106.54
Equat. Guinea	430	0	0	0	0	0	0	0	0	0	0	0	4.84	96.85
Eritrea	4100	0	0	0	0	0	0	0	0	165.27	0	0	4.99	75.64
Estonia	1400	0	0	0	0	0	0	0	0	474.51	0	0	42.52	120.97
Ethiopia	64300	2.89	0.04	0	0	0	0	0.03	0	359.95	0	0	5.08	92.88
Falkland Isl.	0	0	0	0	0	0	0	0	0	0	0	0	24.21	96.85
Fiji	730	23	0	0	0	0	0	0	0	0	0	0	26.97	111.57
Finland	5180	323.44	0	0	0.05	0	0	2.3	11.54	1659.77	0	4.62	55.79	100.48
France	60430	126.58	0.27	0	0.02	0.55	1.08	0.3	39.82	202.03	5.78	3.83	65.18	108.72
Fr. Guiana	130	0	0	0	0	0	0	0	0	0	0	0	24.21	96.85
Fr. Polynesia	210	0	0	0	0	0	0	0	0	0	0	0	40	98.16
Gabon	1200	66.43	0	0	0	0	0	0	0	1018.62	0	0	17.14	107.02
Gambia	690	0	0	0	0	0	0	0	0	0	0	0	5.67	114.14
Georgia	5000	132.86	0	0	0	0	0	0	0	18.6	0	0	19.23	97.63
Germany	82170	30.22	0	0.16	0.14	1.47	0	22.3	28.47	79.49	2.64	9.01	50.12	116.95

Ghana	19300	39.24	0	0	0	0	0	0	0	366.24	0	0	5.81	124.84
Gibraltar	30	0	0	0	0	0	0	0	0	0	0	0	4.84	121.07
Greece	10560	40.01	0	0.25	0	12.46	0	7.8	0	118.77	0	0.13	41.02	138.4
Greenland	50	200	0	0	0	0	0	0	0	0	0	0	48.43	96.85
Grenada	90	0	0	0	0	0	0	0	0	0	0	0	32.64	101.21
Guadeloupe	440	0	0	0	0	0	0	0	0	0	0	0	24.21	96.85
Guam	120	0	0	0	0	0	0	0	0	0	0	0	14.53	96.85
Guatemala	11400	23.31	0	0	0	0	0	0	0	454.54	0	0	10.07	95.06
Guinea	4050	6	0	0	0	0	0	0	0	0	0	0	3.87	110.07
Guinea Bissau	920	0	0	0	0	0	0	0	0	0	0	0	7.85	105.18
Guyana	870	0	0	0	0	0	0	0	0	0	0	0	20.29	104.75
Haiti	8000	3.32	0	0	0	0	0	0	0	252.44	0	0	6.44	93.12
Honduras	6400	39.44	0	0	0	0	0	0	0	276.11	0	0	16.66	99.32
HongKong	6800	0	0	0	0	0	0	0	0	9.77	0	0	38.74	96.85
Hungary	10020	1.99	0	0.66	0	0	0	0	2.65	47.21	0	0	53.9	113.56
Iceland	280	2595.6	623.51	2984.7	0	0	0	0	4.75	0	0	0	67.65	94.19
India	1015900	8.37	0	0	0	0.17	0	0.4	0	263.64	0	2.3	9.39	108.18
Indonesia	210400	4.93	1 44	0	0	0	0	0.02	0	299.95	0	0	5.67	134.87
IranIR	63700	6.67	0	0	0	0	0	0.06	0	16.48	0	0	13.03	128.04
Iraq	23300	2.85	0	0	0	0	0	0.08	0	1.71	0	0	4.26	102.13
Ireland	3790	25.59	0	0	0	0	0	9.7	0	47.68	0	9.82	54.38	120.58
Israel	6200	0.2	0	0	0.007	127.72	0	0.08	0	0	0	0	31.96	140.53
Italy	57730	87.5	6.7	0	0.03	0.25	0	2.2	7.69	36.71	0	2.97	45.28	132.01
Ivory Coast	16000	12.46	0	0	0	0	0	0	0	350.43	0	0	4.75	120.68
Jamaica	2600	5.11	0	0	0	0	0	0	0	245.29	0	0	18.98	111.43
Japan	126920	78.55	3.01	2.33	0.25	8.41	0	0.34	10.56	47.64	0	0	27.55	106.2
Jordan	4900	0.3	0	0	0	17.62	0	0.08	0	0	0	0	15.79	117.34

Continued

Table 2.1-1 Year 2000 renewable energy use (W/cap.), cf. Figs. 2.1-2–2.1-14—cont'd

COUNTRY W/cap. or number	2000 Pop./1000	HYDRO	GEOTH POWR	GEOTH HEAT	PV	SOLAR HEAT	TIDAL	WIND	BIO-RES	BIO-SOLID	BIO-LIQ	BIO-GAS	ANIM. FOOD	VEG. FOOD
Kazakhstan	14900	57.96	0	0	0	0	0	0.04	0	6.24	0	0	31.23	113.61
Kenya	30100	4.86	1.63	0	0	0	0	0	0	519.98	0	0	11.33	83.78
Kiribati	60	0	0	0	0	0	0	0	0	0	0	0	18.84	124.36
Korea	47280	9.7	0	0	0.008	1.18	0	0.06	41.84	4.83	0	1.1	22.47	127.31
Korea DPR	22300	109.03	0	0	0	0	0	0	0	59.58	0	0	5.96	99.81
Kuwait	2000	0	0	0	0	0	0	0	0	0	0	0	34.53	117.14
Kyrgyzstan	4900	319.96	0	0	0	0	0	0.02	0	0	0	0	26.73	112.3
Laos	2960	13	0	0	0	0	0	0	0	0	0	0	7.7	102.08
Latvia	2400	132.86	0	0	0	0	0	0	0	548.06	0	0	33.32	104.94
Lebanon	4300	12.36	0	0	0	2.16	0	0	0	40.17	0	0	19.56	133.22
Lesotho	1980	0.4	0	0	0	0	0	0	0	0	0	0	4.84	106.54
Liberia	2510	7	0	0	0	0	0	0	0	0	0	0	3.24	97.29
Libya ArabJam	5300	0	0	0	0	0	0	0.04	0	35.1	0	0	17.82	142.23
Liechtenstein	30	0	0	0	0	0	0	0	0	0	0	0	48.43	106.59
Lithuania	3700	10.77	0	0	0	3.59	0	0	0	226.23	0	0	34.09	113.12
Luxembourg	440	30.2	0	0	0	0	0	4.09	81.53	48.31	0	3.02	54.29	124.94
Macedonia FY	2000	66.43	1.99	0	0	0	0	0	0	139.51	0	0	24.21	121.31
Madagascar	8460	5	0	0	0	0	0	0	0	0	0	0	9.59	87.65
Malawi	9860	0	0	0	0	0	0	0	0	0	0	0	2.37	103.2
Malaysia	23300	34.21	0	0	0	0	0	0	0	144.27	0	0	27.41	113.95
Maldives	270	0	0	0	0	0	0	0	0	0	0	0	31.82	93.7
Mali	8070	1.1	0	0	0	0	0	0	0	0	0	0	4.84	106.25
Malta	400	0	0	0	0	0	0	0	0	0	0	0	44.21	127.36
Marshall Isl.	50	0	0	0	0	0	0	0	0	0	0	0	24.21	96.85

Martinique	450	0	0	0	0	0	0	0	0	0	0	0	24.21	96.85
Mauritania	2250	5	0	0	0	0	0	0	0	0	0	0	15.84	107.02
Mauritius	1220	11	0	0	0	0	0	0	0	0	0	0	20.68	123.87
Mexico	97220	38.94	6.94	0	0.014	0.59	0	0.02	0	109.81	0	0.08	28.23	125.04
Micronesia	0	0	0	0	0	0	0	0	0	0	0	0	24.21	96.85
Moldova Rep	4300	3.09	0	0	0	0	0	0	0	18.54	0	0	18.98	114.87
Monaco	30	0	0	0	0	0	0	0	0	0	0	0	45.28	108.72
Mongolia	2310	0.08	0	0	0	0	0	0	0	0	0	0	45.67	50.27
Morocco	28700	2.78	0	0	0	0.28	0	0.56	0	20.37	0	0	10.36	133.17
Mozambique	17700	45.04	0	0	0	0	0	0	0	496.18	0	0	2.32	90.99
Myanmar	47700	4.46	0	0	0	0	0	0	0	255.7	0	0.03	6	131.62
Namibia	1800	88.58	0	0	0	0	0	0	0	125.48	0	0	12.88	115.4
Nauru	10	0	0	0	0	0	0	0	0	0	0	0	19.37	96.85
Nepal	23000	8.09	0	0	0	0	0	0	0	388.77	0	3.7	7.75	110.22
Netherlands	15920	1	0	0	0.08	1.34	0	8.9	41.81	28.38	0	11.02	57.38	102.18
New Caledonia	170	94	0	0	0	0	0	0	0	0	0	0	31.72	101.02
New Zealand	3830	734.39	82.56	164.43	0	0	0	2.74	0	286.89	0	10.41	52.59	104.89
Nicaragua	5100	5.21	3.13	0	0	0	0	0	0	369.94	0	0	8.77	99.08
Niger	8650	0	0	0	0	0	0	0	0	0	0	0	5.42	95.74
Nigeria	126900	5.23	0	0	0	0	0	0	0	757.29	0	0	4.21	133.8
Niue	0	0	0	0	0	0	0	0	0	0	0	0	14.53	96.85
N Mariana Isl.	30	0	0	0	0	0	0	0	0	0	0	0	14.53	96.85
Norway	4490	3603.3	0	0	0.13	0	0	0.87	36.69	354.8	0	1.48	56.13	109.15
Oman	2400	0	0	0	0	0	0	0	0	0	0	0	33.9	96.85
Pakistan	138100	14.24	0	0	0	0	0	0	0	231.09	0	0.01	20.77	97.97
Palau Islands	20	0	0	0	0	0	0	0	0	0	0	0	19.37	96.85
Panama	2900	123.7	0	0	0	0	0	0	0	210.75	0	0	28.09	92.4

Continued

Table 2.1-1 Year 2000 renewable energy use (W/cap.), cf. Figs. 2.1-2–2.1-14—cont'd

COUNTRY W/cap. or number	2000 Pop./1000	HYDRO	GEOTH POWR	GEOTH HEAT	PV	SOLAR HEAT	TIDAL	WIND	BIO-RES	BIO-SOLID	BIO-LIQ	BIO-GAS	ANIM. FOOD	VEG. FOOD
Papua N Guin.	4420	8	0	0	0	0	0	0	0	0	0	0	10.12	95.25
Paraguay	5500	1111.2	0	0	0	0	0	0	0	553.2	0	0	28.09	94.58
Peru	25700	71.86	0	0	0	2.74	0	0	0	115.29	0	0	16.8	110.31
Philippines	75600	11.78	17.57	0	0	0	0	0.02	0	167.66	0	0.01	17.14	98.06
Poland	38650	6.22	0	0	0	0	0	0.1	0	123.31	0.7	1.03	43.1	120.39
Portugal	10010	129.28	0.92	0.13	0.009	2.39	0	3.3	23.1	249.14	0	0.13	51.19	128.72
Puerto Rico	4940	4.6	0	0	0	0	0	0	0	0	0	0	48.43	121.07
Qatar	600	0	0	0	0	0	0	0	0	0	0	0	38.74	96.85
Reunion	650	38	0	0	0	0	0	0	0	0	0	0	29.06	96.85
Romania	22400	75.33	0.12	0	0	0	0	0	0	169.05	0	0	32.74	125.81
Russia Fed	185500	108	0.04	0	0	0	0	0.02	0	0	0	0	31.38	109.88
Rwanda	145600	128.76	0.05	0	0	0	0	0	0	63.97	0	0	2.47	98.11
Saint Lucia	150	0	0	0	0	0	0	0	0	0	0	0	32.11	105.33
San Marino	30	0	0	0	0	0	0	0	0	0	0	0	45.04	132.01
S Tome&Princ.	150	0	0	0	0	0	0	0	0	0	0	0	4.41	111.33
Saudi Arabia	20700	0	0	0	0	0	0	0.04	0	0	0	0	21.6	117.63
Senegal	9500	0	0	0	0	0	0	0	0	240.55	0	0	9.64	99.66
Seychelles	80	0	0	0	0	0	0	0	0	0	0	0	23.97	93.8
Sierra Leone	3480	0.2	0	0	0	0	0	0	0	0	0	0	3.15	87.12
Singapore	4000	0	0	0	0	21.26	0	0	0	0	0	0	29.06	96.85
Slovak R	5400	99.89	0	0	0	0	0	0.04	0	19.68	0	0	38.16	113.56
Slovenia	2000	219.23	0.66	0	0	0	0	0	0	305.59	0	0	45.08	108.33
Solomon IsL.	350	0	0	0	0	0	0	0	0	0	0	0	10.27	100
Somalia	10510	0	0	0	0	0	0	0	0	0	0	0	29.88	48.96

South Africa	42800	3.73	0	0	0	0	0	0	0	393.32	0	0	17.92	121.84
Spain	39930	81.19	0	0.2	0.02	1.03	0	21.3	7.75	130.47	0	3.79	44.41	117.92
Sri Lanka	19400	18.49	0	0	0	0	0	0	0	291.75	0	0.1	7.51	108.91
St Kitts&Nevis	50	0	0	0	0	0	0	0	0	0	0	0	36.17	93.85
St Vincent&Gr.	120	0	0	0	0	0	0	0	0	0	0	0	22.23	102.66
Sudan	31100	4.27	0	0	0	0	0	0	0	602.37	0	0	22.08	91.62
Suriname	420	166	0	0	0	0	0	0	0	0	0	0	17.68	110.75
Swaziland	700	12	0	0	0	0	0	0	0	0	0	0	18.89	107.99
Sweden	8870	1017.1	0	0	0.03	0.75	0	8.97	60.22	1155.48	0	4.19	49.83	100.78
Switzerland	7190	585.42	0	16.82	0.21	4.25	0	0.12	140.26	92.76	0	11.64	52.88	106.59
Syria Arab Rep	16200	65.61	0	0	0	0	0	0.08	0	0	0	0	19.85	127.26
Taiwan Teipei	22200	45.49	0	0	0	0	0	0.02	0	0.6	0	0	29.06	96.85
Tajikistan	6200	255.01	0	0	0	0	0	0	0	0	0	0	7.17	76.13
Tanzania UR	33700	7.49	0	0	0	0	0	0	0	567.73	0	0	5.91	86.39
Thailand	60700	11.38	0	0	0	0	0	0.04	0	312.13	0	0.15	13.75	107.6
Togo	4500	7	0	0	0	0	0	0	0	307.06	0	0	3.87	108.91
Tonga	100	0	0	0	0	0	0	0	0	0	0	0	29.06	96.85
Trinidad&Tob.	1300	0	0	0	0	0	0	0	0	30.66	0	0	21.16	113.32
Tunisia	9600	1.38	0	0	0	0.28	0	0.34	0	171.62	0	0	16.8	142.95
Turkey	66840	52.8	0.13	3.56	0	5.21	0	0.09	0	128.31	0	0.1	18.21	147.22
Turkmenistan	5200	0.4	0	0	0	0	0	0	0	0	0	0	21.94	107.6
Turks&Caicos I	10	0	0	0	0	0	0	0	0	0	0	0	29.06	96.85
Tuvalu	10	0	0	0	0	0	0	0	0	0	0	0	29.06	96.85
Uganda	16670	0	0	0	0	0	0	0	0	0	0	0	6.83	107.41
Ukraine	48500	26.85	0	0	0	0.03	0	0.04	0	7.12	0	0	28.33	110.7
U. Arab Emir.	2900	0	0	0	0	0	0	0	0	0	0	0	38.16	116.47
U. Kingdom	59760	9.76	0	0.02	0.003	0.24	0	2.13	6.23	18.65	0	17.88	48.52	112.93

Continued

Table 2.1-1 Year 2000 renewable energy use (W/cap.), cf. Figs. 2.1-2–2.1-14—cont'd

COUNTRY W/cap. or number	2000 Pop./1000	HYDRO	GEOTH POWR	GEOTH HEAT	PV	SOLAR HEAT	TIDAL	WIND	BIO-RES	BIO-SOLID	BIO-LIQ	BIO-GAS	ANIM. FOOD	VEG. FOOD
United States	275420	103.04	6.09	2.5	0.05	7.32	0	2.84	34.29	290.69	15.55	15.91	50.51	132.16
Uruguay	3300	245.6	0	0	0	0	0	0	0	169.1	0	0	46.88	92.49
US Virgin Isl.	100	0	0	0	0	0	0	0	0	0	0	0	43.58	121.07
Uzbekistan	24800	27.32	0	0	0	0	0	0	0	0	0	0	20.97	93.85
Vanuatu	150	0	0	0	0	0	0	0	0	0	0	0	23.15	102.13
Vatican City	0	0	0	0	0	0	0	0	0	0	0	0	45.04	132.01
Venezuela	24200	296.47	0	0	0	0	0	0	0	29.65	0	0	17.19	92.06
Vietnam	78500	21.16	0	0	0	0	0	0.02	0	383.02	0	0.02	13.17	111.91
W Sahara	90	0	0	0	0	0	0	0	0	0	0	0	15.84	96.85
W Samoa	190	0	0	0	0	0	0	0	0	0	0	0	29.06	96.85
Yemen	17500	0	0	0	0	0	0	0	0	6.07	0	0	6.59	92.11
Serbia Monten	10600	130.36	0	0	0	0	0	0	0	31.34	0	0	48.23	76.22
Zambia	10100	88.14	0	0	0	0	0	0	0	674.84	0	0	4.5	88.09
Zimbabwe	12600	29.53	0	0	0	0	0	0	0	589.45	0	0	6.83	95.64
COUNTRY	POPULAT	HYDRO	GEO-P	GEO-H	PV	SOL-H	TIDAL	WIND	WASTE	B-SOLID	B-LIQ	B-GAS	ANIM.	VEG.

These are some of the external conditions for our planet, spelt out in energy units. It is a little more difficult to obtain reliable estimates for the amount of energy residing within the Earth itself. The kinetic energy of molecular motion, i.e. heat energy, is of the order of 5×10^{30} J. This estimate represents the total heat energy, relative to the absolute zero temperature. It is extrapolated from the value 4×10^{30} J for the heat energy in the interior of the Earth relative to the average surface temperature of 287 K.

The materials forming the Earth carry further energy, in addition to the heat energy corresponding to their temperature. About 10^{21} J is, on average, present as kinetic energy in the atmospheric and oceanic circulation, and the potential energy of the continental height-relief, relative to sea-level, is about 2×10^{25} J, taking into account density variations in the crust (Goguel, 1976). Much larger amounts of energy are involved in the chemical and nuclear bindings, which determine the state and structure of matter. The carbon compounds of biological material provide an example of chemical energy. During earlier periods of the Earth's history, fossilisation of biological material created the deposits of coal, oil and natural gas, of which at least 10^{23} J is presently believed to be recoverable in a form suitable for fuel uses. Current standing crops of biomass correspond to an average of 1.5×10^{22} J.

Nuclear energy may be released in large quantities from nuclear reactions, such as fission of heavy nuclei or fusion of light nuclei. Except for spontaneously fissioning nuclear isotopes in the Earth's crust, which release about 4×10^{20} J y^{-1}, an initial amount of energy must be provided in order to get the energy-releasing fission or fusion processes going. Set-ups for explosive release of nuclear energy involving both types of processes are used for military purposes. As further discussed in section 3.7.3, only the fission process has as yet been demonstrated as a basis for controlled energy supply systems, and, with necessary additional improvements in the technology of fast breeder reactors, recoverable resources of nuclear fuels are estimated to be of the order of 10^{24} J. If fusion of deuterium nuclei to form helium nuclei could be made viable on the basis of deuterium present in sea water, this resource alone would amount to more than 10^{31} J.

Energy conversion processes depleting certain materials of the Earth may be said to constitute irreversible processes. This is often true from a practical point of view, even if the reverse process may be theoretically possible.

The terms "energy use", "spending energy", etc., which are commonly used in energy literature as well as in everyday language, are of course imprecise expressions describing energy conversion processes. Such processes are in most cases associated with an increase in entropy. The entropy is a property of a system, which quantifies the "quality" of the energy contained in the system. The system may be, for example, an amount of fuel, a mass of air in motion, or the entire Earth–atmosphere system.

The entropy change for a process (e.g. an energy conversion process), which brings the system from a state 1 to a state 2, is defined by

$$\Delta S = \int_{T_1}^{T_2} T^{-1} \mathrm{d}Q \tag{2.1.1}$$

where the integral is over successive infinitesimal and reversible process steps (not necessarily related to the real process, which may not be reversible), during which an amount of heat dQ is transferred from a reservoir of temperature T to the system. The imagined reservoirs may not exist in the real process, but the initial and final states of the system must have well-defined temperatures T_1 and T_2 in order for (2.1.1) to be applicable.

Conversion of certain forms of energy, such as electrical or mechanical energy, among themselves may, in principle, not change entropy, but in practice some fraction of the energy always gets converted into heat. The energy processes characteristic of man's activities on Earth involve a series of successive conversion processes, usually ending with all the converted energy in the form of heat, radiated to space or released to the atmosphere, from which radiation of heat energy into space also takes place. The temperatures involved (the T_2) are typically 200–300 K.

Stored energy of any form, which may be converted to heat that is ultimately lost to space, could then be called a "non-renewable energy resource". The term a "renewable energy resource" is used for energy flows, which are replenished at the same rate as they are "used". The prime renewable energy resource is thus solar radiation intercepted by the Earth, because the Earth (i.e. the Earth–atmosphere system) re-radiates to space an amount of heat equal to the amount of solar radiation received. To utilise solar energy thus means converting it in a way convenient for man, but the net result is the same as if man had not interfered, i.e. ultimately to convert solar radiation into heat radiated to space. Such usage may involve a delay in returning the heat, either as a part of man's conversion scheme or by a natural process. For this reason energy stores, which are part of the natural process of converting solar energy into heat re-radiation, are also considered as "renewable energy resources".

"Renewable energy" is not narrowly defined here, and it may be taken to include the usage of any energy storage reservoir which is being "refilled" at rates comparable to that of extraction.

The amount of solar energy intercepted by the Earth and hence the amount of energy flowing in the "solar energy cycle" (from incident radiation flux via reflection, absorption and re-radiation to heat flux away from the Earth) is about 5.4×10^{24} J per year.

Energy fluxes of other than solar origin which occur naturally at the surface of the Earth are numerically much smaller. For example, the heat flux from the interior of the Earth through the surface is about 9.5×10^{20} J y^{-1}, and the energy dissipated in connection with the slowing down of the Earth's rotation (due to tidal attraction by other masses in the solar system) is of the order of 10^{20} J y^{-1}.

2.1.2.1 Man's energy history

The minimum energy requirement of man may be taken as the amount of "exchangeable" chemical energy that can be associated with the amount of food necessary to maintain life processes for someone performing a minimum of work and not losing weight. This minimum depends on the temperature of the surroundings, but for an adult man is generally considered to lie in the region of 60–90 W on average for extended periods, corresponding to $(6 \text{ to } 8) \times 10^6$ J day^{-1}. The total life requirements are, of course, more than energy, comprising adequate supplies of water, nutrients, etc.

In order to perform any (muscle) work not purely vegetative, additional energy must be supplied in the form of food, or energy stored in the body will become depleted. The efficiency in converting stored energy into work typically ranges from 5–50%, with the lower efficiencies being associated with activities involving large fractions of static conversion (e.g. carrying a weight, which requires the conversion of body energy even if the weight is not being moved). The percentage complementary to the efficiency is released as various forms of heat energy.

The maximum average rate of food energy intake that a human being can continue for extended periods is about 330 W, and the maximum average rate at which work can be delivered for extended periods is of the order of 100 W (Spitzer, 1954). During work periods, the "man-power" output level may be 300–400 W, and the maximum power which can be delivered by an adult male for a period of about a minute is roughly 2000 W.

Although it is not certain that the rates of energy conversion by the human body have remained constant during the evolution of man, it may be reasonable to assume that the average amount of "muscle power" used by the earliest members of the genus *Homo*, which evidence suggests lived some 4×10^6 years ago in Africa (Leakey, 1975), was of the order of 25 W.

The total energy flux received by an individual man in a food gathering or hunting society is then the sum of the energy in the food, averaging say 125 W, and the absorbed flux of radiation and heat from the surroundings, which may reach considerably larger values, but is highly dependent on clothing, climate and the nature of the surroundings (cf. e.g. Budyko, 1974). The outgoing energy flux again consists of heat and radiation fluxes, turnover of organic material, plus the amount of energy converted into work. For growing individuals, the net flux is positive and the mass of biological material increases, but also for adult individuals with zero net energy flux, new biomass continues to be produced to replace "respiration losses".

Man has successively developed new activities, which have allowed him to gain access to larger amounts of energy. Solar energy may have been used for drying purposes, and as soon as fires became available a number of activities based on firewood energy may have started, including heating, food preparation and process heat for tool making. The earliest evidence for fires used in connection with dwellings is from Hungary, 350 000–400 000 years ago (H. Becker, 1977, personal communication).

A good fire in open air, using some 10–50 kg of firewood per hour, may convert energy at a rate of 10^4–10^5 W, whereas indoor fires are likely to have been limited to about 10^3 W. Several persons would presumably share a fire, and it would probably not burn continuously at such a power level, but rather would be re-lit when required, e.g. from glowing embers. It is thus difficult to estimate the average fire energy per person, but it would hardly exceed 100 W in primitive societies. The efficiency of delivering energy for the desired task is quite low, in particular for open-air fires.

The next jump in energy utilisation is generally considered to have been associated with the taming of wild animals to form livestock and the introduction of agriculture. These revolutions have been dated to about 10^4 years ago for the Near East region (cf. e.g. DuRy, 1969), but may have been developed in other regions at about the same time, e.g. in Thailand and Peru (Pringle, 1998). This time corresponds to the ending of the last ice age, which may have caused changes in the low-latitude climate, including altered precipitation rates. The introduction of livestock would have promoted the tendency to settle at a given place (or vice versa), increasing in turn the requirement for food beyond the capacity of a hunting society. Agriculture was based at first on wild varieties of wheat, for example, and it is believed that artificial irrigation was necessary at many of the sites where evidence of agriculture (various tools) has been found. The power for water transport and, later, pumping would then be derived from suitable draught animals in the livestock pool, as a substitute for man's own muscle power. The transition from a hunting to an agricultural society, often called the *Neolithic* or "new stone age", occurred several thousand years later in the temperate zones of northern America and Europe.

The creation of cultures of growing size and level of sophistication, leading to the formation of large cities, for example, at the rivers Euphrates, Tigris and the Nile, about 5000 years ago, witnessed a growing use of energy for ploughing, irrigation, grinding and transport (of food supplies and of materials, e.g. in connection with buildings and monuments), as well as the harvest of solar energy through agricultural crops. It is not known exactly how much of the physical work was performed by men and how much by animals, but it is likely that another 100–200 W was added to the average energy usage per capita in the most developed regions.

It is also important to bear in mind that there must have been large differences in energy use, both between different societies and between individuals within a given society. Throughout man's history (the meaning of "history" not being restricted to imply the presence of written records) there have been individuals whose access to energy was largely limited to that converted by their own bodies. Large regions in Asia and Africa today have an average energy spending per person, which is only a few hundred watts above the muscle power level (with firewood as an important source). This means that parts of the population today use no more energy than the average person during the Neolithic period.

The energy sources that have emerged so far are direct solar radiation, environmental heat, animal biomass, as well as primary (plant) biomass in the form of food and later as firewood, plus mechanical work from the muscle power of animals. In the Near East, oil was used for lighting, and bitumen had non-energy uses. Boat travel in the open sea (the Mediterranean) is believed to have started over 9000 years ago (Jacobsen, 1973), and there is evidence of wind energy utilisation by means of sails in Egypt about 4500 years ago (Digby, 1954). Per person, wind energy may not at this time have contributed a significant proportion of the total energy use in the Mediterranean region, but later, when trade became more developed (from about 4000 years ago), the total amount of energy spent on transportation on land and at sea constituted a less negligible share (maybe a few per cent) of the total amount of energy spent in the "developed regions" of the world at the time.

The building of houses in many cases implied the creation of a required indoor climate with utilisation of solar energy. In the low-latitude regions, structures of high heat capacities were employed in order to smooth out day-to-night temperature variations, and in many cases the houses were built partly underground, and the evaporation of soil moisture was utilised to create cool environments for living (during hot periods) and food storage (Bahadori, 1977). In regions with a colder climate, a number of insulating* building materials (e.g. roofs made of straw) were employed to reduce heat losses, and heat production not involving fires was increased by keeping livestock within the living area of the houses, so as to benefit from their respirational heat release.

Water mills and windmills (e.g. the vertical axis panemone type probably derived from waterwheels, or the sail-wing type presumably copied from sail-ships) also played a role from a certain stage in development. The earliest mention of windmills in actual use is from India about 2400 years ago (Wulff, 1966). Considering its low efficiency and overall size, it is unlikely that wind power has at any time accounted for a large proportion of the average energy use. On the other hand, windmills and water mills offered the only alternative to muscle power for high-quality (i.e. low-entropy) mechanical energy, until the invention of the steam engine.

The industrial revolution 200–300 years ago was connected with placing at the disposal of man amounts of power capable of producing work far beyond his own muscle power. However, at that time firewood was barely a renewable resource in the developed regions of the world, despite quite extensive programmes to plant new forests to compensate for usage. The increase in energy usage made possible by the growing industrialisation did not really accelerate, therefore, before large amounts of coal became available as fuel. In the 20th century, the large growth in energy consumption was made possible by the availability of inexpensive fossil fuels: coal, natural gas and oil.

An outline of the possible development in energy usage up to the present is presented in Figs. 2.1-15–2.1-17. Only over the past century or two have reliable world-wide data on energy usage been recorded, and even for this period the data comprise mainly direct use of commercial fuels, supplemented with incomplete information on biomass and other renewables. One reason for this is that it is more difficult to specify the remaining energy use, because e.g. solar collectors are often not individually monitored, local biomass use is not quantified in energy units, environmental heat gains vary from day to day, and so on. In Figs. 2.1-15–2.1-17, which are anyway only indicative, fuels are included in terms of their gross energy value, independently of end-use efficiency. The use of renewable energy flows, on the other hand, is given as an estimated net energy at the primary conversion stage, i.e. the energy in food intake rather than the total amount of energy absorbed by the plants or the total biomass of plants and animals. The environmental energy contribution to maintaining man's body temperature as well as the regulation of indoor climate by the choice of materials and building systems ("passive energy systems") are excluded.

* The term "insulating" is taken to include suppression of convective heat transfer.

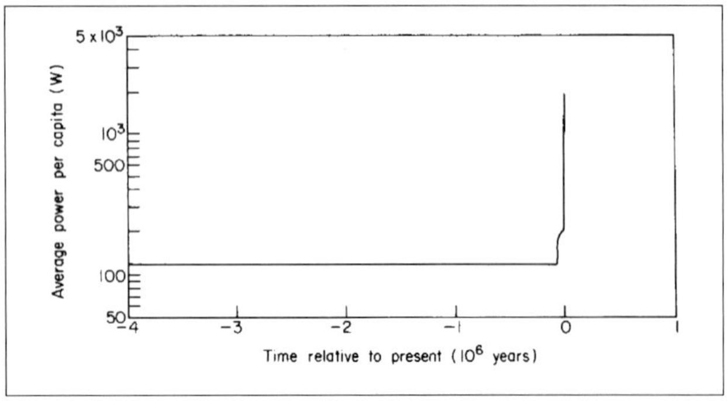

Figure 2.1-15 Trends in average rate of energy conversion per capita, not including fluxes associated with the local thermal environment.

Figure 2.1-15 shows the trend in average rate of energy conversion per capita, on a linear time scale, and Fig. 2.1-16 shows the same trend on a logarithmic time scale, extending backwards from the year 2000. Figure 2.1-16 also indicates the estimated spread in energy usage, with the upper curve representing the societies with highest energy use, at a given time, and the lower curve representing the societies with the lowest energy use. These curves, which do not reflect any great degree of accuracy, do not represent rigorous limits, and values outside the interval may certainly be appropriate for individuals of a given society – the very rich or the very poor.

The energy conversion rate corresponding to food only has been taken as 125 W throughout the time interval. The increase in energy usage from about -10^5 y is associated with access to fire. The amount of energy derived from fires depends on whether fires were used only for cooking or also for heating. The choice of the average curve further rests on the assumption that between -7×10^4 and -10^4 y (i.e. during the latest ice age, cf. Fig. 2.91) about half of the world population used fires for heating purposes.

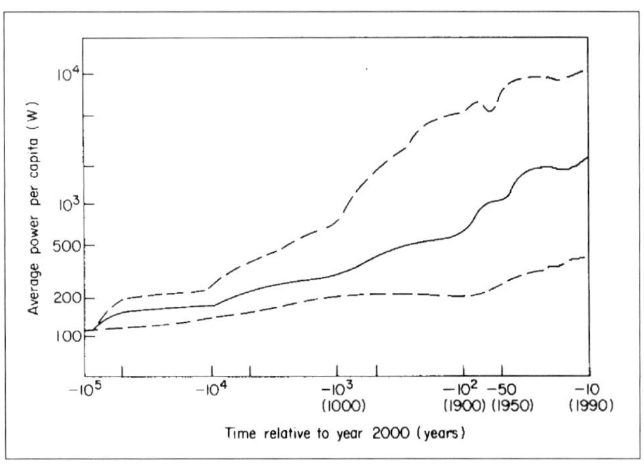

Figure 2.1-16 Trends in average rate (solid line) of energy conversion per capita, not including fluxes associated with the local thermal environment (same as Fig. 2.1-1, but on a logarithmic time scale). Dashed lines indicate the corresponding trends for the societies, which at a given time have the highest and the lowest average energy usage. For the more recent period, data from Darmstadter *et al.* (1971) and European Commission (1997) have been used, in a smoothed form.

In the time interval -10^4 to -10^3 y, human settlements developed into a variety of societies, some of which had a very high degree of organisation and urbanisation. The increase in energy usage was mainly associated with more systematic heating and cooking practices, with tool production (e.g. weapons) and with transportation (e.g. by riding or by draught animals). With increasing population density, materials which previously had been available in the immediate natural surroundings had to be transported from far away, or substitutes had to be manufactured; either way, additional energy had to be spent. In several of the societies in question, mechanical work was performed not only by animals but also by human slaves, so that the average per capita energy usage was less affected. The trends of the curves also reflect the differences in development characterising different geographical regions. Simultaneously with the culmination of the civilisations in Mesopotamia and Egypt, northern Europe and northern America entered the Neolithic period, with warm climatic conditions quite different from those of the preceding several thousand years.

During the last 1000 years, the increasing energy usage was, in part, due to the shift in population distribution towards higher latitudes, and to overall increased requirements for space heating in such regions (the "little ice age"). It should also be mentioned that the efficiency of converting the energy of firewood (supplemented by animal dung and later by peat) into useful heat for cooking, craft work, hot water and space heating was quite low, for example in 16th-century Europe, but gradually improved as the 20th century approached (Bjørnholm, 1976). During the period 1500–1900, the curves are a result of this feature (in particular the early high maximum value attained for the most affluent societies) combined with increased energy demand (e.g. larger proportions of the population acquiring energy-demanding habits or lifestyles, such as taking hot baths, drinking hot beverages, washing clothes in hot water). The development in the last century is dominated by the energy consumption of the industrialised countries (industrial process heat, transportation, increased room temperature, refrigeration, lighting, etc.). During this period, the top curve in Fig. 2.1-16 represents the extravagant energy use of an average American, while the lowest curve represents the average energy use in the poor regions of Africa or India, including non-commercial fuels such as cow dung and stray wood (which used to be absent from official statistics, as first noted by Makhijani, 1977).

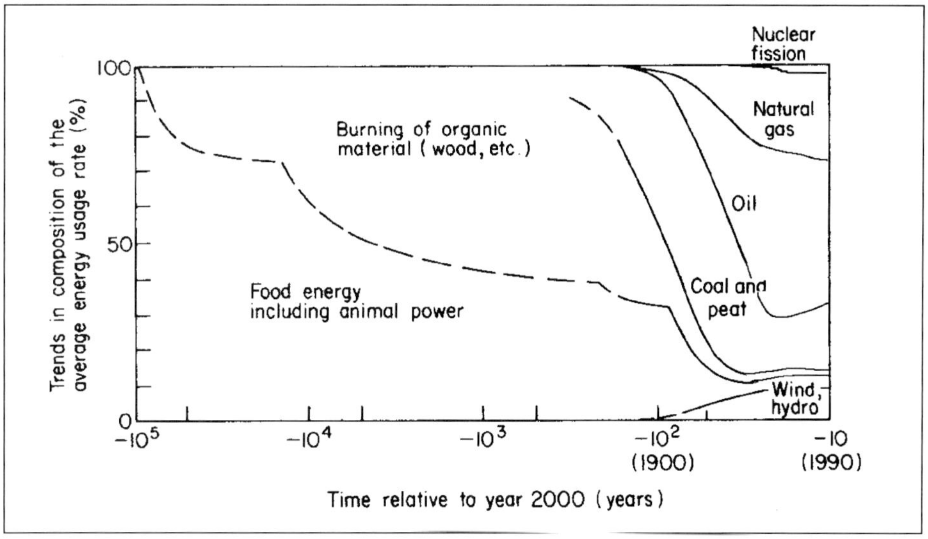

Figure 2.1-17 Trends in the distribution on different types of energy resources of the average rate of energy use. The most recent period is based on smoothed data from Darmstadter *et al.* (1971) and European Commission (1997), and the basis for the estimates pertaining to earlier periods is explained in the text. Needless to say, such estimates should be regarded as very tentative, and the definition of average use is itself uncertain, particularly for the early periods (e.g. the 20% contribution from fires 50 000 years ago depends sensitively on the fraction of the world population living in regions where space heating was desirable).

In Fig. 2.1-17, a sketch of the distribution of the energy consumption on different sources of energy is attempted. Again, only for the past century or two have actual data been used. The shape of the curve describing the diminishing share of food energy starting about 10^5 years ago is again dependent on the picture of emerging cultures and geographical distribution of the population, outlined above. It is clear, however, that the energy basis for human societies has been renewable energy sources until quite recently. Whether all the wood usage should be counted as renewable is debatable. Early agricultural practice (e.g. in northern Europe) involved burning forest areas for farming purposes and repeating the process in a new area after a few years, as the crop yield diminished owing to nutrient deficiency of the soil. Most forests not being converted into permanent agricultural land survived this exploitation, owing to the low population density and the stability of the soils originating from glacier deposits. Similar overuse, or overgrazing by livestock, would be (and was in fact) disastrous in low-latitude regions with a very shallow soil layer, which would simply be eroded away if the vegetation cover was removed. Replantation of forests has been common in northern Europe during the last few centuries, but the strongly increasing demand for wood over the last century (not only for fuel purposes), as well as construction work associated with urbanisation, has led to an actual decrease in forest area in most parts of the world.

From the middle of the 19th century, the non-renewable fossil fuels have rapidly increased their share of the total energy usage, to the present 80–90%. In the beginning, fossil fuels replaced wood, but they soon became the basis for exponential growth in energy use, associated with a number of novel energy-demanding activities. During the same period, usage of hydropower has increased, and recently nuclear fission power passed the 1% level. Growth has been interrupted by wars and periods of economic recession. The high dependence on non-renewable energy sources has developed over a very short period of time. The briefness of this era compared with the history of man on Earth stands out clearly on the linear scale used in Fig. 2.1-15.

2.1.3 The energy future and the role of renewable energy

Figure 2.1-16 shows a very large difference between the energy use of the leading countries or leading persons within countries, as compared with the least energy-using inhabitants of poor countries. This feature is currently changing, as the level of global interaction increases and every world citizen becomes aware of the kind of lifestyle that is "possible". However, the current development does not seem to indicate a diminishing ratio of the energy use of those using most and those using least energy. This is also true of other commodities related to living standard.

Energy use and resource depletion does not, of course, constitute the primary goals of any society or individual within a society. For example, average Europeans or

Japanese use about half as much energy as the average North American, but have a living standard which certainly is not lower than that of the North American citizens. This underlines the fact that the living standard and welfare depends on having primary (food, shelter, relations) as well as secondary standards of individual preference fulfilled and that this can be done in different ways with different implications for energy use.

The relationship between economic activities and social welfare has been debated for a considerable period of time, as has the possibility of physical limits to growth in material exploitation of the resources of a finite planet. The answer of conventional economists to this is that the inventiveness of man will lead to substitution of the materials threatened by exhaustion with others, in an ever-ongoing process. Recognising the finiteness of fossil and nuclear energy sources, this leads to the general prediction that renewable energy sources must take over at some stage, and the only debate is on how soon this will happen.

Most current geologists believe that oil and natural gas production will peak sometime in the next two decades. After that prices are bound to rise, thereby easing the introduction of alternative energy sources. Accepting a higher price of energy, it is also implied that energy must be used more efficiently, in order to prevent the belief that the higher energy cost slows down the development of human welfare.

This development in energy use is linked to another problem that may serve to accelerate the energy transition, namely, the increased awareness of the negative implications of environmental impacts of energy production and use. Early man was capable of causing environmental disturbance only on a very local scale. However, extensive burning of forests, for example, to provide land for agriculture, which would later be abandoned when overexploitation diminished the crop or grazing yields, may have been instrumental in creating the desert and semidesert regions presently found at low latitudes (Bryson, 1971). This is already an important example of a possibly man-made climatic change. Recently, man has reached a technological level enabling him to convert energy at rates which can be maintained over extended areas and which are no longer small compared to the energy fluxes of solar origin that are responsible for the climate.

The average heat flux of anthropogenic origin (i.e. from fossil fuels) in an industrial and urban area such as the Los Angeles Basin (about 10^{10} m^2) was estimated in 1970 to be 7 Wm^{-2} (Lees, 1970). The global average value around 1970 was 0.015 Wm^{-2}, and the average solar flux absorbed by the Earth-atmosphere system is 240 Wm^{-2}. For comparison, a forest fire, which would burn down an area of fertile, tropical forests in one week, would release a heat flux of about 1000 Wm^{-2}. Yet the average heat flux from forest fires in all continental regions, the average being over several years, is less than the average anthropogenic heat flux. The nuclear weapons arsenal built up during the last 50 years is somewhere in the range of 10^4–10^5 megatons (Feld, 1976), with the higher figure corresponding to about 4.4×10^{20} J. If these weapons were detonated within a 24-hour interval, the average energy flux would be 5×10^{15} W, and if the target area were 10^{12} m^2, the average heat flux would be 5000 Wm^{-2}. The destructive effects would not be confined to those related to the immediate energy release. Radioactive contamination of the environment would cause additional death and decay, and would establish further mechanisms for climatic disturbance (e.g. destruction of the stratospheric ozone shield), in addition to the difficulty presented to human survival as the dominant species on our planet.

It has been suggested that fusion energy constitutes an alternative to renewable energy as the long-term solution. However, it is not clear at present whether fusion energy on Earth will ever become a feasible and practical source of controlled energy supply. It will create nuclear waste in amounts similar to those of fission technologies and will counteract the development towards decentralised technologies characterising the present trend. It is probably an exaggeration to imagine that the introduction of one kind of energy technology rather than another will determine or solve such institutional problems. What may be true, though, is that certain types of technology are more suitable for societies organised in a particular way and that the kind of technology imagined in connection with the use of certain renewable energy resources fits well both with the needs of sophisticated, decentralised societies based upon information technology and with the needs of the presently underprivileged regions.

Science and technology literature contains a range of suggestions for handling future energy demands. In the past, some of the technologies thus brought forward as "technically feasible" have actually been developed to commercial viability, and others not, for a variety of reasons. Renewable energy has over the last decades passed from the level of technical feasibility to a level of cautious introduction into the marketplace and not least into long-term government planning. One reason for the slow penetration is that some influential funding institutions, including the European Commission, have continued to use a large fraction of their R&D funds, as well as loan and aid money, on fission and fusion, ignoring the unattractiveness of the long-range radioactive waste problems and hoping to obtain short-term industry advantages in export of outdated fission technology to former East-block and developing nations. If funds were wholeheartedly aimed at a rapid transition from the fossil to the renewable era, the progress could be much faster. This has been demonstrated by a number of recent scenario studies, some of which is described elsewhere (Sorensen, 2004). The general question of who controls technology development has been discussed by Elliott and Elliott (1976) and by Sørensen (1983; 2001a). During recent decades, a number of "grassroot" movements have advocated use of renewable energy, and it can be hoped that these preferences are preserved as that generation of people make their way into decision-making positions.

Renewable energy sources are typically characterised by a theoretical maximum rate at which energy may be extracted in a "renewable" mode, i.e. the rate at which new energy is arriving or flowing into the reservoirs associated with many of the renewable energy flows. In some cases, the additional loop on a given renewable energy cycle, caused by man's utilisation of the source, will by itself modify the rate at which new energy is arriving (for instance, utilisation of temperature differences in the oceans may alter surface evaporation rates and the velocities of ocean currents; in both cases the mechanisms for establishing the temperature differences may become altered). The geothermal energy flux from the interior of the Earth is not a renewable resource, since the main part of the flux is associated with a cooling of the interior. On the other hand, it is a very small fraction of the heat which is lost per year (2.4×10^{-10}), so for practical purposes geothermal energy behaves as a renewable resource. Only in the case of over-exploitation, which has characterised some geothermal steam projects, renewability is not ensured.

In Chapter 3.2, the nature and origin of renewable energy sources are discussed in what may resemble an odyssey through the sciences of astrophysics, atmospheric physics and chemistry, oceanography and geophysics. The importance of connecting all the pieces into an interlocking, overall picture becomes evident when the possible environmental impact of extended use of the renewable energy sources in the service of mankind is investigated elsewhere (Sorensen, 2004).

Chapter 3.2 provides, for each renewable energy source, an estimate of the size of the resource, defined as the maximum rate of energy extraction, which on an annual average basis will become renewed, independently of whether it is possible to extract such energy by known devices. Also issues of power density and variability are discussed in this chapter.

Chapter 4.1 opens with some general features of energy conversion devices and then describes a number of examples of energy conversion equipment suitable for specific renewable energy sources.

Chapter 13.1 gives an overview of various methods of energy transport and storage, which, together with the energy conversion devices, will form the ingredients for total energy supply systems discussed in Chapter 6.

The modelling of the performance of individual renewable energy devices as well as whole systems and, finally, scenarios for the global use of renewable energy, with consideration of both spatial and temporal constraints in matching demand and supply is described elsewhere (Sorensen, 2004).

Renewable energy resources are placed in the framework of current economic thinking, as a preliminary effort to quantify some of the considerations which should be made in constructing a viable energy supply system. Indirect economic factors can be considered, which leads to the description of the methodology of life-cycle analysis, which together with the scenario technique constitutes the package for an up-to-date economic analysis.

Renewable energy research and development areas most important for the near future are discussed elsewhere (Sorensen, 2004).

2.1.4 Suggested topics for discussion

2.1.4.1

Find out what has happened since the year 2000 to the amounts of renewable energy used, based upon available statistical data, either globally or for your region.

2.1.4.2

Order the prices given on page 32 according to the type of energy (heat, fuels for industry or transportation, electricity) and the type of customers (wholesale, retail). Try to find out what has happened to these prices since the year 2000 (e.g. using the annually updated sources such as OECD, IEA or Danish Energy Agency). Have the relative prices changed?

Section Three

Alternate energy sources

Chapter 3.1

3.1

Alternate energy sources

One form of energy can be converted into another form with a certain loss of efficiency. The various forms of energy currently available and the methods to generate them appear below. Some of these energy forms are renewable, and others are nonrenewable, as outlined in Fig. 3.1-1. The nonrenewable energy forms have been practiced traditionally for several hundreds of years and are currently expected to disappear completely in a few hundred years or less.

- Heat
  - by burning fossil fuels
  - solar radiation
  - warm air, water, subsurface water, and ocean
  - nuclear energy
  - earth's core (hot springs)
  - electricity passing through wires
- Light
  - sun
  - fluorescent and incandescent lightbulbs
  - LED
  - laser
  - burning fuels
- Electricity
  - photovoltaic
  - dynamo generators
  - batteries
  - hydrogen fuel cells
  - static (friction, lightning)
- Radio waves
  - radio transmitters
- Mechanical
  - hydroelectric facilities
- Sound
  - vibrating surfaces (microwave)

The two most common forms of energy used by humans are heat and electricity.

3.1.1 Greenhouse gases

When sunlight strikes the earth's surface, some of it is reflected toward space as infrared radiation (heat). Many chemical compounds found in the earth's atmosphere allow sunlight to enter the atmosphere freely, while certain gases and vapors absorb this infrared radiation and trap the heat in the atmosphere (these are known as *greenhouse gases*). Over a period of time, the cumulative effect of these greenhouse gases is a slow increase in the earth's surface temperature. Some of the gases found in nature, such as water vapor, carbon dioxide, methane, and nitrous oxide, and others, which are exclusively human made, like gases used for aerosols and chemicals such as chlorofluorocarbons (CFCs), exhibit greenhouse properties.

During the past 20 years, about three quarters of human made carbon dioxide emissions were due to the burning of fossil fuels. Based on 2004 data, next to the United States, China is the biggest producer of greenhouse gases (Jia, 2004). China mainly relies on coal and oil for energy. Between 1996 and 2003, oil imports in China increased from 20 to 90 million tons. Energy consumption in China is expected to continue rising significantly as it aims to quadruple its GDP by 2020. According to China, in 1994 it emitted 2.6 billion tons of carbon dioxide, 34.3 million tons of methane, and 850,000 tons of nitrous oxide. In the United States, greenhouse gas emissions arise mostly due to energy requirements (which represent 82% of total U.S. human made greenhouse gas emissions) (Energy Information Administration, Office of Integrated Analysis & Forecasting, 2002).

Renewable Energy Focus Handbook 2009; ISBN: 9780123747051

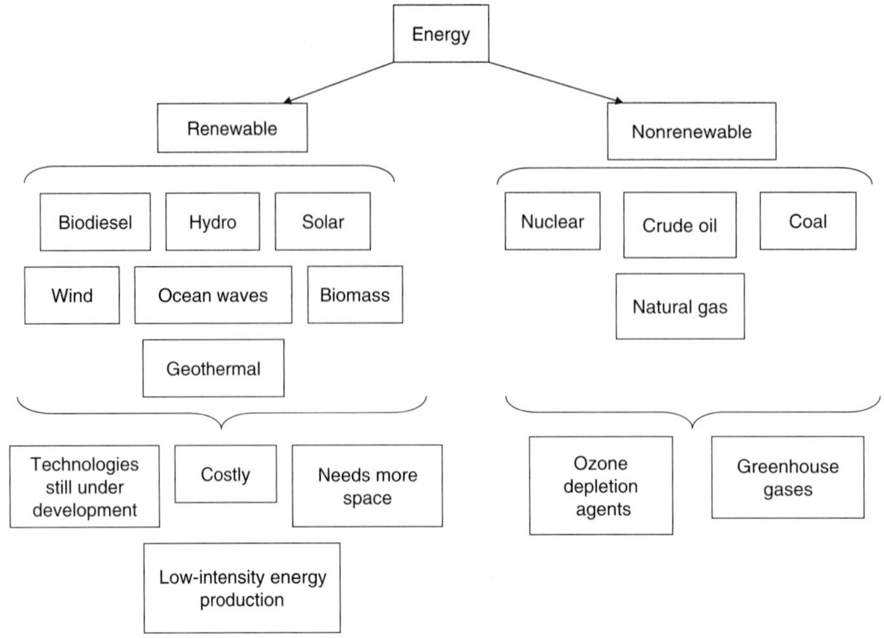

Figure 3.1-1 Energy sources.

Energy requirements include fuel used for electricity generation and that used for heating and cooling needs of homes and offices. The United States currently produces about 25% of global carbon dioxide emissions from burning fossil fuels but is projected to lower its carbon intensity by 2025 and remain below the world average (National Energy Information Centre, EIA, Washington, DC (www.eia.doe.gov/environment. html).

The Kyoto Protocol is an international treaty bringing many of the world's developed nations together in an effort to limit greenhouse gas emissions and reduce the effects of global warming. Russia ratified the treaty in February 2005, while, as of this publication, the United States and Australia were among the few countries that had not ratified it. All companies operating in ratifying countries must comply with the protocol regardless of where they are based. Six specific gases that contribute toward global warming have been identified: carbon dioxide, methane, nitrous oxide, hydrofluorocarbons (HFC), perfluorocarbons (PFC), and sulfur hexafluoride (SHF). The global warming potential of these gases with respect to CO_2 are 21, 310, 140–11,700, 6500–9200, and 23,900, respectively (Dore et al., 2003). Although carbon dioxide has the lowest global warming potential potency, it is released in far greater amounts than any of the other gases due to human activities and is responsible for 82% of all the global warming caused in the United Kingdom. The half-life of some man-made chemicals is so long that they persist longer than the natural global warming gases.

A *sink* is a reservoir that uptakes a chemical element or compound from another part of its cycle. For example, soil and trees tend to act as natural sinks for carbon, as billions of tons of carbon each year in the form of CO_2 are absorbed by oceans, soils, and trees. The EU-15 countries had agreed to cut, by 2012, 8% of the 1990 values, but the data collected in late 2006 by the European Commission predict that the values will be only 0.6% below the base year levels by 2010. Worse still, Austria, Belgium, Denmark, Ireland, Italy, Portugal, and Spain may even exceed their individual limits.

The next most prominent greenhouse gas is methane (9%), which comes from landfills, coal mines, oil and gas operations, and agriculture. Nitrous oxide (5% of total emissions) is emitted from burning fossil fuels and from the use of nitrogenous fertilizers and industrial processes. Manmade gases such as HFC, PFC, and SHF (2% of total emissions) are released as byproducts from industrial processes and through leakage from cooling systems.

World carbon dioxide emissions are expected to increase by 1.9% annually between 2001 and 2025. Much of the increase in these emissions is expected to occur in developing nations such as China and India (which will be above the world average at 2.7% annually between 2001 and 2025). Data collected from Antarctic ice cores, before the industrial emissions began, led to the conclusion that atmospheric CO_2 levels were about 280 μL/L. The concentrations stayed between 260 and 280 μL/L throughout the 10,000 years preceding the beginning of the years of industrial emissions. Since the beginning of the Industrial Revolution (c.1850), the concentrations of many of the greenhouse gases have increased. Most of the increase in carbon dioxide occurred after 1945. The current amount of CO_2 is 364 ppm

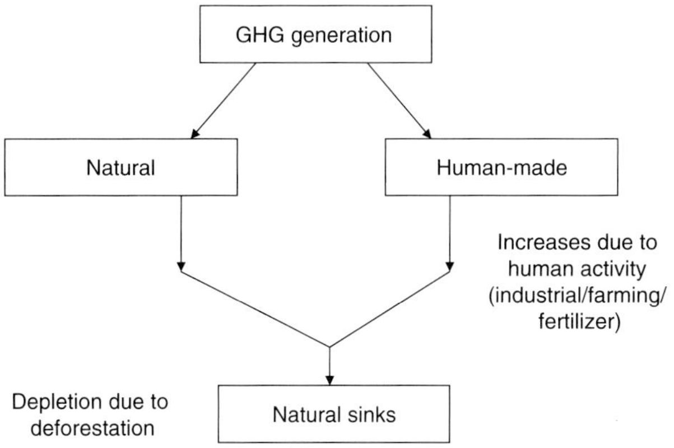

Figure 3.1-2 Greenhouse gas (GHG) generation and consumption.

(increase of 31% over pre-industrial era), and CH_4 is 1745 ppb (increase of 150% over pre-industrial era).

Human activities have raised the levels of greenhouse gases primarily by releasing carbon dioxide, methane, and other gases (see Fig. 3.1-2). These activities include burning of fossil fuels and deforestation, leading to higher carbon dioxide concentrations; livestock and paddy rice farming, land use and wetland changes, pipeline losses, and covered, vented, landfill emissions, all of which lead to higher methane release; use of CFCs in refrigeration systems; use of halons in fire extinguishers; and various manufacturing processes. Water vapor is a natural greenhouse gas that accounts for the largest percentage of the greenhouse effect. Water vapor levels fluctuate regionally, but in general humans do not have a direct effect on its levels. The current amounts of CFC-11, CFC-12, CFC-113, CCl_4, and HCFC-22 are 268, 533, 84, 102, and 69 ppt, respectively (IPCC, 1998).

The concentration of greenhouse gases in the atmosphere keeps changing dynamically due to several reactions and processes, such as (Archer, 2005; Caldeira and Wickett, 2005):

1. condensation and precipitation of water vapor from the atmosphere,
2. oxidation of methane by hydroxyl radical and water vapor,
3. mixing and interchange of atmospheric gases into oceans or other regions,
4. chemical reactions of gases in atmosphere with other compartments of the planet, such as reduction in CO_2 amount due to photosynthesis of plants or, after dissolving in the oceans, formation of carbonic acid and bicarbonate and carbonate ions,
5. dissociation of halocarbons by UV light releasing Cl^* and F^* as free radicals in the stratosphere with harmful effects on ozone (such as "ozone depletion" or "ozone hole"),
6. dissociative ionization reactions caused by high-energy cosmic rays or lightning discharges (e.g., lightning forms N^* atoms from N_2, and the former then reacts with O_2 to form NO_2).

A material balance for the accumulation of CO_2 in the environment consists of CO_2 generation due to natural causes and human activities. The depletion of CO_2 is due to its absorption in the natural sinks, as shown in the following model:

$$\frac{Vd[CO_2]}{dt} = r^h{}_G + r^n{}_G - r_S,$$

$r^h{}_G$ = rate of generation of CO_2 due to human activity
= [rate of burning of fossil fuels][f_1] + Σ [rate of production of industrial chemicals/ manufacturing operations][f_2],
$r^n{}_G$ = rate of generation of CO_2 due to natural causes,
r_S = rate of consumption of CO_2 due to absorption in various sinks
= Σ[amount of various sinks] [g_1],
V = volume of the biosphere,
f_1, f_2 = stoichiometry of CO_2 production,
g_1 = efficiency of absorption of CO_2.

The various sinks for CO_2 absorption have been depleting because of deforestation (leading to a decrease in rate of CO_2 absorption), while the rate of manufacturing operations has been increasing. This leads to an overall increase in CO_2 concentration ([CO_2]) with time.

3.1.1.1 Lifetime of greenhouse chemicals

Recovery from a large input of atmospheric CO_2 from burning fossil fuels will result in an effective lifetime of tens of thousands of years (Archer, 2005). Methane has an atmospheric lifetime of 12 to 15 years. The methane is degraded to water and CO_2 by chemical reactions in the atmosphere. Nitrous oxide has an atmospheric lifetime of 120 years, while CFC-12 has an atmospheric lifetime of 100 years. HCFC-22 has an atmospheric lifetime of 12.1 years, and tetrafluoromethane has an atmospheric lifetime of 50,000 years. Sulfur hexafluoride has an atmospheric lifetime of 3200 years.

3.1.1.2 Oil

Crude oil production (measured in thousands of barrels per day) in various countries (Nov. 2006 data) (Development report on the oil markets in 2006, Ministry of Energy, State of Kuwait (www.moo.gov.kw)) is given in Table 3.1-1.

Table 3.1-1 Crude oil production

Algeria	1,343
Indonesia	872
Iran	3,803
Kuwait	2,383
Iraq	2,002
Libya	1,713
Nigeria	2,258
Qatar	802
Saudi Arabia	8,857
UAE	2,448
Venezuela	2,488

In 1956, Hubbert proposed that over time crude oil production in a country would follow a bell-shaped curve, with a peak production at a particular time and a decrease with further time. Nine major large oil-producing countries, including the United States, Great Britain, Venezuela, and Norway, had reached their peak global oil production in 1998, and their production volume has been slowly decreasing since then. Factors that influence the peak date include worldwide recession, military or political factors, etc. The cumulative depletion among this group is now about 1.5 million barrels per day (Mbpd). Thus, presently the current demand for oil is growing at about 3.5%, which is 82 Mbpd worldwide. The peak production is estimated to be above 90 Mbpd. But the International Energy Agency (IEA) as well as the OECD are optimistic, and authorities claim that production will be 110 Mbpd within the next 10 to 14 years. This is based on the assumptions that significant future discoveries will be made and that technologies for the use of nonconventional sources such as shale oil and oil sands will become competitive and hence will be practiced.

3.1.1.3 Solar energy

Solar energy can be converted directly into heat by passive or active systems. The passive systems use a thermo siphon and have no pumps. The thermo siphon operates only when the fluid is hot. Other space heating systems use a thermal diode to achieve similar effects. Passive solar water distillers may rely on capillary action to pump water. Active solar systems use additional equipment such as circulation pumps, air blowers, or tracking systems that aim the solar arrays or collectors at the sun. These mechanisms are typically powered by electricity. A wide range of power technologies exist to convert the solar energy. A few of these include:

1. Photovoltaic cell produces electricity directly from solar energy.
2. Hydroelectric power stations produce indirect solar power.
3. Sunlight is concentrated onto a thermal collector, and the surface is heated up. The heat is carried away by a fluid.
4. Sunlight strikes a solar sail on a spacecraft and is converted directly into a force on the sail, which causes motion of the craft.
5. Sunlight strikes a light mill and causes the vanes to rotate.
6. Sunlight is focused on an externally mounted reflective channel that conducts sunlight into building interiors to supplement lighting.

Photovoltaic (PV) devices use semiconducting materials to convert sunlight directly into electricity. Solar radiation, which is nearly constant outside the earth's atmosphere, varies with changing atmospheric conditions (clouds and dust) and the changing position of the earth relative to the sun. Of the total solar energy received, 19% is absorbed by the atmosphere, while clouds reflect 35% of the total energy. The peak power received at sea level is 1000 W/m^2. For example, in North America the average power of the solar radiation lies somewhere between 125 and 375 W/m^2, meaning at a rate between 3 and 9 $kWh/m^2/day$. Photovoltaic panels currently have an efficiency of 15%, and, hence, a solar panel delivers 19 to 56 W/m^2, or 0.45–1.35 $kWh/m^2/day$ (annual day and night average). A 173-m^2 photovoltaic system in the 30-year lifetime of the system is estimated to prevent 2100 lb of NO_x, 6100 lb of SO_x, and 756 tons of carbon dioxide that will be produced if oil is used to produce the same amount of energy. The average lowest retail cost of a large solar panel declined from $7.50 to $4 per watt between 1990 and 2005. The cost of producing electricity from solar radiation is still not yet competitive.

According to an April 2000 article in the *Electric Power Research Institute (EPRI) Journal*, photovoltaic arrays in a geostationary earth orbit at an altitude of 22,300 miles would receive eight times the sunlight that is received at earth's surface. Such arrays would be unaffected by cloud cover, atmospheric dust, or the earth's day–night cycle. A drawback to concentrated sunlight is that it is hot. If not converted into electricity, radiation that is focused turns into heat and can damage the arrays. Current research is directed toward studying ways to capture waste heat and convert it to electricity by means of thermal voltaic processes and special coatings on the

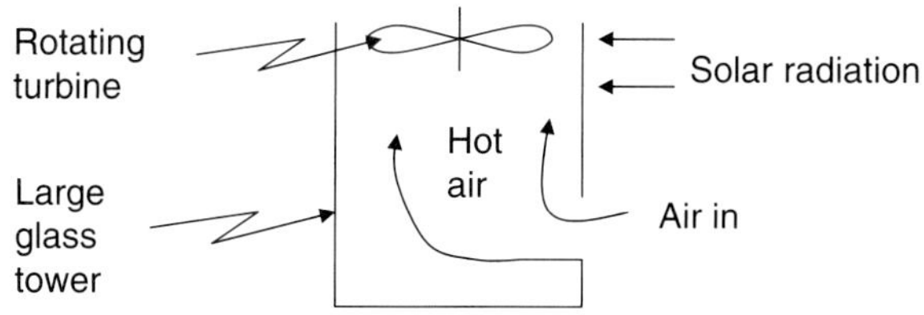

Figure 3.1-3 Solar updraft tower.

mirrors and lenses that can reject portions of the sun's spectrum that PV arrays do not use, thereby reducing excess heat. Another approach is to convert stored solar energy to microwave radiation and beam it down to a combination rectifier-antenna (rectenna), located in an isolated area. The rectenna could convert the microwave energy to direct current power.

3.1.1.4 Solar updraft tower

Figure 3.1-3 shows a low-tech solar thermal power plant where air passes under a very large agricultural glass house (between 2 and 30 km in diameter) that is heated by the sun; it is then channeled upwards toward a convection tower. It then rises naturally and is used to drive turbines, which generate electricity. An energy tower (see Fig. 3.1-4) is an alternative proposal for the solar updraft tower and is driven by spraying water at the top of the tower. Evaporation of water causes a downdraft by cooling the air, thereby increasing its density, driving wind turbines at the bottom of the tower. It requires a hot, arid climate and large quantities of water (sea water may be used for this purpose) but does not require the large glass house of the solar updraft tower.

A solar pond is a low-cost approach to harvesting solar energy. The pond has three layers of water: the top layer with a low salt content; an intermediate layer with a salt gradient, which sets up a density gradient that prevents heat exchange by natural convection in the water; and a bottom layer that has a high salt content, which can reach a temperature of 90°C. The different densities of the layers prevent convection currents. The heat trapped in the salty bottom layer can be used for different purposes, such as heating of buildings, industrial processes, or generating electricity.

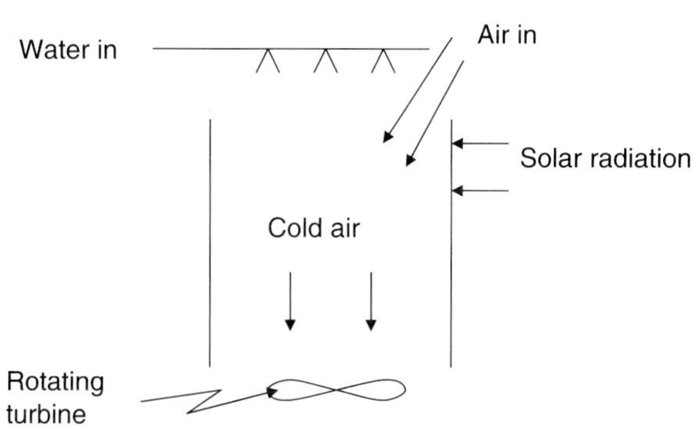

Figure 3.1-4 Energy tower.

Solar chemical refers to a number of possible processes that harness solar energy by absorbing sunlight in a chemical reaction in a way similar to photosynthesis in plants but without using living organisms. Metals, such as zinc, have been shown to drive photoelectrolysis of water. Transition metal compounds, such as titania, titanates, niobates, and tantalates, exhibit very low efficiency photoelectrolysis of water. Man-made photosynthesis reactions could also convert solar energy and CO_2 into useful chemicals.

The world's largest solar power plant, which is located in the Mojave Desert, in California, consists of 4 km^2 of solar reflectors. This plant produces 90% of the world's commercially produced solar power. The total world peak power of installed solar panels was around 5300 MW as of the end of 2005 (International Energy Agency, 2006). The largest solar plant, SEGS in California, produces 350 MW, and the largest nuclear power plants each generate more than 1000 MW per year.

3.1.1.5 Ocean waves

Since water is about a thousand times heavier than air, even a slow-flowing stream of water can yield great amounts of energy. Wind causes waves on the surface of the ocean and on lakes as it transfers some of its energy to the water, through friction between the air molecules and the water molecules. Stronger winds cause larger waves. Wave energy is another form of alternative and sustainable energy. The up and down motion of the waves can be captured to generate power. An experimental wave energy machine installed off the coast of Scotland can generate enough electricity for about 400 homes on a nearby island. The machine has an oscillating chamber made of concrete. As the waves come inside the chamber, the air pocket inside this chamber is compressed. This compressed air is forced through a small hole onto a turbine that generates electricity.

3.1.1.6 Hydroelectricity

Canada generated 61% of its electricity supply from hydroelectricity in 1999, mostly with large dams (Renewable Energy in Canada; Conference Board of Canada, 2003). Hydroelectric generation does not produce significant greenhouse gases, but it does have other major environmental impacts. The reservoirs often destroy nearby habitat by submerging vast areas of highly productive forest and wildlife habitat. The dams also damage freshwater ecosystems by blocking the movement of fish and other organisms. Pollution from mercury and other contaminants is a problem in many reservoirs in northern Canada. Large dams are also known to cause earthquakes. Hydroelectric power now supplies about 715,000 MWe, or 19% of the world's electricity. Hydroelectric power can be far less expensive than electricity generated from fossil fuel or nuclear energy.

3.1.1.7 Wind

In 2005 the U.S. wind energy industry installed 2431 MW of new wind power capacity in 22 states, making the total U.S. wind power capacity 9149 MW (Wind & Hydropower Technologies Program; U.S. Department of Energy; Energy Efficiency and Renewable Energy). Commercial wind turbines are now installed in 30 U.S. states, producing enough electricity annually to equal the power used by 2.3 million U.S. households. Wind power developers invested more than $3 billion in new wind turbines in 2005. The AWEA predicted that installations in 2006 would approach 3000 MW of new wind power capacity (U.S. Department of Energy; Energy Efficiency and Renewable Energy, 2006). Texas now challenges California's status as the state with the most installed wind power. FPL Energy and GE Energy supply most of the wind turbine in the United States. China, Germany, and India follow in the capacity of installed wind power generation systems. The current installed capacity of wind farms is 59,322 MW (less than 1% of worldwide electricity use). The total installed wind power capacity in 2005 of the top five countries are

1. Germany, 18,428 MW,
2. Spain, 10,941 MW,
3. United States, 9549 MW,
4. India, 5200 MW,
5. Denmark, 3128 MW.

Small-scale turbines that are approximately 2 m in diameter weigh about 16 kg, and produce 900 W are available. Wind strengths vary and hence do not give a continuous, steady flow of power. The wind speeds are seasonal, and wind farms require a large land area, where no other activity could be performed. Wind turbines may harm birds. Wind mills located at a slightly higher altitude could capture more energy than those located at the ground level.

3.1.1.8 Geothermal energy

Geothermal energy is obtained from the earth's internal heat and can be used to generate steam to run a steam turbine, which in turn could generate electricity. Three miles below the earth's surface, the temperature is 100°C, which suffices to boil enough water to run a steam-powered electric power plant.

Instead of drilling three miles beneath the earth's surface, scientists could access this power at *geothermal hotspots*, which are volcanic features found all around the world. At the hotspot the mantle is thin, and excess heat from the interior of the earth is transmitted to the outer crust. A few of the hotspots are the volcanic islands of Hawaii, the mineral deposits and geysers in Yellowstone National Park, or the hot springs in Iceland. Iceland produced 170 MW geothermal power and heated 86% of all houses in the year 2000 through geothermal energy (Ragnarsson, 2000). Geothermal power is generated in over 20 countries at an operational capacity of 8000 MW (equivalent to 17% of its electricity from geothermal sources). These geothermal hotspots can easily be used to generate electricity. Some examples follow:

1. One system consists of pumping hot water into permeable sedimentary hotspots found underground and then using the steam liberated to generate electricity. The used steam is condensed and sent back down to the permeable sedimentary stream.
2. Another system utilizes volcanic magma that is still partly molten at around 650°C to boil water, which generates electricity.
3. The third approach uses hot dry rock, which is just hardened magma, but still is extremely hot. To recover this heat from these rocks, water is circulated through the rock to produce steam.

The first system listed above is not as useful as the other methods because of the acidic nature of the fluids (sulfurous and sulfuric acids) found underground. These acidities would damage the equipment, reducing the economic effectiveness of the system. This is a general problem of geothermal energy systems, making them more expensive than other alternative energy sources.

3.1.1.9 Hydrogen

Hydrogen is not an energy source but an energy carrier. One of the main reasons for switching to hydrogen is to prevent global warming caused by fossil fuels since the energy produced by hydrogen does not produce acid rain, CO_2, dust, or nitrous oxides. Of course, if H_2 is produced from fossil fuel, then the whole purpose is lost. Hydrogen and oxygen can react in a fuel cell to produce electricity and water as their reaction's byproduct. Currently, 96% of hydrogen is made from fossil fuels. Based on 2004 data, in the United States 90% is made from natural gas, with an efficiency of 72%. Only 4% of hydrogen is made from water via electrolysis. Currently, the vast majority of electricity comes from fossil fuels in plants that are 30% efficient and from electrolysis that is 70% efficient, which leads to a 20% efficient process to create one unit of hydrogen energy. Using renewable energy is much

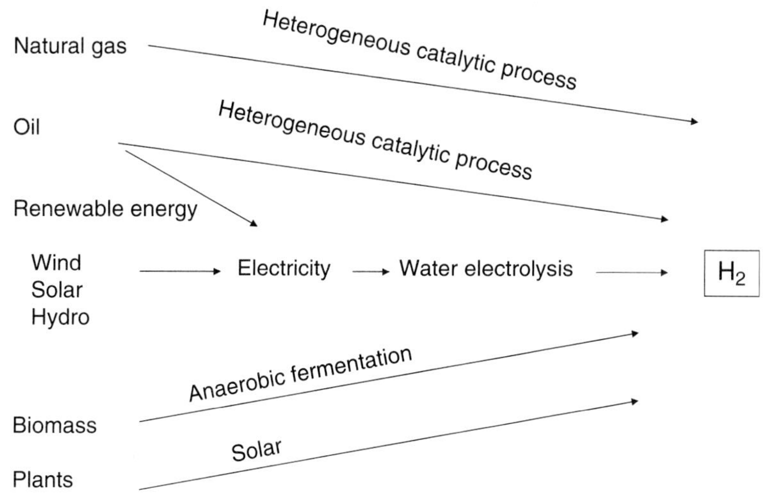

Figure 3.1-5 Various processes for the production of hydrogen.

more useful than using fossil fuel to produce hydrogen. Current wind turbines perform at 30–40% efficiency, producing hydrogen at an overall 25% efficiency. The best solar cells available have an efficiency of 10%, leading to an overall efficiency of 7%. Algae can be used to produce hydrogen at an efficiency of about 0.1% (see Fig. 3.1-5).

Hydrogen can be made from biomass, but this process has several problems:

1. It is a very seasonal process.
2. It contains plenty of moisture and hence requires space to store and energy to dry before gasification.
3. Limited supply of biomass is available for this process.
4. The quantities are not large or consistent enough for large-scale hydrogen production.
5. A large land space is required (cultivated biomass in good soil has a low yield of 10 tons/2.4 acres).
6. The soil will be degraded from erosion, and there will be a loss of fertility if stripped of biomass.
7. Any energy put into the land to grow the biomass, such as fertilizing, planting, and harvesting, will add to the energy costs.
8. The delivery costs to the central power plant are high.
9. The process is not suitable for pure hydrogen production.

To be stored, hydrogen must be compressed and liquefied. To compress hydrogen to 10,000 psi in a multistage compressor, 15% of the energy needs to be utilized. Handling and storage require extreme precautions because the hydrogen is so cold (−423°F), requiring cryogenic support systems. As hydrogen pressure in the storage tank is increased, the volume decreases but the thickness of the tank material increases, adding to cost. Transporting hydrogen or sending it through pipelines also requires a lot of care. Hydrogen also tends to make metal brittle due to the formation of metal hydride. Hydrogen has the lowest ignition point of any fuel, 20 times less than that of gasoline, so it can explode easily due to a spark or static.

The numerous problems associated with fuel cells must be addressed before such technology becomes practical:

1. Fuel cells are heavy. A metal hydride storage system that can hold 5 kg of hydrogen, including the alloy, container, and heat exchangers, would weigh approximately 300 kg, which would lower the vehicle's fuel efficiency.
2. Fuel cells are expensive. In 2003, they cost at least $1 million [U.S. Department of Energy's National Energy Technology Laboratory (NETL) and the Electric Power Research Institute (EPRI)].
3. They are currently not reliable.
4. They need a much less expensive catalyst than platinum.
5. They can clog and lose power if the hydrogen has impurities.
6. They do not last more than 1000 hr.
7. They have yet to achieve a driving range of more than 100 miles. They still cannot compete with electric hybrids like the Toyota Prius, which is already more energy-efficient and lower in CO_2 generation than projected fuel cells.

3.1.1.10 Nuclear power

Nuclear power plants provide about 17% of the world's electricity. There are 400 nuclear power plants around the world, with more than 25% in the United States. France depends more than most countries on nuclear power for electricity, generating about 75% of its electricity from nuclear power. In the United States, nuclear power supplies about 15% of the electricity.

A nuclear reactor uses enriched uranium in the form of *pellets*, with a 1-in. diameter and length. The pellets are arranged into long rods, which are collected together into *bundles*. The bundles are submerged in water inside a pressure vessel. The water acts as a coolant. In order for the reactor to work, the bundle, submerged in water, must be slightly supercritical. Control rods made of a material that absorbs neutrons are inserted into the bundle. Raising and lowering the control rods allow operators to control the rate of the nuclear reaction. To increase the rate of reaction and produce more heat, the rods are raised out of the uranium bundle. To create less heat, the rods are lowered into the uranium bundle. When the rods are lowered completely into the uranium bundle, the reactor shuts down. The uranium bundle heats the water and turns it to steam. The steam drives a steam turbine, which spins a generator to produce power. In some reactors, the steam from the reactor goes through a secondary heat exchanger to convert another loop of water to steam, which drives the turbine. The advantage of this design is that the radioactive water/steam never comes in contact with the turbine. In some other reactors, the coolant fluid in contact with the reactor core is gas (CO_2) or liquid metal (Na, K); these types of reactor allow the core to be operated at higher temperatures. The reactor's pressure vessel is typically housed inside a concrete liner that acts as a radiation shield.

The main disadvantages of nuclear power are the handling and later disposal of the radioactive waste. A safe method for the disposal of radioactive waste, other than burying it deep in the land in concrete bunkers, has yet to be identified. Also, the reactors need several safety protections and safeguards to avoid human exposure to radioactive radiation and leakage of material. While Uranium-235 fission produces no CO_2 gas directly, the processes of mining, milling, refining, moving uranium, and disposing of radioactive waste (ore concentrations) require energy equivalent to the CO_2 gas emissions of a natural gas-fired power station.

3.1.1.11 Biodiesel

Biodiesel is a vegetable oil-based (soy or canola oil) fuel that runs in present-day diesel engines, without any modifications to the hardware. Biodiesel and biodiesel blends can be used in all compression–ignition (CI) engines that were designed to be operated on diesel fuel. It is cheaper than oil, sustainable, and nontoxic; it does not produce acid rain (absence of sulfur); and it does not contribute as much as fossil fuels do to global warming. Studies have shown it reduces engine wear by as much as 30%, primarily because it provides excellent lubrication (Agarwal *et al.*, 2003). Even 2% biodiesel in normal diesel will help achieve this improvement. Biodiesel fuel yields 220% more energy than that required to produce, transport, and distribute it, which is due to the fact that the feedstock crop collects solar energy and transforms it into the biodiesel feedstock oil.

Various biodiesel blends, which include different ratios of biodiesel and diesel from crude oil, can be used in vehicles depending upon the vehicle's requirement and weather conditions. A 20% biodiesel will provide a higher octane rating, superior lubricity, significant emission reductions, and less toxic emissions; will virtually eliminate visible soot emissions; and will have similar fuel consumption, horsepower, and torque. Premium biodiesel is a fuel manufactured from vegetable oils by a transesterification process. Soybean oil is currently the leading source of vegetable oil for biodiesel manufacture in the United States.

Problems with biodiesel are that it is not readily available in large quantities and the amount of NO_X increases by 15%, which contributes to the generation of smog. Another disadvantage is that the viscosity increases at lower ambient temperatures, hence requiring additives for lowering the fuel's gel point.

3.1.2 Renewable energy

Green power describes electricity produced from renewable sources that are less harmful to the environment than fossil fuels. So energy produced from solar, wind power, geothermal, biomass, and small hydroelectric plants is considered green power. *Renewable energy* is an unending source of energy that quickly replenishes itself. Renewable energy does not cause pollution or release toxic substances into the atmosphere; of course, hydroelectric dams cause damage to flora and land area. Some renewable energy systems have environmental problems. Wind turbines can be hazardous to flying birds, while hydroelectric dams can create barriers for migrating fish—a serious problem in the Pacific Northwest, which has seen the destruction of salmon populations. Biomass and biofuels require large amounts of land area. Renewable energy sources provide relatively low-intensity energy, so in order to convert them into useful energy, the collection needs to be distributed over large areas.

Renewable energy comes from an energy resource being replaced by a natural process at a rate that is equal to or faster than the rate at which that resource is being consumed. Renewable energy is a subset of sustainable energy. Most renewable forms of energy, other than geothermal and tidal power, ultimately derive from solar energy. Energy from biomass is derived from plant material and is produced by photosynthesis using the power of the sun. Wind energy derives from winds, which are also generated by the sun's uneven heating of the atmosphere. Hydropower depends on rain, which again depends on the sunlight's power to evaporate water. Even fossil fuels are derived from solar energy since fossil fuels originated from plant material. Renewable energy resources may be used directly or used to create other, more convenient forms of energy. Examples of direct use are solar ovens, geothermal heating, and water and windmills. Biomass refers to any form of plant or animal tissue, including wood, straw, biological waste products such as manure, and other natural materials that contain stored energy. The energy stored in biomass can be released by burning the material directly or by feeding it to microorganisms that use it to make biogas.

The various types of renewable energy are:

1. wind energy,
2. water power,
3. solar energy,
4. geothermal energy,
5. biofuels (including liquid biofuel, solid biomass, and biogas).

3.1.2.1 Biofuel

Biofuel is any fuel with an 80% minimum content by volume of materials derived from living organisms harvested within the 10 years preceding its manufacture. A drawback with biomass is that it needs to be grown, collected, dried, fermented, and burned. All of these steps require resources and an infrastructure. The carbon in biofuels was recently extracted from atmospheric carbon dioxide by the growing plants, so burning it does not result in a net increase of carbon dioxide in the earth's atmosphere. Hence it is considered a renewable source.

Agricultural products, including straw, lumber, manure, sewage, garbage, and food leftovers, are used for the production of bioenergy. Currently, most biofuel is burned to release its stored chemical energy, which is not very efficient. Converting it into electricity using fuel cells is very efficient. Most bioenergy is consumed in developing countries and is used for direct heating. Sweden and Finland supply 17% and 19%, respectively, of their energy needs with bioenergy.

The biomass could be residue from harvesting or crops specifically grown for the task. The biomass could be converted into useful products through the sugar platform. The bulk of the plant material contains cellulose and lignin. The cellulose or hemicellulose is broken down into sugars. Then these sugars can be converted into ethanol or other building block chemicals. Biofuels can be classified as solid, liquid, and gaseous, and several examples exist in each of these classes. Lignin can be burned as a fuel. The second approach involves conversion of biomass to gaseous or liquid fuel by heating it under oxygen-limiting conditions (pyrolysis), which are further converted to useful products. A large number of information resources are available with the U.S. Department of Energy Efficiency and Renewable Energy (http://www1.eere.energy.gov/biomass/for_researchers.html). The Biomass Document Database provides access to most biofuel and many other biomass documents produced by the National Bioenergy Center and its subcontractors since 1980. The National Renewable Energy Laboratory (NREL) Publications Database contains bibliographic information about publications developed or written by NREL staff and subcontractors (http://www.nrel.gov/publications/). The Department of Energy's (DOE) Information Bridge (http://www.osti.gov/bridge/) contains documents and bibliographic citations of DOE research report literature from 1994. The citations relate to physics, chemistry, materials, biology, environmental sciences, energy technologies, engineering, computer and information science, and renewable energy.

3.1.2.2 Solid biomass

Solid forms of biomass are:

- wood,
- straw,
- animal waste,
- crops such as maize, rice, peanuts, and cotton,
- dried, compressed peat.

Certain types of biomass have attracted research and industrial attention, since they are available in very large quantities and have low market value. They are algae, bagasse from sugarcane, dried distiller's grain, firewood, hemp, jatropha, maize (corn), manure, meat and bone meal, peat, rice hulls, silage, stover, and whey.

3.1.2.3 Liquid biomass

A number of liquid forms of biomass can be used as a fuel:

- Bioalcohols.
- Ethanol produced from sugar cane is being used as automotive fuel in Brazil. Ethanol produced from corn is being used as a gasoline additive (oxygenator) in the United States.
- Methanol, which is currently produced from natural gas, can also be produced from biomass.

- An acetone–butanol-ethanol mixture can be produced by anaerobic fermentation. This mixture can be used in existing gasoline engines.
- Biologically produced oils can be used in diesel engines:
  straight vegetable oil,
  waste vegetable oil,
  biodiesel obtained from transesterification of animal fats and vegetable oil.

Liquid biofuel is usually bioalcohol such as ethanol and biodiesel and virgin vegetable oils. E85 is a fuel composed of 85% ethanol and 15% gasoline that is currently being sold to consumers in the United States. The European Union plans to add 5% bioethanol to Europe's petrol by 2010.

3.1.2.4 Gaseous biomass

Forms of gaseous biomass include:

- biomethane produced by the natural decay of garbage or agricultural manure,
- wood gas,
- hydrogen produced by the electrolysis of water,
- gasification, which produces carbon monoxide.

Many organic materials can release gases, due to the metabolism of organic matter by bacteria under anaerobic fermentation. Under high-pressure, high-temperature, and anaerobic conditions, many organic materials such as wood can be gasified to produce gas. Biogas can be produced from current waste streams, such as paper production, sugar production, sewage, and animal waste.

3.1.3 Future sources of renewable energy

A difference in salt concentration exists between sea water and river water. This gradient can be utilized to generate electricity by separating positive and negative ions by ion-specific membranes. Brackish water is produced. It is predicted that one third of the electricity needs in the Netherlands can be covered with this system. Ocean thermal energy conversion (OTEC) uses the temperature difference between the warmer surface of the ocean and the cold lower recesses to employ a cyclic heat engine. Lake-bottom water is constant at 4°C, and this water can be used for cooling fluids that flow through submerged pipes.

3.1.4 Conclusions

Countries that are currently highly dependent on fossil fuel should focus all their efforts toward harnessing renewable energy sources. Although the cost of production is still high, more effort would bring down this value. Tapping renewable energy sources would also decentralize energy supply to the individual household, ending energy supply's status as a political issue.

Different sources (including waste) are being investigated for extracting energy. For example, Ozmotech, a major Australian company, has developed a pyrolysis process under inert-atmosphere to convert 400,000 tons of plastic waste into 350 million liters of diesel per annum (Ecos, 2006).

References

Agarwal, A. K., Bijwe, J., and Das, L. M., Wear assesment in a biodiesel fueled compression ignition engine, *J. of Engng. Gas Turbines and Power*, **125**(3): 820–826, 2003.

Archer, D., Fate of fossil fuel CO_2 in geologic time, *J. Geophys. Res.*, **110**: C09S05, 2005.

Biodiesel for Oregon, Oregon Environmental Council, Portland OR (www.biofuels4oregon.org).

Caldeira, K. and Wickett, M. E., Ocean model predictions of chemistry changes from carbon dioxide emissions to the atmosphere and ocean, *J. Geophys. Res.*, **110**: C09S04, 2005.

Conference Board of Canada, Renewable Energy in Canada, 2003.

Dore, C. J., Goodwin, J. W. L., Watterson *et al.* National Atmospheric Emissions Inventory, U.K., Emissions of Air Pollutants 1970–2003, 2003.

Ecos, Ozmotech's plastic-to-fuel solution in demand, **130**: 5, 2006.

Energy Information Administration, Office of Integrated Analysis & Forecasting, Emissions of Greenhouse Gases in the United States 2001. Washington, D.C., 2002.

Henley, M., Potter, S., Howell, J., and Mankins, J., Wireless power transmission options for space solar power, *EPRI Journal* (Spring): 6–17, 2000.

Hubbert, M. K., *Nuclear Energy and the Fossil Fuels*, Presented before the Spring Meeting of the Southern District, American Petroleum Institute, March 1956.

IPCC, Radiative forcing report from 1994, updated (to 1998) by IPCC TAR, 1998.

Jia H. P., The People's Republic of China—Initial National Communication on Climate Change—2004, United Nations Framework Convention on Climate Change (UNFCCC), 24 Nov. 2004; www.scidev.net

Ragnarsson, A., *Geothermal Development in Iceland*, Proceedings World Geothermal Congress 2000, Kyushu-Tohoku, Japan, May 28–June 10, 2000.

U.S. Department of Energy; Energy Efficiency and Renewable Energy, U.S. Wind Power Industry Tempers Its 2006 Forecast Slightly. EERE Network News, November 2006.

Chapter 3.2

3.2

Energy reserves and renewable energy sources

During industrial expansion, we can rapidly deplete available resources. This has occurred for petroleum on U.S. soil. Proven, recoverable oil reserves in the United States would only power our thirst for oil for 3 years if cut off from oil imports. Technology can meet the challenges of dwindling U.S. oil reserves but only by switching to the abundance of other energy forms and reserves. This energy is available in three forms: fossilized solar energy, nuclear energy, and recent solar energy.

3.2.1 Fossil fuel reserves

The "fossil" designation of certain fuels implies that the fuel energy content originates from prehistoric vegetation or organisms. Fossil fuels are the most commonly used energy source to drive our machines. Unlike wind and sunlight, which are dispersed in low concentrations across the surface of the Earth, commonly used fossil fuels tend to be concentrated at locations near the Earth's surface. Where these are easily accessible, we are able to collect them with great efficiency. Fossil fuel sources include the following:

- Coal
- Petroleum
- Heavy oil
- Oil sands
- Oil shale
- Methane hydrates
- Natural gas.

In Wyoming, some coal seams are 40 feet thick and less than 100 feet underground. In the Middle East, hundreds of barrels per day of crude oil can flow from a single well under its own pressure. Each source provides abundant energy.

Coal,[1] petroleum,[2] and natural gas[3] are accessible fossil fuels and easy to use (see box, "Petroleum and Gas"). By far, they are today's most popular fuels. Figure 3.2-1 summarizes the known accessible reserves of these fuels in the entire world and in the United States. World recoverable reserves for coal, natural gas, and petroleum are 2.5E + 19 (25 billion billion), 6.9E + 19, and 5.3E + 18 Btus.[4] For coal, the total estimated reserves are about a factor of ten higher than the estimate for recoverable reserves.[5]

In the year 2000, the United States consumed 19.7 million barrels of petroleum per day or 3.8E + 16 Btus per year. This consumption would deplete known U.S. petroleum reserves in about 3 years and estimated U.S. petroleum reserves in about 7.6 years. These statistics are summarized in Table 3.2-1.

Petroleum represents about 53% of the total annual U.S. energy consumption.[i] The U.S. total energy consumption would deplete U.S. estimated reserves of petroleum in four years. Natural gas would last 30 years, and coal would last 90 years. Coal can be used for much more than electrical power production (see box, "Strategic technologies"). World natural gas reserves would last 1,000 years if they were used only to meet U.S. consumption. Figure 3.2-1 illustrates the relative magnitude of these reserves.

Estimated energy reserves in heavy oil, oil sands, oil shale, and methane hydrates dwarf known reserves in coal, natural gas, and petroleum. One evolutionary route to form these three reserves includes the advanced

[i] The total U.S. energy consumption is about 7.1E + 16 Btus per year.

Renewable Energy Focus Handbook 2009; ISBN: 9780123747051

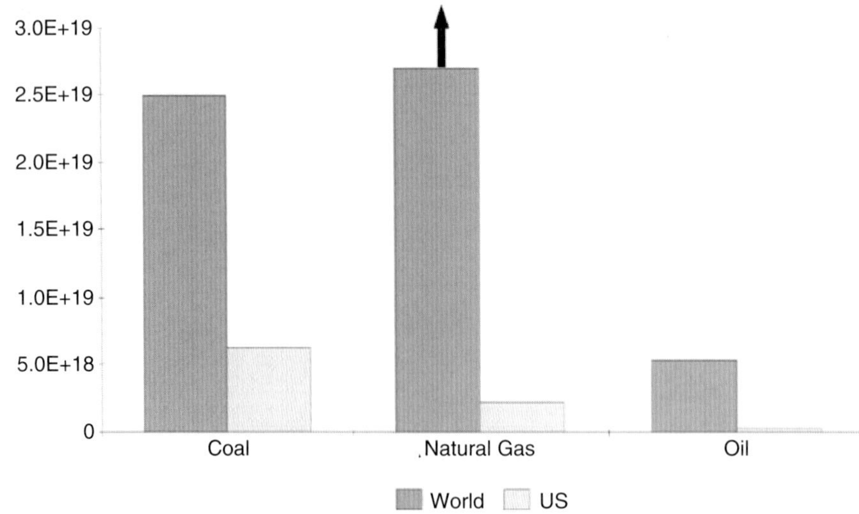

Figure 3.2-1 Summary of world and U.S. fuel reserves in Btus.

Petroleum and gas—from the ground to the refinery

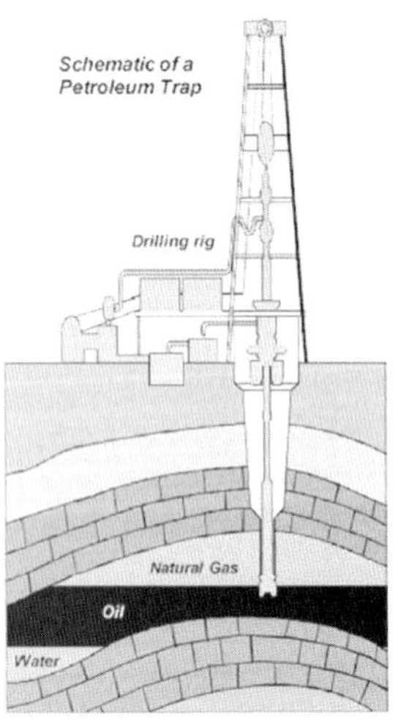

The inserted image[6] illustrates a typical petroleum reservoir and drilling used to recover that petroleum. A rock cap has kept the reserve isolated from atmospheric oxygen for millions of years. Since the petroleum is lighter than water, it floats above water aquifers. Petroleum gases, including natural gas, are the least dense material in the formation and are above the oil. Drilling and inserting a pipe into the petroleum reserve allow recovery. Oil reserves recovered by conventional land drilling applications are typically several hundred feet to about one mile deep.

stages of petroleum decay. Petroleum contains a wide range of hydrocarbons, ranging from the very volatile methane to nonvolatile asphaltines/tars. When petroleum is sealed securely in rock formations, the range of volatility is preserved for millions of years or converted to methane if buried deeper, where it is converted by geothermal heat.

If the rock overburden is fractured, either due to erosion of rocks above the formation or simply due to the weak or porous nature of the rock, the more volatile components of petroleum escape. This leaves less-volatile residues in the forms of heavy oils, oil sand, and oil shale.

Heavy oils are volatile-depleted deposits that will not flow at reservoir conditions but need assistance for recovery. Oil sand liquids are heavier than heavy oils—typically not mobile at reservoir conditions, but heat or solvents can make the oil flow through the porous rock. Oil shales are usually immobile and present in rocks that do not allow oil flow. Unlike the oil in oil sands that can be removed in situ or with low amounts of solvent and heat, the oil in shale tends to be very difficult to remove.

This makes oil shales more difficult to recover than oil sand. Unlike coal, which is a concentrated fossil fuel, oil shale is best characterized as a relatively nonvolatile oil dispersed in a shale. World reserves are estimated to be 600 to 3,000 times world crude oil reserves.[7] Lower estimates specific to the western United States place reserves at two to five times known world oil reserves.[8]

Heavy oil reserves in Venezuela are estimated to be from 0.1 trillion barrels[9] to 1 trillion barrels.[10] These heavy oils are generally easier to recover than oil sands and much easier to recover than oil from oil shale. The United States, Canada, Russia, and Middle East also have heavy oil reserves totaling about 0.3 trillion barrels[11]

Table 3.2-1 Comparison of energy reserves and rates of consumption

Energy description	Amount (Btus)	Amount (as indicated)
World Recoverable Coal Reserves	2.5E 19	
World Recoverable Natural Gas Reserves	6.9E 19	
World Recoverable Petroleum Reserves	5.3E 18	
U.S. Petroleum Consumption (year 2000)	3.8E 16	19.7E 6 barrels
U.S. Petroleum Consumption divided by U.S. Known Petroleum Reserves		3 years
U.S. Total Recoverable Coal Reserves Divided by U.S. Total Energy Consumption		90 years

(lower side of estimates). In all, the heavy oil reserves are estimated to be slightly greater than all the more-easily recovered conventional crude oil reserves.

Surface reserves of oil sands have been mined and converted to gasoline and diesel since 1969 in Alberta, Canada. Production costs are about \$20 per barrel.[12] These supply about 12% of Canada's petroleum needs. The sands are strip-mined and extracted with hot water. Estimated reserves in Alberta are 1.2–1.7 trillion barrels with two open pit mines now operating.[13] Other estimates put oil sand reserves in Canada, Venezuela, and Russia at about 3, 3, and 0.6 trillion barrels respectively. Estimates approximate 90% of the world's heavy oil (and oil sand) to be in Western Canada and Venezuela.[14]

Strategic technologies of the 21st century

A few technologies stand out as logical extensions of current technology that have both economic and strategic value. These technologies generally require greater coordination than first generation technologies, and they are based on adding value to abundant indigenous resources. Uranium reprocessing is a strategic technology for electrical power generation.

Coal and biomass qualify as abundant local resources that could be better utilized. The solid fuel refinery uses these feed stocks to supply an array of conversion and synthesis processes, including electrical power generation, production of liquid fuels, and production of chemicals like ammonia. The synthesis gas pipeline is based on much the same concept as the solid fuel refinery, but in the synthesis gas pipeline a pipe network is fed mixtures of hydrogen and carbon monoxide from several source locations. Likewise, the pipeline is used as a feed stock for chemicals and fuel by many processes. Two technologies stand out to tap into these resources (see "Option 1" and "Option 2" boxes).

Cumulatively, oil sand reserves are six to ten times proven conventional crude oil reserves. (Conventional crude oil reserves are reported at 1 trillion barrels.)

If methane escapes from an underground deposit (due to porous rock or erosion of overburden) and comes in contact with a combination of increased pressure, water, and cold temperatures, methane hydrate is formed. Methane hydrate is ice that contains methane and is stable below the freezing point of water as well as at temperatures slightly warmer than the freezing point of water at high pressure. Conditions are right for the formation of methane hydrates on the sea floor (under a few hundred feet of seawater, where the temperature is relatively constant at the temperature of maximum water density, 4°C; this is true even in the Caribbean) and in the Arctic permafrost. In addition to formation mechanisms involving petroleum decay, methane is commonly formed directly from biomass—both geological and recent biomass methane can form hydrates.

Methane hydrate reserves are not presently recovered. Countries like Japan have great interest in the potential of this technology because of the lack of natural fossil fuel reserves in the country and the large coastal water areas. In the United States, methane hydrates have received the attention of congressional hearings where reserves were estimated at 400 million trillion cubic feet (200,000 trillion cubic feet of gas (Tcf) in reserves under the jurisdiction of the United States). In the most conservative interpretation, these hydrates have enough energy to last maybe 5,000 years.

Natural gas emissions from these reserves occur naturally, so the methane greenhouse effect from this source will occur regardless of whether or not we use the energy. If we burn the natural gas from these emissions, the resulting carbon dioxide would have about one-tenth of the greenhouse effect of the equivalent methane release. The release of methane from methane hydrates on the sea floor may have contributed to the end of many of the ice ages. During ice ages, lower sea levels reduced the pressure on the sea floors. This lower pressure would lead to methane release.[15]

Option 1: synthesis gas pipeline

In the solid fuel refinery, the synthesis gas generation can be separated from the other processes. This represents a potential added cost to recovering the heat from the synthesis, which is not required in a solid fuel refinery. The synthesis gas pipeline has this drawback, plus the expense of the pipeline, but the benefits are many.

Reduced transportation costs (pipeline versus railroad) are an advantage, as well as the ease of disposal of coal ash at the mining location as landfill to replace space created by removing the coal. Sulfur removal from the coal would be easier using this approach. This would allow the reopening and increased use of several minefields across the country that contain high-sulfur coal.

The synthesis gas will allow electrical power generation from coal approaching 50% thermal efficiency, which is about 10% better than direct firing coal. Furthermore, customers are likely to use this option, since they would not have the burden of building a gasification facility. The synthesis gas would have all the advantages of natural gas, but it would presumably be less costly on an equal energy basis.

A synthesis gas pipeline would allow smaller companies and entrepreneurs to enter the energy big business. Opportunities would exist both for production of synthesis gas (potentially from biomass or municipal solid waste) and use of the gas from the pipeline in a large array of processes. The pipeline would make it possible for companies to enter business with smaller investments and easy feedstock acquisition.

The trick to reducing natural methane emissions is to mine and recover those reserves that are most likely to release naturally. Most hydrate mining research involves changing the temperature and pressure at the solid reserve location to cause the methane hydrates to melt or sublime and then to recover the methane that is evolved. Experts at the congressional hearing agreed that *Alaska's North Slope was the most likely candidate for initial research because of its relative easy access (compared to the deep-water Gulf of Mexico) and in-place infrastructure.*[16]

In some instances, natural gas reserves are below ground and in equilibrium with methane hydrate reserves. When the natural gas is recovered, the methane hydrates melt, resupplying the gas in the reserve for easy recovery.

Table 3.2-2 summarizes the energies available in recoverable fuels. All of these fuels are in concentrated deposits, with the exception of uranium. The uranium availability includes recovering uranium from sea waters, which is possible but costly compared to today's prices.

Table 3.2-2 Relative abundance of recoverable fuels

U.S. Petroleum	1/20 ×
World Petroleum	1 ×
World Recoverable Coal	5 ×
World Recoverable Heavy Oil and Oil Sands	10 ×
World Recoverable Natural Gas	15 ×
World Recoverable Oil Shale	>500 ×
World Methane Hydrates	>5,000 ×
World Recoverable Uranium	50,000 ×

The numbers approximate the magnitudes of the different reserves relative to conventional crude oil reserves.

Estimating energy reserves has historically been inaccurate. For example, in the 1980s when the oil-producing countries shifted from a mentality of "creating the perception of oil shortfalls" to setting production quotas based on "countries reported reserves," the reported reserves increased dramatically. Likewise, reported coal reserves decreased by a factor of 10 from 1980[17] to 2002 due to redefinitions of recoverable reserves and the influence of oil companies on U.S. policy and U.S. Department of Energy positions.

Qualitatively, the image portrayed by Table 3.2-2 is correct. The costs of fuels reflect this. Petroleum ($45–$75 per barrel) and natural gas cost $9.00-$15.00 and $6.00–$12.00 per MBtu, respectively.

Coal costs about $1.20–$1.40 per MBtu. Uranium costs $0.62 per MBtu when 3.4% is fissioned or $0.062 per MBtu if a fission of 35% is assumed.

The following uses of these reserves are consistent with recent trends:

- Major oil corporations will likely progressively tap into oil sands and heavy oil as the reserves of petroleum are depleted. The corporations will likely be able to meet petroleum needs for several decades in the progression; however, this progression will

Option 2: solid fuel refinery

A solid fuel refinery capitalizes on the strengths of several processes to overcome the weaknesses of each process to produce much higher final conversion efficiencies. For example, electrical power generation does not effectively use low-temperature steam and flue gases, but it produces large quantities of these gases.

For Fischer-Tropsch synthesis, conversion of the last bit of hydrogen and carbon monoxide to liquid fuels is considerably more expensive than the initial conversion of fresh feed. In a solid fuel refinery, the residual carbon monoxide and hydrogen are suitable for driving a gas turbine to generate electricity. These residual gases left from liquefied fuel production are used for electrical power generation that is more efficient than generation by a coal-fired power plant.

Ammonia is an important chemical produced in a solid fuel refinery because of the energy intensity of the ammonia production process and its value as a fertilizer. Building one big gasifier to feed several processes brings an important economy of scale for this important step.

Solid fuel refineries are a reality in South Africa, as illustrated by the flow diagram. A large array of chemicals are produced. The overall process makes the best use of coal resources for supplying many useful products.

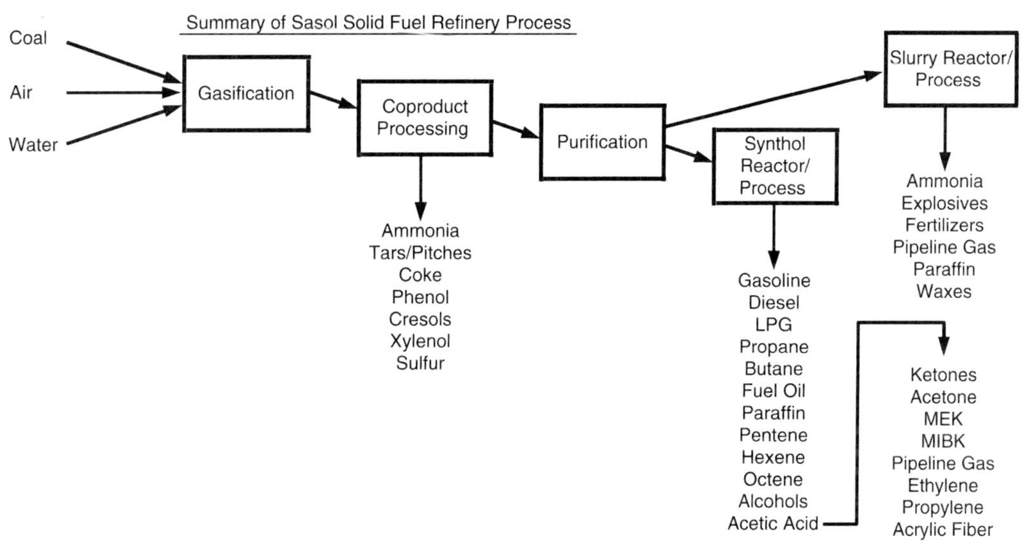

likely occur to maintain corporate profit, will result in huge trade deficits, and may be at the expense of continuous military activity.

- Natural gas use in the United States is likely to follow the course of petroleum. The depletion of U.S. reserves will lead to increasing reliance on imports.
- Coal will continue increased use in electrical power generation. However, its high rate of carbon dioxide generation and limited recoverable reserves will dampen its expansion relative to 1970s estimates that stated centuries of abundance. New technology is likely to bring a greater portion of the total coal reserves into the "recoverable" coal reserves category.
- Energies in oil shale are unlikely to be realized due to the high energy and cost of recovery.
- Methane hydrate recovery is uncertain because not enough is known about safe methods for its recovery, and the extent of reserves of "recoverable" methane hydrates depends on this technology.

Impacting technologies likely to develop in the next 30 years include 100% nuclear fission of nuclear fuel for abundant electricity, hybrid vehicles that are partially rechargeable with grid electricity, and chemicals and fuels made from biomass. Closed cycle nuclear fuel cycles that use all of the energy in uranium (including U-238) include reprocessing of spent nuclear fuel and the concentration of fission products so that the volume of waste is 100 times less than if the fuel rods were directly placed in repositories. These are the technologies that can change the rules, can deliver sustainability, and can promote economic prosperity by eliminating the need to import oil and natural gas.

3.2.2 Cosmic history of fossil energy reserves

Similar to the fossilized bones of a dinosaur indicating its existence in the remote past, fossil fuels are the remains

What is permanence?

The Permanence Scale in Figure 3.2-2 is the binding energy in MeV for each nucleon (a proton or neutron) in the nucleus of an atom. The most stable atoms are those with the highest binding energy per nucleon. The nucleon-binding energy is plotted against the atomic mass number. The maximum binding energy per nucleon occurs at mass number 54, the mass number for iron. As the atomic mass number increases from zero, the binding energy per nucleon also increases because the number of protons (with positive charge) increases, and protons strongly repel each other. The binding energy of the neutron (no charge) holds the nucleus of the atom together. The atomic mass number for each atom is approximately the mass of the protons and neutrons in the nucleus, since the electrons (equal to the number of protons) have a mass that is about 1/1,837 that of a nucleon.

Above a mass number of about 60, the binding energy per nucleon decreases. Visualize the large atom nucleus as many protons try to get away from each other, with a much larger number of neutrons holding that nucleus together. The common isotope of iron-56 (about 92% of the mass of natural iron) has 26 protons and 30 neutrons. The common isotope of uranium-238 (about 99.27% of the mass of natural uranium) has 92 protons and 146 neutrons. The iron nucleus is very stable. The uranium-238 isotope does spontaneously decay, but it will take about 4.5 billion years for one-half of a lump of pure U-238 to decay, so it is practically a stable isotope.

An additional comment: The nucleon-binding energy increases as the atomic mass decreases from 260 to 60. This is the primary reason for the energy release in the fission process that makes nuclear reactors work. Since the forces in the nucleus of large atoms are so carefully balanced, a small energy addition (a low-energy neutron entering the nucleus) will cause the large forces acting between the protons to become unbalanced, and the nucleus comes apart (flies apart), releasing lots of energy. These large atoms are the fuel for nuclear reactors. The strong binding energies of the smaller atoms lead to the permanent end point of the natural isotope decay processes.

of plants and microscopic organisms. Their compositions reveal a history going back hundreds of millions of years; however, the history of the energy in these fuels does not start there. This history goes back to the origin of the universe. An understanding of the dominant role nuclear energy has played in the history of energy helps us understand how nuclear energy will always be part of the energy mix we use.

The arrays of different elements in our planet, the solar system, and the galaxy reveal their history. Hydrogen is the smallest of the atoms assigned an atomic number of one. Physicists tell us that at the birth of the universe, it consisted mostly of hydrogen. Stars converted hydrogen to helium, and supernovas (see the box "Supernovas") generated the larger atoms through atomic fusion.

Atoms are identified based on the number of protons (positively charged subatomic particles). Since both hydrogen and helium are smaller atoms than the nitrogen and oxygen in the air, the Goodyear blimp (filled with helium) and the Hindenburg zeppelin (filled with hydrogen) floated in air.

Helium has 2 protons, lithium has 3 protons, carbon has 6 protons, and oxygen has 8 protons. The number of protons in an atom is referred to as the "atomic number" of the atom. Atoms are named and classified by their atomic number. Atoms having between 1 and 118 protons have been detected and named (see the box "Making New Molecules in the Lab"). The atomic mass is the sum of the mass of neutrons, protons, and electrons in an atom— the atomic mass and the atomic spacing determines the density of materials.

Protons are packed together with neutrons (subatomic particles without a charge) to form an atom nucleus. There are more stable and less stable combinations of these protons and neutrons. Figure 3.2-2 illustrates the permanence of nuclei as a function of the atomic mass number (the atomic mass is the sum of the protons and neutrons). Helium 3—abbreviated He,3—is shown to have a lower permanence than He,4. Two neutrons simply hold the two protons in He,4 together better than one neutron in He,3. In general, the number of neutrons in an atom must be equal to or greater than the number of protons, or that atom will disintegrate into more stable combinations of protons and neutrons.

The wealth of information in Figure 3.2-2 explains much about the chemistry of our planet. For example, why do hydrogen atoms combine to form helium instead of breaking apart to form hydrogen atoms? Figure 3.2-2 illustrates that the "permanence" is greater for He,4 than for hydrogen (H,1). In nuclear reaction processes, atoms tend to move uphill on the curve of Figure 3.2-2 toward more stable states. In Chapter 10, the concept of atomic stability will be discussed in greater detail, and the term "binding energy" will be defined and used in place of "permanence."

To make atomic transitions to more permanent/stable atoms, extreme conditions are necessary (see the box "Supernovas"). On the sun, conditions are sufficiently extreme to allow hydrogen to fuse to more stable, larger molecules. In nuclear fission reactions in a nuclear power plant or in Earth's natural uranium deposits, large molecules break apart to form more stable smaller molecules. Each atomic event is toward more stable

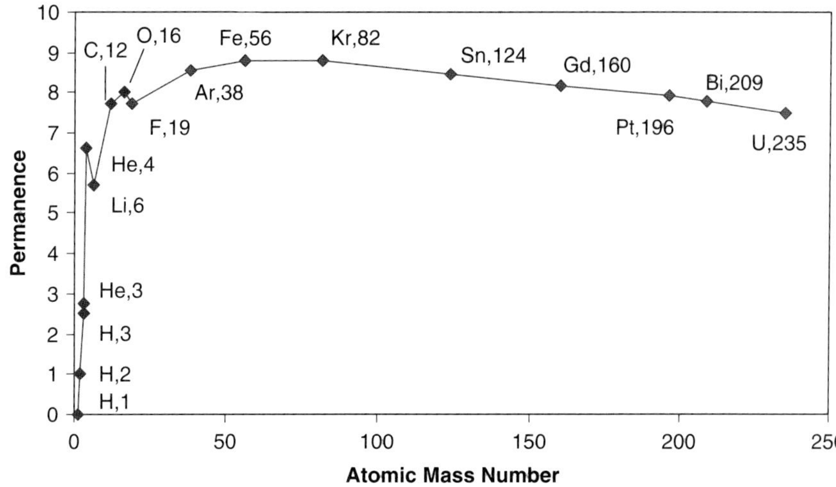

Figure 3.2-2 Impact of atomic mass number on permanence of atoms. H is hydrogen, He is helium, Li is lithium, C is carbon, O is oxygen, F is fluorine, Ar is argon, Fe is iron, Kr is krypton, Sn is tin, Gd is gadolinium, Pu is plutonium, Bi is bismuth, and U is uranium.

combinations of protons and neutrons, and this process releases energy.

Figure 3.2-3 is the starting point for qualitative understanding the history of energy. Nuclear reactions are where the history of energy begins. The story of the history of energy goes something like this: Once upon a time, long ago—about 15 billion years—there was a big bang. From essentially nothingness, in an infinitely small corner of space, protons and helium were formed. Carbon, iron, copper, gold, and the majority of other atoms did not exist.

Unimaginably large quantities of hydrogen and helium clustered together to form stars. The most massive of these stars formed supernovas. Fusion conditions were so intense in these supernovas that atoms of essentially all known atomic numbers were formed. Hence, carbon, oxygen, iron, copper, gold, and the vast array of atoms that form solid objects around us were formed.

Uranium was also formed along with atoms larger than uranium. The largest of these atoms rapidly fell apart to form more stable molecules. Uranium and plutonium have intermediate stability. They could be induced to fall apart but were stable enough to last for billions of years without spontaneous decomposition.

The spinning masses continued to fly outward from the big bang. As time passed, localized masses collected to form galaxies, and within these galaxies, solar systems, stars, planets, asteroids, and comets formed.

It is here where energy and the universe as we know it began to take form. We are just beginning to understand the processes of the stars and supernovas to tap into the vast amounts of binding energy in the atom. The atomic binding energy available in one pound of uranium is equivalent to the chemical binding energy present in 8 million pounds of coal.[ii]

Supernovas—the atomic factories of the universe

Astronomers have observed "lead stars" that produced heavier metals like lead and tungsten. Three have been observed about 1,600 light-years from Earth. To paraphrase a description of the process:

Stars are nuclear "factories" where new elements are made by smashing atomic particles together. Hydrogen atoms fuse to create helium. As stars age and use up their nuclear fuel, helium is fused into carbon.

Carbon, in turn, is fused into oxygen, and the process continues to make heavier elements until a natural limit is reached at iron. To make elements heavier than iron, a different system is needed that adds neutrons to the atomic nuclei. Neutrons are a kind of atomic "ballast" that carry no electric charge.

Scientists believe there are two places where this can occur: inside very massive stars when they explode as supernovas and more commonly, in normal stars right at the end of their lives before they burn out.

3.2.3 Nuclear energy

In principle, nuclear energy is available in all elements smaller than iron through nuclear fusion and all elements larger than iron through nuclear fission. Iron is on the top

[ii] 12,000 tons per day = 365 days/year × 40%/30%/(750 kg × 1 ton/~1,000 kg).

From hydrogen to helium

The most abundant atom in the universe is hydrogen. Hydrogen is the fuel of the stars. The diagram illustrates how four protons interact to produce a new stable molecule: helium. Considerable energy is released in the process—about 25 MeV of energy for every helium atom formed (570 million Btus per gram helium formed).

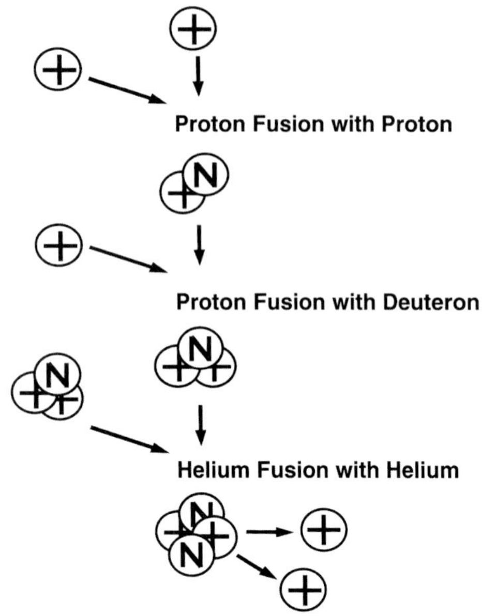

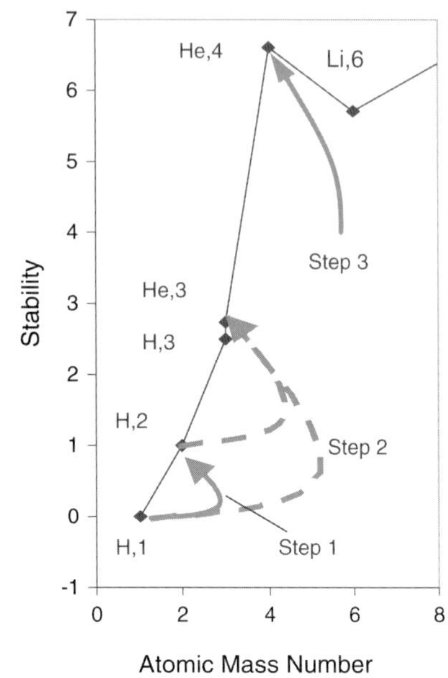

of the permanence curve so it is one of the most abundant elements. While all other atoms can degrade to iron, iron does not degrade. When most of the nuclear energy of the universe is expended, it will consist mostly of iron and elements of similar atomic number (all being stable).

In general, the largest atoms are the most likely—given enough time—to undergo nuclear decays such as the release of an alpha particle (a helium atom). Atoms larger than uranium have undergone fission to the extent that they can no longer be found on Earth. The amount of U-238 on Earth today is slightly less than half of what was present at Earth's formation. The amount of U-235 on Earth today is less than 1% of what was present at Earth's formation.

Of interest to us is the ability to perform these nuclear processes in a controlled and safe manner, because the nuclear binding energy can be used to produce electricity. The energy released as protons, neutrons, and atoms combines and rearranges in the progression to higher "binding energy." We are able to use nuclear fission on a practical/commercial scale with one naturally occurring element: uranium. We could perform fission on elements larger than uranium, but these are not readily available. In the H-bomb, we have demonstrated an ability to tap the energy of fusion for massive destruction, but use of fusion for domestic energy production is much more difficult. Practical nuclear fusion methods are an area of active research.

The only practical nuclear energy sources today are nuclear fission of uranium in nuclear reactors and the recovery of geothermal heat produced by nuclear decay under the surface of Earth (occurring continuously). Uranium is the primary fuel for both of these processes.

At 18.7 times the density of water, uranium is the heaviest of all the naturally occurring elements (the lightest is hydrogen; iron is 7.7 times the density of water). All elements (as defined by the number of protons in the nucleus) occur in slightly differing forms known as *isotopes*. These different forms are caused by the varying number of neutrons packed with the 92 protons in uranium's nucleus. Uranium has 16 isotopes, only two are stable. Most (99.3%) of natural uranium is composed of uranium-238 (U-238, 238 is the sum of neutrons and protons) and U-235, about 0.71% of natural uranium.

U-235 is slightly less stable than U-238 and when enriched to 3% to 8% can be made to release heat continuously in a nuclear reactor. Enriched to 90% U-235 and the sudden release of large amounts of energy becomes a nuclear bomb. We have mastered the technology to perform both of these processes. The U-238 decays

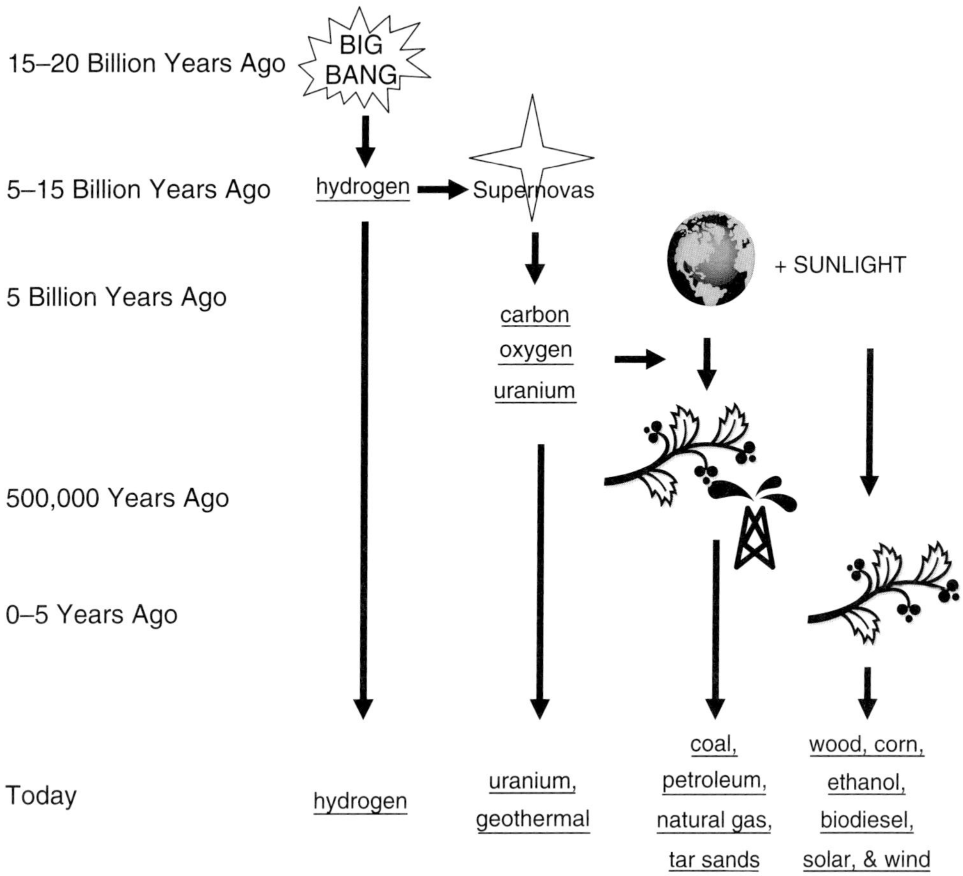

Figure 3.2-3 The history of energy.

slowly with about 2 kilograms of uranium decaying to 1 kilogram of uranium and slightly less than 1 kilogram of fission products in about 4.5 billion years. About half of the U-238 present when Earth was formed (and >99% of the U-235) has decayed, keeping the Earth's interior a molten metal core.

U-235 decays faster than U-238. We are able to induce the fission of U-235 by bombarding it with neutrons. When one neutron enters the U-235 nucleus to form U-236, it breaks apart almost instantly because it is unstable. It breaks apart to form the nucleus of smaller atoms plus two or three neutrons. These two or three neutrons can collide with U-235 to produce another fission in a sustained chain reaction.

Nuclear fission occurs when the nucleus of an atom captures a neutron and breaks apart expelling two or three neutrons. The U-235 continuously undergoes fission (fission half life of 1.8E 18 years, alpha-decay half life of 6.8E 8 years) that proceeds slowly because there is so little U-235 in the metal; most of the emitted neutrons are lost with only a few producing fission. The exception was the natural nuclear reactor that formed in Oklo, Gabon (Africa), about 2 billion years ago. This occurred when the concentration of U-235 in ore at Oklo was high enough to cause a chain fission of the U-235 leading to lower-than-normal U-235 concentrations and trace plutonium in the deposits today.

Our use of nuclear fission to make a bomb is based on an uncontrolled chain reaction. A neutron chain reaction results when, for example, two of the neutrons produced by U-235 fission produce two new fission events. This will occur when nearly pure U-235 is formed into a sphere that contains a critical mass: about 60 kilograms of metal. Then in each interval of 10 billionths of a second, the number of fission events grows from 1, 2, 4,... 64,... 1,024, 2,048,... as illustrated by Figure 3.2-4. The transition from very few fission events to an uncountable number occurs in less than a microsecond. The enormous energy released in this microsecond is the source of the incredible explosive power of a nuclear fission bomb.

This escalating chain reaction is to be distinguished from the controlled steady-state process as depicted by Figure 3.2-5. In a controlled steady-state process, a nearly constant rate of fission occurs (rather than a rapidly increasing rate) with a resulting constant release of energy.

The first nuclear bomb used in war exploded over Hiroshima, Japan, was a U-235 bomb. Two hemispheres containing half of the critical mass are slammed together

Making new molecules in the lab

On small scales, scientists are able to make new molecules similar to the way in which supernovas combine two smaller molecules to form a larger molecule. As illustrated in the figure, krypton and lead combine to form a compound nucleus. This nucleus is very unstable and rapidly degrades to a more stable atom indicated as Element 118.

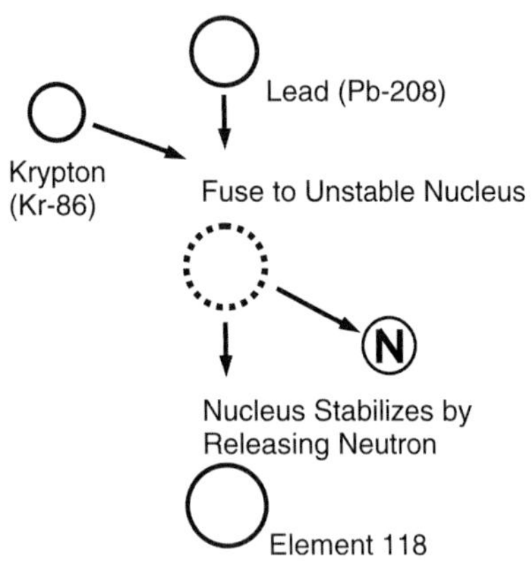

This work was performed at the Lawrence Berkeley National Lab. During this synthesis, two atoms are joined to actually form a less stable molecule. This is possible in a particle accelerator that puts kinetic energy (high speed) into the krypton. This high speed provides the extra energy needed to fuse the nuclei. The atomic rearrangement resulting in the release of a neutron helps lock in a final element that is stable. High atomic number atoms are unstable and not found in nature.

with conventional explosive charges. In the resulting nuclear explosion, about 2% of the U-235 mass underwent fission. Everything else in the bomb was instantly vaporized. The fireball and the explosion shock wave incinerated and leveled a vast area of Hiroshima. This is the legacy of nuclear energy that indelibly etched fear into the minds of world citizens. The second explosion at Nagasaki was a plutonium bomb, followed by the development and testing of even more powerful and fearsome nuclear weapons during the Cold War period, adding to this legacy of fear.

For a nuclear bomb, the rapid chain reaction depicted by Figure 3.2-4 is competing with the tendency for the melting/vaporizing uranium to rapidly splat over the surroundings. This "splatting" action tends to stop the chain reaction by separation of small pieces of uranium. Weapons' grade U-235 is typically at least 80% U-235; higher purities give increased release of the nuclear energy (more fission and less splatting).

The enormous energy available from U-235 in a very small space led U.S. naval technologists to consider using nuclear energy to power submarines. The task is to configure the nuclear fuel (U-235 and U-238) so that exactly one of the neutrons produced by a U-235 fission produces one new fission. The shape of the reactor core and control rods (that absorb neutrons) combines to serve as a "throttle" to match the energy release to load. The thermal energy produces steam that propels the vessel and provides electric power. All of this technology development was done with private industrial firms under contract by the military and was classified "top secret."

The industrial firms that built the nuclear reactors for the military also built steam turbines and generators for electric power stations. The first nuclear reactor built to produce electric power for domestic consumption was put into service in Shipingsport, Ohio, in 1957, just 15 years after the "Top Secret" Manhattan Project was assembled to build a nuclear weapon. This represents a remarkable technological achievement. Today, modern nuclear reactors produce electricity based on technology similar to that used in the submarines.

In nuclear reactors and the bomb, neutron sources can be used to supplement the neutrons created by the natural decay of U-235 to create greater control in attaining criticality. The chain reaction is started by inserting some beryllium mixed with polonium, radium, or another alpha-emitter. Alpha particles from the decay cause the release of neutrons from the beryllium as it turns to carbon-12.

3.2.3.1 Reserves

Uranium reserves are difficult to estimate; however, an estimate can be readily made on the energy in the spent rods from U.S. nuclear power generation. Current nuclear technology uses 3.4% of the uranium in the fuel, leaving 96.6% of the uranium unused. The amount of nuclear fuel in spent nuclear fuel rods in U.S. nuclear facilities has an energy content comparable to the entire recoverable U.S. coal reserve.[iii] The depleted uranium created during the fabrication of the initial nuclear fuel rods has about four times as much energy as that remaining in the spent nuclear fuel rods. Combined, this stockpiled uranium in the United States has the capacity

[iii] Assuming 30 years of spent fuel at the current rate, this translates to 75 years of capacity to meet all U.S. energy needs at the present rate of consumption with near-zero generation of greenhouse gases.

Modern nuclear reactors in the United States

Modern nuclear power plants use a pressurized water reactor to produce the thermal energy to produce the steam to drive a turbine and generate electricity. The fuel is 3% to 4% U-235 enriched uranium oxide pellets sealed in tubes that are held in racks in the reactor pressure vessel. This maintains the geometry of the reactor core. The water that removes the heat from the core leaves the reactor at about 320°C, and it would boil at a pressure of about 70 atmospheres (850 psi). The pressure in the reactor vessel is held at 150 atmospheres (2,250 psi), so it never boils. This hot water is pumped to a heat exchanger, where the steam to drive the turbines is produced. The high-pressure reactor cooling water will always contain small amounts of radioactive chemicals produced by the neutrons in the reactor. This radioactivity never gets to the steam turbine where it would make it difficult to perform maintenance on the turbine and steam-handling equipment.

Large pressurized water reactors produce about 3,900 megawatts of thermal energy to produce about 1,000 megawatts of electric power. The reactor core contains about 100 tons of nuclear fuel. Each of the nuclear fuel racks has places where control rods can be inserted. The control rods are made of an alloy that contains boron. Boron metal absorbs neutrons, so with these rods in position, there will not be enough neutrons to initiate a chain reaction. When all of the fuel bundles are in position and the lid of the pressure vessel is sealed, the water containing boric acid fills the pressure vessel. The control rods are withdrawn, and the boron water solution still absorbs the neutrons from U-235 fission. As the water circulates, boric acid is slowly removed from the water and the neutron production rate increases; the water temperature and pressure are closely monitored. When the neutron production rate produces the rated thermal power of the reactor, the boron concentration in the water is held constant. As the fuel ages through its life cycle, the boron in the water is reduced to maintain constant power output.

If there is an emergency that requires a power shutdown, the control rods drop into the reactor core by gravity. The control rods quickly absorb neutrons, and fission power generation stops. The radioactive fission products in the fuel still generate lots of heat, as these isotopes spontaneously decay after fission stops. Water circulation must continue for several hours to remove this radioactive decay heat.

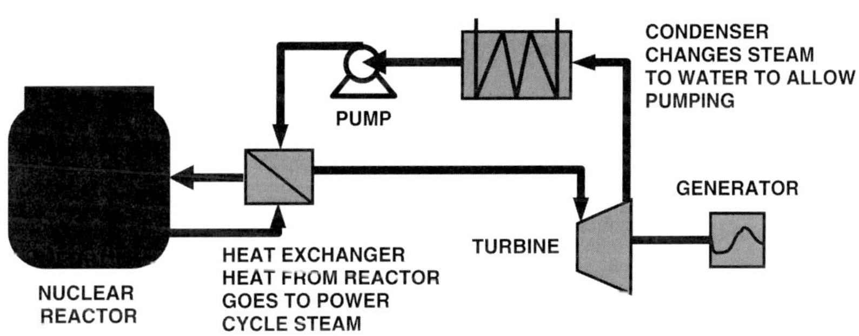

to meet all of the U.S. energy needs with near-zero greenhouse gas emissions for the next 250 years.[iv]

3.2.3.2 Reprocessing technology

The 250 years of capacity from uranium that has already been mined will require reprocessing. A typical spent nuclear fuel rod in the United States contains about 3.4% fission products, 0.75%–1% unused U-235, 0.9% Pu-239, 94.5% U-238, and trace amounts of atoms having atomic masses greater than U-235 (referred to as transuranic elements).

Not only would reprocessing tap this 250 years of energy available from stockpiled uranium, the additional nuclear waste would be the fission products produced by the recycled uranium fuel.

Reprocessing involves removing the 3.4% that is fission products and enriching the U-235 and/or Pu-239 to meet the "specifications" of nuclear reactor fuel. The "specifications" depend on the nuclear reactor design. Nuclear reactors and fuel-handling procedures can be designed that allow nuclear fuel specifications to be met at lower costs than current reprocessing practice in France. For comparative purposes, the cost of coal, U.S. nuclear fuel

[iv] 250,000 tons of spent fuel were in storage in 2001 worldwide, with waste inventories increasing about 12,000 tons per year. About 3,000 tons of spent fuel are reprocessed in France.

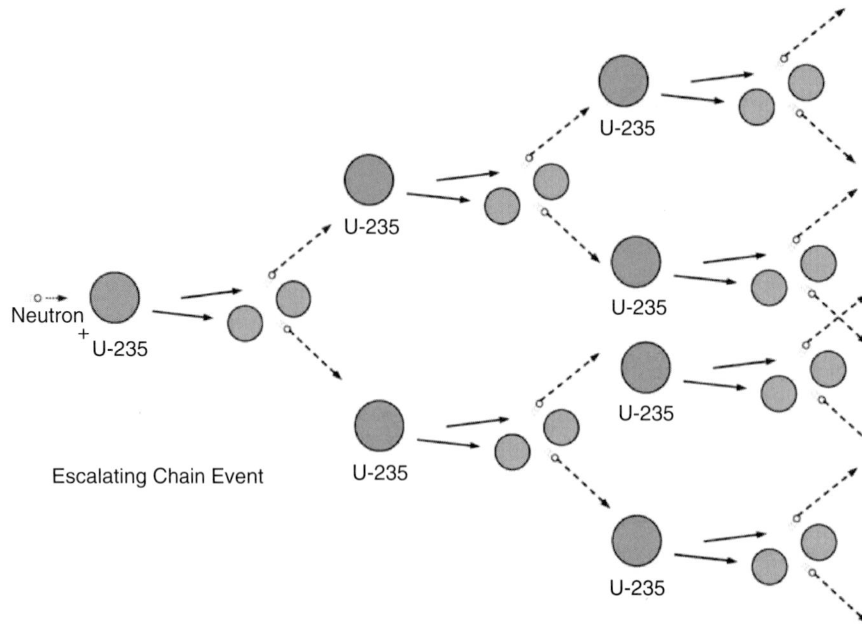

Figure 3.2-4 Escalating chain reaction such as in a nuclear bomb.

from mined uranium, and French reprocessed fuel is about 1.05, 0.68, and 0.90 ¢ per kWh or electricity produced.

From 33% to 40% of the energy produced in nuclear power plants today originates from U-238 and is released by the Pu-239 formed and subsequently fissions: For every three parts of U-235 entering the reactor, about two parts of U-235 plus Pu-239 leave the reactor and, to date, remain stored at the power plant site. All of the uranium, the two parts U-235 and Pu-239 are the target of reprocessing technology. To tap this 250 year stockpile of fuel, new fast-neutron reactors could be put in place that produce more Pu-239 than the combined U-235 and Pu-239 in the original fuel.

Decades of commercial nuclear power provide stockpiles of spent fuel rods, billions of dollars collected on a 0.1 cent per kWh tax levied, and retained to process the spent fuel rods. A remarkable safety history for U.S. designed reactors is set against a costly history of regulations that limit the ability of the technology to advance. These circumstances provide opportunity or perpetual problems, depending on the decisions made to use (or not) nuclear power.

Figure 3.2-6 summarizes the accumulation of spent fuel currently being stored on site at the nuclear power plants in the United States.

The United States uses about 98 GW of electrical power generating capacity from nuclear energy. The construction and startup of most of these plants occurred between 1973 and 1989. In 2007 the inventory of spent nuclear fuel will correspond to about 30 years of operation at current generation rates of the nuclear power infrastructure. Figures 3.2-6 approximates the total spent fuel inventories and cumulative inventories of U-235 and Pu-239 under two different scenarios. Figure 3.2-6 illustrates that reprocessing is the key to decreasing Pu-239 and U-235 inventories and ending the accumulation of spent fuel nuclear reactor sites.

If reprocessing would have initiated in 2005 to meet all current nuclear power plants, the current inventories, along with the Pu-239 that is generated as part of PWR

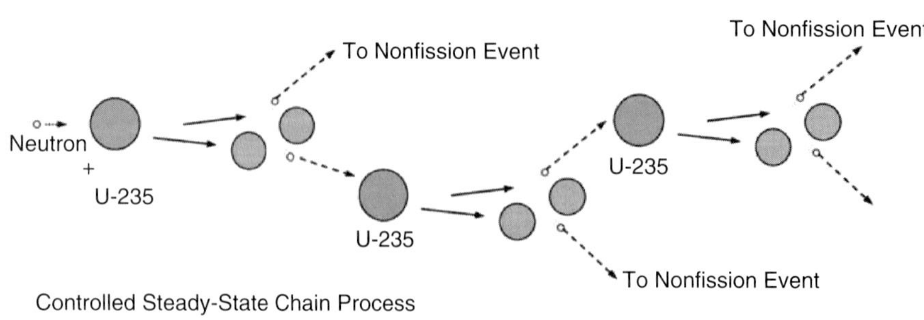

Figure 3.2-5 Controlled steady-state chain nuclear fission such as in a nuclear reactor.

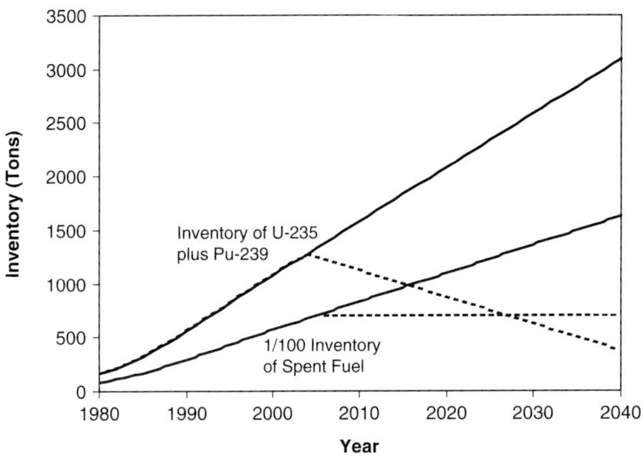

Figure 3.2-6 Approximate inventory of commercial spent nuclear fuel and fissionable isotopes having weapons potential (Pu-239 and U-235). The solid lines are for continued operation without reprocessing, and the dashed lines are for reprocessing (starting in 2005) to meet the needs of current nuclear capacity.

operation, would provide sufficient Pu-239 to operate at existing capacity through 2045. If in 2005 the demand for Pu-239 and U-235 increased threefold (~300 GW capacity), the current inventories would last until 2019. This does not include the use of Pu-239 and U-235 in weapons' inventories, or depend on fast neutron reactor technology to convert the much greater inventories of U-238 into Pu-239 fuel. This partly explains trends of discontinuing breeder reactor research and operation. Breeder reactors will not be needed for some time.

Fast-neutron reactor technology would allow nuclear reactors to meet all energy needs for the next 250 years without generating additional radioactive materials and without mining additional uranium. The potential of this technology should not be ignored.

3.2.3.3 Geothermal

Geothermal energy is heat that is released from the continuous nuclear decay occurring in uranium that is distributed throughout the Earth. The Earth's core has remained hot for billions of years for two reasons: (1) thousands of feet of the Earth's crust provide a good insulation that hinders the loss of heat to surrounding space; and (2) heavier elements (like uranium) tend to be more concentrated toward Earth's center, where these elements undergo natural radioactive decay releasing heat.

The warmer the geothermal heat source, the more useful the energy. For most locations, higher temperatures are located several thousand feet under the surface, and the cost of accessing them is too great compared to alternatives. At the center of the Earth, some 3,700 miles below the surface, temperatures reach 9,000°F, and metals and rocks are liquid.

At locations such as Yellowstone Park and Iceland, useful geothermal heat is available a few hundred feet under the surface or at the surface (hot springs and geysers). Even at these locations the costs of the underground pipe network necessary to create an electrical power plant is a capital-intensive facility. On a case-by-case basis, geothermal heating has been economical. Much of Iceland's residential and commercial heating is provided by geothermal energy (see box).

Some manufacturers refer to the use of groundwater or pipes buried in the ground used in combination with a heat pump as geothermal furnaces. These furnaces do not use geothermal heat. Rather, the large mass of the Earth simply acts as energy storage to take in heat during the summer and give up heat during the winter.

3.2.4 Recent solar energy

3.2.4.1 Use of sunlight

Solar energy provides the most cost effective means to reduce heating costs and can be used to directly produce electricity. Both can be cost effective, depending on the local cost of conventional electrical energy alternatives.

Solar heating is the most commonly used and least expensive use of sunlight. Building location, orientation, and window location can be used to displace auxiliary heating such as a natural gas furnace. Windows located on the south side of a northern hemisphere building will bring in sunlight to heat during the winter. A strategically located tree or well-designed roof overhang can block the sunlight during the summer. The number and placement of windows will vary, based on design preference. Aesthetics, solar functionality, and nonsolar functionality (siding on a building) are available for building construction. New building designs are available that provide cost-effective combinations for solar systems.

Solar water heating systems are the next most popular use of solar energy. They use solar heat to reduce the consumption of natural gas or electricity to heat water. Clarence Kemp is known as the father of solar energy in the United States. He patented the first commercial Climax Solar Water Heater. This and competing systems sold about 15,000 units in Florida and California by 1937.

In 1941, between 25,000 and 60,000 were in use in the United States, with 80% of the new homes in Miami having solar hot water heaters. Use outside the United States has developed, especially in regions where electricity costs are high and the climate is warm.

When confronted with the oil boycott and subsequent oil supply incentives, Israel proceeded with a major initiative to use solar water heaters. More than 90% of Israeli households owned solar water heaters at the start of the 21st century. Solar water heaters are also quite popular in

Geothermal heating in Iceland

The first trial wells for hot water were sunk by two pioneers of the natural sciences in Iceland, Eggert Olafsson and Bjarni Palsson, at Thvottalaugar in Reykjavik and in Krisuvik on the southwest peninsula in 1755–1756. Additional wells were sunk at Thvottalaugar in 1928 through 1930 in search of hot water for space heating. They yielded 14 liters per second at a temperature of 87°C, which in November 1930 was piped three kilometers to Austurbacjarskoli, a school in Reykjavik that was the first building to be heated by geothermal water. Soon after, more public buildings in that area of the city as well as about 60 private houses were connected to the geothermal pipeline from Thvottalaugar.

The results of this district heating project were so encouraging that other geothermal fields began to be explored in the vicinity of Reykjavik. Wells were sunk at Reykir and Reykjahbd in Mosfellssveit, by Laugavegur (a main street in Reykjavik), and by Ellidaar, the salmon river flowing at that time outside the city but now well within its eastern limits. Results of this exploration were good. A total of 52 wells in these areas are now producing 2,400 liters per second of water at a temperature of 62–132°C.

Hitaveita Reykjavikur (Reykjavik District Heating) supplies Reykjavik and several neighboring communities with geothermal water. There are about 150,000 inhabitants in that area, living in about 35,000 houses. This is way over half the population of Iceland. Total harnessed power of the utility's geothermal fields, including the Nesjavellir plant, amounts to 660 MWt, and its distribution system carries an annual flow of 55 million cubic meters of water.

Australia. At sunny locations where electricity is expensive and where natural gas is not available, solar water heating is a good option. It is easy to store water so the solar energy collected during the day is available at night.

Considerable research has been conducted using mirrors to focus sunlight and generate the high temperatures required to produce steam for electrical power generation. To date, most of these systems are too costly. Alternatively, the direct conversion of sunlight to electricity is popular in niche markets, and new technology is poised to expand this use.

In the 1950s, Bell Laboratory scientists made the first practical photovoltaic solar cell. Today, photovoltaic technology is widely used on flat screen computer monitors and for producing electrical power for electric devices in remote locations. These remote devices include highway signs that cannot be easily connected to grid electricity and small electrical devices like handheld calculators.

Table 3.2-3 U.S electricity power production in 1999 in billions of kilowatt hours

Coal	1,884.3	50.8%
Nuclear	728.3	19.6%
Natural gas	556.2	15.0%
Hydroelectric	319.5	8.6%
Petroleum	123.6	3.3%
Wood	37.6	1.0%
MSW	20.2	0.5%
Geothermal	16.8	0.5%
Wind	4.5	0.12%
Solar	0.8	0.02%
Total	3,711.8	

http://www.eia.doe.gov/pub/oil_gas/petroleum/analysis_publications/oil_market_basics/default.htm. "Electrical Energy Consumption."

Solar energy is usually not used for power generation in competition with grid electricity. In some locations, photovoltaic cells on roofs provide an alternative for enthusiasts where consumer electrical prices are above $0.10 per kWh. Small solar roof units show a better payback to meet individual electrical needs than commercial units designed to sell power to the electrical grid. While consumers will usually pay more than $0.08 per kWh for electricity, when selling electricity to the grid one typically receives less than $0.04 per kWh.

The south facing walls and roof sections of every building in the United States are potentially useful solar receivers. Materials having both aesthetic and solar function are generally not available today, but such systems will likely be developed. From this perspective, there is great potential for solar energy to replace a portion of grid electrical power. At 0.8 billion kWh in 1999, solar electrical power on the grid provided about 0.02% of the electrical energy production (see Table 3.2-3).

3.2.4.2 Hydroelectric

Water stored in high-elevation lakes or held by dams creates high-pressure water at the bottom of the dam. The energy stored in the high-pressure water can be converted to shaft work using hydroelectric turbines to produce electricity. Most of the good dam sites in the

United States have been developed, so this is a relatively mature industry. At 319.5 billion kWh in 1999, hydroelectric power on the grid provided 8.6% of the electrical energy production. Environmentalists oppose dam construction in the Northwest, and there is active lobbying to remove dams on the Columbia River.

3.2.4.3 Wind energy

There have been over 8 million wind turbines installed since the 1860s in the United States. Wind energy is one of the oldest and most widely used forms of power. Traditionally, the shaft work was used to pump water or mill grain. In the 1930s, an infant electrical power industry was driven out of business by policies favoring the distribution of fossil fuel electricity.

Between 1976 and 1985, over 5,500 small electric utility units (1–25 kW) were installed on homes and remote locations in response to high oil prices. The installation and use dwindled when government subsidies ended in 1985. More recently, wind farms have been installed to provide grid electrical power. Between 1981 and 1990, approximately 17,000 machines were installed with output ranging from 20 to 350 kilowatts with a total capacity of 1,700 megawatts.

The price of electrical power from wind has decreased from more than $1.00 per kWh in 1978 to about $0.05 per kWh in 1998, with costs projected as low as $0.025 per kWh (for large wind farms). At $0.025 per kWh, wind power competes with the fuel costs for most fossil fuel power plants. Projections aside, for most locations in 2002, a low price for wind power was $0.045 per kWh, but some wind farms need $0.06 per kWh to operate at a profit. The primary issue for more widespread use of wind power is not the cost of the wind turbines. Issues are (1) high maintenance costs because of the large number of wind turbines needed to generate as much power as a typical coal-fired power plant,(2) environmental impact (noise pollution and poor aesthetics), and (3) dependable power on demand. (The wind doesn't always blow.)

The dependability issue is the ability of wind power to supply continuous electrical power. The ability of a facility to provide electricity is characterized by its capacity factor. This factor is the actual energy supplied by a wind turbine compared to the theoretical power supplied if it operated continuously at its design capacity. Wind power suffers from low-capacity factors because of the lack of wind at night and the lack of power demand when the wind is blowing. Capacity factors for wind farms range from 0.20 to 0.35 compared to 0.5 for fossil fuel plants, 0.6 for some new gas turbine plants, and 0.85 for nuclear power.

One of the less obvious opportunities for electrical power supply is energy storage. Storing wind energy as it is available for use during peak demand times will increase the value of the wind energy and would increase capacity factors. This could increase the value of wind energy from a wind farm by a factor of three or more. Such an increase in the value of wind energy would change the economic outlook of wind power from marginal to profitable.

Wind power generated 4.5 billion kWh in 1999 or 0.12% of the electrical energy production in the United States.

3.2.4.4 Biomass

Energy storage limits the utility for both wind power and solar energy. Nature's way for storing solar energy is biomass. Biomass is biological material: vegetation, grass, wood, corn, palm oil, and similar plant material. Time, the absence of oxygen, and compaction (promoted by Earth overburdens) convert biomass to coal, petroleum, or other geological variations of these materials.

Wood has been used through the ages to produce heat. Today, wood supplies heat and is used to generate electrical power, corn is converted to ethanol, and vegetable oils are converted to biodiesel. Unlike fossil fuels, biomass is not available in reserves that have accumulated for years. Rather, biomass grows each year and must be harvested and used quickly to get quality fuel. The supply must be renewed every year.

The availability of biomass is reported as estimated amounts produced per year. A wide range of biomass types, as well as different terrain and climate, control biomass availability. The supply of biomass tends to increase with increasing prices. Table 3.2-4 summarizes example prices and availability of solid biomass (not including fats and oils).

Solid biomass is used for energy five different ways: (1) burning for heat or electrical power generation; (2) conversion to ethanol; (3) pyrolysis; (4) gasification; and (5) anaerobic (without oxygen) methane production (landfill gas) The high cost of biomass makes conversion to liquid fuels and use as chemical feed stocks the best applications for most of these renewable energy sources. Direct combustion and anaerobic methane are handled on a case-by-case basis, where they are generally profitable if the biomass has already been collected for other reasons. For quality liquid fuel production, two technologies stand out: ethanol production and gasification for Fischer-Tropsch fuel production. When including oil seed crops (e.g. soybeans and rapeseeds), a third option of biodiesel is also becoming quite attractive.

3.2.5 Ethanol and biodiesel from agricultural commodities

Table 3.2-4 shows the number of gallons of ethanol that can be produced from the most common forms of

Table 3.2-4 U.S. estimate of supplies and production potential of ethanol from biomass.[A] Except for reliable corn numbers, conversions are optimistic

	Price $/ dry ton	Quantity million dry tons/yr	Conversion gallons ethanol/ton	Ethanol equivalent millions of gallons/yr	Cost of feedstock/ gallon ethanol
$2.40 bu (56 lb) Corn, U.S.[B,C]	85.7	280	89	24,920	$0.96
Refuse Derived Waste	15	80	80	6,400	$0.19
Wood Waste *Cheap*	30	10	110	1,100	$0.27
Wood Waste *Expensive*	45	80	110	8,800	$0.41
Switchgrass *Cheap*	25	5	95	475	$0.26
Switchgrass *Expensive*	45	250	95	23,750	$0.47

Estimated per-ton yields of ethanol from corn, sorghum, refuse-derived fuel, wood waste, and stitchgrass are 89, 86, 80, 110, and 95, respectively. Corn has 56 pounds per bushel with an assumed price of $2.40 per bushel ($85.71/ton) with an annual U.S. production estimate of 10 billion bushels or 280 million tons.

[A] K. Shaine *et al.*, "Modeling the Penetration of the Biomass-Ethanol Industry and Its Future Benefits," September 18, 1996, Bioenergy '96, Opryland Hotel, Nashville, TN.

[B] R. M. Tshiteya, *Conversion Technologies: Biomass to Ethanol.* Golden, CO: National Renewable Energy Laboratory, September 1992, pp. 3–5.

[C] http://www.usda.gov/nass/pubs/trackrec/track02a.htm#corn. Corn production U.S.

biomass. The corn data of Table 3.2-4 are important points of reference. Corn is the largest commodity crop in the United States and provides high yields of dense biomass. While the price per ton of corn is almost twice the price of large-volume switchgrass and wood, the corn prices and volumes are current, while the other biomass prices are based on estimates that may be optimistic.

Dried distiller grains are a by-product sold as a high-protein, high-fat cattle feed when producing ethanol from corn. Over half of the corn costs are recovered by the sale of this by-product. It is because of these by-products that corn is used more than other biomass crops (e.g. sugar cane). Other biomass materials may actually have a higher yield of ethanol per acre, but they do not have the valuable by-products.

Corn is the most commonly used biomass for producing ethanol. The production process consists of adding yeast, enzymes, and nutrients to the ground starchy part of corn to produce a beer. The beer contains from 4% to 18% ethanol, which is concentrated by distillation that is similar to the distillation used to produce whiskey. The final fuel ethanol must contain very little water for use as motor fuel. In gasoline, the water is an insoluble phase that extracts the ethanol and settles to the bottom of the tank. If this water gets into the engine, the engine will stall.

About 90 gallons of ethanol are produced from one ton of corn. The production cost is $1.05 to $1.25 per gallon. Ethanol has about two-thirds the energy content of gasoline, so these prices translate to $1.57 to $1.85 per equivalent gasoline gallon. This is more than gasoline that is produced for about $1.30 per gallon[v] before the $0.42 (average) motor fuel tax is added.

Estimates of gasoline used in U.S. cars, vans, pickup trucks, and SUVs are about 130 billion gallons of gasoline per year. (These numbers agree with the motor gasoline consumption of 3.05 billion barrels reported elsewhere.) About 500 million prime, corn-producing acres would be required for ethanol to replace all of the gasoline. This is about one-quarter of the land in the lower 48 states. The lower 48 states have about 590 million acres of grassland, 650 million acres of forest, and 460 million acres of croplands (most is not prime acreage).

[v] $50 per barrel divided by 42 gallons per barrel plus a refining cost.

If all of the current corn crop were converted to ethanol, this would replace about 17 billion gallons of gasoline—less than 15% of our current consumption. Estimates of dedicating acreage for ethanol production equivalent (yield-based) to current gasoline consumption would require nine times the acreage used for the current U.S. corn crop. This approach is not realistic. However, if hybrid vehicle technology doubles fuel economy and the electrical power grid further reduces gasoline consumption to about 60 billion gallons, substantial ethanol replacement of gasoline is possible. Use of perennial crops would be a necessary component of large-volume ethanol replacement for gasoline.

Corn is an annual crop and must be planted each year. For this reason, costs for mass production of wood and grasses are potentially lower than for corn. In the late 20th century, corn-to-ethanol production facilities dominated the biomass-to-ethanol industry. This was due to (1) less expensive conversion technologies for starch-to-ethanol compared to cellulose-to-ethanol required for wood or grasses and (2) generally ambitious farmer-investors who viewed this technology as stabilizing their core farming business and providing a return on the ethanol production plant investment. State governments usually provide tax credits for investment dollars to build the ethanol plants, and there is a federal subsidy for fuel grade ethanol from biomass.

Because of lower feedstock costs (see Table 3.2-4), wood-to-ethanol and grass-to-ethanol technologies could provide lower ethanol costs—projections are as low as $0.90 per equivalent gasoline gallon. Research focus has recently been placed on cellulose-to-ethanol. The cost of cellulose-to-ethanol has improved from more costly to about the same as corn-to-ethanol technology. Based on present trends, cellulose-to-ethanol technology could dominate ethanol expansion in the 21st century. It would require large tracts of land dedicated to cellulose production.

The current world production of oils and fats is about 240 billion pounds per year (32.8 billion gallons, 0.78 billion barrels), with production capacity doubling about every 14 years. This compares to a total U.S. consumption of crude oil of 7.1 billion barrels per year of which 1.35 billion barrels is distillate fuel oil (data for the year 2000). With proper quality control, biodiesel can be used in place of fuel oil (including diesel) with little or no equipment modification. Untapped, large regions of Australia, Colombia, and Indonesia could produce more palm oil. This can be converted to biodiesel that has 92% of the energy per gallon as diesel fuel from petroleum. This biodiesel can be used in the diesel engine fleet without costly engine modifications.

Table 3.2-5 Comparison of annual U.S. gasoline and diesel consumption versus ethanol and biodiesel production capabilities

Gasoline Consumption (billions of gallons per year)	130
Distillate Fuel Oil (including diesel) Consumption	57
Ethanol from Corn [equivalent gasoline gallons]	25 [17]
Biodiesel from Soybeans [equivalent diesel gallons]	4.25 [3.8]

In the United States, ethanol is the predominant fuel produced from farm commodities (mostly from corn and sorghum), while in Europe, biodiesel is the predominant fuel produced from farm commodities (mostly from rapeseed). In the United States, biodiesel is produced predominantly from waste grease (mostly from restaurants and rendering facilities) and from soybeans.

In the United States, approximately 30% of crop area is planted to corn, 28% to soybeans, and 23% to wheat. For soybeans this translates to about 73 million acres (29.55 million hectares) or about 2.8 billion bushels (76.2 million metric tons). Soybeans are 18%–20% oil by weight, and if all of the U.S. soybean oil production were converted[vi] to biodiesel, it would yield about 4.25 billion gallons of biodiesel per year. Typical high yields of soybeans are about 40 bushels per acre (2.7 tons per hectare), which translates to about 61 gallons per acre. By comparison, 200 bushels per acre of corn can be converted to 520 gallons of ethanol per acre.

Table 3.2-5 compares the consumption of gasoline and diesel to the potential to produce ethanol and biodiesel from U.S. corn and soybeans. If all the starch in corn and all the oil in soybeans were converted to fuel, it would only displace the energy contained in 21 billion gallons of the 187 billion gallons of gasoline and diesel consumed in the United States. Thus, the combined soybean and corn production consumes 58% of the U.S. crop area planted each year. It is clear that farm commodities alone cannot displace petroleum oil for transportations fuels. At best, ethanol and biodiesel production is only part of the solution. U.S. biodiesel production in 2005 was about 0.03 billion gallons per year compared to distillate fuel oil consumption of 57 billion gallons per year.

Converting corn and soybean oil to fuel is advantageous because the huge fuel market can absorb all excess crops and stabilize the price at a higher level. In addition, in times of crop failure, the corn and soybeans that

[vi] 76.2 million metric tons of beans is about 14.5 billion kilograms of soybean oil. This translates to about 16 billion liters, using a density of 0.9 g/cc or about 4.25 billion gallons per year.

normally would be used by the fuel market could be diverted to the feed market. The benefits of using soybeans in the fuel market can be further advanced by plant science technology to develop high-oil content soybeans.

Soybeans sell for about $0.125 per pound, while soybean oil typically sells for about twice that ($0.25 per lb). The meal sells for slightly less than the bean at about $0.11 per pound. Genetic engineering that would double the oil content of soybeans (e.g., 36%–40%) would make the bean, on the average, more valuable. In addition, the corresponding 25% reduction in the meal content would reduce the supply of the meal and increase the value of the meal. At a density of 0.879 g/cc, there are about 7.35 lbs of biodiesel per gallon. A price of $0.25 per lb corresponds to $1.84 per gallon; $0.125 per lb to $0.92 per gallon.

Fuel production from corn and soybean oil would preferably be sustainable without agricultural subsidies of any kind (none for ethanol use, biodiesel use, farming, or not farming). A strategy thus emerges that can increase the value of farm commodities, decrease oil imports, decrease the value of oil imports, and put U.S. agriculture on a path of sustainability without government subsidies. To be successful, this strategy would need the following components:

1. **Develop better oil-producing crops.**
   - Promote genetic engineering of soybeans to double oil content and reduce saturated fat content (saturated fats cause biodiesel to plug fuel filters at moderate temperatures).
   - Promote the establishment of energy crops like the Chinese tallow tree in the South that can produce eight times as much vegetable oil per acre as soybeans.
2. **Pave the future for more widespread use of diesel engines and fuel cells.**
   - Promote plug-in HEV technology that uses electricity and higher fuel efficiency to displace 80% of gasoline consumption. Apply direct-use ethanol fuel cells for much of the remaining automobile transportation energy needs.
   - Continue to improve diesel engines and use of biodiesel and ethanol in diesel engines. Fuel cells will not be able to compete with diesel engines in trucking and farm applications for at least a couple of decades.
3. **Pass antitrust laws that are enforced at the border.**
   - If the oil-exporting countries allow the price of petroleum to go up astronomically, do not allow subsequent price decreases to bankrupt new alternative fuel facilities.
4. **Fix the dysfunctional U.S. tax structure.**
   - Restructure federal and state taxes to substantially eliminate personal and corporate income taxes and replace the tax revenue with consumption taxes (e.g., 50%) on imports and domestic products. This would increase the price of diesel (red diesel, no highway tax).
   - Treat farm use of ethanol and biodiesel as an internal use of a farm product, and, therefore, no consumption tax would be applied.

Increased use of oil crops would include use of rapeseed in drier northern climates (rapeseed contains about 35% oil) and use of Chinese tallow trees in the South. Chinese tallow trees are capable of producing eight times as much oil per acre as soybeans. If Chinese tallow trees were planted in an acreage half that of soybeans and the oil content of soybeans were doubled, 17–20 billion gallons of diesel could be replaced by biodiesel allowing continued use of soybean oil in food applications. This volume of biodiesel production would cover all agricultural applications and allow the imports to be optional.

Chinese tallow trees are one of the fastest-growing trees. In addition to producing oil crops, clippings and old trees could be used for ethanol production. Chinese tallow tree orchards could readily become the largest agriculture crop in the United States. High-protein animal feed is an additional potential by-product.

The plug-in HEV technology would displace about 104 billion gallons per year of gasoline with electricity and increase efficiency. The electricity could be made available from the reprocessed spent nuclear fuel and adding advanced technology nuclear reactors. About half of the remaining 26 billion gallons of gasoline could be displaced with ethanol and half with continued use of gasoline.

In this strategy, up to 55 billion gallons of annual diesel and gasoline consumption would still need to be met with fossil fuel sources. These could be met with petroleum, coal-derived liquid fuels (like Fischer-Tropsch fuels), and Canadian oil sand fuels. Increase of electric trains for freight could displace much of the 55 billion gallons. It would be a buyer's market for liquid fossil fuels.

These proposed technologies were cost effective and sustainable at $55 per barrel of crude oil and a tax strategy that equally taxed domestic and imported products (the consumption tax). A variety of continued strategies could assure that the United States did not have to import petroleum and that farmers could achieve higher-value nonfood uses for their products.

The consumption tax is emerging as a preferred way to end the stress on U.S. manufacturing with domestic taxes that are not applied to imports. All the technology to replace petroleum is demonstrated and cost-effective

Table 3.2-6 Current and projected costs and production of U.S. ethanol and bio-diesel

	Application	Billions of gallons per year	Pricing
2005			
Ethanol[liv]	Octane Enhancer, Oxygenate for CO Nonattainment	3.4	$1.20–$1.50
Biodiesel	Primarily as 2% to 20% Additive in Bus Fleets and Farm Applications	0.03	$1.30–$2.30 per gallon
2015 (2025)			
Ethanol	Direct-Use Ethanol Fuel Cells, Octane Enhancer, Oxygenate for CO Nonattainment Areas	10 (20)	$1.50 per gallon
Biodiesel	Predominant Farm Fuel, 50% Market Share in South, Fleets	5 (15)	$2.10 for farm application (soybean and rapeseed based, no tax); $2.25 for South (beef tallow tree based, including consumption tax)

[liv] 2004 Gasoline Price Increases: An Analysis Summary Prepared by Renewable Fuels Association. (http://www.ethanolrfa.org/, March 2004.)

with the possible exception of low-temperature direct-use ethanol fuel cells. Intermediate temperature PEM fuel cells should make direct use of ethanol cost effective in ten years.

To use prices based on 2005, Table 3.2-6 summarizes the uses, volumes, and prices of ethanol production and compares these to a likely scenario if a consumption tax is implemented and antitrust laws are enforced at the border. Table 3.2-7 summarizes the impact of the federal incentive of 5.4 ¢ per gallon tax exemption that goes to blenders placing 10% ethanol in gasoline—this is applied against the 18.4 ¢ federal excise tax on gasoline. The federal tax incentive is paid to the blenders, so if the blender pays $1.25 per gallon for ethanol, the federal government provides a reduction ($0.54 per gallon of ethanol blended to 10% ethanol in gasoline) in the highway taxes that are paid by the blender.

The price of petroleum fuels relative to the price of vegetable oil is an important factor that will impact sustainable biodiesel and ethanol production. Based on $55 per barrel petroleum and a consumption tax strategy that would tax imports the same as domestic production, there is a basis for developing a biodiesel and ethanol industry that can be sustainable and compete with $2.25 per gallon diesel (includes consumption tax, excludes highway tax).

Table 3.2-7 Example cost of ethanol and impact of tax credit

Example Ethanol Wholesale Price	123 ¢/gallon
Alcohol Fuel Tax Incentive	54 ¢/gallon
Effective Ethanol Price	69 ¢/gallon
Effective Ethanol Price for Energy in 1 Gallon Gasoline	103 ¢
Gasoline Wholesale Price ($55/barrel crude oil, $0.14/gallon refining cost)	145 ¢/gallon

The price of vegetable oil ranges from $0.92 to $1.84, depending on whether the oil is priced at the same value as the soybean or a premium price is received for the oil component of the bean. In principle, higher-oil seed crops could sustainably provide vegetable oil at $1.50 per gallon while maintaining a premium value (more on a per-pound basis than for soybeans) for the bean. At a reasonable $0.40 per pound processing cost, a biodiesel cost of $1.90 per gallon would have been sustainable. Prices as low as $1.70 per gallon may be attainable with catalyst development. This gives $2.25 per gallon for diesel. With the 50% consumption tax, imported oil would have to be below $45 per barrel for the price of the diesel to be less than $1.90 per gallon.

Subsidies of $0.54 per gallon for ethanol and $1.00 per gallon for biodiesel approximately compensate for all the U.S. taxes collected on agriculture and processing that go toward the production of these fuels (essentially no U.S. taxes are applied on imported petroleum). These incentives were appropriate in view of U.S. tax strategies in 2005. A continued use of these incentives rather than a consumption tax would have at least three drawbacks: (1) many citizens will perceive the incentive as political favor rather than a mechanism to allow the fuels to compete fairly with imported petroleum, (2) the incentives require periodic renewal and can be eliminated when production becomes high enough, and (3) the incentives currently do not apply to other technologies such as plug-in HEV technology that would help realize the true value of ethanol to replace petroleum.

If large-scale Chinese tallow tree farming were to occur, the farming should be profitable at oil prices as low as $1.30 per gallon ($0.90 per gallon for the oil plus $0.40 for processing). At these prices, the biodiesel would compete in the trucking fuel industry where a 50% consumption tax would take that to $1.95 per gallon—considerably less than petroleum at $2.25 per gallon.

For ethanol use in advanced plug-in HEVs, the technology is demonstrated and cost effective with the possible exception of low-temperature direct-use ethanol fuel cells. Intermediate temperature PEM fuel cells should make direct use of ethanol cost-effective in less than 10 years.

Consumption taxes will not represent an additional tax burden on U.S. consumers if properly implemented; however, the taxes will be more apparent. The average consumer will undoubtedly welcome the absence of income taxes. However, when the price tag on that new $30,000 pickup truck (in 2005) became $45,000, there would be some distress. The implementation should be gradual to allow consumers to become accustomed to paying taxes, at the point of sale rather than on income. An initial phase of eliminating corporate income taxes, about a 5% reduction in all personal income taxes and a 20%–30% consumption tax, would be a good start. The price of a pickup truck should not increase by 20%–30%, because the lack of corporate income taxes should allow the $30,000 pickup to first have a price decrease to about $28,000 and a total price of about $34,000 when the tax is applied.

The United States is poised to level the playing field between imported and domestic production through use of consumption taxes. This correction to a current tax structure is much needed. If and when this transition happens, U.S. farmers would do well to support the transition and to make sure that ethanol and biodiesel used in agricultural applications would be free of this consumption tax. It should be considered an internal transaction.

3.2.6 Emergence of nuclear power

In the pursuit of sustainable energy, nuclear power emerges for four reasons:

1. On a Btu basis, nuclear fuel is the least expensive, and it is economically sustainable. Nuclear fuel has the potential to be ten times less expensive than any alternative (less than $0.10 per MBtu).
2. Nuclear fuel is the most readily available fuel. It is stockpiled at commercial reactors in the form of spent fuel.
3. Nuclear fuel is the most abundant. Enough has already been mined, processed, and stored in the United States to supply all energy needs for centuries.
4. There is no technically available alternative to give sustainable energy supply for current and projected energy demand.

This last point is emphasized in this chapter. It is impractical to try to replace transportation fuels with biomass, let alone nontransportation energy expenditures. The limited availability of petroleum is already inciting military conflict to keep the oil flowing, not to mention the contribution of the trade deficit drag on the U.S. economy. The imports of natural gas are growing rapidly, and at prices greater than $6 per MBtu, it is too expensive for use for electrical power generation.

Coal will be important for decades to produce electrical power and for centuries as a feedstock to the chemical industry. However, coal is already used for about 50% of electric power production (see Table 3.2-3). Nuclear energy is less expensive on a fuel basis. A more rigorous comparison of electrical power costs for coal versus nuclear is found in *Renewable Energy*, Sorensen (2004).

References

1. http://www.eia.doe.gov/fueloverview.html.
2. http://www.eia.doe.gov/fueloverview.html.
3. http://www.eia.doe.gov/fueloverview.html.
4. Ibid. A–Z.
5. *Kirk-Othmer Encyclopedia of Chemical Technology*. New York: John Wiley & Sons, 1980, Vol. 12, p. 326.
6. From U.S. DOE Public Domain. http://www.eia.doe.gov/pub/oil_gas/petroleum/analysis_publications/oil_market_basics/Supply_Petroleum_Trap.htm, 2003.
7. *Kirk-Othmer Encyclopedia of Chemical Technology*. New York: John Wiley & Sons, 1980, Vol. 12, p. 326.
8. *Kirk-Othmer Encyclopedia of Chemical Technology*. New York: John Wiley & Sons, 1977, Vol. 11, p. 326.
9. S. M. Farouq Ali, Heavy Crude Oil Recovery, in Ender Okandan, *The Hague*. Boston: Martinus Nijhoff Publisher, 1984, p. 50.
10. *Kirk-Othmer Encyclopedia of Chemical Technology*. New York: John Wiley & Sons, 1980, Vol. 11, p. 887.
11. Ibid.
12. http://www.petroleumworld.com/SF021906.htm.

13. Canada's Oil Sands and Heavy Oil. Petroleum Communication Foundation, Canadian Centre for Energy Information, Alberta. See http://www.centreforenergy.com/EE-OS.asp.
14. Evolution of Canada's Oil Gas Industry. Canadian Centre for Energy Information, Alberta, Canada, 2004. See http://www.centreforenergy.com/EE-OS.asp.
15. http://www.agiweb.org/legis105/ch4.html.
16. J. E. Mielke, Methane Hydrates: Energy Prospect or Natural Hazard? CRS Report to Congress. Order Code RS20050, February 14, 2000.
17. *Kirk-Othmer Encyclopedia of Chemical Technology.* New York: John Wiley & Sons, 1980, Vol. 12, p. 326.

Chapter 3.3

3.3

The individual energy sources

3.3.1 Solar radiation

An assessment of the "magnitude" of solar radiation as an energy source will depend on the geographical location, including local conditions such as cloudiness, turbidity, etc. In *Renewable Energy* Sorensen (2004) a number of features of the radiation flux at a horizontal plane are described, such as spectral distribution, direct and scattered parts, geographical variations, and dependence on time, from annual to diurnal variations at a given location. The seasonal variation in solar radiation on a horizontal plane is shown in Fig. 3.3-1.

For actual applications, it is often necessary to estimate the amount of radiation received by tilted or complexly shaped devices, and it is useful to look at relations which allow relevant information to be extracted from some basic measured quantities. For instance, radiation data often exist only for a horizontal plane, and a relation is therefore needed to predict the radiation flux on an arbitrarily inclined surface. In regions at high latitudes, directing solar devices towards the Equator at fairly high tilt angles actually gives an increase in incident energy relative to horizontally placed collectors.

Only the incoming radiation will be discussed in detail in this section, since the outgoing flux may be modified by the specific type of energy conversion device considered. In fact, such a modification is usually the very idea of the device. A description of some individual solar conversion devices will be taken up in Chapter 4 of *Renewable Energy* (2004), and their potential yield in Chapter 6, in terms of combined demand and supply scenarios.

3.3.1.1 Direct radiation

The inclination of a surface, e.g. a plane of unit area, may be described by two angles. The tilt angle, s, is the angle between vertical (zenith) and the normal to the surface, and the azimuth angle, γ, is the angle between the southward direction and the direction of the projection of the normal to the surface onto the horizontal plane; γ is counted positive towards east [in analogy to the hour angle]. In analogy to the expression at the top of the atmosphere, the amount of direct radiation reaching the inclined surface characterised by (s, γ) may be written

$$D_{s,\gamma} = S_N \cos\theta, \tag{3.3.1}$$

where S_N is the "normal radiation", i.e. the solar radiation from the direction to the Sun. The normal radiation is entirely "direct" radiation, according to the definition of "direct" and "scattered" radiation in Chapter 2 of *Renewable Energy* (2004). The angle θ is the angle between the direction to the Sun and the normal to the surface specified by s and γ. The geometrical relation between θ and the time-dependent coordinates of the Sun, declination δ and hour angle ω, is

$$\cos\theta = (SC - CSC)\sin\delta + (SS\sin\omega + (CC + SSC)\cos\omega)\cos\delta, \tag{3.3.2}$$

where the time-independent constants are given in terms of the latitude ϕ and the parameters (s, γ) specifying the inclination of the surface,

$$\begin{aligned} SC &= \sin\phi\cos s, \\ CSC &= \cos\phi\sin s\cos\gamma, \\ SS &= \sin s\sin\gamma, \\ CC &= \cos\phi\cos s, \\ SSC &= \sin\phi\sin s\cos\gamma. \end{aligned} \tag{3.3.3}$$

For a horizontal surface ($s = 0$), θ equals the zenith angle z of the Sun, and (3.3.2) reduces to the expression

Renewable Energy Focus Handbook 2009; ISBN: 9780123747051

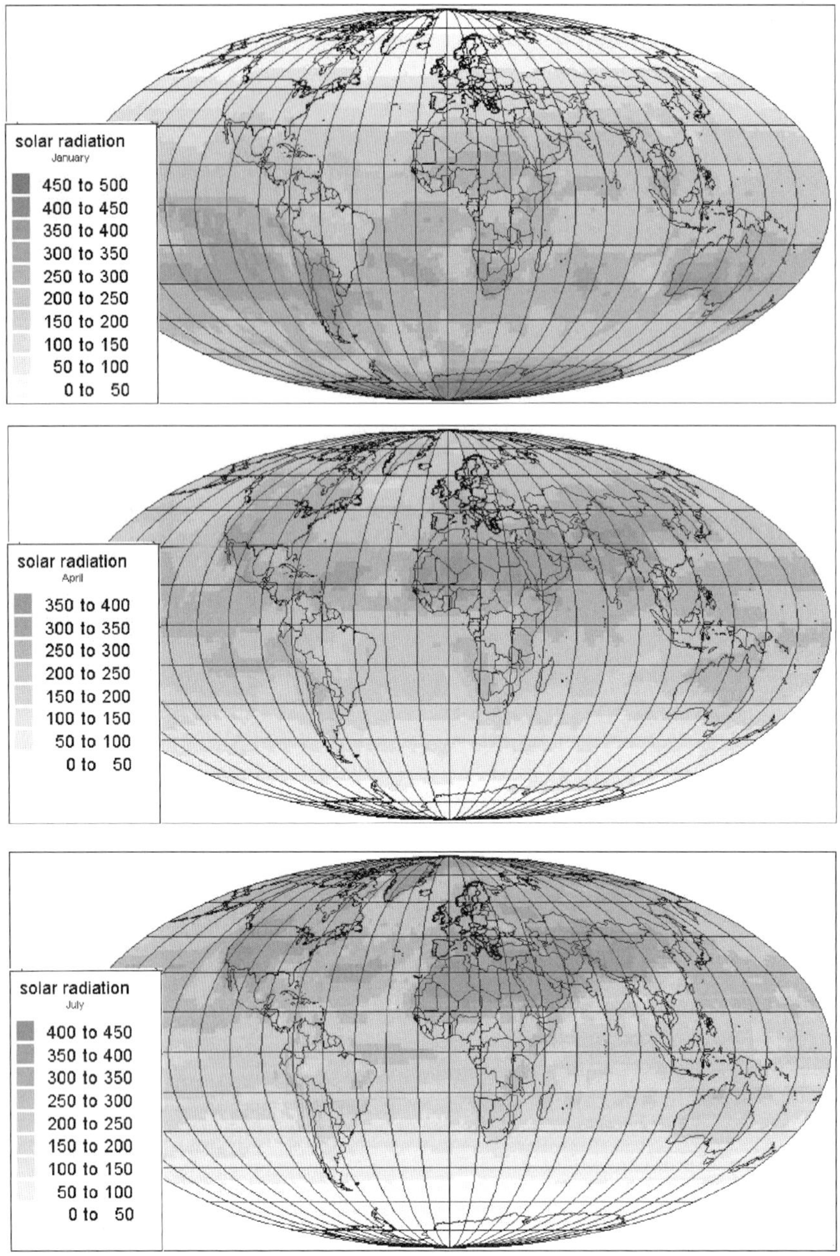

Figure 3.3-1a,b,c Average short-wavelength solar radiation on a horizontal plane at the Earth's surface (W m^{-2}), for the months of January (*a*, above), April (*b*, middle) and July (*c*, bottom) of 1997 (NCEP-NCAR, 1998).

for cos z given in section 2.2.1 of *Renewable Energy* (2004). Instead of describing the direction to the Sun by δ and ω, the solar altitude $h = \frac{1}{2}\pi - z$ and azimuth Az (conventionally measured as positive towards the west, in contrast to the hour angle) may be introduced. The two sets of coordinates are related by

$$\begin{aligned} \sin h &= \sin\delta \sin\phi + \cos\delta \cos\phi \cos\omega, \\ \sin Az \cos h &= -\cos\delta \sin\omega. \end{aligned} \qquad (3.3.4)$$

In the height–azimuth coordinate system, (3.3.2) may be written

$$\cos\theta = \sin h \cos s + \cos h \sin s \cos(Az + \gamma) \qquad (3.3.5)$$

Again, the sign conventions for Az and γ are opposite, so the argument of the last cosine factor in (3.3.5) is really the difference between the azimuth of the Sun and the projection of the normal to the surface considered. If cos θ found from (3.3.2) or (3.3.5) is negative, it means that the Sun is shining on the "rear" side of the surface considered. Usually, the surface is to be regarded as "one-sided", and cos θ may then be replaced by zero, whenever it assumes a negative value, in order that the calculated radiation flux (3.3.1) becomes properly zero.

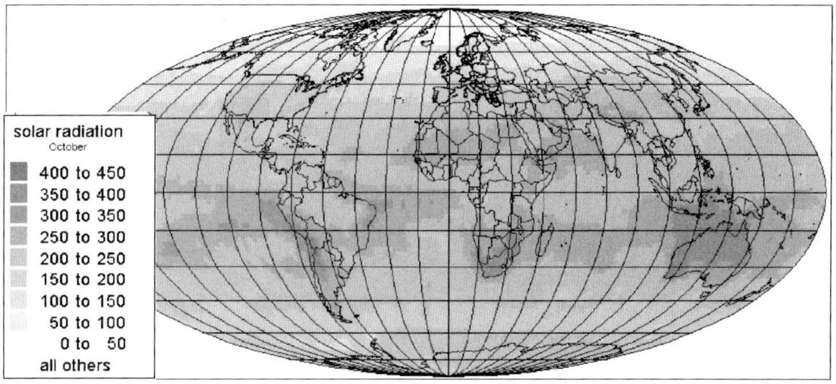

Figure 3.3-1d Average short-wavelength solar radiation on a horizontal plane at the Earth's surface (W m^{-2}), for the month of October of 1997 (NCEP-NCAR, 1998).

If data giving the direct radiation flux D on a horizontal plane are available, and the direct flux impinging on an inclined surface is wanted, the above relations imply that

$$D_{s,\gamma} = D \cos\theta / \cos z.$$

It is evident that care should be taken in applying this relation when the Sun is near the horizon.

More reliable radiation fluxes may be obtained if the normal incidence radiation flux S_N is measured (as function of time) and (3.3.1) is used directly. S_N is itself a function of zenith distance z, as well as a function of the state of the atmosphere, including ozone mixing ratio, water vapour mixing ratio, aerosol and dust content, and cloud cover. The dependence on zenith angle is primarily a question of the path that the radiation has taken through the atmosphere. This path-length is shortest when the Sun is in zenith (often denoted "air mass one" for a clear sky) and increases with z, being quite large when the Sun is near the horizon and the path is curved due to diffraction. The extinction in the atmosphere is normally reduced at elevated locations (or low-pressure regions, notably mountain areas), in which case the effective air mass may become less than one. Figure 3.3-2 gives some typical variations of S_N with zenith angle for zero, small and heavy particle load ("turbidity"). Underlying assumptions are: cloudless sky, water vapour content throughout a vertical column equal to 0.02 $m^3\,m^{-2}$, standard sea-level pressure and mean distance to the Sun (Robinson, 1966).

Since a complete knowledge of the state of the atmosphere is needed in order to calculate S_N, such a calculation would have to be coupled to the equations of state and motion discussed in Chapter 2, *Renewable Energy* (2004). Only some average behaviour may be described without doing this, and if, for example, hourly values of S_N are required in order to predict the performance of a particular solar energy conversion device, it would be better to use measured values of S_N (which are becoming available for selected locations throughout the world, cf. e.g. Turner, 1974) or values deduced from measurements for horizontal surfaces. Measuring techniques are discussed by Coulson (1975), among others.

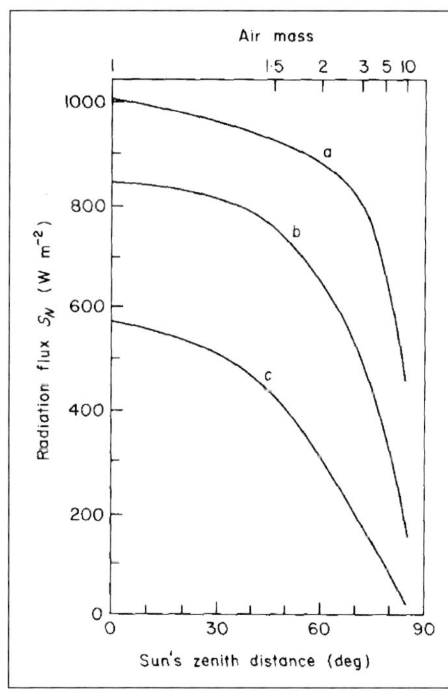

Figure 3.3-2 Normal incidence radiation as a function of zenith angle for a cloudless sky with different particle content: (a), hypothetical atmosphere with zero turbidity [$B = 0$, no aerosols but still water vapour and molecular (Rayleigh) scattering]; (b), clear sky ($B = 0.01$); (c), atmosphere with heavy aerosol pollution ($B = 0.4$). B is the turbidity coefficient B_λ (defined in the text) averaged over wavelengths. Sea-level pressure (10^5 N m^{-2}), a water content of 0.02 $m^3\,m^{-2}$, an ozone content of 0.0034 $m^3\,m^{-2}$ (both referred to standard temperature and pressure) and mean Earth–Sun distance have been assumed. At the top, an approximate air mass scale is provided, relative to the one for a vertical path from sea-level (based on Robinson, 1966).

Attempts to parametrise the solar radiation received at the ground are usually made for the global radiation (direct plus scattered, and for tilted planes also radiation reflected onto the surface), rather than separately for normal incidence and scattered radiation.

Dependence on turbidity and cloud cover

The variability of S_N due to turbidity may be demonstrated by noting the range of S_N values implied by extreme high or low turbidities, in Fig. 3.3-2, and by considering the spread in mean daily turbidity values, an example of which is shown in Fig. 3.3-3. The turbidity coefficient B_λ for a given wavelength may be defined through an attenuation expression of the form

$$S_N(\lambda)/E_{0+}^{sw}(\lambda) = 10^{-m_r(\tau_\lambda^s+\tau_\lambda^a+B_\lambda)}.$$

Here m_r is the relative air mass (optical path-length in the atmosphere), τ_λ^s describes the scattering on air molecules and τ_λ^a is the absorption by ozone. In terms of the cross sections $\sigma_s(\lambda)$, $\sigma_a(\lambda)$ and $\sigma_p(\lambda)$ for scattering on air molecules, ozone absorption and attenuation by particles, one has

$$-m_r\tau_\lambda^s = \log_{10}\{\exp(\int \sigma_s(\lambda)ds)\},$$

and similarly for τ_λ^a and B_λ. Figure 3.3-3 gives both daily and monthly means for the year 1972 and for a wavelength of 5×10^{-7} m (Bilton *et al.*, 1974). The data suggest that it is unlikely to be possible to find simple, analytical expressions for the detailed variation of the turbidity or for the solar fluxes, which depend on turbidity.

Another major factor determining the amount and distribution of different types of fluxes is the cloud cover. Both cloud distribution and cloud type are of importance. For the direct radiation flux the important questions are whether the path-line is obscured or not, and, if it is, how much attenuation is caused by the particular type of cloud. Figure 3.3-4 shows the total and scattered flux on a horizontal plane (and by subtraction the direct), for a clear sky and three different types of clouds, as a function of the zenith angle of the Sun. The cloud classification represents only a general indication of category. It is evident in this example that altocumulus and stratus clouds are almost entirely impermeable for direct radiation, whereas cirrus clouds allow the penetration of roughly half the direct radiation flux. Meteorological observations containing records of sunshine (indicating whether or not the direction to the Sun is obscured) and of cloud cover (as a percentage and identifying the cloud types and their estimated height) may allow a fairly reliable estimate of the fraction of direct radiation reaching a plane of given orientation.

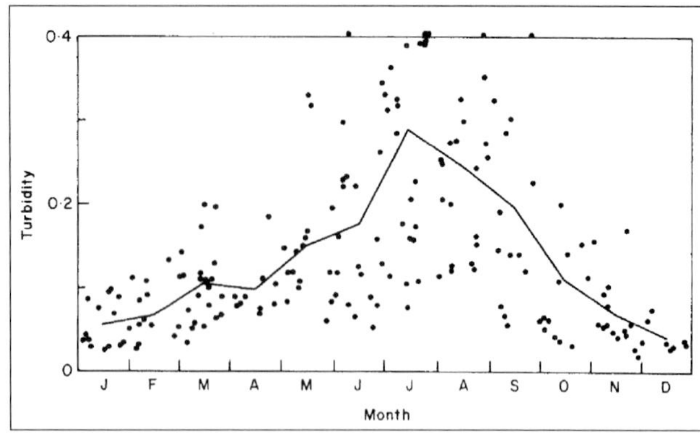

Figure 3.3-3 Turbidity coefficient B_λ at the wavelength $\lambda = 5 \times 10^{-7}$ m, as measured during 1972 at Raleigh (North Carolina) by Bilton *et al.* (1974). Daily means are indicated by dots, monthly means by the curve. Only days with a cloudless sky in the direction of the Sun have been included. Data points above $B_\lambda = 0.4$ are indicated at 0.4 (cf. the high turbidity curve of Fig. 3.3-2).

3.3.1.2 Scattered radiation

The scattered radiation for a clear sky may be estimated from knowledge of the composition of the atmosphere. The addition of radiation scattered from clouds requires knowledge of the distribution and type of clouds, and the accuracy with which a calculation of the scattered flux on a given plane can be made is rather limited. Even with a clear sky, the agreement between calculated and observed fluxes is not absolute.

Assuming that the distribution of intensity, $S_{h,Az}^{scatt.}$, for scattered radiation as a function of the directional coordinates (h, Az) [or (δ, ω)] is known, then the total flux of scattered radiation reaching a plane tilted at an angle s and directed azimuthally, at an angle γ away from south, may be written

$$d_{s,\lambda} = \int S_{h,Az}^{scatt.} \cos\theta \, d\Omega = \int_0^{\pi/2} d(Az) \int_{h_{min}(Az)}^{\pi/2} dh\, S_{h,Az}^{scatt.} \cos\theta(h, Az) \cos h. \quad (3.3.6)$$

Here $h_{min}(Az)$ is the smallest height angle, for a given azimuth, for which the direction defined by (h, Az) is on the "front" side of the inclined plane. The unit solid angle is $d\Omega = \sin z\, dz\, d\,(Az) = -\cos h\, dh\, d\,(Az)$. For a horizontal plane $\theta(h, Az) = z = ½\pi - h$. If the scattered radiation is isotropic,

$$S_{h,Az}^{scatt.} = \text{constant} = S^{scatt.},$$

then the scattered radiation flux on a horizontal plane becomes

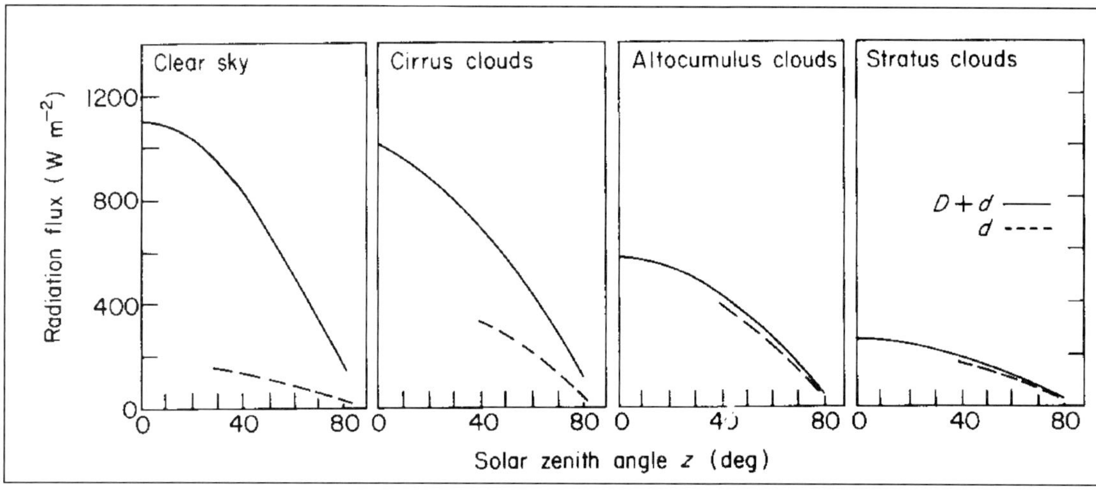

Figure 3.3-4 The influence of clouds on the radiation flux reaching a horizontal plane (total short-wavelength flux, $D + d$, and scattered flux alone, d) (based on measurements by J.Millard and J.Arvesen, as quoted in Turner, 1974).

$$d = \pi S^{scatt.}, \tag{3.3.7}$$

and the scattered radiation flux on an arbitrarily inclined surface may be written

$$d_{s,\gamma} = d \cos^2 (s/2), \tag{3.3.8}$$

an expression that can also be derived by considering the fraction of the sky "seen" by the tilted surface.

A realistic distribution of scattered radiation intensity is not isotropic, and as mentioned earlier it is not even constant for a cloudless sky and fixed position of the Sun, but depends on the momentary state of the atmosphere. Figure 3.3-5 shows the result of measurements for a cloudless sky (Kondratyev and Fedorova, 1976), indicating that the assumption of isotropy would be particularly poor for planes inclined directly towards or away from the Sun (Az equal to 0 or π relative to the solar azimuth). Robinson (1966) notes, from observations, that the main differences between observed distributions of scattered radiation and an isotropic one are (a) increased intensity for directions close to that of the Sun and (b) increased intensity near the horizon. In particular, the increased intensity in directions near the Sun is very pronounced, as is also evident from Fig. 3.3-5, and Robinson suggests that about 25% of d, the scattered radiation on a horizontal plane, should be subtracted from d and added to the direct radiation, before the calculation of scattered radiation on an inclined surface is performed using the isotropic model. However, such a prescription would not be generally valid, because the increased intensity in directions near the Sun is a function of the turbidity of the atmosphere as well as of cloud cover.

It is evident from Fig. 3.3-4 that the effect of clouds generally is to diminish direct radiation and increase scattered radiation, although not in the same proportion. Cirrus clouds and in particular altocumulus clouds substantially increase the scattered flux on a horizontal plane.

The radiation scattered by clouds is not isotropically distributed. Figure 3.3-6 gives the luminance distribution, i.e. the intensity of total radiation, along the great circle containing zenith as well as the direction to the Sun, the height angle of which was 20° at the time of measurement. For complete cloud cover the luminance distribution is entirely due to scattered

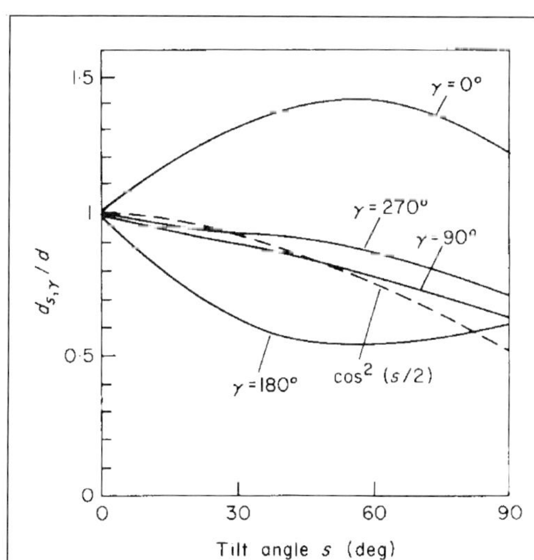

Figure 3.3-5 Ratio between scattered radiation flux on an inclined surface (tilt angle s, azimuth angle γ) and scattered radiation flux on a horizontal surface for a cloudless sky with solar height 15° (based on Kondratyev and Fedorova, 1976).

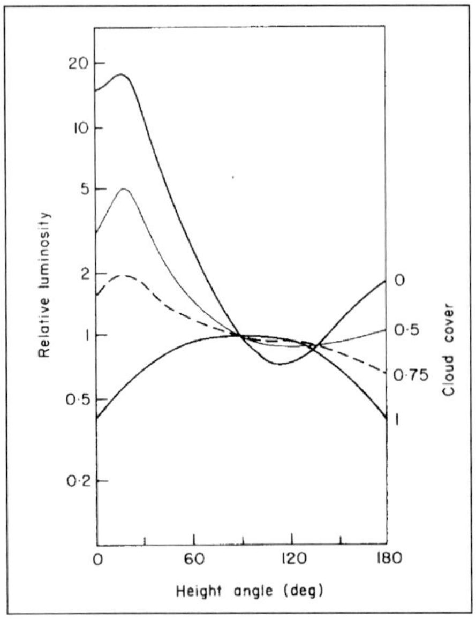

Figure 3.3-6 Luminance distribution along a great circle containing zenith as well as the direction towards the Sun, as a function of cloud cover. The distributions have been normalised at $h = 90°$. The data are based on several measurements with solar height 20° and mean turbidity, performed by Tonne and Normann (1960).

radiation from the clouds, and it is seen to be maximum at zenith and falling to 0.4 times the zenith value at the horizon.

3.3.1.3 Total short-wavelength radiation

For inclined surfaces or surfaces surrounded by elevated structures, the total short-wavelength (wavelengths below, say, 3×10^{-6} m) radiation flux comprises not only direct and scattered radiation, but also radiation reflected from the ground or from the surroundings onto the surface considered.

Reflected radiation

The reflected radiation from a given area of the surroundings may be described in terms of an intensity distribution, which depends on the physical nature of the area in question, as well as on the incoming radiation on that area. If the area is specularly reflecting, and the incoming radiation is from a single direction, then there is also a single direction of outgoing, reflected radiation, a direction which may be calculated from the law of specular reflection (polar angle unchanged, azimuth angle changed by π). Whether the reflection is specular or not may not only depend on the fixed properties of the particular area, but also on the wavelength and polarisation of the incoming radiation.

The extreme opposite of specular reflection is completely diffuse reflection, for which by definition the reflected intensity is isotropic over the hemisphere bordered by the plane tangential to the surface at the point considered, no matter what the distribution of incoming intensity. The total (hemispherical) amount of reflected radiation is in this case equal to the total incoming radiation flux times the albedo a of the area in question,

$$R = aE_+,$$

where, for horizontal surfaces, $E_+ = D + d$ (for short-wavelength radiation).

In general, the reflection is neither completely specular nor completely diffuse. In this case, the reflected intensity in a given direction, e.g. specified by height angle and azimuth, $S_{\Omega r}^{refl.}$, depends on the distribution of incident radiation intensities $S_{\Omega_r}^{inc.}$,*

$$S_{\Omega_r}^{refl.} = \int_{hemisphere} \rho_2(\Omega_r, \Omega_i) S_{\Omega_i}^{inc.} \cos\theta\,(\Omega_i)\,d\Omega_i. \tag{3.3.9}$$

Here $\rho_2(\Omega_r,\Omega_i)$ is called the bi-angular reflectance (Duffie and Beckman, 1974; these authors include an extra factor π in the definition of ρ_2). Dividing by the average incoming intensity,

$$\int S_{\Omega_i}^{inc.} \cos\theta(\Omega_i)\,d\Omega_i \Big/ \int \cos\theta\,(\Omega_i)\,d\Omega_i = E_+/\pi,$$

a reflectance depending only on one set of angles (of reflected radiation) is defined,

$$\rho_1\,(\Omega\,r) = \pi\,S_{\Omega_r}^{refl.}/E_+. \tag{3.3.10}$$

In general, ρ_1 is not a property of the reflecting surface, since it depends on incoming radiation, but if the incoming radiation is isotrophic (diffuse or black-body radiation), the incoming intensity can be divided out in (3.3.10).

* The intensity S_Ω is here defined as the energy flux passing through an infinitesimal area into an infinitesimal solid angle in the direction specified by Ω (e.g. h and Az). The infinitesimal area is perpendicular to the direction Ω, and the dimension of S_Ω is energy per unit time, unit area and unit solid angle. The energy flux passing through a unit area is found by multiplying S_Ω by $\cos\theta$, with θ being the angle between the direction Ω and the normal to the plane, and by $d\Omega$, and then integrating over the hemisphere above or below the unit area considered (giving directional fluxes E_- or E_+). These definitions are consistent with those used in sections 2.1 and 2.2 of *Renewable Energy* (2004), remembering that the incident solar radiation represents a limiting case of infinitesimal solid angle for the definition of intensity, because the rays from the Sun are treated as parallel, and recalling that several of the energy fluxes considered in section 2.2 of *Renewable Energy* (2004) were global averages.

The total hemispherical flux of reflected radiation may be found by integration of (3.3.9),

$$R = \int_{hemisphere} S^{refl.}_{\Omega_r} \cos\theta\,(\Omega_r)\, d\Omega_r, \tag{3.3.11}$$

and the corresponding hemispherical reflectance,

$$\rho = R/E_+ = a, \tag{3.3.12}$$

is equal to the albedo defined above.

All of the above relations have been written without reference to wavelength, but they are valid for each wavelength λ, as well as for appropriately integrated quantities.

In considering the amount of reflected radiation reaching an inclined surface, the surrounding surfaces capable of reflecting radiation are often approximated by an infinite, horizontal plane. If the reflected radiation is further assumed to be isotropic of intensity $S^{refl.}$, then the reflected radiation flux received by the inclined surface is

$$R_{s,\gamma} = \pi\, S^{refl.} \sin^2(s/2). \tag{3.3.13}$$

The factor $\sin^2(s/2)$ represents the fraction of the hemisphere above the inclined surface from which reflected radiation is received. This fraction is evidently complementary to the fraction of the hemisphere from which scattered radiation is received and, therefore, equal to $1 - \cos^2(s/2)$ from (3.3.8). The albedo (3.3.12) may be introduced, noting from (3.3.11) that $R = \pi S^{refl.}$ for isotropic reflection,

$$R_{s,\gamma} = aE_+ \sin^2(s/2). \tag{3.3.14}$$

For short-wavelength radiation, the flux E_+ on a horizontal plane equals $D + d$, the sum of direct and scattered short-wavelength fluxes.

If the geometry of the reflecting surroundings is more complicated (than a plane), or if the reflected intensity is not isotropic, the calculation of the amount of reflected radiation reaching a given inclined surface involves an integration over the hemisphere seen by the inclined surface. Thus, for each direction of light reflected onto the inclined plane, a contribution to $S^{refl.}_{\Omega_r}$ is included from the first unobscured point on the line of sight capable of reflecting radiation. If semi-transparent objects (such as a water basin) are present in the surroundings, the integration becomes three dimensional, and the refraction and transmission properties of the partly opaque objects must be considered.

Figure 3.3-7 shows an example of the enhancement of indirect radiation that may result from increasing albedo of the surroundings – in this case it is due to snow cover in winter (Kondratyev and Fedorova, 1976). The rejected radiation on the inclined surface reaches a maximum value for the tilt angle $s = 90°$ (vertical).

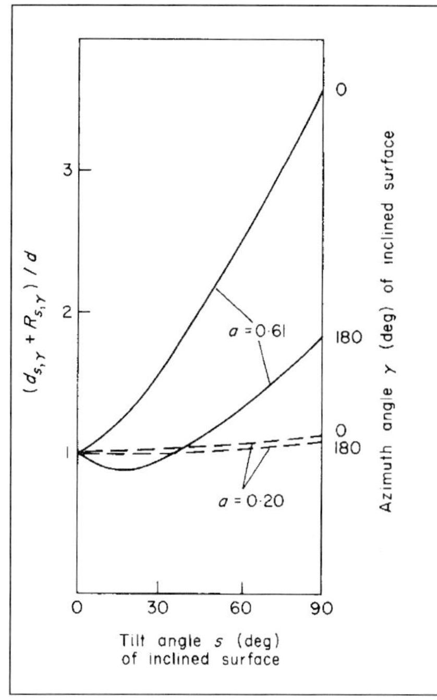

Figure 3.3-7 The ratio of scattered plus reflected radiation flux on inclined surfaces to that on a horizontal surface (scattered flux d only) for a clear sky with the Sun at height 34°. Measurements corresponding to two different ground albedos are depicted, based on summer and winter (snow-covered ground) conditions at a location in the Soviet Union (Kondratyev and Fedorova, 1976).

Average behaviour of total short-wavelength radiation

The sum of direct, scattered and reflected radiation fluxes constitutes the total short-wavelength (sw) radiation flux. The total short-wavelength flux on a horizontal surface is sometimes referred to as the "global radiation", i.e. $D + d$ (if there are no elevated structures to reflect radiation onto the horizontal surface). For an inclined surface of tilt angle s and azimuth γ, the total short-wavelength flux may be written

$$E^{sw}_{s,\gamma} = D_{s,\gamma} + d_{s,\gamma} + R_{s,\gamma}, \tag{3.3.15}$$

with the components given by (3.3.1), (3.3.6) or (3.3.8), and (3.3.14) or a generalisation of it. The subscript "+" on E^{sw} has been left out, since the direction of the flux is clear from the values of s and γ (the E_- flux would generally require $s > 90°$).

Examples of the influence of clouds, and of solar zenith angle, on global radiation $E^{sw}_{s=0\ \gamma}$ have been given in Fig. 3.3-4. Figure 3.3-8 illustrates the influence of cloud cover on the daily sum of total short-wavelength

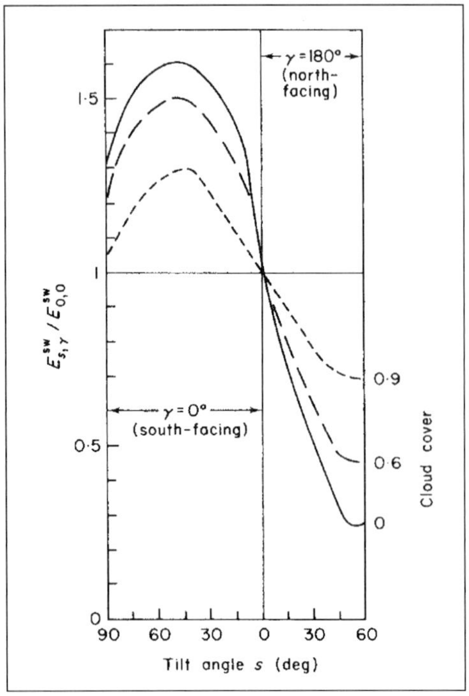

Figure 3.3-8 Daily sums of total short-wavelength radiation on an inclined plane, relative to that of a horizontal plane, for different cloud cover and as a function of tilt angle for north- and south-facing surfaces (based on measurements in the Soviet Union, Kondratyev and Fedorova, 1976).

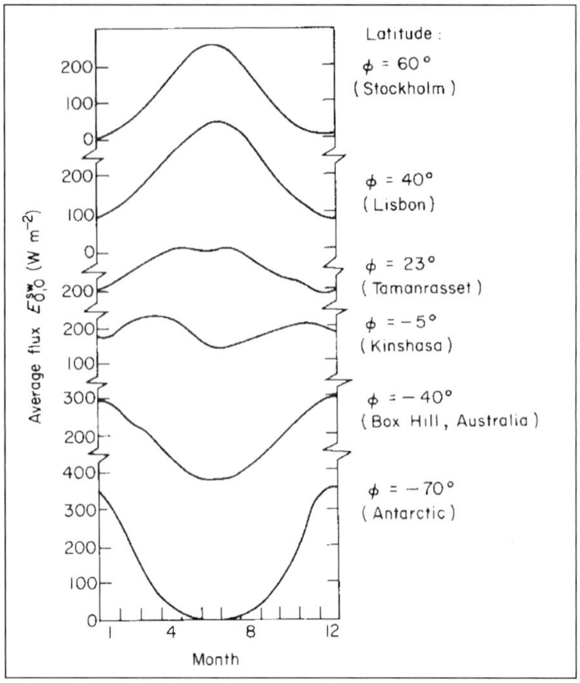

Figure 3.3-9 Smoothed variation with seasons of the average daily flux (24-h average) of total short-wavelength radiation on a horizontal plane for selected geographical locations (based on Flack and Morikofer, 1964).

radiation received by inclined surfaces, relative to that received by a horizontal surface. For south-facing slopes the radiation decreases with increasing cloud cover, but for north-facing slopes the opposite takes place.

Monthly averages of total short-wavelength radiation for different geographical locations are shown in Figs 3.3-9 and 3.3-10 for a horizontal surface. In Fig. 3.3-10, two sets of data are compared, each representing a pair of locations with the same latitude. The two at $\phi \approx 43°$N correspond to a coastal and a continental site, but the radiation, averaged for each month, is nearly the same. The other pair of locations are at 35–36°N. Albuquerque has a desert climate, while Tokyo is near the ocean. Here the radiation patterns are vastly different during summer by as much as a factor two. The summer solar radiation in Tokyo is also smaller than in both the 43°N sites, presumably due to the influence of urbanisation and a high natural frequency of cloud coverage. Other average properties of radiation fluxes on horizontal surfaces were considered elsewhere, notably for a latitude 53°N location.

Turning now to inclined surfaces, Fig. 3.3-11 shows the total short-wavelength radiation on a vertical plane facing south, west, east and north for a location at $\phi = 56°$N. The monthly means have been calculated from the hourly data of the Danish reference year (Andersen *et al.*, 1974), using the isotropic approximation (3.8) and (3.14) with an assumed albedo $a = 0.2$. The Danish reference year consists of selected meteorological data exhibiting "typical" fluctuations. This is achieved by selecting monthly

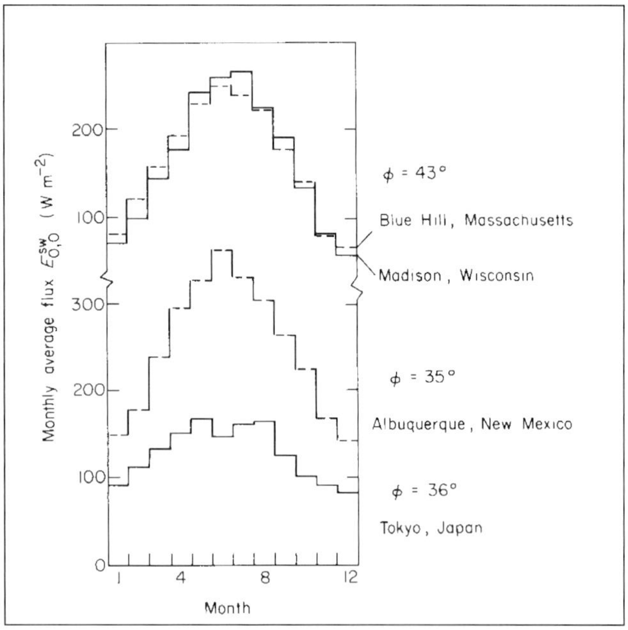

Figure 3.3-10 Monthly average flux on a horizontal plane for pairs of locations with similar latitude but different climate [based on data from NOAA (US National Oceanic and Atmospheric Administration), quoted from Duffie and Beckman, 1974].

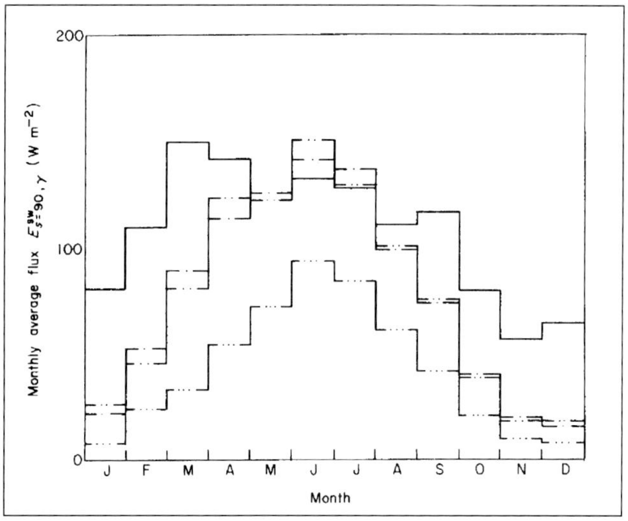

Figure 3.3-11 Monthly average flux on vertical surface with different azimuthal orientation, based on the Danish reference year (latitude 56°) and an assumed ground albedo of 0.2 (0.9 with snow cover). Key:—— $\gamma = 0°$ (south); —·—·— $\gamma = -90°$ (west); —··—··— $\gamma = 90°$ (east); —— ··· —— $\gamma = 180°$ (north). This and the following calculations assume that the scattered radiation is isotropic.

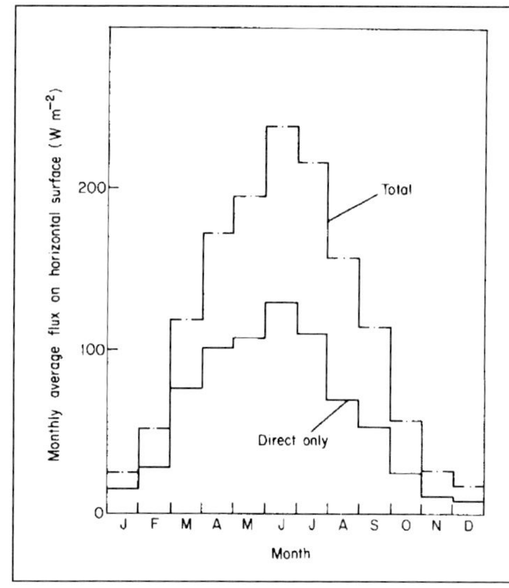

Figure 3.3-12 Monthly average short-wavelength flux on horizontal surface, and the direct part, D, based on the Danish reference year, $\phi = 56°N$ (Andersen *et al.*, 1974).

sequences of actual data, with monthly averages of each of the meteorological variables in the vicinity of the 30-year mean. The different sequences, which make up the reference year, have been taken from different years. The variables pertaining to solar radiation are global radiation ($D + d$), normal incidence radiation (S_N) and scattered radiation on a horizontal plane (d). Only global radiation has been measured over extended periods of time, but the two other variables have been constructed in such a way that the three variables, of which only two are independent, become reasonably consistent. Several other countries are in the process of constructing similar "reference years", which will allow an easy intercomparison of, for example, performance calculations of solar energy devices or building insulation prescriptions. Monthly averages of the basic data of the Danish reference year, $D + d$ and D, are shown in Fig. 3.3-12. Reference years have subsequently been constructed for a number of other locations (European Commission, 1985).

Figure 3.3-13 shows the composition of the total short-wavelength flux on a vertical, south-facing surface, in terms of direct, scattered and reflected radiation. It has been assumed that the albedo changes from 0.2 to 0.9 when the ground is covered by snow (a piece of information also furnished by the reference year). Snow cover is in this way responsible for the relative maximum in reflected flux for February. The proportion of direct radiation is substantially higher during winter for this vertical surface than for the horizontal surface (Fig. 3.3-12). Figure 3.3-14 gives the variation of the yearly average fluxes with the tilt angle s, still for a south-facing surface ($\gamma = 0$). The variation of monthly averages of the total flux with tilt angle for a south-facing surface is given in Fig. 3.3-15.

According to Fig. 3.3-14, the maximum yearly short-wavelength radiation in Denmark ($\phi = 56°N$) is obtained on a south-facing surface tilted about 40°, but the maximum is broad. The maximum direct average flux is obtained at a tilt angle closer to ϕ, which is clear because at the top of the atmosphere the maximum

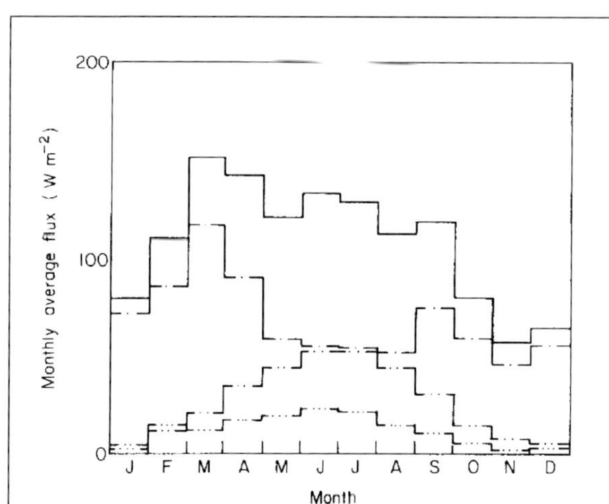

Figure 3.3-13 Components of monthly average fluxes on vertical, south-facing surface. Key: —— total short-wavelength radiation; —·—·— direct radiation $D_{s,\gamma}$; —··—··— scattered radiation $d_{s,\gamma}$; — ··· — reflected radiation $R_{s,\gamma}$ (in all cases $\gamma = 0$). The calculation is based on the Danish reference year, $\phi = 56°N$, and an assumed albedo of 0.2 (0.9 with snow cover).

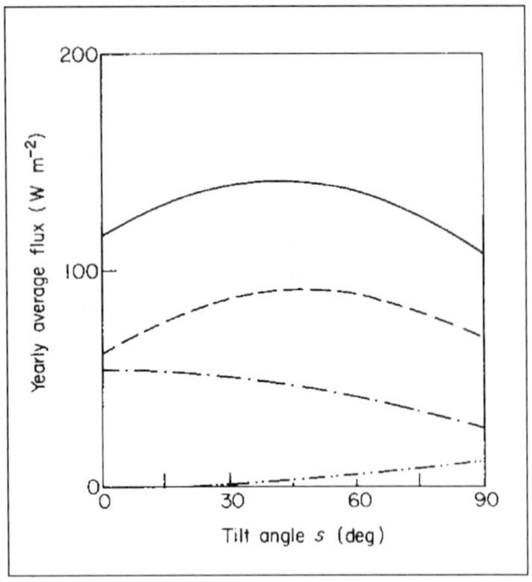

Figure 3.3-14 Components of yearly average fluxes on a vertical, south-facing surface. Key: —— total short-wavelength radiation; ------ direct radiation $D_{s,\gamma}$; –·–·–·– scattered radiation d_s,γ; –··–··– reflected radiation $R_{s,\gamma}$ (in all cases $\gamma = 0$). The calculation is based on the Danish reference year, $\phi = 56°$N, and an assumed albedo of 0.2 (0.9 with snow cover).

would be attained for s equal to the latitude plus or minus the Sun's declination, the extremes of which are about ±23° at summer and winter solstices. From Fig. 3.3-15, one can see that the most constant radiation over the year is obtained for $s = 90°$. The December maximum value is for $s = 75°$, but the solar radiation on an $s = 90°$ surface is only 1–2% smaller.

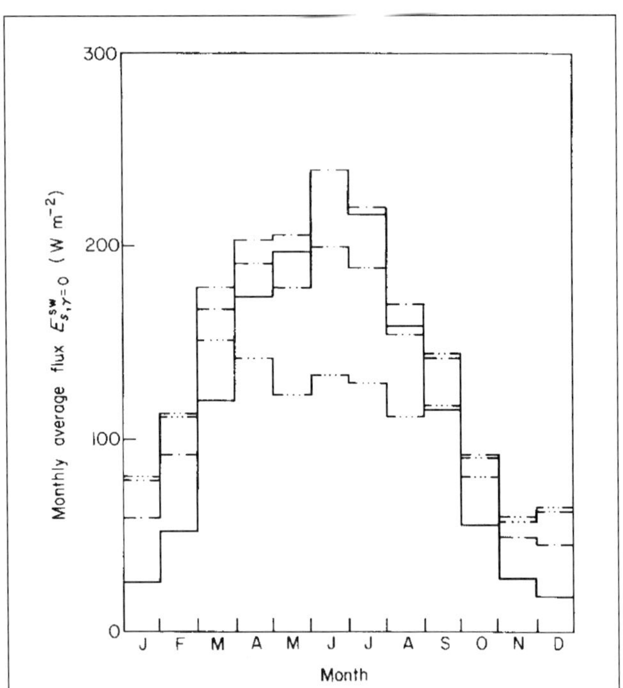

Figure 3.3-15 Monthly average of total short-wavelength radiation on inclined, south-facing surfaces, based on the Danish reference year, $\phi = 56°$N. Key: —— tilt angle $s = 0$; –·–· $s = 30°$; –··–·· $s = 60°$; –···– $s = 90°$.

3.3.1.4 Long-wavelength radiation

The long-wavelength radiation reaching a plane situated at the Earth's surface may be quite substantial, in fact, exceeding the short-wavelength radiation when averaged over 24 h. However, the outgoing long-wavelength radiation is (again on average) still larger, so that the net long-wavelength flux is directed away from the surface of the Earth. The ground and most objects at the Earth's surface emit long-wavelength (lw) radiation approximately as a black body, while the long-wavelength radiation from the atmosphere often deviates substantially from any black-body spectral distribution.

In general, the emission and absorption of radiation by a body may be described by the spectral and directional emittance, $\varepsilon\lambda(\Omega)$, and the corresponding absorptance, $\alpha_\lambda(\Omega)$. These quantities are related to $e(v)$ and $K(V)$, which may also depend on the direction Ω, by

$$e(v,\ \Omega) = \varepsilon_\lambda(\Omega)\ \mathrm{d}F^{Planck}/\mathrm{d}v, \qquad k(v,\ \Omega) = \alpha_\lambda\ (\Omega)$$

for corresponding values of frequency v and wavelength λ.

Thus $\varepsilon_\lambda(\Omega)$ is the emission of the body, as a fraction of the black-body radiation for emission in the direction Ω, and $\alpha_\lambda(\Omega)$ is the fraction of incident flux, again from the direction specified by Ω, which is absorbed by the body.

Where a body is in complete thermodynamic equilibrium with its surroundings (or more specifically those with which it exchanges radiation), Kirchoff's law is valid (see e.g. Siegel and Howell, 1972),

$$\varepsilon_\lambda(\Omega) = \alpha_\lambda(\Omega). \tag{3.3.16}$$

However, since both $\varepsilon_\lambda(\Omega)$ and $\alpha_\lambda(\Omega)$ are properties of the body, they are independent of the surroundings, and hence (3.3.16) is also valid away from equilibrium. This does not apply if wavelength averages are considered. Then $\varepsilon(\Omega)$ is still a property of the surface, but $\alpha(\Omega)$ depends on the spectral composition of the incoming radiation, as follows from the definition of $\alpha(\Omega)$,

$$\alpha(\Omega) = \int \alpha_\lambda(\Omega)\ S_\lambda^{inc}\ (\Omega)\ \mathrm{d}\lambda \Big/ \int S_\lambda^{inc}\ (\Omega)\ \mathrm{d}\lambda.$$

Further averaging over directions yields the hemispherical emittance ε and absorptance α, the former still being a property of the surface, the latter not.

Consider now a surface of temperature T_s and orientation angles (s, γ), which emit long-wavelength radiation in such a way that the hemispherical energy flux is

$$E_{-}^{lw,emission} = \varepsilon^{lw}\,\sigma\,T_s^4, \tag{3.3.17}$$

where σ is Stefan's constant (5.7×10^{-8} W m^{-2} K^{-4}). For pure black-body radiation from a surface in thermal equilibrium with the temperature T_s (forming an isolated system, e.g. a cavity emitter), ε^{lw} equals unity.

The assumption that $\varepsilon_\lambda = \varepsilon$ (emissivity independent of wavelength) is referred to as the "grey-body" assumption. It implies that the energy absorbed from the environment can be described by a single absorptivity $\alpha^{lw} = \varepsilon^{lw}$. Further, if the long-wavelength radiation from the environment is described as black-body radiation corresponding to an effective temperature T_e, the incoming radiation absorbed by the surface may be written

$$E_{+}^{lw,abs.} = \varepsilon^{lw}\,\sigma\,T_e^4. \tag{3.3.18}$$

If the part of the environmental flux reflected by the surface is denoted R $(= \rho^{lw}\,\sigma\,T_e^4)$, then the total incoming long-wavelength radiation flux is $E_+^{lw} = R + E_+^{lw,abs}$, and the total outgoing flux is $E_-^{lw} = R + E_-^{lw,emission}$, and thus the net long-wavelength flux is

$$E^{lw} = \varepsilon^{lw}\,\sigma(T_e^4 - T_s^4). \tag{3.3.19}$$

The reflection from the environment of temperature T_e back onto the surface of temperature T_s has not been considered, implying an assumption regarding the "smallness" of the surface considered as compared with the effective surface of the environment. Most surfaces that may be contemplated at the surface of the Earth (water, ice, grass, clay, glass, concrete, paints, etc.) have long wavelength emissivities ε^{lw} close to unity (typically about 0.95). Materials such as iron and aluminium with non-polished surfaces have low long-wavelength emissivities (about 0.2), but often their temperature T_s is higher than the average temperature at the Earth's surface (due to high absorptivity for short-wavelength radiation), so (3.3.19) may still be a fair approximation, if T_s is chosen as an average temperature of physical surfaces. If the surface is part of an energy-collecting device, a performance evaluation will require the use of the actual temperature T_s with its variations.

The deviations of the long-wavelength radiation received from the environment from that of a black-body are more serious, leading both to a change in wavelength dependence and to a non-isotropic directional dependence. Decisive in determining the characteristics of this radiation component is the average distance, for each direction, to the point at which the long-wavelength radiation is emitted. Since the main absorbers in the long-wavelength frequency region are water vapour and CO_2, and since the atmospheric content of water vapour is the most variable of these, then one may expect the long-wavelength flux to be primarily a function of the water vapour content m_v (Kondratyev and Podolskaya, 1953).

At wavelengths with few absorption bands the points of emission may be several kilometres away, and owing to the temperature variation through the atmosphere, one expects this component of T_e to be 20–30 K below the ambient temperature T_a at the surface. As humidity increases, the average emission distance diminishes, and the effective temperature T_e becomes closer to T_a. The temperature of the surface itself, T_s, is equal to or larger than T_a, depending on the absorptive properties of the surface. This is also true for physical surfaces in the surroundings, and therefore the "environment" seen by an inclined surface generally comprises partly a fraction of the sky with an effective temperature T_e below T_a and partly a fraction of the ground, possibly with various other structures, which has an effective temperature of T_e above T_a.

Empirical evidence for inclined surfaces

Figure 3.3-16 shows the directional dependence of long-wavelength radiation from the environment, averaged over wavelengths, for very dry and humid atmospheres. The measurements on which the figure is based (Oetjen *et al.*, 1960) show that the black-body approximation is only valid for directions close to the horizon (presumably implying a short average distance to points of emission), whereas the spectral intensity exhibits deeper and deeper minima, in particular around $\lambda = 10^{-5}$ m, as the direction approaches zenith. Relative to the black-body radiation at ambient temperature, T_a, the directional flux in Fig. 3.3-16 starts at unity at the horizon, but drops to 79 and 56%, respectively, for the humid (Florida Beach region) and the dry atmospheres (Colorado mountainous region). If an effective black-body temperature is ascribed, although the spectral distributions are not Planckian, the reduced T_e is 94.3 and 86.5% of T_a, corresponding to temperatures 27.5 and 37.9 K below ambient temperature. Although it is not clear whether they are in fact, Meinel and Meinel (1976) suggest that the two curves in Fig. 3.3-16 be used as limiting cases, supplemented with black-body emissions from ground and surrounding structures, which are characterised by ε^{lw} between 1.0 and 1.1, for the calculation of net long-wavelength radiation on inclined surfaces.

In calculations of the performance of flat-plate solar collectors it has been customary to use an effective environmental temperature about 6 K less than the ambient temperature T_a (Duffie and Beckman, 1974; Meinel and Meinel, 1976), but since the surfaces

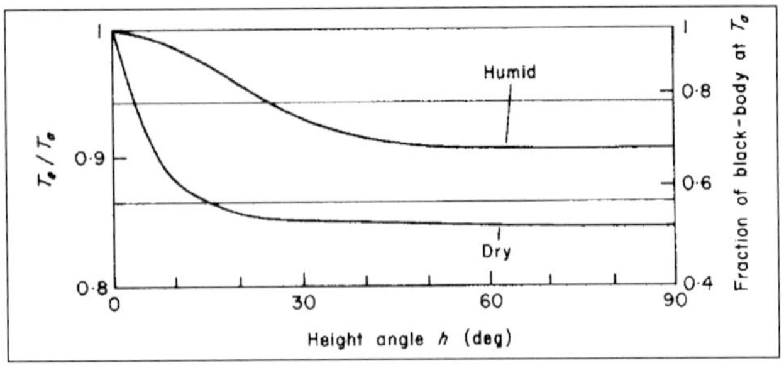

Figure 3.3-16 Variation in incoming long-wavelength radiation with height above the horizon (flat ground). The scale on the left is the ratio between the effective temperature of the long-wavelength radiation received and the ambient air temperature, whereas the scale on the right gives the average flux relative to that of black-body radiation at ambient temperature. The two curves represent measurements for a very humid and a very dry atmosphere; performed by Oetjen *et al.* (1960) and quoted by Meinel and Meinel (1976).

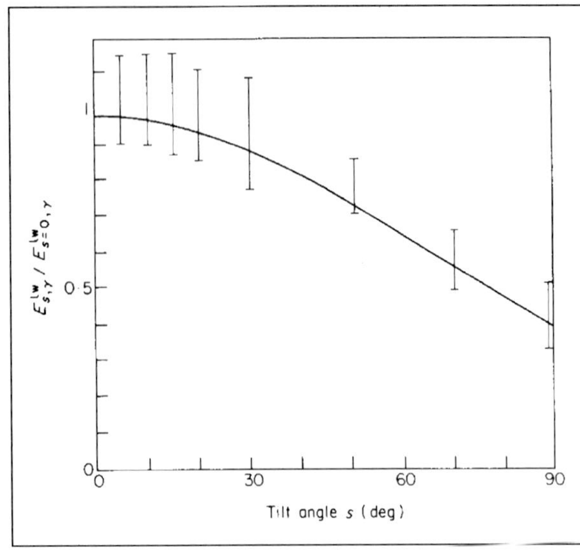

Figure 3.3-17 Ratio of net long-wavelength radiation on inclined surfaces and the corresponding flux for a horizontal surface. The error bars represent the spread of measured values. Although measurements for different azimuth angles γ have been aggregated, the position of data points within the error bars is not noticeably correlated with γ. The curve has been calculated with assumptions mentioned in the text (based on Kondratyev and Fedorova, 1976).

considered in this context are usually placed with tilt angles in the range 45–90°, a fraction of the hemisphere "seen" by the surface will be the ground (and buildings, trees, etc.). Thus, the effective temperature of the long-wavelength radiation received will be of the form

$$T_e = C\,T_{e,atmosphere} + (1 - C)\,T_{e,ground}, \tag{3.3.20}$$

where C is at most $\cos^2(s/2)$, corresponding to the case of an infinitely extended, horizontal plane in front of the inclined surface considered. For this reason T_e will generally not be as much below T_a as indicated by Fig. 3.3-16, but it is hard to see how it could be as high as $T_a - 6$, since $T_{e,ground}$ is rarely more than a few degrees above T_a. Silverstein (1976) estimates values of $T_e - T_a$ equal to –20 K for horizontal surfaces and –7 K for vertical surfaces.

Figure 3.3-17 shows measured values of the ratio of net long-wavelength radiation (3.3.19) on a black-painted inclined surface to that on the same surface in horizontal position. The quantity varying with angle of inclination is the effective black-body temperature (which could alternatively be calculated on the basis of curves such as those given in Fig. 3.3-16), $T_{e,s,\gamma}$, so that Fig. 3.3-17 can be interpreted as giving the ratio

$$(T^4_{e,s,\gamma} - T^4_s)/(T^4_{e,0,0} - T^4_s).$$

No significant differences are found between the data points for $\gamma = 0°$, $\pm 90°$ and 180°, which are available for each value of the tilt angle s. The data are in agreement with the vertical to horizontal temperature difference $T_{e,s=90°} - T_{e,0}$ found by Silverstein and the calculation of Kondratyev and Podolskaya (1953) for an assumed atmospheric water mixing ratio of about 0.02 $m^3\ m^{-2}$ vertical column. However, the accuracy is limited, and it would be valuable to expand the experimental activity in order to arrive at more precise determinations of the effective environmental temperature under different conditions.

3.3.1.5 Variability of solar radiation

The fact that the amount of solar radiation received at a given location at the Earth's surface varies with time has been implicit in several of the data discussed in the preceding sections. The variability and part-time absence of solar radiation due the Earth's rotation (diurnal cycle) and orbital motion (seasonal cycle, depending on latitude) are well known and simple to describe, and so the emphasis here will be on a description of the less simple influence of the state of the atmosphere, cloud cover, etc.

Most of the data presented in section 3.1.1 have been in the form of averages over substantial lengths of time (e.g. a month), although some results were discussed in terms of instantaneous values, such as intensity as a function of solar zenith angle. In order to demonstrate the time structure of radiation quantities more clearly, Fig. 3.3-18 shows hourly averages of normal incidence radiation, S_N, as well as scattered radiation on a horizontal

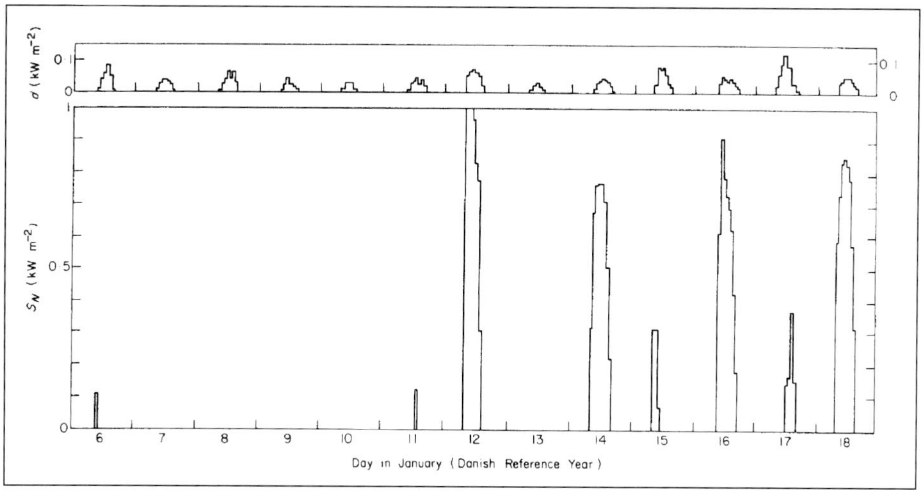

Figure 3.3-18 Hourly values of the normal incidence flux, S_N, and the scattered flux, d, on a horizontal plane for the Danish referencc year, ϕ — 56°N. Thirteen consecutive winter days are shown.

surface, d, hour by hour over a 13-day period of the Danish reference year (latitude 56°N). For this period in January, several days without any direct (or normal incidence) radiation appear consecutively, and the scattered radiation is quite low. On the other hand, very clear days occasionally occur in winter, as witnessed by the normal flux received on 12 January in Fig. 3.3-18. Fluctuations within each hour are also present, as can be seen from the example shown in Fig. 3.3-19. The data are for two consecutive days at a latitude of about 39°N (Goddard Space Flight Center in Maryland; Thekaekara, 1976), collected at 4-s intervals. The figure conveys the picture of intense fluctuations, particularly on a partially cloudy day.

The variability of incident solar energy may be displayed in various ways. Figure 3.3-20 gives the frequency distributions for the daily integrated radiation energy received by a horizontal plane for two different locations.

At Mauna Loa (Hawaii, at a height of 3400 m), the frequency curve peaks at about 6 kWh m^{-2} d^{-1}, while at Argonne (Illinois) the maximum is broad and extends from zero to about 2.5 kWh m^{-2} d^{-1}. The results are shown for two sampling periods: one year and several years. One sees that individual years may exhibit frequency distributions quite different from long-term averages.

Figure 3.3-21 gives, for the same two locations, some impression of the higher order structure of the fluctuations by trying to indicate the frequency of poor solar radiation days lumped together. For selected daily energy sums, the figure shows, for one year as well as for the average of several years, the number of occurrences of n consecutive days of daily solar radiation below the selected value. At Mauna Loa, the solar radiation falls below 1.742 kWh m^{-2} d^{-1} only one day a year and never falls below 1.161 kWh m^{-2} d^{-1}, while at Argonne, 11 consecutive days with solar radiation below 1.742 kWh m^{-2} d^{-1} on average occur once every year and isolated days with this condition occur nine times a year.

Geographical distribution of solar power

The geographical distribution of average incoming (shortwave) solar radiation on a horizontal plane is shown in Figs. 3.3-1a–d for each of the four seasons. The data exhibit considerable dependence on conditions of cloud

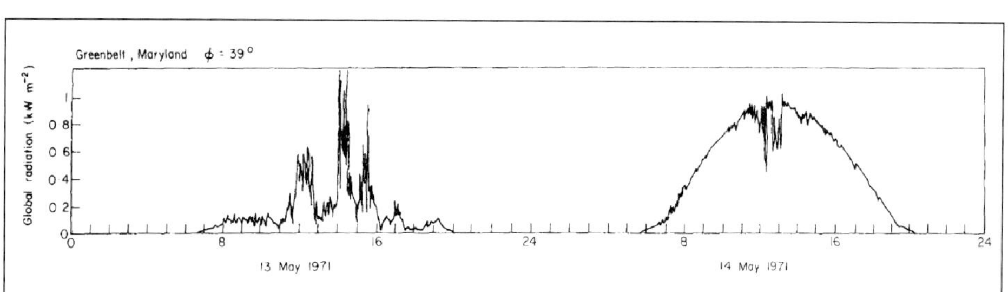

Figure 3.3-19 Two days of continuous record of total short-wavelength flux on a horizontal plane (based on Thekaekara, 1976).

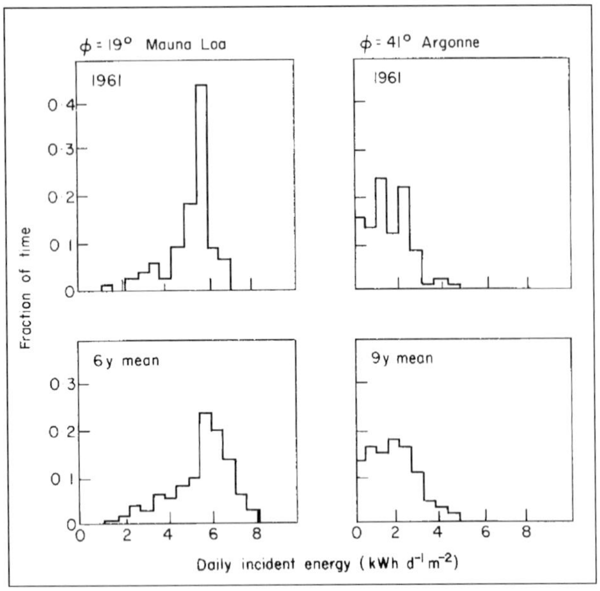

Figure 3.3-20 Frequency distribution of daily sums of total short-wavelength radiation flux on a horizontal plane for Mauna Loa (Hawaii, $\phi = 19°$N, elevation of measurement location: 3400 m) and Argonne (Illinois, $\phi = 41°$N). The upper curves are for 1961, while the frequency distributions shown in the lower row are based on several years of observation (based on Machta, 1976).

cover and other variables influencing the disposition of radiation on its way from the top to the bottom of the atmosphere. These data form the basis for further analysis in terms of suitability of the power flux for energy conversion in thermal and electricity-producing devices (Chapter 4.1). Methods for estimating solar radiation on inclined surfaces from the horizontal data will be introduced in Chapter 6.1, along with an appraisal of the fraction of the solar resource that may be considered of practical use after consideration of environmental and area use constraints.

Power duration curves

The accumulated frequency distribution is called a "power duration curve", since it gives the fraction of time during which the energy flux exceeds a given value E, as a function of E. Figure 3.3-22a,b gives power duration curves for total and for direct radiation on a horizontal surface and two southward inclined surfaces for a location at latitude $\phi = 56°$N (Danish reference year; cf. European Commission, 1985). Since data for an entire year have been used, it is not surprising that the energy flux is non-zero for approximately 50% of the time. The largest fluxes, above 800 W m^{-2}, are obtained for only a few hours a year, with the largest number of hours with high incident flux being for a surface inclined about 45° (cf. Fig. 3.3-14). The shape of the power duration curve may be used directly to predict the performance of a solar energy device, if this is known to be sensitive only to fluxes above a certain minimum value, say, 300 W m^{-2}. The power duration curves for direct radiation alone are shown on the right. They are relevant for devices sensitive only to direct radiation, such as most focusing collectors.

Figure 3.3-22c gives the power duration curves for normal incidence radiation, S_N, and for scattered radiation on a horizontal plane, d. The normal incidence curve is of interest for fully tracking devices, i.e. devices which are being continuously moved in order to face the direction of the Sun. By comparing Fig. 3.3-22c with Fig. 3.3-22a,b, it can be seen that the normal incidence curve lies substantially above the direct radiation curve for an optimum but fixed inclination ($s = 45°$). The power duration curve for normal incidence radiation is not above the curve for total radiation at $s = 45°$, but it would be if the scattered radiation received by the tracking plane normal to the direction of the Sun were added to the normal incidence (direct by definition) radiation. The power duration curve for scattered radiation alone is also shown in Fig. 3.3-22c for a horizontal plane. The maximum scattered flux is about 350 W m^{-2}, much higher than the fluxes received during the winter days shown in Fig. 3.3-18.

3.3.1.6 Wind

It follows from the discussion in *Renewable Energy*, 2004 (Ch. 2) that the kinetic energy content of the atmosphere, on average, equals about seven days of kinetic energy production or dissipation, also assuming average rates. Utilising wind energy means installing a device which converts part of the kinetic energy in the atmosphere to, say, mechanical useful energy; i.e. the device draws primarily on the energy stored in the atmospheric circulation and not on the energy flow into the general circulation (1200 TW). This is also clear from the fact that the production of kinetic energy, i.e. the conversion of available potential energy into kinetic energy does not take place to any appreciable extent near the Earth's surface. Newell *et al.* (1969) give height–latitude distributions of the production of zonal kinetic energy for different seasons. According to these, the regions of greatest kinetic energy, the zonal jet-streams at mid-latitudes and a height of about 12 km, are maintained from conversion of zonal available potential energy in the very same regions.

One of the most important problems to resolve in connection with any future large-scale energy extraction from the surface boundary region of the atmosphere's wind energy storage is that of the nature of the mechanism for restoring the kinetic energy reservoir in the presence of man-made energy extraction devices. To

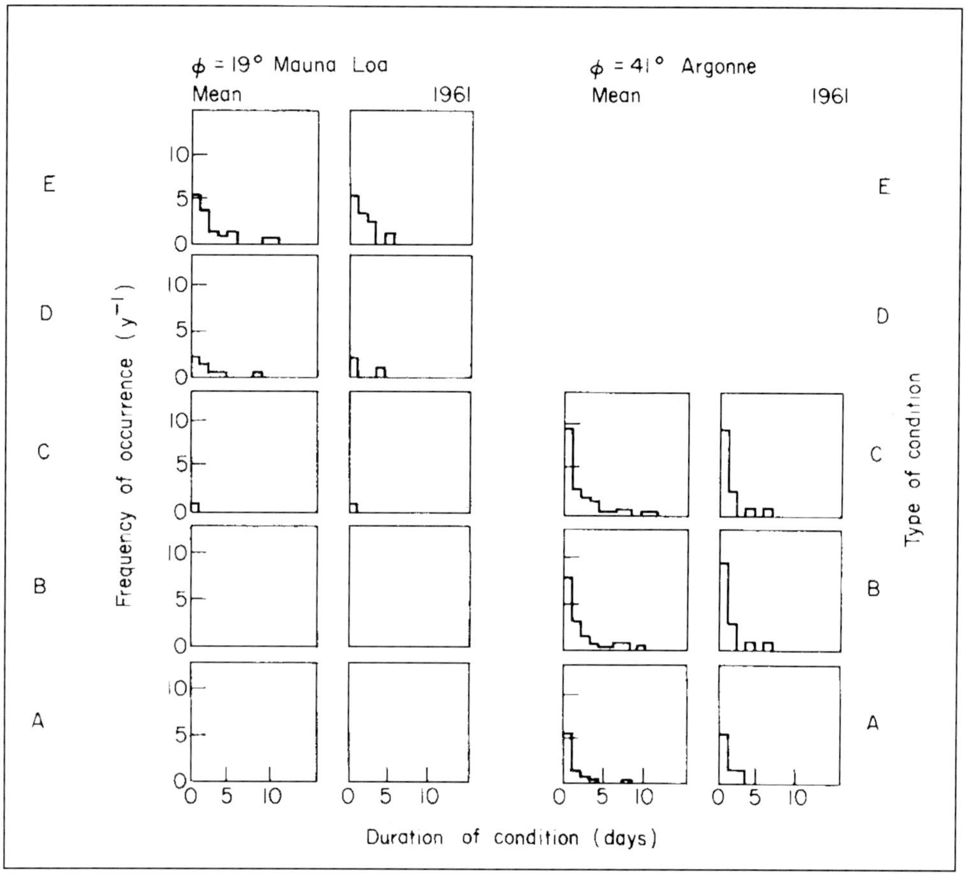

Figure 3.3-21 Occurrence of rows of consecutive days with daily sums of radiation on a horizontal plane below a certain value for two different locations. Five conditions are considered in rows from bottom to top: radiation in Wh m^{-2} d^{-1} below 581 (A), below 1161 (B), below 1742 (C), below 3483 (D), and below 5225 (E). The histograms give the number of times per year, for which one of the conditions A–E persists during the number of days indicated on the abscissa. The two sets of columns correspond to mean frequencies for a number of years of observation (cf. Fig. 3.3-20) and to 1961 alone (based on Machta, 1976).

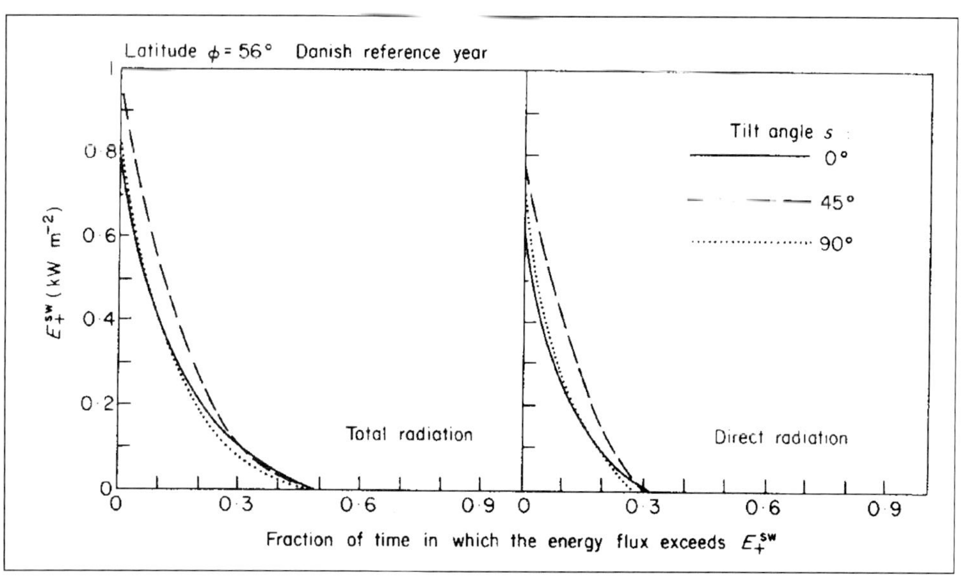

Figure 3.3-22a,b One-year power duration curves of total (*a:* left) and direct (*b:* right-hand side) short-wavelength radiation on south-facing surfaces of three different inclinations, based on the Danish reference year, $\phi = 56°$N.

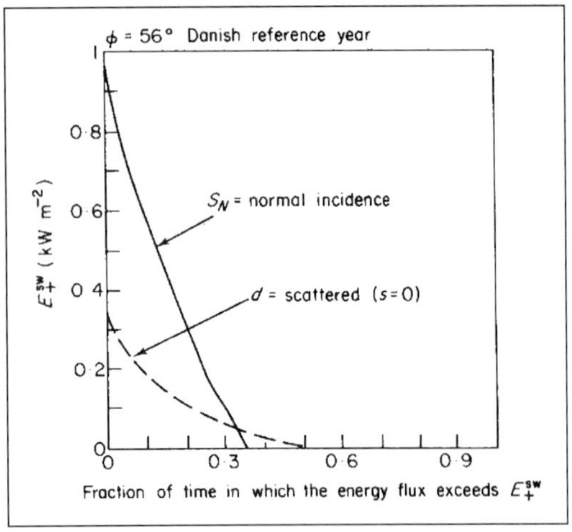

Figure 3.3-22c One-year power duration curves of normal incidence radiation alone, and of scattered radiation alone, based on the Danish reference year (ϕ = 56°N). The normal incidence curve would correspond to the radiation received by a fully tracking instrument. The curve for scattered radiation is for a horizontal surface.

a substantial extent, this mechanism is known from the study of natural processes, by which energy is being removed from the kinetic energy storage in a given region of the atmosphere. The store of energy is replenished by the aforementioned conversion processes, by which kinetic energy is formed from the much larger reservoir of available potential energy (four times larger than the kinetic energy reservoir). The available potential energy is created by temperature and pressure differences, which in turn are formed by the solar radiation flux and the associated heat fluxes.

3.3.1.7 Wind velocities

The horizontal wind profile

The horizontal components of the wind velocity are typically two orders of magnitude larger than the vertical component, but the horizontal components may vary a great deal with height under the influence of frictional and impact forces on the ground and the geostrophic wind above, which is governed by the Earth's rotation. A simple model for the variation of horizontal velocity with height in the lower (Prandtl) planetary boundary layer, assuming an adiabatic lapse rate [i.e. temperature gradient] and a flat, homogeneous surface.

The zero point of the velocity profile is often left as a parameter, so that it is:

$$u = \kappa^{-1} (\tau/\rho)^{1/2} \log((z + z_0 - d_0)/z_0) \qquad (3.3.21)$$

The zero displacement, d_0, allows the zero point of the profile to be determined independently of the roughness length z_0, both being properties of the surface. This is convenient, for example, for woods and cities with roughness lengths from about 0.3 m to several metres, for which the velocity profile "starts" at an elevation of maybe 10 or 50 m above the ground. For fixed z_0 and d_0, a family of profiles is described by (3.3.21), depending on the one parameter $(\tau/\rho)^{1/2}$, called the friction velocity. For a given location it is essentially determined by the geostrophic wind speed, still assuming an adiabatic ("neutral") temperature profile.

If the atmosphere is not neutral, the model leading to (3.3.21) has to be modified. In fact, (2.33) in Chapter 2, *Renewable Energy* (2004) may be viewed as a special case in a more general theory, due to Monin and Obukhov (1954) (see also Monin and Yaglom, 1965), according to which the wind speed may be expressed as

$$u(z) = \kappa^{-1} (\tau/\rho)^{1/2} (f(z/L) - f(z_0/L)).$$

Only in the case of an adiabatic lapse rate is the function f equal to a logarithm. This case is characterised by $z \ll L$, where the parameter L describing the atmosphere's degree of stability may be written

$$L = c_P\, \rho\, (\tau/\rho)^{3/2}\, T_0/(g\, \kappa\, Q_0). \qquad (3.3.22)$$

L is called the Monin–Obukhov length, g is the acceleration of gravity and T_0 and Q_0 are the temperature and heat flux at the surface. Thus, L is a measure of the transport of heat near the surface, which again determines the stability of the atmosphere. If the heat flux is sufficiently large and directed from ground to atmosphere (Q_0 and thus L negative), then transport of heat and matter is likely to be highly convective and unstable, so that there will be rapid mixing and therefore less variation of wind speed with height. On the other hand, if Q_0 and L are positive, the atmosphere becomes stable, mixing processes become weak and stratification becomes pronounced, corresponding to almost laminar motion with increasing wind speed upwards.

The connection between L and the temperature gradient becomes clear in the convective case, in which one can prove that z/L equals the Richardson number,

$$Ri = g\frac{1}{\theta}\frac{\partial\theta}{\partial z}\Big/\left(\frac{\partial u}{\partial z}\right)^2, \qquad (3.3.23)$$

where θ is the potential temperature. From the definition,

$$\frac{\partial\theta}{\partial z} = \frac{\theta}{T}\left(\frac{\partial T}{\partial z} - \frac{c_P - c_V}{c_P}\frac{T}{P}\frac{\partial P}{\partial z}\right),$$

is a direct measure of the deviation from adiabaticity. The adiabatic atmosphere is characterised by $Ri = 0$. The identity of z/L and Ri is easy to derive, but the linear relationship $l = kz$ may not be valid in general. Still, the identity may be maintained, if the potential temperature in (3.3.23) is regarded as a suitably defined average.

Figure 3.3-23 gives an example of calculated velocity profiles for neutral, stable and unstable atmospheres, assuming fixed values of z_0 (0.01 m, a value typical of grass surfaces) and $(\tau/\rho)^{1/2} = 0.5$ m s^{-1} (based on Frost, 1975). Many meteorological stations keep records of stability class, based on either observed turbulence, Ri or simply $\partial T/\partial z$. Thus, it is often possible with a certain amount of confidence to extrapolate wind speed measurements taken at one height (e.g. the meteorological standard of 10 m) to other heights that are interesting from the point of view of wind energy extraction.

The wind profiles also show that a wind energy conversion device spanning several tens of metres and placed in the planetary boundary layer is likely to experience substantially different wind velocities at different ends of the device (e.g. rotor blades).

The simple parametrisations of wind profiles considered above can be only expected to be valid over flat and homogeneous terrain. The presence of obstacles such as hills, vegetation of varying height and building structures may greatly alter the profiles and will often create regions of strong turbulence, where no simple average velocity distribution will give an adequate description. If an obstacle such as a hilltop is sufficiently smooth, however,

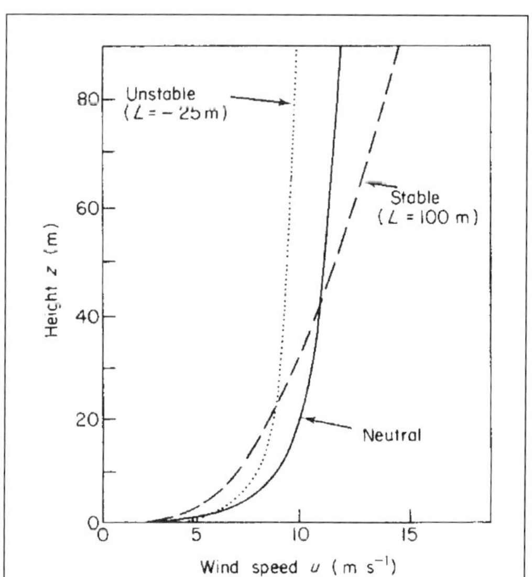

Figure 3.3-23 Wind speed profiles (i.e. variation with height) for a flat terrain characterised by a roughness length of 0.01 m and for a friction velocity of $(\tau/\rho)^{1/2} = 0.5$ m s^{-1}. The three different curves correspond to an atmosphere which is neutral (has an adiabatic lapse rate), stable or unstable. The stability is expressed in terms of the Monin–Obukhov length (based on Frost, 1975).

and the atmospheric lapse rate neutral or stable, the possibility exists of increasing the wind speed in a regular fashion at a given height where an energy collecting device may be placed. An example of this is shown in Fig. 3.3-24, based on a calculation by Frost *et al.* (1974). The elliptical hill shape extends infinitely in the direction perpendicular to the plane depicted. However, most natural hill shapes do not make a site more suitable for energy extraction. Another possibility of enhanced wind flow at certain heights exists in connection with certain shapes of valleys, which may act like a shroud to concentrate the intensity of the wind field, while still preserving its laminar features (see e.g. Frost, 1975). Again, if the shape of the valley is not perfect, regions of strong turbulence may be created and the suitability for energy extraction diminishes.

Only in relatively few cases can the local topography be used to enhance the kinetic energy of the wind at heights suitable for extraction. In the majority of cases, the optimum site selection will be one of flat terrain and smoothest possible surface in the directions of the prevailing winds. Since the roughness length over water is typically of the order 10^{-3} m, the best sites for wind energy conversion will usually be coastal sites with large fetch regions over water.

Wind speed data

Average wind speeds give only a very rough idea of the power in the wind (which is proportional to u^3) or the kinetic energy in the wind (which is proportional to u^3), owing to the fluctuating part of the velocity, which makes the average of the cube of u different from the cube of the average, etc. Whether the extra power of the positive speed excursions can be made useful depends on the response of the particular energy conversion device to such fluctuations.

For many locations, only average wind speed data are available, and since it is usually true in a qualitative way that the power which can be extracted increases with increasing average wind speed, a few features of average wind speed behaviour will be described.

In comparing wind speed data for different sites, the general wind speed–height relation (such as the ones shown in Fig. 3.3-23) should be kept in mind, along with the fact that roughness length, average friction velocity and statistics of the occurrence of different stability classes are also site-dependent quantities.

Figure 3.3-25 shows the seasonal variation of wind speed, based on monthly average values, for selected locations. Except for Risø and Toronto, the sites represent near optimum wind conditions for their respective regions. Not all parts of the Earth are as favoured with winds as the ones in Fig. 3.3-25. Figure 3.3-26 shows the variation of average wind speed throughout the day at

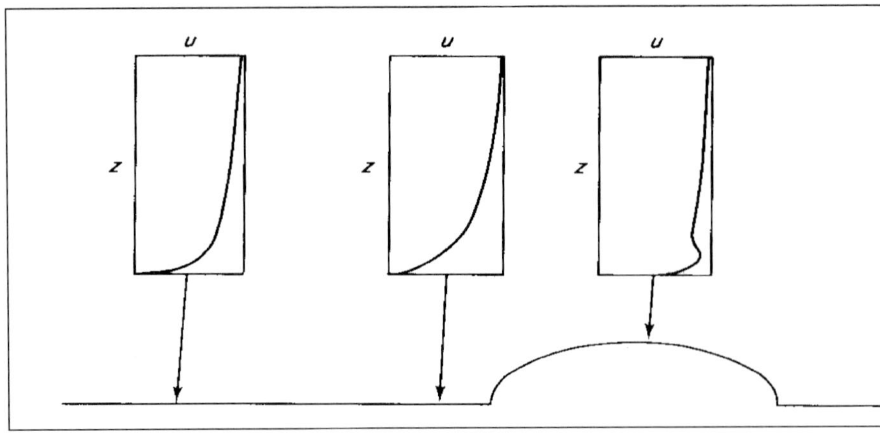

Figure 3.3-24 Wind speed profiles for a wind directed perpendicular to an elliptically shaped hill for two locations upwind, plus at the top of the hill. The profile far away from the disturbance corresponds to a surface of roughness length of 0.005 m, a friction velocity of 0.4 m s^{-1} and a neutral atmosphere (based on Frost *et al.*, 1974).

two shoreline sites and one maritime site in Singapore. The average wind speed at the latter site is slightly over 3 m s^{-1}, and there is little seasonal variation. At the two land-based observational stations, the overall average speed is only around 2 m s^{-1}, with very little gain when the height increases from 10 to 65 m.

Global wind speed data were shown in Chapter 2 of *Renewable Energy* (2004). Figure 3.3-27 shows the levels of power in the wind at potential hub height for wind turbines (about 70 m), on a seasonal base, using data from NCEP-NCAR (1998). These data have been made more consistent by a re-analysis method (Kalnay *et al.* 1996), in which general circulation models have been run to improve predictions in areas of little or poor data. The construction of power in the wind estimates from wind speed data is explained in section 6.2.5. It involves an averaging of circulation model data from height levels of 1000 and 925 mb (expressed as pressure levels), plus an empirical way of relating the average of the third power of the wind speed (the power) to the third power of the average wind speed. This procedure is necessarily very approximative. The main observation is that the highest power levels are found over oceans, including some regions close to shorelines. This points to the possibility of obtaining viable extraction of wind power from off-shore plants located at suitable shallow depths. It is also seen that seasonal variations are significant, particularly over large continents.

Figure 3.3-28 shows an example of measured wind profiles as a function of the hour of the day, averaged over a summer and a winter month. The tendency to instability due to turbulent heat transfer is seen during summer day hours (cf. the discussion above). Changes in wind direction with height have been measured (e.g. by Petersen,

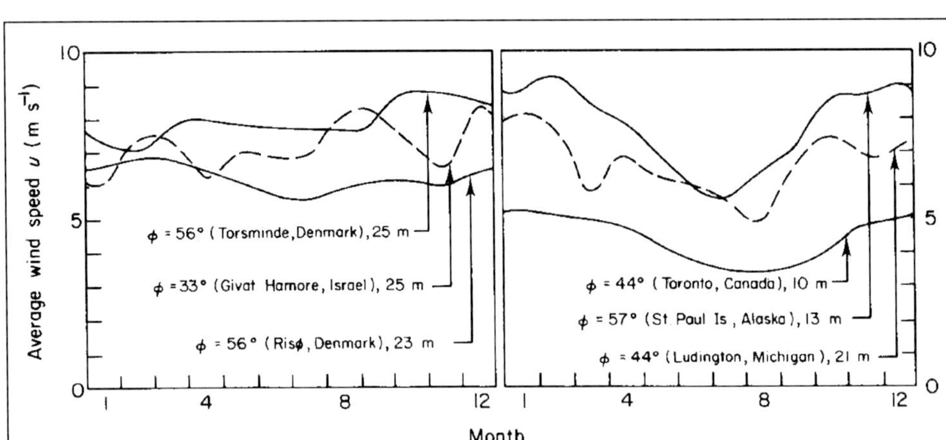

Figure 3.3-25 Seasonal variation of average wind speed for selected locations and heights. Torsminde data for 1961 from Jensen (1962), Givat Hamore data Nov. 1959 to Oct. 1960 from Frenkiel (1964), Risø 1958–1967 data from Petersen (1974), Toronto 1955–1972 data from Brown and Warne (1975), St. Paul Island 1943–1971 data from Wentink (1976), and the Ludington 1970,1972 data from Asmussen (1975).

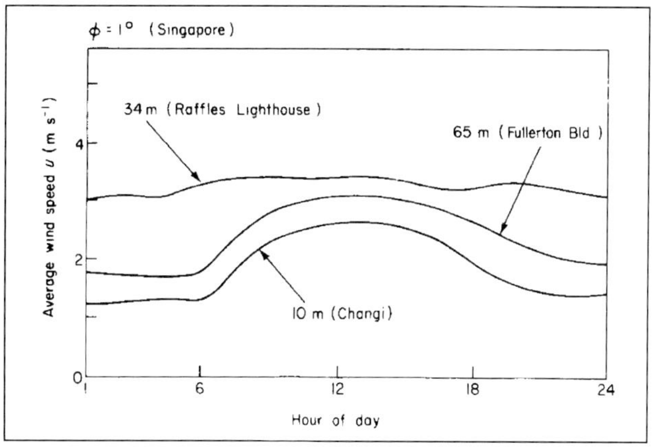

Figure 3.3-26 Variation of average wind speed with hour of the day for three Singapore stations. The data cover the period 1970–1974 (based on Nathan *et al.*, 1976)

1974). In the upper part of the planetary boundary layer this is simply the Ekman spiral.

The continuation of the velocity profile above the Prandtl boundary layer may be important in connection with studies of the mechanisms of restoring kinetic energy lost to an energy extraction device (e.g. investigations of optimum spacing between individual devices). Extraction of energy in the upper troposphere does not appear attractive at the present technological level, but the possibility cannot be excluded. Figure 3.3-29 shows the 10-year average wind profile at Risø ($\phi = 56°$) matched to a profile stretching into the stratosphere, based on German measurements. The average wind speed near the tropopause compares well with the zonal means.

3.3.1.8 Kinetic energy in the wind

In analogy to Fig. 3.3-29, the height variation of the kinetic energy density is obtained in the form shown in Fig. 3.3-30. The density variation is indicated in Fig. 3.3-29, and the hourly data from Risø, Denmark, have been used to construct the kinetic energy at heights of 7–123 m, the average values of which form the lower part of the curve in Fig. 3.3-30. The fluctuations in wind speed diminish with height (cf. Fig. 3.3-36), and at z = 123 m the difference between the annual average of u^2 and the square of the average u is only about 1%. For this reason, the average wind speeds of Fig. 3.3-29 were used directly to form the average kinetic energy densities for the upper part of the figure. The kinetic energy density peaks some 3 km below the peak wind speed, owing to the diminishing air density. The curve from Fig. 3.3-30 has been redrawn in Fig. 3.3-31, on non-logarithmic scales. This illustrates the bend of the curve at a height of 1–2 km, where the near-logarithmic increase of wind speed with height is replaced by a much stronger increase.

The kinetic energy density curve of Fig. 3.3.31 may readily be integrated to give the accumulated amount of kinetic energy below a given height, as shown in Fig. 3.3-32. This curve illustrates the advantage of extending the sampling area of a wind energy-collecting device into the higher regions of the atmosphere. The asymptotic limit of the accumulation curve is close to 2×10^6 J m^{-2} of vertical column. This order of magnitude corresponds to that of the estimate given in Fig. 2.50 in Chapter 2, *Renewable Energy* (2004), of 8×10^5 J m^{-2} as a global average of zonal kinetic energy, plus a similar amount of eddy kinetic energy. The eddy kinetic energy, which is primarily that of large-scale eddies (any deviation from zonal means), may, in part, be included in the curve based on Fig. 3.3-29, which gives mean wind speed and not zonal mean. Since the data are for a latitude of 50–56°N, it may also be assumed that the kinetic energy is above the global average.

3.3.1.9 Power in the wind

The energy flux passing through an arbitrarily oriented surface exposed to the wind is obtained by multiplying the kinetic energy density by $\boldsymbol{v}\cdot\boldsymbol{n}$, where $\boldsymbol{v}$ is the wind velocity and $\boldsymbol{n}$ is the normal to the surface. The energy flux (power) may then be written

$$E = \tfrac{1}{2}\,\rho\,u^3 \cos\theta, \qquad (3.3.24)$$

where θ is the angle between $\boldsymbol{v}$ and $\boldsymbol{n}$. If the vertical component of the wind velocity can be neglected, as well as the short-term fluctuations [termed v'] then the flux through a vertical plane perpendicular to the direction of the wind becomes

$$E = \tfrac{1}{2}\,\bar{\rho}\,(V^*)^3, \qquad (3.3.25)$$

where V^* is the average, horizontal wind velocity corresponding to a given time-averaging interval. If the wind speed or direction cannot be regarded as constant over the time interval under consideration, then an average flux (average power) may be defined by

$$E = \frac{1}{\Delta t}\int_{t_1}^{t_1+\Delta t} \tfrac{1}{2}\,\rho\,v^3 \cos\theta \, dt_2, \qquad (3.3.26)$$

where both v and θ (and in principle ρ) may depend on the time integrand t_2. A situation often met in practice is one of an energy-collecting device, which is able to follow some of the changes in wind direction, but not the very rapid ones. Such a "yaw" mechanism can be built into the

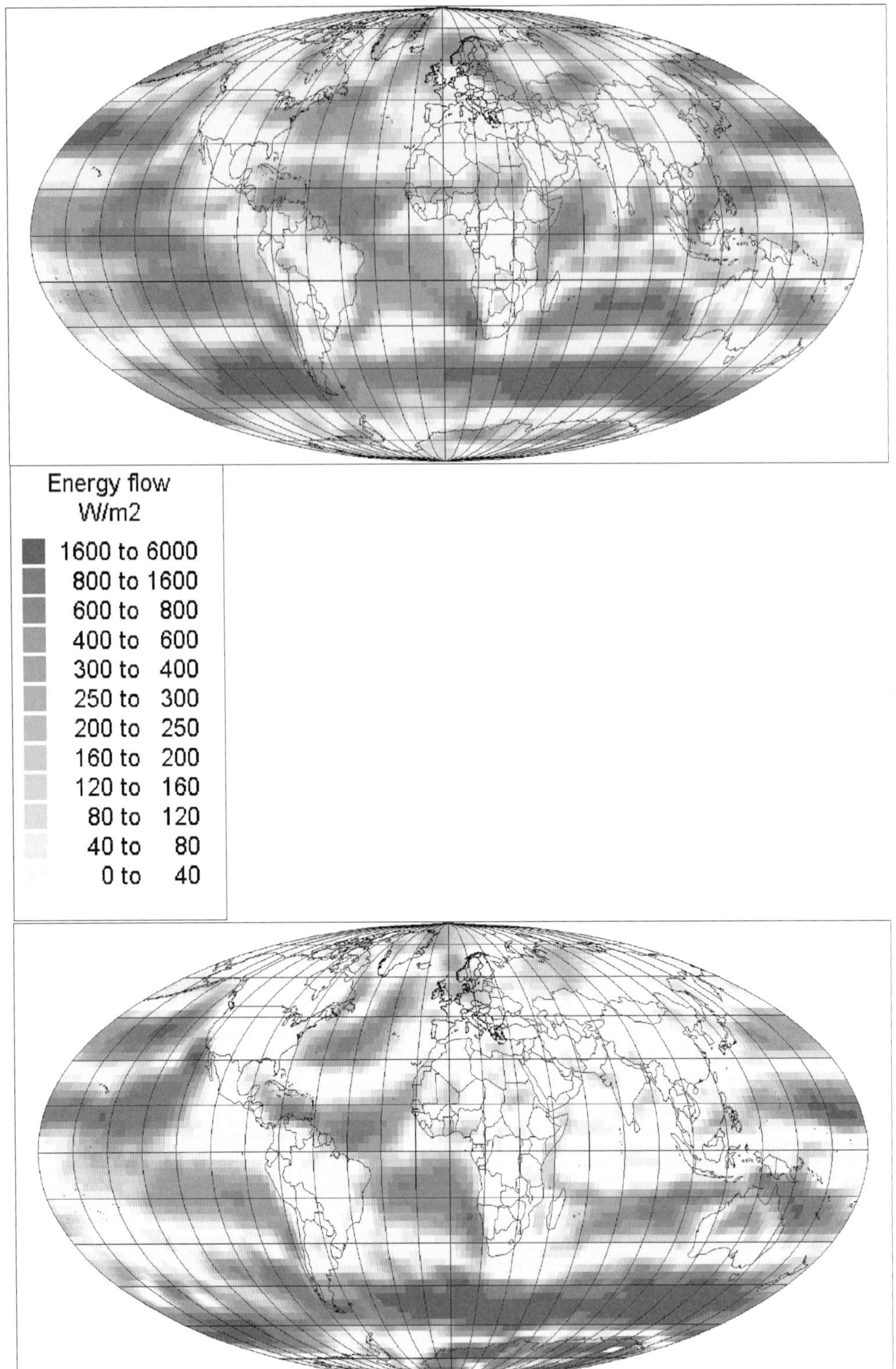

Figure 3.3-27a,b Maps of wind power regimes for January and April 1997, based on NCEP-NCAR (1998). The power levels are estimated for a height of 70 m above ground, presently a typical hub height for wind turbines. (These and following area-based geographical information system (GIS) maps are from Sørensen and Meibom (1998), copyright B. Sørensen).

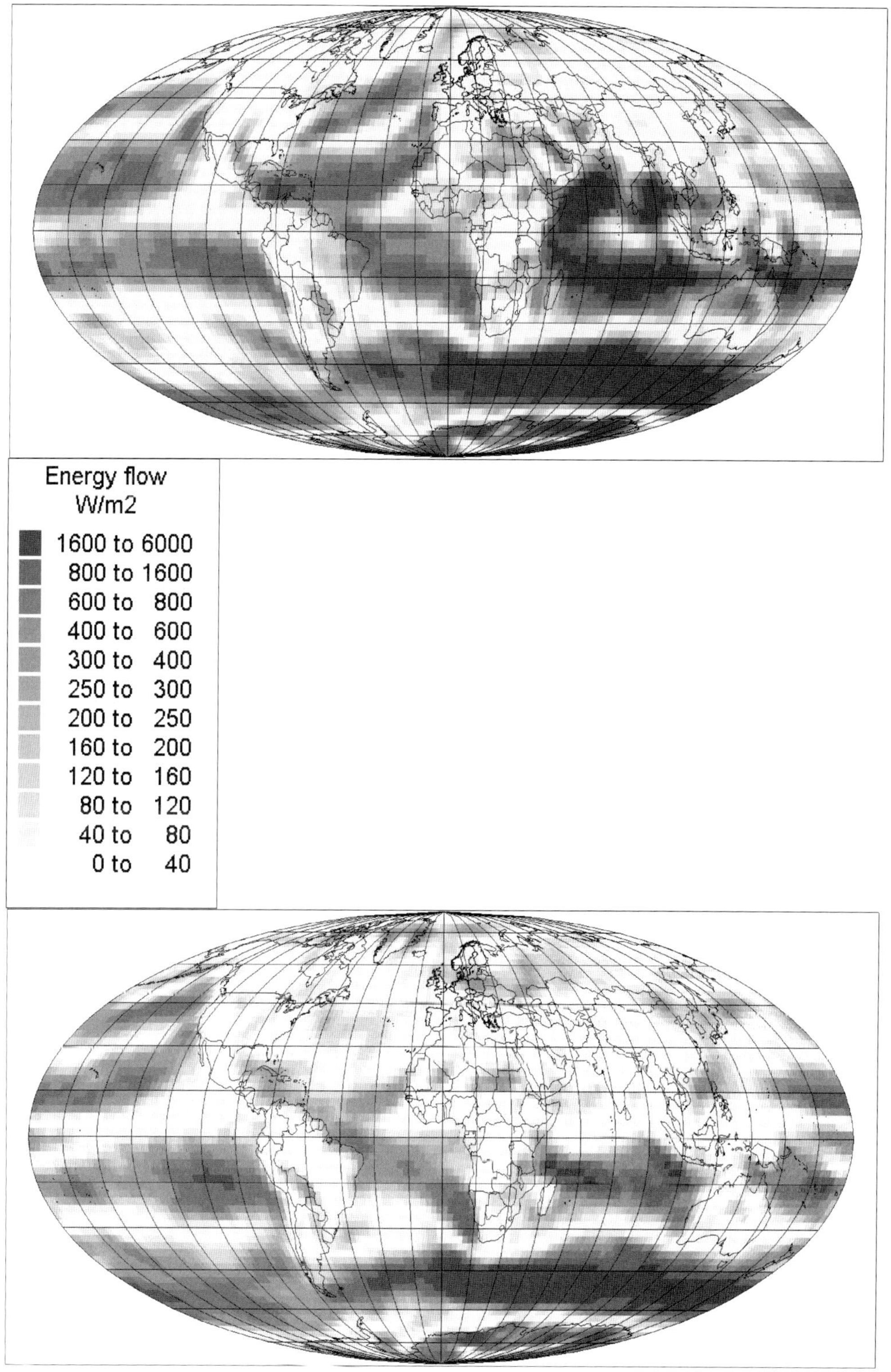

Figure 3.3-27c,d Maps of wind power regimes for July and October 1997, based on NCEP-NCAR (1998). The power levels are estimated for a height of 70 m above ground, presently a typical hub height for wind turbines.

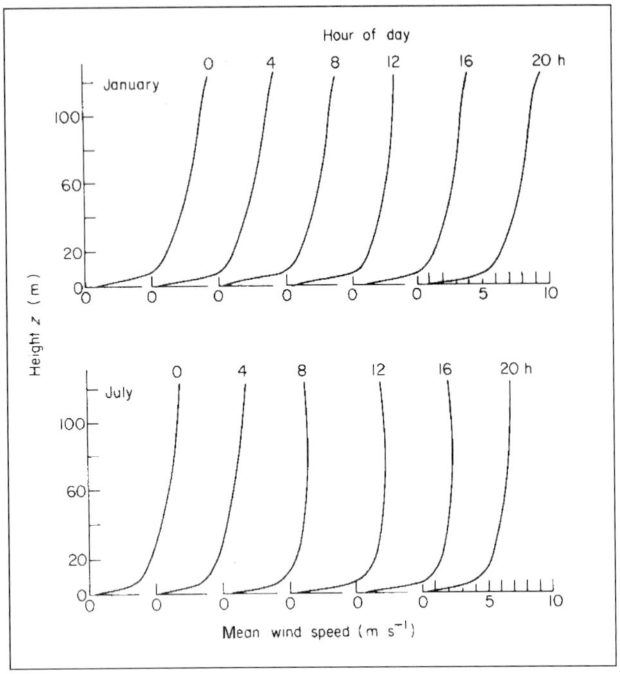

Figure 3.3-28 Variation of wind speed height profile with the hour of day at Risø, Denmark (56°N). Each curve has been averaged over all data from the period 1958–1967 for one winter or summer month (based on data from Petersen, 1974).

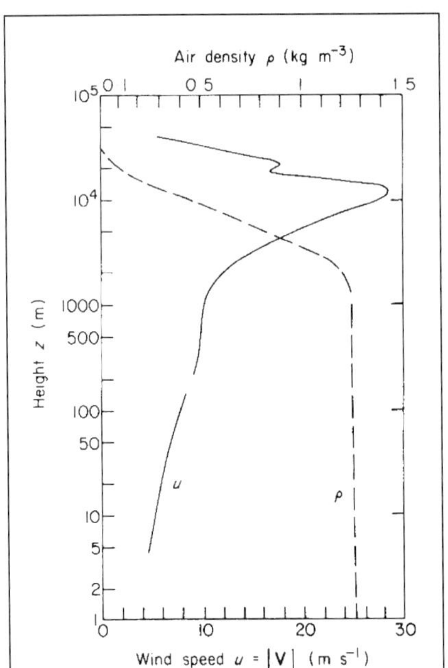

Figure 3.3-29 Annual average wind speed vs height profile at latitudes 50–56°N. The lower part of the curve is based on Risø data (Petersen, 1974), and the upper part is based on a set of curves derived from German measurements (Hütter, 1976). The one matching the lower part of the curve has been selected. Also indicated (dashed line) is an average height profile of the air density, used in the construction of Figs. 3.3-30–3.3-33.

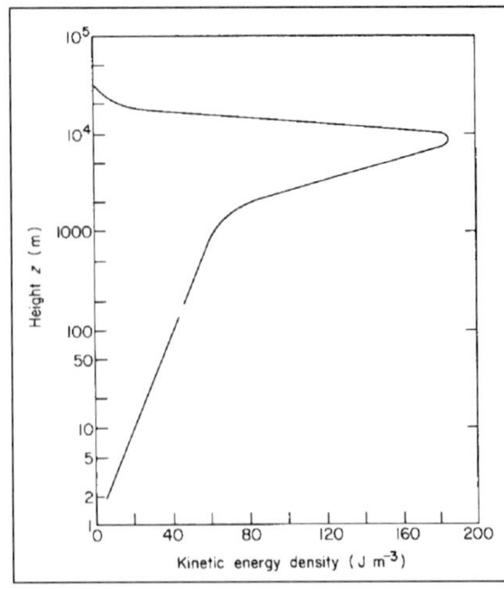

Figure 3.3-30 Annual average height profile of kinetic energy density at latitudes 50-56°N. The lower part of the curve is based on hourly data from Risø (Petersen, 1974), while the upper part has been calculated from the wind speeds of Fig. 3.3-29, neglecting any fluctuations. The contribution from the vertical component of the wind speed has been considered negligible.

prescription for defining θ, including the effect of a finite response time (i.e. so that the energy-collecting surface is being moved towards the direction the wind had slightly earlier).

Most wind data are available in a form where short-term variations have been smoothed out. For example, the Risø data used in the lower part of Fig. 3.3-29 are 10-min averages centred around every hour. If such data are used to construct figures for the power in the wind, as in Figs. 3.3-33 and 3.3-34 and the duration curves in Figs. 3.3-39 and 3.3-40, then two sources of error are introduced. One is that random excursions away from the average wind speed will imply a level of power which on average is larger than that obtained from the average wind speed [owing to the cubic dependence in (3.3.24)]. The other is that owing to changes in wind direction the

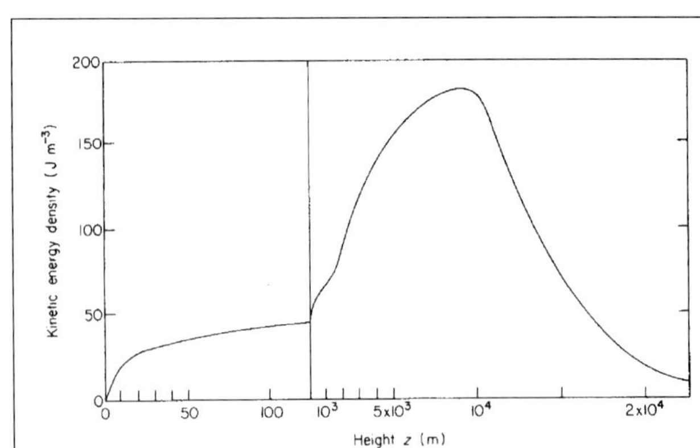

Figure 3.3-31 Same as Fig. 3.3-30, on a non-logarithmic scale (or rather two segments each with their linear scale).

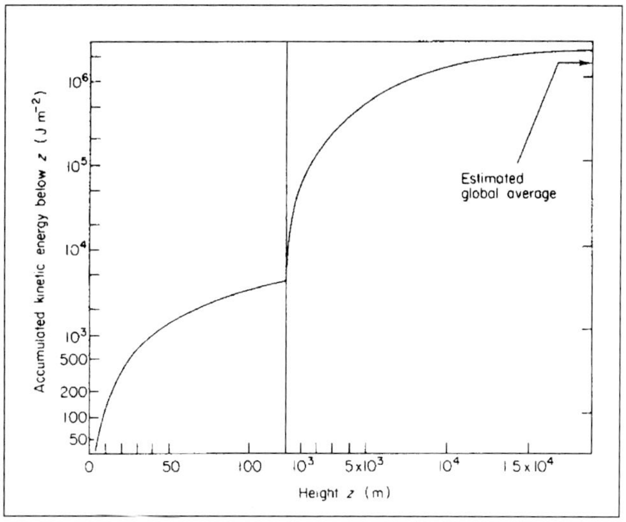

Figure 3.3-32 Accumulated kinetic energy below a given height, for latitudes around 50–56°N, obtained by integration of the curve in Fig. 3.3-31.

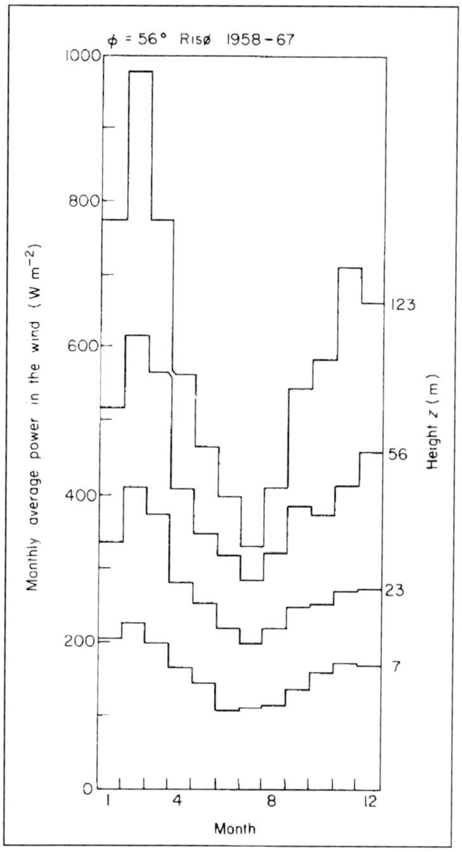

Figure 3.3-34 Monthly average power in the wind at Risø, Denmark, for different heights. Calculation based on 10 years of data (Petersen, 1974).

flux calculated from the average wind speed will not be passing through the same surface during the entire averaging period, but rather through surfaces of varying orientation. This second effect will tend to make the predicted power level an overestimate for any surface of fixed orientation.

Jensen (1962) overcame both of these difficulties by basing his measurements of power in the wind on a statistical ensemble of instantaneous measurements and by mounting unidirectional sensors on a yawing device which was chosen to respond to changing wind directions in a manner similar to that of the wind energy generators being considered. However, it is not believed that monthly average power or power duration curves will be substantially affected by using average data such as the ones from Risø (note that the ergodic hypothesis in section 2.3.3 (Renewable Energy, 2004), relating the time averages to the statistical ensemble means, is not necessarily valid if the time averages are formed only over one 10-min interval for each hour).

In the upper part of Fig. 3.3-33 the power has been calculated from the average wind speeds of Fig. 3.3-29, using (3.3.25). The seasonal variations at lower heights are shown in Fig. 3.3-34. It is clear that the amount of seasonal variation increases with height. This indicates that the upper part of Fig. 3.3-33 may well underestimate the average of summer and winter energy fluxes.

Additional discussion of the power in the wind, at different geographical locations, may be found in section 3.2. The overall global distribution of power in the wind at different seasons is shown in Fig. 3.3-27.

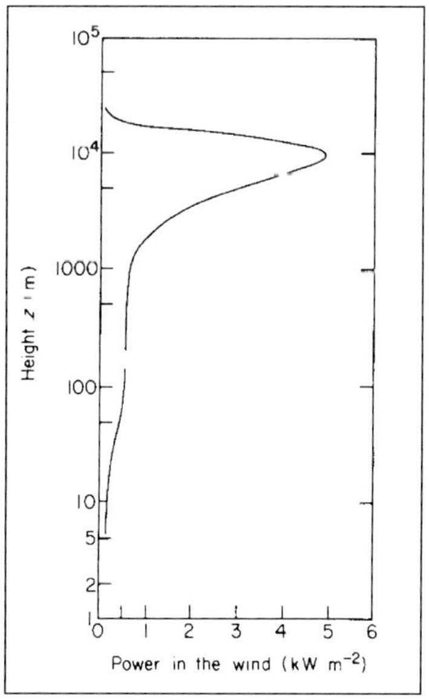

Figure 3.3-33 Annual average height profile of power in the wind, at latitudes 50–56°N. The lower part of the curve has been calculated from the Risø data of Petersen (1974), and the upper part is from the wind speeds of Fig. 3.3-29, neglecting any fluctuations.

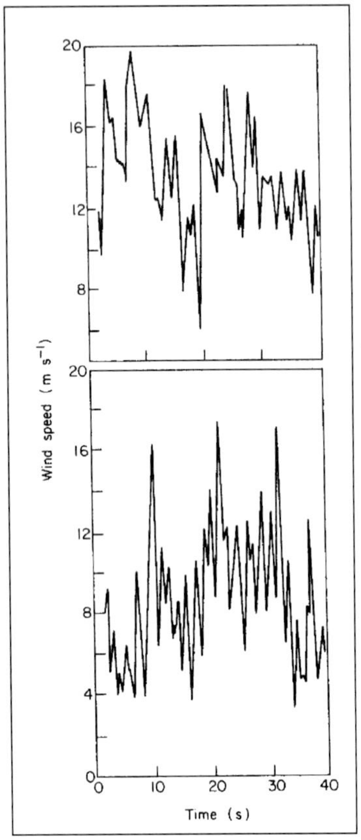

Figure 3.3-35 Short-term variation in wind speed, from simultaneous records at two locations 90 m apart (measuring height probably in the range 5–10 m) (based on Banas and Sullivan, 1976).

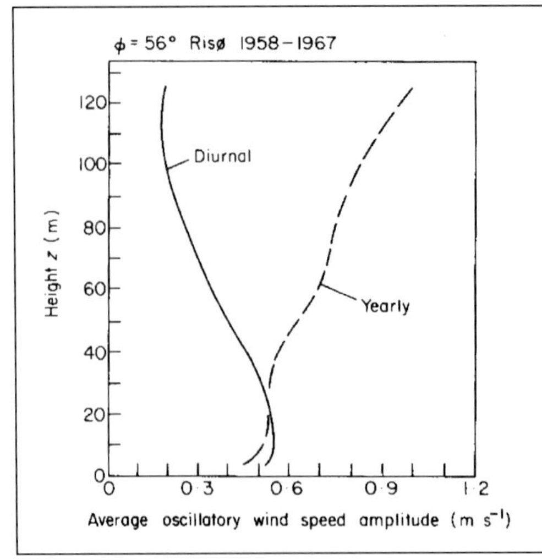

Figure 3.3-36 Height dependence of diurnal and yearly amplitude of oscillation of the wind speed at Risø. The average estimate is based on a Fourier decomposition of the data with no smoothing (Petersen, 1974).

3.3.1.10 Variability in wind power

An example of short-term variations in wind speed at low height is given in Fig. 3.3-35. These fluctuations correspond to the region of frequencies above the spectral gap of Fig. 2.110. The occurrences of wind "gusts", during which the wind speed may double or drop to half the original value over a fraction of a second, are clearly of importance for the design of wind energy converters. On the other hand, the comparison made in Fig. 3.3-35 between two simultaneous measurements at distances separated horizontally by 90 m shows that little spatial correlation is present between the short-term fluctuations. Such fluctuations would thus be smoothed out by a wind energy conversion system, which comprises an array of separate units dispersed over a sufficiently large area.

The trends in amplitudes of diurnal and yearly wind speed variations are displayed in Fig. 3.3-36, as functions of height (Petersen, 1974). Such amplitudes are generally site-dependent, as one can deduce, for example, from Figs. 3.3-26 and 3.3-28 for diurnal variations and from Fig. 3.3-25 for seasonal variations. The diurnal amplitude in Fig. 3.3-26 diminishes with height, while the yearly amplitude increases with height. This is probably a quite general phenomenon when approaching the geostrophic wind, but the altitude dependence may depend on geographical position and local topography. At some locations, the diurnal cycle shows up as a 24-h

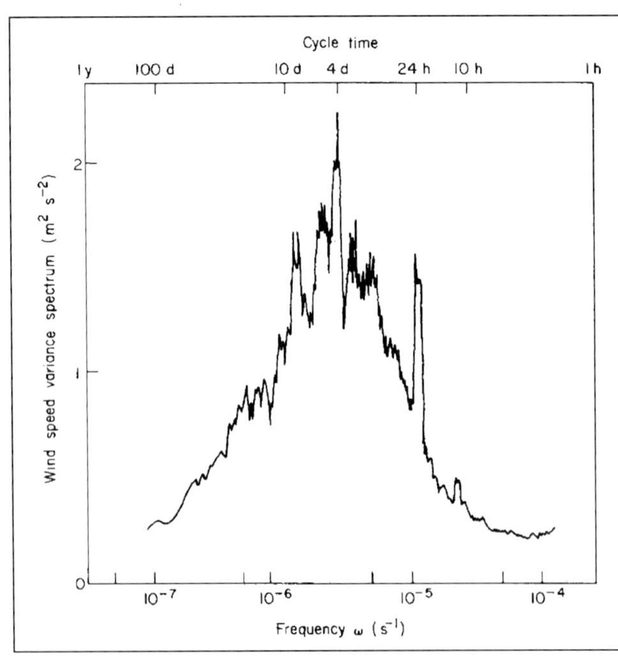

Figure 3.3-37 Variance spectrum of horzontal wind speeds at Risø ($\phi = 56°$N) at a height of 56 m (based on 10 years of observation; Petersen, 1974). The spectrum is smoothed by using frequency intervals of finite length.

peak in a Fourier decomposition of the wind speed data. This is the case at Risø, Denmark, as seen from Fig. 3.3-37, while the peak is much weaker at the lower height. The growth in seasonal amplitude, with height, is presumably determined by processes taking place at greater height as well as by seasonal variations in atmospheric stability, etc.

The wind speed variance spectrum covers a frequency interval between the yearly period and the spectral gap. In addition to the 24-h periodicity, the amplitude of which diminishes with increasing height, the figure exhibits a group of spectral peaks with periods in the range 3–10 days. At the selected height of 56 m, the 4-day peak is the most pronounced, but moving up to 123 m, the peak period around 8 days is more marked (Petersen, 1974). It is likely that these peaks correspond to the passage time of typical meso-scale front and pressure systems.

In analysing the variability of wind speed and power during a certain period (e.g. month or a year), the measured data are conveniently arranged in the forms of frequency distributions and power duration curves, much in the same manner as discussed in section 3.1.5. Figure 3.3-38 gives the 1-year frequency distribution of wind speeds at two Danish locations for a height of about 50 m. The wind speed frequency distribution (dashed curve) at Gedser (near the Baltic Sea) has two maxima, probably associated with winds from the sea and winds approaching from the land side (of greater roughness). However, the corresponding frequency distribution of power (top curve) does not exhibit the lower peak, as a result of the cubic dependence on wind speed. At the Risø site, only one pronounced maximum is present in the wind speed distribution. Several irregularities in the power distribution (which are not preserved from year to year) bear witness to irregular landscapes with different roughness lengths, in different directions from the meteorological tower.

Despite the quite different appearance of the wind speed frequency distribution, the power distribution for the two Danish sites peaks at roughly the same wind speed, between 10 and 11 m s^{-1}.

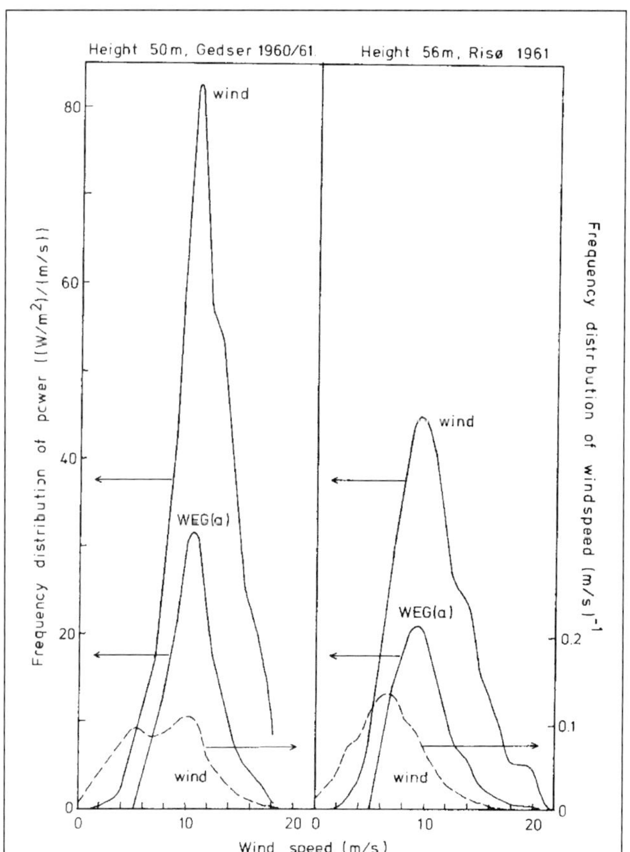

Figure 3.3-38 Frequency distribution of wind speed (right-hand scale) and of power for a height of about 50 m at two Danish sites. The middle curves give frequency distributions for the output of a typical wind energy generator (Sørensen, 1978a).

Power duration curves

On the basis of the frequency distributions of the wind speeds (or alternatively that of power in the wind), the power duration curves can be constructed, giving the percentage of time when the power exceeds a given value. Figures 3.3-39 and 3.3-40 give a number of such curves, based on periods of a year or more. In Fig. 3.3-39, power duration curves are given for four US sites which have been used or are being considered for wind energy conversion and for one of the very low-wind Singapore sites.

In Fig. 3.3-40, power duration curves are given for the two different Danish sites considered in Fig. 3.3-38, as well as for a site on the west coast of Sweden, at three different heights. These three curves for the same site have a similar shape, but in general the shape of the power duration curves in Figs. 3.3-39 and 3.3-40 depends on location. Although the Swedish Ringhals curves have non-negligible time percentages with very large power, the Danish Gedser site has the advantage of a greater frequency of medium power levels. This is not independent of conversion efficiency, which typically has a maximum as function of wind speed.

3.3.2 Ocean waves

The order of magnitude of the total energy in wave motion is about 10^{-3} of the total kinetic energy in the atmospheric wind systems. The wave energy of about 10 kJ m^{-2} found as an annual average in the North Atlantic Ocean corresponds to the accumulated wind energy up to a height of about 200 m, according to

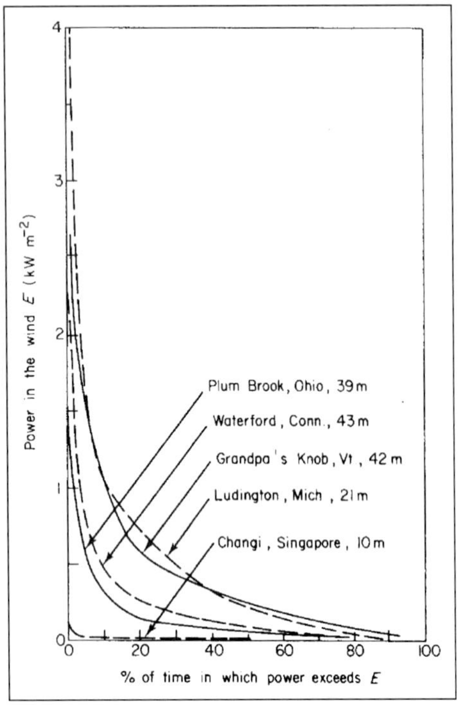

Figure 3.3-39 One-year duration curves of power in the wind, for a number of locations and heights [Swanson *et al.*, 1975 (Plum Brook and Grandpa's Knob); Coste, 1976 (Waterford); Asmussen, 1975 (Ludington); Nathan *et al.*, 1976 (Changi)].

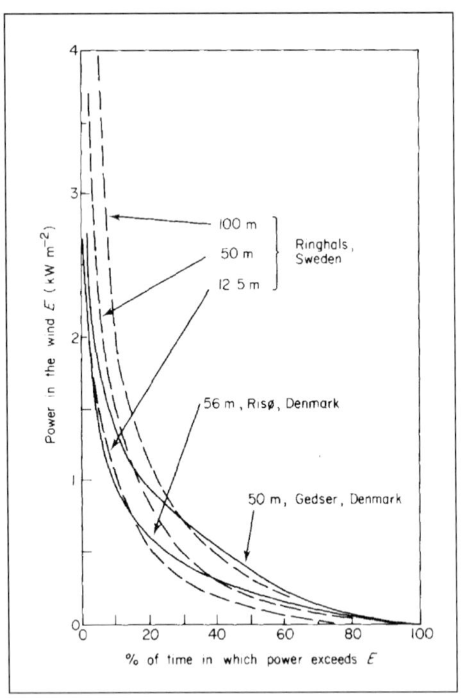

Figure 3.3-40 One-year duration curves of power in the wind at Scandinavian sites (including different heights at the Ringhals site) [based on data from Ljungström, 1975 (Ringhals); Petersen, 1974 (Risø); Jensen, 1962 (Gedser)].

Fig. 3.3-32. This implies that, although the amount of energy stored in waves is much smaller than the amount stored in wind, the wave energy may still be equivalent to the height-integrated fraction of wind energy accessible for practical use, at least at the current level of technology.

From the tentative estimates found elsewhere, the average turnover time for the energy in wave motion in the open ocean may be of the order of a few days. This time is consistent with average dissipation mechanisms, such as internal friction and weak wave–wave interactions, plus shoreline dissipation modes. The input of energy by the wind, on the other hand, seems to be an intermittent process which for extended intervals of time involves only slow transfer of energy between waves and turbulent wind components, or vice versa, and between wind and wave fields propagating in different directions (any angle from 0 to 2π). However, large amounts of energy may be transferred from wind to waves during short periods of time (i.e. "short periods" compared with the average turnover time). This suggests that the energy storage in waves may be varying more smoothly than the storage in wind (both waves and wind represent short-term stored solar energy, rather than primary energy flows. The wave fields exhibit short-term fluctuations, which may be regarded as random. On a medium time scale, the characteristics of the creation and dissipation mechanisms may make the wave energy a more "dependable" energy source than wind energy, but on a seasonal scale, the variations in wind and wave energy are expected to follow each other.

3.3.2.1 Wave spectra

The energy spectrum $F(k)$ of a random wave field has been defined in *Renewable Energy* (2004). Since the wavelength (or wave number k) is very difficult to observe directly, it is convenient instead to express the spectrum in terms of the frequency,

$$v = \omega(k)/2\pi = 1/T.$$

The frequency is obtained from the period, T, which for a harmonic wave equals the zero-crossing period, i.e. the time interval between successive passages of the wave surface through the zero (average) height, in the same direction. The spectral distribution of energy, or "energy spectrum" when expressed in terms of frequency, $F_1(v) = 2\pi\, F_1(\omega)$, is usually normalised to the total energy (Barnett and Kenyon, 1975), whereas the wavenumber-dependent spectrum $F(k)$ is normalised to the average potential energy. Thus,

$$\int F_1(\omega)\, d\omega = W^{total} = 2\int F(k)\, dk.$$

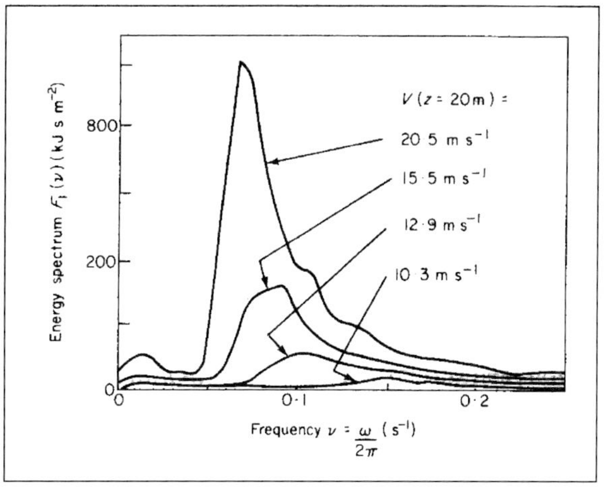

Figure 3.3-41 Energy spectrum of waves estimated to be "fully developed" for the Atlantic Ocean (data have been grouped according to the wind speed at a height of 20 m) (based on Moskowitz, 1964).

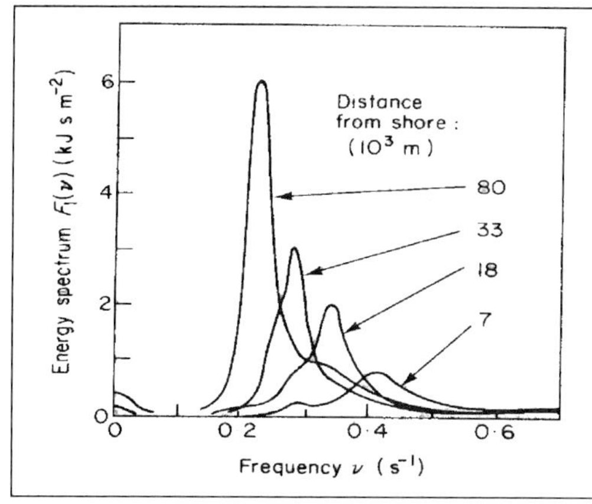

Figure 3.3-42 Fetch-limited energy spectrum of waves in the southern part of the North Sea. The wind is blowing from the continental shore (westward from Helgoland) (based on Hasselmann *et al.*, 1973).

Figure 3.3-41 shows a set of measured energy spectra, F_1 based on data from the Atlantic Ocean (Moskowitz, 1964). The wave fields selected were judged to correspond to "fully developed waves", and data corresponding to the same wind speed at a height of 20 m were averaged in order to provide the spectra shown in the figure. It is seen that the spectral intensity increases and the frequency corresponding to the peak intensity decreases, with increasing wind speed.

Based on the similarity theory of Monin and Obukhov (1954), Kitaigorodskii (1970) suggested that F_I (ω) $g^2 \rho_w^{-1} V^{-5}$ (V being the wind speed) would be a universal function of $\omega V g^{-1}$, with both quantities being dimensionless. Based on the data shown in Fig. 3.3-41, which approximately satisfies Kitaigorodskii's hypothesis, Moskowitz (1964) suggested the following analytical form for the energy spectrum of fully developed gravity waves:

$$F_1(\omega) = 8.1 \times 10^{-3}\, \rho_w\, g^3\, \omega^{-5} \exp(-0.74 \times (V\{z = 20\,\mathrm{m}\}w/g)^{-4}). \quad (3.3.27)$$

The usefulness of this relation is limited by its sensitivity to the wind speed at a single height and by the difficulty of determining whether a given wave field is "fully developed" or not.

If the wave field is "fetch-limited", i.e. if the wind has only been able to act over a limited length, then the energy spectrum will peak at a higher frequency, and the intensity will be lower, as shown in Fig. 3.3-42. Hasselmann *et al.* (1973) have generalised (3.3.27) to situations in which the wave field is not necessarily fully developed, according to such data.

The position of the spectral peak will move downwards as a function of time owing to non-linear wave–wave interactions, i.e. interactions between different spectral components of the wave field (Hasselmann, 1962). This behaviour is clearly seen in the laboratory experiments of Mitsuyasu (1968), from which Fig. 3.3-43 shows an example of the time derivative of the energy spectrum, $\partial F_1(\nu)/\partial t$. Energy is transferred from middle to smaller frequencies. In other experiments, some transfer is also taking place in the direction of larger frequencies. Such a transfer is barely present in Fig. 3.3-43. The shape of the rate-of-transfer curve is similar to the one found in later experiments for real sea (Hasselmann *et al.*, 1973), and both are in substantial agreement with the non-linear theory of Hasselmann.

Observations of ocean waves often exist in the form of statistics on the occurrences of given combinations of wave height and period. The wave height may be expressed as the significant height, H_s, or the root mean square displacement of the wave surface, H_{rms}. For a harmonic wave, these quantities are related to the surface amplitude, a, by

$$a^2 = 2\,(H_{rms})^2 = H_s^2/8.$$

The period is usually given by the observed zero-crossing period. Figures 3.3-44 and 3.3-45 give examples of such measured frequency distributions, based on a year's observation at the North Atlantic Station "India" (59°N, 19°W, the data quoted by Salter, 1974) and at the North Sea Station "Vyl" (55.5°N, 5.0°E, based on data discussed elsewhere). The probability of finding periods between 8 and 10 s and wave heights, a, between 0.3 and 1.5 m is quite substantial at the site "India", of the order of 0.3. Also the probability of being within a band of

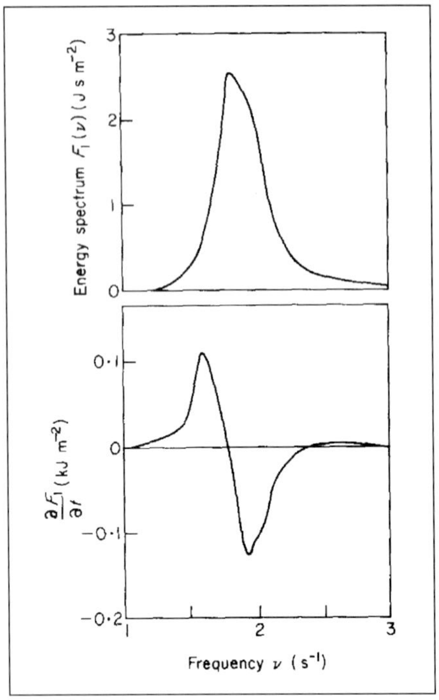

Figure 3.3-43 Simultaneous spectra of energy (top) and its time derivative (based on laboratory experiments by Mitsuyasu, 1968).

zero-crossing periods with a width of about 2 s and a centre which moves slowly upwards with increasing wave height is nearly unity (the odd values characterising

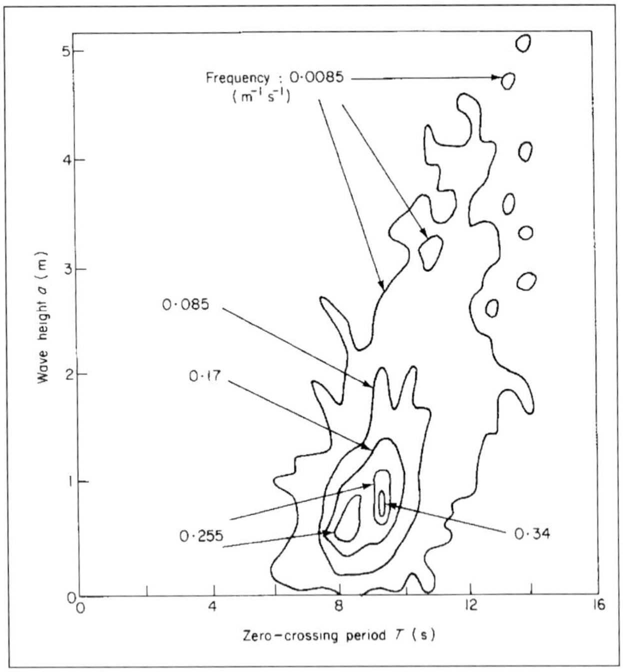

Figure 3.3-44 Frequency distribution of wave heights and zero-crossing periods for Station India (59°N, 19°W) in the North Atlantic [based on one year of observation by Draper and Squire (1967), quoted by Salter, 1974].

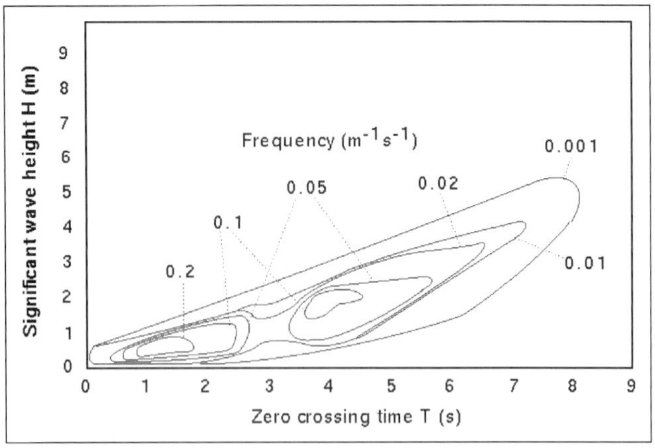

Figure 3.3-45 Frequency distribution of significant wave heights and zero-crossing periods for Station Vyl (55°N,5°E) in the North Sea (based on one year of observation by Nielsen, 1977).

the contour lines of Fig. 3.3-44 are due to the original definition of sampling intervals in terms of feet and H_s).

3.3.2.2 Power in the waves

The power of a harmonic wave, i.e. the energy flux per unit length of wave crest, passing through a plane perpendicular to the direction of propagation

$$P = \rho_w\, g\, U_w\, a^2/4 = \rho_w\, g^2\, T\, a^2\, /\, (8\pi)$$
$$= \rho_w\, g^2\, a^2\, /\, (4\omega). \qquad (3.3.28)$$

For the spectral distribution of energy given by the function F_1 (section 3.3.1), each component must be multiplied by the group velocity $\partial\omega(k)/\partial k$. Taking $\omega = (gk)^{1/2}$ for ocean gravity waves, the group velocity becomes $g/(2\omega)$, and the power becomes

$$P = \int \frac{\partial \omega}{\partial k} F_1\,(\omega)\, d\omega = ½\, g \int \frac{F_1\,(\omega)}{\omega}\, d\omega. \qquad (3.3.29)$$

Based on observed energy spectra, F_1, as a function of frequency or period $T = 2\pi/\omega$, the power distribution [(3.3.29) before integration] and total power may be constructed. Figure 3.3-46 gives the power distribution at the North Atlantic station also considered in Fig. 3.3-44. Using data for an extended period, the average distribution (labelled "year") has been calculated and, taking out data for the periods December–February and June–August, the curves labelled "winter" and "summer" have been constructed (Mollison *et al.*, 1976).

Compared with Fig. 3.3-34, for example, it is seen that the seasonal variations in wave power at the open ocean site considered in Fig. 3.3-46 are quite substantial and equivalent to the seasonal variations in wind power at considerable height (maybe around 100 m, considering

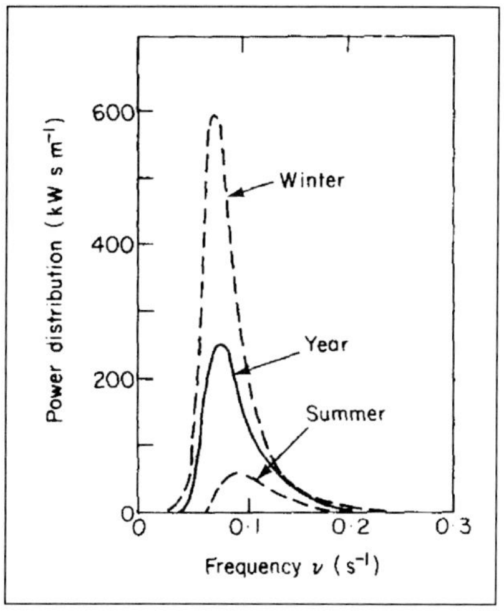

Figure 3.3-46 Frequency distribution of wave power, based on one year of observations (full line) or summer or winter data only (dashed lines), for Station India (59°N, 19°W) in the North Atlantic. The yearly average power is 91 kW m^{-1} (based on Mollison *et al.*, 1976).

that the roughness length over the ocean is smaller than at the continental site considered in Fig. 3.3-36).

Figure 3.3-47 summarises some of the data available on the geographical distribution of wave power. The figures given are yearly mean power at a specific location, and most of the sites chosen are fairly near to a coast, although still in open ocean. The proximity of a shore is considered important in connection with the potential for energy extraction from wave motion. Whether such a condition will be maintained depends on the development of suitable transmission methods for long-range energy transfer.

In Fig. 3.3-48 a more detailed map of wave power availability for the north European region is shown, based on the iso-power lines estimated in initial assessments of wave power in the United Kingdom, supplemented with estimates based on data for the waters surrounding Denmark. One might note the rapid decrease in power when passing the Hebrides in approaching the Scottish coast from the Atlantic and also when moving southwards through the North Sea.

The variability of wave power may be described in terms similar to those used for wind energy. Figure 3.3-49 shows the power duration curves for Station India in the North Atlantic, based on all year or the summer or winter periods only, as in Fig. 3.3-46. Again, the occurrence of periods of low and high power depends markedly on seasonal changes.

Waves in a climatic context

As in the case of wind energy, little is known about the possible impact of large-scale energy extraction from wave power. One may argue that the total amount of energy involved is so small compared to the energy exchanged by atmospheric processes that any climatic consequence is unlikely. However, the exchange of latent

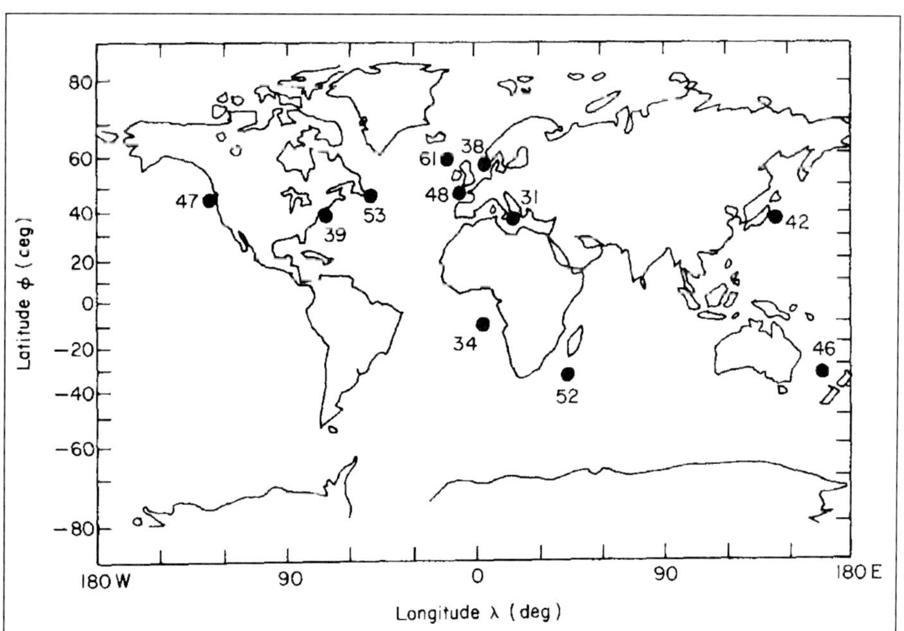

Figure 3.3-47 Annual average wave power (in kW m^{-1}) for selected sites (based on United Kingdom Energy Technology Support Unit, 1976).

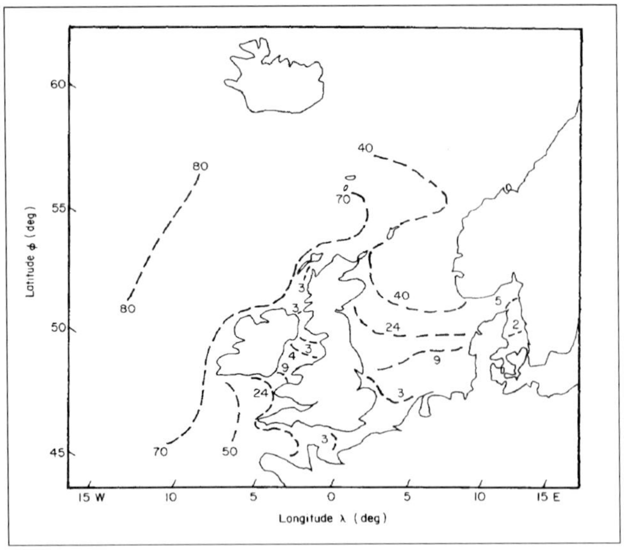

Figure 3.3-48 Contours of estimated equal annual average wave power in the North Sea, adjacent straits and a part of the North Atlantic Ocean (unit: kW m^{-1}) (based on United Kingdom Energy Technology Support Unit, 1976).

and sensible heat, as well as matter, between the ocean and the atmosphere may, to a large extent, depend on the presence of waves. In particular, the rates of transfer may be high in the presence of breaking waves, and the extraction of energy from the wave motion may prevent the waves from developing the wave profile to the point of breaking. Thus, a study of the environmental implications of wave energy utilisation, which seems to lend itself very naturally to computer simulation techniques, should be undertaken in connection with proposed energy extraction schemes. As suggested in previous sections, this is likely to be required for any large-scale use of renewable energy, despite the intuitive feeling that such energy sources are "non-polluting". Yet the limited evidence available, mostly deriving from analogies to natural processes, does suggest that the results of more detailed analyses will be that renewable energy flows and stores can be utilised in quantities exceeding present technological capability, without worry about environmental or general climatic disturbances.

3.3.3 Water flows and tides

3.3.3.1 Ocean currents

The maximum speed in the centre of the Gulf Stream is about 2 m s^{-1}, corresponding to an energy density $\frac{1}{2}\,\rho_w V_w^2 = 2$ kJ m^{-3} and a power of $\frac{1}{2}\,\rho_w V_w^3 = 4$ kW m^{-2}. This power level, for example, approaches that of wave power at reasonably good sites, taken as power per metre of wave crest rather than per square metre perpendicular to the flow, as used in the above case. However, high average speed of currents combined with stable direction is found only in a few places. Figure 3.3-50 shows the isotachs in a cross section of the Gulf Stream, derived from a single set of measurements (in June 1938; Neumann, 1956). The maximum current speed is found

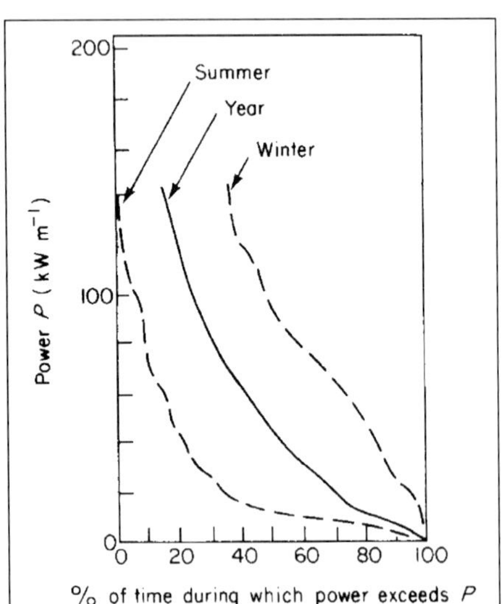

Figure 3.3-49 Time duration curves for wave power at Station India (59°N, 19°W) in the North Atlantic for the whole year (solid line) or only summer or winter (dashed lines) (based on Mollison *et al.*, 1976).

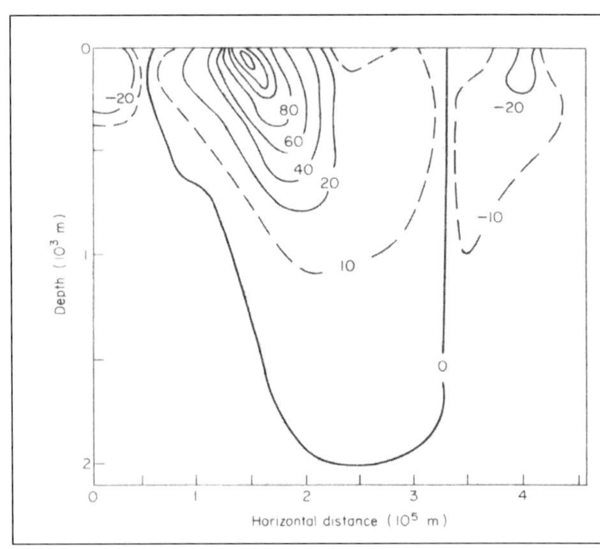

Figure 3.3-50 Contours of equal speed (in 10^{-2} m s^{-1}) along a cross section through the Gulf Stream from Bermuda to the US coast. The measurements were performed over a limited period of time (in June 1938, by Neumann, 1956), but they are consistent with other measurements in the same region (e.g. Florida Strait measurements by Brooks and Niiler, 1977).

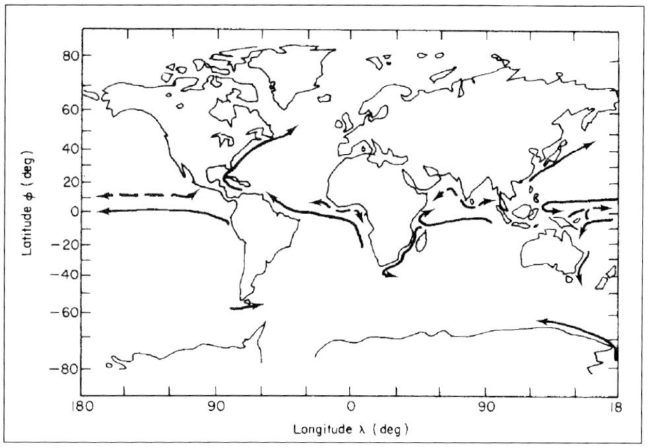

Figure 3.3-51 Indication of the location of strong surface currents (approximately defined as having average speeds above 0.5 m s^{-1}). Currents drawn with dashed lines are characterised by seasonal changes in direction (based on Defant, 1961).

at a depth of 100–200 m, some 300 km from the coast. The isotachs expand and become less regular further north, when the Gulf Stream moves further away from the coast (Niiler, 1977).

Even for a strong current like the Gulf Stream, the compass direction at the surface has its most frequent value only slightly over 50% of the time (Royal Dutch Meteorological Institute, as quoted by Neumann and Pierson, 1966), and the power will, on average, deviate from that evaluated on the basis of average current speeds (as is the case for wind or waves) owing to the inequality of $<V^3>$ and $<V>^3$.

The geographical distribution of the strongest surface currents is indicated in Fig. 3.3-51. The surface currents in the Atlantic Ocean, along with the horizontal currents at three different depths, are sketched in Fig. 3.3-52. The current speeds indicated are generally decreasing downwards, and the preferred directions are not the same at different depths. The apparent "collision" of oppositely directed currents, e.g. along the continental shelf of Central America at a depth of around 4 km, conceals the vertical motion which takes place in most of these cases. Figure 3.3-53 gives a vertical cross section with outlined current directions. This figure shows how the oppositely directed waters "slide" above and below each other. The coldest water is flowing along the bottom, while the warmer water from the North Atlantic is sliding above the cold water from the Antarctic.

Variability in current power

The particular topography of coastal regions may have an important influence on currents. As an example, water forced through a narrow strait may acquire substantial speeds. The strait Storebælt ("Great Belt") between two Danish isles, which provides an outlet from the Baltic Sea, may serve as an illustration. It is not extremely narrow (roughly 20 km at the measurement site), but narrow enough to exhibit only two current directions, separated by 180°. The currents may seem fairly steady, except for periods when the direction is changing between north-going and south-going velocities, but when the energy flux is calculated from the third powers of the current speeds, the fluctuations turn out to be substantial. Figure 3.3-54, which gives the power at 3-h

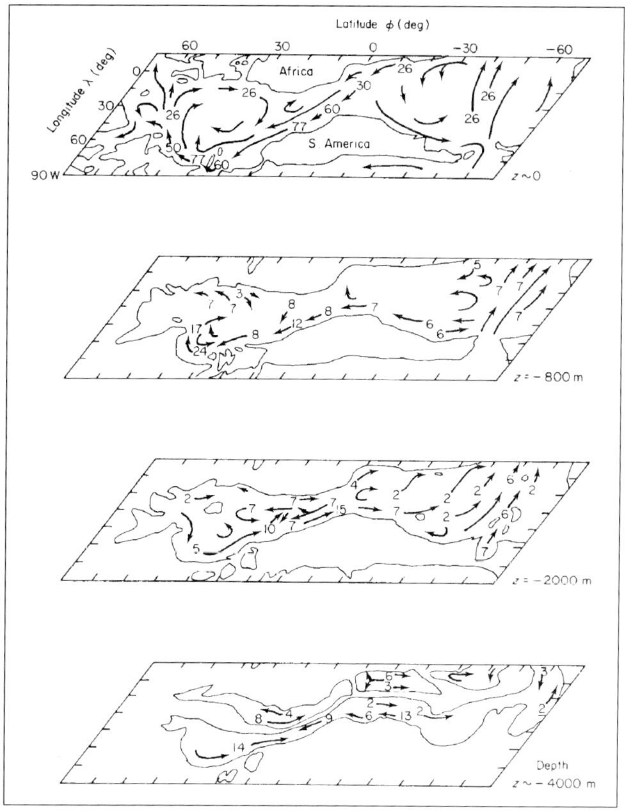

Figure 3.3-52 Indication of average horizontal current speeds (in 10^{-2} m s^{-1}) in the Atlantic Ocean for different depths (based on Defant, 1961).

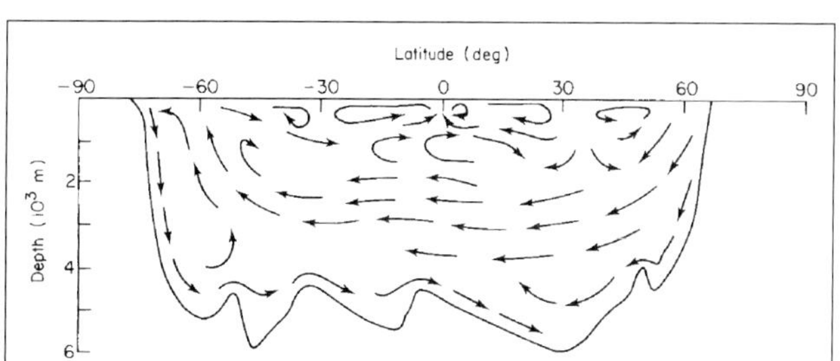

Figure 3.3-53 Indication of average current directions within a vertical cross section of the Atlantic Ocean (based on Neumann and Pierson, 1966).

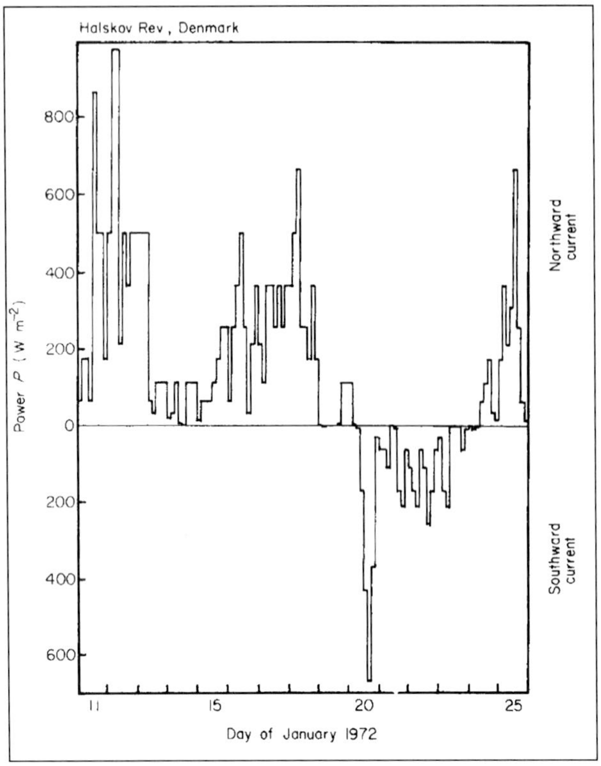

Figure 3.3-54 Power of surface current, based on observations made at 3-h intervals, for Halskov Rev, Denmark, during a 15-day period in 1972 (the strait is narrow enough to exhibit only two opposite current directions) (based on Danish Meteorological Institute, 1973).

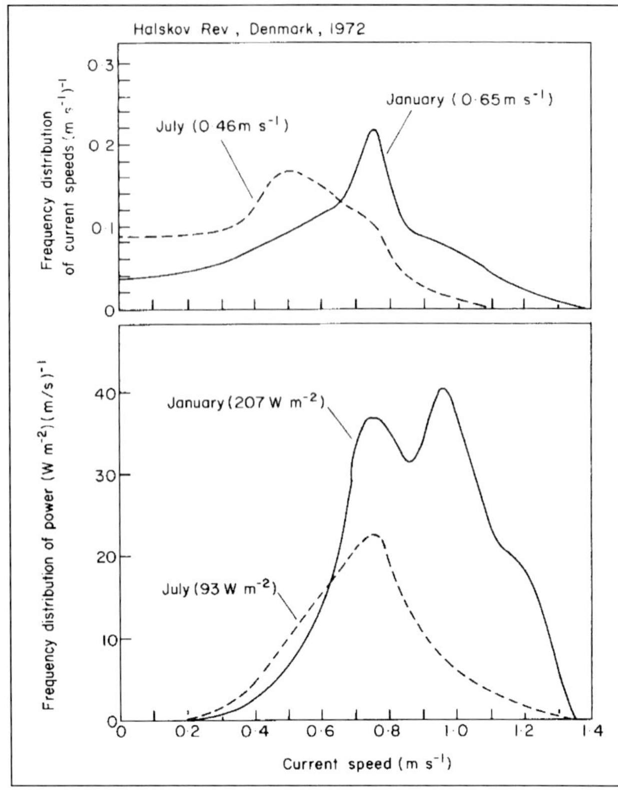

Figure 3.3-56 Frequency distributions of current speeds and power, based on a summer and a winter month (full and dashed curves, respectively), for Halskov Rev, Denmark. The data (taken from Danish Meteorological Institute, 1973) have been smoothed in calculating the distributions. The monthly average speed and power are indicated on the figure (in parentheses).

intervals, during two weeks of January 1972, clearly illustrates this.

Figure 3.3-55 shows, again for the Halskov Rev position in Storebælt, the variation in current speed with the hour of the day, based on one-month averages. A 12-h periodicity may be discerned, at least during January.

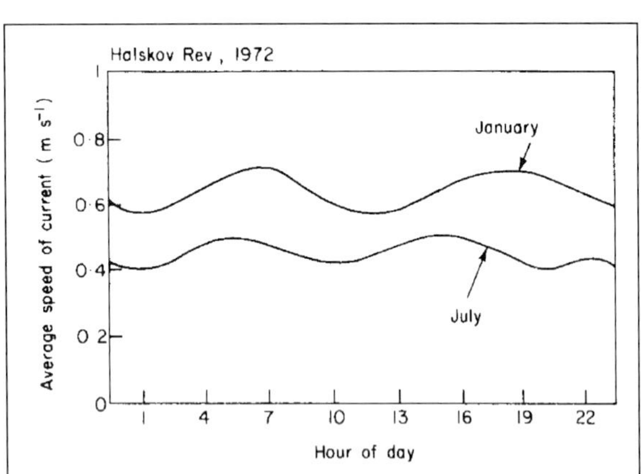

Figure 3.3-55 Dependence of average current speed on the hour of the day, for a summer and a winter month, for Halskov Rev, Denmark (based on Danish Meteorological Institute, 1973).

This period is smaller than the one likely to be found in the open sea due to the motion of water particles in stationary circles under the influence of the Coriolis force and having a period equal to 12 h divided by sin ϕ (ϕ is the latitude).

In Fig. 3.3-56, the frequency distributions of current speed and power are shown for a summer and a winter month. These curves are useful in estimating the performance of an energy extraction device, and they can be used, for example, to construct the power duration curve of the current motion, as shown in Fig. 3.3-57. This is the power duration curve of the currents themselves. That of an energy extracting device will have to be folded with the efficiency function of the device.

The peak in the frequency distribution of power is at a lower current speed for July than for January, and the average power is 93 W m^{-2} in July, as compared with 207 W m^{-2} in January (the corresponding kinetic energy densities are 138 and 247 J m^{-3}). This indicates that the fluctuations around the average values have a substantial effect on the energy and, in particular, on the power, because from the average current speeds, 0.46 m s^{-1}

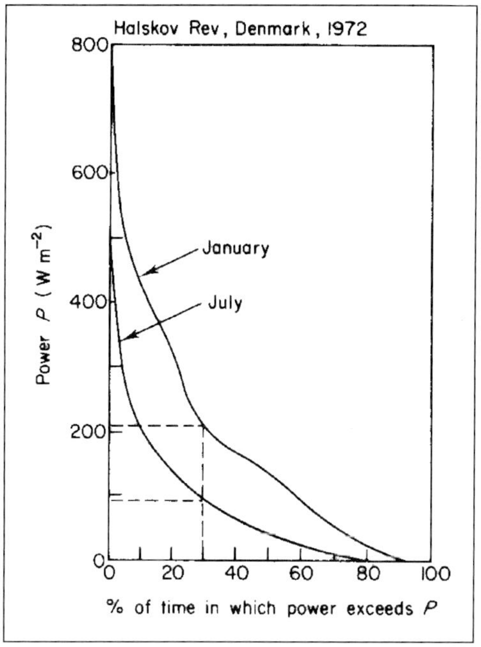

Figure 3.3-57 Power duration curves for power in currents at Halskov Rev, Denmark, based on a summer and a winter month of 1972. Thin, dashed lines indicate the monthly average power and the percentage of time during which it is available.

(July) and 0.65 m s^{-1} (January), the calculated kinetic energy densities would have been 108 and 211 J m^{-3}, and the power would have taken the values 50 and 134 W m^{-2}.

Few locations, whether in coastal regions or in open oceans, have higher average current speeds than the Danish location considered in Figs. 3.3-54–3.3-57, so that average power levels in the range 100–200 W m^{-2} are likely to be more representative than the 4000 W m^{-2} found (at least at the time of the measurement reported in Fig. 3.3-50) in the core of the Gulf Stream. This means that at many locations the power in currents is no greater than that found in the wind at quite low heights (cf. Fig. 3.3-34). Also, the seasonal variations and fluctuations are similar, which is not unexpected for wind-driven currents. For the currents at greater depths this may not hold true, partly because of the smoothing due to a long turnover time and partly because not all the deep sea motion is wind driven, but may also be associated with temperature and salinity gradients, as discussed in *Renewable Energy* (2004).

The power duration curves in Fig. 3.3-57 may be compared to those of wind power shown in Figs 3.3-39 and 3.3-40. The fraction of time in which the monthly average power is available in the Halskov Rev surface current is 0.3 in both January and July. The overall average power of about 150 W m^{-2} is available for about 45% of the time in January, but only 17% of the time in July.

3.3.3.2 River flows, hydropower and elevated water storage

The kinetic energy of water flowing in rivers or other streams constitutes an energy source very similar to that of ocean currents. However, rather than being primarily wind driven or caused by differences in the state of the water masses themselves, the river flows are part of the hydrological cycle. Water vapour evaporated into the atmosphere is transported and eventually condensed. It reaches the ground as water or ice, at the elevation of the particular location. Thus, the primary form of energy is potential. In the case of ice and snow, a melting process (using solar energy) is usually necessary before the potential energy of elevation can start to transform into kinetic energy. The origin of many streams and rivers is precisely the ice melting process, although they are joined by ground water flows along their subsequent route. The area from which a given river derives its input of surface run-off, melt-off and ground water is called its "drainage basin".

The flow of water in a river may be regulated by means of dam building, if suitable reservoirs exist or can be formed. In this way the potential energy of water stored at an elevation can be transformed into kinetic energy (e.g. driving a turbine) at the times most convenient with respect to utilisation.

An estimate of the potential hydro-energy of a given geographical region could in principle be obtained by hypothetically assuming that all precipitation was retained at the altitude of the local terrain and multiplying the gravitational potential *mg* by the height above sea-level. The annual precipitation over land amounts to about 1.1×10^{17} kg of water, and taking the average elevation of the land area as 840 m (Sverdrup et al., 1942), the annually accumulated potential energy would amount to 9×10^{20} J, corresponding to a mean energy flux (hydropower) of 2.9×10^{13} W.

Collection of precipitation is not usually performed as part of hydropower utilisation; but rather the natural processes associated with soil moisture and vegetation are allowed to proceed, leading to a considerable re-evaporation and some transfer to deeper lying ground water which eventually reaches the oceans without passing through the rivers. The actual run-off from rivers and overground run-off from polar ice caps comprise, on average, only about 0.36 of the precipitation over land, and the height determining the potential energy may be lower than the average land altitude, namely, that given by the height at which the water enters a river flow (from ground water or directly). The minimum size of stream or river which can be considered useful for energy extraction is, of course, a matter of technology, but these general considerations would seem to place an upper limit on the hydroenergy available of roughly 3×10^{20} J y^{-1}, corresponding to a power average of below 10^{13} W (10 TW).

If, instead of using average precipitation and evaporation rates together with average elevation, the geographical variations of these quantities are included, the result is also close to 3×10^{20} J y^{-1} or 10^{13} W. These figures are derived from the integral over all land areas,

$$\frac{dW^{pot}}{dt} = \int_{land\,area} (r - e)\, gz\, dA,$$

where r and e are the rates of precipitation and evaporation (mass of water, per unit of area and time), g is the gravitational acceleration and z is the height above sea level. The observed annual mean precipitation and evaporation rates quoted by Holloway and Manabe (1971) were used in the numerical estimate.

Geographical distribution of hydropower resources

A different estimate of hydropower potential is furnished by counts of actual rivers with known or assumed water transport and falling height. According to such an estimate by the World Energy Conference (1974,1995), the installed or installable hydro-generation capacity resource at average flow conditions may, in principle, amount to 1.2×10^{12} W, for both large installations and smaller ones and down to "micro-hydro" installations of around 1 MW. On the other hand, it is unlikely that environmental and other considerations will allow the utilisation of all the water resources included in the estimate. The World Energy Conference (1995) estimates 626 GW as a realistic reserve (including an already installed capacity producing, on average, 70 GW).

Figure 3.3-58 gives an idea of the geographical distribution of the hydropower resources on a national basis. The largest remaining resources are in South America. The figures correspond to average flow conditions, and the seasonal variations in flow are very different for different regions. For example, in Zaire practically all the reserves would be available year round, whereas in the USA only 30% can be counted on during 95% of the year.

Figure 3.3-59 gives seasonal variations (for two years) in the flow into the existing hydropower reservoirs in Norway, a country where the primary filling of reservoirs is associated with the melting of snow and ice during the late spring and early summer months.

Environmental impact

The environmental impact of non-regulated hydrogeneration of power is mainly associated with preventing the migration of fish and other biota across the turbine area, but the building of dams in connection with large hydro facilities may have an even more profound influence on the ecology of the region, in addition to introducing accident risks. For large reservoirs, there has been serious destruction of natural landscapes and dislocation of populations living in areas to be flooded. There are ways to avoid some of the problems. Modular construction, where the water is cascaded through several smaller reservoirs, has been used, e.g. in Switzerland, with a substantial reduction in the area modified as a result. The reservoirs need not be constructed in direct connection with the generating plants, but can be separate installations placed in optimum locations, with a two-way turbine that uses excess electric production from other regions to pump water up into a high-lying reservoir. When other generating facilities cannot meet demand, the water is then led back through the turbines. This means that although the water cycle may be unchanged on an annual average basis, considerable seasonal modifications of the hydrological cycle may be involved. The influence of such modifications on the vegetation and climate of the region below the reservoir, which would otherwise receive a water flow at a different time, has to be studied in each individual case. The same may be true for the upper region, for example, owing to increased evaporation in the presence of a full reservoir.

Although these modifications are local, they can influence the ecosystems with serious consequences for man. An example is provided by the building of the Aswan Dam in Egypt, which has allowed water snails to migrate from the Nile delta to the upstream areas. The water snails may carry parasitic worms causing schistosomiasis, and this disease has actually spread from the delta region to Upper Egypt since the building of the dam (Hayes, 1977).

It is unlikely that hydropower utilisation will ever be able to produce changes in the seasonal hydrological cycle, which could have global consequences, but no detailed investigation has yet been made. Such a study could proceed along the same lines as the investigation of the influence of deforestation.

3.3.3.3 Tides

The average rate of dissipation of tidal energy, as estimated from the slowing down of the Earth's rotation, is about 3×10^{12} W. Of this, about a third can be accounted for by frictional losses in definite shallow sea regions, bays and estuaries, according to Munk and MacDonald (1960).

In order to understand the concentration of tidal energy in certain coastal regions, a dynamic picture of water motion under the influence of tidal forces must be considered. The equations of motion for the oceans must be generalised to include the acceleration due to tidal

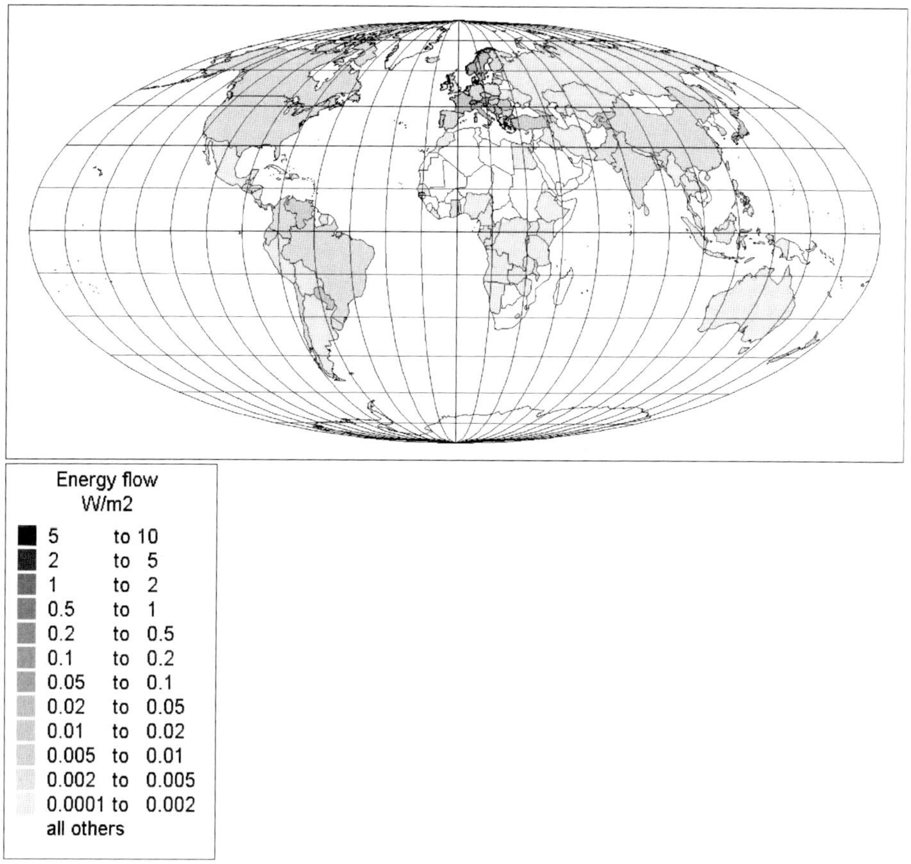

Figure 3.3-58 Hydropower potential average production (W m^{-2}), including existing plants, plants under construction, and planned and contemplated installations, both for large- and small-scale power schemes (based on data from World Energy Council, 1995).

attraction, i.e. F^{tidal}/m where an approximate expression for the tidal force is given in (2.67) in *Renewable Energy* (2004). Numerical solutions to these equations (see e.g. Nihoul, 1977) show the existence of a complicated pattern of interfering tidal waves, exhibiting in some points zero amplitude (nodes) and in other regions deviations from the average water level far exceeding the "equilibrium tides" of a fraction of a metre. These features are in agreement with observed tides, an example of which is shown in Fig. 3.3-60. Newer data based on

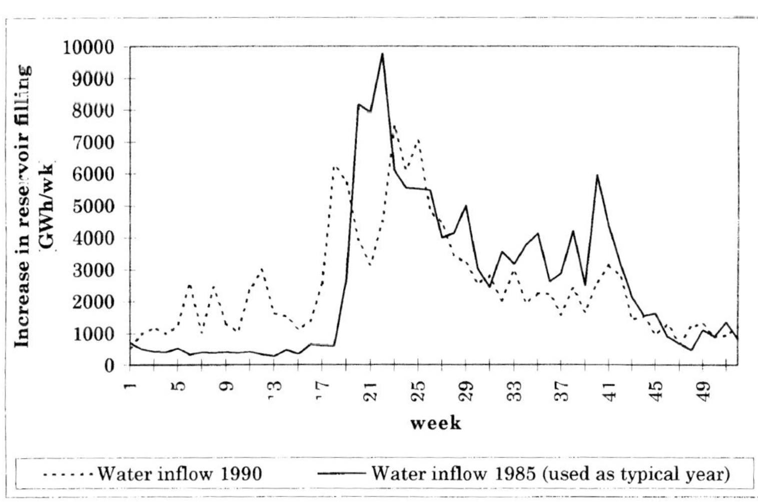

Figure 3.3-59 Seasonal variation in the power associated with the water flow into Norwegian hydropower reservoirs for a typical year (1985) and a "dry year" (1990) (based on Andersen, 1997; Meibom *et al.*, 1999).

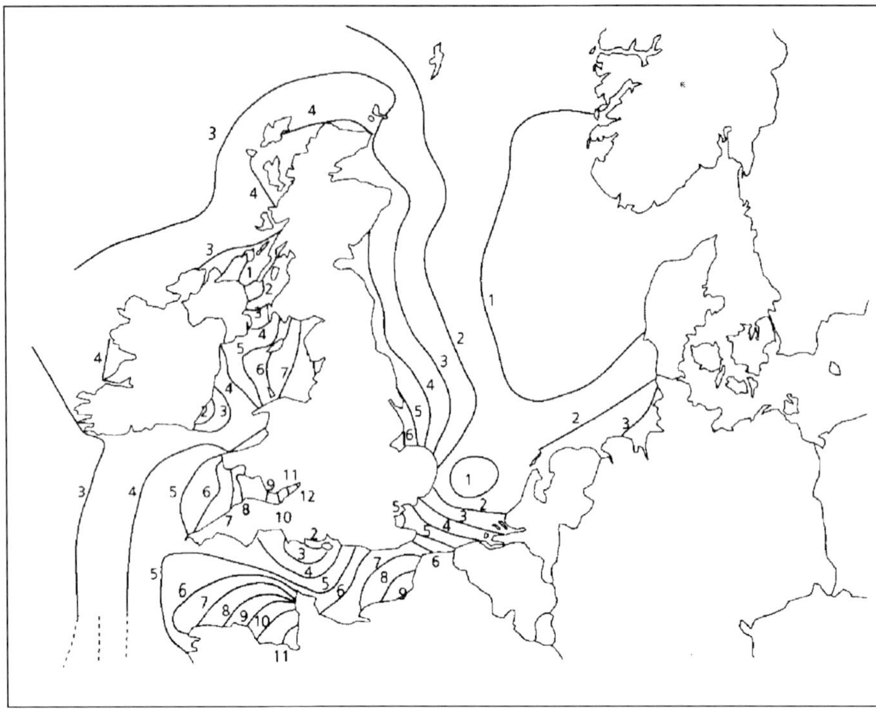

Figure 3.3-60 Tidal range *H* (difference between highest and lowest level in m) of the semidiurnal lunar tides in the North Sea and adjacent straits (based on Kiilerich, 1965; Cavanagh *et al.*, 1993).

satellite measurements can be followed in near-real time on the internet (NASA, 2004).

The enhancement of tidal amplitudes in certain bays and inlets can be understood in terms of resonant waves. Approximating the inlet by canal of constant depth h, the phase velocity of the tidal waves is $U_t = (gh)^{1/2}$ (Wehausen and Laitone, 1960), and the wavelength is

$$\lambda_t = T_t U_t,$$

where T_t is the period. For the most important tidal wave, T_t equals half a lunar day, and the condition for resonance may be written

$$L = i\,\lambda_t/4 = 0.25\,i\,T_t\,g^{1/2}\,h^{1/2}, \qquad (3.3.30)$$

where i is an integer, so that the length L of the inlet is a multiple of a quarter wavelength. For $i = 1$, the resonance condition becomes $L = 34973h^{1/2}$ (L and h in metres). Bays and inlets satisfying this condition are likely to have high tidal ranges, with the range being the difference between highest and lowest water level. An example of this is the Severn inlet near Bristol in the UK, as seen from Fig. 3.3-60. Cavanagh *et al.* (1993) estimate the total European potential to be 54 GW or about 100 TWh y^{-1}, of which 90% is in France and the UK.

The tides at a given location cannot be expected to have a simple periodicity, but rather are characterised by a superposition of components with different periods, the most important of which being equal to one or one-half lunar or solar day. As a function of time, the relative phases of the different components change, leading to regularly changing amplitudes, of which two different patterns are shown in Fig. 3.3-61. The upper one is dominated by the half-day period, while the lower one is dominated by the full-day period.

If the water level at high tide, averaged over an area A, is raised by an amount H over the level at low tide, then the potential energy involved is

$$W^{tidal} = \rho_w H\,A\,g\,H,$$

and the maximum power which could be extracted or dissipated would, as an average over a tidal period T_t, is

$$P^{tidal} = \rho_w\,g\,A\,H^2\,T_t^{-1}. \qquad (3.3.31)$$

Based on measured tidal ranges, and on an estimate of the area A of bay or inlet which could reasonably be enclosed by a barrage with the purpose of utilising the energy flow (3.3.31), a number of favourable sites have been identified, as shown in Fig. 3.3-62. These include only localities with considerable concentration of tidal energy, considering that if the tidal range decreases, the area to be enclosed in order to obtain the same power quadratically increases, and the length of barrage will

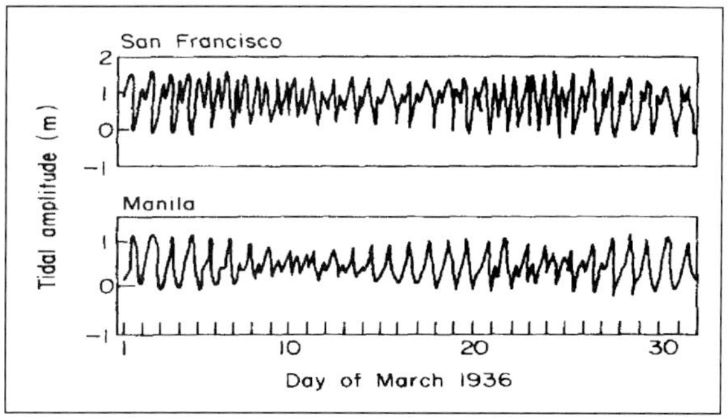

Figure 3.3-61 Examples of the time-development of tidal amplitudes at two different locations for the same month of 1936 (based on Defant, 1961).

have to be correspondingly greater. For the same reason, sites of adequate tidal range, but no suitable bay which could be enclosed by a reasonably small length of barrage have been excluded. Of course, the term "reasonable" rests on some kind of economic judgement, which may be valid only under given circumstances. It is estimated that 2–3 GW may be extracted in Europe, half of which at costs in the range 10–20 euro-cents (or US cents) per kWh (Cavanagh *et al.*, 1993, using costing methodology of Baker, 1987), and 20–50 GW in North America (Sorensen and MacLennan, 1974; Bay of Fundy Tidal Power Review Board, 1977).

Environmental impacts may arise from utilisation of tidal power. When the La Rance tidal plant was built in the 1960s, the upper estuary was drained for water during a two-year period, a procedure that would hardly be considered environmentally acceptable today. Alternative building methods using caissons or diaphragms exist, but in all cases the construction times are long and careful measures have to be taken to protect the biosphere (e.g. when stirring up mud from the estuary seabed). Generally, the coastal environment is affected by the building and operation of tidal power plants, both during construction and to a lesser extent during operation, depending on the layout (fish bypasses etc., as known from hydropower schemes). Some fish species may be killed in the turbines, and the interference with the periodic motion of bottom sand may lead to permanent siltation problems (and it has at la Rance).

The total estimated power of about 120 GW at the best sites throughout the world may be larger than what can be considered economic, but smaller than the amount of tidal energy actually available. It is still 12% of the above-mentioned estimate of the total dissipation of tidal energy in the vicinity of land, and it is unlikely that all the coastal sites yielding a total of 1000 GW would be suitable for utilisation, so the order of magnitude is likely to be correct. The 2000 GW remaining relative to the tidal power derived from astronomical data (Munk and MacDonald, 1960) presumably becomes dissipated in the open ocean.

The maximal tidal power believed to be accessible, as well as the total resource estimate, is about 10% of the corresponding figures for hydropower. Tidal variations are coupled to river run-off and sea level rise due to global greenhouse warming (Miller and Douglas, 2004).

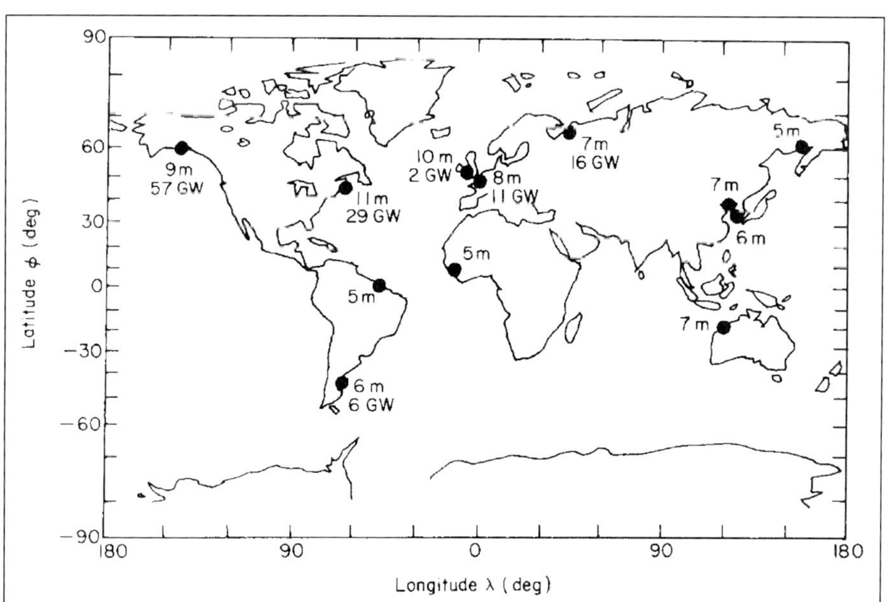

Figure 3.3-62 Tidal range for selected locations and the estimated average tidal power for each bay in a number of cases where utilisation has been proposed (based on King Hubbert, 1969; Gray and Gashus, 1972; Sorensen and MacLennan, 1974).

3.3.4 Heat flows and stored heat

A large fraction of the incoming solar radiation is stored as heat near the surface of the Earth. On average 47% is absorbed by the oceans and continents. A more detailed picture shows that 38% is absorbed by the oceans, 9% by the continents and 24% by the atmosphere. Chapter 2 in Renewable Energy (2004) dealt with some of the ways in which this energy could be distributed and eventually dissipated. Other sections dealt with the incoming radiation itself and with the kinetic energy in atmospheric and oceanic motion derived from the solar input by a number of physical processes. Storage in terms of potential energy has also been considered in connection with the processes of lifting water to a higher elevation, either in waves or in evaporated water, which may later condense and precipitate at a higher topographical level. Still, the energy involved in such kinetic and potential energy-creating processes is much less that the latent and sensible heat fluxes associated with the evaporation and condensation processes themselves and with the conversion of short-wavelength radiation to stored heat. The heat will be re-radiated as long-wavelength radiation in situations of equilibrium, but the temperature of the medium responsible for the absorption will rise, and in some cases the temperature gradients between the absorbing regions and other regions (deep soil, deep ocean, etc.), which do not themselves absorb solar radiation, cause the establishment of significant heat flows.

Utilisation of heat flows and stored heat may be direct if the desired temperature of use is no higher than that of the flow or storage. If this is not so, two reservoirs of different temperature may be used to establish a thermodynamic cycle yielding a certain amount of work, limited by the second law of thermodynamics. An alternative conversion scheme makes use of the heat pump principle by expending work added from the outside. Such conversion methods will be considered in Chapter 4.1, while the focus here will be on identifying those heat sources, which look most suitable for utilisation. The solar energy stores and flows will be surveyed in section 3.3.5.1, whereas section 3.3.5.2 will deal with the energy stored in the interior of the Earth and the corresponding geothermal flows.

3.3.4.1 Solar-derived heat sources

The ability of land and sea to act as a solar energy absorber and further as an energy store, to a degree determined by the local heat capacity and the processes, which may carry heat away, is of profound importance for the entire biosphere. For example, food intake accounts only for 25–30% of man's total income of energy during a summer day in central Europe (Budyko, 1974). The rest is provided by the body's absorption of solar energy. The biological processes that characterise the present biosphere have vanishing rates if the temperature departs from a rather narrow range of about 270–320 K. The present forms of life thus depend on the greenhouse effect maintaining temperatures at the Earth's surface in this range, at least during a fraction of the year ("growing season"), and it is hard to imagine life on the frozen "white Earth" which would result if the absorption processes were minimised.

Utilisation of heat stores and flows may then be defined as uses in addition to the benefits of the "natural" thermal regime of the Earth, but often no sharp division between "natural" and "artificial" uses can be drawn. For this reason, an assessment of the "magnitude" of the resource is somewhat arbitrary, and in each case it must be specified what is included in the resource base.

From total energy absorption rates, and it is clear that the oceans are by far the most significant energy accumulators. The distributions of yearly average temperatures along cross sections of the main oceans are shown in Figs 2.63–2.65.

The potential of a given heat storage for heat pump usage depends on two temperatures: that of the storage and the required usage temperature. This implies that no energy "amount" can be associated with the energy source, and a discussion of heat pump energy conversion will therefore be deferred until Chapter 4.1.

For utilisation without addition of high-grade mechanical or electric energy, two reservoirs of different temperature must be identified. In the oceans, the presence of a number of temperature gradients exist, most noticeable at latitudes below 50° and depths below 1000–2000 m. Near the Equator, the temperature gradients are largest over the first few hundred metres, and they are seasonally stable, in contrast to the gradients in regions away from the Equator, which are largest during summer and approach zero during winter. At still higher latitudes, there may be ice cover during a part or all of the year. The temperature at the lower boundary of the ice, i.e. at the water–ice interface, is quite constantly equal to 271.2 K (the freezing point of sea water), and in this case a stable temperature difference can be expected between the water and the atmosphere, particularly during winter.

It follows that over half the solar energy absorbed by the oceans is used to evaporate water. Of the remaining part some will be transferred to the atmosphere as sensible heat convection, but most will eventually be re-radiated to the atmosphere (and maybe to space) as long-wavelength radiation. The length of time during which the absorbed energy will stay in the oceans, before being used in one of the above ways, determines the temperature regimes. At the ocean surface, residence times of a fraction of a day will be associated with the diurnal temperature variations, while residence times normally increase with depth and reach values of several hundred years in the deepest regions.

If oceanic temperature gradients are used to extract energy from the surface region, a cooling of this region may result, unless the currents in the region are such that the cooling primarily takes place elsewhere. A prediction of the climatic impact of heat extraction from the oceans thus demands a dynamic investigation, such as the one offered by the general circulation models discussed elsewhere. Attention should be paid to the energy loss from the ocean's surface by evaporation, which would decrease at the places with lowered temperatures. The extraction of energy from the oceans may for this reason lead to an increase of the total downward energy flux through the surface, as noted by Zener (1973). In addition to possible climatic effects, oceanic energy extraction may have an impact on the ecology, partly because of the temperature change and partly because of other changes introduced by the mixing processes associated with at least some energy conversion schemes (e.g. changing the distribution of nutrients in the water). Assuming a slight downward increase in the concentration of carbon compounds in sea water, it has also been suggested that artificial upwelling of deep sea water (used as cooling water in some of the conversion schemes discussed in Chapter 4.1) could increase the CO_2 transfer from ocean to atmosphere (Williams, 1975).

The power in ocean thermal gradients

In order to isolate the important parameters for assessing the potential for oceanic heat conversion, Fig. 3.3-63 gives a closer view of the temperature gradients in a number of cases with large and stable gradients. For the locations at the Equator, seasonal changes are minute, owing to the small variations in solar input. For the locations in Florida Strait, the profiles are stable, except for the upper 50–100 m, owing to the transport of warm water originating in the tropical regions by the Gulf Stream. The heat energy content relative to some reference point of temperature T_{ref} is $c_V(T - T_{ref})$, but only a fraction of this can be converted to mechanical work, and the direct use of water at temperatures of about 25°C above ambient is limited, at least in the region near the Equator. Aiming at production of mechanical work or electricity, the heat value must be multiplied with the thermodynamic efficiency (cf. Chapter 4.1), i.e. the maximum fraction of the heat at temperature T which can be converted to work. This efficiency, which would be obtained by a hypothetical conversion device

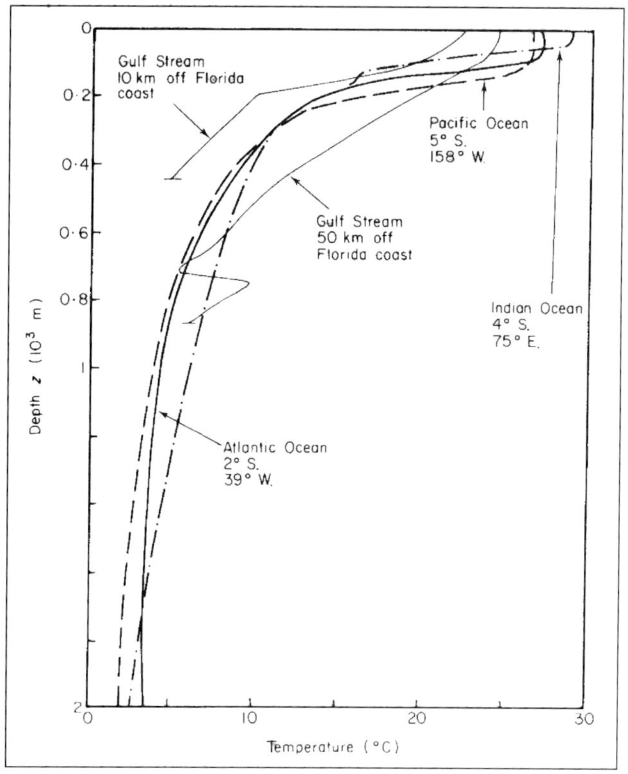

Figure 3.3-63 Temperature profiles for equatorial sites in the major oceans, as well as for the Gulf Stream at the Florida Strait. The profiles are derived from time-limited observations and may not represent annual averages (based on Neumann and Pierson, 1966; Sverdrup *et al.*, 1942).

operating in accordance with an ideal Carnot cycle, is (cf. section 4.1.1)

$$\eta_{Carnot} = (T - T_{ref})/T, \qquad (3.3.32)$$

where the temperatures should be in K.

Taking the reference temperature (the "cold side") as T_{ref} = 6°C, corresponding to depths of 380 m (Gulf Stream, 10 km from coast), 680 m (Gulf Stream, 50 km from coast), 660 m (Atlantic Ocean), 630 m (Pacific Ocean) and 1000 m (Indian Ocean) (see Fig. 3.3-63), one obtains the maximum energy which can be extracted from each cubic metre of water at temperature T (as a function of depth), in the form of mechanical work or electricity,

$$\eta_{Carnot}\, c_V(T - T_{ref}).$$

This quantity is shown in Fig. 3.3-64 for the sites considered in the preceding figure. It is seen that, owing to the quadratic dependence on the temperature difference, the work density is high only for the upper 100 m of the tropical oceans. Near the coast of Florida the work density drops more slowly with depth, owing to the

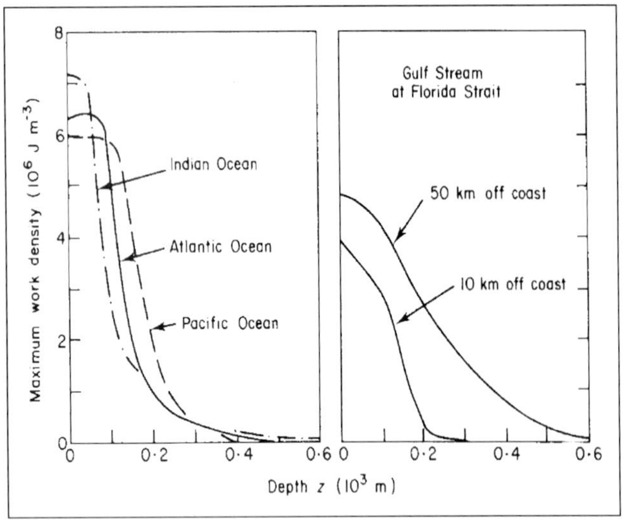

Figure 3.3-64 Maximum work density for the locations of Fig. 3.3-63, corresponding to an ideal Carnot machine extracting work from the temperature difference between the temperature at a given depth and a cold reservoir at constant temperature 6°C (the depth at which such cooling water might be collected can be found from Fig. 3.3-63).

presence of the core of the warm Gulf Stream current (cf. Fig. 3.3-50). Below the current the work density drops with a different slope, or rather different slopes, because of the rather complex pattern of weaker currents and counter-currents of different temperatures.

The power which could be extracted is obtained by multiplying the work density by the transfer coefficient of the conversion device, i.e. the volume of water from which a maximum energy prescribed by the Carnot value can be extracted per second. This quantity is also a function of T and T_{ref}, and losses relative to the ideal Carnot process are included in terms of a larger "effective" volume necessary for extracting a given amount of work. If the device were an ideal Carnot machine and the amount of water processed were determined by the natural flow through the device, then at a flow rate of 1 m s^{-1} (typical of the Gulf Stream current) the power extracted would be 3–4 MW m^{-2} facing the current flow direction, according to Fig. 3.3-64. In practice, as the treatment in Chapter 4 will show, only a fraction of this power will be obtained.

An indication of the global distribution of locations with large ocean temperature gradients may be obtained from Fig. 3.3-65, which gives the average temperature difference between the surface water and the water at a depth of 200 m for the time of the year when this difference is smallest. As mentioned above, the seasonal variation is very small in the equatorial region, whereas the summer temperature difference is far greater than the winter temperature difference at mid- and high latitudes, as seen from Fig. 3.3-66.

The availability of currents, which could promote the flow through a temperature gradient-utilising device, may be estimated from the material in section 3.3.4.1, but it should be borne in mind that the currents are warm only after passage through the equatorial region. It then follows from Fig. 3.3-51 that most currents in the Northern Hemisphere are warm, while most currents in the Southern Hemisphere originate in the Antarctic region and are cold.

Temperature gradients in upper soil and air

The temperature gradient due to absorption of solar radiation in continental soil or rock has a diurnal and a seasonal component. Transport processes in soil or rock are primarily by conduction rather than by mass motion (ground water does move a little, but speeds are negligible compared with the currents in the oceans), which implies that the regions which are heated are much smaller (typically depths of up to 0–7 m in the diurnal cycle and up to 15 m in the seasonal cycle). The seasonal cycle may be practically absent for locations in the neighbourhood of the Equator. The picture of the diurnal cycle is very similar except for scale. It follows that the potential for extracting mechanical work or electricity from solar-derived soil or rock temperature gradients is very small. Direct use of the heat is usually excluded as well, because the temperatures in soil or rock are mostly similar to or slightly lower than that of the air. An exception may be dry rock, which can reach temperatures substantially above the ambient after some hours of strong solar exposure. Extraction of heat from soil or rock by means of the heat pump principle is possible and will be discussed in the next chapter.

Temperature gradients in the atmosphere are rather small but of stable sign (except for the layer near the ground) up until the tropopause. At higher altitudes the gradient changes sign a few times. The temperature gradient over the first few hundred metres above ground is largely determined by the sign of the net total radiation flux. This exhibits a daily cycle characterised by downward net flux during daylight and upward net flux at night.

The density of air also decreases with height, and the heat capacity is small (e.g. compared with that of water), so the potential for energy conversion based on atmospheric heat is small per unit of volume. Energy production using the atmospheric temperature gradient would require the passage of very large quantities of air through an energy extraction device, and the cold air inlet would have to be placed at an elevation of the order of kilometres above the warm air inlet. However, heat pump use of air near ground level is possible, just as it is for solar energy stored in the upper soil.

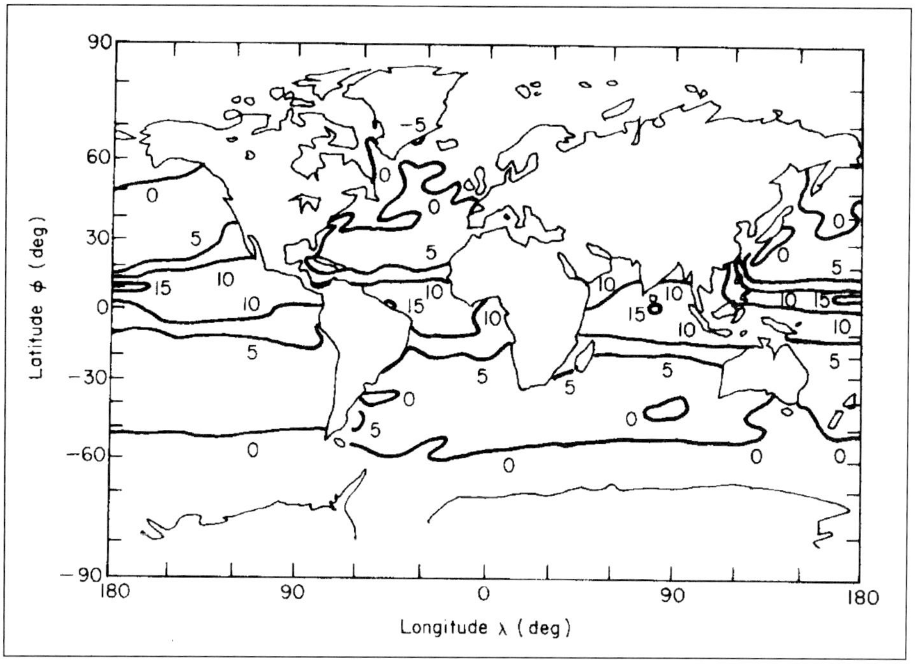

Figure 3.3-65 Average minimum temperature difference (°C) between sea surface and a depth of 200 m. "Average minimum" means that a smoothed curve for the seasonal variations in temperature difference has been used to determine the minimum (based on Putnam, 1953).

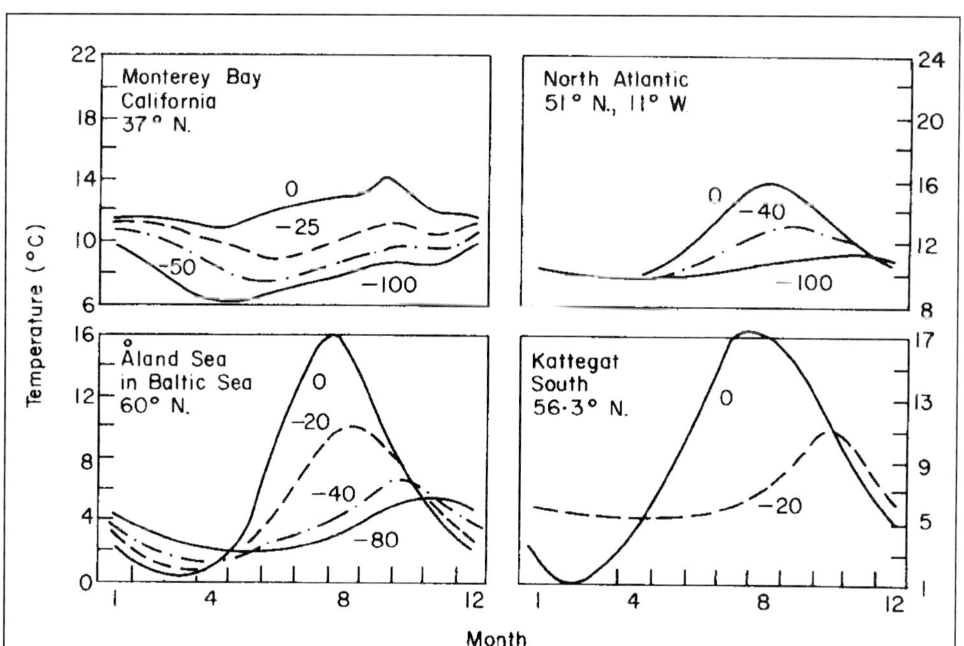

Figure 3.3-66 Trends of seasonal variations in water temperature at different depths (indicated in metres below surface) for selected locations in open ocean and near shore (based on Sverdrup *et al.*, 1942; Neumann and Pierson, 1966; Danish Meteorological Institute, 1973).

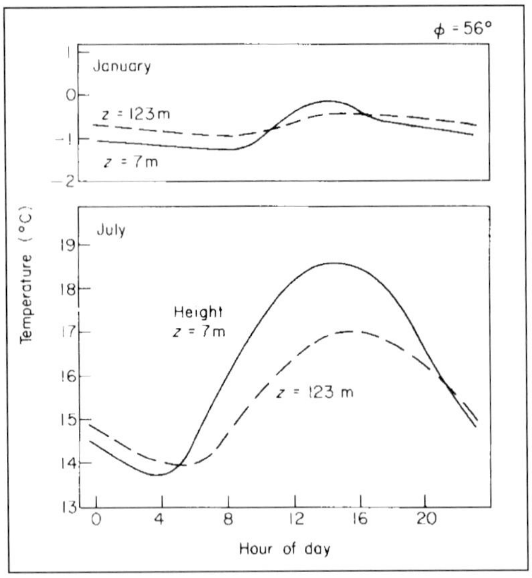

Figure 3.3-67 Variations of air temperature for two heights over ground and two months (January and July), as a function of the hour of the day. Ten years of observations from the meteorological tower at Risø, Denmark (Petersen, 1974) have been used to form the average trends.

3.3.4.2 Geothermal flows and stored energy

Heat is created in some parts of the interior of the Earth as a result of radioactive disintegrations of atomic nuclei. In addition, the material of the Earth is in the process of cooling down from an initial high temperature or as a result of heat released within the interior by condensation and possibly other physical and chemical processes.

Regions of particularly high heat flow

Superimposed on a smoothly varying heat flow from the interior of the Earth towards the surface are several abnormal flow regions. The subsurface temperature exhibits variations associated with the magnitude of the heat flow and the heat capacity, or in general the thermal properties, of the material. The presence of water (or steam) and other molten material is important for the transportation and concentration of heat. It is the vapour-dominated concentrations of geothermal energy which have attracted most attention as potential energy extraction sources, up until the present. However, the occurrence of hot springs or underground steam reservoirs is limited to a very few locations. Reservoirs containing superheated water (brine) are probably more common, but are hard to detect from general geological data.

Some reservoirs attract special interest owing to chemical processes in brine–methane mixtures which release heat and increase the pressure, while the conductivity of such mixtures is low (so-called "geopressurised systems", cf. Rowley, 1977). Dry rock accumulation of heat is probably the most common type of geothermal heat reservoir (i.e. storage with temperature above average), but it does not lend itself so directly to utilisation because of the difficulties in establishing sufficiently large heat transfer surfaces. A number of high-temperature reservoirs are present in connection with volcanic systems in the form of lava pools and magma chambers.

Whereas the smoothly varying average heat flow from the interior of the Earth can be considered a renewable energy resource (see below), the reservoirs of abnormal temperature are not necessarily large in relation to possible usage and do not get renewed at rates comparable to possible extraction rates. It is estimated that electric power generation by use of geothermal steam at the locations where such steam is available will be possible only over a period of about 50 years. The total amount of geothermal heat stored in water or steam to a depth of 10 000 m is estimated to be 4×10^{21} J, of which some (1–2) $\times\ 10^{20}$ J is capable of generating steam of above 200°C (World Energy Conference, 1974). The same source estimates the total amount of energy stored in dry rocks to a depth of 10 km as around 10^{27} J. The steam of temperatures above 200°C represents an average power of 240×10^9 W for a period of 50 years.

It follows from the above that most of the abnormal geothermal heat reservoirs must be considered as non-renewable resources on the same footing as fossil and fissile deposits. However, the average geothermal heat flow also corresponds to temperature gradients between the surface and reachable interior of the Earth, which can be used for energy extraction, at least if suitable transfer mechanisms can be established. These may be associated with water flow in regions of high permeability, such that water which has been cooled by an energy extraction device may regain the temperature of the surrounding layer in a relatively short time.

The origin of geothermal heat

The radioactive elements mainly responsible for the geothermal heat production at present are ^{235}U (decay rate $9.7 \times 10^{-10}\ y^{-1}$), ^{238}U (decay rate $1.5 \times 10^{-10}\ y^{-1}$), ^{232}Th (decay rate $5.0 \times 10^{-11}\ y^{-1}$) and ^{40}K (decay rate $5.3 \times 10^{-10}\ y^{-1}$). These isotopes are present in different concentrations in different geological formations. They are more abundant in the granite-containing continental shields than in the ocean floors. Most of the material containing the radioactive elements is concentrated in the upper part of the Earth's crust. In the lower half of the crust (total depth around 40 km), the radiogenic heat production is believed to be fairly constant at a value of about $2 \times 10^{-7}\ Wm^{-3}$ (Pollack and Chapman, 1977).

The rate of radiogenic heat production at the top of the continental crust is typically at least ten times higher, but it decreases with depth and reaches the lower crust value approximately half way through the crust. Little is known about the radioactivity in the mantle (occupying the volume between the core and the crust) and in the core (the radius of which is about half the total radius), but from models of their composition it is believed that very little radioactive heat production takes place in the mantle or core.

From the radioactive decay of the above-mentioned isotopes, it can be deduced that the temperature of an average piece of rock near the surface of the crust must have decreased some 900°C during the last 4.5×10^9 years (i.e. since the Sun entered the main burning sequence; Goguel, 1976). At present, the radiogenic heat production accounts for an estimated 40% of the continental average heat flow at the surface. The rest, as well as most of the heat flow at the oceanic bottoms, may then be due to cooling associated with expenditure of stored heat.

In order to assess the nature and origin of the heat stored in the interior of the Earth, a model of the creation and evolution of the planet must be formulated. The presence of heavy elements in the Earth's crust would be explained if the material forming the crust was originally produced during one or more supernovae outbursts. It is also consistent to assume that this material was formed over a prolonged period, with the last contribution occurring some 10^8 years before the condensation of the material forming the Earth (Schramm, 1974).

A plausible model of the formation of the planetary system assumes an initial nebula of dust and gases (including the supernova-produced heavy elements), the temperature of which would be increasing towards the centre. The Sun might then be formed by gravitational collapse. The matter not incorporated into the "protosun" would be slowly cooling, and parts of it would condense into definite chemical compounds. The planets would be formed by gravitational accretion of matter, at about the same time as or slightly earlier than the formation of the Sun (Reeves, 1975).

One hypothesis is that the temperature at a given distance from the protosun would be constant during the formation of planets like the Earth ("equilibrium condensation model", see e.g. Lewis, 1974). As temperature declines, a sequence of condensation processes occur: at 1600 K oxides of calcium, aluminium, etc.; at 1300 K nickel–iron alloys; at 1200 K enstatite ($MgSiO_3$); at 1000 K alkali-metal-containing minerals (feldspar, etc.); at 680 K troilite (FeS). The remaining iron would be progressively oxidised and at about 275 K water ice would be formed. The assumption that the Earth was formed at a constant temperature of about 600 K would then imply that it would have a composition characteristic of the condensation state at that temperature.

Other models can be formulated which would also be relatively consistent with the knowledge of the (initial) composition of the Earth's interior. If, for example, the planet formation process was slow compared to the cooling of the nebula, then different layers of the planet would have compositions characteristic of different temperatures and different stages of condensation.

It is also possible that the creation of the Sun and the organisation of the primordial nebula did not follow the scheme outlined above, but were, for example, the results of more violent events such as the close passage of a supernova star.

If a constant temperature of about 600 K during the formation of the Earth is accepted, then the subsequent differentiation into crust, mantle and core (each of the two latter having two subdivisions) is a matter of gravitational settling, provided that the interior was initially in the fluid phase or that it became molten as a result of the decay of radioactive isotopes (which as mentioned must have been much more abundant at that time, partly because of the exponential decay of the isotopes present today and partly owing to now absent short-lived isotopes). A crust would have formed very quickly as a result of radiational cooling, but the presence of the crust would then have strongly suppressed further heat loss from the interior. The central region may have been colder than the mantle during the period of core formation by gravitational settling (Vollmer, 1977).

The gravitational settling itself is a process by which gravitational (potential) energy is transformed into other energy forms. It is believed that most of this energy emerged as heat, and only a smaller fraction as chemical energy or elastic energy. If the differentiation process took place over a short time interval, the release of 26×10^{30} J would have heated material of heat capacity similar to that of crustal rocks to a temperature of about 5000 K. If the differentiation was a prolonged process, then the accompanying heat would to a substantial extent have been radiated into space, and the temperature would never have been very high (Goguel, 1976). The present temperature increases from about 300 K at the surface to perhaps 4000 K in the centre of the core. During the past, convective processes as well as formation of steam plumes in the mantle are likely to have been more intense than at present. The present heat flow between the mantle and the crust is about 25×10^{12} W, while the heat flow at the surface of the crust is about 30×10^{12} W (Chapman and Pollack, 1975; Pollack and Chapman, 1977).

If heat conduction and convective processes of all scales can be approximately described by the diffusion, supplemented with a heat source term describing the radioactive decay processes, in analogy to the general transport equation, then the equation describing the cooling of the Earth may be written

$$dT/dt = \text{div}\,(K\,\text{grad}\,T) + S/\rho \qquad (3.3.33)$$

where K is the effective diffusion coefficient and S is the radiogenic heat production divided by the local heat capacity C. The rate of heat flow is given by

$$E^{sens} = \lambda\,\partial T / \partial r \qquad (3.3.34)$$

where the thermal conductivity λ equals KC. If the temperature distribution depends only on the radial co-ordinate r, and K is regarded as a constant, (3.3.33) reduces to

$$dT/dt = K\,(\partial^2 T/\partial r^2 + 2\,r^{-1}\,\partial T / \partial r) + S/\rho \qquad (3.3.35)$$

The simplified version is likely to be useful only for limited regions. The diffusion coefficient K is on average of the order of 10^{-6} m^2 s^{-1} in the crust and mantle, but increases in regions allowing transfer by convection. The thermal conductivity λ of a material depends on temperature and pressure. It averages about 3 W m^{-1} K^{-1} in crust and mantle, but may be much higher in the metallic regions of the core. The total amount of heat stored in the Earth (relative to the average surface temperature) may be estimated from $\int_{volume} \lambda\,K^{-1}(T - T_s)\,dx$, which has been evaluated as 4×10^{30} J (World Energy Conference, 1974). Using the average heat outflow of 3×10^{13} W, the relative loss of stored energy presently amounts to 2.4×10^{-10} per year. This is the basis for treating the geothermal heat flow on the same footing as the truly renewable energy resources. Also, in the case of solar energy, the energy flow will not remain unchanged for periods of the order billions of years, but will slowly increase until the Sun leaves the main sequence of stars (at which time the radiation will increase dramatically).

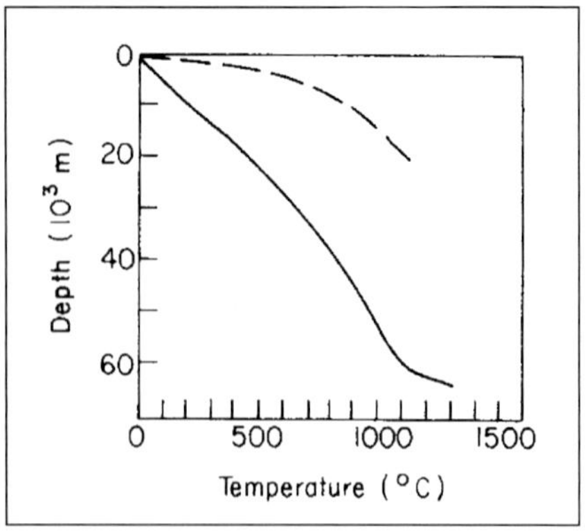

Figure 3.3-68 Temperature as a function of depth for a young and an old geological formation, representing average extremes of temperature profiles due to the smoothly varying part of the geothermal heat flow. The young formation (dashed line) corresponds to a mid-ocean ridge, and the old formation (solid line) corresponds to a Precambrian continental shield. The kink in the solid line may be associated with the onset of convective processes in the mantle (based on a model calculation, adjusted to observations, by MacGregor and Basu, 1974).

Distribution of the smoothly varying part of the heat flow

The important quantities for evaluating the potential of a given region for utilisation of the geothermal heat flow are the magnitude of the flow and its accessibility.

The temperature gradient and heat flow are usually large in young geological formations and, in particular, at the mid-oceanic ridges, which serve as spreading centres for the mass motion associated with continental drift. In the continental shields, the gradient and flow are much smaller, as illustrated in Fig. 3.3-68. Along sections from the mid-oceanic ridges to the continental shields the gradient may gradually change from one to the other of the two types depicted in Fig. 3.3-68. In some regions the temperature gradients change much more abruptly (e.g. in going from the Fenno-Scandian Shield in Norway to the Danish Embayment; cf. Balling, 1976), implying in some cases a high heat flux for land areas (e.g. northwestern Mexico, cf. Smith, 1974).

Maps of the geographical distribution of heat flow, both at the surface of the crust and at the surface of the mantle, have been prepared by Chapman and Pollack (1975; Pollack and Chapman, 1977) by supplementing available data with estimates based on tectonic setting and, for oceanic regions, the age of the ocean floor. The results, shown in Figs 3.3-69 and 3.3-70, are contours of a representation in terms of spherical harmonic functions of latitude and longitude, $Y_{lm}(\phi, \lambda)$, of maximum degree $l = 12$. The advantage of this type of analysis is that oscillations of wavelength smaller than about 3000 km will be suppressed, so that the maps are presumably describing the smoothly varying average flow without perturbations from abnormal flow regions. A comparison between the calculation shown in Fig. 3.3-69 and one in which the model predictions have been used also in regions where measurements do exist, has convinced Chapman and Pollack that little change will result from future improvements in the data coverage. The map showing the mantle flow, shown in Fig. 3.3-70, is obtained by subtracting the contribution from the radiogenic heat production in the crust from the surface flow map. To do this, Pollack and Chapman have used a model for the continental regions in which the heat production decreases exponentially from its local value at the surface of the crust,

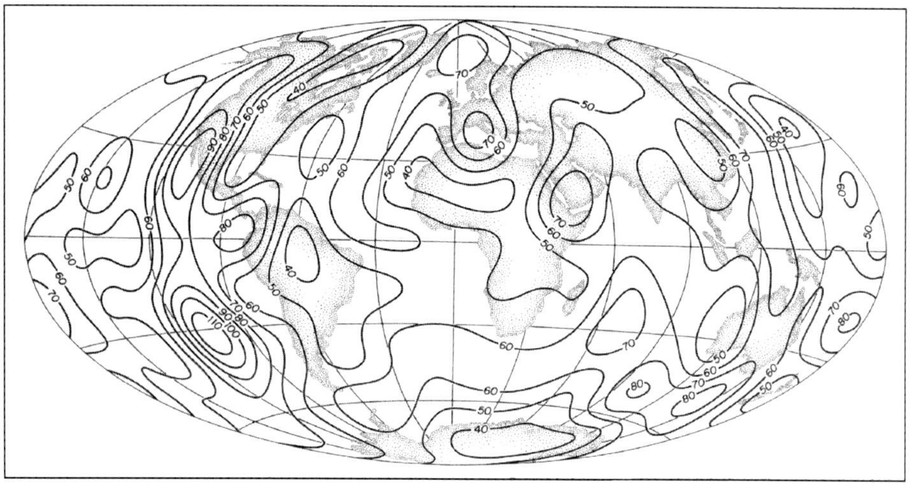

Figure 3.3-69 Surface heat flow contours (in 10^{-3} W m^{-2}) for the smoothly varying part of the geothermal flux, calculated on the basis of available data supplemented with a theoretical model. From H. Pollack and D. Chapman (1977), *Earth and Planetary Science Letters* **34**,174–184, copyright by Elsevier Scientific Publ. Co., Amsterdam.

$$S = S_s \exp(-z/b),$$

where $b = 8.5$ km (if not measured), until it reaches the constant value $CS = 2.1 \times 10^{-7}$ W m^{-3} assumed to prevail throughout the lower crust. For oceanic regions, the difference between mantle and surface heat flows is estimated on the basis of the cooling of the oceanic crust (assumed to behave like a 6.5-km thick layer of basalt) with a fixed lower boundary temperature of 1200°C. The time during which such cooling has taken place (the "age" of the sea floor) is determined from the present surface heat flow (Pollack and Chapman, 1977).

It is evident that the mantle heat flow is very regular, with low continental values and increasing heat flow when approaching the oceanic ridges, particularly in the southern Pacific Ocean. The surface heat flow is considerably more irregular, in a manner determined by the composition of the crust at a given location, but the mean fluctuation around the average value of 5.9×10^{-2} W m^{-2} is only 22%. The surface heat flow has similar values if evaluated separately for continents and oceans (5.3 and 6.2×10^{-2} W m^{-2}), while the contrast is larger for the upper mantle flow (2.8 and 5.7×10^{-2} W m^{-2}, respectively).

The usefulness of a given flow for purposes of energy extraction depends on the possibilities for establishing a heat transfer zone of sufficient flux. As mentioned earlier, the most appealing heat transfer method is by circulating water. In this case the rate of extraction

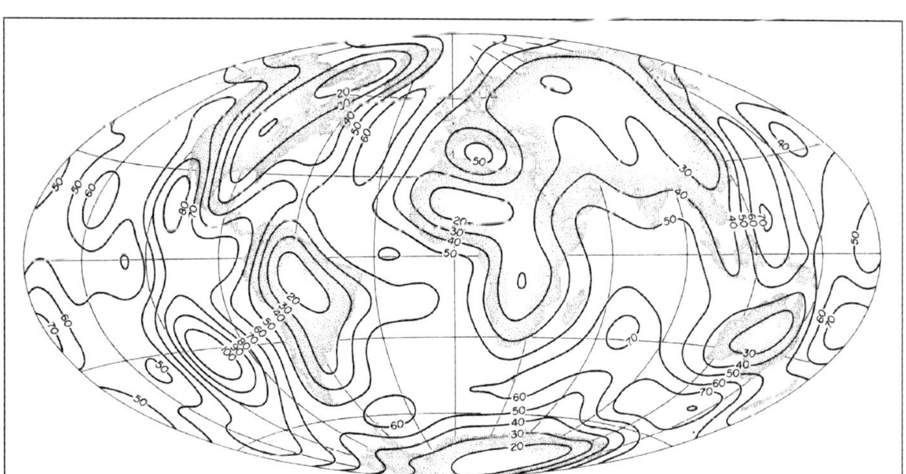

Figure 3.3-70 Heat flow contours at the top of the mantle (10^{-3} W m^{-2}), calculated from the surface fluxes shown in Fig. 3.3-69 by subtracting the contribution originating in the crust, for which a model has been constructed (see text). Ffrom H. Pollack and D. Chapman (1977), *Earth and Planetary Science Letters* **34**, 174–184, copyright by Elsevier Scientific Publ. Co., Amsterdam.

depends on the permeability of the geological material (defined as the fluid speed through the material, for a specified pressure gradient and fluid viscosity). The pressure is also of importance in determining whether hot water produced at a depth will rise to the top of a drill-hole unaided or will have to be pumped to the surface. High permeability is likely to be found in sand-containing sedimentary deposits, while granite and gneiss rock have very low porosity and permeability. Schemes for hydraulic fractionation or granulation caused by detonation of explosives have been proposed, with the purpose of establishing a sufficiently large heat transfer area, so that the vast areas of hot rock formations could become suited for extraction of geothermal energy, for example, by circulating water through such man-made regions of porous material.

3.3.5 Biological conversion and storage of energy

The amount of solar energy stored in the biosphere in the form of standing crop biomass (plants, animals) is roughly 1.5×10^{22} J (see section 2.4.1), and the average rate of production is 1.33×10^{14} W or 0.26 W m^{-2}. The residing biomass per unit area is small in oceans and large in tropical forests. The average rate of production on the land surface is 7.6×10^{13} W or 0.51 W m^{-2} (Odum, 1972), and over 0.999 of the standing crop is found on land.

Still larger amounts of dead organic matter are present on the Earth in various degrees of fossilation, expressed in carbon mass units rather than in energy units; 1 kg carbon roughly corresponds to 41.8×10^6 J). Thus, the fossil deposits (varieties of coal, oil and natural gas) are estimated to be of the order of 6×10^{23} J. This may be an underestimate, in view of the floating distinction between low-concentration energy resources (peat, shale, tar sand, etc.) and non-specific carbon-containing deposits. A maximum of about 1×10^{23} J is believed to be recoverable with technology that can be imagined at present, suggesting that the fossil energy resources are not very large compared with the annual biomass production. On the other hand, more sophisticated methods are generally required to convert fresh biomass into useful energy than is the case for the most pure and concentrated fossil fuel deposits. For bio-energy sources which have been in use for a long time, such as firewood and straw, an important requirement is often drying, which may be achieved by further use of direct solar radiation.

It should not be forgotten that plants and derived forms of biomass serve man in essential ways other than as potential energy sources, namely, as food, as raw material for construction and – in the case of green plants – as producers of atmospheric oxygen. The food aspect is partially an indirect energy usage, because man and other animals convert the energy stored in plants by metabolic processes, which furnish the energy for the life processes. It is also more than that, acting as a source of nutrients, vitamins, etc., which are required for reasons other than their energy content.

An assessment of the possible utilisation of bio-energy for purposes other than those associated with the life-cycles themselves must take into account the food requirements as well as other tasks performed by vegetation and animal stock, e.g. prevention of soil erosion, conservation of diversity of species and balanced ecological systems. Only energy uses which can be implemented in harmony with (maybe even in combination with) these other requirements can be considered acceptable. Although plant systems convert only a small fraction of the incident solar radiation, plans for energy usage should also consider the requirements necessary to avoid any adverse climatic impact. The management of extended areas of vegetation does hold possibilities of interfering with climate, e.g. as a result of the strong influence of vegetation on the water cycle (soil moisture, evaporation, etc.). Examples of such relations and actual climatic changes caused by removal of vegetation, overgrazing, can be found in *Renewable Energy* (2004), Chapter 2.

Before the question of plant productivity in different geographical regions and under different conditions is considered, a brief survey of the biochemistry of photosynthetic processes in green plants will be outlined.

3.3.5.1 Photosynthesis

Mechanism of green plant photosynthesis

The cells of green plants contain a large number of chloroplasts in their cytoplasm. The interior of a chloroplast contains a liquid, the stroma, rich in dissolved proteins, in which floats a network of double membranes, the thylakoids. The thylakoid membranes form a closed space, containing chlorophyll molecules in a range of slightly different forms, as well as specific proteins important for the photosynthetic processes. In these membranes the photo-induced dissociation of water takes place

$$2H_2O + \text{light} \rightarrow 4e^- + 4H^+ + O_2, \qquad (3.3.36)$$

and the electrons are transferred, through a number of intermediaries, from the internal to the external side of the membrane. The transport system is depicted in Fig. 3.3-71, where the ordinate represents the redox potential [the difference in redox potential between two states roughly equals the amount of energy, which is released or (if negative) which has to be added, in order to go from one state to the other].

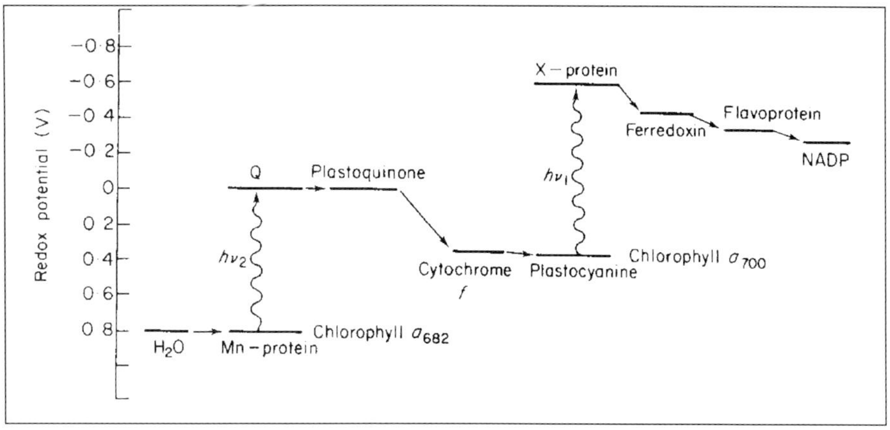

Figure 3.3-71 Model of the pathway of electron transfer in green plant photosynthesis (based on Trebst, 1974).

The first step in the reaction is the absorption of solar radiation by a particular chlorophyll pigment (a_{680}). The process involves the catalytic action of an enzyme, identified as a manganese complex, which is supposed to trap water molecules and become ionised as a result of the solar energy absorption. Thereby, the molecules (of unknown structure) denoted Q in Fig. 3.3-71 become negatively charged, i.e. electrons have been transferred from the water–manganese complex to the Q-molecules of considerably more negative redox potential (Kok et al., 1975; for a recently updated review, see the treatment given in Chapter 2 of Sørensen, 2004a).

The photosensitive material involved in this step is called "photosystem II". It contains the chlorophyll pigment a_{680} responsible for the electron transfer to the Q-molecules, where 680 is the approximate wavelength (in 10^{-9} m), for which the light absorptance peaks. It is believed that the initial absorption of solar radiation may be affected by any of the chlorophyll pigments present in the thylakoid and that a number of energy transfers take place before the particular pigment, which will transport an electron from H_2O to Q, receives the energy (Seliger and McElroy, 1965).

The gross absorptance as a function of wavelength is shown in Fig. 3.3-72 for green plants and, for comparison, a purple bacterium. The green plant spectrum exhibits a peak just above 400×10^{-9} m and another one a little below 700×10^{-9} m. The bacterial spectrum is quite different with a pronounced peak in the infrared region around 900×10^{-9} m. The positions of the peaks are somewhat different for the different types of chlorophyll (a, b, c,...), and they also move if the chlorophyll is isolated from its cellular surroundings. For a given chlorophyll type, different pigments with different absorption peak positions result from incorporation into different chemical environments (notably bonding to proteins).

Returning to Fig. 3.3-71, the electrons are transferred from the plastoquinone to cytochrome f by a redox reaction, which transfers a corresponding number of protons (H^+) across the thylakoid membrane. The electrons are further transferred to plastocyanine, which leaves them to be transported to another molecule of unknown structure, the X-protein. The energy required for this process is provided by a second photo-excitation step, in what is called "photosystem I", by means of the chlorophyll pigment a_{700} (spectral peak approximately at 700×10^{-9} m).

The electrons are now delivered to the outside region where they are picked up by ferredoxin, a small protein molecule situated at the outer side of the thylakoid membrane system. Ferredoxin may enter a number of different reactions, of which a very important one transfers the electrons via flavoprotein to

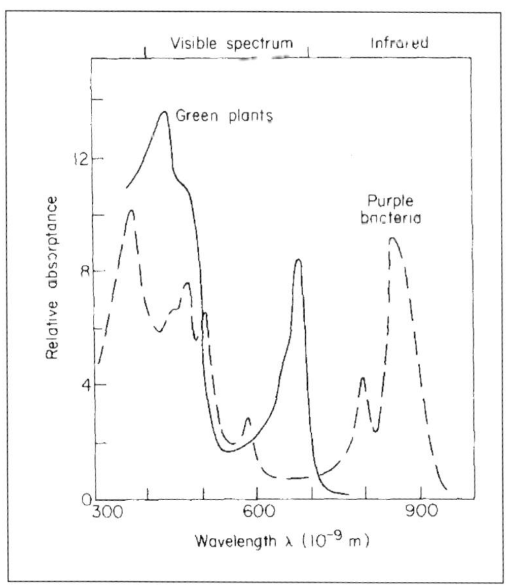

Figure 3.3-72 Spectrum of relative absorptance for a green plant and for a purple bacterium (based on Clayton, 1965).

nicotinamide–adenine dinucleotide phosphate (NADP), which reacts with protons to form $NADPH_2$, the basis for the carbon dioxide assimilation. The protons formed by (3.3.36) and by the cytochrome *f*–plastoquinone reduction are both formed inside the membrane, but they penetrate the thylakoid membrane by a diffusion process, regulated by the reaction (3.3.38) described below. The $NADPH_2$-forming reaction may then be written

$$2NADP + 4e^- + 4H^+ \rightarrow 2NADPH_2, \qquad (3.3.37)$$

which describes the possible fate of electrons and protons formed by the process (3.3.36), after transport by the sequence shown in Fig. 3.3-71 and proton diffusion through the membrane.

The transport of protons from outside the thylakoid membrane to its inside space, by means of the plastoquinone–cytochrome *f* redox cycle, creates an acidity (pH) gradient, which in turn furnishes the energy necessary for the phosphorylation of adenosine diphosphate (ADP) into adenosine triphosphate (ATP),

$$ADP^{3-} + H^+ + HPO_4^{2-} \rightarrow ATP^{4-} + H_2O. \qquad (3.3.38)$$

This process, which involves the action of an enzyme, stores energy in the form of ATP (4.8×10^{-20} J per molecule), which may later be used to fuel energy-demanding processes, e.g. in the chloroplast stroma or in the cytoplasm outside the chloroplast (see e.g. Douce and Joyard, 1977). It is the analogue of the energy-stocking processes taking place in any aerobic cell of plants and animals (i.e. cells using oxygen), as a result of the degradation of food (saccharides, lipids and proteins), and associated with expenditure of oxygen (the respiratory chain, the Krebs cycle; see e.g. Volfin, 1971). The membrane system involved in the case of food metabolism is the mitochondrion.

By means of the energy stored in ATP and the high reducing potential of $NADPH_2$, the CO_2 assimilation process may take place in the stroma of the chloroplasts independently of the presence or absence of solar radiation. The carbon atoms are incorporated into glyceraldehyde 3-phosphate, which forms the basis for synthesis of glucose and starch. The gross formula for the process leading to the synthesis of glyceraldehyde 3-phosphate in the chloroplasts (the Benson, Bassham and Calvin cycle; cf. Douce and Joyard, 1977) is

$$\begin{aligned} &3CO_2 + 9ATP^{4-} + 6NADPH_2 + 5H_2O \\ &\rightarrow POCH_2CH(OH)CHO^{2-} + 9ADP^{3-} \\ &+ 8HPO_4^{2-} + 6NADP + 9H^+. \end{aligned} \qquad (3.3.39)$$

The reactions (3.3.36) to (3.3.39)may be summarised:

$$\begin{aligned} &3CO_2 + 2H_2O + HPO_4^{2-} + \text{light} \\ &\rightarrow POCH_2CH(OH)CHO^{2-} + 3O_2. \end{aligned} \qquad (3.3.40)$$

Going one step further and including the synthesis of glucose, the classical equation of photosynthesis is obtained,

$$6H_2O + 6CO_2 + 4.66 \times 10^{-18} J \rightarrow C_6H_{12}O_6 + 6O_2, \qquad (3.3.41)$$

where 4.66×10^{-18} J is the net energy to be added by solar radiation.

Efficiency of conversion

The basis for energy utilisation has traditionally been the undifferentiated biomass produced. Other possibilities would include the direct dissociation of water by the action of sunlight (photolysis), or the formation of hydrogen rather than $NADPH_2$, after the initial oxygen formation (3.3.36) and electron transport to ferredoxin (Benemann and Weare, 1974; Mitsui and Kumazawa, 1977),

$$4e^- + 4H^+ \rightarrow 2H_2. \qquad (3.3.42)$$

In both cases the problem is to control the recombination process,

$$2H_2 + O_2 \rightarrow 2H_2O + 9.47 \times 10^{-19} J, \qquad (3.3.43)$$

so that this energy-releasing process takes place where desired and not immediately on the formation of hydrogen. The plants accomplish this by means of the thylakoid membrane and later by the chloroplast external membrane. Man-made processes may attempt to copy the membrane principle in various ways (Broda, 1975; Calvin, 1974, 1977; cf. Chapter 4.1), but it is not clear that present suggestions will allow separation of hydrogen and oxygen on a large scale. An advantage in this connection may lie in using the green plants to perform the first step (3.3.36) and transport electrons and protons (hydrogen ions) through their membranes, but to prevent the energy from being too deeply trapped in organic matter, for example, by adding a strongly reducing agent to ferredoxin, as well as suitable enzymes, in order to accomplish the reaction (3.3.42) at this stage.

The maximum theoretical efficiency of this process is the ratio between the heat release 9.47×10^{-19} J in (3.3.43) and the solar energy input required. The latter depends on the absorption properties of the plant and its chlorophyll molecules (cf. Fig. 3.3-72), as well as on the number of light quanta required for each molecular transformation (3.3.36). The energy, E, of each light quantum of wavelength λ may be found from

$$\lambda E_\lambda = hc = 1.986 \times 10^{-25}\ \text{J m}.$$

The minimum number of quanta required to transport one electron as depicted in Fig. 3.3-71 is two, one of wavelength 680×10^{-9} m and the other of 700×10^{-9} m (although these need not be the quanta originally absorbed). Since (3.3.36) requires the transport of four electrons, the minimum requirement would be 8 quanta with a total energy of 2.3×10^{-18} J. Experimental estimates of the number of quanta needed typically give values between 8 and 10.

The efficiency of the photosynthetic process containing only the steps (3.3.36) and (3.3.42) may be written

$$\eta' = \eta_\lambda\ \eta_{geom}\ \eta_{chem}, \tag{3.3.44}$$

where η_λ is the fraction of the frequencies in the solar spectrum (depending on cloud cover, etc.) that is useful in the photosynthetic process, η_{geom} is the geometrical efficiency of passing the incoming radiation to the chlorophyll sites (depending on penetration depth in leaves, on reflectance from outer and inner surfaces and on absorption by other constituents of the leaves), and η_{chem} is the efficiency of the photochemical reactions, the maximum value of which is given by

$$\eta_{chem} \leq 9.47 \times 10^{-19}/2.30 \times 10^{-18} = 0.41.$$

This efficiency pertains to the amount of internal heat produced by (3.3.43). Only a part of this can be converted into useful work. This part is obtained by replacing in (3.3.43) the enthalpy change 9.47×10^{-19} J by the change in free energy, $\Delta G = 7.87 \times 10^{-19}$ J (cf. section 4.1.1). In this way the efficiency of the photochemical reaction becomes

$$\eta_{chem,free} \leq 0.34.$$

The efficiency η_λ associated with the chlorophyll absorption spectrum typically lies in the range 0.4–0.5 (Berezin and Varfolomeev, 1976), and the geometrical efficiency η_{geom} may be around 0.8 (for the leaf of a green plant, not in the reduction associated with the penetration of radiation through other vegetation, e.g. in a forest environment).

The overall maximum efficiency of about $\eta' \approx 0.14$ found here for a hypothetical hydrogen (H_2)-producing system is valid also for actual green plants which assimilate CO_2. Over extended periods of time, the biomass production efficiency will have to incorporate still another factor, η_{resp}, expressing the respiration losses associated with the life-cycle of the plant,

$$\eta = \eta'\ \eta_{resp}. \tag{3.3.45}$$

The respirative energy losses emerge as heat and evaporated water, at rates depending on temperature and wind speed (Gates, 1968). The value of η_{resp} is 0.4–0.5 for land plants and somewhat larger for aquatic plants and algae.

Actual plants may get close to the theoretical maximum efficiency, although the average plant does not. For the blue-green alga *Anacystis nidulans*, Goedheer and Hammans (1975) report an energy efficiency of $\eta_{chem} \approx 0.30$, based on 36 h of growth, including 6 h of irradiation, with a generous supply of nitrogen and other nutrients, as well as a CO_2-enriched atmosphere (i.e. 73% of the maximum energy efficiency calculated above).

It is estimated that, on average, each CO_2 molecule in the atmosphere becomes assimilated in a plant once every 200 years and that each O_2 molecule in the atmosphere is "renewed" through a plant once every 2000 years (Seliger and McElroy, 1965).

Bacterial photosynthesis

Several bacteria use solar radiation to dissociate a compound of the general form H_2X, with a net reaction scheme of the form

$$CO_2 + 2H_2X + \text{light} \rightarrow (CH_2O) + H_2O + 2X, \tag{3.3.46}$$

(Van Niel, 1941). Here (CH_2O) should be understood not as free formaldehyde, but as part of a general carbohydrate compound in analogy to the more precise equation (3.3.40). Actually, (3.3.46) was proposed to be valid for both green plant and bacterial photosynthesis, but there is no detailed analogy, since the bacterial photosynthesis has been found to take place in a single step, resembling the photosystem I of the green plant two-step process. Other photo-induced reactions do take place in bacteria, connected with the ATP formation, which in this case is not a side-product of the primary photosynthetic process.

The compound H_2X may be H_2S (sulphur bacteria), ethanol C_2H_5OH (fermentation bacteria), etc. Most photosynthetic bacteria are capable of absorbing light in the infrared region (wavelength $800–1000 \times 10^{-9}$ m). The role which the NADP-$NADPH_2$ cycle (3.3.37) and (3.3.39) plays for green plant photosynthesis is played by NAD (nicotinamide-adenine dinucleotide)–$NADH_2$ for photosynthetic bacteria. The redox potential of $NADH_2$ is more or less the same as that of the initial compounds, e.g. H_2S, so practically no energy is stored in the process of bacterial photosynthesis (Hind and Olson, 1968).

3.3.5.2 Productivity in different environments

Ecological systems

The gross primary production of a plant or, in general, of an ecological system is the rate at which solar energy is assimilated, i.e. the total amount of energy produced by photosynthesis. The net primary production, on the other hand, is the difference between gross primary production and respiration. The respiration processes involve an increase in redox potential (oxidation), either by consumption of oxygen [aerobic respiration, the net result being equivalent to (3.3.43)] or by the action of some other agent of oxidation (anaerobic respiration; if the oxidant is an organic compound, the respiration process is called fermentation).

The primary producers are a part of an ecological system. Figure 3.3-73 gives a schematic view of the energy and matter flow through such a system. The primary chain comprises the primary producers capable of carrying out photosynthesis, plant-eating organisms (herbivores) and a chain of successive carnivorous predators (carnivores), of which some may also be plant eating (like man). Each compartment in the chain is called a trophic level, and the photosynthetic organisms are called autotrophs, while the consuming organisms further along the chain are called heterotrophs. Over short periods of time, withering and death of autotrophs can be neglected, and the net primary production available to the heterotrophic part of the community equals the gross primary production less the respiration of the autotrophs. Over longer periods, respiration, predation and death must be considered in order to describe the amounts of biomass in each compartment ("standing crop"). Generally, the biomass diminishes along the chain, but owing to the different average lifetime, which is often longest for the highest trophic levels of the food chain, the biomass at a given time may be maximal for, say, the second or third member of the chain. Also, the seasonal dependence of primary production in most geographical regions, in conjunction with an often short lifetime of the autotrophs, leads to a biomass distribution with much more long-range stability in the higher levels. The stability depends on the age of the ecosystem. A young system has fewer species and is more sensitive to external disturbances (climate variations, immigration of new predator species, etc.) which may destroy the food basis for the higher trophic levels. Old ecosystems are characterised by higher diversity and, hence, are better equipped to circumvent sudden changes in external conditions, at least for a while.

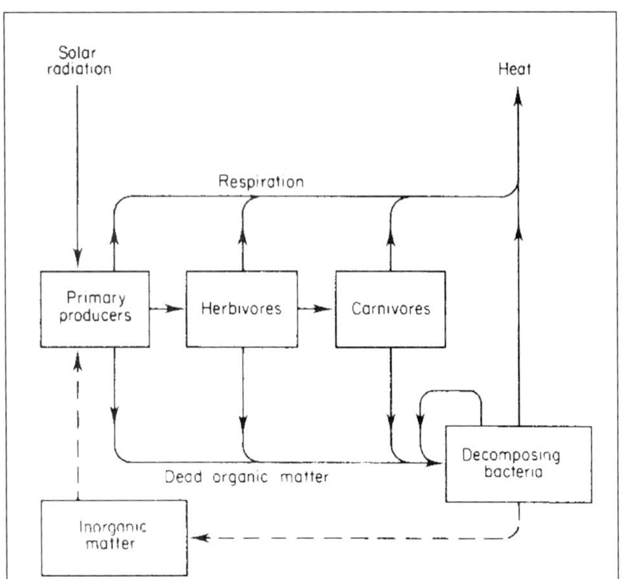

Figure 3.3-73 Model of an ecological system exhibiting flows of energy (solid lines, which in most cases also indicate the flow of organic matter) and of inorganic matter (dashed lines; inorganic matter includes CO_2, nutrients, etc.). Exchange of heat between compartments and their environment is not shown, neither is the possible presence of photosynthetic bacteria in the system. Some of the flows may be "delayed", e.g. man's harvesting of food crops and wood, which may go into storage rather than being consumed immediately (i.e. transferred to the next compartment along the chain, "herbivores" in this example).

The dead organic matter may precipitate towards the sea floor or detritus layer of the land surface, where an environment suitable for sustaining a number of decomposing bacteria evolves. Thereby nutrients are transformed back to the inorganic form necessary for renewed uptake by autotrophs.

It follows from Fig. 3.3-73 that the flows of energy, which could be used for energy extraction serving human society, diminish along the food chain. This does not negate the fact that the concentration of energy flow may be higher and perhaps easier to utilise at one of the higher levels, as compared with the primary producers. For example, livestock leave a large fraction of their respiration heat and manure within their sheds, a spatial region much smaller than the size of a grazing range which may yield the same energy value.

Important factors in determining the net productivity of a given autotrophic system are solar radiation, temperature regime, water availability, climate in general, carbon dioxide and nutrient access, as well as the nature of the entire ecosystem, density and structure of plants in the neighbourhood, predation and harvesting by heterotrophs including man. The growth in biomass would be exponential if optimal conditions for photosynthetic production could be maintained indefinitely. Actual growth is stopped when one or more of the conditions can no longer be met; the factor(s) responsible for halting the exponential growth is(are) called the limiting factor(s). A mature ecosystem may reach a stable situation in which the net production (and hence growth) of the community as a whole is zero.

Limiting factors

The effective area of a plant which is exposed to solar radiation depends on shadow effects caused by other vegetation, on the shape and arrangements of leaves or more generally of light-absorbing pigments. This, together with the loss caused by scattering and absorption on the leaf surface or other parts of the plant not involved in the light accumulation, accounts for the maximum geometrical efficiency in (3.3.44) being around 0.8. The effective area of fully developed terrestrial plant systems (forests, erect-leaved cereal crops) is large enough to intercept practically all incident radiation (Loomis and Gerakis, 1975), and most of the losses under such circumstances are due to reflection.

The availability of solar radiation itself is, of course, of basic importance. The geographical distribution of radiation has been discussed in connection with, section 3.3.1. The corresponding distributions of productivity for typical coastal and oceanic regions of the North Atlantic are shown in Fig. 3.3-74. Total autotrophic plusheterotrophic biomass at these sites was found to be 167×10^3 J m^{-2} (coastal region) and 8×10^3 J m^{-2} (open ocean) (Odum, 1972).

Too high a solar intensity may diminish efficiency, if the chlorophyll pigments are unable to absorb it all or if the radiation destroys the chlorophyll. Also, the state of the photosynthetic systems as well as of the further energy conversion and transport system of the plant may play a role. Efficiency may be lowered by damage or ageing of such components (Wassink, 1975).

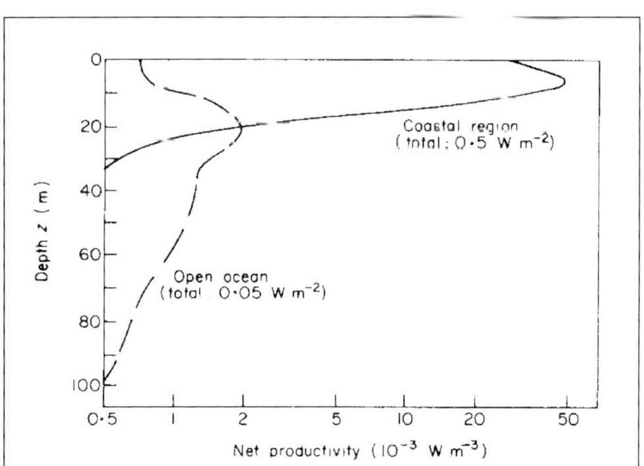

Figure 3.3-74 Net primary production as function of depth for two locations in the North Atlantic. The corresponding average biomasses (standing crops) are 167 kJ m^{-2} (coastal region) and 8 kJ m^{-2} (open ocean) (based on Odum, 1972).

If changes do not occur too rapidly, many plants are able to adapt to external conditions of light, humidity etc., to a certain extent. The pathway of carbon synthesis described in section 3.3.6.1 involves a molecule (3.3.40) with three carbon atoms (C_3-pathway). Some plants are able to follow a different pathway, in which four carbon atoms are synthesised (C_4-pathway), notably as malic acid ($HO_2CCH{:}CHCO_2H$). In this case less CO_2 is needed in order to synthesise e.g. sugar, so that plant growth may occur more rapidly than for the C_3-plants, at least in some environments.

The temperature plays an important role, both the temperature of air and soil for terrestial plants and the temperature of water for aquatic plants. The life cycles of aquatic plankton and algae are of relatively short duration, and for some species the biomass production is confined to a few weeks of the year. Many terrestrial plants are dormant during cold periods, and they are able to withstand low temperatures during winter. If, however, sudden frost occurs in the middle of a growth period, the tolerance is often low and damage may result. The number of days during the year with temperatures above certain values, depending on species, may be defined as the growing season, a concept which has been used extensively in estimates of average properties of productivity. A phenomenological treatment of the influence of thermal conditions on plant and animal life may be found in the work of Budyko (1974).

Figure 3.3-75 shows the trends of gross primary production of an evergreen oak forest, along with temperature and average solar radiation. The dashed line gives the net production of the community as a whole (trees and canopy), i.e. the surplus or growth rate, which is seen to be high most of the year except in summer and with peaks in spring and autumn. This is an example of a growing ecosystem. If the intervention of man can be achieved without changing the trophic relationships within the community, it would appear that a harvest (energy extraction) corresponding to the net production could be sustained, leaving a stable ecosystem with no growth. In this way man would enter as an additional loop in the system depicted in Fig. 3.3-73, and the ecosystem would provide a truly renewable energy source.

When the standing crop biomass is small, only a small fraction of the solar radiation is intercepted. Thus, the productivity initially increases with increasing biomass until a turning point of optimal growth conditions is reached, as seen in Fig. 3.3-76. After this point, the productivity slowly decreases towards equilibrium (zero).

Water flows through all plants at average rates much higher than the rate at which water is dissociated by the first photosynthetic step (3.3.36). Extra water is required for the oxidation of the manganese enzyme involved in this process, as well as for the cyclic ADT–ATP

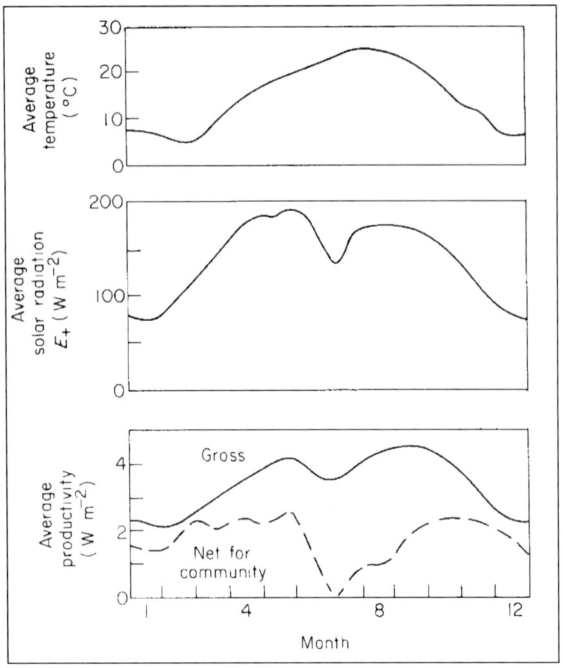

Figure 3.3-75 Trends of average temperature, solar radiation, gross and net production of biomass for an evergreen oak forest community at Minamata, Kyushu, Japan. The curves have been constructed from data collected in 1971/1972, and the conversion factor 1 kg dry matter = 18.8 × 10^6 J has been used (based on Kira, 1975).

and NADP–NADPH$_2$ processes, but these amounts of water are not consumed. The same is true for other metabolic processes in the plant, and water also performs physical tasks such as keeping the surfaces of membranes wet and transporting dissolved substances.

The overall photosynthetic process (3.3.41) requires one water molecule for each carbon atom assimilated, or 1.5 kg of water for each 1 kg of carbon or for each 42 × 10^6 J of biomass, corresponding to 2–3 kg of dry matter. The wet weight is typically in the range 4–10 times the dry weight, implying that the total water content of the plant is at least 4 times the amount fixed by photosynthesis and typically much higher.

The transport of water through a terrestrial plant is achieved by some kind of free-energy gradient, often referred to as the "water potential" [cf. (3.49) and e.g. Canny (1977)]. The corresponding force may thus be of a more general form than that associated with osmotic pressure. Loss of water by transpiration can be viewed as a local decrease in water potential near the area where water is leaving the plant. This means that a potential gradient has been established, which will attract water from other parts of the plant. Thereby the potential gradient moves through the plant and builds up in the root system, which then attracts water from the surrounding soil. If the water that can be drawn from the soil is insufficient, the plant will try to decrease its transpiration by contracting the stomata (openings at leaf surfaces). If the water deficit produces a strong and prolonged stress on the plant, the stomata will become tightly closed, in which case a subsequent situation of water saturation does not immediately lead to a complete resumption of the rate of photosynthesis at the previous level (as illustrated in Fig. 3.3-77).

The turnover of water associated with a biomass production of 1 kg of dry matter is quoted as lying in the range 200–900 kg of er, with values around 500 kg being typical for grass and cereal crops (Geiger, 1961; cf. also Penman, 1970, which quotes 400 kg for food crops).

As an upper limit on the flow of water available to terrestrial plants, one may take the total land precipitation rate of 1.1 × 10^{17} kg of water per year

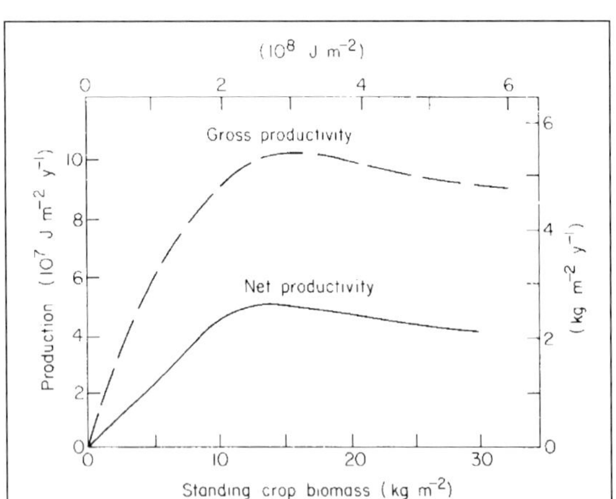

Figure 3.3-76 Trends of gross and net productivity as functions of standing crop biomass for a fir species (*Abies sachalinensis*) (based on Kira, 1975).

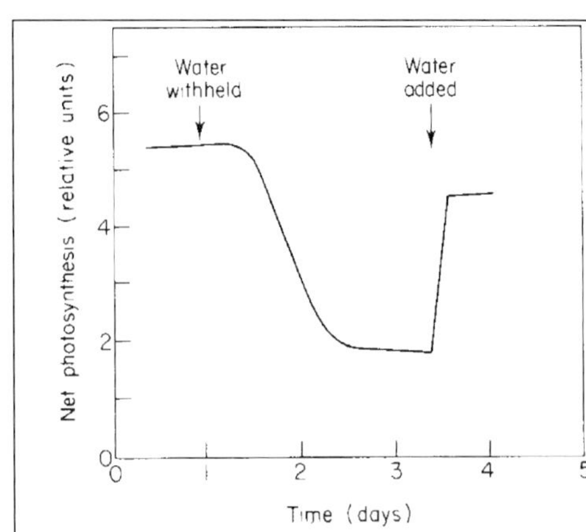

Figure 3.3-77 Effect on productivity of withholding water for 2.5 days. The experiment was performed on sunflower leaves under conditions of generous light (based on Slavik, 1975).

(Fig. 2.61). Assuming the requirement of the plants to be 500 kg of water per kg of dry matter per year, one deduces a maximum sustainable production of 2.2×10^{14} kg of dry matter per year, or in energy units 4.0×10^{21} J y^{-1} (1.3×10^{14} W).

Since the total terrestrial gross primary production is presently 7.6×10^{13} W (Odum, 1972), and the net primary production may amount to about half this value, it is clear that a large fraction of the evapotranspiration (6.7×10^{16} kg of water per year) over land is due to plants. Accordingly, a substantial increase in terrestrial plant production would have to involve increased recycling of water on time scales shorter than one year. One might consider achieving this by artificial irrigation, using some of the water run-off from land areas (3.9×10^{16} kg of water per year). However, about 70% of the water diverted to food crops is not immediately returned to the run-off streams. In view of the many other important functions performed by the rivers and streams responsible for the run-off, this suggests that there may be little room for expanding terrestrial biomass production on a world-wide scale if fresh-water resources alone are used for irrigation.

It would at least appear that schemes for ensuring quick recycling of the water added are as important as expansion of the irrigation potential and that really significant increases in biomass production must involve oceanic water resources, either by marine biomass production or by use of (eventually distilled) sea water for irrigation. Some terrestrial plants actually accept sea water irrigation, with some reduction in conversion efficiency (Epstein and Norlyn, 1977).

If the availability of CO_2 is the only limiting factor, it is reasonable to assume that the productivity is proportional to the amount of CO_2 entering the plant. The concentration X of CO_2 at the photosynthetic sites may be related to the atmospheric CO_2-concentration X_{atm}, e.g. by a linear relation

$$X = f X_{atm},$$

where f depends on properties of the plant (inside pressure, opening of stomata, etc.), as well as on the CO_2 eddy diffusion parameter in the air surrounding the plant. The atmospheric concentration, X_{atm}, decreases with height when no photosynthesis takes place (night condition). However, during daytime in low wind conditions, the CO_2 concentration exhibits a minimum half way through the canopy of vegetation (Saeki, 1975).

If the proportionality factor between production P and CO_2 internal concentration X is g, when no other limiting factors are encountered, then the rate of primary production as function of incident radiation, E_+, may be written

$$\frac{1}{P} = \frac{1}{\eta' E_+} + \frac{1}{gX'} \quad (3.3.47)$$

where η' (3.3.44) is the efficiency at low incident radiation, when light is the only limiting factor. In (3.3.47), temperature requirements are supposed to be satisfied, and the water and nutrient supply are supposed to be adequate.

The state of the photosynthetic system within the plant may give rise to different efficiencies η_{chem} under different conditions. Basically, this can be understood in terms of competing pathways for the chain of molecular reactions involved. In the electron transport chain, shown in Fig. 3.3-71, the Q to plastoquinone electron transport in photosystem II may receive competition by fluorescent de-excitation of the Q-molecule. A similar de-excitation may happen for the X-protein in photosystem I. In both cases, the energy captured by the chlorophyll pigments will become lost from the point of view of biomass production. This may be summarised by stating that under conditions which are not optimal more than 8 light quanta are required for each CO_2 molecule assimilated (the excess energy being lost as heat or radiation).

In aquatic environments, the rate of photosynthesis may depend on such factors as salinity and pH-value (acidity). In Sweden, it has been suggested that decreased productivity of coniferous forests is associated with increasing acidity of precipitation (and hence of soil) (Swedish Ministries of Foreign Affairs and of Agriculture, 1971).

Last, but not least, the role of nutrients as a limiting factor should be mentioned. Deprivation of nutrients decreases and eventually halts production. Transfer of plants to nutrient-free environments has shown effects of decreasing magnitude resulting from the deprivation of nitrogen, phosphorus and potassium in that order.

It follows that an average of 1.4×10^{-3} kg of nitrogen is fixed for each kg of carbon fixed in the terrestrial biosphere, in which there is about 0.0343 kg of N (kg of C)$^{-1}$ (the standing crop biomass shows a different relationship between N and C, owing to different turnover times). The amount of nitrogen with which the soil must be enriched in order to increase productivity, if N is a limiting factor, may be much larger, owing to the nature of the pathways of uptake. Delwicke (1970) quotes experiments in which 0.76 kg of N was needed in order to increase production by 1 kg of dry matter.

In evaluating the net energy yield of cultivated land or sea, non-solar energy subsidies must also be considered. These may be in the form of machinery and fuels for running the machinery (ploughing and harvesting tools). In present day practices of terrestrial agriculture, the largest energy subsidy in areas of intense farming is in the manufacture of fertilisers. In regions demanding irrigation, water pumping may expend substantial amounts of energy, whereas the amount of machinery employed typically is large only in regions of intense farming, so that the corresponding energy subsidy remains a fraction of that spent on fertiliser. If little machinery is used, more man-power is needed, combined with power provided by draught animals. Also, transport of harvested crops costs energy, particularly for remote fishing, and in the case of food crops, processing and packaging may in industrialised regions account for the largest part of the total energy subsidy. Food derived from animals also involves less efficient utilisation of primary energy, as discussed in connection with Fig. 3.3-73.

In the case of "energy crops", the size of energy subsidies in fertilisers may be of particular interest, in order to determine whether the extra primary production, i.e. solar energy utilisation, exceeds the energy subsidy. Using present-day manufacturing techniques, including the mining of phosphate rock, the energy needed to produce 1 kg of nitrogen in the form of fertiliser is about 10^8 J, the energy needed for 1 kg of phosphorus is about 1.4×10^7 J, and the energy for 1 kg of potassium is about 9×10^6 J (Steinhart and Steinhart, 1974; Blaxter, 1974). As an example of fertiliser usage in intense farming practice, the average use of fertiliser in Denmark in 1965–1966 amounted to 0.0114 kg of N per m^2, 0.003 58 kg of P per m^2 and 0.0109 kg of K per m^2 (Danish Statistical Survey, 1968), which with the above conversion figures correspond to a total subsidy of 1.3×10^6 J m^{-2} y^{-1} (0.04 W m^{-2}). The harvested yield averaged 0.5 kg of dry matter per m^2 and year. For the most common crop (barley), the net primary production would be about twice the amount harvested or 2×10^7 J m^{-2} y^{-1}, and the fertiliser energy subsidy would thus be 6.5%. By 1971, the fertiliser energy consumption had risen by 23%, but the harvest yield was unchanged per unit area (Danish Statistical Survey, 1972). This implies that, because of this farming practice, nutrients in a form suitable for uptake are being depleted from the soil in such a way that increasing amounts of fertiliser must be added in order to maintain a constant productivity.

A proposed scheme for "energy plantations" yielding biomass at the rate of about 15×10^7 J m^{-2} y^{-1} (e.g. based on sunflower) under (southern) US conditions anticipates the use of 4×10^6 J m^{-2} y^{-1} for fertilisers and 2×10^6 J m^{-2} y^{-1} for irrigation and other machinery, a total energy subsidy of 5% (Alich and Inman, 1976).

Productivity data

Data on productivity of different species under different climatic conditions are plentiful, but often hard to summarise because of the differences in techniques used and ways of representing the results. Most controlled experiments are performed under optimal conditions, and they do not reflect the average productivity of the particular plant and geographical region. On the other hand, statistical information for entire regions or countries does not usually convey the exact conditions of growth. In many cases, the productivity of a given crop depends on the history of previous uses of the land. For natural ecosystems there are other factors which make it difficult to compare data from different sources. The conditions are not always sufficiently well described by climatic zone and solar radiation data. For instance, the geology of the upper soil may vary over very short distances. Altogether, it is not reasonable to give more than broad ranges or limits inside which the productivity of a given plant in a given environment will normally lie, not excluding exceptional values outside the range given.

In order to partially eliminate the dependence on the amount of solar radiation, Fig. 3.3-78 indicates such ranges for the overall efficiency (3.3.45), representing the ratio between net primary production and total solar radiation. In this way, for example, the woods in different climatic regions (boreal coniferous, deciduous, temperate broad-leaved evergreen and subtropical or tropical rain forests) become represented by a fairly narrow range of efficiencies, despite the large variations in absolute productivity.

The range of efficiencies of natural plants goes from practically nothing to a little over 2%, the highest values being reached for tropical rain forests and algal cultures of coral reefs. Cultivated crops (terrestrial or marine) may reach some 4–5% under optimal conditions and nutrient subsidy. The optimum efficiency from the theoretical discussion of equations (3.3.44) and (3.3.45) is roughly

$$\eta = \eta_\lambda \, \eta_{geom} \, \eta_{chem} \, \eta_{resp} \cdot\cdot 0.5 \times 0.8 \times 0.4 \times 0.6 \\ = 0.1. \qquad (3.3.48)$$

Each of the conditions is difficult to achieve in practice, at least for extended periods. As mentioned earlier, mature ecosystems are characterised by diversity, which may not be reconcilable with maximum efficiency of primary production, and young ecosystems, such as non-perennial crops, need a growing period (cf. Fig. 3.3-76) before they reach full productivity, and thus their average efficiency on an area basis and over prolonged periods of time cannot be optimal.

Figure 3.3-79 gives a recent estimate of the geographical distribution of potential biomass production. These are net primary production data derived from the "Terrestrial Ecosystem Model (TEM)" of the Woods Hole group (Melillo and Helfrich, 1998; Raich *et al.*, 1991; Melillo *et al.*, 1993; McGuire *et al.*, 1997; Sellers

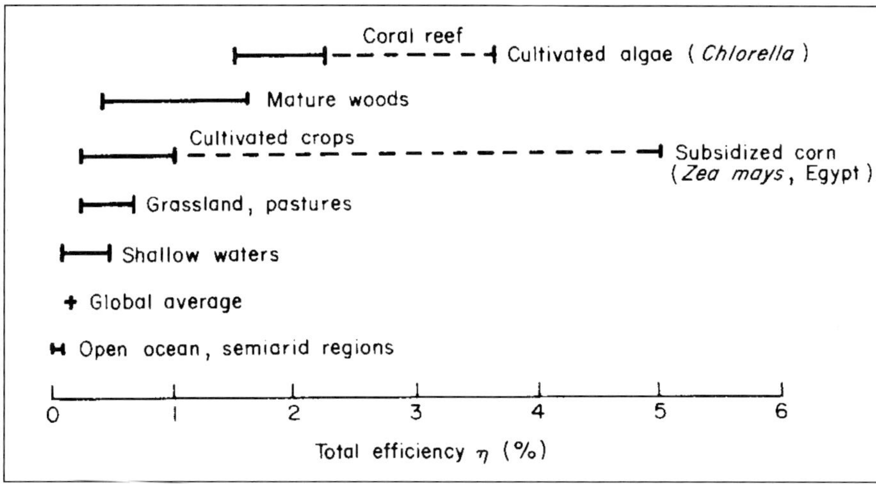

Figure 3.3-78 Intervals of typical values of total photosynthetic efficiency for different plants and communities (solid lines), as well as intervals of possible improvement for subsidised and optimised cultures (dashed lines). For non-perennial plants, the efficiency is an average over the growing season (constructed on the basis of information from Kira, 1975; Caldwell, 1975; Loomis and Gerakis, 1975).

et al., 1997). The assumption is that a mature ecosystem of natural vegetation has developed, and the model takes into account solar radiation, moisture, temperature, as well as access to water and nutrients. Not included here is global warming (increased CO_2), which could induce increased primary production in a fairly complex pattern and change the borders of natural vegetation zones, sometimes by several hundred kilometres (IPCC,

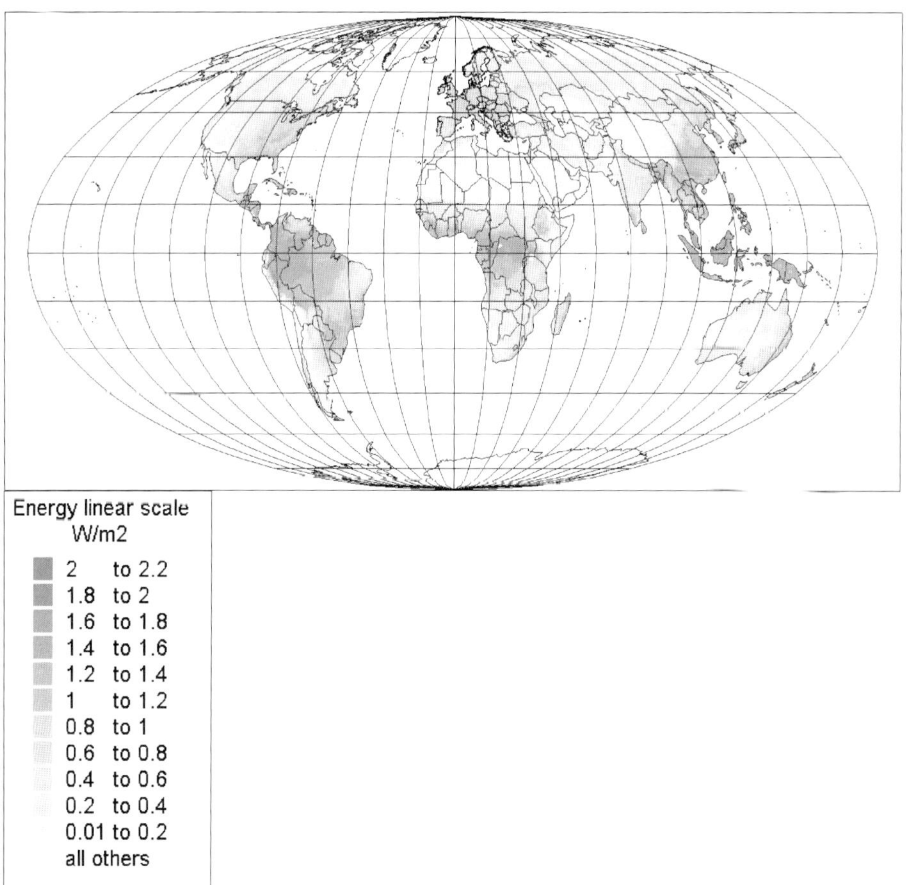

Figure 3.3-79 Annual average energy content in W m^{-2} of potential net biomass production in mature ecosystems (based on Melillo and Helfrich, 1998).

1996b). Use of these biomass data for energy scenarios is illustrated elsewhere. Seasonal variations of actual biomass land coverage, as deduced from satellite-based sensors, are regularly posted on the Internet (NOAA, 1998).

3.3.6 Other energy sources

The preceding parts of this chapter have been concerned with renewable energy flows and stores based on solar radiation, gravitation between celestial bodies, and the mechanical energy involved in atmospheric and oceanic circulation, as well as a number of (sensible or latent) heat sources, the origin of which were solar or geological. It is believed that the most important such sources have been listed, but there could among those omitted be some that still have a certain amount of significance.

In the description of heat sources, in particular, the list is far from complete, considering that any difference in temperature constitutes a potential source of energy. It may not be practical to attempt to extract energy from the difference in air or soil temperature between the arctic and equatorial regions, but the corresponding difference in water temperature is, in fact, the basis for the temperature gradients in some oceanic regions (discussed in section 3.3.5.1), owing to the water transport by currents. One case in which a stable temperature difference might be usable in ways similar to those proposed for ocean temperature gradients is offered by the difference between the arctic air temperature and the water temperature at the base of the ice sheets covering arctic oceans. The average air temperature is about −25°C for the period October–March, but the temperature at the ice–water interface is necessarily close to 0°C (although not quite zero because of salinity- and pressure-induced depreciation of the freezing point, cf. Fig. 3.3-66).

It is natural to ask whether other energy forms present in the regime of the Earth, such as electromagnetic, earthquake, chemical (of which bio-energy has been treated in section 3.3.6) or nuclear energy, can be considered as energy resources and to what extent they are renewable. The rest of this section will briefly deal with examples of such energy forms.

3.3.6.1 Atmospheric electricity

The Earth is surrounded by a magnetic field, and from a height of about 80 km a substantial fraction of atmospheric gases is in an ionised state. It is thus clear that electromagnetic forces play an important role in the upper atmosphere. Manifestations of electromagnetic energy in the lower parts of the atmosphere are well

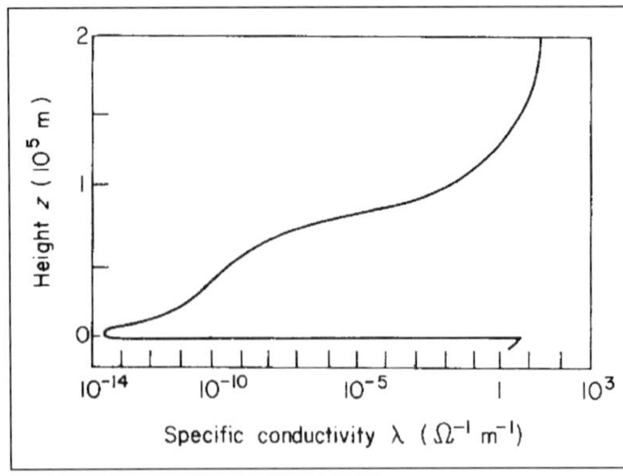

Figure 3.3-80 Electrical conductivity of the atmosphere at noon on a clear day as a function of height (at about 10^5 m height, a substantial night-time reduction in λ takes place) (based on Webb, 1976).

known in the form of lightning. Speculations on the possibility of extracting some of the electrical energy contained in a thunderstorm have appeared. An estimation of the amounts of energy involved requires knowledge of the electrical properties of the atmosphere.

In Fig. 3.3-80, calculated values of the electrical conductivity of the atmosphere are shown as a function of height for the lowest 200 km (Webb, 1976). It is shown that the conductivity is low in the troposphere, increases through the stratosphere and increases very sharply upon entering the ionosphere. Strong horizontal current systems are found in the region of high conductivity at a height of around 100 km (the "dynamo currents"). Vertical currents directed towards the ground are 6–7 orders of magnitude smaller for situations of fair weather, averaging about 10^{-12} A m^{-2} in the troposphere. Winds may play a role in generating the horizontal currents, but the energy input for the strong dynamo currents observed at mid-latitudes is believed to derive mainly from absorption of ultraviolet solar radiation by ozone (Webb, 1976).

The fair weather downward current of 10^{-12} A m^{-2} builds up a negative charge on the surface of the Earth, the total value of which on average is around 10^5 C. Locally, in the thunderstorm regions, the current is reversed and much more intense. Together these currents constitute a closed circuit, the flow through which is 500–1000 A, but in such a way that the downward path is dispersed over most of the atmosphere, while the return current in the form of lightning is concentrated in a few regions (apart from being intermittent, causing time variations in the charge at the Earth's surface). The power P associated with the downward current may be found from

$$P = \int I^2 \lambda^{-1} \, dz,$$

where I ($\approx 10^{-12}$ A m^{-2}) is the average current and $\lambda(z)$ is the conductivity function (e.g. Fig. 3.3-80 [at noon]), or more simply from

$$P = IV = I\,Q/C,$$

where C is the capacitance of the Earth relative to infinity (V the corresponding potential difference) and Q ($\approx 10^5$ C) is its charge. The capacitance of a sphere with the Earth's radius r_s (6.4×10^6 m) is $C = 4\pi\varepsilon_0 r_s = 7 \times 10^{-4}$ F (ε_0 being the dielectric constant for vacuum), and thus the average power becomes $P \approx 6 \times 10^{-5}$ W m^{-2}.

The energy stored in the charge Q of the Earth's surface, relative to a situation in which the charges were all moved to infinity, is

$$W_{\infty}^{electric} = \tfrac{1}{2}\,Q^2/C \approx 3 \times 10^{12}\ \text{J}.$$

In "practice", the charges could only be moved up to the ionosphere (potential difference less than 10^6 V), reducing the value of the stored energy by almost two orders of magnitude. In any case, the charge of the Earth, as well as the trophospheric current system, constitutes energy sources of very small magnitude, even if the more concentrated return flux of lightning could be utilised (multiplying the average power estimate by the area of the Earth, one obtains a total power of 3×10^{10} W).

3.3.6.2 Salinity differences

Useful chemical energy may be defined as energy that can be released through exothermic chemical reactions. In general, chemical energy is associated with chemical bindings of electrons, atoms and molecules. The bindings may involve overlapping electron wavefunctions of two or more atoms, attraction between ionised atoms or molecules, and long-range electromagnetic fields created by the motion of charged particles (notably electrons). In all cases, the physical interaction involved is the Coulomb force. Examples of chemical energy connected with molecular binding structure have been given in section 3.3.6 (bio-energy sources, including fossil fuels).

The organisation of atoms or molecules in regular lattice structures represents another manifestation of chemical bindings. Some substances possess different crystalline forms, which may exist under given external conditions. In addition to the possibility of different solid phases, phase changes associated with transitions among solid, liquid and gas phases all represent different levels of latent energy. Examples in which such latent energy differences have been considered as potential energy sources have been mentioned, for example, in section 3.3.4.2.

Solutions represent another form of chemical energy, relative to the pure solvent. The free energy of a substance with components $i = 1,2,\ldots$, there being n_i mol of the ith component, may be written

$$G = \sum_i n_i \mu_i, \tag{3.3.49}$$

where μ_i is called the "chemical potential" of component i. For a solution, μ_i can be expressed in the form (see e.g. Maron and Prutton, 1959)

$$\mu_i = \mu_i^0 + \mathscr{R}\,T \log (f_i x_i), \tag{3.3.50}$$

where $\mathscr{R}$ is the gas constant (8.3 J K^{-1} mol^{-1}), T is the temperature (K) and $x_i = n_i/(\Sigma_j n_j)$ the mole fraction. μ_i^0 is the chemical potential that would correspond to $x_i = 1$ at the given pressure P and temperature T, and f_i is the "activity coefficient", an empirical constant which approaches unity for "ideal solutions", an example of which is the solvent of a very dilute solution (whereas, in general, f_i cannot be expected to approach unity for the dissolved component of a dilute solution).

It follows from (3.3.49) and (3.3.50) that a solution represents a lower chemical energy than does the pure solvent. The most common solution present in large amounts on the Earth is saline ocean water. Relative to this, pure or fresh water such as river run-off represent an elevated energy level. In addition, there are salinity differences within the oceans.

Taking the average ocean salinity as about 33×10^{-3} (mass fraction), and regarding this entirely as ionised NaCl, $n_{Na+} = n_{Cl-}$ becomes about 0.56×10^3 mol and $n_{water} = 53.7 \times 10^3$ mol, considering a volume of 1 m^3. The chemical potential of ocean water, μ, relative to that of fresh water, μ^0, is then from (3.3.50)

$$\mu - \mu^0 = \mathscr{R}\,T \log x_{water} \approx -2\,\mathscr{R}\,T n_{Na^+}/n_{water}.$$

Consider now a membrane which is permeable for pure water but impermeable for salt (i.e. for Na^+ and Cl^- ions) as indicated in Fig. 3.3-81. On one side of the membrane, there is pure (fresh) water, on the other side saline (ocean) water. Fresh water will flow through the membrane, trying to equalise the chemical potentials μ^0 and μ initially prevailing on each side. If the ocean can be considered as infinite and being rapidly mixed, then n_{Na^+} will remain fixed, also in the vicinity of the membrane. In this case each m^3 of fresh water penetrating the membrane and becoming mixed will release an amount of energy, which from (3.3.49) is

$$\delta G = \sum_i (n_i\,\delta\mu_i + \mu_i\,\delta n_i) \approx n_{water}\,(\mu^0 - \mu) \approx 2\,\mathscr{R}\,T n_{Na^+}. \tag{3.3.51}$$

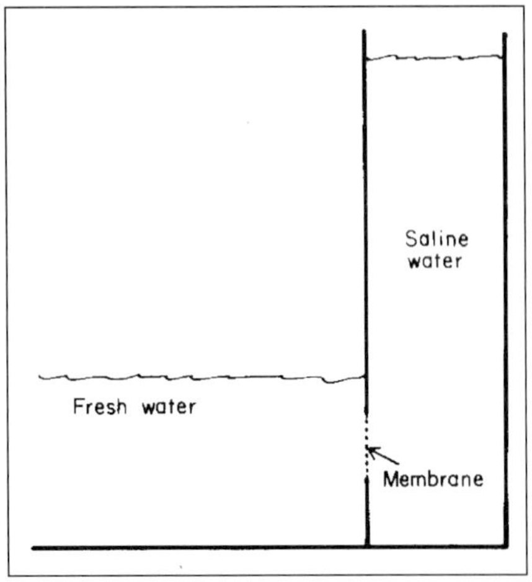

Figure 3.3-81 Schematic picture of an osmotic pump. In order to mix the fresh water penetrating the semi-permeable membrane in the direction towards the right and to maintain the salinity in the salt water compartment, new saline water would have to be pumped into the salt water compartment, and water motion near the membrane would have to be ensured.

For a temperature $T \approx 285$ K (considered fixed), $\delta G \approx 2.65 \times 10^6$ J. The power corresponding to a fresh-water flow of 1 $m^3 s^{-1}$ is thus 2.65×10^6 W (cf. Norman, 1974). The world-wide run-off of about 4×10^{13} $m^3 y^{-1}$ would thus correspond to an average power of around 3×10^{12} W.

The arrangement schematically shown in Fig. 3.3-81 is called an osmotic pump. The flow of pure water into the tube will ideally raise the water level in the tube, until the pressure of the water head balances the force derived from the difference in chemical energy. The magnitude of this "osmotic pressure", P^{osm}, relative to the atmospheric pressure P_0 on the fresh water surface, is found from the thermodynamic relation

$$V\, dP - S\, dT = \sum\nolimits_i n_i\, d\mu_i,$$

where V is the volume, S is the entropy and T is the temperature. Assuming that the process will not change the temperature (i.e. considering the ocean a large reservoir of fixed temperature), insertion of (3.3.51) yields

$$P^{osm} = \delta P \approx n_{water}\, V^{-1}\, \delta\mu_{water} \approx 2\mathcal{R}\, T\, n_{Na^+}\, V^{-1}. \qquad (3.3.52)$$

Inserting the numerical values of the example above, $P^{osm} = 2.65 \times 10^6$ N m^{-2}, corresponding to a water-head some 250 m above the fresh water surface. If the assumption of fixed mole fraction of salt in the tube is to be realised, it would presumably be necessary to pump saline water into the tube. The energy spent for pumping, however, would be mostly recoverable, since it also adds to the height of the water-head, which may be used to generate electricity as in a hydropower plant.

An alternative way of releasing the free energy difference between solutions and pure solvents is possible when the dissolved substance is ionised (the solution is then called electrolytic). In this case direct conversion to electricity is possible, as further discussed in Chapter 4.1.

3.3.6.3 Nuclear energy

The atomic nuclei (consisting of protons and neutrons) carry the bulk of the mass of matter on Earth as well as in the known part of the present universe. By a nucleus is usually understood a bound system of Z protons and N neutrons (unbound systems, resonances, may be formed under laboratory conditions and they are observed in cosmic ray showers for short periods of time). Such a nucleus contains an amount of nuclear binding energy, given by

$$E_{Z,N} - (NM_n + ZM_p)\, c^2 = -B,$$

with the difference between the actual energy $E_{Z,N}$ of the bound system and the energy corresponding to the sum of the masses of the protons and neutrons if these were separated from each other. It thus costs energy to separate all the nucleons, and this is due to the attractive nature of most nuclear forces.

However, if the binding energy B (i.e. the above energy difference with opposite sign) is evaluated per nucleon, one obtains a maximum around ^{56}Fe with lower values for both lighter and heavier nuclei. Figure 3.3-82 shows the trends of $-B/A$, where $A = Z + N$ is the nucleon number. For each A, only the most tightly bound nucleus $(Z, A - Z)$ has been included, and only doubly even nuclei have been included (if Z or N is odd, the binding energy is about 1 MeV lower).

This implies that nuclear material away from the iron region could gain binding energy (and thereby release nuclear energy) if the protons and neutrons could be re-structured to form ^{56}Fe. The reason why this does not happen spontaneously for all matter on Earth is that potential barriers separate the present states of nuclei from that of the most stable nucleus and that very few nuclei are able to penetrate these barriers at the temperatures prevailing on Earth.

A few nuclei do spontaneously transform to more tightly bound systems (the natural radioactive nuclei mentioned in section 3.3.5.2), but the rate at which they penetrate the corresponding "barriers" is low, since otherwise they would no longer be present on Earth now, some 5×10^9 years after they were formed. As mentioned in section 3.3.5.2, these nuclei are responsible for 40% of the average heat flow at the surface of continents

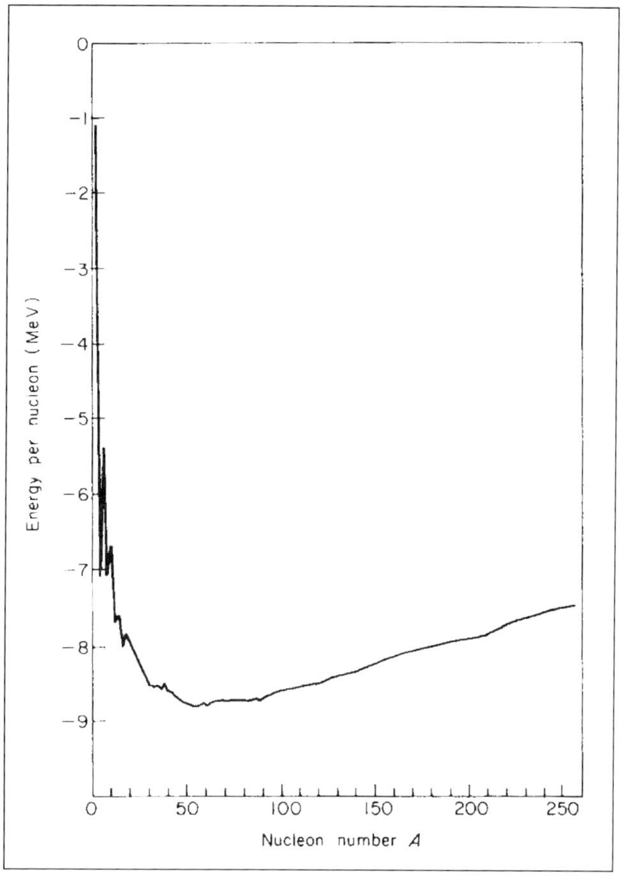

Figure 3.3-82 Trends of nuclear binding energy of bound nuclei (taken per nucleon) relative to a hypothetical situation in which all nucleons are separated (see text) (based on Bohr and Mottelson, 1969).

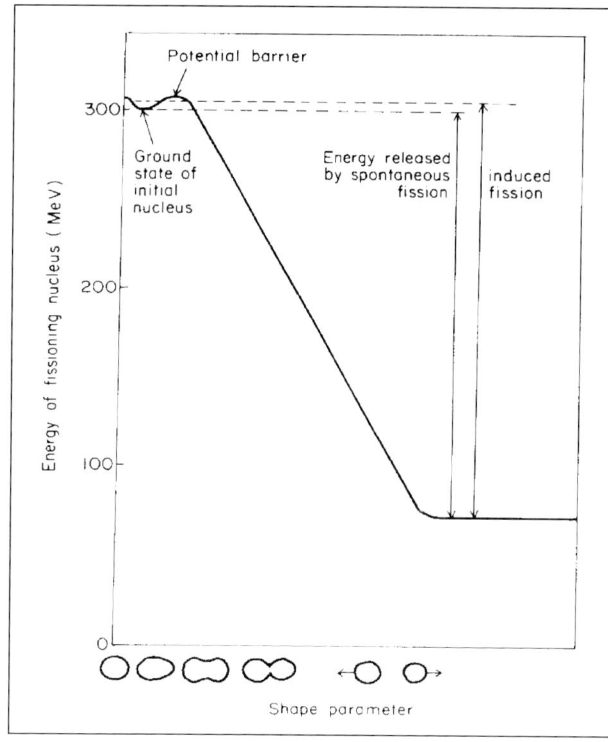

Figure 3.3-83 Schematic view of a nuclear fission process described by a "one-dimensional" shape parameter (axially symmetric shapes are indicated along the lower boundary of the figure). The zero point of the energy scale on the ordinate corresponds to the energy of a (fractional) number of ^{56}Fe nuclei with mass identical to that of the fissioning nucleus.

and contribute almost nothing to the heat flow at the ocean floors.

As schematically illustrated in Fig. 3.3-83, the barrier which a nucleus of $A \sim 240$ must penetrate in order to fission is typically a very small fraction (2–3%) of the energy released by the fission process. This barrier has to be penetrated by quantum tunnelling. The width of the barrier depends on the state of the initial nucleus. It is smaller if the nucleus is in an excited state rather than in its ground state. Some heavy isotopes, such as ^{235}U, easily absorb neutrons, forming a compound system with dramatically reduced fission barrier and hence with dramatically increased fission probability. This process is called "induced fission", and it implies that by adding a very small amount of energy to a "fissile" element such as ^{235}U, by slow-neutron bombardment, a fission energy of some 200 MeV can be released, mostly in the form of kinetic energy of the products, but a few per cent usually occurring as delayed radioactivity of unstable fragments. An example of binary fission, i.e. with two end-nuclei plus a number of excess neutrons (which can induce further fission reactions), is

$$^{235}_{92}U + n \rightarrow ^{142}_{54}Xe + ^{92}_{38}Sr + 2n.$$

The probability of finding an asymmetrical mass distribution of the fragments is often larger than the probability of symmetric fission. The reason why further fission processes yielding nuclei in the $^{56}{}_{26}Fe$ region do not occur is the high barrier against fission for nuclei with $Z^2/A \cdot\cdot 30$ and the low probability of direct fission of, say, uranium into three fragments closer to the iron region (plus a larger number of neutrons).

The amount of recoverable fissile material in the Earth's crust would correspond to a fission energy of about 10^{22} J (Ion, 1975). However, heavy elements other than those fissioning under slow-neutron bombardment can be made fissile by bombardment with more energetic neutrons or some other suitable nuclear reaction, spending an amount of energy much smaller than that which may later be released by fission. An example is $^{238}{}_{92}U$, which may absorb a neutron and emit two electrons, thus forming $^{239}{}_{94}Pu$, which is a "fissile" element, capable of absorbing neutrons to form $^{240}{}_{94}Pu$, the fission cross section of which is appreciable. Including resources by which fissile material may be "bred" is this way, a higher value may be attributed to the recoverable energy resource associated with nuclear fission. This value depends both on the availability of resource material such as ^{238}U (the isotope ratio ^{238}U to ^{235}U is practically the same for all geological formations of terrestrial origin) and on the possibility of constructing devices with a high "breeding ratio". Some estimates indicate a 60-fold increase over the energy value of the "non-breeder" resource (World Energy Conference, 1974).

As indicated in Fig. 3.3-82, the energy gain per nucleon which could be released by fusion of light elements is several times larger than that released by fission of heavy elements. A number of possible fusion processes have been mentioned in Chapter 2 of *Renewable Energy* (2004). These reactions take place in stars at high temperature. On Earth, they have been demonstrated in the form of explosive weapons, with the temperature and pressure requirements being provided by explosive fission reactions in a blanket around the elements to undergo fusion. Controlled energy production by fusion reactions such as $^2H+^3H$ or $^2H+^2H$ is under investigation, but at present the necessary temperatures and confinement requirements have not been met. Theoretically, the fusion processes of elements with mass below the iron region are, of course, highly exenergetic, and energy must be added only to ensure a sufficient collision frequency. In practice, it may take a while to realise devices for energy extraction for which the net energy gain is at all positive.

In theory, the nuclear fusion energy resources are much larger than the fission resources, but in any case only a fraction of the total nuclear energy on Earth (relative to an all-iron state) could ever become energy resources on a habitable planet. In principle, the nuclear energy resources are clearly non-renewable, and the recoverable amounts of fissionable elements do not appear large compared with the possible requirements of man during the kind of time span for which he may hope to inhabit planet Earth. However, it cannot be denied that the fusion resources might sustain man's energy expenditure for a long time, if, for example, the reaction chain entirely based on naturally occurring deuterium could be realised [cf. the cosmological reactions occurring some 300 s after the singularity in the Big Bang theory at temperatures between 10^9 and 10^8 K]:

$$
\begin{aligned}
&{}^2_1\text{H} + {}^2_1\text{H} \rightarrow {}^1_1\text{H} + {}^3_1\text{H} + 3.25\ \text{MeV},\\
&{}^2_1\text{H} + {}^2_1\text{H} \rightarrow {}^1_0n + {}^3_2\text{He} + 4.0\ \text{MeV},\\
&{}^2_1\text{H} + {}^3_1\text{H} \rightarrow {}^1_0n + {}^4_2\text{He} + 17.6\ \text{MeV},\\
&{}^2_1\text{H} + {}^3_2\text{He} \rightarrow {}^1_1\text{H} + {}^4_2\text{He} + 18.3\ \text{MeV}.
\end{aligned}
$$

The abundance of ^2H in sea water is about 34×10^{-6} (mass fraction). The potential nuclear energy released by one of the deuterium-to-helium fusion chains is thus 10^{13} J m^{-3} of sea water, or over 10^{31} J for all the oceans (an energy storage equivalent to the entire thermal energy stored in the interior of the Earth, cf. section 3.3.5.2).

The prime environmental concern over utilisation of fission or fusion energy is the inherent change in the radioactive environment. The fragments formed by fission reactions cover a wide range of nuclear isotopes, most of which are unstable and emit nuclear radiation of gamma or particle type (see e.g. Holdren, 1974). Also, the fusion devices are likely to produce large amounts of radioactivity, because the high-energy particles present in the reaction region may escape and experience frequent collisions with the materials forming the "walls", the confinement, and thereby induce nuclear reactions. The range of radioactive elements formed by fusion reactions can be partially optimised from an environmental point of view (minimising the production of the most biologically hazardous isotopes) by choosing appropriate materials for the confinement, but the choice of materials is also limited by the temperature and stability requirements (see e.g. Post and Ribe, 1974).

Large-scale implementation of fission- or fusion-based energy conversion schemes will raise the question of whether it will be possible safely to manage and confine the radioactive "wastes" that, if released to the general environment, could cause acute accidents and long-range alterations in the radiological environment to which man is presently adapted. These dangers exist equally for the use of nuclear reactions in explosive weapons, in addition to the destructive effect of the explosion itself, and in this case no attempt is even made to confine or control the radioactive material.

In addition to possible health hazards, the use of nuclear energy may also have climatic consequences, e.g. associated with the enhanced heat flow (during nuclear war; cf. Feld, 1976) or with the routine emissions of radioactive material (e.g. ^{85}Kr from the uranium fuel cycle; cf. Boeck, 1976).

In summary, nuclear energy production based on the existing once-through reactors constitutes a parenthesis in history, given the very limited amounts of fissile resources available for this mode of operation, comparable at best to oil reserve availability. Nuclear fusion research has been ongoing for more than 50 years, so far with little success. Commercialisation is still predicted to happen some 50 years into the future, just as it was at any earlier stage.

3.3.7 Suggested topics for discussion

3.3.7.1

Discuss the reflected flux of solar radiation on an inclined plane, in front of which a horizontal mirror of given dimensions has been placed (cf. e.g. Seitel, 1975).

3.3.7.2

Which data on solar radiation, wind, waves, etc. are available in your region (e.g. check with the local meteorological service)?

3.3.7.3

Discuss the cooling of the Earth in terms of simplified models such as (a) a uniform sphere of constant heat capacity and diffusion coefficient, initially at a uniform temperature, (b) adding radiogenic heat production in the crust, by constraining the surface temperature to be the sum of two terms, one entirely due to the heat flow from the interior, and the other term exponentially decaying, such that the drop in 4.5×10^9 years becomes 900 K.

Use present surface heat flow and temperature (average values) as boundary conditions, and discuss the long-range development of the thermal conditions in the interior of the Earth (cf. e.g. Goguel, 1976). Recent models suggest a core temperature of nearly 4000 K (Hofmeister, 1999).

3.3.7.4

Consider a compartment model for biological energy production and transfer within a closed ecosystem of the general form shown in Fig. 3.3-73. Assume, for instance, that the limiting factor governing the rate of primary production is solar radiation, such that the production becomes proportional to a schematic solar input, constant or with only seasonal sinusoidal variations [cf. (3.3.47)]. Think of a food chain, in which you may find data on the rates and biomass levels appearing in Fig. 3.3-73 or some modification of it, and try to determine conditions for stability (no net production of the community as a whole).

If you have a computer at hand, you might try to set up a numerical simulation model for the system, assuming each rate (time derivative of a compartment level) to be a linear combination of all the compartment levels (some coefficients may be zero, of course). In this case you may also be able to follow the growth phase of the system, and attempt to judge whether the parameter values chosen are reasonable (cf. e.g. Odum, 1972, where examples of parameter values may also be found; Patten, 1971,1972; Sørensen, 1975a).

3.3.7.5

Use the current biomass production data (available on the Internet) mentioned in connection with Fig. 3.3-79 to estimate possible bio-energy sources present in your region and their seasonal distribution. Does the seasonality have implications for energy use? Compare your results with the model considered in Chapter 6.2

Section Four

Energy conversion

Chapter 4.1

4.1

Energy conversion processes

4.1.1 Basic principles of energy conversion

A large number of energy conversion processes take place in nature, some of which were described in Sections 2 and 3. Mankind is capable of performing a number of additional energy conversion processes by means of various devices invented during the history of man. Such devices may be classified according to the type of construction used, according to the underlying physical or chemical principle, or according to the forms of energy appearing before and after the action of the device. In this chapter, a survey of conversion methods, which may be suitable for the conversion of renewable energy flows or stored energy, will be given. A discussion of general conversion principles will be made below, followed by an outline of engineering design details for specific energy conversion devices, ordered according to the energy form being converted and the energy form obtained. The collection is necessarily incomplete and involves judgement about the importance of various devices.

4.1.1.1 Conversion between energy forms

For a number of energy forms, Table 4.1-1 lists some examples of energy conversion processes or devices currently in use or contemplated, organised according to the energy form emerging after the conversion. In several cases more than one energy form will emerge as a result of the action of the device, e.g. heat in addition to one of the other energy forms listed. Many devices also perform a number of energy conversion steps, rather than the single ones given in the table. A power plant, for example, may perform the conversion process chain between energy forms: chemical → heat → mechanical → electrical. Diagonal transformations are also possible, such as conversion of mechanical energy into mechanical energy (potential energy of elevated fluid → kinetic energy of flowing fluid → rotational energy of turbine) or of heat into heat at a lower temperature (convection, conduction). A process in which the only change is that heat is transferred from a lower to a higher temperature is forbidden by the second law of thermodynamics. Such transfer can be established if at the same time some high-quality energy is degraded, e.g. by a heat pump (which is listed as a converter of electrical into heat energy in Table 4.1-1, but is further discussed in the heat conversion section, 4.1.6.1).

The efficiency with which a given conversion process can be carried out, i.e. the ratio between the output of the desired energy form and the energy input, depends on the physical and chemical laws governing the process. For the heat engines, which convert heat into work or vice versa, the description of thermodynamic theory may be used in order to avoid the complication of a microscopic description on the molecular level (which is, of course, possible, e.g. on the basis of statistical assumptions). According to thermodynamic theory (again the "second law"), no heat engine can have an efficiency higher than that of a reversible Carnot process, which is depicted in Fig. 4.1-1, in terms of different sets of thermodynamic state variables,

(P, V) = (pressure, volume),
(T, S) = (absolute temperature, entropy),

and

(H, S) = (enthalpy, entropy).

The entropy was defined in (2.1.1), apart from an arbitrary constant fixed by the third law of thermodynamics (Nernst's law), which states that S can be taken as zero

Renewable Energy Focus Handbook 2009; ISBN: 9780123747051

Table 4.1-1 Examples of energy conversion processes listed according to the initial energy form and one particular converted energy form (the one primarily wanted)

Initial energy form	Converted energy form				
	Chemical	**Radiant**	**Electrical**	**Mechanical**	**Heat**
Nuclear					Reactor
Chemical			Fuel cell, battery discharge		Burner, boiler
Radiant	Photolysis		Photovoltaic cell		Absorber
Electrical	Electrolysis, battery charging	Lamp, laser		Electric motor	Resistance, heat pump
Mechanical			Electric generator, MHD	Turbines	Friction, churning
Heat			Thermionic & thermoelectric generators	Thermodynamic engines	Convector, radiator, heat pipe

at zero absolute temperature ($T = 0$). The enthalpy H is defined by

$$H = U + PV, \tag{4.1.1}$$

in terms of P, V and the internal energy U of the system. According to the first law of thermodynamics, U is a state variable given by

$$\Delta U = \int dQ + \int dW, \tag{4.1.2}$$

in terms of the amounts of heat and work added to the system [Q and W are not state variables, and the individual integrals in (4.1.2) depend on the paths of integration]. The equation (4.1.2) determines U up to an arbitrary constant, the zero point of the energy scale. Using the definition (2.1.1),

$$dQ = T\,dS$$

and

$$dW = -P dV,$$

both of which are valid only for reversible processes. The following relations are found among the differentials:

$$dU = T\,dS - P\,dV,$$

$$dH = T\,dS + V\,dP. \tag{4.1.3}$$

These relations are often assumed to have general validity.

If chemical reactions occur in the system, additional terms $\mu_i\,dn_i$, should be added on the right-hand side of both relations (4.1.3), in terms of the chemical potentials, which were discussed briefly in section 3.3.7.2.

For a cyclic process such as the one shown in Fig. 4.1-1, $\int dU = 0$ upon returning to the initial locus in one of the diagrams, and thus according to (4.1.3) $\int T\,dS = \int P\,dV$. This means that the area enclosed by the path of the cyclic process in either the (P, V) or the (T, S) diagram equals the work $-W$ performed by the system

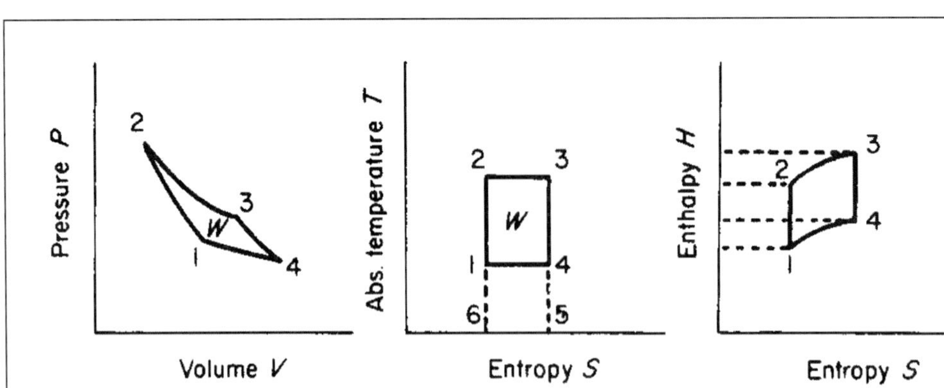

Figure 4.1-1 The cyclic Carnot process in different representations. Traversing the cycle in the direction 1→2→3→4 leads to the conversion of a certain amount of heat into work (see text for details).

during one cycle (in the direction of increasing numbers on Fig. 4.1-1).

The amount of heat added to the system during the isothermal process 2→3 is $\Delta Q_{23} = T(S_3 - S_2)$, if the constant temperature is denoted T. The heat added in the other isothermal process, 4→1, at a temperature T_{ref}, is $\Delta Q_{41} = -T_{ref}(S_3 - S_2)$. It follows from the (T, S) diagram that $\Delta Q_{23} + \Delta Q_{41} = -W$. The efficiency by which the Carnot process converts heat available at temperature T into work, when a reference temperature of T_{ref} is available, is then

$$\eta = \frac{-W}{\Delta Q_{23}} = \frac{T - T_{ref}}{T}. \tag{4.1.4}$$

The Carnot cycle (Fig. 4.1-1) consists of four steps: 1→2, adiabatic compression (no heat exchange with the surroundings, i.e. $dQ = 0$ and $dS = 0$); 2→3, heat drawn reversibly from the surroundings at constant temperature (the amount of heat transfer ΔQ_{23} is given by the area enclosed by the path 2-3-5-6-2 in the (T, S)-diagram); 3→4, adiabatic expansion; and 4→1, heat given away to the surroundings by a reversible process at constant temperature [$|\Delta Q_{41}|$ equal to the area of the path 4-5-6-1-4 in the (T, S)-diagram].

The (H, S)-diagram is an example of a representation in which energy differences can be read directly on the ordinate, rather than being represented by an area.

It requires long periods of time to perform the steps involved in the Carnot cycle in a way that approaches reversibility. As time is important for man (the goal of the energy conversion process being power rather than just an amount of energy), irreversible processes are deliberately introduced into the thermodynamic cycles of actual conversion devices. The thermodynamics of irreversible processes are described below using a practical approximation, which will be referred to in several of the examples to follow.

Irreversible thermodynamics

The degree of irreversibility is measured in terms of the rate of energy dissipation,

$$D = T\, dS/dt, \tag{4.1.5}$$

where dS/dt is the entropy production of the system while held at the constant temperature T (i.e. T may be thought of as the temperature of a large heat reservoir, with which the system is in contact). In order to describe the nature of the dissipation process, the concept of "free energy" may be introduced (cf. Callen, 1960; Prigogine, 1968).

The free energy, G, of a system is defined as the maximum work that can be drawn from the system under conditions where the exchange of work is the only interaction between the system and its surroundings. A system of this kind is said to be in thermodynamic equilibrium if its free energy is zero.

Consider now a system divided into two subsystems, a small one with extensive variables (i.e. variables proportional to the size of the system) U, S, V, etc. and a large one with intensive variables T_{ref}, P_{ref}, etc., which is initially in thermodynamic equilibrium. The terms "small system" and "large system" are meant to imply that the intensive variables of the large system (but not its extensive variables U_{ref}, S_{ref}, etc.) can be regarded as constant, regardless of the processes by which the entire system approaches equilibrium.

This implies that the intensive variables of the small system, which may not even be defined during the process, approach those of the large system when the combined system approaches equilibrium. The free energy, or maximum work, is found by considering a reversible process between the initial state and the equilibrium. It equals the difference between the initial internal energy, $U_{init} = U + U_{ref}$, and the final internal energy, U_{eq}, or it may be written (all in terms of initial state variables) as

$$G = U - T_{ref}S + P_{ref}V, \tag{4.1.6}$$

plus terms of the form $\Sigma \mu_{i,ref} n_i$ if chemical reactions are involved, and similar generalisations in case of electromagnetic interactions, etc.

If the entire system is closed it develops spontaneously towards equilibrium through internal, irreversible processes, with a rate of free energy change

$$\frac{dG}{dt} = \frac{d}{dt}(U_{init} - U_{eq}(t)) = \left(\frac{\partial}{\partial S(t)} U_{eq}(t)\right)\frac{dS(t)}{dt},$$

assuming that the entropy is the only variable. $S(t)$ is the entropy at time t of the entire system, and $U_{eq}(t)$ is the internal energy that would be possessed by a hypothetical equilibrium state defined by the actual state variables at time t, i.e. $S(t)$ etc. For any of these equilibrium states, $\partial U_{eq}(t)/\partial S(t)$ equals T_{ref} according to (4.1.3), and by comparison with (4.1.5) it is seen that the rate of dissipation can be identified with the loss of free energy, as well as with the increase in entropy,

$$D = -dG/dt = T_{ref}dS(t)/dt. \tag{4.1.7}$$

For systems met in practice, there will often be constraints preventing the system from reaching the absolute equilibrium state of zero free energy. For instance, the small system considered above may be separated from the large one by walls keeping the volume V constant. In such cases the available free energy (i.e. the maximum amount of useful work that can be extracted) becomes the absolute amount of free energy, (4.1.6), minus the free energy of the relative equilibrium which the combined system can be made to approach in the presence of the constraint. If the extensive variables in the constrained equilibrium state are denoted U^0, S^0, V^0, etc., then the available free energy becomes

$$\Delta G = (U - U^0) - T_{ref}(S - S^0) + P_{ref}(V - V^0), \quad (4.1.8)$$

eventually with the additions involving chemical potentials, etc. In the form (4.1.6) or (4.1.8), G is called the Gibbs potential. If the small system is constrained by walls, so that the volume cannot be changed, the free energy reduces to the Helmholtz potential $U - TS$, and if the small system is constrained so that it is incapable of exchanging heat, the free energy reduces to the enthalpy (4.1.1). The corresponding forms of (4.1.8) give the maximum work that can be obtained from a thermodynamic system with the given constraints.

A description of the course of an actual process as a function of time requires knowledge of "equations of motion" for the extensive variables, i.e. equations which relate the currents such as

$$\begin{aligned} J_s &= \mathrm{d}S/\mathrm{d}t \text{ (entropy flow rate) or} \\ J_Q &= \mathrm{d}Q/\mathrm{d}t \text{ (heat flow rate),} \\ J_m &= \mathrm{d}m/\mathrm{d}t \text{ (mass flow rate) or} \\ J_\theta &= \mathrm{d}\theta/\mathrm{d}t \text{ (angular velocity),} \\ J_q &= \mathrm{d}q/\mathrm{d}t \\ &= I \text{ (charge flow rate or electrical current), etc.} \end{aligned} \quad (4.1.9)$$

to the (generalised) forces of the system. As a first approximation, the relation between the currents and the forces may be taken as linear (Onsager, 1931),

$$\boldsymbol{J}_i = \Sigma_j L_{ij} \boldsymbol{F}_j. \quad (4.1.10)$$

The direction of each flow component is $\boldsymbol{J}_i/J_i$. The arbitrariness in choosing the generalised forces is reduced by requiring, as did Onsager, that the dissipation be given by

$$D = -\mathrm{d}G/\mathrm{d}t = \Sigma_i \boldsymbol{J}_i \cdot \boldsymbol{F}_i. \quad (4.1.11)$$

Examples of the linear relationships (4.1.10) are Ohm's law, stating that the electric current J_q is proportional to the gradient of the electric potential ($F_q \propto$ grad ϕ), and Fourier's law for heat conduction or diffusion, stating that the heat flow rate $E^{sens} = J_Q$ is proportional to the gradient of the temperature.

Considering the isothermal expansion process required in the Carnot cycle (Fig. 4.1-1), heat must be flowing to the system at a rate $J_Q = dQ/dt$, with $J_Q = LF_Q$ according to (4.1.10) in its simplest form. Using (4.1.11), the energy dissipation takes the form

$$D = T\mathrm{d}S/\mathrm{d}t = J_Q F_Q = L^{-1} J_Q^2.$$

For a finite time Δt, the entropy increase becomes

$$\Delta S = (\mathrm{d}S/\mathrm{d}t)\Delta t = (LT)^{-1} J_Q^2 \Delta t = (LT\Delta t)^{-1} (\Delta Q)^2,$$

so that in order to transfer a finite amount of heat ΔQ, the product $\Delta S\ \Delta t$ must equal the quantity $(LT)^{-1}\ (\Delta Q)^2$. In order that the process approaches reversibility, as the ideal Carnot cycle should, ΔS must approach zero, which is seen to imply that Δt approaches infinity. This qualifies the statement made in the beginning of this subsection that, in order to go through a thermodynamic engine cycle in a finite time, one has to give up reversibility and accept a finite amount of energy dissipation and an efficiency which is smaller than the ideal one (4.1.4).

Efficiency of an energy conversion device

A schematic picture of an energy conversion device is shown in Fig. 4.1-2, sufficiently general to cover most types of converters in practical use (Angrist, 1976; Osterle, 1964). There is a mass flow into the device and another one out from it, as well as an incoming and outgoing heat flow. The work output may be in the form of electric or rotating shaft power.

It may be assumed that the converter is in a steady state, implying that the incoming and outgoing mass flows are identical and that the entropy of the device itself is constant, i.e. that all entropy created is being carried away by the outgoing flows.

From the first law of thermodynamics, the power extracted, E, equals the net energy input,

$$E = J_{Q,in} - J_{Q,out} + J_m(w_{in} - w_{out}). \quad (4.1.12)$$

The magnitude of the currents is given by (4.1.9), and their conventional signs may be inferred from Fig. 4.1-2. The specific energy content of the incoming mass flow, w_{in}, and of the outgoing mass flow, w_{out}, are the sums of potential energy, kinetic energy and enthalpy. The significance of the enthalpy to represent the thermodynamic energy of a stationary flow is established by Bernoulli's theorem (Pippard, 1966). It states that for a stationary flow, if heat conduction can be neglected, the enthalpy is constant along a streamline. For the uniform mass flows assumed for the device in Fig. 4.1-2, the specific enthalpy, h, thus becomes a property of the flow, in analogy with the kinetic energy of motion and, for example, the geopotential energy,

$$w = w^{pot} + w^{kin} + h. \quad (4.1.13)$$

The power output may be written

$$E = -\boldsymbol{J}_\theta \cdot \boldsymbol{F}_\theta - \boldsymbol{J}_q \cdot \boldsymbol{F}_{q'} \quad (4.1.14)$$

with the magnitude of currents given by (4.1.9) and the generalised forces given by

$$\begin{aligned} \boldsymbol{F}_\theta &= \int \boldsymbol{r} \times \mathrm{d}\boldsymbol{F}_{mech}(\boldsymbol{r}) \quad \text{(torque),} \\ \boldsymbol{F}_q &= -\mathrm{grad}(\phi) \quad \text{(electric field)} \end{aligned} \quad (4.1.15)$$

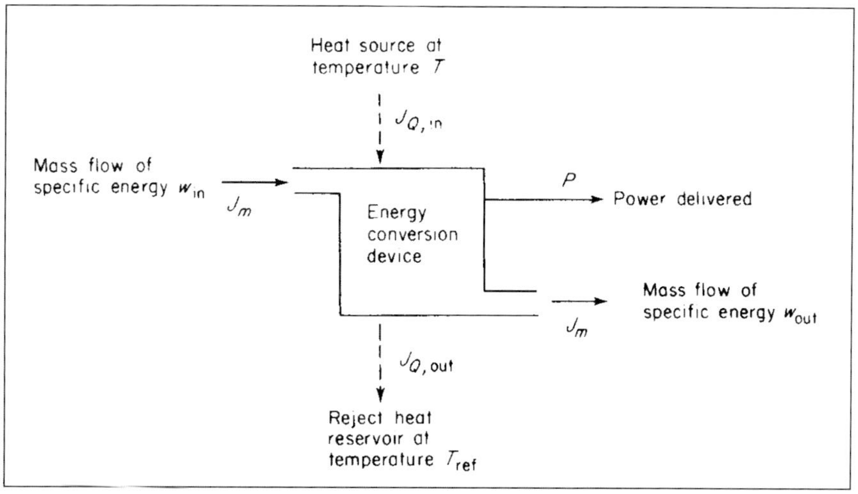

Figure 4.1-2 Schematic picture of an energy conversion device with a steady–state mass flow. The sign convention is different from the one used in (4.1.2), where all fluxes into the system were taken as positive.

corresponding to a mechanical torque and an electric potential gradient. The rate of entropy creation, i.e. the rate of entropy increase in the surroundings of the conversion device (as mentioned, the entropy inside the device is constant in the steady-state model), is

$$dS/dt = (T_{ref})^{-1} J_{Q,out} - T^{-1} J_{Q,in} + J_m (S_{m,out} - S_{m,in}),$$

where $s_{m,in}$ is the specific entropy of the mass (fluid, gas, etc.) flowing into the device, and $s_{m,out}$ is the specific entropy of the outgoing mass flow. $J_{Q,out}$ may be eliminated by use of (4.1.12), and the rate of dissipation obtained from (4.1.7),

$$\begin{aligned} D = T_{ref} dS/dt &= J_{Q,in}(1 - T_{ref}/T) \\ &+ J_m(w_{in} - w_{out} - T_{ref}(s_{m,in} - s_{m,out})) \\ &- E = \max(E) - E. \end{aligned} \tag{4.1.16}$$

The maximum possible work (obtained for $dS/dt = 0$) is seen to consist of a Carnot term (closed cycle, i.e. no external flows) plus a term proportional to the mass flow. The dissipation (4.1.16) is brought in the Onsager form (4.1.11),

$$D = J_{Q,in} F_{Q,in} + J_m F_m + \boldsymbol{J}_\theta \cdot \boldsymbol{F}_\theta + \boldsymbol{J}_q \cdot \boldsymbol{F}_{q'} \tag{4.1.17}$$

by defining generalised forces

$$F_{Q,in} = 1 - T_{ref}/T,$$

$$F_m = w_{in} - w_{out} - T_{ref}(s_{m,in} - s_{m,out}) \tag{4.1.18}$$

in addition to those of (4.1.15).

The efficiency with which the heat and mass flow into the device is converted to power is, in analogy to (4.1.4),

$$\eta = \frac{E}{J_{Q,in} + J_m w_{in}}, \tag{4.1.19}$$

where the expression (4.1.16) may be inserted for E. This efficiency is sometimes referred to as the "first law efficiency", because it only deals with the amounts of energy input and output in the desired form and not with the "quality" of the energy input related to that of the energy output.

In order to include reference to the energy quality, in the sense of the second law of thermodynamics, account must be taken of the changes in entropy taking place in connection with the heat and mass flows through the conversion device. This is accomplished by the "second law efficiency", which for power-generating devices is defined by

$$\eta^{(2.law)} = \frac{E}{\max(E)} = -\frac{\mathbf{J}_\theta \cdot \mathbf{F}_\theta + \mathbf{J}_q \cdot \mathbf{F}_q}{J_{Q,in} F_{Q,in} + J_m F_m}, \tag{4.1.20}$$

where the second expression is valid specifically for the device considered in Fig. 4.1-2, while the first expression is of general applicability, when $\max(E)$ is taken as the maximum rate of work extraction permitted by the second law of thermodynamics. It should be noted that $\max(E)$ depends not only on the system and the controlled energy inputs, but also on the state of the surroundings.

Conversion devices for which the desired energy form is not work may be treated in a way analogous to the example in Fig. 4.1-2. In the form (4.1.17), no distinction is made between input and output of the different energy forms. Taking, for example, electrical power as input (sign change), output may be obtained in the form of heat or in the form of a mass stream. The efficiency expressions (4.1.19) and (4.1.20) must be altered, placing the actual input terms in the denominator and the actual output terms in the numerator. If the desired output energy form is denoted W, the second law efficiency can be written in the general form

$$\eta^{(2.law)} = W/\max(W). \qquad (4.1.21)$$

For conversion processes based on other principles than those considered in the thermodynamic description of phenomena, alternative efficiencies could be defined by (4.1.21), with max(W) calculated under consideration of the non-thermodynamic types of constraint. In such cases, the name "second law efficiency" would have to be modified.

4.1.1.2 Thermodynamic engine cycles

A number of thermodynamic cycles, i.e. (closed) paths in a representation of conjugate variables, have been demonstrated in practice. They offer examples of the compromises made in modifying the "prototype" Carnot cycle into a cycle that can be traversed in a finite amount of time. Each cycle can be used to convert heat into work, but in traditional uses the source of heat has mostly been the combustion of fuels, i.e. an initial energy conversion process, by which high-grade chemical energy is degraded to heat at a certain temperature, associated with a certain entropy production.

Figure 4.1-3 shows a number of engine cycles in (P, V)-, (T, S)- and (H, S)-diagrams corresponding to Fig. 4.1-1.

The working substance of the Brayton cycle is a gas, which is adiabatically compressed in step 1-2 and expanded in step 3-4. The remaining two steps take place at constant pressure (isobars), and heat is added in step 2-3. The useful work is extracted during the adiabatic expansion 3-4, and the simple efficiency is thus equal to the enthalpy difference $H_3 - H_4$ divided by the total input $H_3 - H_1$. Examples of devices operating on the Brayton cycle are gas turbines and jet engines. In these cases, the cycle is usually not closed, since the gas is exhausted at point 4 and step 4-1 is thus absent. The somewhat contradictory name given to such processes is "open cycles".

The Otto cycle, presently used in a large number of automobile engines, differs from the Brayton cycle in that steps 2-3 and 4-1 (if the cycle is closed) are carried out at constant volume (isochores) rather than at constant pressure.

The Diesel cycle (e.g. common in ship and lorry engines) has step 2-3 as isobar and step 4-1 as isochore, while the two remaining steps are approximately adiabates. The actual designs of the machines, involving turbine wheels or piston-holding cylinders, etc., may be found in engineering textbooks (such as Hütte, 1954).

Closer to the Carnot ideal is the Stirling cycle, involving two isochores (1-2 and 3-4) and two isotherms.

The Ericsson cycle has been developed with the purpose of using hot air as the working fluid. It consists of

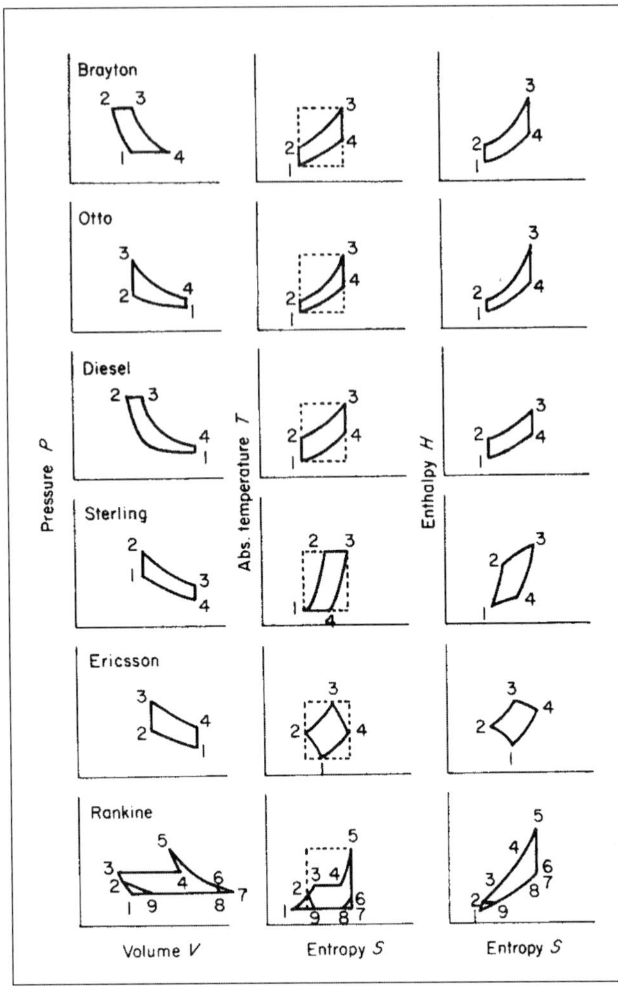

Figure 4.1-3 Examples of thermodynamic cycles in different representations. For comparison, the Carnot cycle is indicated in the (P, S)-diagram (dashed lines). Further descriptions of the individual cycles are given in the text (cf. section 4.1.2 for an alternative version of the Ericsson cycle).

two isochores (2-3 and 4-1) and two curves somewhere between isotherms and adiabates (cf. e.g. Meinel and Meinel, 1976).

The last cycle depicted in Fig. 4.1-3 is the Rankine cycle, the appearance of which is more complicated owing to the presence of two phases of the working fluid. Step 1-2-3 describes the heating of the fluid to its boiling point. Step 3-4 corresponds to the evaporation of the fluid, with both fluid and gaseous phases being present. It is an isotherm as well as an isobar. Step 4-5 represents the superheating of the gas, followed by an adiabatic expansion step 5-7. These two steps are sometimes repeated one or more times, with the superheating taking place at gradually lowered pressure, after each step of expansion to saturation. Finally, step 7-1 again involves mixed phases with condensation at constant pressure and temperature. The condensation often does not start until a temperature below that of saturation is reached. Useful work is extracted during the expansion step 5-7, so the simple efficiency equals the enthalpy difference $H_5 - H_7$ divided by the total input $H_6 - H_1$. The second law efficiency is obtained by dividing the simple efficiency by the Carnot value (4.1.4), for $T = T_5$ and $T_{ref} = T_7$.

Thermodynamic cycles such as those of Figs. 4.1-1 and 4.1-3 may be traversed in the opposite direction, thus using the work input to create a low temperature T_{ref} (cooling, refrigeration; T being the temperature of the surroundings) or to create a temperature T higher than that (T_{ref}) of the surroundings (heat pumping). In this case step 7-5 of the Rankine cycle is a compression (8-6-5 if the gas experiences superheating). After cooling (5-4), the gas condenses at the constant temperature T (4-3), and the fluid is expanded, often by passage through a nozzle. The passage through the nozzle is considered to take place at constant enthalpy (2-9), but this step may be preceded by undercooling (3-2). Finally, step 9-8 (or 9-7) corresponds to evaporation at the constant temperature T_{ref}.

For a cooling device the simple efficiency is the ratio of the heat removed from the surroundings, $H_7 - H_9$, and the work input, $H_5 - H_7$, whereas for a heat pump it is the ratio of the heat delivered, $H_5 - H_2$, and the work input. Such efficiencies are often called "coefficients of performance" (COP), and the second law efficiency may be found by dividing the COP by the corresponding quantity ε_{Carnot} for the ideal Carnot cycle (cf. Fig. 4.1-1),

$$\varepsilon^{cooling}_{Carnot} = \frac{\Delta Q_{14}}{W} = \frac{T_{ref}}{T - T_{ref}}, \tag{4.1.22a}$$

$$\varepsilon^{heatpump}_{Carnot} = \frac{\Delta Q_{32}}{W} = \frac{T_{ref}}{T - T_{ref}}. \tag{4.1.22b}$$

In practice, the compression work $H_5 - H_7$ (for the Rankine cycle in Fig. 4.1-3) may be less than the energy input to the compressor, thus further reducing the COP and the second law efficiency, relative to the primary source of high-quality energy.

4.1.1.3 Thermoelectric and thermionic conversion

If the high-quality energy form desired is electricity, and the initial energy is in the form of heat, there is a possibility of utilising direct conversion processes, rather than first using a thermodynamic engine to create mechanical work and then in a second conversion step using an electricity generator.

Thermoelectric generators

One direct conversion process makes use of the thermoelectric effect associated with heating the junction of two different conducting materials, e.g. metals or semiconductors. If a stable electric current, I, passes across the junction between the two conductors A and B, in an arrangement of the type depicted in Fig. 4.1-4, then quantum electron theory requires that the Fermi energy level (which may be regarded as a chemical potential μ_i) is the same in the two materials ($\mu_A = \mu_B$). If the spectrum of electron quantum states is different in the two materials, the crossing of negatively charged electrons or positively charged "holes" (electron vacancies) will not preserve the statistical distribution of electrons around the Fermi level,

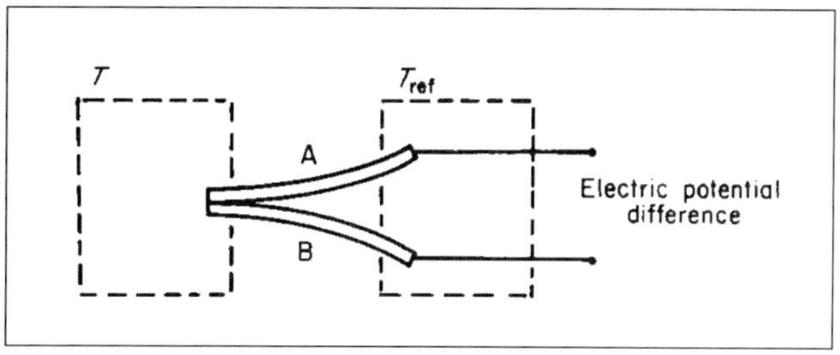

Figure 4.1-4 Schematic picture of a thermoelectric generator (thermocouple). The rods A and B are made of different materials (metals or better p- and n-type semiconductors).

$$f(E) = (\exp((E - \mu_i)/kT) + 1)^{-1}, \tag{4.1.23}$$

with E being the electron energy and k being Boltzmann's constant. The altered distribution may imply a shift towards a lower or a higher temperature, such that the maintenance of the current may require addition or removal of heat. Correspondingly, heating the junction will increase or decrease the electric current. The first case represents a thermoelectric generator, and the voltage across the external connections (Fig. 4.1.4) receives a term in addition to the ohmic term associated with the internal resistance R_{int} of the rods A and B,

$$\Delta\phi = -IR_{\text{int}} + \int_{T_{ref}}^{T} \alpha \mathrm{d}T',$$

The coefficient α is called the Seebeck coefficient. It is the sum of the Seebeck coefficients for the two materials A and B, and it may be expressed in terms of the quantum statistical properties of the materials (Angrist, 1976). If α is assumed independent of temperature in the range from T_{ref} to T, then the generalised electrical force (4.1.15) may be written

$$F_q = R_{int}J_q - \alpha T F_{Q,in}, \tag{4.1.24}$$

where J_q and $F_{q,in}$ are given in (4.1.9) and (4.1.18).

Considering the thermoelectric generator (Fig. 4.1-4) as a particular example of the conversion device shown in Fig. 4.1-2, with no mass flows, the dissipation (4.1.11) may be written

$$D = J_Q F_Q + J_q F_q.$$

In the linear approximation (4.1.10), the flows are of the form

$$J_Q = L_{QQ}F_Q + L_{Qq}F_q,$$
$$J_q = L_{qQ}F_Q + L_{qq}F_q,$$

with $L_{Qq} = L_{qQ}$ because of microscopic reversibility (Onsager, 1931). Considering F_Q and J_q (Carnot factor and electric current) as the "controllable" variables, one may solve for F_q and J_Q, obtaining F_q in the form (4.1.24) with $F_Q = F_{Q,in}$ and

$$L_{qq} = (R_{int})^{-1}; \quad L_{qQ} = L_{Qq} = \alpha T/R_{int}.$$

The equation for J_Q takes the form

$$J_Q = CTF_Q + \alpha TJ_q, \tag{4.1.25}$$

where the conductance C is given by

$$C = (L_{QQ}L_{qq} - L_{Qq}L_{qQ})/(L_{qq}T).$$

Using (4.1.24) and (4.1.25), the dissipation may be written

$$D = CTF_Q^2 + R_{int}J_q^2, \tag{4.1.26}$$

and the simple efficiency (4.1.19) may be written

$$\eta = \frac{-J_qF_q}{J_Q} = \frac{-J_q(R_{int}J_q - \alpha TF_Q)}{CTF_Q + \alpha TJ_q}. \tag{4.1.27}$$

If the reservoir temperatures T and T_{ref} are maintained at a constant value, F_Q can be regarded as fixed, and the maximum efficiency can be found by variation of J_q. The efficiency (4.1.27) has an extremum at

$$J_q = \frac{CF_Q}{\alpha}\left(\left(1 + \frac{\alpha^2 T}{R_{int}C}\right)^{1/2} - 1\right), \tag{4.1.28}$$

corresponding to a maximum value

$$\max(\eta) = F_Q\frac{(1 + ZT)^{1/2} - 1}{(1 + ZT)^{1/2} + 1}, \tag{4.1.29}$$

with $Z = \alpha^2(R_{int}C)^{-1}$. Equation (4.1.29) is accurate only if the linear approximation (4.1.10) is valid. The maximum second law efficiency is obtained from (4.1.29) by division by F_Q [cf. (4.1.20)].

The efficiencies are seen to increase with temperature, as well as with Z. Z is largest for certain materials (A and B in Fig. 4.1-4) of semiconductor structure and small for metals as well as for insulators. Although R_{int} is small for metals and large for insulators, the same is true for the Seebeck coefficient α, which appears squared. C is larger for metals than for insulators. Together, these features combine to produce a peak in Z in the semiconductor region. Typical values of Z are about 2×10^{-3} $(\mathrm{K})^{-1}$ at $T = 300$ K (Angrist, 1976). The two materials A and B may be taken as a p-type and an n-type semiconductor, which have Seebeck coefficients of opposite signs, so that their contributions add coherently for a configuration of the kind shown in Fig. 4.1-4.

Thermionic generators

Thermionic converters consist of two conductor plates separated by vacuum or by a plasma. The plates are maintained at different temperatures. One, the emitter, is at a temperature T large enough to allow a substantial emission of electrons into the space between the plates due to the thermal statistical spread in electron energy (4.1.23). The electrons (e.g. of a metal emitter) move in a potential field characterised by a barrier at the surface of the plate. The shape of this barrier is usually such that the probability of an electron penetrating it is small until a critical temperature, after which it increases rapidly ("red-glowing" metals). The other plate is maintained at a lower temperature T_{ref}. In order not to have a build-up of space charge between the emitter and the collector, atoms of a substance such as caesium may be introduced in this area. These atoms become ionised near the hot emitter (they give away electrons to make up for the electron deficit in the emitter material), and for a given caesium pressure the positive ions exactly neutralise the space charges of the travelling electrons. At the collector surface, recombination of caesium ions takes place.

The layout of the emitter design must allow the transfer of large quantities of heat to a small area in order to maximise the electron current responsible for creating the electric voltage difference across the emitter–collector system, which may be utilised through an external load circuit. This heat transfer can be accomplished by a so-called "heat pipe" – a fluid-containing pipe which allows the fluid to evaporate in one chamber when heat is added. The vapour then travels to the other end of the pipe, condenses and gives off the latent heat of evaporation to the surroundings, whereafter it returns to the first chamber through capillary channels, under the influence of surface tension forces.

The description of the thermionic generator in terms of the model converter shown in Fig. 4.1-2 is very similar to that of the thermoelectric generator. With the two temperatures T and T_{ref} defined above, the generalised force F_Q is defined. The electrical output current, J_q, is equal to the emitter current, provided that back-emission from the collector at temperature T_{ref} can be neglected and provided that the positive-ion current in the intermediate space is negligible in comparison with the electron current. If the space charges are saturated, the ratio between ion and electron currents is simply the inverse of the square root of the mass ratio, and the positive-ion current will be a fraction of a per cent of the electron current. According to quantum statistics, the emission current (and hence J_q) may be written

$$J_Q = AT^2\exp(-e\phi_e/(kT)), \quad (4.1.30)$$

where ϕ_e is the electric potential of the emitter, $e\phi_e$ is the potential barrier of the surface in energy units, and A is a constant (Angrist, 1976). Neglecting heat conduction losses in plates and the intermediate space, as well as light emission, the heat $J_{Q,in}$ to be supplied to keep the emitter at the elevated temperature T equals the energy carried away by the electrons emitted,

$$J_{Q,in} = J_q(\phi_e + \delta + 2kT/e), \quad (4.1.31)$$

where the three terms in brackets represent the surface barrier, the barrier effectively seen by an electron due to the space charge in the intermediate space, and the original average kinetic energy of the electrons at temperature T (divided by e), respectively.

Finally, neglecting internal resistance in plates and wires, the generalised electrical force equals the difference between the potential ϕ_e and the corresponding potential for the collector ϕ_c,

$$-F_q = \phi_c - \phi_e, \quad (4.1.32)$$

with insertion of the above expressions (4.1.30) to (4.1.32). Alternatively, these expressions may be linearised in the form (4.1.10) and the efficiency calculated exactly as in the case of the thermoelectric device. It is clear, however, that a linear approximation to (4.1.30), for example, would be very poor.

4.1.1.4 Turbines and other flow-driven converters

A turbine is a device delivering rotational shaft power on the basis of some other type of mechanical energy. If the temperature of the surroundings is regarded as fixed, the simple model in Fig. 4.1-2 allows the energy dissipation (4.1.17) to be written

$$D = J_m F_m + J_\theta F_\theta, \quad (4.1.33)$$

since from (4.1.18) $F_{Q,in}$ is zero, and no electrical output has been considered in this conversion step. The output variables are the angular velocity of the shaft, J_θ (4.1.9), and the torque acting on the system, F_θ (4.1.15), while the input variables are the mass flow rate, J_m (4.1.9), and the generalised force is F_m given in (4.1.18). The specific energy contents w_{in} and w_{out} are of the form (4.1.13), corresponding to e.g. the geopotential energy of a given water head,

$$w_{in}^{pot} = g\Delta z, \qquad w_{out}^{pot} = 0, \quad (4.1.34)$$

the kinetic energy of the working fluid,

$$w_{in}^{kin} = ½u_{in}^2, \qquad w_{out}^{kin} = ½u_{out}^2, \quad (4.1.35)$$

and the enthalpy connected with the pressure changes,

$$h_{in} = P_{in}/\rho_{in}, \qquad h_{out} = P_{out}/\rho_{out}, \quad (4.1.36)$$

where the internal energy term in (4.1.1), assumed constant, has been left out, and the specific volume has been expressed in terms of the fluid densities ρ_{in} and ρ_{out} at input and output.

If a linear model of the Onsager type (4.1.10) is adopted for J_m and J_θ and these equations are solved for J_m and F_θ, one obtains

$$J_m = L_{m\theta}J_\theta/L_{\theta\theta} + (L_{mm} - L_{m\theta}L_{\theta m}/L_{\theta\theta})F_m,$$
$$-F_\theta = -J_\theta/L_{\theta\theta} + L_{\theta m}F_m/L_{\theta\theta}. \quad (4.1.37)$$

The coefficients may be interpreted as follows: $L_{m\theta}/L_{\theta\theta}$ is the mass of fluid displaced by the turbine during one radian of revolution, $(L_{mm} - L_{m\theta}\,L_{\theta m}/L_{\theta\theta})$ is a "leakage factor" associated with fluid getting through the turbine without contributing to the shaft power, and finally, $L_{\theta\theta}^{-1}$ represents the friction losses. Insertion into (4.1.33) gives the linear approximation for the dissipation,

$$D = (L_{mm} - L_{m\theta}L_{\theta m}/L_{\theta\theta})(F_m)^2 + (J_\theta)^2/L_{\theta\theta}. \quad (4.1.38)$$

An ideal conversion process may be approached if no heat is exchanged with the surroundings, in which case (4.1.19) and (4.1.12) give the simple efficiency

$$\eta = (w_{in} - w_{out})/w_{in}. \quad (4.1.39)$$

The second law efficiency in this case is, from (4.1.20), (4.1.14) and (4.1.12),

$$\eta^{(2.\ law)} = (w_{in} - w_{out})/(w_{in} - w_{out} - T_{ref}(s_{m,in} - s_{m,out})). \quad (4.1.40)$$

The second law efficiency becomes unity if no entropy change takes place in the mass stream. The first law efficiency (4.1.39) may approach unity if only potential energy change of the form (4.1.34) is involved. In this case $w_{out} = 0$, and the fluid velocity, density and pressure are the same before and after the turbine. Hydroelectric generators approach this limit if working entirely on a static water head. Overshot waterwheels may operate in this way, and so may the more advanced turbine constructions, if the potential to kinetic energy conversion (in penstocks) and pressure build-up (in the nozzle of a Pelton turbine and in the inlet tube of many Francis turbine installations) are regarded as "internal" to the device (cf. section 4.4.1.5). However, if there is a change in velocity or pressure across the converter, the analysis must take this into account, and it is no longer obvious whether the first law efficiency may approach unity or not.

Free stream flow turbines

Consider, for example, a free stream flow passing horizontally through a converter. In this case, the potential energy (4.1.34) does not change and may be left out. The pressure may vary near the converting device, but far behind and far ahead of the device the pressure is the same if the stream flow is free. Thus,

$$w = w^{kin} = ½(u_x^2 + u_y^2 + u_z^2) = ½\boldsymbol{u} \cdot \boldsymbol{u},$$

and

$$w_{in} - w_{out} = ½(\boldsymbol{u}_{in} - \boldsymbol{u}_{out}) \cdot (\boldsymbol{u}_{in} + \boldsymbol{u}_{out}). \quad (4.1.41)$$

This expression and hence the efficiency would be maximum if $\boldsymbol{u}_{out}$ could be made zero. However, the conservation of the mass flow J_m requires that u_{in} and u_{out} satisfy an additional relationship. For a pure, homogeneous streamline flow along the x-axis, the rate of mass flow is

$$J_m = \rho\, A_{in}\, u_{x,in} = \rho\, A_{out}\, u_{x,out}, \quad (4.1.42)$$

in terms of areas A_{in} and A_{out} enclosing the same streamlines, before and after the passage through the conversion device. In a more general situation, assuming rotational symmetry around the x-axis, there may have been induced a radial as well as a circular flow component by the device. This situation is illustrated in Fig. 4.1-5. It will be further discussed in section 4.4.1.3, and the only case treated here is the simple one in which the radial and tangential components of the velocity field, u_r and u_t, which may be induced by the conversion device, can be neglected.

The axial force ("thrust") acting on the converter equals the momentum change,

$$F_x = J_m(u_{x,in} - u_{x,out}). \quad (4.1.43)$$

If the flow velocity in the converter is denoted u, an effective area of conversion, A, may be defined by

$$J_m = \rho A u_x. \quad (4.1.44)$$

according to the continuity equation (4.1.42). Dividing (4.1.43) by ρA, one obtains the specific energy transfer from the mass flow to the converter, within the conversion area A. This should equal the expression (4.1.41) for the change in specific energy, specialised to the case of homogeneous flows $\boldsymbol{u}_{in}$ and $\boldsymbol{u}_{out}$ along the x-axis,

$$u_x(u_{x,in} - u_{x,out}) = ½\,(u_{x,in} + u_{x,out})(u_{x,in} - u_{x,out})$$

or

$$u_x = ½\,(u_{x,in} + u_{x,out}) \quad (4.1.45)$$

The physical principle behind this equality is simply energy conservation, and the assumptions so far have been the absence of heat exchange [so that the energy change (4.1.12) becomes proportional to the kinetic energy difference (4.1.41)] and the absence of induced rotation (so that only x-components of the velocity needs to be considered). On both sides of the converter, Bernoulli's equation is valid, stating that the specific energy is constant along a streamline with ϕ = constant defining the streamlines). Far from the converter, the pressures are equal but the velocities are different, while the velocity just in front of or behind the converter may be taken as u_x, implying a pressure drop across the converter,

$$\Delta P = ½\,\rho\,(u_{x,in} + u_{x,out})\,(u_{x,in} - u_{x,out}) \quad (4.1.46)$$

The area enclosing a given streamline field increases in a continuous manner across the converter at the same time as the fluid velocity continuously decreases. The

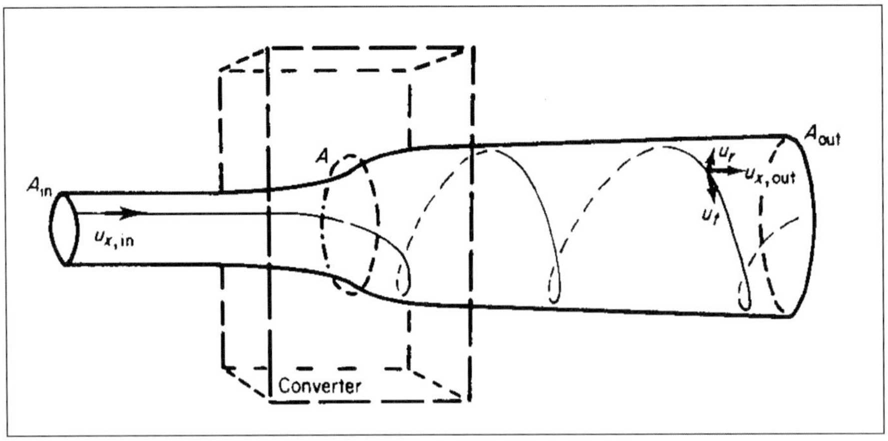

Figure 4.1-5 Schematic picture of a free stream flow converter or turbine. The incoming flow is a uniform streamline flow in the x-direction, while the outgoing flow is allowed to have a radial and a tangential component. The diagram indicates how a streamline may be transformed into an expanding helix by the device. The effective area of the converter, A, is defined in (4.1.44).

pressure, on the other hand, rises above the ambient pressure in front of the converter, then discontinuously drops to a value below the ambient one, and finally increases towards the ambient pressure again, behind ("in the wake of") the converter.

It is customary (see e.g. Wilson and Lissaman, 1974) to define an "axial interference factor", a, by

$$u_x = u_{x,in}(1 - a) \quad (4.1.47)$$

in which case (4.1.45) implies that $u_{x,out} = u_{x,in}(1 - 2a)$. With this, the power output of the conversion device can be written

$$E = J_m(w_{in} - w_{out}) = \rho A(u_{x,in})^3 2a(1 - a)^2, \quad (4.1.48)$$

and the efficiency can be written

$$\eta = E/(J_m w_{in}) = 4a(1 - a) \quad (4.1.49)$$

It is seen that the maximum value of η is unity, obtained for $a = ½$, corresponding to $u_{x,out} = 0$. The continuity equation (4.1.42) then implies an infinite area A_{out}, and it will clearly be difficult to defend the assumption of no induced radial motion.

In fact, for a free stream device of this type, the efficiency (4.1.49) is of little relevance since the input flux may not be independent of the details of the device. The input area A_{in}, from which streamlines would connect with a fixed converter area A, could conceivably be changed by altering the construction of the converter. It is therefore more appropriate to ask for the maximum power output for fixed A, as well as fixed input velocity $u_{x,in}$, this being equivalent to maximising the "power coefficient" defined by

$$C_p = E/(½\,\rho\,A\,(u_{x,in})^3) = 4a\,(1 - a)^2. \quad (4.1.50)$$

The maximum value is obtained for $a = 1/3$, yielding $C_p = 16/27$ and $u_{x,out} = u_{x,in}/3$. The areas are $A_{in} = (1 - a)\,A = 2/3\,A$ and $A_{out} = (1 - a)A/(1 - 2a) = 2A$, so in this case it is not unlikely that it may be a reasonable approximation to neglect the radial velocity component in the far wake.

The maximum found above for C_p is only a true upper limit with the assumptions made. By discarding the assumption of irrotational flow, it becomes possible for the converter to induce a velocity field, for which rot($\boldsymbol{u}$) is no longer zero. It has been shown that if the additional field is in the form of a vortex ring around the converter region, so that it does not contribute to the far wake, then it is possible to exceed the upper limit power coefficient 16/27 found above (cf. section 4.1.3.4).

Magnetohydrodynamic converters

For the turbines considered above, it was explicitly assumed that no heat was added. Other flow-type converters are designed to receive heat during the process. An example of this is the gas turbine, which was described in section 4.4.1.2 from the point of view of thermodynamics. The gas turbine (Brayton cycle) allows the conversion of heat into shaft power, but it may equally well be viewed as the successive conversion of heat into kinetic energy of flow and of the kinetic energy of flow into shaft power.

The magnetohydrodynamic converter is another device converting heat into work, but delivering the work directly as electrical power without intermediate steps of mechanical shaft power. The advantage is not in avoiding the shaft power to electrical power conversion, which can be done with small losses, but rather in avoiding a construction with moving parts, thereby permitting higher working temperatures and higher efficiency. The heat added is used to ionise a gas, and this conducting gas ("plasma") is allowed to move through an expanding duct, upon which an external magnetic field $\boldsymbol{B}$ is acting. The motion of the gas is sustained by a pressure drop between the chamber where heat is added and the open end of the expanding duct. The charged particles of velocity u in the plasma are subjected to a Lorentz force

$$\boldsymbol{F} = \boldsymbol{\rho}_{el}\,\boldsymbol{u} \times \boldsymbol{B}, \quad (4.1.51)$$

where the direction of this induced force is perpendicular to $\boldsymbol{B}$ and $\boldsymbol{u}$, but opposite for the positive atoms and for the negatively charged electrons. Since the mass of an electron is much smaller than that of an atom, the net induced current will be in the direction given by a negative value of $\boldsymbol{\rho}_{el}$. Assuming a linear relationship between the induced current $\boldsymbol{J}_{ind}$ and the induced electric field $\boldsymbol{E}_{ind} = \boldsymbol{F}/\boldsymbol{\rho}_{el} = \boldsymbol{u} \times \boldsymbol{B}$, the induced current may be written

$$\boldsymbol{J}_{ind} = \sigma \boldsymbol{u} \times \boldsymbol{B}, \quad (4.1.52)$$

where $\boldsymbol{\sigma}$ is the electrical conductivity of the plasma. This outlines the mechanism by which the magnetohydrodynamic (MHD) generator converts kinetic energy of moving charges into electrical power associated with the induced current $\boldsymbol{J}_{ind}$ across the turbine. A more detailed treatment must take into account the contributions to the force (4.1.51) on the charges, which arise from the induced velocity component $\boldsymbol{J}_{ind}/\boldsymbol{\rho}_{el}$, as well as the effect of variations (if any) in the flow velocity $\boldsymbol{u}$ through the generator stage (see e.g. Angrist, 1976).

The generator part of the MHD generator has an efficiency determined by the net power output after subtraction of the power needed for maintaining the magnetic field $\boldsymbol{B}$. Only the gross power output can be considered as given by (4.1.12). Material considerations require that the turbine be cooled, so in addition to power output there is a heat output in the form of a coolant flow, as well as the outgoing flow of cooled gas. The temperature of the outflowing gas is still high (otherwise recombination of ions would inhibit the functioning of the converter), and the MHD stage is envisaged as being followed by one or more conventional turbine stages. It is believed that the total power generation in all stages could be made to exceed that of a conversion system based entirely on turbines with moving parts, for the same heat input.

Very high temperatures are required for the ionisation to be accomplished thermally. The ionisation process can be enhanced in various ways. One is to "seed" the gas with suitable metal dust (sodium, potassium, caesium, etc.), for which case working MHD machines operating at temperatures around 2500 K have been demonstrated (Hammond *et al.*, 1973). If the heat source is fossil fuel, and particularly if it is coal with a high sulphur content, the seeding has the advantage of removing practically all the sulphur from the exhaust gases (the seeding metals are rather easily retrieved and must anyway be recycled for economic reasons).

4.1.1.5 Photovoltaic conversion

Conversion of radiant energy (light quanta) into electrical energy can be achieved with the use of semiconductor materials, for which the electron excitation caused by impinging light quanta has a strongly enhancing effect on the conductivity.

It is not sufficient, however, that electrons are excited and are able to move more freely, if there is no force to make them move. Such a force would arise from the presence of a gradient of electrical potential, such as the one found in a *p–n* junction of doped semiconductor materials (a *p–n* junction is a junction of a *p*-type and an *n*-type semiconductor, as further described below). A *p–n* junction provides an electrical field which will cause the electrons excited by radiation (such as solar) to move in the direction from the *p*-type to the *n*-type material and cause the vacancies (holes) left by the excited electrons to move in the opposite direction. If the electrons and holes reach the respective edges of the semiconductor material, the device is capable of delivering electrical power to an external circuit. The motion of electrons or holes receives competition from recombination processes (electrons being recaptured into vacancies), making such factors as overall dimensions and electron mobility in the material used of importance.

The *p–n* junction

An essential constituent of photovoltaic cells is the *p–n* junction. A refresher on the semiconductor physics needed for understanding the *p–n* junction is given in section 4.4. A at the end of this chapter. When a *p*-type and an n-type semiconductor are joined so that they acquire a common surface, they are said to form a *p–n* junction. This will initially cause electrons to flow in the *n* to *p* direction because the electron density in the conduction band is higher in *n*-type than in *p*-type material and because the hole density in the valence band is higher in the *p*-type than in the *n*-type material (the electron flow in the valence band can also be described as a flow of positive holes in the direction *p* to *n*).

This electron flow builds up a surplus of positive charge in the *n*-type material and a surplus of negative charge in the *p*-type material, in the neighbourhood of the junction (mainly restricted to distances from the junction of the order of the mean travelling distance before recombination of an electron or a hole in the respective materials). These surplus charges form a dipole layer, associated with which is an electrostatic potential difference, which will tend to hinder any further unidirectional electron flow. Finally, an equilibrium is reached in which the potential difference is such that no net transfer of electrons takes place.

Another way of stating the equilibrium condition is in terms of the Fermi energy. Originally, the Fermi energies of the *p*- and *n*-type materials, μ_p and μ_n, are different, but at equilibrium $\mu_p = \mu_n$. This is illustrated in Fig. 4.1-6, and it is seen that the change in the relative positions of the conduction (or valence) bands in the two types of material must equal the electron charge, $-e$, times the equilibrium electrostatic potential.

The number of electrons in the conduction band may be written

$$n_c = \int_{E_c}^{E'_c} n'(E) f(E) \mathrm{d}E \tag{4.1.53}$$

where E_c and E'_c are the lower and upper energy limit of the conduction band, $n'(E)$ is the number of quantum states per unit energy interval (and, for example, per unit volume of material, if the electron number per unit volume is desired), and finally, $f(E)$ is the Fermi–Dirac distribution (4.1.23). If the electrons in the conduction band are regarded as free, elementary quantum mechanics gives (see e.g. Shockley, 1950)

$$n'(E) = 4\pi h^{-3} (2m)^{3/2} E^{1/2}, \tag{4.1.54}$$

where h is Planck's constant and m is the electron mass. The corrections for electrons moving in condensed matter, rather than being free, may to a first approximation be included by replacing the electron mass by an "effective" value.

If the Fermi energy is not close to the conduction band,

$$E_c - \mu >> kT,$$

the Fermi–Dirac distribution (4.1.23) may be replaced by the Boltzmann distribution,

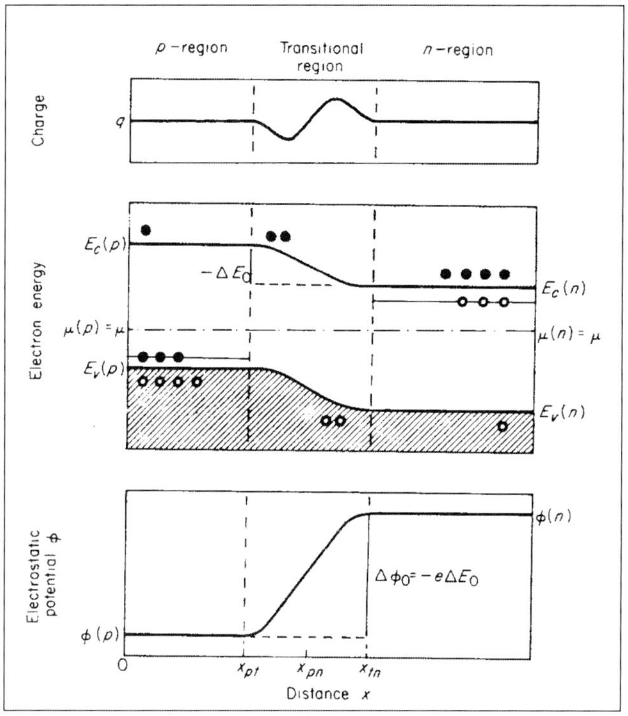

Figure 4.1-6 Schematic picture of the properties of a p–n junction in an equilibrium condition. The x-direction is perpendicular to the junction (all properties are assumed to be homogeneous in the y- and z-directions). The charge (top) is the sum of electron charges in the conduction band and positive hole charges in the valence band, plus charge excess or defect associated with the acceptor and donor levels. In the electron energy diagram (middle), the abundance of minority charge carriers (closed circles for electrons, open circles for holes) is schematically illustrated. The properties are further discussed in the text.

$$f_B(E) = \exp(-(E - \mu)/kT). \qquad (4.1.55)$$

Evaluating the integral, (4.1.53) then gives an expression of the form

$$n_c = N_c \exp(-(E_c - \mu)/kT). \qquad (4.1.56)$$

The number of holes in the valence band is found in an analogous way,

$$n_v = N_v \exp(-(\mu - E_v)/kT), \qquad (4.1.57)$$

where E_v is the upper limit energy of the valence band.

The equilibrium currents in a p–n junction such as the one illustrated in Fig. 4.1-6 can now be calculated. Considering first the electron currents in the conduction band, the electrons thermally excited into the conduction band in the p-region can freely flow into the n-type materials. The corresponding current, $I_0^-(p)$, may be considered proportional to the number of electrons in the conduction band in the p-region, $n_c(p)$, given by (4.1.56),

$$I_0^-(p) = \alpha N_c \exp(-(E_c(p) - \mu(p))/kT), \qquad (4.1.58)$$

where the constant α depends on electron mobility in the material and on the electrostatic potential gradient, grad ϕ. The electrons excited into the conduction band in the n-type region will have to climb the potential barrier in order to move into the p-region. The fraction of electrons capable of doing this is given by a Boltzmann factor of the form (4.1.56), but with the additional energy barrier $\Delta E_0 = -\Delta\phi_0/e$ ($-e$ being the electron charge),

$$n_c(n) = N_c \exp(-(E_c(n) - \mu(n) - \Delta E_0)/kT).$$

Using $-\Delta E_0 = E_c(p) - E_c(n)$ (cf. Fig. 4.1-6) and considering the current $I_0^-(n)$ as being proportional to $n_c(n)$, the corresponding current may be written

$$I_0^-(n) = \alpha' N_c \exp(-(E_c(n) - \mu(n))/kT), \qquad (4.1.59)$$

where α' depends on the diffusion parameter and on the relative change in electron density, $n_c^{-1}\mathrm{grad}(n_c)$, considering the electron motion against the electrostatic potential as a diffusion process. The statistical mechanical condition for thermal equilibrium demands that $\alpha = -\alpha'$ (Einstein, 1905), so (4.1.58) and (4.1.59) show that the net electron current,

$$I_0^- = I_0^-(p) + I_0^-(n),$$

becomes zero precisely when

$$\mu(p) = \mu(n),$$

which is then the condition for thermal equilibrium. The same is true for the hole current,

$$I_0^+ = I_0^+(p) + I_0^+(n).$$

If an external voltage source is applied to the p–n junction in such a way that the n-type terminal receives an additional electrostatic potential $\Delta\phi_{ext}$ relative to the p-type terminal, then the junction is no longer in thermal equilibrium, and the Fermi energy in the p-region is no longer equal to that of the n-region, but satisfies

$$\mu(p) - \mu(n) = e^{-1}\Delta\phi_{ext} = \Delta E_{ext} \qquad (4.1.60)$$

if the Boltzmann distributions of electrons and of holes are to maintain their shapes in both p- and n-regions. Similarly $E_c(p) - E_c(n) = -(\Delta E_0 + \Delta E_{ext})$, and assuming that the proportionality factors in (4.1.58) and (4.1.59) still bear the relationship $\alpha = -\alpha'$ in the presence of the external potential, the currents are connected by the expression

$$I^-(n) = -I^-(p)\exp(\Delta E_{ext}/kT).$$

The net electron current in the conduction band then becomes

$$I^- = I^-(n) + I^-(p) = -I^-(p)(\exp(\Delta E_{ext}/kT) - 1). \qquad (4.1.61)$$

For a positive $\Delta\phi_{ext}$, the potential barrier which electrons in the n-region conduction band (see Fig. 4.1-6) have to climb increases and the current $I^-(n)$ decreases exponentially (ΔE_{ext} negative, "reverse bias"). In this case, the net current I^- approaches a saturation value equal to $I^-(p)$, according to (4.1.61).

For negative $\Delta\phi_{ext}$, (positive ΔE_{ext}, "forward bias"), the current $I^-(n)$ increases exponentially with the external potential. In both cases $I^-(p)$ is assumed to remain practically unchanged, when the external potential of one or the other sign is applied, considering that $I^-(p)$ is primarily limited by the number of electrons excited into the conduction band in the p-type material, a number which is assumed to be small in comparison with the conduction band electrons in the n-type material (cf. Figs 4.1-6).

The contributions to the hole current, I^+, behave similarly to those of the electron current, and the total current I across a p–n junction with an external potential $\Delta\phi_{ext} = -e\,\Delta E_{ext}$ may be written

$$I = I^- + I^+ = -I(p)(\exp(\Delta E_{ext}/kT) - 1). \quad (4.1.62)$$

The relationship between current and potential is called the "characteristic" of the device, and the relation (4.1.62) for the p–n junction is illustrated in Fig. 4.1-7 by the curve labelled "no light". The constant saturation current $I(p)$ is sometimes referred to as the "dark current".

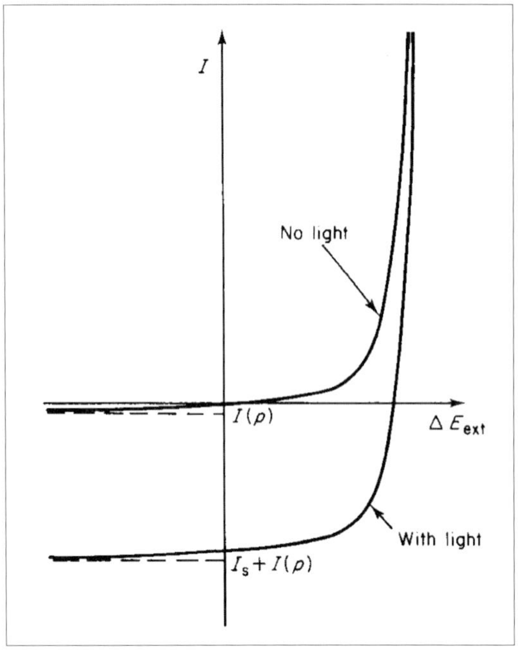

Figure 4.1-7 Characteristics (i.e. current as a function of external voltage) of a p–n junction, in the dark and with applied light. The magnitude of the short-circuit current, I_s, is a function of light intensity and spectral distribution.

Solar cells

A p–n junction may be utilised to convert solar radiation energy into electric power. A solar cell is formed by shaping the junction in such a way that, for example, the p-type material can be reached by incident solar radiation, e.g. by placing a thin layer of p-type material on top of a piece of n-type semiconductor. In the dark and with no external voltage, the net current across the junction is zero, as was shown in the previous subsection, i.e. the intrinsic potential difference $\Delta\phi_0$ is unable to perform external work.

However, when irradiated with light quanta of an energy $E_{light} = h\nu = hc/\lambda$ (h is Planck's constant, c is the velocity of light and ν and λ are the frequency and wavelength of radiation), which is larger than the energy difference between the conduction and valence band for the p-type material,

$$E_{light} \geq E_c(p) - E_v(p),$$

then electrons may be photo-excited from the valence band into the conduction band. The absorption of light quanta produces as many holes in the valence band of the p-type material as electrons in the conduction band. Since in the dark there are many fewer electrons in the p-type conduction band than holes in the valence band, a dramatic increase in the number of conduction band electrons can take place without significantly altering the number of holes in the valence band. If the excess electrons are sufficiently close to the junction to be able to reach it by diffusion before recombining with a hole, then the current in this direction exceeds $I_0^-(p)$ of (4.1.58) by an amount I_s, which is the net current through the junction in case of a short-circuited external connection from the n-type to the p-type material. The photo-induced current is not altered if there is a finite potential drop in the external circuit, since the relation between the current (4.1.62) and the external potential drop $e\,\Delta E_{ext}$ was derived with reference only to the changes in the n-region.

An alternative n–p type of solar cell may consist of a thin n-type layer exposed to solar radiation on top of a p-type base. In this case, the excess holes in the n-type valence band produce the photo-induced current I_s.

The total current in the case of light being absorbed in the p-type material and with an external potential drop is then

$$I = I_s - I(p)(\exp(-\Delta\phi_{ext}/kT) - 1). \quad (4.1.63)$$

The short-circuit current I_s depends on the amount of incident light with frequencies sufficient to excite electrons into the conduction band, on the fraction of this light actually being absorbed, and on the conditions for transporting the excess electrons created in the conduction band, in competition with electron–hole recombination processes. I_s may be written as the sum of a conduction and a diffusion type current, both related to the number of excess electrons in the conduction band, n_c^{ind}, induced by the absorption of light,

$$I_s = e\left(m_c E_e n_c^{ind} + k_c\, dn_c^{ind}/dx\right), \quad (4.1.64)$$

where e is the numerical value of the electron charge (1.6×10^{-19} C), m_c is the mobility of conduction band electrons [e.g. 0.12 $m^2\ V^{-1}\ s^{-1}$ for silicon (Loferski, 1956), the dependence on the degree of doping being displayed in Fig. 4.1-8], E_e is the local electrical field, k_c is the diffusion constant [e.g. $k_c = 10^{-3}\ m^2\ s^{-1}$ (Loferski, 1956)], and x is the depth below the solar cell surface, assumed to be the only significant co-ordinate (as in Fig. 4.1-6).

The excess electron number density, $n_c^{ind}(x)$, at a depth x, changes when additional electrons are photo-excited, when electrons are carried away from x by the current I_s, and when electrons recombine with holes,

$$\frac{\partial n_c^{ind}(x)}{\partial t} = \int \sigma(\nu) n_{ph}(\nu) \exp(-\sigma(\nu)x) d\nu + \frac{1}{e}\frac{\partial I_s}{\partial x} - n_c^{ind}(x)\frac{1}{\tau_c}. \quad (4.1.65)$$

Here $\sigma(\nu)$ is the cross section for absorption of light quanta ("photons") in the p-type material, and $n_{ph}(\nu)$ is the number of photons at the cell surface ($x = 0$) per unit time and unit interval of frequency ν. The absorption cross section is zero for photon energies below the semiconductor energy gap, $h\nu < E_c(p) - E_v(p)$, i.e. the material is transparent to such light. The most energetic light quanta in visible light could theoretically excite more than one electron per photon (e.g. 2–3 in Si with an energy gap slightly over 1 eV), but the probability for exciting just one electron to a higher energy is higher, and such a process is usually followed by a transfer of energy to other degrees of freedom (e.g. lattice vibrations and ultimately heat), as the excited electron approaches the lower part of the conduction band, or as the hole left by the electron de-excites from a deep level to the upper valence band. Thus, in practice the quantum efficiency (number of electron–hole pairs per photon) hardly exceeds one.

The last parameter introduced in (4.1.65), τ_c, is the average lifetime of an electron excited into the conduction band, before recombination [τ_c may lie in the interval 10^{-11} to 10^{-7}, with 10^{-9} being a typical value (Wolf, 1963)]. The lifetime τ_c is connected to the cross section for recombination, σ_c, and to the mean free path l_c of electrons in the conduction band by

$$l_c = \sigma_c^{-1} = v_c \tau_c N_a,$$

where v_c is the average thermal velocity of the electrons, $v_c = (2kT/m)^{1/2}$ (m being the electron mass, k being Boltzmann's constant and T being the absolute temperature) and N_a is the number of recombination centres ("acceptor impurities").

The boundary conditions for solving (4.1.65) may be taken as the absence of excess minority carriers (electrons or holes) at the junction $x = x_{pn}$,

$$n_c^{ind}(x_{pn}) = 0,$$

and a prescribed (material dependent) excess electron gradient at the surface $x = 0$. This gradient, $(dn_c^{ind}/dx)\,|_{x=0}$, is often expressed in terms of a surface recombination velocity, s_c, through (4.1.64) by writing the left-hand side

$$I_s = s_c n_c^{ind}(0).$$

Typical values of s_c are of the order of 10^3 m s^{-1} (Wolf, 1963, 1971).

For n–p type solar cells, expressions analogous to the above can be used.

Once $n_c^{ind}(x)$ has been found, I_s can be calculated. Figure 4.1-7 shows an example of the total current through a p–n junction, as a function of applied voltage but for a fixed rate of incoming solar radiation on the p-type surface. The short-circuit current I_s increases

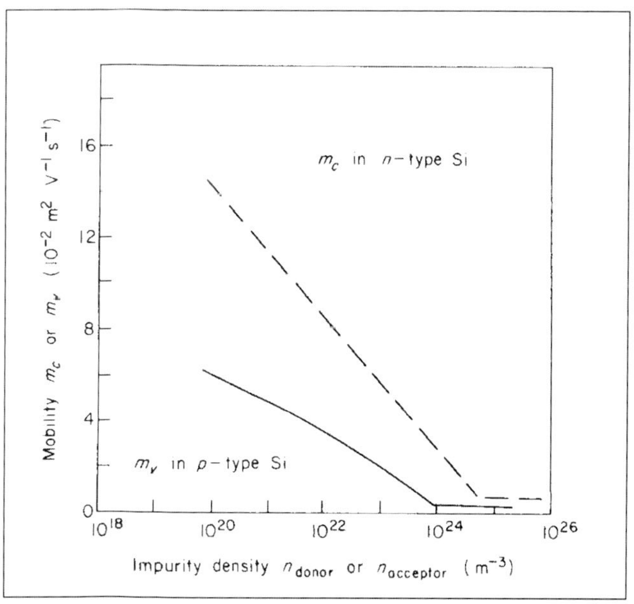

Figure 4.1-8 Mobility of minority carriers in Si at room temperature (about 300 K), extrapolated from measurements (Wolf, 1963). The mobility plotted is the "conduction mobility", equal to the conductivity divided by the number of minority carriers and by the electron charge. The attenuation of the flow of actual carriers by recombination effects (trapping) is not considered.

The drop in free energy (4.1.71) is usually considered to be mainly associated with the reaction (4.1.73), expressing G in terms of a chemical potential (3.3.49), e.g. of the H^+ ions dissolved in the electrolyte. Writing the chemical potential μ as Faraday's constant times a potential ϕ, the free energy for n moles of hydrogen ions is

$$G = n\mu = n\mathcal{F}\phi = nN_A e\phi. \quad (4.1.74)$$

When the hydrogen ions "disappear" at the positive electrode according to the reaction (4.1.73), this chemical free energy is converted into the electrical energy (4.1.70) or (4.1.71), and since the numbers of electrons and hydrogen ions in (4.1.73) are equal, $n = n_e$, the chemical potential μ is given by

$$\mu = \mathcal{F}\phi = \mathcal{F}\Delta\phi_{ext}. \quad (4.1.75)$$

Here ϕ is the quantity usually referred to as the electromotive force (e.m.f.) of the cell, or "standard reversible potential" of the cell, if taken at standard atmospheric pressure and temperature. From the value of ΔG quoted above, corresponding to -2.37×10^5 J per mole of H_2O formed, the cell e.m.f. becomes

$$\phi = -\Delta G/n\mathcal{F} = 1.23\text{V}, \quad (4.1.76)$$

with $n = 2$ since there are two H^+ ions for each molecule of H_2O formed. The chemical potential (4.1.75) may be parametrised in the form (3.3.50), and the cell e.m.f. may thus be expressed in terms of the properties of the reactants and the electrolyte [including the empirical activity coefficients appearing in (3.3.50) as a result of generalising the expression obtained from the definition of the free energy, (4.1.6), assuming P, V and T to be related by the ideal gas law, $PV = \mathcal{R}T$, valid for one mole of an ideal gas (cf. e.g. Angrist, 1976)].

The efficiency of a fuel cell is the ratio between the electrical power output (4.1.70) and the total energy lost from the fuel. However, it is possible to exchange heat with the surroundings, and the energy lost from the fuel may thus be different from ΔG. For an ideal (reversible) process, the heat added to the system is

$$\Delta Q = T\Delta S = \Delta H - \Delta G,$$

and the efficiency of the ideal process thus becomes

$$\eta^{ideal} = -\Delta G/(-\Delta G - \Delta Q) = \Delta G/\Delta H. \quad (4.1.77)$$

For the hydrogen–oxygen fuel cell considered above, the enthalpy change during the two processes (4.1.72) and (4.1.73) is $\Delta H = -9.5 \times 10^{-19}$ J or -2.86×10^5 J per mole of H_2O formed, and the ideal efficiency is

$$\eta^{ideal} = 0.83.$$

There are reactions with positive entropy change, such as $2C+O_2 \rightarrow 2CO$, which may be used to cool the surroundings and at the same time create electric power with an ideal efficiency above one (1.24 for CO formation).

In actual fuel cells, a number of factors tend to diminish the power output. They may be expressed in terms of "expenditure" of cell potential fractions on processes not contributing to the external potential,

$$\Delta\phi_{ext} = \phi - \phi_1 - \phi_2 - \phi_3 - \cdots,$$

where each of the terms $-\phi_i$ corresponds to a specific loss mechanism. Examples of loss mechanisms are blocking of pores in the porous electrodes, e.g. by piling up of the water formed at the positive electrode in the process (4.1.73), internal resistance of the cell (heat loss) and the building up of potential barriers at or near the electrolyte–electrode interfaces. Most of these mechanisms limit the reaction rates and thus tend to place a limit on the current of ions that may flow through the cell. There will be a limiting current, I_L, beyond which it will not be possible to draw any more ions through the electrolyte, because of either the finite diffusion constant in the electrolyte, if the ion transport is dominated by diffusion, or the finite effective surface of the electrodes at which the ions are formed. Figure 4.1-11 illustrates the change in $\Delta\phi_{ext}$ as a function of current,

$$I = \Delta\phi_{ext}R_{ext},$$

expressed as the difference between potential functions at each of the electrodes, $\Delta\phi_{ext} = \phi_c - \phi_a$. This representation gives some indication of whether the loss mechanisms are connected with the positive or negative electrode processes, and it is seen that the largest fraction

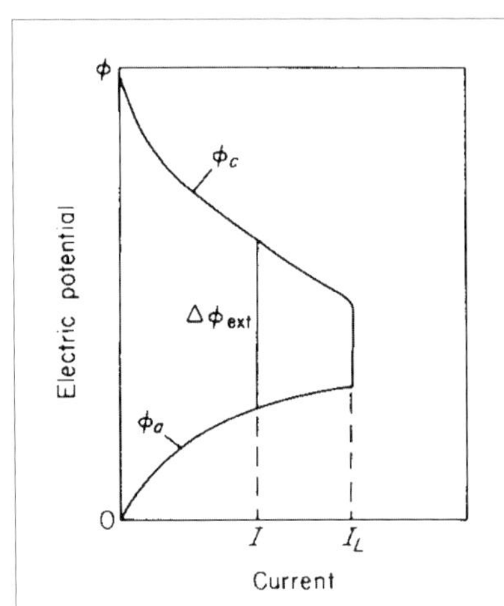

Figure 4.1-11 Fuel cell negative electrode potential ϕ_a and positive electrode potential ϕ_c as a function of current. The main cause of the diminishing potential difference $\Delta\phi_{ext}$ for increasing current is at first incomplete electrocatalysis at the electrodes, for larger currents also ohmic losses in the electrolyte solution, and finally a lack of ion transport (based on Bockris and Shrinivasan, 1969).

of the losses is connected with the more complex positive electrode reactions in this example. For other types of fuel cell it may be negative ions that travel through the electrolyte, with corresponding changes in characteristics.

It follows from diagrams of the type shown in Fig. 4.1-11 that there will be an optimum current, usually lower than I_L, for which the power output will be maximum,

$$\max(E) = I^{opt}\Delta\phi_{ext}^{opt}.$$

Dividing by the rate at which fuel energy ΔH is added to the system in a steady-state situation maintaining the current I^{opt}, one obtains the actual efficiency of conversion maximum,

$$\max(\eta) = I^{opt}\Delta\phi_{ext}^{opt}/(\mathrm{d}H/\mathrm{d}t). \qquad (4.1.78)$$

The potential losses within the cell may be described in terms of an internal resistance, the existence of which implies a non-zero energy dissipation, if energy is to be extracted in a finite time. This is the same fundamental type of loss as that encountered in the case of solar cells (4.1.67) and in the general thermodynamic theory described in section 4.1.1.

Other electrochemical conversion schemes

A driven-cell conversion based on the dissociation of water may be accomplished by electrolysis, using components similar to those of the hydrogen–oxygen fuel cell to perform the inverse reactions. Thus, the efficiency of an ideal conversion may reach 1.20, according to (4.1.78), implying that the electrolysis process draws heat from the surroundings.

By combining a fuel cell with an electrolysis unit, a regenerative system has been obtained, and if hydrogen and oxygen can be stored, an energy storage system or a "battery" results. The electric energy required for the electrolysis need not come from the fuel cell, but may be the result of an intermittent energy conversion process (e.g. wind turbine, solar photovoltaic cell).

Direct application of radiant energy to the electrodes of an electrochemical cell has been suggested, aiming at the achievement of the driven-cell process (e.g. dissociation of water) without having to supply electric energy. The electrodes could be made of suitable *p*- and *n*-type semiconductors, and the presence of photo-induced electron–hole excitations might modify the electrode potentials ϕ_a and ϕ_c in such a way that the driven-cell reactions become thermodynamically allowed. Owing to the electrochemical losses discussed above, additional energy would have to be provided by the solar radiation.

If the radiation-induced electrode processes have a low efficiency, the overall efficiency may still be higher for photovoltaic conversion of solar energy into electricity followed by conventional electrolysis (Manassen *et al.*, 1976). In recent years, several such photoelectrochemical devices have been constructed. Some of these will be described below in section 4.2.3.

4.1.2 Conversion of solar radiation

4.1.2.1 Heat generation

Conversion of solar energy to heat requires a light-absorbing material, a *collector*, which is able to distribute the absorbed radiant energy over internal degrees of freedom associated with kinetic energy of motion at the molecular level (e.g. lattice vibrations in the case of a solid). The Earth and its atmosphere are examples of such collectors. Absorption of solar energy will raise the temperature of the collector or transfer energy to a reservoir, if the collector is connected to one. The collector will also emit radiation, and it may lose heat energy by conduction and convection processes. The frequency spectrum of the emitted radiation will correspond to the Planck spectrum for the collector temperature T_c, if the collector is in a state allowing the definition of a thermodynamic temperature.

Man-made collectors may try to achieve a large absorption by minimising reflection and transmission and to achieve small losses, e.g. by operating the collector at temperatures not much above ambient air temperatures or, if higher load temperatures are required, by reducing the heat loss rates, e.g. by suitable transparent covers and insulation.

One may distinguish between "passive" and "active" systems, according to whether energy is specifically added (pumps, etc.) in order to bring the collector heat gain to the load areas or not. A passive system need not be characterised by the absence of definite heat flow paths between collectors and load areas, but such flows should be "natural", i.e. they should not depend on other energy inputs provided by man. There may be borderline cases in which the term "natural circulation" would be difficult to define.

Examples of passive solar heat systems are ordinary windows in buildings which transmit a large fraction of the solar radiation (if the angle of incidence is small, i.e. if the direction of the incident light does not make a small angle with the pane). The room behind the window may, to a large extent, act like a black body, absorb practically all of the radiation transmitted through the window and re-emit only a small fraction again to the outside (provided that the total window area of the room is not large).

Another kind of passive solar heat system uses the heat capacity of walls facing the sun during the daytime. The walls absorb radiation and accumulate it (the heat

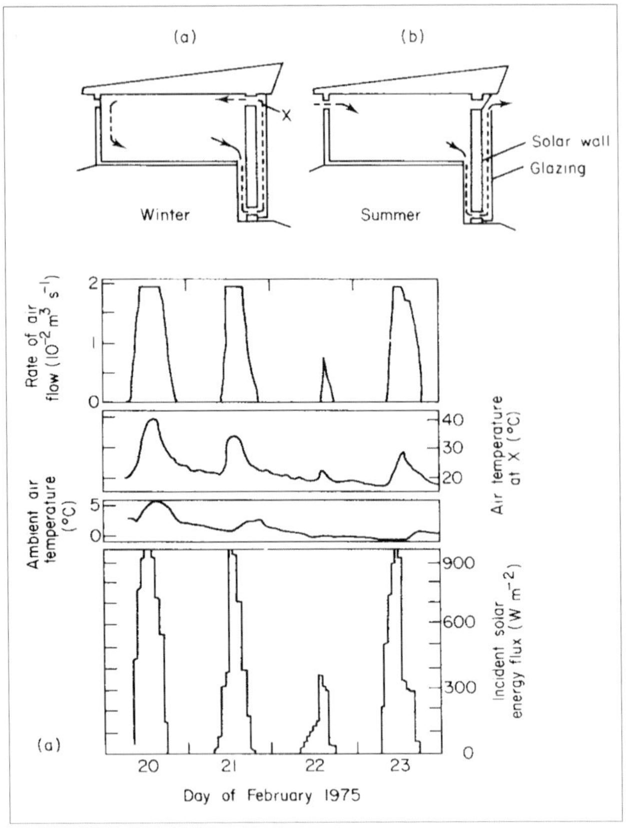

Figure 4.1-12 Solar wall type passive heating and cooling system. Top cross sections show air flows during (a) winter and (b) summer. Curves below show, for a selected period of a few days operation, the air flow-rate and temperature [at the location X in (a)], ambient temperature outside the house, and solar radiation. The inside wall temperature remains above 20°C during the period considered. The house is situated at 43°N latitude (based on Trombe, 1973; Stambolis, 1976).

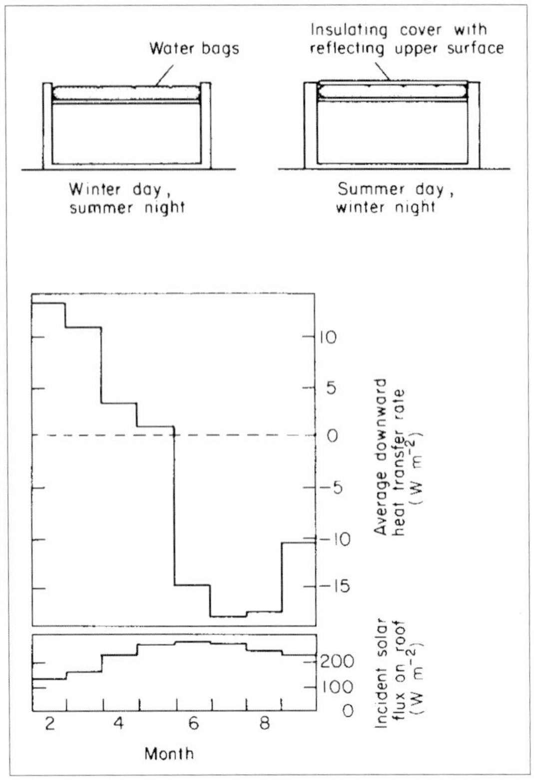

Figure 4.1-13 Solar roof-type passive heating and cooling system. The top cross sections show operation of cover panel at day and night, during summer and winter. Below are heat transfer rates through the roof (positive downward) and solar radiation, both averaged month by month during an 8-month period. The house is situated at 35°N latitude (based on Hay and Yellot, 1972; Stambolis, 1976).

capacity of building materials increases roughly in proportion to mass), and at night they lose heat to their colder surroundings, including the inside area which is thus heated. The wall's heat capacity also serves to cool the building during at least the first part of the daytime, if the wall temperature after the night's cooling off is lower than the next day's ambient temperature. More elaborate versions of such solar wall systems, directing the natural convection according to conditions (night/day, summer/winter), are shown in Figs. 4.1-12 and 4.1-13. These systems require a little "active" help in opening and closing vents, shutters or covers. Being dependent on a daily cycle, these systems are of most interest in climatic regions where the daily solar input is substantial also during the cold season.

Greenhouses are also passive solar-collecting systems, as are water heaters based on water bags (e.g. placed on roofs) or on flat-plate collectors (see below) delivering heat to water, which is transferred by natural circulation to a storage tank lying higher than the collector. A saline pond can play the role of the collector, with a heat exchanger at the bottom of the pond transferring the collected energy to a working fluid, which can be circulated to the load area by natural or forced circulation. The solar pond itself (see Fig. 4.1-14) contains water with a high content of dissolved salts, causing the formation of a salinity and density gradient, preventing the mixing of surface and bottom water. The water absorbs some solar radiation, but if the pond is shallow most of the absorption takes place at the bottom, thereby creating a stable temperature gradient increasing towards the bottom, because heat is transferred upwards only by slow processes.

Flat-plate collectors

The term "flat-plate collector" is used for absorbers with a generally flat appearance, although the collecting surface need not be flat in detail (it might have V-shaped carvings or even focusing substructure). The side of the collector facing the sun may have a cover system (e.g. one

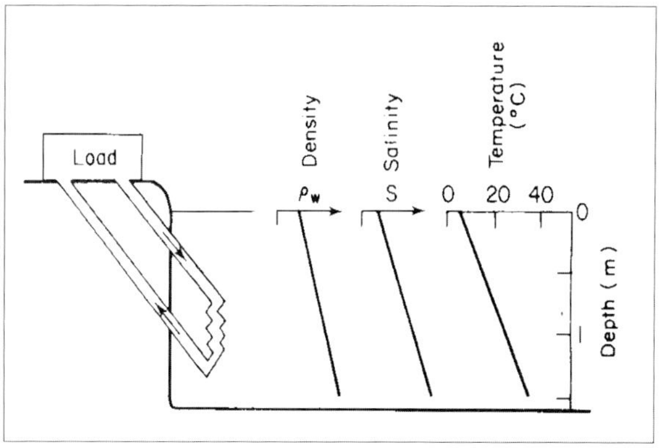

Figure 4.1-14 Schematic cross section through a solar pond (cf. Tabor, 1967). The temperature profile may have a secondary peak within the top hundredths of a metre, due to absorption of ultra-violet radiation in this layer, and depending on the re-radiation conditions (effective sky temperature relative to temperature of pond surface layer).

or more layers of glass), and a mass flow J_m^c of some fluid (e.g. water or air) passes the absorber and is supposed to carry the heat, which is transferred from the absorber plate to the mass flow, to the load area (i.e. the place of actual usage) or to some temporary energy storage. The general layout is shown in Fig. 4.1-15.

The absorber is characterised by an absorptance, $\alpha_\lambda(\Omega)$, which may depend on wavelength and on the direction of incident light (cf. section 3.1.4). In many situations, it may be assumed in a first approximation that the absorptance is independent of direction and of wavelength ("grey surface"). For a black painted surface, this α may be around 0.95, but, if the surface is structureless, the assumption that α is independent of direction breaks down when the angle between the normal to the surface and the direction of incidence exceeds about 60°. Towards 90°, $\alpha_\lambda(\Omega)$ actually approaches zero (Duffie and Beckman, 1974). According to (3.3.16), the emittance ε equals the absorptance (both are properties of the surface in the grey-surface approximation), and high absorptance thus implies high emittance for all wavelengths, including those characterising the thermal emission from the absorber of temperature T_c. In order to reduce this loss, use is made of surface treatments causing the emittance to assume two distinct values, a high one for the short wavelengths of the solar spectrum (implying a high solar absorptance α^{sw}) and a low one (α^{lw}) for the longer wavelengths characterising the emission from typical absorber temperatures, assumed to have a spectrum approximately equal to the black-body spectrum for that temperature. Such surfaces are called *selective surfaces*, and their wavelength-dependent absorptance/emittance may resemble that shown in Fig. 4.1-16, exhibiting the regions that can be characterised by $\alpha^{sw} \approx 0.95$ and $\alpha^{lw} \approx 0.1$. A further discussion

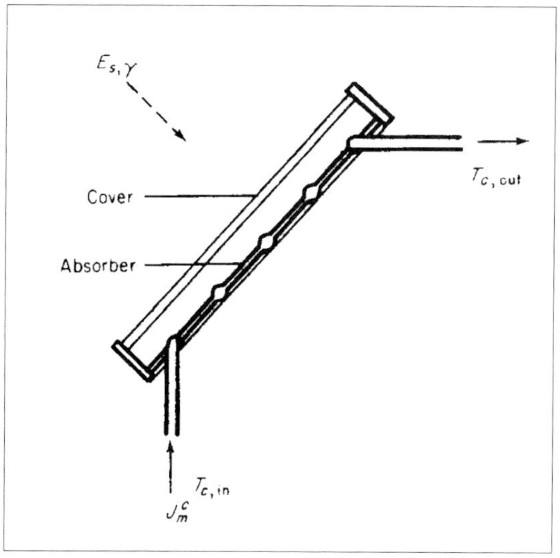

Figure 4.1-15 Example of flat-plate solar collector. The performance may not be given simply by the net incident radiation flux $E_{s,\gamma}$ (s is tilt angle and γ is azimuth angle), because the transmission–absorption product of the cover system may be different for different components of the incident radiation.

of selective surface technologies may be found in Meinel and Meinel (1976).

As indicated in Fig. 4.1-15, the absorber may be covered by material that reduces heat losses and at the same time transmits most of the incoming radiation. The cover may consist, for example, of one or more layers of glass. The transmittance of glass depends on the type of glass and in particular on minority constituents such as Fe_2O_3. Figure 4.1-17 gives an example of the wavelength dependence of the transmittance, τ_λ, through ordinary window glass, defined as the fraction of incident light which is neither reflected [cf. (3.3.12)] nor absorbed

$$\tau_\lambda = 1 - \rho_\lambda - \alpha_\lambda. \tag{4.1.79}$$

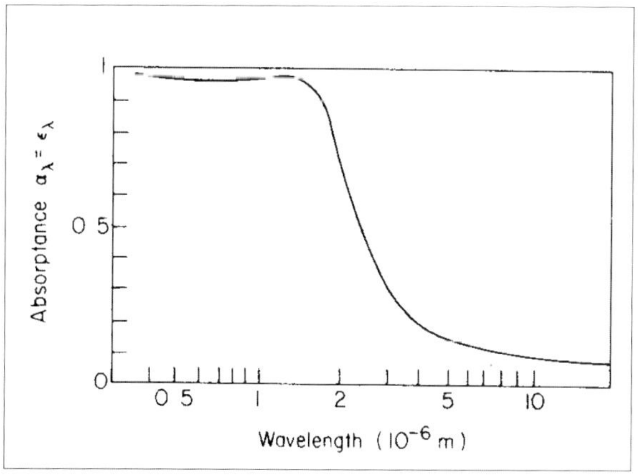

Figure 4.1-16 Spectral absorptance (or emittance) of a commercial "blackchrome" type of selective surface (based on Masterson and Seraphin, 1975).

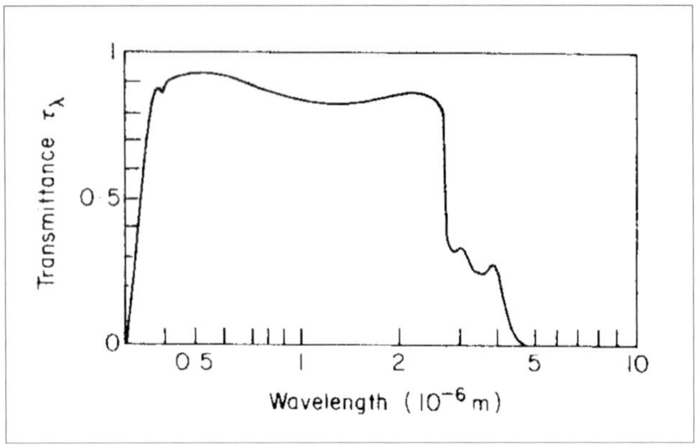

Figure 4.1-17 Spectral transmittance, $(1 - \rho_\lambda - \alpha_\lambda)$, of 2.8×10^{-3} m thick glass layer with a Fe_2O_3 content of 0.1 (based on Dietz, 1954).

For a collector of the type considered, multiple reflections may take place between the absorber and the cover system as well as within the cover system layers, if there are more than one. The reflections are mostly specular (angle of exit equal to angle of incidence) at glass surfaces, but mostly diffuse (hemispherically isotropic) at the absorber surface. Thus, the amount of radiation absorbed by the absorber plate, in general, cannot be calculated from knowledge of the total incoming flux but must be calculated for each incident direction from which direct, scattered or reflected light is received, as well as for each wavelength. The total absorbed flux is then obtained by integration.

In many cases a sufficiently good approximation can be obtained by considering only the two wavelength intervals, denoted "short" and "long" wavelengths above (the dividing wavelength being about 3×10^{-6} m), in view of the near-constancy of the absorptance and transmittance of the relevant materials (cf. Figs. 4.1-16 and 4.1-17). The net short-wavelength energy gain by the collector may then be written

$$E_c^{sw} = A \int E_{c+}^{sw}(\Omega) P^{t.a.}(\Omega)\, \mathrm{d}\Omega, \qquad (4.1.80)$$

in terms of the incoming flux from a given direction Ω, $E_{c+}^{sw}(\Omega)$, and the "transmission–absorption product", $P^{t.a.}(\Omega)$, describing the fraction of incident short-wavelength radiation from the direction Ω, which gets transmitted through the cover system and gets absorbed by the absorber plate, the area of which is denoted A. It has been assumed that A serves as a simple proportionality constant, implying that incident radiation as well as transmission–absorption properties are uniform over the entire collector area. For a cover system consisting of N layers, e.g. of glass, characterised by a refraction index relative to air, n (~1.5 for glass), and an extinction coefficient, x, such that the reduction in intensity of radiation from traversing a distance d through the material is $\exp(-xd)$, the total thickness of each cover layer being L, then $P^{t.a.}(\Omega)$ may be approximately written (see e.g. Duffie and Beckman, 1974)

$$P^{t.a.}(\Omega) = \frac{1-\rho}{1+(2N-1)\rho} e^{-xNL\cos\theta} \times \frac{\alpha^{sw}}{1-(1-\alpha^{sw})\rho_d(N)}, \qquad (4.1.81)$$

where θ is the polar angle between the incident direction Ω and the normal to the collector surface, and ρ is the reflectance of one cover layer, given by the Fresnel formula in the case of unpolarised light

$$\rho = \tfrac{1}{2}\left(\frac{\sin^2(\theta' - \theta)}{\sin^2(\theta' + \theta)} + \frac{\tan^2(\theta' - \theta)}{\tan^2(\theta' + \theta)}\right),$$

with the polar angle of refraction, θ', given by

$$\theta' = \mathrm{Arc\ sin}\left(\frac{\sin\theta}{n}\right).$$

The factorisation of (4.1.81) into a factor describing the multiple reflection through the N-layer cover system in the absence of absorption, times a factor describing the attenuation of the intensity from passage through glass, and a final factor describing the absorption in the absorber plate, after multiple reflections back and forth between the plate and cover system, is a valid approximation for most practical applications, owing to the smallness of the extinction product xL. The final factor in (4.1.81) may be rewritten as

$$\frac{\alpha^{s\omega}}{1-(1-\alpha^{s\omega})\rho_d} = \alpha^{s\omega} \sum_{i=1}^{\infty} ((1-\alpha^{s\omega})\rho_d)^i,$$

revealing that it is the sum of terms corresponding to i reflections back and forth between the plate and cover (considering only the lower cover surface), and assuming the non-absorbed fraction, $(1-\alpha^{sw})$, from each impact on the absorber plate to be diffusely reflected from the cover inside surface with a reflectance ρ_d different from that of specular reflection, ρ. In order to make up for processes in which light reflected back from the absorber gets transmitted to and reflected from cover layers other than the lowest one, and then eventually reaches back to become absorbed by the absorber plate, effective values of ρ_d depending on N may be used, as indicated in (4.1.81). Duffie and Beckman (1974) suggest that the diffuse reflectance ρ_d may be approximated by the specular reflectance for an incident polar angle of 60°, using values of 0.16, 0.24, 0.29 and 0.32 for $N = 1$, 2, 3 and 4.

For the direct part (3.3.1) of the incident radiation flux (3.3.15), only one direction Ω is permitted [given by

(3.3.2)], and the integration in (4.1.81) should be left out. For specular reflection on to the collector, the same is true in simple cases (cf. Seitel, 1975), but for complicated geometry of the specularly reflecting objects in the surroundings more than one direction may have to be taken into account. In general cases of reflection on to the collector, as well as for scattered radiation from the atmosphere [see (3.3.6)], the integration in (4.1.80) would generally have to be kept, but approximations such as multiplying the total scattered *or* reflected flux at the outside collector surface, (3.3.6) or (3.3.8) and (3.3.11) or (3.3.14), by the transmission–absorption product for an "average" angle of incidence, such as $\theta = 60°$, are sometimes used, permitting the total short-wavelength gain by the collector plate to be written

$$E_c^{sw} = A(D_{s,\gamma}P^{t.a.}(\theta) + (d_{s,\gamma} + R_{s,\gamma}) < P^{t.a.} >), \tag{4.1.82}$$

e.g. with $< P^{t.a.} >= P^{t.a.}(\theta = 60°)$.

The net long-wavelength energy gain by the collector is the difference between the long-wavelength radiation received from the surroundings, and the long-wavelength radiation emitted by the collector system. Considering the temperature T_c of the absorber plate as constant, the thermal radiation from the plate is of the form (3.3.17), and the net gain of an absorber without cover would be

$$E_c^{lw}(N = 0) = A\varepsilon_i^{lw}\sigma(T_e^4 - T_c^4), \tag{4.1.83}$$

where the temperature T_e of the environment, in general, is lower than the ambient air temperature, T_a, as discussed in section 3.3.1.4.

Finally, sensible heat may be exchanged between the collector and its surroundings through conduction and convection processes. The back and edge heat transfers are usually expressed in the same way as heat losses from buildings,

$$E_{back}^{sens} = -A\,U_{back}(T_c - T_b), \tag{4.1.84}$$

where U_{back} is a constant depending on the insulating properties of the materials separating the back side of the absorber plate from the environment of temperature T_b, which may be equal to the ambient temperature, T_a, if the collector is mounted freely or equal to the indoor temperature, T_L, if the collector forms an integral part of a building wall or roof. U_{back} may be assumed to include the "edge" heat losses from the sides of the collector, mainly of conduction type.

The net exchange of sensible energy at the front of the collector depends on the number of cover layers and on the wind speed, V, since the convection of heat is largely limited by the rate at which warm air at the top of the front cover is removed. In the absence of cover, the front (or top) exchange of heat may be written

$$E_{back}^{sens}(N = 0) = -Af_1(T_c - T_a), \tag{4.1.85}$$

where f_1 is a polynomial in V with empirical coefficients, as suggested by Duffie and Beckman (1991),

$$f_1 = max(5.0;\ 3095\,V[\mathrm{m\,s^{-1}}]^{0.6})\ [\mathrm{W\,K^{-1}\,m^{-2}}]. \tag{4.1.86}$$

With glass covers, the expressions (4.1.83) and (4.1.85) must be modified, for example, by explicitly considering the heat transfer equations connecting each layer and solving for steady-state solutions with constant (but different) temperatures of each layer. Klein (1975) has parametrised these solutions in an explicit form, given by

$$E_c^{lw} = \frac{A\sigma(T_e^4 - T_c^4)}{(\varepsilon_p^{lw} + c_1N(1-\varepsilon_p^{lw}))^{-1} + N(2+(f_2-1)/N) - N)/\varepsilon_g^{lw}}, \tag{4.1.87}$$

where ε_p^{lw} and ε_g^{lw} are the long-wavelength emittances of the absorber plate and of the cover glasses, $c_1 = 0.05$ and

$$f_2 = (1 + 0.089\,f_1 - 0.1166\,f_1\varepsilon_p^{lw})(1 + 0.07866N), \tag{4.1.88}$$

with f_1 inserted from (4.1.86) using the units indicated (Duffie and Beckman, 1991). The relation replacing (4.1.85) is

$$E_{top}^{sens} = -A\left(T_c - T_a\right) \cdot \left\{\left(\frac{NT_C}{f_3}f_4^{-C_2} + f_1^{-1}\right)^{-1} + \sigma(T_c + T_e)(T_c^2 + T_e^2)\left((\varepsilon_p^{lw} + 0.00591Nf_1)^{-1} + \frac{2N + f_2 - 1 + 0.133\varepsilon_p^{lw}}{\varepsilon_g^{lw}} - N\right)^{-1}\right\}. \tag{4.1.89}$$

Here $c_2 = 0.43(1 - 100/T_c)$ and

$$f_3 = 520[\mathrm{K}](1 - 0.000051s^2), \tag{4.1.90}$$

with s being the tilt angle of the collector or 70°, whichever is smaller. Further,

$$f_4 = (T_c - T_a)/(N + f_2), \tag{4.1.91}$$

unless this expression becomes smaller than 1, in which case f_4 should rather be equal to 1. All temperatures in (4.1.87) and (4.1.89) should be inserted in K. The parametrisations are clearly phenomenological, and substantial changes between editions of the Duffie and Beckman book, notably affecting winter results, emphasise the *ad hoc* nature of the procedure.

The total net energy gain of the absorber plate of the collector is the sum of the contributions specified above,

$$E_c^{gain} = E_c^{sw} + E_c^{lw} + E_{top}^{sens} + E_{back}^{sens}. \tag{4.1.92}$$

Stalled and operating collector

If the mass flow $J_m^c = dm/dt$ through the solar collector is zero, the entire energy gain (4.1.92) will be used to raise the temperature of the collector. This situation of a "stalled collector" is of considerable interest for determining the maximum temperature which the materials used in constructing the collector must be able to withstand, for example, in a situation of pump failure or complete loss of circulating fluid. Denoting the heat capacity of the "dry" collector C', taken per unit area of collector, the equation determining the time development of the stalled-collector plate temperature becomes

$$AC' \, dT_c/dt = E_c^{sens}(T_c). \tag{4.1.93}$$

As T_c rises, the negative terms in (4.1.92) increase, and if the incident energy can be regarded as constant, an equilibrium situation will result in which the heat losses exactly balance the gains and the temperature thus remains constant,

$$E_c^{gain}(T_{c,max}) = 0. \tag{4.1.94}$$

The determination of $T_{c,max}$ and the magnitude of the individual terms in (4.1.92), as a function of T_c, are illustrated in Fig. 4.1-18 for a high incoming flux E_c^{sw}, environmental temperatures $T_e = T_a = T_L = 300$ K, and no wind, based on the explicit expressions (4.1.84)–(4.1.92). Two collectors are considered, one with selective surface ($\varepsilon_p^{lw} = 0.11$) and one with non-selective black surface ($\varepsilon_p^{lw} = \alpha^{sw} = 0.94$). Both have two layers of cover glass. The corresponding equilibrium temperatures are approximately $T_{c,max} = 550$ K and $T_{c,max} = 480$ K. If wind were present, the convective heat loss (4.1.89) would be much higher and the maximum temperature correspondingly lower.

The time required to reach a temperature T_c close to $T_{c,max}$ can now be calculated from (4.1.93) by integration, as illustrated in Fig. 4.1-19. While the maximum temperature is independent of the heat capacity of the collector, the time scales linearly with C'. The value $C' = 10^4$ J m^{-2} K^{-1} used in Fig. 4.1-19 corresponds to a fairly light collector (e.g. absorber plate of extruded aluminium), in which case the asymptotic temperature region is reached after about 1 h of constant high incoming radiation. The shortness of this period makes the assumption of a constant E_c^{sw} acceptable. The amount of time needed to reach operating temperatures of about 50°C is roughly 7 min. It would double if the ambient temperature were at the freezing point of water and would further increase if the radiation were lower than the 1000 W m^{-2} assumed in the example used in Figs. 4.1-18 and 4.1-19. A short response time of the collector is also of relevance in situations of variable radiation (e.g. caused by frequent cloud passage).

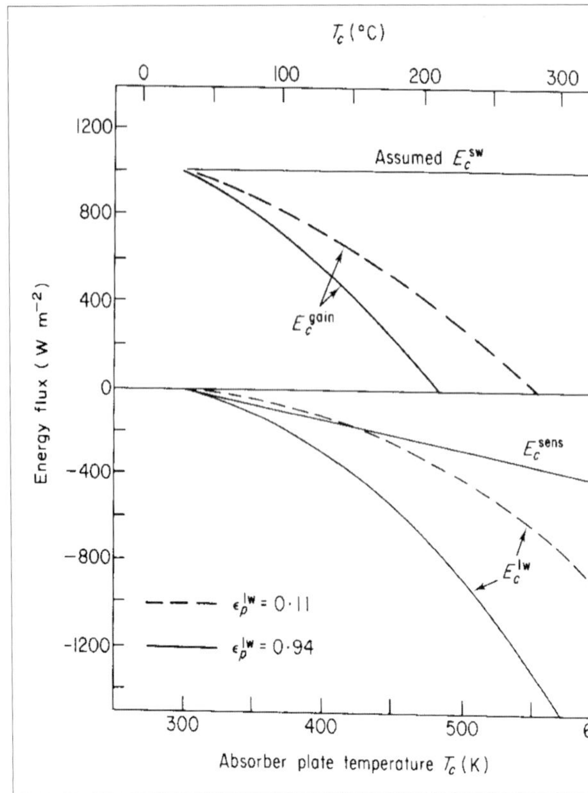

Figure 4.1-18 Energy flux components for a stalled collector, as a function of plate temperature T_c for a selective and a non-selective absorber surface (dashed and solid lines). There are two cover glasses of emittance $\varepsilon_g^{lw} = 0.88$, there is no wind (the curves are sensitive to wind speed changes), and all environmental temperatures have been taken as 300 K. The plate absorptance is $\alpha^{sw} = 0.94$, and the collector is placed horizontally. The curve E_c^{sens} represents the sum of fluxes reaching the top and back side of the collector (both positive towards the collector).

In case of an operating collector, J_m^c is no longer zero and a fluid carries energy away from the absorber (cf. Fig. 4.1-15). The temperature of the outgoing fluid, $T_{c,out}$, will in this case be higher than that of the incoming fluid, $T_{c,in}$, and it is no longer consistent to assume that the plate temperature T_c is uniform over the plate. The change in plate temperature must be at least as large as the change in fluid temperature, between the points where the fluid is receiving its first and last heat transfer contribution from the absorber plate. Because of the non-linear relation between plate temperature and heat loss, it is not correct to evaluate the losses at the mean plate temperature, T_c. Still, this is often done in approximate calculations, choosing

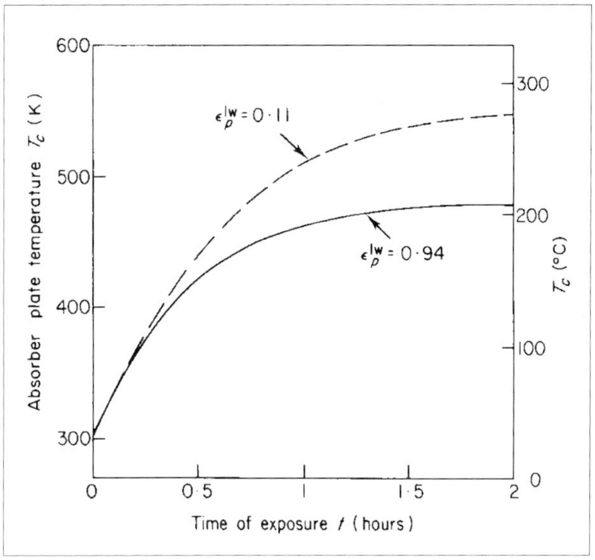

Figure 4.1-19 Plate temperature for a stalled, dry collector as a function of time. The incident flux is constant E_c^{sw} = 1000 W m^{-2}, and the heat capacity C' in (4.1.93) has been taken as 10^4 J m^{-2} K^{-1}, corresponding to a fairly lightweight construction. Other parameters are as in Fig. 4.1.1-18. About 7 min are required to reach operating temperatures (some 50 K above ambient). The time scale is proportional to C', so results for other values of C' are easily inferred.

$$\overline{T}_c = ½(T_{c,in} + T_{c,out}), \qquad (4.1.95)$$

and assuming further that the average fluid temperature is also the average plate temperature. Incomplete heat transfer between absorber plate and fluid channels, as well as the non-linearity of loss functions, may be simulated by adding a constant term to (4.1.95), or by using a factor larger than ½. The transfer of heat to the fluid may be effectuated by passing the fluid along the absorber plate (e.g. used in most collectors with air as a working fluid) or by the fluid's passage through pipes in highly heat-conducting contact with the absorber plate (most often used when the working fluid is water, oil, etc.). In the latter case, the non-uniform heat extraction constitutes a further cause of non-uniform plate temperature (the temperature gradients towards the pipe locations being, of course, the reason for obtaining a heat flow from any plate position to the relatively small area of contact with the fluid-carrying pipes).

If the fluid of fixed inlet temperature $T_{c,in}$ is allowed to pass only once through the collector (e.g. for hot water production), and if the simplifying assumptions mentioned above are made, then the equation replacing (4.1.93) for an operating collector will be

$$\overline{E}_c^{gain}(T_c) = AC'\, d\overline{T}_c/dt + J_m^c C_p^c(T_{c,out} - T_{c,in}), \qquad (4.1.96)$$

or in a steady state situation just

$$E_c^{gain}(\overline{T}_c) = J_m^c C_p^c(T_{c,out} - T_{c,in}).$$

C_p^c is the heat capacity of the fluid flowing through the collector. This equation determines $T_{c,out}$, but the left-hand side depends on $T_{c,out}$, through (4.1.95) or some generalisation of this relation between the flow inlet and outlet temperatures and the effective average plate temperature. Therefore, it is conveniently solved by iteration,

$$T_{c,out}^{(i+1)} = T_{c,in} + E_c^{gain}(½(T_{c,in} + T_{c,out}^{(i)}))/(J_m^c C_p^c). \qquad (4.1.97)$$

A better approximation for the transfer of heat from the absorber plate to the fluid channels may be obtained along the lines described below for the general problem of heat exchange encountered in several places with application of the solar collectors in heat supply systems, for example, involving a heat storage which is not part of the collector flow circuit.

Heat exchange

A situation like the one depicted in Fig. 4.1-20 is often encountered in energy supply systems. A fluid is passing through a reservoir of temperature T_3, thereby changing the fluid temperature from T_1 to T_2. In order to determine T_2 in terms of T_1 and T_3, in a situation where the change in T_3 is much smaller than the change from T_1 to T_2, the incremental temperature change of the fluid by travelling a short distance dx through the pipe system is related to the amount of heat transferred to the reservoir, assumed to depend linearly on the temperature difference,

$$J_m C_p^{fluid} dT^{fluid}/dx = h'(T_3 - T^{fluid}).$$

Integrating from T_1 at the inlet ($x = x_1$) gives

$$T^{fluid}(x) - T_3 = (T_1 - T_3)\exp(-h'(x - x_1)/(J_m C_p^{fluid})), \qquad (4.1.98)$$

where h' is the heat transfer per unit time from a unit length of the pipe for a temperature difference of one

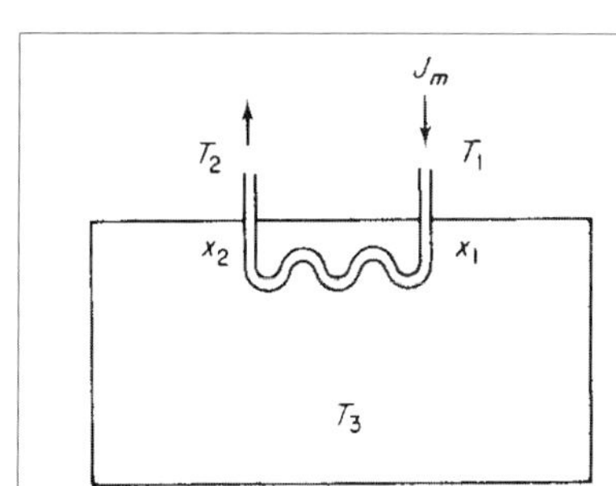

Figure 4.1-20 Heat exchanger, an idealised example of a well-mixed T_3-reservoir.

unit. The heat transfer coefficient for the entire heat exchanger is

$$h = \int_{x_1}^{x_2} h' \mathrm{d}x,$$

which is sometimes written $h = U_h A_h$, with U_h being the transfer coefficient per unit area of pipe wall and A_h being the effective area of the heat exchanger. For $x = x_2$, (4.1.98) becomes (upon re-ordering)

$$(T_1 - T_2) = (T_1 - T_3)(1 - \exp(-h/(J_m C_p^{fluid}))). \tag{4.1.99}$$

Flat-plate collector with heat storage

The general layout of a flat-plate solar collector with heat storage is shown in Fig. 4.1-21. There are two heat exchangers on the collector side and two on the load side. If the actual number is less, for example if the storage is in terms of a fluid that can be circulated directly through the collector, then the corresponding heat transfer coefficient [h in (4.1.99)] should be taken as infinitely large. While the relationship between temperatures at every heat exchanger may be taken from (4.1.99), the net energy transfer to the storage is associated with a change in the storage temperature (the average value of which is $\overline{T}_s$), which may be obtained from the equation accounting for the relevant energy fluxes, assuming the storage temperature to be uniform,

$$SC^s \, \mathrm{d}T_s/\mathrm{d}t = J_m^c C_p^c (T_{c,out} - T_{c,in}) - J_m^L C_p^L (T_{L,in} - T_{L,out}) - h_s(\overline{T}_s - T_0). \tag{4.1.100}$$

Here S is the storage volume (or mass), C^s is the heat capacity on a volume (or mass) basis of the material used for storage (assuming for simplicity storage in the form of sensible heat), superscripts c and L distinguish the flows in the collector and the load circuits, and finally, T_0 is the temperature of the surroundings of the storage volume (e.g. soil in the case of underground storage), to which the storage is losing energy through its walls. Equation

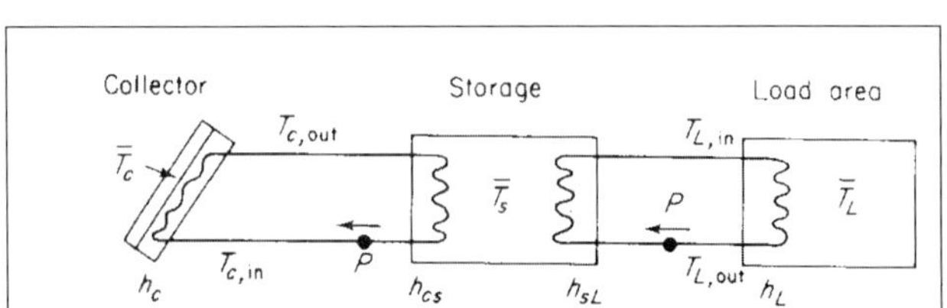

Figure 4.1-21 Solar heating system with storage. h denotes a heat exchanger, P denotes a circulation pump, and the Ts are temperatures.

(4.1.100) must be supplemented by (4.1.96) and a similar equation for the load area, e.g. of the form

$$LC^L \, \mathrm{d}T_L/\mathrm{d}t = J_m^L C_p^L (T_{L,in} - T_{L,out}) - h_L(\overline{T}_L - T_0), \tag{4.1.101}$$

assuming that the load area is a building of volume (or mass) L and average temperature $\overline{T}_L$, which may lose energy to surroundings of ambient air temperature T_a. If the load is in the form of water heating, it may be described by a term of the same form as the first one on the right-hand side of (4.1.101), but with $T_{L,out}$ being the water inlet temperature and $T_{L,in}$ being the water temperature of the water supplied by the system for usage.

Together with the four heat exchanger equations (4.1.99) with h_c, h_{cs}, h_{sL} and h_L (see Fig. 4.1.-21), (4.1.96), (4.1.100) and (4.1.101) constitute seven equations for determining the seven temperatures appearing in Fig. 4.1-21. If the load circuit is not activated, the four equations for $T_{c,in}$, $T_{c,out}$, $\overline{T}_s$ and $\overline{T}_c$ remain. Instead of the heat exchanger equation (4.1.99) for the collector, which for $h_c \to \infty$ gives $T_{c,out} = \overline{T}_c$ owing to the assumption that the reservoir of temperature $\overline{T}_c$ is large compared with the amounts of heat exchanged, relations such as (4.1.95) may be used, or a more elaborate calculation taking into account the geometry of the collector and fluid pipes or channels may be performed.

The efficiency of the collector itself may be defined as

$$\eta_c = J_m^c C_p^c (T_{c,out} - T_{c,in})/E_{s,\gamma}^{s\omega}, \tag{4.1.102}$$

where the numerator may be obtained from (4.1.96) and the denominator from (3.3.15). Only if the heat capacity of the collector is neglected, or if the collector is operating at a constant temperature, can the collector heat output (to the fluid) be taken as the collector gain $E_c^{gain}(T_c)$ from (4.1.92) directly.

If an energy store is used in connection with the collector, the relevant efficiency is that of delivering the heat to the storage, which is formally the same as (4.1.102). In practice, however, the introduction of an energy storage may completely alter the efficiency by changing $T_{c,in}$, which again causes a change in $T_{c,out}$. When the storage temperature, $\overline{T}_s$, is low, the average collector plate temperature $\overline{T}_c$ will be fairly low and the loss terms correspondingly small. Since $T_{c,in}$ will always be higher than T_s (but not always higher than $\overline{T}_s$, since often the fluid circuit is made to leave the storage at a place of low temperature, if T_s is not uniform), the accumulation of stored energy and associated increases in T_s and $T_{c,in}$ will cause the collector loss terms to increase, implying a diminishing efficiency. Examples of this behaviour will be given in Chapter 6.2, in the discussion of simulations of complete solar heating systems.

Another important factor in determining the efficiency of energy collection is the rate of fluid circulation through the collector circuit, J_m^c. Figure 4.1-22 gives an example of the performance of the same system with two different values of J_m^c. With the low fluid velocity, the exit temperature from the collector, $T_{c,out}$, is about 45°C above the inlet temperature, but the energy transferred to the storage is much less than in the case of a higher fluid velocity, causing a smaller difference between $T_{c,out}$ and $T_{c,in}$, but larger net collector gain, owing to the smaller loss terms. The optimum fluid flow rate for a given heat exchange capability h_{cs} (assuming the heat transfer at the collector to be nearly perfect) may be found by calculating the collection efficiency as a function of J_m^c or, as a more generally applicable criterion, by calculating the coverage of the actual energy demand

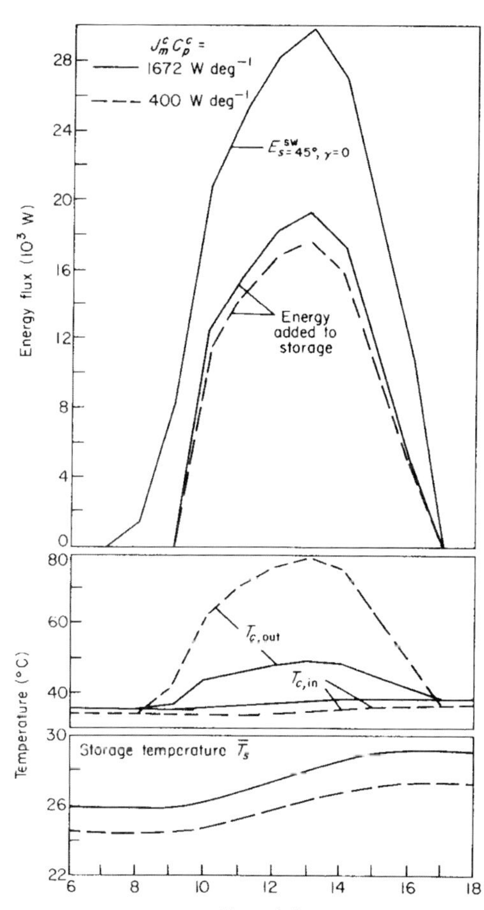

Figure 4.1-22 Single-day performance of a solar heating system, based on a simulation model using the Danish Reference Year (latitude 56°N). The day is 8 February, the collector area $A = 40$ m^2, and the storage volume $S = 25$ m^3 (water equivalent). The collector has one glass cover layer and a selective surface ($\varepsilon_c^{lw} = 0.11$), and the heat transfer coefficient $h_{cs} = 4000$ W K^{-1}. Two different flow rates are considered for the collector circuit, $J_m^c C_p^c = 1672$ or 400 W K^{-1} corresponding to 0.4 or 0.096 kg of water per second.

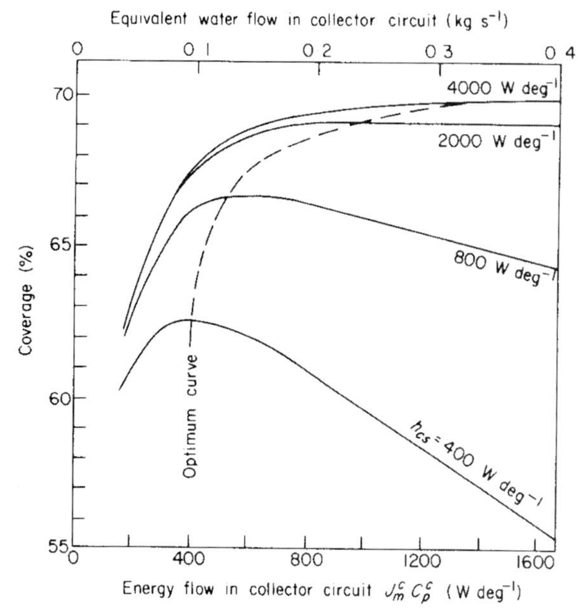

Figure 4.1-23 Annual average coverage of heating and hot water load for a one-family dwelling (average load 1.91 kW) under conditions set by the Danish Reference Year (56°N) (cf. section 6.2.1) as a function of water flow in collector circuit and heat exchanger transfer coefficient. The collector is as described in the caption to Fig. 4.1-22 and placed on a south-facing roof with tilt angle $s = 45°$.

which is sought covered by the solar heat system. If a given solar heat system is considered, the percentage of the total heat load over an extended period, such as a year, may be evaluated as function of J_m^c, as done in Fig. 4.1-23.

If 100% coverage is aimed at, the parameter to be optimised may be taken as the minimum size (e.g. collector area A) which will allow the full load to be covered at any time. The example shown in Fig. 4.1-23 relates to the heating and hot water requirements of a one-family dwelling at 56°N latitude. For each value of the heat exchange coefficient h_{cs} between the collector circuit and the storage, the optimum flow rate is determined. With increasing h_{cs}, the region of "acceptable" values of J_m^c becomes larger, the increased J_m^c indicated for large h_{cs} does not significantly improve the performance of the system, and it will soon become uneconomical due to the energy spent in pumping the fluid through the circuit (the energy required for the pump is negligible for low fluid speeds, but the pipe resistance increases with the fluid speed, and the resistance through the heat exchanger increases when h_{cs} does). The results of Fig. 4.1-23 may then be interpreted as suggesting a modest increase in fluid velocity, from about 0.1 to about 0.2 kg (or litre) s^{-1}, when the heat exchange coefficient is increased from 400 to 4000 W K^{-1} (for the particular system considered).

The optimisation problem outlined in this example is typical of the approach to the concept of an "efficiency" for the total solar heat system, including storage and load sections. Because the time distribution of the load is, in general, barely correlated with that of the solar energy collection (in fact, it is more "anti-correlated" with it), an instantaneous efficiency defined as the ratio between the amount of energy delivered to the load area and the amount of incident solar energy is meaningless. A meaningful efficiency may be defined as the corresponding ratio of average values, taken over a sufficient length of time to accommodate the gross periodicity of solar radiation and of load, i.e. normally a year

$$\bar{\eta}_{system} = \bar{c}/\bar{E}^{sw}_{s,\gamma}, \qquad (4.1.103)$$

where c is the average load covered by the solar energy system and $\bar{E}^{sw}_{s,\gamma}$ is the average incident short-wavelength radiation on the surface of the collector (its cover if any). Conventionally, the long-wavelength or other heat contributions to the incident energy flux are not included in the efficiency calculations (4.1.103) or (4.1.102).

Concentrating collectors and tracking systems

Various advanced versions of the flat-plate collector are conceivable in order to increase the amount of absorbed energy and decrease the amount of heat lost. The incident average flux may be increased by replacing the fixed installation (of tilt and azimuth angles s and γ) by a movable one, which can be made to track the direction towards the Sun. Fully tracking (i.e. in both height and azimuth angle) devices may be mounted in a manner similar to the one used for (star-)tracking astronomical instruments, and incompletely tracking systems (e.g. only following the azimuth angle of the Sun) might combine a simpler design with some improvement in energy gain. An idea of the gain made possible by tracking the Sun can be obtained by comparing Fig. 3.3-19 with Fig. 3.3-20. The scattered and (diffusely) rejected fluxes do not change much on average by going from a fixed to a tracking collector, but the direct part (3.3.1) becomes replaced by the normal incidence flux S_N. For a flat-plate collector, the maximum average gain from using a fully tracking system (less than a factor 2) would rarely justify the extra cost of the tracking system which, at least at present, exceeds the cost of doubling the collector area.

Other possible improvements of the flat-plate collector system include coating the cover layers to reduce reflection losses on the outside and increase reflection (back on to the absorber) on the inside surfaces. The mainly convective heat loss E^{sens}_{top} may be reduced by installation of shields that diminish the wind velocity outside the top cover layer. The strong dependence of convection losses on wind speed is clear from (4.1.89). The convective losses may be more radically diminished by evacuating the space between the absorber plate and the cover system. In vacuum, heat transport can take place only by radiation. The evacuation of the substantial space between absorber and cover typical of collector designs of the type shown in Fig. 4.1-15 is not practical, and systems using evacuation-type insulation would be built in a different way, e.g. with units of evacuated cylinders or spheres containing the absorbers and, possibly, concentrating equipment.

Focusing, or more generally concentrating, devices constitute a separate class of solar collectors. These are necessary if very high temperatures are required, but may, in principle, also be considered for heating purposes involving modest temperature rises over the ambient. For a flat-plate collector without any of the special loss-reducing features mentioned above, the net gain is a rapidly decreasing function of the plate temperature, as seen in Fig. 4.1-18. Since the losses are proportional to the absorber area, there is a special interest in reducing this area relative to the collector area A determining the incident radiation energy.

One extreme is a point-focusing device, such as the parabolic reflectors in Fig. 4.1-24 or the lenses in Fig. 4.1-26. If the absorber is placed in the focus on the symmetry axis and its dimension corresponds to that of the image of the Sun when the Sun is in the direction of the symmetry axis ("optical axis"), then direct radiation will be accepted, but only as long as the optical axis is made to track the Sun. Imperfections and, in the case of lenses, the wavelength dependence of the refraction index may further enlarge the image dimensions. The absorber dimension would have to be chosen so that it covered the bulk of the image under most conditions.

The total short-wavelength radiation received by the absorber, E^{sw}_c, may be written in the same form as for the flat-plate collector, (4.1.80) or (4.1.82), but with a transmission–absorption product, $P^{t.a.}(\Omega)$, suitable for the focusing geometry rather than (4.1.81). If reflections and re-absorptions at the absorber surface are neglected, the transmission–absorption product for a focusing device may be written as the product of the absorptance α^{sw} of the absorber surface and a transmission function $t^{sw}(\Omega)$, which expresses the fraction of the incident

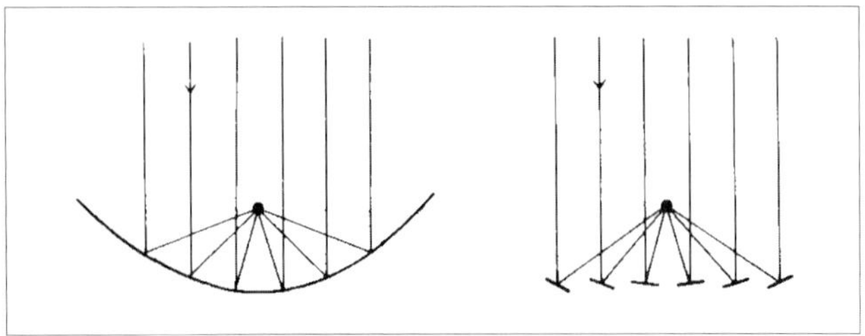

Figure 4.1-24 Parabolic- and Fresnel-type reflectors, e.g. for use in heliostats and tower absorbers, respectively.

short-wavelength radiation from the direction Ω that reaches the absorber surface (directly or after one or more reflections or refractions within the device),

$$p^{t.a.}(\Omega) = \alpha^{sw} t^{sw}(\Omega). \qquad (4.1.104)$$

For the parabolic reflector (Fig. 4.1-24) and the lens (Fig. 4.1-26), the idealised transmission function is zero, unless the absorber is placed exactly in the focal point corresponding to incoming radiation from the direction Ω. Thus, either the collector has to fully track the Sun or the absorber has to be moved to the focal point corresponding to a given direction to the Sun. For scattered radiation, which by definition does not come from the direction of the Sun (the same being normally the case for reflected radiation), $t^{sw}(\Omega)$ is zero, and such radiation will not be available for these devices. As a compromise, the acceptance angle may be increased at the expense of a high concentration ratio. Some scattered and reflected radiation will be accepted, and complete tracking may not be necessary.

Figure 4.1-25 gives a number of examples of such devices. The inclined "booster" mirrors along the rim of a flat-plate collector (Fig. 4.1-25a) result in a modest increase in concentration for incident angles not too far from the normal to the absorber plate, but a decrease (due to shadows) for larger incident angles. The V- or cone-shaped collector with the absorber in the shape of a wall or an upright line (cylinder) in the middle (Fig. 4.1-25b) has an acceptance angle depending on the distance from the convergence point of the cone. At this point, the acceptance angle equals the opening angle, but at points further out along the cone sides the acceptance angle is smaller (depending on the length of the absorber line, or wall height) relative to that of the cone. A fairly high concentration can be achieved with a cusp construction (Fig. 4.1-25c), comprising two parabolic sections, each of which has its focus at the point of intersection between the opposite parabolic section and the absorber. Figure 4.1-25d shows a "trapping" device consisting of a cylinder (or sphere) with a window, and an absorber displaced from the centre. The dimensions of the absorber are such that a large fraction of the radiation passing through the window will be absorbed after (multiple) reflections on the reflecting inside walls of the cylinder. Also, for the long-wavelength radiation emitted by the absorber, the reflecting walls will lead a substantial fraction back to the absorber, so that the loss-to-gain ratio is improved, even though there is no actual concentration (if the dimension of the absorber is as large as that of the window).

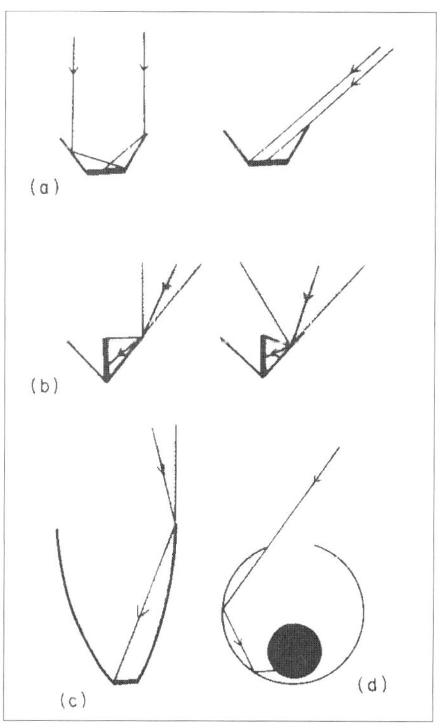

Figure 4.1-25 Examples of concentrating collectors.

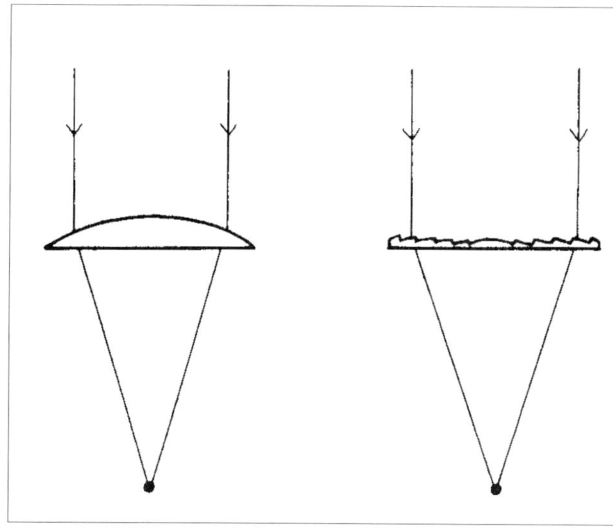

Figure 4.1-26 Simple Fresnel-type lenses.

Energy collection from focusing systems

A survey of a number of optically focusing systems for solar energy collection may be found in Meinel and Meinel (1976).

In the absence of cover systems, the long-wavelength emission from a focusing system may be written in the form (4.1.83),

$$-E_c^{lw} = A_a \varepsilon^{lw} \sigma (T_c^4 - T_e^4), \qquad (4.1.105)$$

where A_a is the absorber area.

Similarly, the sensible heat exchange with the surroundings of ambient temperature T_a may be given in analogy to (4.1.84) and (4.1.85),

$$E_c^{sens} = -A_a U(T_c - T_a), \tag{4.1.106}$$

where u may depend on wind speed in analogy to (4.1.86). The collector gain is now given by $E_c^{gain} = E_c^{sw} + E_c^{lw} + E_c^{sens}$, where the first term is proportional to A, while the two last terms are proportional to A_a. Assuming now a fluid flow J_m^c through the absorber, the expression for determination of the amount of energy extracted becomes [cf. (4.1.96)]

$$\begin{aligned} J_m^c C_p^c(T_{c,out} - T_{c,in}) &= E_c^{gain}(\overline{T}_c) - A_a C' \mathrm{d}\overline{T}_c/\mathrm{d}t \\ &= E_c^{sw} + E_c^{lw}(\overline{T}_c) + E_c^{sens}(\overline{T}_c) \\ &\quad - A_a C' \mathrm{d}\overline{T}_c/\mathrm{d}t, \end{aligned} \tag{4.1.107}$$

where $A_a C'$ is the heat capacity of the absorber and the relation between $\overline{T}_c$ and $T_{c,out}$ may be of the form (4.1.97) or (4.1.99). Inserting (4.1.82) with use of (4.1.104), (4.1.105) and (4.1.106) into (4.1.107), it is apparent that the main additional parameters specifying the focusing system are the area concentration ratio,

$$X = A/A_a, \tag{4.1.108}$$

and the energy flux concentration ratio,

$$C^{flux} = At^{SW}(\Omega)/A_i, \tag{4.1.108'}$$

where t^{SW} is given in (4.1.104) and A_i is the area of the actual image (which may be different from A_a). Knowledge of A_i is important for determining the proper size of the absorber, but it does not appear in the temperature equation (4.1.107), except indirectly through the calculation of the transmission function t^{SW}.

For a stalled collector, $J_m^c = 0$ and $\mathrm{d}\overline{T}_c/\mathrm{d}t = 0$ give the maximum absorber temperature $T_{c,max}$. If this temperature is high, and the wind speed is low, the convective heat loss (4.1.108) is negligible in comparison with (4.1.105), yielding

$$\sigma(T_{c,max}^4 - T_c^4) = X\, D\, \alpha^{sw} \varepsilon^{lw},$$

where D is the incident short-wavelength flux in the direction of the Sun [cf. (4.1.82)], assuming the device accepts only the flux from this direction and assuming perfect transmission to the absorber ($t^{SW} = 1$). With the same conditions, the best performance of an operating collector is obtained from (4.1.108′), assuming a steady-state situation. The left-hand side of (4.1.108′) is the amount of energy extracted from the collector per unit of time, E^{extr}:

$$E^{extr} = A\alpha^{SW} D - A_a \varepsilon^{lw} \sigma(\overline{T}_c^4 - T_c^4), \tag{4.1.109}$$

and the efficiency corresponding to the idealised assumptions

$$\eta_c^{ideal} = \frac{E^{extr}}{AD} = \alpha^{SW}\left(1 - \frac{\varepsilon^{lw}\sigma}{X\alpha^{SW}D}(\overline{T}_c^4 - T_e^4)\right). \tag{4.1.110}$$

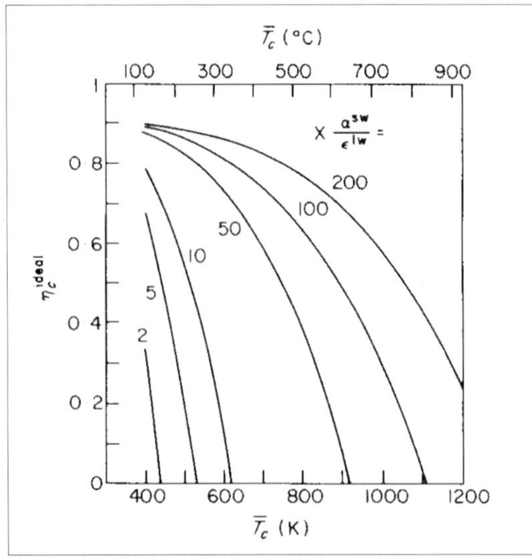

Figure 4.1-27 Ideal collector efficiency for concentrating collectors, evaluated for a direct radiation incident flux of $D = 800$ W m^{-2}, an environmental temperature $T_c = 300$ K and $\alpha^{SW} = 0.9$.

This relation is illustrated in Fig. 4.1-27, assuming $D = 800$ W m^{-2}, $T_c = 300$ K and $\alpha^{SW} = 0.9$ independent of absorber temperature. It is clear that the use of a selective absorber surface (large ratio $\alpha^{SW}/\varepsilon^{lw}$) allows a higher efficiency to be obtained at the same temperature and that the efficiency drops rapidly with decreasing light intensity D. In case the total short-wavelength flux on the collector is substantially different from the direct component, the actual efficiency relative to the incident flux $E_{s,\gamma}^{SW}$ of (3.3.15),

$$\eta_c = E^{extr}/E_{s,\gamma}^{SW},$$

may become far less than the "ideal" efficiency (4.1.110).

4.1.2.2 Applications for cooling, pumping, etc.

A variety of applications of solar energy thermal conversion have been considered, in addition to heat use. Living comfort and food preservation require cooling in many places. As in the case of heating, the desired temperature difference from the ambient is usually only about 10–30°C. Passive systems such as those depicted in Figs. 4.1-12 and 4.1-13 may supply adequate cooling in the climatic regions for which they were designed. Radiative cooling is very dependent on the clearness of the night sky (cf. section 3.3.1.4). In desert regions with a clear night sky, the difference between the winter ambient temperature and the effective temperature of the night sky has been used to freeze water. In Iran, the ambient winter temperature is typically a few degrees Celcius above the freezing point of water, but the temperature T_e of the night sky is below 0°C. In previous centuries, ice (for use in the palaces of the rich) was produced from shallow ponds surrounded by walls to screen out the daytime winter Sun, and the ice produced during the night in such natural icemakers ("yakhchal", see Bahadori, 1977) could be stored for months in very deep (say, 15 m) underground containers.

Many places characterised by hot days and cool nights have taken advantage of the heat capacity of thick walls or other structures to smooth out the diurnal temperature variations. If additional cooling is required, active solar cooling systems may be employed, and a "cold storage" may be introduced in order to cover the cooling need independent of the variations in solar radiation, in analogy to the "hot storage" of solar heating systems.

The solar cooling system may consist of flat-plate collectors delivering the absorbed energy to a "hot storage", which is then used to drive an absorption cooling cycle (Fig. 4.1-28), drawing heat from the "cool storage", to which the load areas are connected (see e.g. Wilbur and Mancini, 1976). In principle, only one kind of storage is necessary, but with both hot and cold storage the system can simultaneously cover both heating needs (e.g. hot water) and cooling needs (e.g. air conditioning).

The absorption cooling cycle (Fig. 4.1-28) is achieved by means of an absorbent–refrigerant mix, such as $LiBr–H_2O$ or $H_2O–NH_3$. The lithium–bromide–water mix is more suitable for flat-plate solar collector systems, giving higher efficiency than the water–ammonia mix for the temperatures characteristic of flat-plate collectors. LiBr is hygroscopic, i.e. it can mix with water in any ratio. The solar heat is used in a "generator" to vaporise some water from the mix. This vapour is led to a condenser unit, using a coolant flow, and is then expanded to regain a gaseous phase, whereby it draws heat from the area to be cooled, and is returned to an "absorber" unit. Here it becomes absorbed in the $LiBr–H_2O$ mix with the help of a coolant flow. The coolant inlet temperature would usually be the ambient one, and the same coolant passes through the absorber and the condenser. The coolant exit temperature would then be above ambient, and the coolant cycle could be made closed by exchanging the excess heat with the surroundings (e.g. in a "cooling tower"). The refrigerant-rich mix in the absorber is pumped back to the generator and is made up for by recycling refrigerant-poor mix from the generator to the absorber (e.g. entering through a spray system). In order not to waste solar heat or coolant flow, these two streams of absorbent–refrigerant mix are made to exchange heat in a heat exchanger. The usefulness of a given absorbent–refrigerant pair is determined by the temperature dependence of vaporisation and absorption processes.

In dry climates a very simple method of cooling is to spray water into an air-stream (evaporative cooling). If the humidity of the air should remain unchanged, the air has first to be dried (spending energy) and then cooled by evaporating water into it, until the original humidity is again reached.

In principle, cooling by means of solar energy may also be achieved by first converting the solar radiation to electricity by one of the methods described in section 4.1.2.3 and then using the electricity to drive a thermodynamic cooling cycle, such as the Rankine cycle in Fig. 4.1-3.

The same applies if the desired energy form is work, as in the case of pumping water from a lower to a higher reservoir (e.g. for irrigation of agricultural land). In practice, however, the thermodynamic cycles discussed in section 4.1.1.2 are used directly to make a pump produce mechanical work on the basis of the heat obtained from a solar collector. Figure 4.1-29 shows two

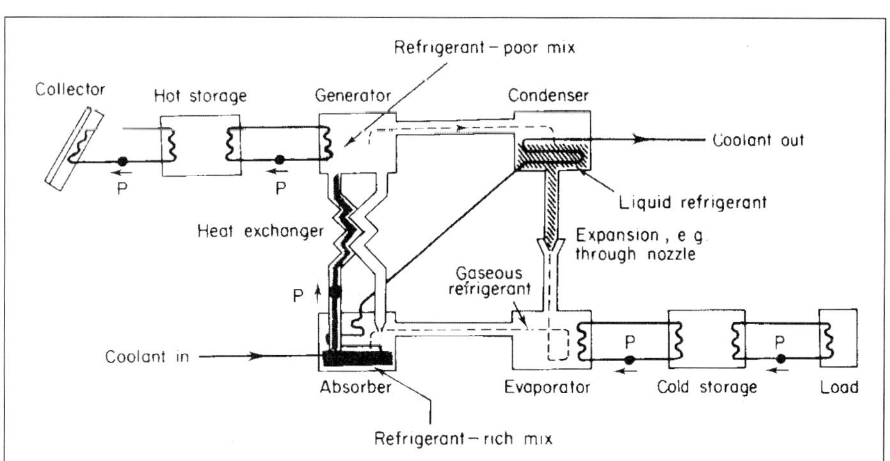

Figure 4.1-28 Solar absorption cooling system using a mix of absorbent (e.g. LiBr) and refrigerant (e.g. H_2O). P denotes pump. The coolant may be recycled after having rejected heat to the environment in a cooling tower.

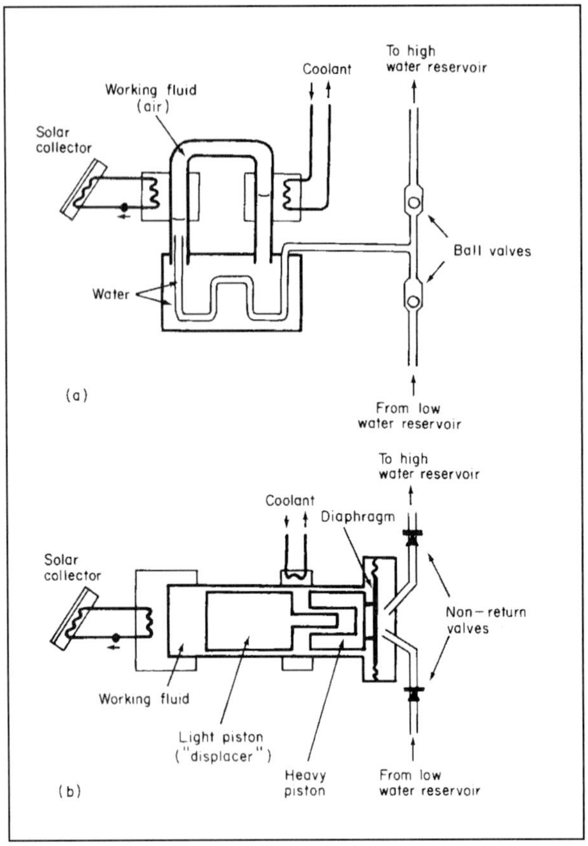

Figure 4.1-29 Solar water pumps based on Stirling cycles. In (a), air and water should not mix. The ball valves produce cyclic operation. In (b), the light and heavy pistons should be out of phase by a half cycle (based on West, 1974; Beale 1976).

types of device based on the Stirling cycle, using air or another gas as a working fluid (cf. Fig. 4.1-3). In the upper part of the figure, two ball valves ensure an oscillatory, pumping water motion, maintained by the tendency of the temperature gradient in the working fluid (air) to move air from the cold to the hot region. The lower part of the figure shows a "free piston Stirling engine", in which the oscillatory behaviour is brought about by the presence of two pistons of different mass, the heavy one being delayed by half an oscillatory period with respect to the lighter one. The actual water pumping is made by the "membrane-movements" of a diaphragm, but if the power level is sufficient the diaphragm may be replaced by a piston-type pump. For Stirling cycle operation, the working fluid is taken as a gas. This is an efficient cycle at higher temperatures (e.g. with use of a focusing solar collector), but if the heat provided by the (flat-plate) solar collector is less than about 100°C, a larger efficiency may be obtained by using a two-phase working fluid (corresponding to one of the Rankine cycles shown in Fig. 4.1-3).

If focusing solar collectors are used, the diaphragm may be replaced by a more efficient piston pump.

Two Rankine-type solar pumps, based on a fluid-to-gas and gas-to-fluid cycle, are outlined in Fig. 4.1-30. The one shown in the upper part of the figure is based on a cyclic process. The addition of heat evaporates water in the container on the left, causing water to be pumped through the upper one-way valve. When the vapour reaches the bottom of the U-tube, all of it moves to the container on the right and condenses. New water is drawn from the bottom reservoir, and the pumping cycle starts all over again. This pump is intended for wells of shallow depth, below 10 m (Boldt, 1978).

The lower part of Fig. 4.1-30 shows a pump operated by a conventional Rankine engine, expanding the working fluid to gas phase passing through a turbine, and then condensing the gas using the pumped water as coolant, before returning the working fluid to the heat exchanger receiving solar absorbed heat. For all the pumps based on a single thermodynamic cycle, the maximum efficiency

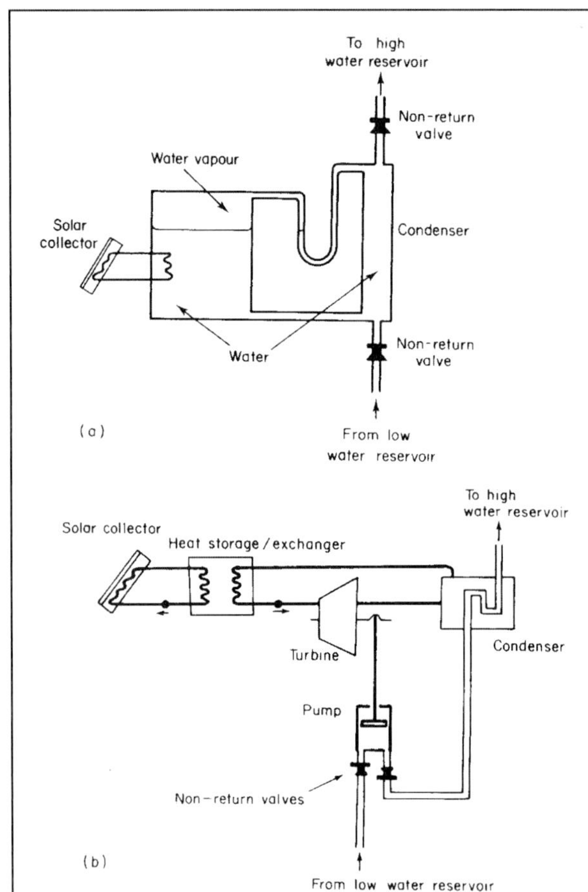

Figure 4.1-30 Solar water pumps based on Rankine cycles. In (a), the formation of vapour, its transfer through the U-tube and condensation produces a cyclic pressure–suction variation, which draws water from the lower reservoir and delivers water to the high reservoir (based on Boldt, 1978). In (b), a turbine-type pump is shown, the working fluid of which may be an organic compound (cf. Meinel and Meinel, 1976).

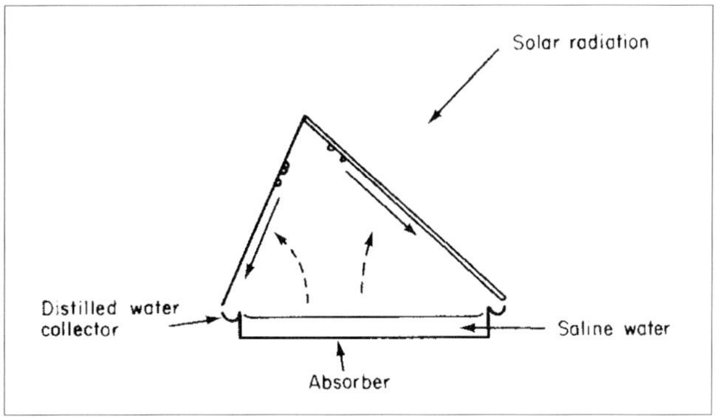

Figure 4.1-31 Solar still.

(which cannot be reached in a finite time) is given by (4.1.4) and the actual efficiency is given by an expression of the form (4.1.19).

Distillation of saline water (e.g. sea water) or impure well water may be obtained by a solar still, a simple example of which is shown in Fig. 4.1-31. The side of the cover system facing the Sun is transparent, whereas the part sloping in the opposite direction is highly reflective and thus remains at ambient temperature. It therefore provides a cold surface for condensation of water evaporated from the saline water surface (kept at elevated temperature due to the solar collector part of the system). In more advanced systems, some of the heat released by condensation (about 2.3×10^6 J kg^{-1}, which is also the energy required for the vaporisation) is recovered and used to heat more saline water (Talbert *et al.*, 1970).

4.1.2.3 Solar electricity generation

Conversion of solar radiation into electric energy may be achieved either in two steps by, for example, first converting radiation to heat, as described in section 4.1.2.1, and then heat into electricity, using one of the methods described in section 4.1.1.2 or 4.1.1.3, or alternatively by direct conversion of radiation into electricity, using the photovoltaic conversion scheme discussed in section 4.1.6. Two-step conversion using chemical energy rather than heat as the intermediate energy form is possible, for instance, by means of the photogalvanic conversion scheme briefly mentioned in section 4.1.6. Examples of the layout of systems capable of performing such conversions are given below, and their possible performance is discussed.

Photo-thermoelectric converters

A two-step conversion device may be of the general form shown in Fig. 4.1-32. The solar collector may be of the flat-plate type or may perform a certain measure of concentration, requiring partial or full tracking of the Sun. The second step is performed by a thermodynamic engine cycle, for example, a Rankine cycle with expansion through a turbine as indicated in the figure. Owing to the vaporisation of the engine-cycle working fluid in the heat exchanger linking the two conversion steps, the heat exchange performance does not follow the simple description leading to (4.1.99). The temperature behaviour in the heat exchanger is more like the type shown in Fig. 4.1-33. The fluid in the collector circuit experiences a uniform temperature decrease as a function of the distance travelled through the heat exchanger, from x_1 to x_2. The working fluid entering the heat exchanger at x_2 is heated to boiling point, after which further heat exchange is used to evaporate the working fluid (and perhaps superheat the gas), so that the temperature curve becomes nearly flat after a certain point.

The thermodynamic engine cycle (the right-hand side of Fig. 4.1-32) can be described by the method outlined in section 4.1.1, with the efficiency of heat to electricity conversion given by (4.1.19) [limited by the ideal Carnot process efficiency (4.1.4)],

$$\eta_w = E^{electric}/J^{w}_{Q,in} \leq (T_{w,in} - T_{w,out})/T_{w,in},$$

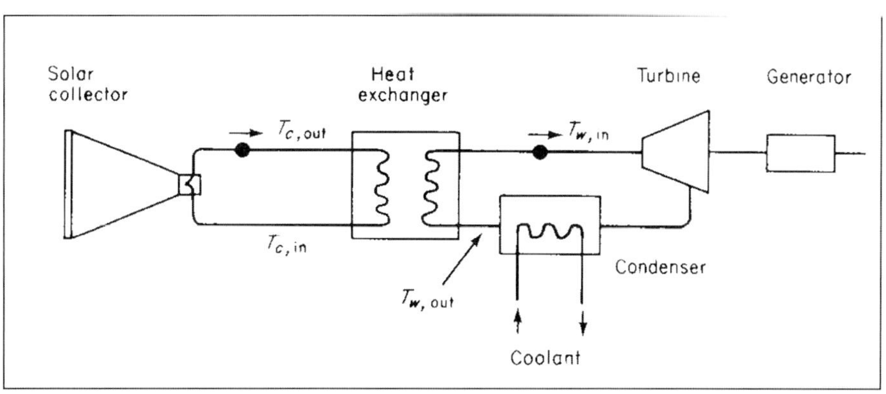

Figure 4.1-32 Photo-thermoelectric generator shown as based on concentrating solar collectors.

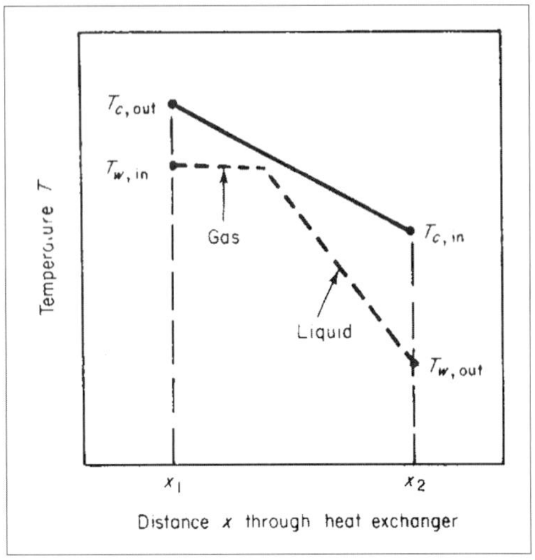

Figure 4.1-33 Temperature behaviour in heat exchanger for two-phase working fluid of temperature T_w and the fluid in the collector circuit of temperature T_c (based on Athey, 1976).

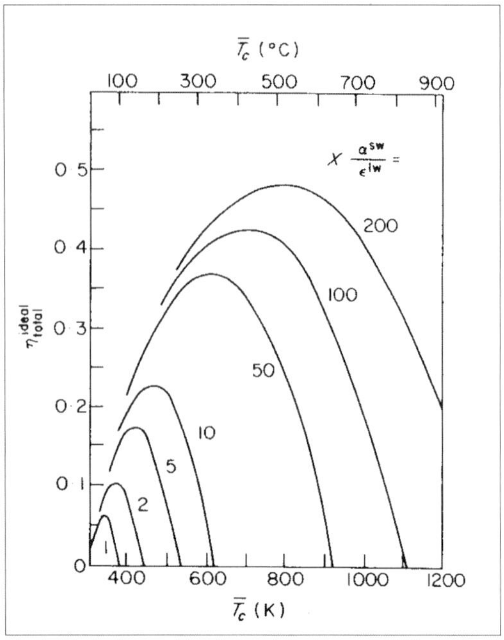

Figure 4.1-34 Ideal overall efficiency of photo-thermoelectric converter, based on the assumptions underlying Figs. 4.1.27 and 4.1.33 and with the Rankine cycle efficiency replaced by an ideal Carnot cycle efficiency.

where $E^{electric} = -J_q F_q$ is the electric power output [cf. (4.1.14)] and

$$J^{w}_{Q,in} = J^{c}_{m} C^{c}_{p} (T_{c,out} - T_{c,in})$$

is the rate of heat transfer from the collector circuit to working fluid circuit in the heat exchanger. The right-hand side is given by (4.1.107) for a concentrating solar collector and by (4.1.96) and (4.1.92) for a flat-plate collector. The overall conversion efficiency is the product of the efficiency η_c of the collector system and η_w,

$$\eta = \eta_c \eta_w.$$

The determination of the four temperatures $T_{c,in}$, $T_{c,out}$, $T_{w,in}$ and $T_{w,out}$ (cf. Fig. 4.1-32) requires an equation for the collector performance, for the heat transfer to the collector fluid, for the heat transfer in the heat exchanger and for the processes involving the working fluid of the thermodynamic cycle. If the collector performance is assumed to depend only on an average collector temperature, $\overline{T}_c$ [e.g. given by (4.1.95)], and if $T_{w,in}$ is considered to be roughly equal to $\overline{T}_c$ (Fig. 4.1-33 indicates that this may be a fair first-order approximation), then the upper limit for η_ω is approximately

$$\eta_\omega = (\overline{T}_c - T_{w,out})/\overline{T}_c.$$

Taking $T_{w,out}$ as 300 K (determined by the coolant flow through the condenser in Fig. 4.1-32), and replacing η_c by the idealised value (4.1.110), the overall efficiency may be estimated as shown in Fig. 4.1-34, as a function of $\overline{T}_c$. The assumptions for η_c^{ideal} are as in Fig. 4.1-27: no convective losses (this may be important for flat-plate collectors or collectors with a modest concentration factor) and an incident radiation flux of 800 W m^{-2} reaching the absorber (for strongly focusing collectors this also has to be direct radiation). This is multiplied by η_w^{ideal} (the Carnot limit) to yield η_{total}^{ideal} depicted in Fig. 4.1-34, with the further assumptions regarding the temperature averaging and temperature drops in the heat exchanger which allowed the introduction of $\overline{T}_c$ as the principal variable both in η_w^{ideal} and in η_c^{ideal}.

In realistic cases, the radiation reaching the absorber of a focusing collector is perhaps half of the total incident flux $E^{SW}_{s,\gamma}$ and η_w is maybe 60% of the Carnot value, i.e.

$$\eta_{total} \approx 0.3\, \eta^{ideal}_{total}.$$

This estimate may also be valid for small concentration values (or rather small values of the parameter $X\alpha^{sw}/\varepsilon^{lw}$ characterising the curves in Figs 4.1-34 and 4.1-27), since the increased fraction of $E^{sw}_{s,\gamma}$ being absorbed is compensated for by high convective heat losses. Thus, the values of η^{ideal}_{total} between 6 and 48%, obtained in Fig. 4.1-34 for suitable choices of $\overline{T}_c$ (this can be adjusted by altering the fluid flow rate J^c_m), may correspond to realistic efficiencies of 2–15%, for values of $X\alpha^{sw}/\varepsilon^{lw}$ increasing from 1 to 200.

The shape of the curves in Fig. 4.1-34 is brought about by the increase in η^{ideal}_w as a function of $\overline{T}_c$, counteracted by the accelerated decrease in η^{ideal}_c as a function of $\overline{T}_c$, which is seen in Fig. 4.1-27.

Photo-thermoelectric conversion systems based on flat-plate collectors have been discussed, for example, by Athey (1976); systems based on solar collectors of the type shown in Fig. 4.1-25d have been considered by

Meinel and Meinel (1972). They estimate that a high temperature on the absorber can be obtained by use of selective surfaces, evacuated tubes (e.g. made of glass with a reflecting inner surface except for the window shown in Fig. 4.1-25d), and a crude Fresnel lens (see Fig. 4.1-26) concentrating the incoming radiation on the tube window. Molten sodium is suggested as the collector fluid. Finally, fully tracking systems based on the concept shown on the right-hand side of Fig. 4.1-24 ("power towers") have been suggested, e.g. by Teplyakov and Aparisi (1976) and by Hildebrandt and Vant-Hull (1977).

Photovoltaic converters

A photovoltaic converter consists of a number of solar cells connected in a suitable way, plus eventually some auxiliary equipment such as focusing devices in front of the cells and tracking systems. The maximum efficiency of a solar cell is given by the ratio of the maximum power output (4.1.67) and the incident radiation flux,

$$\max(\eta) = \max(E)/E^{sw}_{s,\gamma}. \quad (4.1.111)$$

It is smaller than unity for a number of reasons. Firstly, as discussed in section 4.1.1.5, radiation of frequency below the semiconductor band gap is not absorbed. Secondly, according to (4.1.69), the excess energy of radiation with frequencies above the semiconductor band gap is not available to the photovoltaic conversion process. This loss would be small if the solar spectrum was peaked across the band gap, but most semiconductor gaps only correspond to a limited part of the broad solar spectrum.

Thirdly, as seen e.g. in Fig. 4.1-9, the maximum power output is less than the maximum current times the maximum voltage. This reduction in voltage, necessary in order to get a finite current, is analogous to the necessity for a finite dissipation term in a thermodynamic engine in order to get energy out in a finite time (cf. section 4.1.1). The expression (4.1.67) does not fit measured power values in detail, and it has been suggested that a second exponential term be added in the current–voltage relation (4.1.63), of similar form but with $\Delta\phi_{ext}$ replaced by $\frac{1}{2}\Delta\phi_{ext}$ (Sah *et al.*,1957). The origin of such a term is thermal generation and recombination of carriers in the vicinity of the junction. Fourthly, the external potential (times the electron charge) has a maximum value (4.1.66), which is smaller than the semiconductor gap $E_c(p) - E_c(n)$, since it equals the difference in Fermi level between the *p*- and *n*-regions [cf. (4.1.60)]. This loss may be diminished by increasing the number of impurities (and hence the number of minority carriers) in both the *p*- and the *n*-regions. However, if these are increased above a certain level, the increased recombination probability offsets the voltage gain.

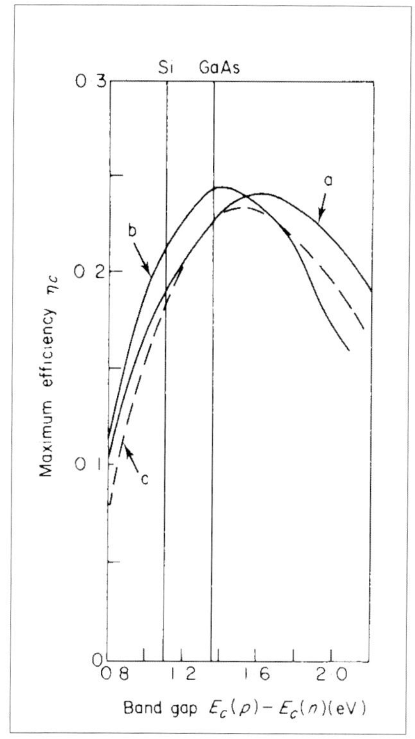

Figure 4.1-35 Early calculation of maximum efficiency for simple *p*-on-*n* solar cell; (a) outside the Earth's atmosphere (E^{sw} = 1350 W m^{-2}); (b) at the Earth's surface under standard conditions (E^{sw} = 890 W m^{-2}, air mass one, water content 0.02 m^3 m^{-2} and major absorption bands included in the calculation); (c) overcast condition (E^{sw} = 120 W m^{-2}) (based on Loferski, 1956).

Figure 4.1-35 shows an early example of calculated maximum efficiency as a function of the semiconductor band gap, including the above-mentioned losses, for radiation conditions corresponding to the top of the Earth's atmosphere (a), for a clear sky day at the Earth's surface with average atmospheric absorption and scattering conditions (b), and for a situation with cloud-covered sky (c). It is shown that for common solar cell materials such as Si or GaAs, the maximum efficiency is larger for the spectral composition of clear-day solar radiation at ground level than for the solar spectrum not disturbed by the Earth's atmosphere. The performance for scattered solar radiation (overcast case) is not substantially impaired.

The wavelength dependence of the collection efficiency (i.e. the efficiency including the first two loss terms discussed above, but not the last two), is shown in Fig. 4.1-36 for a simple silicon cell consisting of a thin (5 × 10^{-7} m) *p*-layer on top of an *n*-layer base (of thickness 4.5 × 10^{-4} m). The total efficiency curve is based on measurements, whereas the separate contributions from the *p*- and *n*-layers have been calculated (Wolf, 1963). Short wavelengths are absorbed in the *p*-layer and give rise to a current of electrons directed towards the junction, but the bulk of the solar wavelengths are not absorbed until

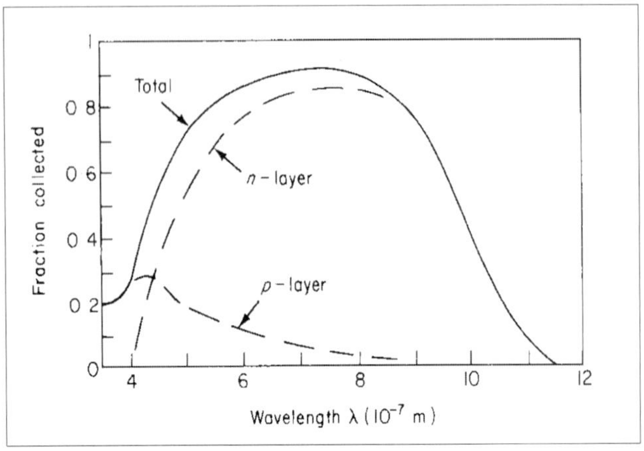

Figure 4.1-36 Spectral collection efficiency for a simple p-on-n solar cell. The curve labelled "total" is based on measurements, and the individual contributions from the p-layer (thickness 5×10^{-7} m) and the n-layer have been calculated. The total thickness is about 4.5×10^{-4} m (based on Wolf, 1963).

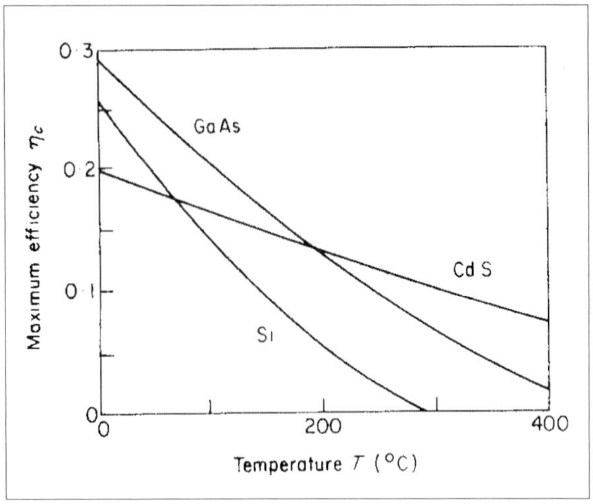

Figure 4.1-37 Calculated temperature dependence of solar cell efficiency (based on Wysocki and Rappaport, 1960).

the photons reach the n-layer base. They give rise to a hole current towards the junction. The efficiency is not independent of cell temperature, as indicated in Fig. 4.1-37. The currently dominating silicon material exhibits an absolute loss in efficiency of 0.4–0.5% for each °C of temperature rise. Figure 4.1-38 adds more recent data on the temperature dependence of different types of solar cells, based upon measurements.

The data shown in Fig. 4.1-38 has, for each type of solar cell, been normalised to a typical absolute efficiency for current commercial or near-commercial versions of the type of device in question. The early theoretical calculations (Wysocki and Rappaport, 1960) are largely confirmed by current measurements, and the mechanisms are thus well understood, at least for conventional photovoltaic devices. The temperature dependence is chiefly due to band-gap effects, which explains why the slope of the crystalline silicon (c-Si) and multicrystalline silicon (m-Si) are identical (Yamamoto *et al.* 1999). In other words, the grain boundaries do not give rise to additional temperature effects. Cd-S cells have a lower

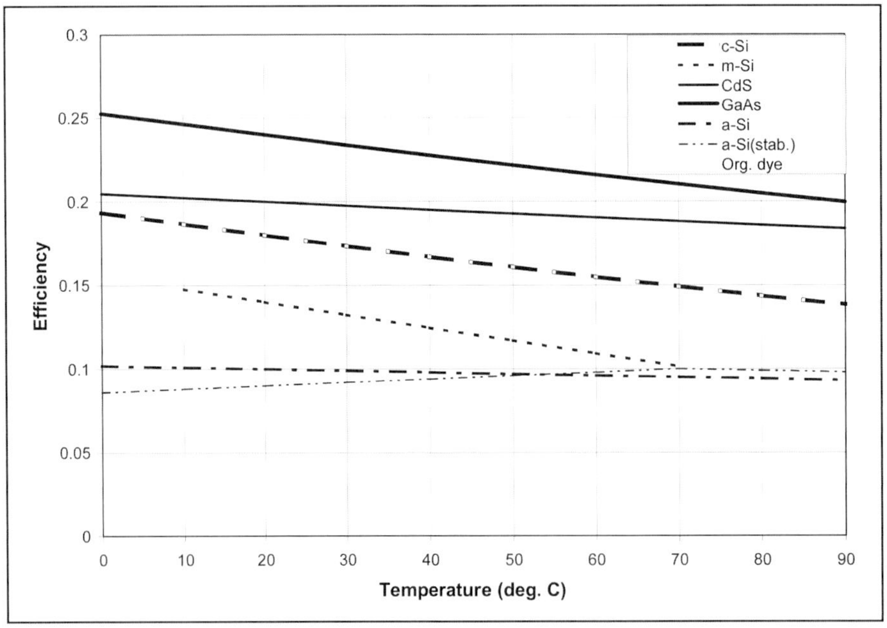

Figure 4.1-38 Solar cell efficiency as function of operating temperature, normalised to a typical 25°C efficiency for each cell type. From Sørensen (2000b), based on Ricaud (1999), Dutta *et al.* (1992), Wysocky and Rappaport (1960), Yamamoto *et al.* (1999), Rijnberg *et al.* (1998).

but still significant temperature gradient, whereas the temperature effect for amorphous silicon cells and organic dye-sensitised TiO_2 cells is very small.

The temperature effect is negative with increasing working temperature for all devices except two: the organic cells show a maximum near 40°C (Rijnberg *et al.*, 1998) and the amorphous silicon-hydrogen cells (a-Si) show a reversal of temperature trends after annealing (Dutta *et al.*, 1992). This positive temperature coefficient only persists until the un-degraded efficiency is reached, and it requires annealing as opposed to light soaking treatment, which causes the development of a stronger negative temperature coefficient. The modest temperature dependence is conveniently modelled by a power expansion of the efficiency,

$$\eta = \eta(298\text{K}) + a(T - 298\text{K}) + b(T - 298\text{K})^2. \quad (4.1.112)$$

The operating temperature dependence of the solar energy to electricity conversion efficiency suggests that cooling the cell by extracting heat may improve the electric performance of the cell and thereby pay for some of the extra expense of the heat extraction equipment. This is further discussed in a case study in section 6.3.1. Typical operating temperatures for un-cooled cells are about 50°C. Fig. 4.1-38 shows that improvement is indeed obtained for e.g. crystalline or multicrystalline silicon photovoltaic (PV) cells, but not notably for dye-sensitised cells or amorphous PV cells. On the other hand, in order to make use of the heat it should preferably be collected at higher temperatures, which would indicate that the best solutions are those with little operating temperature effect of the electricity yields. This trade-off is further illustrated in the simulation models of section 6.3.1.

In cells currently produced, each loss factor is carefully minimised, and the resulting efficiencies have in creased over time, for all types of solar cells, as discussed below.

Monocrystalline silicon cells

The photovoltaic cell principles described in section 4.1.1.5 and above form the basis for monocrystalline cells, which are cells constructed from single crystals, usually in the form of ingots sliced into a number of cells.

A number of improvements have brought the cell efficiency of state-of-the-art monocrystalline silicon cells up to about 25%. The light capture is improved through trapping structures that minimise reflection in directions not benefiting the collection area and by back–side designs reflecting light rays back into the active areas (see e.g. Fig. 4.1-39). The doping degree is altered near electrodes (n^+ and p^+ areas), and a thin oxide layer further helps to prevent electrons reaching the surface rather than the electrode (this process being termed "passivation"). Further, top electrodes may be buried in order not to produce shadowing effects for the incoming light (Green, 1992).

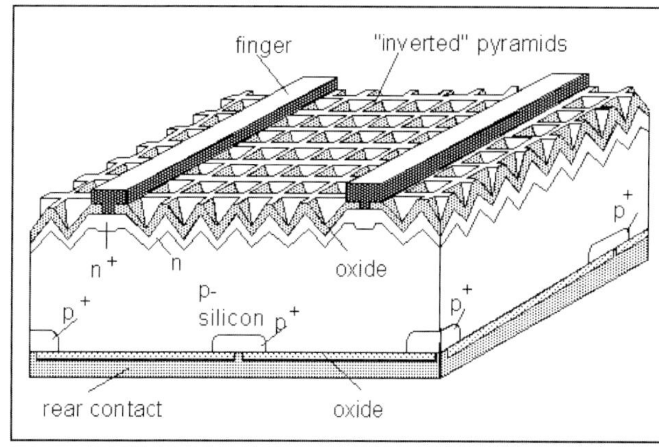

Figure 4.1-39 Structure of a monocrystalline silicon cell with passivated emitter and a locally diffused rear structure ("PERL"), used to obtain a 23% module efficiency (from Green et al., 1998, used with permission).

Figure 4.1-40 shows the measured characteristics, i.e. current as function of voltage, for a cell of the type shown in Fig. 4.1-39.

Simulation of the light trapping and electron transport processes in one, two or three dimensions has helped in selecting the best geometry and degree of doping (Basore, 1991; Müller *et al.*, 1992). Figure 4.1-41 gives an example of one-dimensional simulation of the variation in cell performance as function of doping degree for a cell of the kind shown in Fig. 4.1-39.

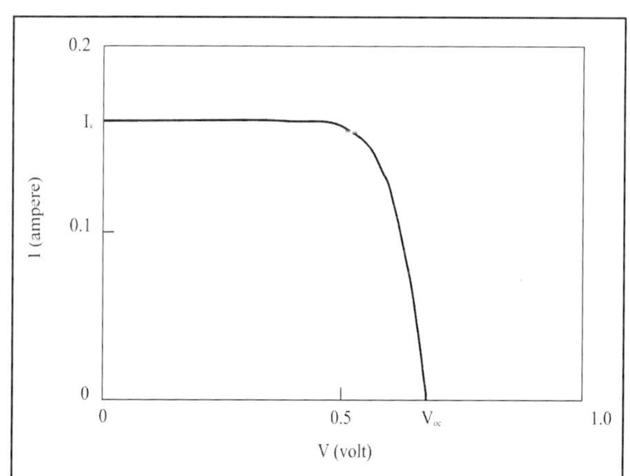

Figure 4.1-40 Current-voltage curve for PERL cell similar to that of Fig. 4.1-39 (but with cell efficiency 22%), as measured at the Sandia Laboratories (USA) at 1006 W m^{-2} airmass 1.5 simulated radiation, for a 4 cm^2 cell. Key findings are V_{OC} = 696 mV (4.1.66), I_S = 160 mA (4.1.64) and a fill factor of 0.79 (Wenham *et al.*, 1995b). The fill factor is the ratio between the area under the curve and $I_S V_{OC}$.

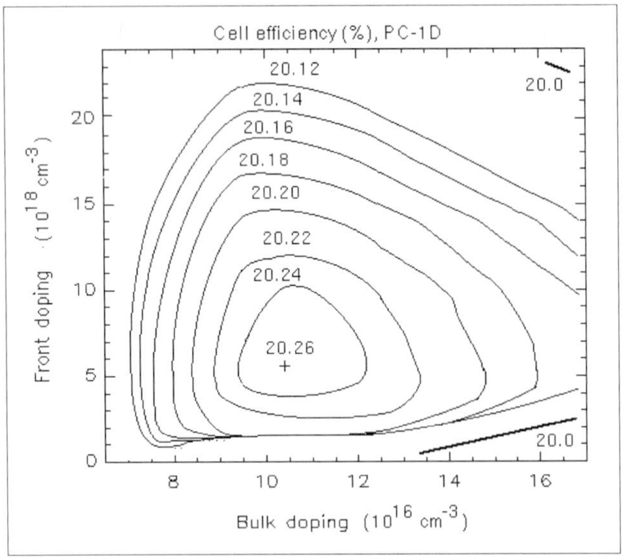

Figure 4.1-41 Calculated efficiency as function of doping parameters for a simplified silicon cell of the type depicted in Fig. 4.1-39. The one-dimensional finite–element model used is described in the text (Sørensen, 1994c).

The model takes into account the 10 μm facet depth and uses a curved one-dimensional path through the cell. The two most important doping parameters (impurities per unit volume) are the uniform bulk doping of the *p*-material and the *n*-doping at the front, assumed to fall off as an error function, thereby simulating the average behaviour of near-electrode sites and sites away from electrodes. Transport and recombination are calculated in a finite-element model (Basore, 1991). The back–side doping is kept at $2 \times 10^{19}\ cm^{-3}$.

Multicrystalline cells

Another technology for producing solar cells uses multicrystalline (sometimes referred to as "polycrystalline") materials, instead of the single-crystal materials. Multicrystalline materials consist of small domains or grains of crystalline material, randomly oriented relative to each other. The crystal grains in multicrystalline materials sustain conductivity in the same way as single crystals do, but the transport of electrons across grain boundaries induces losses, reduces conductivity and thus makes the cells less efficient. On the other hand, they can be produced by simpler methods than those needed for monocrystals, e.g. by evaporating suitable coatings onto a substrate. This field is in rapid development, as it is becoming possible to deposit only a few atomic layers onto a substrate, and with suitable techniques (such as using magnetic fields to align grains) it may soon be possible in this way to form near-monocrystalline layers without having to grow crystals.

It was initially believed that the additional losses at grain boundaries would necessarily make the efficiency of multicrystalline cells substantially lower than what could be obtained by crystalline materials. Actually, the difference has narrowed as a result of better understanding of the options for optimising performance of complex cell structures. One problem has been the damage inflicted upon multicrystalline cells by attempting to copy to them some of the efficiency-improving techniques that have worked well for monocrystalline cells (surface texturing, rear passivation by oxide layers). Yet, etching of inverted pyramids on the surface of multicrystalline cells has improved efficiency considerably (Stock *et al.*, 1996), and recently, less damaging honeycomb texture patterns have brought the efficiency up to 20% (Zhao *et al.*, 1998). This is a trend likely to induce the long-predicted change from expensive ingot-grown monocrystalline cell materials to deposition techniques for multicrystalline materials on suitable backings, much more suited for mass production and price reduction efforts. However, the development away from single-crystalline solar cell materials is slower than anticipated because of the higher maturity of the crystalline industry processes. Figure 4.1-42 shows the structure of the 20% efficient multicrystalline cell, obtaining over 90% absorption of incoming radiation. The advantages of thin-film multicrystalline solar cells over monocrystalline ones would on basic principles seem to more than compensate for the remaining 5% efficiency difference. This does not exclude that crystalline and multicrystalline technologies will continue to co-exist in the marketplace for a while.

Stacked cells

Instead of basing a solar cell on just a single *p–n* junction, it is possible to stack several identical or different cells on top of each other. The use of different cell materials aims

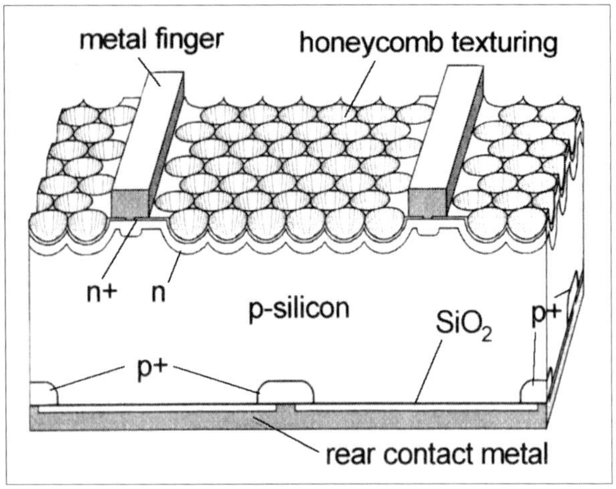

Figure 4.1-42 Texturing used to raise multicrystalline cell efficiency to 20% in a PERL-type cell (from Zhao *et al.*, 1998, used with permission).

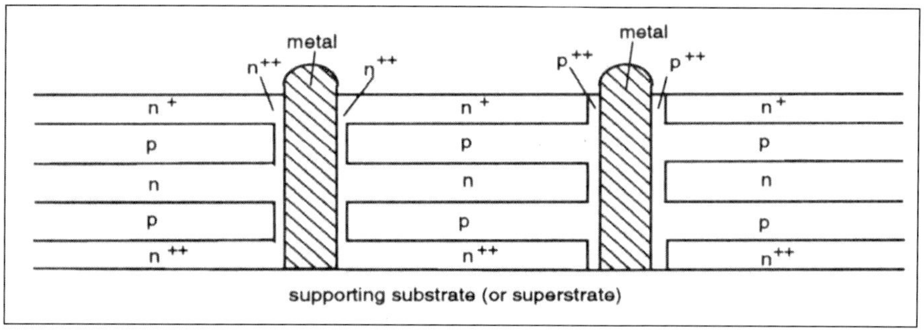

Figure 4.1-43 Concept of a multilayer thin-film silicon solar cell (Wenham *et al.*, 1995a).

at capturing a wider range of frequencies than possible with a single junction. In this case materials of different band gaps will be stacked (Yazawa *et al.*, 1996; Takamoto *et al.*, 1997). In the case of stacked identical cells, the aim is to be able to use lower quality material (e.g. the mentioned thinly sprayed crystalline cells in contrast to ingot-grown ones) and still get an acceptable overall efficiency by stacking several layers of low individual efficiency (Wenham *et al.*, 1995a). This concept is illustrated in Fig. 4.1-43, where a calculation performed for low-quality multicrystalline silicon finds a broad maximum of efficiency for about six layers (Green, 1994).

Figure 4.1-44 shows the industrial process that may be used for producing such multilayer cells, using vapour deposition to deposit each layer and laser grooving to make room for electrodes. Module assembly uses automatic series connection achieved by the layout shown in Fig. 4.1-45. With six junctions of low-quality material (grain sizes above 3 μm), a cell efficiency of 15.2% has been measured (for a small specimen; Sproul *et al.*, 1995). The dependence on the number of layers is weak for the optimum bulk doping concentration of about 10^{18} cm^{-3} and a total thickness of around 10 μm. Efforts to scale the processes up to industrial scale production levels have not succeeded so far, due to difficulties in controlling the transport of electrons from the layers of low-quality silicon to the electrodes.

Amorphous cells

Amorphous semiconductor materials exhibit properties of interest for solar cell applications. While elemental amorphous silicon has a fairly uniform energy

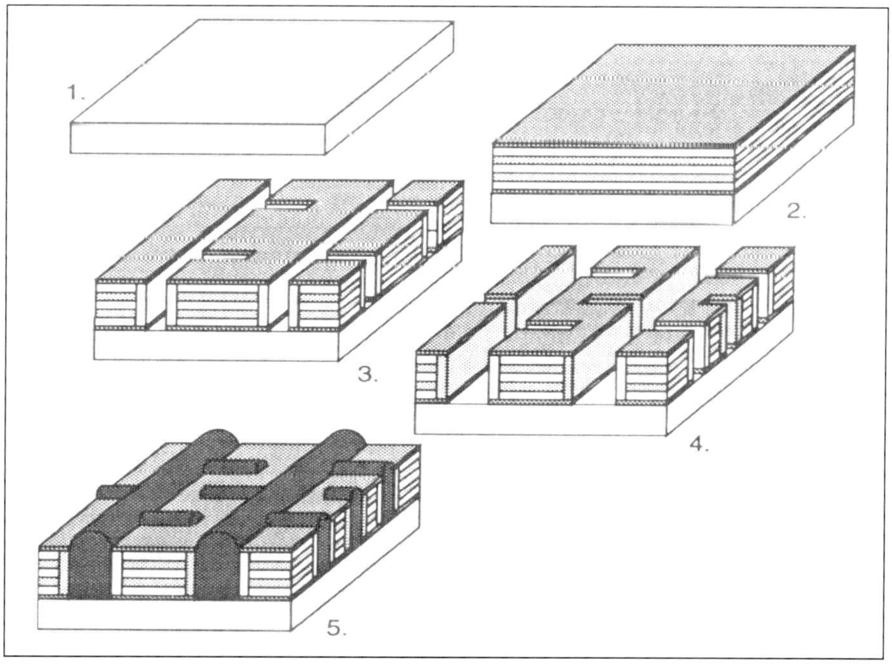

Figure 4.1-44 Production steps: deposition, grooving and metallisation (from Sproul *et al.*, 1995, used with permission).

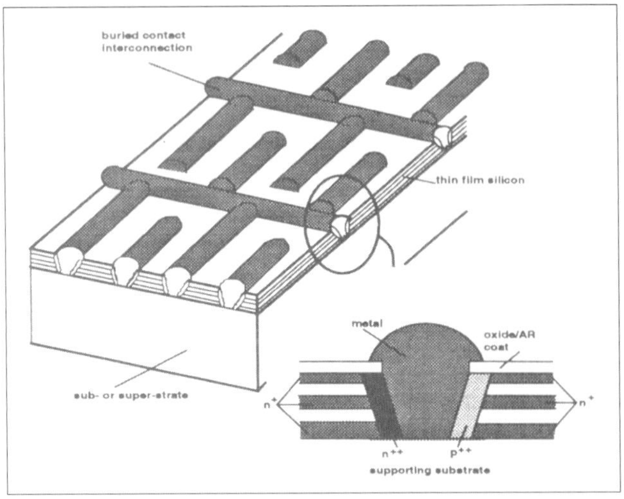

Figure 4.1-45 Design of cells for easy electrical interconnection between adjacent cells in a module (from Wenham *et al.*, 1995a, used with permission).

distribution of electron levels, composite materials have been constructed which exhibit a pronounced energy gap, i.e. an interval of energy essentially without any states, as in a crystal. Spear and Le Comber (1975) first produced such an amorphous material, which was later proved to be a silicon–hydrogen alloy, with the distribution of energy states shown in Fig. 4.1-46 as curve a. Ovshinsky (1978) produced a silicon–fluorine–hydrogen alloy with further depression of gap states (Fig. 4.1-46, curve b). The gap is about 1.6 eV wide and thus should be more favourable with respect to the solar energy spectrum than the 1.1-eV gap of crystalline silicon. Furthermore, doping (introduction of boron or phosphorus atoms) has proved possible, so that *p*-and *n*-type amorphous semiconductors can readily be made, and a certain amount of "engineering" of materials with exactly the desired properties with regard to gap structure, doping efficiency, conductivity, temperature sensitivity and structural stability (lifetime) can be performed.

A theoretical description of the gap occurrence and electron conduction in amorphous materials was first presented by Street and Mott (1975) and followed up by Fritsche (1977) and Pfister and Scher (1977). The basis is the occurrence in amorphous material of defects of particular affinity to attract or reject electrons (e.g. lone pair sites), and the transport of electrons is thought of as a quasi-random "hopping" between sites, some of which are capable of "trapping" an electron for a shorter or longer period of time. Abundance of broken or "dangling" bonds may give rise to serious trapping problems (low conductivity), and the success obtained by incorporating hydrogen seems due to its occupying and neutralising such sites.

Not very long after the theoretical description of amorphous solar cells, Japanese scientists succeeded in creating designs of such cells that were suited for industrial production (Hamakawa *et al.*, 1981) and soon found a market in powering calculators and similar small-scale devices, where the cell cost was relatively unimportant. A typical structure of a commercial amorphous cell is illustrated in Fig. 4.1-47. Band gaps in the range from 1.0 to 3.6 eV can be engineered with different silicon alloys (SiGe, Si, SiC), and such cells may be stacked to obtain a broader frequency acceptance (Ichi-kawa, 1993; Hamakawa, 1998).

However, the simplest version has just one type of material: an intrinsic layer of an a-Si:H compound is the main area of light absorption, and adjacent *p*- and *n*-type

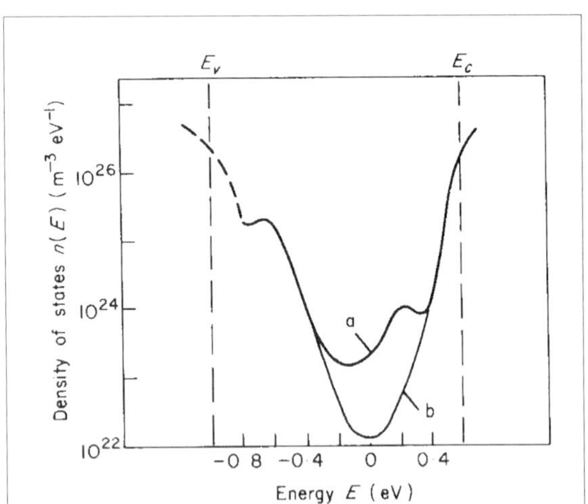

Figure 4.1-46 Density of electron states as a function of excitation energy for silicon-based amorphous materials: (a) silicon–hydrogen alloy; (b) silicium–fluorine–hydrogen alloy (based on Spear and Le Comber, 1975; Ovshinsky, 1978).

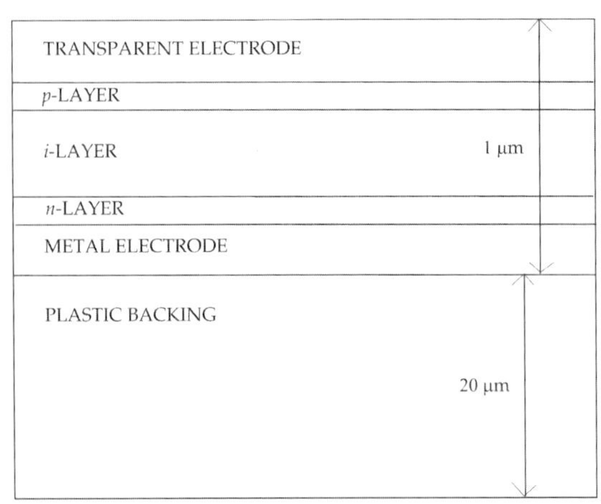

Figure 4.1-47 Structure of the most common a-Si cell. Several *pin*-layers may be stacked.

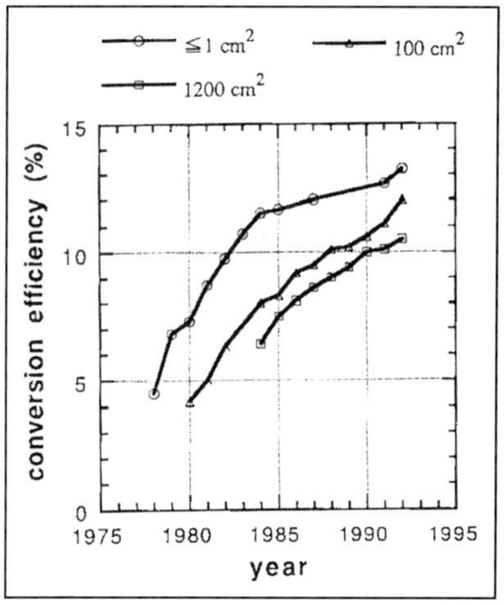

Figure 4.1-48 Development in a-Si solar cell efficiency for different cell area (before degradation; Sakai, 1993).

layers ensure the transport to the electrodes, of which the front one is made of a transparent material. The whole structure is less than 1 μm thick and is deposited onto a plastic backing material. Maximum efficiencies of around 13% have been demonstrated (Fig. 4.1-48), but one problem has persisted: because the structure of the material is without order, bombardment with light quanta may push atoms around, and the material degrades with time. Current practice is to degrade commercial cells before they leave the factory, thereby decreasing the efficiency by some 20%, but in return obtaining reasonable stability over a 10-year period under average solar radiation conditions (Sakai, 1993). Several layers of *p*-, *i*- and *n*-layers may be stacked, and the highest efficiency is obtained by replacing the amorphous *n*-layers by a multicrystalline pure Si or silicon compound layer (Ichikawa, 1993; Ma *et al.*, 1995).

Other materials and other thin-film cells

Use of materials from the chemical groups III and V, such as GaAs, CdS and CdTe, instead of silicon allows better engineering of band gaps in crystalline solar cells to suit particular purposes and brings forward new properties suitable for certain tasks (notably space applications and use in concentrating collectors). Important considerations in selecting materials include temperature dependence where crystalline silicon cell efficiency drops rather fast with the increasing temperature likely to prevail (despite possible active cooling) for cells operating at high levels of solar radiation (see Figs. 4.1-37 and 4.1-38).

The GaAs band gap of 1.43 eV is well suited for the solar spectrum, and with a tandem cell of GaAs plus $GaInP_2$ an efficiency of over 30% has been reached (Bertness *et al.*, 1994; Deb, 1998). At the moment these cells are expensive and are mainly used in space. However, thin-film versions may be developed as in the Si case, as they already have for CIS cells (copper-indium-diselenide). The highest efficiency obtained so far is about 17% for a Cu (In,Ga) Se_2 structure (Tuttle *et al.*, 1996).

Among a range of nontraditional designs of solar cell is the use of spherical droplets. This idea grew out of an idea of reusing scrap material from the microelectronics industry, but it also may increase acceptance of light from different directions and reduce reflection that otherwise would have to be dealt with by surface texturing (Maag, 1993; Drewes, 2003; ATS, 2003).

Organic and other photoelectrochemical solar cells

Photoelectrochemistry is an area of confluence between solar cell technology, discussed here, and battery or fuel cell technology, discussed in section 4.1.7 below. Organic solar cells are a special kind of photoelectrochemical (PEC) devices that try to take advantage of inexpensive organic materials as opposed to the more expensive metals and doped semiconducting materials used in the photovoltaic devices discussed above. Suitable organic materials can, as discussed for biological photosynthesis in section 3.3.6, trap sunlight and convert radiation into other forms of energy. It has been attempted to copy this process in various ways. Calvin (1974) considered a double membrane that would separate the ionised reactants of a photo-excitation process,

$$A + h\nu \rightarrow A^*; \quad A^* + B \rightarrow A^+ + B^-.$$

In addition, a transport system is needed to get the ions to an electrode. No practical version of this idea has been produced. The same is the case for a concept aimed at both removing CO_2 from the atmosphere and at the same time producing methanol (Jensen and Sørensen, 1984, pp. 217–218). The absorption of solar radiation is used to fix atmospheric carbon dioxide to a ruthenium complex [Ru], which is then heated with water steam,

$$[Ru]CO_2 + 2H_2O + 713\ kJ\ mol^{-1} \rightarrow [Ru] + CH_3OH + \tfrac{1}{2}O_2.$$

One scheme that has been realised is the attachment of a ruthenium complex as a dye to TiO_2, which may then transport electrons formed by photoexcitation in the Ru-complex to an electrode. The process is similar to the dye-excitation processes used in conventional photographic prints and was first proposed by Moser (1887). He called the dye substance enhancing the absorption of solar radiation above what can be achieved by the TiO_2 a "sensitiser". The Ru-complex is restored by a redox

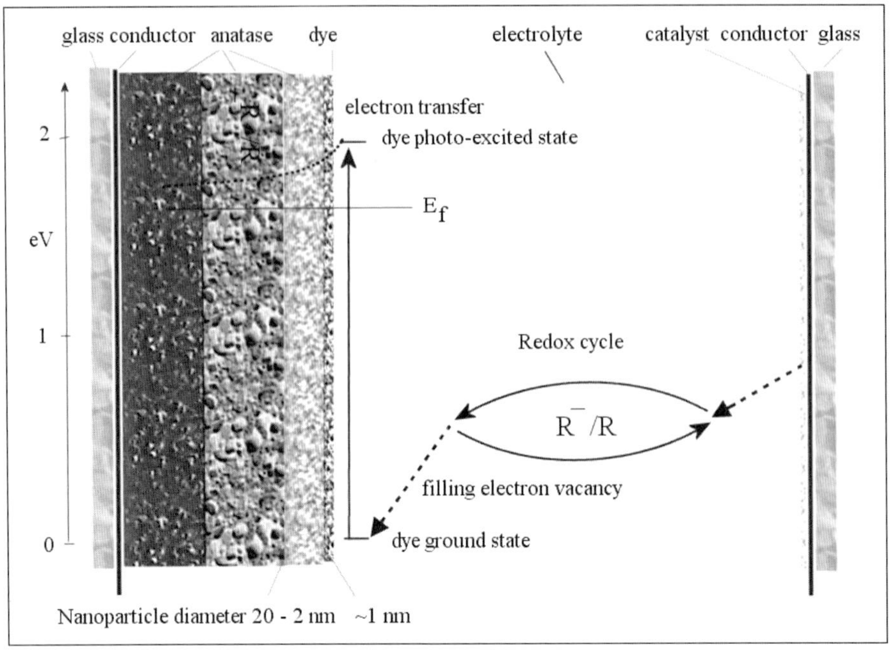

Figure 4.1-49 Layout of photoelectrochemical solar cell with indication of energy levels (E_f is the Fermi level of the semiconductor material). Solar radiation is absorbed in a dye layer, creating an excited electron state, from which an electron is transferred to the semiconductor at left and replenished from the counter electrode through a redox cycle in an electrolyte to the right (Sørensen, 2003a).

process in an electrolyte joining the other electrode, with use of a platinum catalyst as indicated in Fig. 4.1-49. Because a monolayer of even a very efficiently absorbing dye will absorb less than 1% of incoming solar radiation, the dye absorption layer is made three dimensional by adhering the dye to a nanostructured TiO_2 network of nodules, as first demonstrated by Tsubomura *et al.* (1976). They also introduced the liquid electrolyte, capable of penetrating into the cavities in the nanostructured sensitised titanium dioxide and providing the necessary contact for transfer of the replacement electron to the dye. The effective absorption surface may be increased by three orders of magnitude and provides an overall cell efficiency to about 10%. It should still be possible for the TiO_2 nodules to transfer an absorbed electron to the back electrode through a series of transport processes.

The material presently favoured for the anode nanoparticles is TiO_2 in the form of anatase (Fig. 4.1-50). Compared to the other forms of titanium dioxide (rutile and brookite), anatase better accommodates the dye molecules and forms nodules rather than flakes. The large-side dimension of the unit cell shown in Fig. 4.1-50 is about 0.2 nm. Several other semiconductor materials have been investigated, but so far anatase has shown the best overall properties and is used fairly universally. Electron transport through the anatase layers follows conventional solid state physics, except for the issue of nodule coherence. A simple modelling effort has successfully described the transport as random walk, rather than being governed by hopping models (Nelson *et al.*, 2001).

On the cathode side, a redox couple is used to supply the electron to replace the one being excited in the dye and transferred to the anatase before it decays back to the dye ground state. The electrolyte is typically acetonitrile (C_2H_3N), and the redox couple is iodine/tri-iodine (I^-/I^-_3), which has been used rather exclusively since the work of Tsubomura *et al.* (1976). This does not

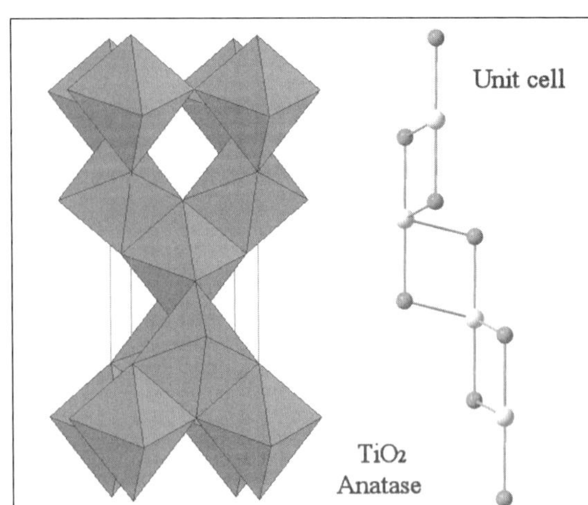

Figure 4.1-50 Anatase structure (first determined by Horn *et al.*, 1970) and unit cell (Sørensen, 2003a).

seem ideal, as the difference between the anatase Fermi level and the I^-/I^-_3 chemical potential, which determines the cell open circuit voltage, is only about 0.9 eV, as compared with typical dye excitation energies of 1.8 eV. Many efforts have been directed at finding more appropriate redox shuttle systems, but so far none has shown overall properties making it preferable to the (I^-/I^-_3) couple (Wolfbauer, 1999). Electrolyte and redox couple integrity and lifetimes are of concern. A comparison to batteries is appropriate, and battery lifetimes are rarely as long as desired for solar cells that may be incorporated directly into building components and structures.

A further energy loss takes place at the cathode, where application of a catalyst is required in order to obtain the desired rate of electron transfer from electrode to electrolyte. As in batteries and fuel cells, traditionally preferred catalysts are based on platinum, but alternatives are under investigation. Generally speaking, the use of liquid electrolytes and catalysts is undesirable, and the much slower electron transfer through the electrolyte and its redox couple (as compared with the semiconductor transport) is likely to be the overall limiting factor for current in the device. However, the reason for this choice is also obvious. The cell is produced by deposition of anatase layers on the anode and subsequent annealing, processes requiring temperatures well over 100°C. The dye is then applied, either by a soaking or by a flushing process, creating the huge intrinsic surface for solar collection. Typical dye melting points are 80–100°C, so applying a second semiconductor material (if one with appropriate properties could be found) from the other side at appropriate temperatures would destroy the cell. More gentle application not requiring high temperatures is not likely to allow the surface contact area between dye and semiconductor to be equally large on both sides.

An alternative would be to find another material not requiring high temperatures for penetrating into the cavity structure of the initial semiconductor plus dye layers. Possible candidates would be conducting polymers or the ion-carrying polymers used in fuel cells. Actual achievements of 2–3% energy conversion efficiency have been obtained with two types of polymer systems. One uses a gel network polymer as electrolyte (Ren *et al.*, 2001). The other is a type of plastic solar cell, where the already known ability of ^{60}C-molecules to absorb solar radiation (Sariciftci *et al.*, 1992) is used to create a fairly large absorption area of ^{60}C sensitiser imbedded in a suitable polymer (Shaheen *et al.*, 2001; Yu *et al.*, 1995).

The choice of sensitiser is ideally based upon fulfilment of requirements including at least the following:

- high absorption capability over the range of spectral frequencies characteristic of sunlight;
- energetically suitable excited states;
- good attachment to semiconductor nanoparticles, which ensures rapid electron transfer (in competition with de-excitation and back-transfer from semiconductor surface to dye sensitiser);
- easily accepting replacement electron from electrolyte;
- dye lifetime consistent with stipulated device life.

The search for optimised sensitisers has usually focused on a particular family of molecules. For example, one group (O'Regan and Grätzel, 1991; Nazeeruddin *et al.*, 1993, 2001; Shklover *et al.*, 1998) has looked at metal complexes based on ruthenium polypyridines, meticulously synthesising one variant after the other, adding rings, thiocyanate ligands and carboxylate groups in different combinations. The size of the molecule, in combination with its excitation spectrum, determines the frequencies of solar radiation that can be absorbed and the associated cross sections. The "black dye" (1 ruthenium atom, 3 pyridine rings, 3 thiocyanate ligands and 3 carboxylate groups) has led to the currently highest overall conversion efficiency of 10% for laboratory cells (area about 10^{-4} m^2). For comparison, an efficiency of 5% is claimed for a large cell (of the order of 1 m^2) in industrial production (STI, 2002). An earlier favourite was the "N3 dye" (Ru, 2 bipyridine rings, 2 thiocyanate ligands and 4 carboxylate groups). It is particularly willing to transfer an excited electron to an anatase surface, a fact that has been attributed to its attachment to the anatase surface by two carboxylate binding sites at approximately the same spacing as the "indents" in one anatase surface. However, the light absorption stops below 800 nm, implying smaller efficiency for many potential real-life collector sites. Figure 4.1-51 compares

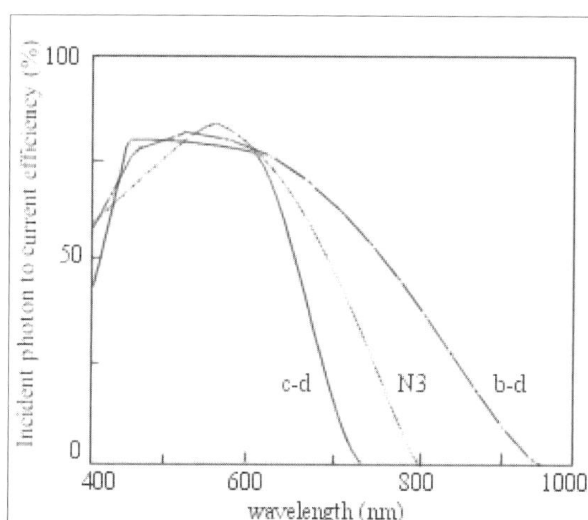

Figure 4.1-51 Spectral sensitivity [c-d: coumarin derivative, N3 and b-d (black dye) ruthenium complexes] (based upon Hara *et al.*, 2001; Nazeeruddin *et al.*, 2001).

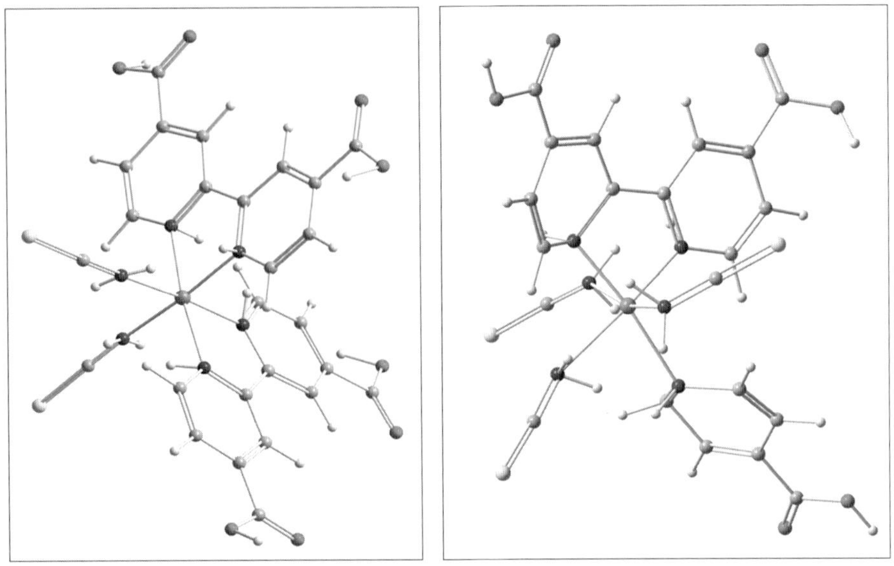

Figure 4.1-52 The structure of N3 ($RuS_2O_8N_6C_{26}H_{24}$, below left) and b-d ($RuS_3O_6N_6C_{21}H_{22}$, below right) ruthenium sensitisers synthesised by Nazee-ruddin *et al.* (1993; 2001). A modest structure optimisation has been performed (Sørensen, 2004b).

spectral sensitivities of the two dyes mentioned above, plus the coumarin-derivative organic dye considered in the following.

Figure 4.1-52 gives the molecular structure of the purple N3 and the black ruthenium dye (Nazeeruddin *et al.*, 1993, 2001). Several variants have been studied (see e.g. Shklover *et al.*, 1998; Zakeeruddin *et al.*, 1997).

The structure of modified sensitiser molecules is often roughly given by that of known components plus some general rules of thumb. Measurements of spectra such as the nuclear magnetic resonance (NMR) spectra will help determine the positions of specific atoms (e.g. hydrogen atoms), but not always in a unique way. The addition of new features to a dye molecule in order to enhance solar absorption or help electron transfer out of the molecule may give rise to structures not amenable to simple guesses, e.g. due to isomerism or for other reasons such as energy surfaces in configuration space with more than one minimum.

Quantum mechanical modelling of the molecular structure* will lead to theoretical predictions of likely structures, as outcomes of optimisation studies. These involve following a path of steepest descent in the potential energy surface, eventually using second-order derivatives in order to stabilise the sizes of jumps made for each of the iterations (Sørensen, 2003a).

Figure 4.1-53 shows the result of a theoretical optimisation (Sørensen, 2001a; 2003a) for a modification of a coumarin dye, being part of a family of molecules studied by Hara *et al.* (2001) as potential inexpensive, purely organic candidates for dyes to use with TiO_2 semiconductor nanoparticles in solar cells.

Figure 4.1-54 shows the molecular orbits near the Fermi level for the coumarin-derivative dye, obtained from a quantum mechanical Hartree-Fock self-consistent field calculation of the best ground state configuration (using software of Frisch *et al.*, 1998 and including a basis of 641 molecular states or about 3 for each physical orbit). It is seen that moving an electron from the highest occupied orbit (HOMO) to the lowest unoccupied one (LUMO) involves reducing the electron density in the coumarin part of the molecule and increasing it along and particularly at the end of the "arm" attached to the molecule. However, at this level of approximation, the energy difference (LUMO minus HOMO) is still nearly 8 eV, as opposed to an experimental value of 1.8 eV (Hara *et al.*, 2001). The second unoccupied orbit (Fig. 4.1-54 bottom) is quite different in nature.

In order to estimate more realistically the energy of the first excited state of the coumarin-derivative dye molecule, a number of calculations have been performed (Sørensen, 2003a). Interactions not included in the self-consistent ground state calculation may be added in a time-dependent Hartree-Foch calculation for one or more excited states (TDHF; Casida *et al.*, 1998), and further improvement is obtained by performing the

*So-called *ab initio* calculations in quantum chemistry involve solving the Schrödinger equation under several simplifying assumptions: nuclear motion is neglected, the basis functions used are linear combinations of Gaussian functions for each atom, and far fewer than needed for completeness. Interactions are first treated by the Hartree-Fock method, implying that each electron is assumed to move in the mean field from all the other particles. Higher electron correlations are added in a perturbative way or by the phenomenological density functional theory.

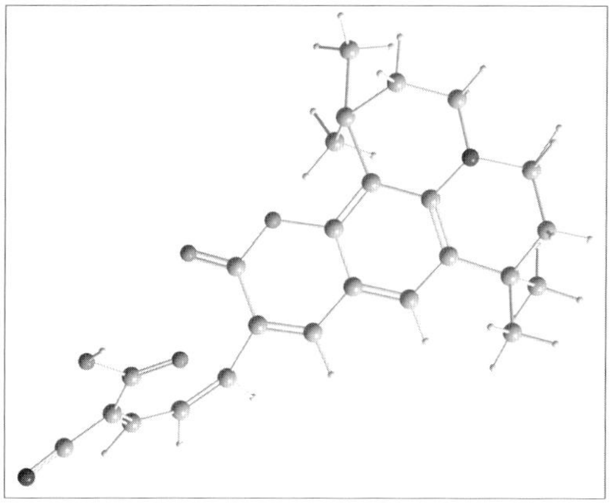

Figure 4.1-53 Optimised structure of coumarin-derivative organic dye (gross formula $H_{26}C_{25}N_2O_4$) synthesised by Hara *et al.* (2001) and yielding overall efficiencies of around 6% when used in PEC solar cells (Sørensen, 2003a).

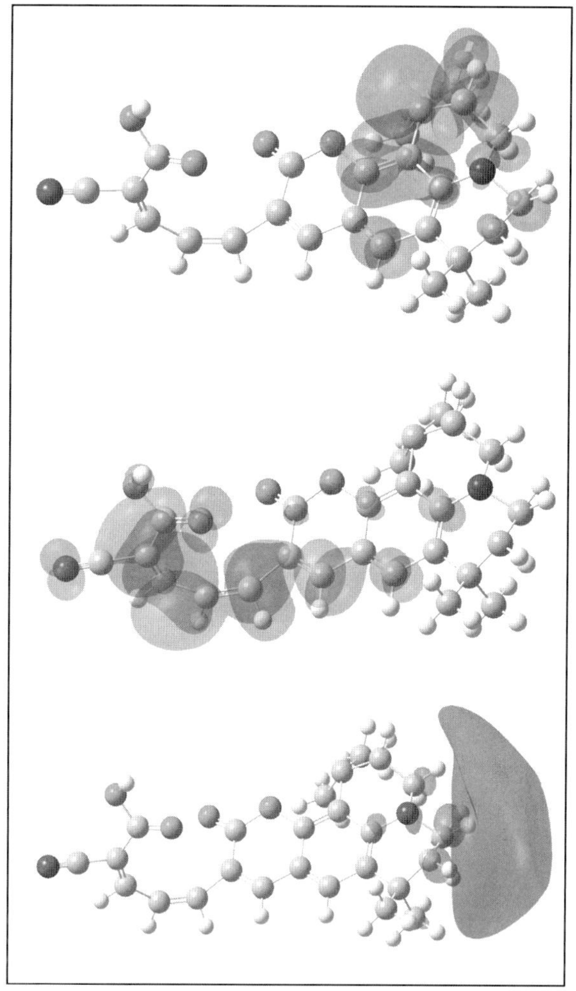

Figure 4.1-54 Electron density of HOMO (top), LUMO (middle) and second unoccupied molecular orbital (bottom) for coumarin-derivative dye, based on self-consistent field (SCF) calculation (Sørensen, 2001a, 2003a).

TDHF calculation on top of ground state calculations including further interaction, such as including exchange forces in a density functional method (Kohn and Sham, 1965; Becke, 1993). Figure 4.1-55 shows the excitation energy of the lowest spin-0 and spin-1 excited states using successive improvements in the sophistication of the interactions included and in the number of basis states used for the calculations (Sørensen, 2003b).

The fact that the first excited state comes down from an initial 8 eV molecular orbital energy difference to the observed 1.8 eV indicates that it must contain a substantial amount of correlations, or in other words that this excitation must comprise collective involvement of a large number of electrons.

This is borne out in Fig. 4.1-56, showing the electron density difference between the first singlet excited state of the coumarin-derivative, and the ground state, using the large basis TDHF calculation. The excited state is made up by a dozen significant molecular orbital (MO) excitation pairs. The density difference shows the expected effect, already surmised from the HF ground state MO's, that the excitation moves electron density from the coumarin core to the peripheral arm added to the molecule. It would then be natural to assume that this is from where the transfer to the anatase surface, to which the dye adheres, takes place. This interpretation is supported by the large dipole moment found in the calculations (13.7 debyes).

The precise attachment of the dye to the anatase surface might be investigated by a combined optimisation of the dye plus a chunk of semiconductor, with the distance and rotation angles of the dye relative to the surface as parameters (cf. Fig. 4.1-57). A recent study of a similar material has revealed the nature of surface distortion and surface Ti and O molecules (Erdman *et al.*, 2002). The distortion only penetrates one layer down, in contrast to the over 10 nm thick space charge regions in surfaces of solids. The conclusion from the present calculation is that both dye and surface are modified, with the dye "arm" being bent to better attach to the surface, and the surface atoms to accommodate the dye particle (but even in the absence of the dye, the surface of the lattice structure would be different from the regular interior used as starting point for the optimisation).

A number of further "handles" on the quantum chemical calculations is the comparison of measured spectra (IR, NMR, etc.) with those predicted by the calculations. The understanding of complex molecules has progressed rapidly as a result of the combination of spectral studies in the laboratory with quantum model calculations.

The photoelectrochemical dye and nanostructure technique has several applications beyond the possibility of forming solar cells. Among these are smart windows

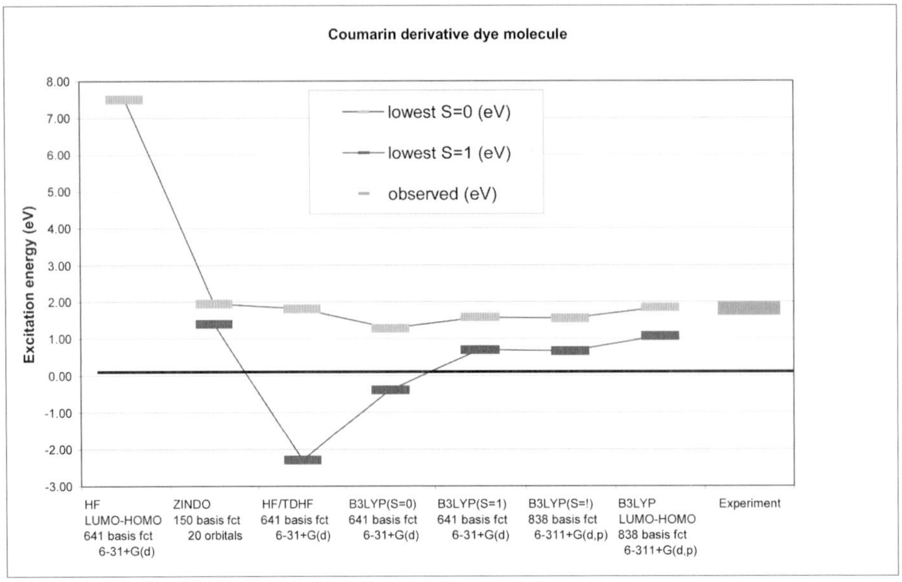

Figure 4.1-55 Measured (right-hand side) and calculated excitation energies of first excited states of spin 0 or 1 for the coumarin-derivative molecule shown in Fig. 4.1-53. The abscissa indicates acronyms for the type of calculation made, with complexity increasing from left to right (except the last, which by comparison with the first column shows the correlations built into the molecular orbits calculated with inclusion of exchange forces). It is seen that two of the calculations give unphysical states of negative energy, indicating the fragile nature of the approximations that effectively use other electron interactions for the excited state calculation than for the ground state calculation (because at a realistic level of approximation, the same approach will not yield both). The problem is solved in the subsequent columns by using basis states more appropriate for the spin-1 calculations (Sørensen, 2003b).

Figure 4.1-56 Calculated electron density difference between the first excited singlet and ground state of coumarin-derivative dye (top: positive values enhanced; below: negative values enhanced; Sørensen, 2001a, 2003a).

(Bechinger *et al.*, 1996; Granqvist *et al.*, 1998; STI, 2002), energy storage (Hauch et al., 2001; Krasovec *et al.*, 2001), environmental monitors (Kamat and Vinodgopal, 1998; Yartym *et al.*, 2001), hydrogen production (Khaselev and Turner, 1998; Luzzi, 1999; Mathew *et al.*, 2001), computer and TV screens also suitable for outdoor use (using the dye as light emitter rather than absorber; Rajeswaran *et al.*, 2000; Tang, 2001; Müller *et al.*, 2003), and three-dimensional data storage (Dwayne-Miller *et al.*, 2001).

Module construction

Individual cells based on monocrystalline materials have areas of typically 100 cm^2 (limited by techniques for growing ingots of monocrystalline material). Multicrystalline and photoelectrochemical cells, where semiconductor material is deposited on a backing template (usually glass), may have larger cell size, and for amorphous cells, there are essentially no limitations. Amorphous cells have been produced on rolls of flexible plastic backing materials, with widths of 1–2 m and rolls of any length. The same may also become possible for other thin-film types, such as spray-deposited multicrystalline materials.

It is customary to assemble cells into modules by a mixture of parallel and series connections, so that the resulting voltages become suitable for standard electricity handling equipment, such as inverters transforming the

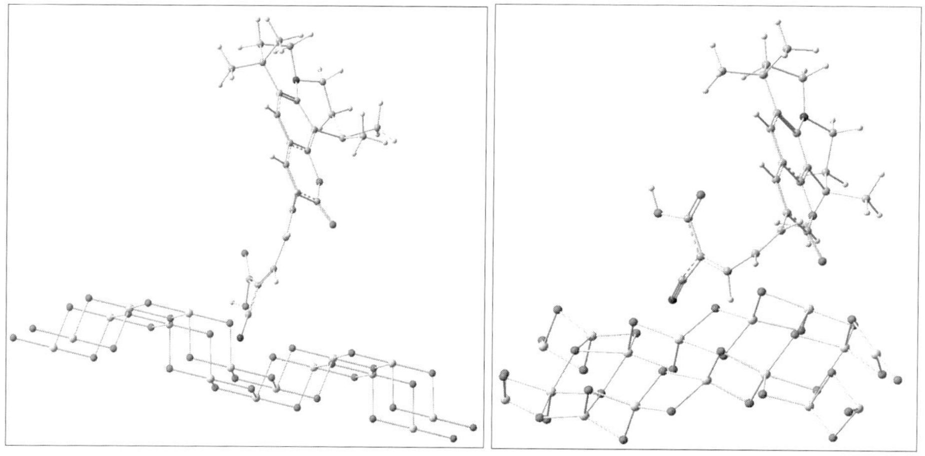

Figure 4.1-57 Attachment of coumarin-derivative dye to and modification of anatase surface. Left: initial guess, right: optimised (the depth of TiO_2 surface layer distortion cannot be determined by modelling just the top layer; Sørensen, 2003a, 2004b).

DC currents into AC currents of grid specifications and quality. Alternatively, microprocessor inverters may be integrated into each module or even into each cell in order to minimise transport losses. In recent years, specific inverters optimised for solar cell applications have been produced, with an inverter efficiency increase from 90% to some 98% (IEA, 1999).

The solar cell technology is then characterised by two main solar radiation conversion efficiencies: the efficiency of each cell and the efficiency of the entire module sold to the customer. The latter is currently about 5% lower than the former, notably because of phase mismatch between the individual cell current components, but this does not need to be so, and the difference between the two efficiencies is expected to diminish in the future.

Optical subsystem and concentrators

As indicated by the devices shown in Figs. 4.1-39 and 4.1.42, optical manipulation of incoming solar radiation is in use for non-concentrating solar cells. They serve the purpose of reducing the reflection on the surface to below a few per cent over the wavelength interval of interest for direct and scattered solar radiation, as seen from Fig. 4.1-58. The collection efficiency is far better than the one shown on Fig. 4.1-36, exceeding 90% for wavelengths between 0.4 and 1.05 μm, and exceeding 95% in the interval 0.65–1.03 μm.

For large-factor concentration of light onto a photovoltaic cell much smaller than the aperture of the concentrator, the principles mentioned for thermal systems (towards the end of section 4.2.1) apply unchanged. Most of these concentrators are designed to focus the light onto a very small area, and thus tracking the Sun (in two directions) is essential, with the implications that scattered radiation largely cannot be used and that the expense and maintenance of non-stationary components have to be accepted.

One may think that abandoning very high concentration would allow the construction of concentrators capable of accepting light from most directions (including scattered light). However, this is not so easy. Devices that accept all angles have a concentration factor of unity (no concentration), and even if the acceptance angular interval is diminished to, say, 0–60°, which would be suitable because light from directions with a larger angle is anyway reduced by the incident cosine factor, only a very small concentration can be obtained (examples such as the design by Trombe are discussed in Meinel and Meinel, 1976).

The difficulty may be illustrated by a simple two-dimensional ray-tracing model of an (arbitrarily shaped)

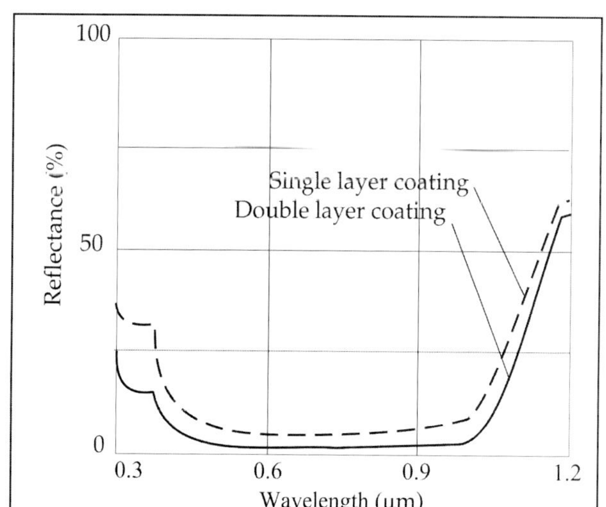

Figure 4.1-58 Front reflection for cell of the type shown in Fig. 4.1-39 (single layer anti-reflection coating) and the same with double layer coating (Zhao *et al.*, 1995).

absorber, i.e. the PV cell, sitting inside an arbitrarily shaped concentrator (trough) with reflecting inner sides, and possibly with a cover glass at the top, that could accommodate some texturing or lens structure. The one component that unfortunately is not available is a semi-transparent cover that fully transmits solar radiation from one side and fully reflects the same wavelengths of radiation hitting the other side. With such a cover, nearly 100% of the radiation coming from any direction could reach the absorber, the only exception being rays being cyclically reflected into paths never reaching the absorber.

Figure 4.1-59 shows some ray-tracing experiments for a catenary-shaped concentrator holding a flat absorber at the bottom, with a concentration factor X = 7 (ratio of absorber and aperture areas, here lines). The cover has an inverted pyramid shape. This is routinely used for flat-plate devices (see Fig. 4.1-39), with the purpose of extending the path of the light rays inside the semi-conductor material so that absorption is more likely. For the concentrator, the purpose is only to make it more likely that some of the rays reach the absorber, by changing their direction from one that would not have led to impact onto the absorber. However, at the same time, some rays that otherwise would have reached the absorber may now become diverted. Thus, not all incident angles benefit from the concentrator, and it has to be decided whether the desired acceptance interval is

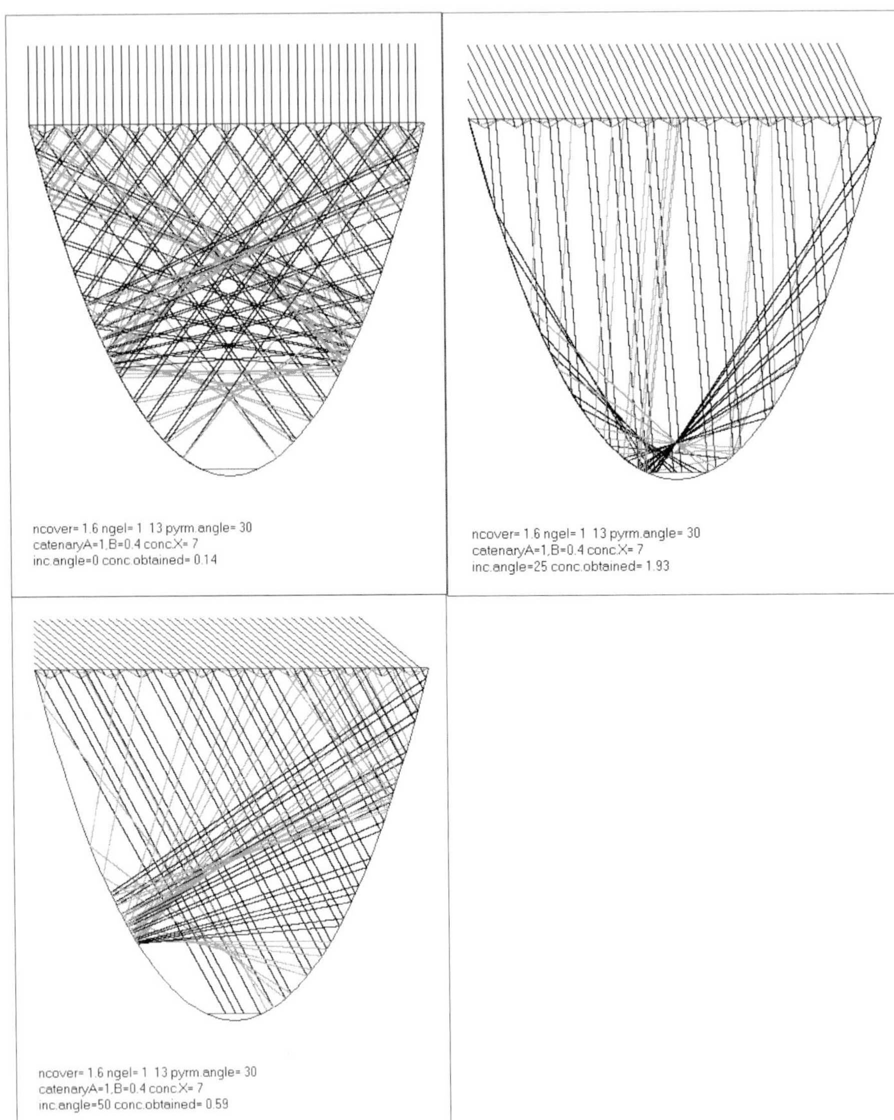

Figure 4.1-59 Two-dimensional ray-tracing through a concentrator with glass cover textured with 30° inverted pyramids, for incident angles 0°, 25° and 50°. Only for angles angles around 25° does the absorber at the bottom receive more light than would have hit it if there had been no concentrator at all.

broad (say, 0–60°) or can be narrowed, e.g. to 0–30°. As illustrated, the concentration of rays towards the centre absorber is paid for by up to half of the incident rays being reflected upwards after suffering total reflection on the lower inside of the cover glass.

The result of varying some of the parameters is shown in Fig. 4.1-60. An inverted pyramid angle of 0° corresponds to a flat-plate glass cover, and the figure shows that in this situation the concentration is limited to incident angles below 20°, going from a factor of four to unity. With increasing pyramid angle, the maximum concentration is moved towards higher incident angles, but its absolute value declines. The average concentration factor over the ranges of incident angles occurring in practice for a fixed collector is not even above unity, meaning that a PV cell without the concentrating device would have performed better.

Another idea is to direct the cover pyramids upwards, but to avoid "stretching out" the rays again at the lower glass boundary by having the entire trough filled with the refractive substance (which might be a gel or plastic material, textured at the upper surface). This is illustrated in Figs. 4.1-61a–c for an X = 5 concentrator. For a deep catenary-shaped collector, all rays up to 10° may be led to the absorber, but in this case (Figs. 4.1-61a and 4.1-62), incident angles above 25° are not accepted. To get reasonable acceptance, pyramid angles around 30° would again have to be chosen, and Figs. 4.1-61a–c show that the concentration penalty in trying to extend the acceptance interval to angles above 35° (by decreasing the depth of the concentrator, see Fig. 4.1-63) is severe. The best choice is still the deep catenary shape, with maximum concentration between 2 and 3 and average concentration over the interesting interval of incident angles (say, to 60°) no more than about 1.5. This is better than the inverted pyramid structure in Fig. 4.1-59, but it is still doubtful whether the expense of the concentrator is warranted. The collector efficiency as measured relative to total aperture area is the ratio of absorber intensity and X, i.e. considerably less than the cell efficiency.

Generation 24 h a day at constant radiation intensity could be achieved by placing the photovoltaic device perpendicular to the Sun at the top of the Earth's atmosphere. If it were placed in a geosynchronous orbit, it has been suggested that the power generated might be transmitted to a fixed location on the surface of the Earth after conversion to microwave frequency (see e.g. Glaser, 1977).

Use of residual energy

Given that practical efficiencies of photovoltaic devices are in the range of 10–30%, it is natural to think of putting the remaining solar energy to work. This has first been achieved for amorphous cells, which have been integrated into window panes, such that at least a part of the energy not absorbed is passed through to the room behind the pane. Of course, conductors and other non-transparent features reduce the transmittance somewhat. The same should be possible for other thin-film photovoltaic materials, including the emerging multicrystalline silicon cells.

Another possibility, particularly for PV panels not serving as windows, is to convert the solar energy not

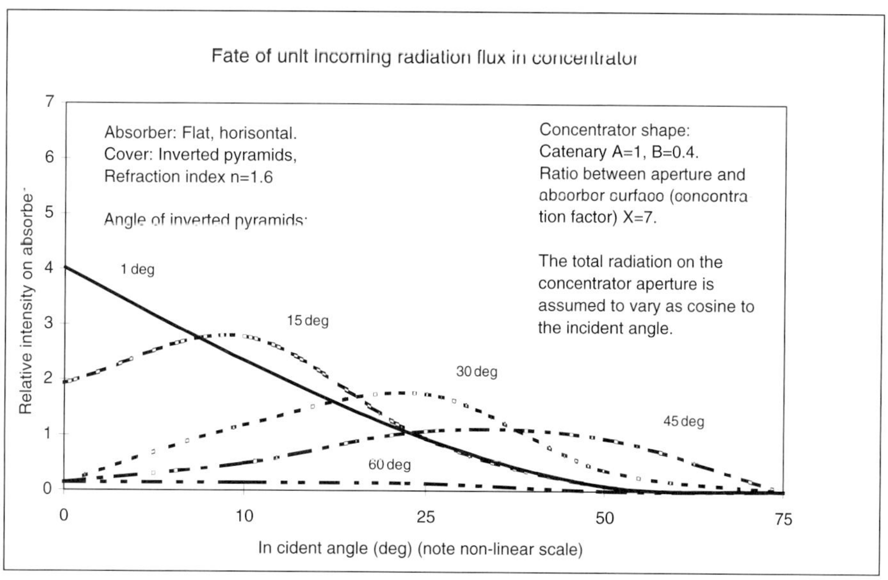

Figure 4.1-60 Intensity of rays on the absorber surface relative to that experienced in the absence of the concentrator (times cosine of incident angle, in order to reflect the different intensity of rays from different directions) for the X = 7 concentrator with a refracting, inverted pyramid cover as illustrated in Fig. 4.1-59.

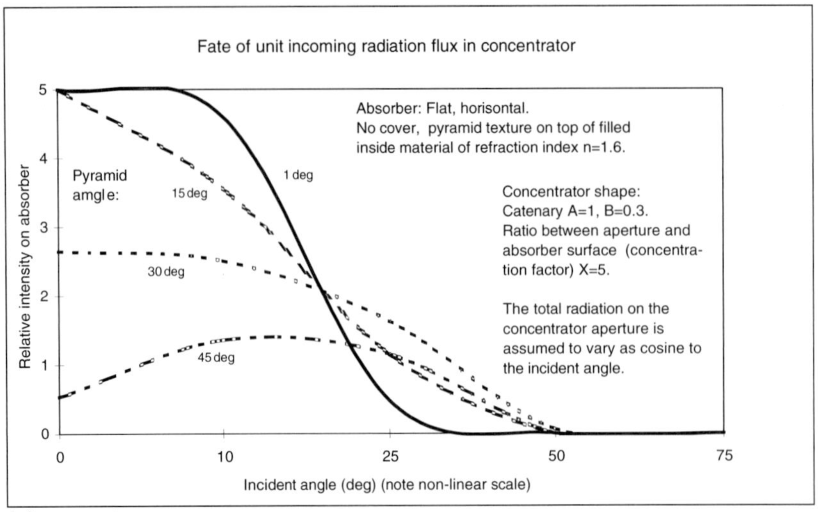

(a)

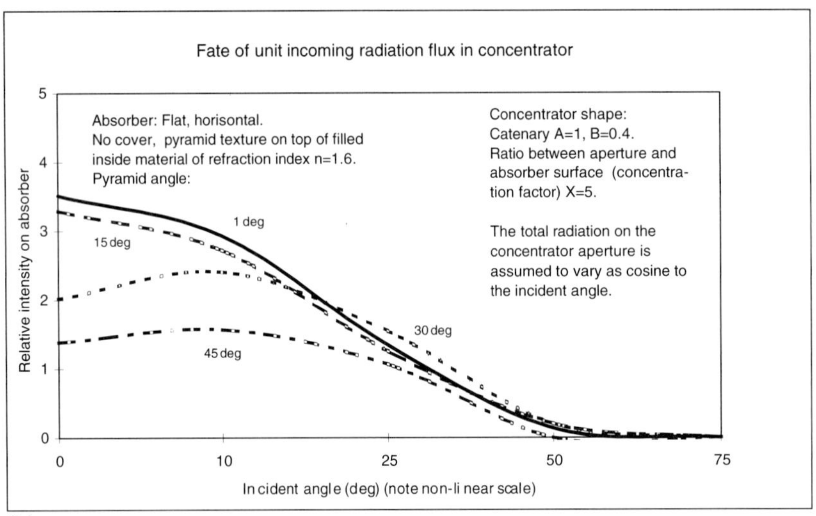

(b)

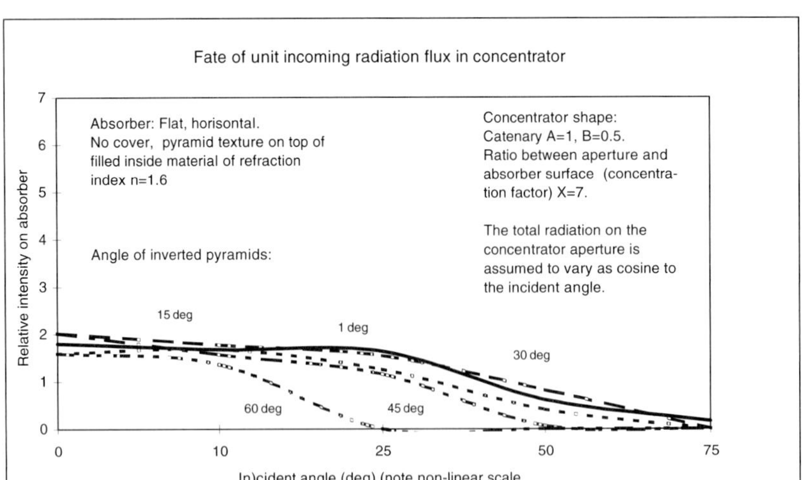

(c)

Figure 4.1-61a-c Intensity of rays on the absorber surface relative to that experienced in the absence of the concentrator (times cosine of incident angle, in order to reflect the different intensity of rays from different directions) for different depths of the $X = 5$ concentrator without cover, but with a refracting inside material having pyramid texture on top, as illustrated in Fig. 4.1-63.

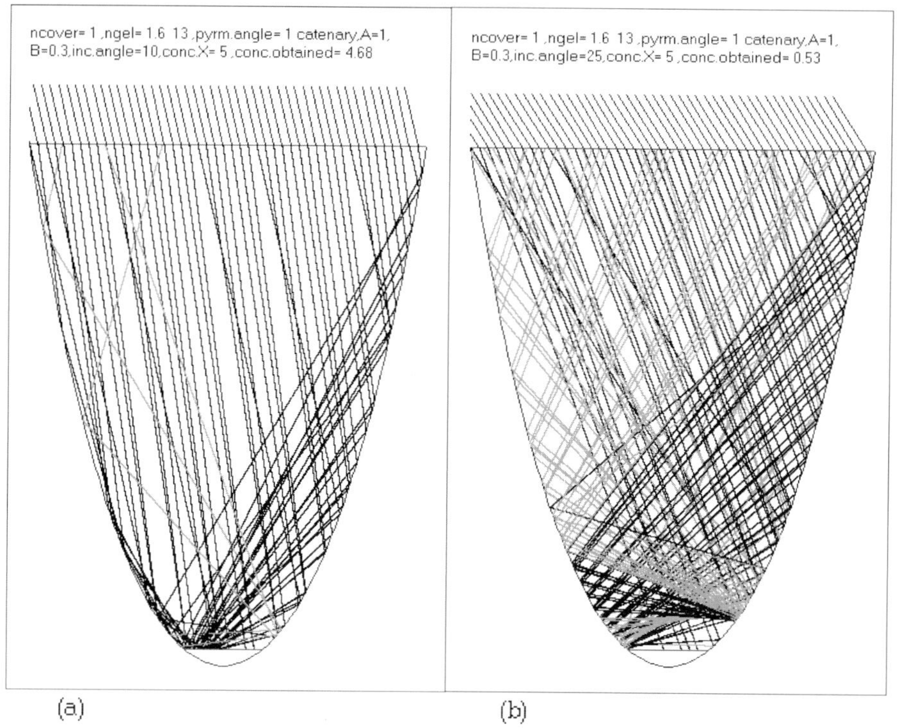

Figure 4.1-62a,b Two-dimensional model of concentrator without cover but filled with a refracting material with slight pyramid texture at top. Incident angles are 10° and 25°.

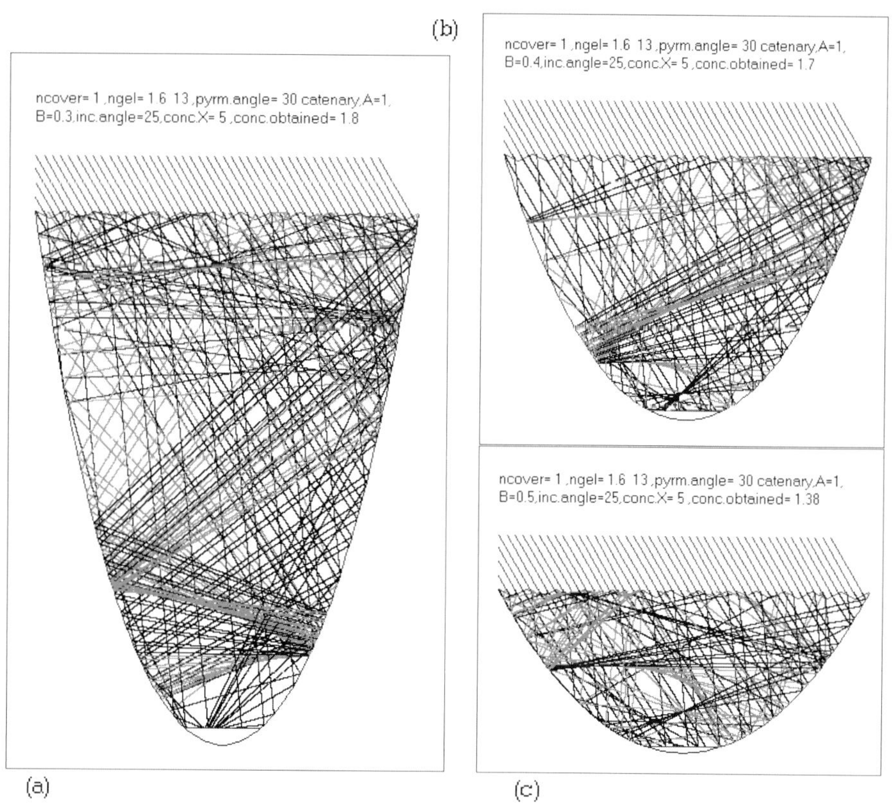

Figure 4.1-63a–c Two-dimensional ray-tracing through a concentrator without cover, but filled with a refracting material, having its top pyramid textured, for an incident angle of 25° and three catenary depths.

giving rise to electricity into heat (as some of it in actuality already is). One may think of cooling the modules of cells by running pipes filled with water or another suitable substance along their bottom side, carrying the heat to a thermal store, or alternatively by an air flow above the collector (but below a transparent cover layer). The energy available for this purpose is the incoming radiation energy minus the reflected and the converted part. Reflection from the back side of the semiconductor material, aimed at increasing the path-length of the photons in the material, could be chosen to optimise the value of the combined heat and power production, rather than only the power production. Examples of combined cycle solar panels are modelled in section 6.3.

For concentrator cells, active cooling is needed in any case, because of the temperature dependence of the characteristics of the photovoltaic process (cf. Fig. 4.1-37).

The maximum electrical efficiency of a single junction photovoltaic device implied by semiconductor physics is around 40% (see discussion earlier in section 4.1.2.3). If reflections can be minimised and many different light absorption materials are stacked, the theoretical Carnot efficiency (4.1.4) for the temperature of the Sun relative to a reference temperature at the surface of the Earth, i.e. about 95%, may be approached. Considerations of power flow optimisation diminish this limit further. Honsberg *et al.* (2001) and Green (2001) find a thermodynamical value of some 87% for an infinite stack of cells, to be used as a starting point for discussing the losses deriving from semiconductor physical arguments.

For a solar cell with electric efficiency below 40%, the possible associated heat gain is of the order of magnitude another 40% (cf. the simulation results in section 6.3.1).

4.1.3 Conversion of wind energy

4.1.3.1 Conversion of wind flow

Conversion of wind energy into linear motion of a body has been utilised extensively, particularly for transportation across water surfaces. A large sail-ship of the type used in the 19th century would have converted wind energy at peak rates of a quarter of a megawatt or more.

The force on a sail or a wing (i.e. profiles of negligible or finite thickness) may be broken down into a component in the direction of the undisturbed wind (drag) and a component perpendicular to the undisturbed wind direction (lift). When referring to an undisturbed wind direction it is assumed that a uniform wind field is modified in a region around the sail or the wing, but that beyond a certain distance such modifications can be disregarded.

In order to determine the force components, Euler's equations may be used. If viscous and external forces are neglected, and the flow is assumed to be irrotational (so that Bernoulli's equation is valid) and steady (so that the time-derivative of the velocity potential vanishes), then the force on a segment of the airfoil (sail or wing) may be written

$$\frac{d\boldsymbol{F}}{dz} = \oint_C P\boldsymbol{n}ds = -\tfrac{1}{2}\rho \oint_C (v\cdot v)\boldsymbol{n}ds.$$

Here dz is the segment length (cf. Fig. 4.1-64), C is a closed contour containing the airfoil profile, $\boldsymbol{n}$ is a unit vector normal to the contour [in the (x, y)-plane] and ds is the path-length increment, directed along the tangent to the contour, still in the (x, y)-plane. Taking advantage of the fact that the wind velocity v approaches a homogeneous field $\boldsymbol{W}$ (assumed to be along the x-axis) far from the airfoil, the contour integral may be reduced and evaluated (e.g. along a circular path),

$$d\boldsymbol{F}/dz = \rho W \Gamma \boldsymbol{e}_y, \tag{4.1.113}$$

$$\Gamma = \oint_C v\cdot ds \approx \pi c W \sin\alpha. \tag{4.1.114}$$

Here $\boldsymbol{e}_y$ is a unit vector along the y-axis, c is the airfoil chord length and α is the angle between the airfoil and $\boldsymbol{W}$. In the evaluation of the circulation Γ, it has been assumed that the airfoil is thin and without curvature. In this case c and α are well defined, but in general the circulation depends on the details of the profile, although an expression similar to the right-hand side of (4.1.114) is still valid as a first approximation, for some average chord length and angle of attack. Equation (4.1.113) is

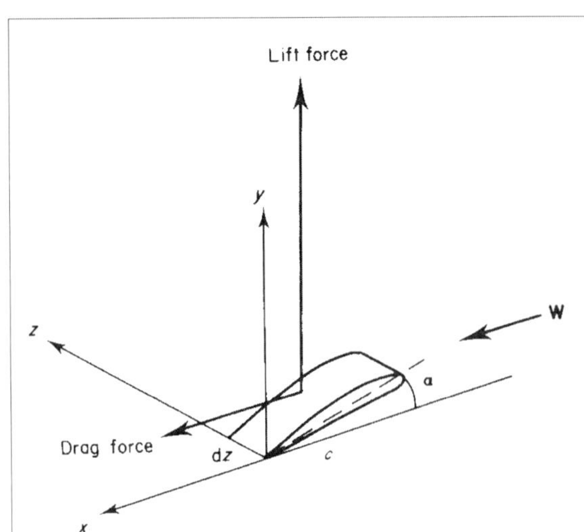

Figure 4.1-64 Forces on an airfoil segment.

known as the theorem of Kutta (1902) and Joukowski (1906).

The expressions (4.1.113) and (4.1.114) are valid in a co-ordinate system fixed relative to the airfoil (Fig. 4.1-64), and if the airfoil is moving with a velocity U, the velocities v and W are to be interpreted as relative ones, so that

$$\boldsymbol{W} = \boldsymbol{u}_{in} - \boldsymbol{U}, \tag{4.1.115}$$

if the undisturbed wind velocity is $\boldsymbol{u}_{in}$.

The assumption that viscous forces may be neglected is responsible for obtaining in (4.1.113) only a lift force, the drag force being zero. Primitive sail-ships, as well as primitive windmills, have been primarily aimed at utilising the drag force. It is possible, however, with suitably constructed airfoils, to make the lift force one or two orders of magnitude larger than the drag force and thereby effectively approach the limit where the viscous forces and hence the drag can be neglected. This usually requires careful "setting" of the airfoil, i.e. careful choice of the angle of attack, α, and in order to study operation at arbitrary conditions the drag component should be retained.

It is customary to describe the drag and lift forces on an airfoil of given shape, as a function of α, in terms of two dimensionless constants, $C_D(\alpha)$ and $C_L(\alpha)$, defined by

$$\begin{aligned} dF_x/dz &= \tfrac{1}{2}\rho C_D W^2 c, \\ dF_y/dz &= \tfrac{1}{2}\rho C_L W^2 c. \end{aligned} \tag{4.1.116}$$

The constants C_D and C_L are not quite independent of the size of the system, which is not unexpected since the viscous forces (friction) in air contribute most to turbulent motion on smaller scales (cf. the discussion in section 2.C). Introducing the Reynolds number,

$$\mathrm{Re} = Wc/\eta,$$

where η is the kinematic viscosity of air as a measure of the ratio between "inertial" and "viscous" forces acting between airfoil and air, the α-dependence of C_D and C_L for fixed Re, as well as the re-dependence for the value of α which gives the highest lift-to-drag ratio, $L/D = C_L/C_D$, may look as shown in Figs 4.1-65 and 4.1-66. The contours of these "high lift" profiles are indicated in Fig. 4.1-65.

Assuming that C_D, C_L and W are constant over the area $A = \int c\, dz$ of the airfoil, the work done by a uniform (except in the vicinity of the airfoil) wind field $\boldsymbol{u}_{in}$ on a device (e.g. a ship) moving with a velocity $\boldsymbol{U}$ can be derived from (4.1.116) and (4.1.115),

$$E = \boldsymbol{F} \cdot \boldsymbol{U}.$$

The angle β between $\boldsymbol{u}_{in}$ and $\boldsymbol{U}$ (see Fig. 4.1-67) may be maintained by a rudder. The power coefficient (4.1.50) becomes

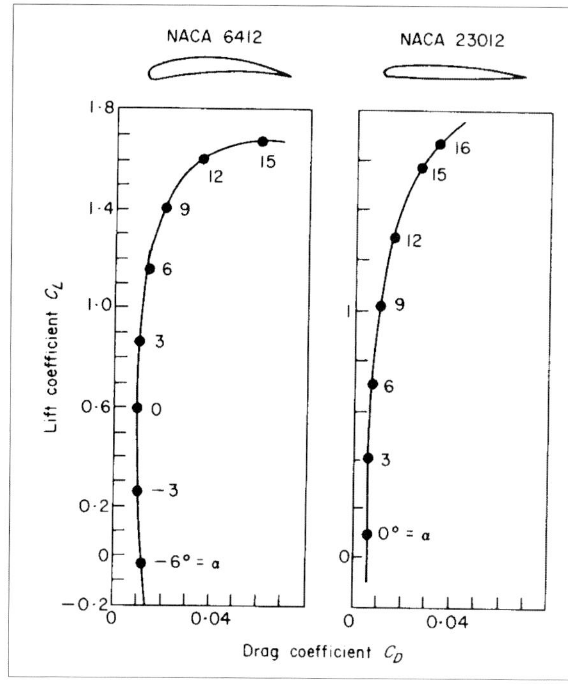

Figure 4.1-65 Lift and drag forces as a function of the angle of attack for two NACA airfoils (National Advisory Committee for Aeronautics; cf. e.g. Betz, 1959). The Reynolds number is Re = 8 × 10^6.

$$C_p = f(C_L \sin\beta - C_D(1 - \sin^2\beta + f^2 - 2f\cos\beta)^{1/2}) \times (1 + f^2 - 2f\cos\beta)^{1/2},$$

with $f = U/u_{in}$. For $C_L = 0$ the maximum C_p is $4C_D/27$, obtained for $f = 1/3$ and $\beta = 0$, whereas the maximum C_p for high lift-to-drag ratios L/D is obtained for β close to $\tfrac{1}{2}\pi$ and f around $2C_L/(3C_D)$. In this case, the maximum

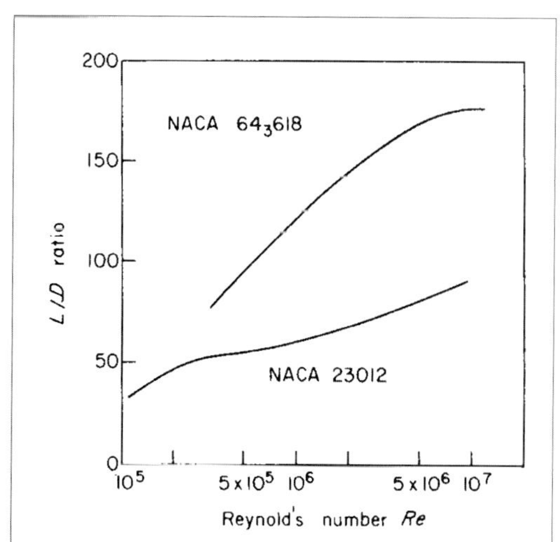

Figure 4.1-66 Reynolds number dependence of the lift-to-drag ratio, defined as the maximum value (as a function of the angle of attack) of the ratio between the lift and drag coefficients C_L and C_D (based on Hütter, 1977).

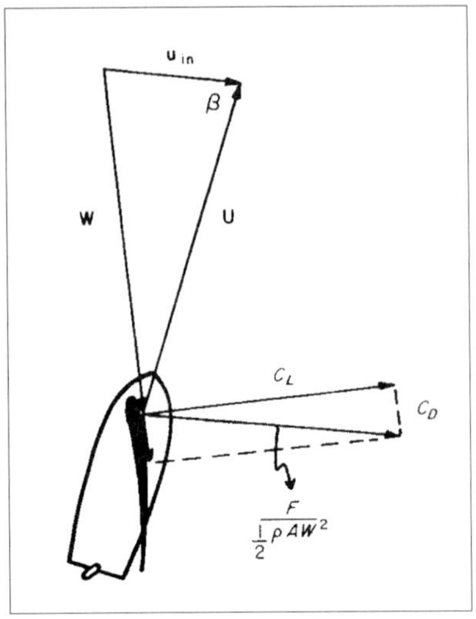

Figure 4.1-67 Velocity and force components for a sail-ship.

C_p may exceed Q by one to two orders of magnitude (Wilson and Lissaman, 1974).

It is quite difficult to maintain the high speeds U required for optimum performance in a linear motion of the airfoil, and it is natural to focus the attention on rotating devices, in case the desired energy form is shaft or electric power and not propulsion. Wind-driven propulsion in the past (mostly at sea) has been restricted to U/u_{in}-values far below the optimum region for high L/D airfoils (owing to friction against the water), and wind-driven propulsion on land or in the air has received little attention.

4.1.3.2 Propeller-type converters

Propellers have been extensively used in aircraft to propel the air in a direction parallel to that of the propeller axis, thereby providing the necessary lift force on the aeroplane wings. Propeller-type rotors are similarly used for windmills, but here the motion of the air (i.e. the wind) makes the propeller, which should be placed with its axis parallel to the wind direction, rotate, thus providing the possibility of power extraction. The propeller consists of a number of blades which are evenly distributed around the axis (cf. Fig. 4.1-68), with each blade having a suitable aerodynamic profile usually designed to produce a high lift force, as discussed in section 4.1.3.1. If there are two or more blades, the symmetrical mounting ensures a symmetrical mass distribution, but if only one blade is used it must be balanced by a counterweight.

Theory of non-interacting streamtubes

In order to describe the performance of a propeller-type rotor, the forces on each element of blade must be calculated, including the forces produced by the direct action of the wind field on each individual element as well as the forces arising as a result of interactions between different elements on the same or on different blades. Since the simple airfoil theory outlined in section 4.1.3.1 deals only with the forces on a given blade segment, in the absence of the other ones and also without the inclusion of "edge effects" from "cutting out" this one segment, it is tempting as a first approximation to treat the different radial segments of each blade as independent. Owing to the symmetrical mounting of blades, the corresponding radial segments of different blades (if more than one) may for a uniform wind field be treated together, considering as indicated in Fig. 4.1-68 an annulus-shaped "streamtube" of flow, bordered by streamlines intersecting the rotor plane at radial distances r and $r+dr$. The approximation of independent contributions from each streamtube, implying that radially induced velocities (u_r in Fig. 4.1-5) are neglected, allows the total shaft power to be expressed as

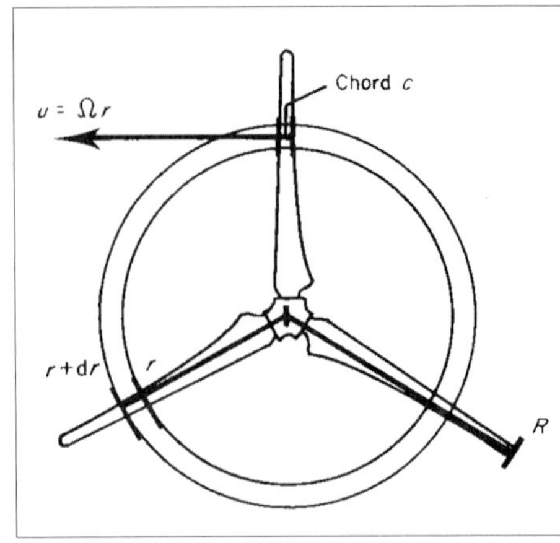

Figure 4.1-68 Definition of the streamtubes and blade integration variable for a propeller-type rotor.

$$E = \int_0^R \frac{dE}{dr} dr,$$

where dE/dr depends only on the conditions of the streamtube at r. Similar sums of independent contributions can in the same order of approximation be used to describe other overall quantities, such as the axial force component T and the torque Q (equal to the power E divided by the angular velocity Ω of the propeller).

In Fig. 4.1-69, a section of a wing profile (blade profile) is seen from a direction perpendicular to the cut. The distance of the blade segment from the axis of rotation is r, and its velocity $r\Omega$ is directed along the y-axis. This defines a coordinate system with a fixed x-axis along the rotor axis and moving y- and z-axes such that the blade segment is fixed relative to this coordinate system.

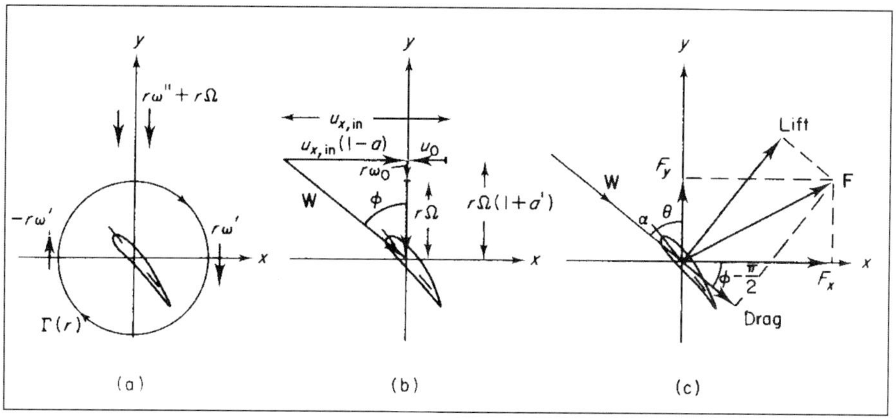

Figure 4.1-69 Velocity and force components of a rotating blade segment in a coordinate system following the rotation (which is in the direction of the y-axis): (a) determination of induced angular velocities; (b) determination of the direction of the apparent wind velocity; (c) determination of force components along x- and y-axes.

In order to utilise the method developed in section 4.1.3.1, the apparent wind velocity W to be used in (4.1.116) must be determined. It is given by (4.1.115) only if the velocity induced by, and remaining in the wake of, the device, $\boldsymbol{u}^{ind} = \boldsymbol{u}_{out} - \boldsymbol{u}_{in}$ (cf. Fig. 4.1-5), is negligible. Since the radial component u_r of $\boldsymbol{u}^{ind}$ has been neglected, $\boldsymbol{u}^{ind}$ has two components, one along the x-axis,

$$u_x^{ind} = u_{x,out} - u_{x,in} = -2au_{x,in}$$

cf. (4.1.47) and one in the tangential direction,

$$u_t^{ind} = \omega^{ind} r = 2a'\Omega r,$$

when expressed in a non-rotating coordinate system (the second equality is defining a quantity a', called the "tangential interference factor").

W is determined by the air velocity components in the rotor plane. From the momentum considerations underlying (4.1.47), the induced x-component u_0 in the rotor plane is seen to be

$$u_0 = \tfrac{1}{2} u_x^{ind} = -au_{x,in}. \quad (4.1.117)$$

In order to determine $r\omega_0$, the induced velocity of rotation of the air in the rotor plane, one may use the following argument (Glauert, 1935). The induced rotation is partly due to the rotation of the wings and partly due to the non-zero circulation (4.1.114) around the blade segment profiles (see Fig. 4.1-69a). This circulation may be considered to imply an induced air velocity component of the same magnitude, $|r\omega'|$, but with opposite direction in front of and behind the rotor plane. If the magnitude of the component induced by the wing rotation is called $r\omega''$ in the fixed coordinate system, it will be $r\omega'' + r\Omega$ and directed along the negative y-axis in the coordinate system following the blade. It has the same sign in front of and behind the rotor. The total y-component of the induced air velocity in front of the rotor is then $-(r\omega'' + r\Omega + r\omega')$, and the tangential component of the induced air velocity in the fixed co-ordinate system is $-(r\omega'' - r\omega')$, still in front of the rotor. But here there should be no induced velocity at all. This follows, for example, from performing a closed line integral of the air velocity v along a circle with radius r and perpendicular to the x-axis. This integral should be zero because the air in front of the rotor has been assumed to be irrotational,

$$0 = \int_{circle\ area} \text{rot}\ v\ dA = \oint v \cdot ds = -2\pi r(r\omega'' - r\omega'),$$

implying $\omega'' = \omega'$. This identity also fixes the total induced tangential air velocity behind the rotor, $r\omega^{ind} = r\omega'' + r\omega' = 2r\omega''$, and in the rotor plane (note that the circulation component $r\omega'$ is here perpendicular to the y-axis),

$$r\omega_0 = r\omega'' = \tfrac{1}{2} r\omega^{ind} = \Omega r a'. \quad (4.1.118)$$

The apparent wind velocity in the co-ordinate system moving with the blade segment, W, is now determined as seen in Fig. 4.1-69b. Its x-component is u_x, given by (4.1.47), and its y-component is obtained by taking (4.1.118) to the rotating coordinate system,

$$W_x = u_x = u_{x,in}(1 - a),$$
$$W_y = -(r\omega_x + r\Omega) = -r\Omega(1 + a'). \quad (4.1.119)$$

The lift and drag forces (Fig. 4.1-69c) are now obtained from (4.1.118) (except that the two components are no longer directed along the coordinate axes), with values of C_D, C_L and c pertaining to the segments of wings at the streamtube intersecting the rotor plane at the distance r from the axis. As indicated in Figs 4.1.69b and c, the angle of attack, α, is the difference between the angle ϕ, determined by

$$\tan \phi = -W_x/W_y, \quad (4.1.120)$$

and the pitch angle θ between blade and rotor plane,

$$\alpha = \phi - \theta. \quad (4.1.121)$$

Referred to the coordinate axes of Fig. 4.1-69, the force components for a single blade segment become

$$F_x = ½\, \rho\, cW^2(C_D(\alpha)\sin\phi + C_L(\alpha)\cos\phi),$$
$$F_y = ½\, \rho\, cW^2(-C_D(\alpha)\cos\phi + C_L(\alpha)\sin\phi), \quad (4.1.122)$$

and the axial force and torque contributions from a streamtube with B individual blades are given by

$$dT/dr = BF(r),$$
$$dQ/dr = BrF_y(r). \quad (4.1.123)$$

Combining (4.1.123) with equations expressing momentum and angular momentum conservation for each of the (assumed non-interacting) streamtubes, a closed set of equations is obtained. Equating the momentum change of the wind in the x-direction to the axial force on the rotor part intersected by an individual streamtube [i.e. (4.1.43) and (4.1.44) with $A = 2\pi r$], one obtains

$$dT/dr = -A\rho\, u_x u_x^{ind} = 2\pi r\rho\, u_{x,in}(1-a)2au_{x,in}, \quad (4.1.124)$$

and similarly equating the torque on the streamtube rotor part to the change in angular momentum of the wind (from zero to $r \times u_t^{ind}$), one gets

$$dQ/dr = A\rho\, u_x u_t^{ind} = 2\pi r^2 \rho\, u_{x,in}(1-a)2a'\Omega r. \quad (4.1.125)$$

Inserting $W = u_{x,in}(1-a)/\sin\phi$ or $W = r\Omega\,(1+a')/\cos\phi$ (cf. Fig. 4.1-69c) as necessary, one obtains a and a' expressed in terms of ϕ by combining (4.1.122)–(4.1.126). Since, on the other hand, ϕ depends on a and a' through (4.1.120) and (4.1.119), an iterative method of solution should be used. Once a consistent set of (a, a', ϕ)-values has been determined as a function of r [using a given blade profile implying known values of θ, c, $C_D(\alpha)$ and $C_L(\alpha)$ as function of r], either (4.1.123) or (4.1.124) and (4.1.125) may be integrated over r to yield the total axial force, total torque or total shaft power $E = \Omega Q$.

One may also determine the contribution of a single streamtube to the power coefficient (4.1.50),

$$C_p(r) = \frac{\Omega dQ/dr}{½\, \rho\, u_{x,in}^3 2\pi r} = 4a'(1-a)\left(\frac{r\Omega}{u_{x,in}}\right)^2. \quad (4.1.126)$$

The design of a rotor may utilise $C_p(r)$ for a given wind speed $u_{x,in}^{design}$ to optimise the choice of blade profile (C_D and C_L), pitch angle (θ) and solidity ($Bc/(\pi r)$) for given angular velocity of rotation (Ω). If the angular velocity Ω is not fixed (as it might be by use of a suitable asynchronous electrical generator, except at start and stop), a dynamic calculation involving $d\Omega/dt$ must be performed. Not all rotor parameters need to be fixed. For example, the pitch angles may be variable, by rotation of the entire wing around a radial axis, such that all pitch angles θ (r) may be modified by an additive constant θ_0. This type of regulation is useful in order to limit the C_p-drop when the wind speed moves away from the design value $u_{x.in}^{design}$. The expressions given above would then give the actual C_p for the actual setting of the pitch angles given by $\theta(r)_{design} + \theta_0$ and the actual wind speed $u_{x,in}$, with all other parameters left unchanged.

In introducing the streamtube expressions (4.1.124) and (4.1.125), it has been assumed that the induced velocities and thus a and a' are constant along the circle periphery of radius r in the rotor plane. As the model used to estimate the magnitude of the induced velocities (Fig. 4.1-69a) is based on being near to a rotor blade, it is expected that the average induced velocities in the rotor plane, u_0 and $r\omega_0$, are smaller than the values calculated by the above expressions, unless the solidity is very large. In practical applications, it is customary to compensate by multiplying a and a' by a common factor, F, less than unity and a function of B, r and ϕ (see e.g. Wilson and Lissaman, 1974).

Furthermore, edge effects associated with the finite length R of the rotor wings have been neglected, as have the "edge" effects at $r = 0$ due to the presence of the axis, transmission machinery, etc. These effects may be described in terms of "trailing vortices" shed from the blade tips and blade roots and moving in helical orbits away from the rotor in its wake. Vorticity is a name for the vector field rot$v$, and the vortices connected with the circulation around the blade profiles (Fig. 4.1-69a) are called "bound vorticity". This can "leave" the blade only at the tip or at the root. The removal of bound vorticity is equivalent to a loss of circulation I (4.1.114) and hence a reduction in the lift force. Its dominant effect is on the tangential interference factor a' in (4.1.125), and so it has the same form and may be treated on the same footing as the corrections due to finite blade number. (Both are often referred to as "tip losses", since the correction due to finite blade number is usually appreciable only near the blade tips, and they may be approximately described by the factor F introduced above.) Other losses may be associated with the "tower shadow", etc.

Model behaviour of power output and matching to load

A calculated overall C_p for a three-bladed, propeller-type wind energy converter is shown in Fig. 4.1-70, as a function of the tip-speed ratio,

$$\lambda = \Omega R/u_{x,in},$$

and for different settings of the overall pitch angle, θ_0. It is clear from (4.1.126) and the equations for determining

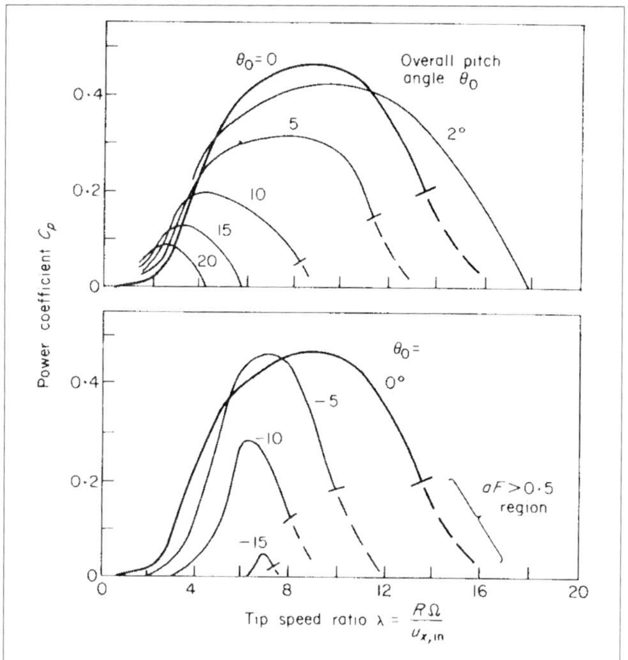

Figure 4.1-70 Dependence of power coefficient on tip-speed ratio for different settings of the overall pitch angle. The calculation is based on the wing geometry shown in Fig. 4.1-71 and the NACA 23012 profile of Fig. 4.1-65.

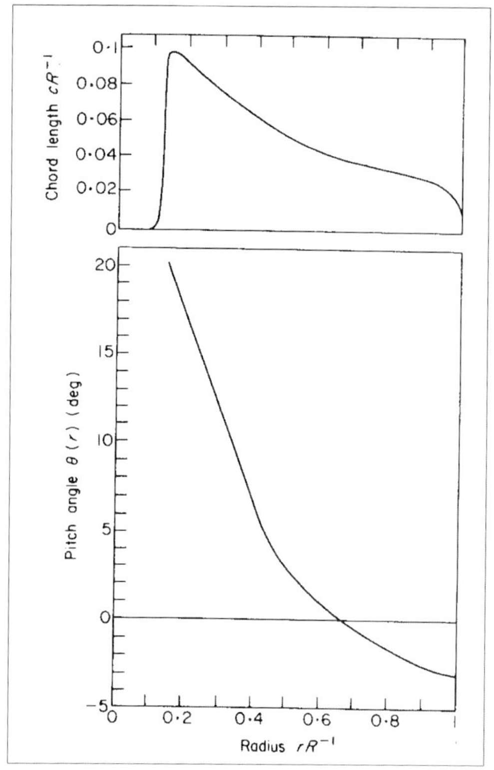

Figure 4.1-71 Chord variation and twist along the blades used for the calculations in this section.

a and a' that C_p depends on the angular velocity Ω and the wind speed $u_{x,in}$ only through the ratio λ. Each blade has been assumed to be an airfoil of the type NACA 23012 (cf. Fig. 4.1-65), with chord c and twist angle θ changing along the blade from root to tip, as indicated in Fig. 4.1-71. A tip-loss factor F has been included in the calculation of sets of corresponding values of a, a' and ϕ for each radial station (each "streamtube"), according to the Prandtl model described by Wilson and Lissaman (1974). The dashed regions in Fig. 4.1-70 correspond to values of the product aF (corresponding to a in the expression without tip loss) larger than 0.5. Since the axial air velocity in the wake is $u_{x,out} = u_{x,in}(1 - 2aF)$ [cf. (4.1.47) with $F = 1$], $aF > 0.5$ implies reversed or re-circulating flow behind the wind energy converter, a possibility which has not been included in the above description. The C_p-values in the dashed regions may thus be inaccurate and presumably overestimated.

Figure 4.1-70 shows that for negative settings of the overall pitch angle θ_0, the C_p distribution on λ-values is narrow, whereas it is broad for positive θ_0 of modest size. For large positive θ_0, the important region of C_p moves to smaller λ-values. The behaviour of $C_p(\lambda)$ in the limit of λ approaching zero is important for operating the wind energy converter at small rotor angular velocities and, in particular, for determining whether the rotor will be self-starting, as discussed in more detail below.

For a given angular speed Ω of the rotor blades, the power coefficient curve specifies the fraction of the power in the wind which is converted, as a function of the wind speed $u_{x,in}$. Multiplying C_p by ½ $\rho A u_{x,in}^3$, where the rotor area is $A = \pi R^2$ (if "coning" is disregarded, i.e. the blades are assumed to be in the plane of rotation), the power transferred to the shaft can be obtained as a function of wind speed, assuming the converter to be oriented ("yawed") such that the rotor plane is perpendicular to the direction of the incoming wind. Figures 4.1-72 and 4.1-73 show such plots of power E, as functions of wind speed and overall pitch angle, for two definite angular velocities ($\Omega = 4.185$ and 2.222 rad s^{-1}, if the length of the wings is $R = 27$ m). For the range of wind speeds encountered in the planetary boundary layer of the atmosphere (e.g. at heights between 25 and 100 m), a maximum power level around 4000 W per average square metre swept by the rotor (10 MW total) is reached for the device with the higher rotational speed (tip speed $R\Omega = 113$ ms^{-1}), whereas about 1000 W m^{-2} is reached by the device with the lower rotational velocity ($R\Omega = 60$ m s^{-1}).

The rotor has been designed to yield a maximum C_p at about $\lambda = 9$ (Fig. 4.1-70), corresponding to $u_{x,in} = 12.6$ and 6.6 m s^{-1} in the two cases. At wind speeds above these values the total power varies less and less strongly, and for negative pitch angles it even starts to decrease with increasing wind speeds in the range characterising "stormy weather". This phenomenon can be used to make the wind energy converter self-regulating, provided

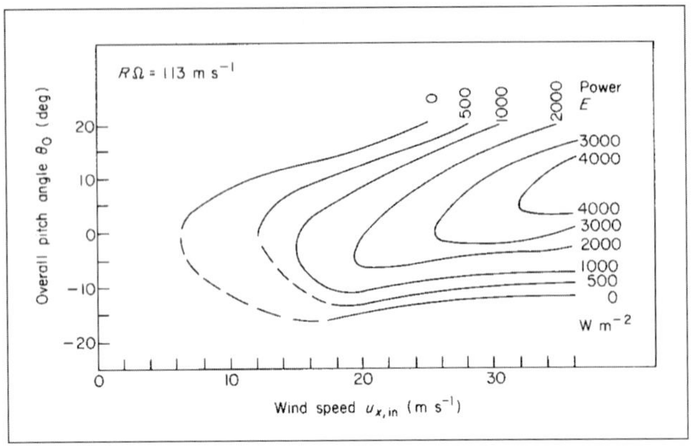

Figure 4.1-72 Power output of the wind energy converter described in Figs 4.1-70 and 4.1-71, for a tip speed of 113 m s^{-1}, as a function of wind speed and overall pitch angle. The actual Reynolds number is in this case about 5 × 10^6 from blade root to tip, in reasonable consistency with the value used in the airfoil data of Fig. 4.1-65. Dashed lines indicate regions where *a* [defined in (4.1.119)] times the tip speed correction, *F*, exceeds 0.5 for some radial stations along the blades.

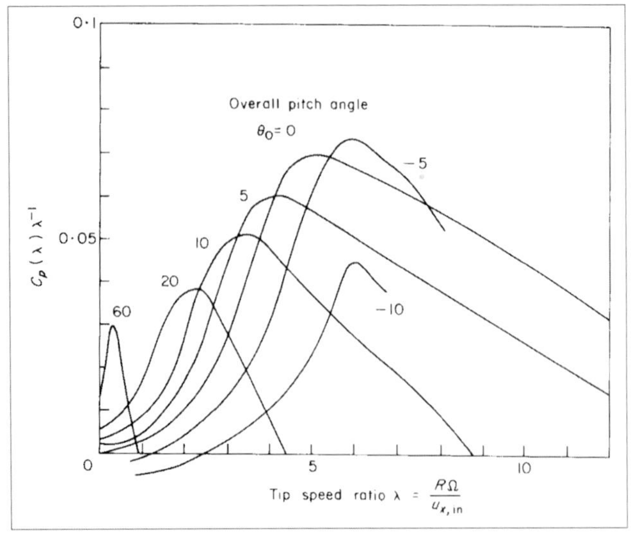

Figure 4.1-74 Variation of power coefficient over tip-speed ratio, as a function of tip-speed ratio, for the converter considered in the previous figures.

that the "flat" or "decreasing" power regions are suitably chosen and that the constancy of the angular velocity at operation is assured by, for example, coupling the shaft to a suitable asynchronous electricity generator (a generator allowing only minute changes in rotational velocity, with the generator angular speed Ωg being fixed relative to the shaft angular speed Ω, e.g. by a constant exchange ratio $n = \Omega g/\Omega$ of a gearbox). Self-regulating wind energy converters with fixed overall pitch angle θ_0 and constant angular velocity have long been used for AC electricity generation (Juul, 1961).

According to its definition, the torque $Q = E/\Omega$ can be written

$$Q = \tfrac{1}{2}\, \rho\, \pi R^3 u_{x,in}^2 C_p(\lambda)/\lambda, \qquad (4.1.127)$$

and Fig. 4.1-74 shows C_p/λ as a function of λ for the same design as the one considered in Fig. 4.1-70. For pitch angles θ_0 less than about −3°, C_p/λ is negative for $\lambda = 0$, whereas the value is positive if θ_0 is above the critical value of about −3°. The corresponding values of the total torque (4.1.127), for a rotor with radius $R = 27$ m and overall pitch angle $\theta_0 = 0°$, are shown in Fig. 4.1-75, as a function of the angular velocity $\Omega = \lambda u_{x,in}/R$ and for different wind speeds. The small dip in Q for low rotational speeds disappears for higher pitch angles θ_0.

The dependence of the torque at $\Omega = 0$ on pitch angle is shown in Fig. 4.1-76 for a few wind speeds. Advantage can be taken of the substantial increase in starting torque with increasing pitch angle, in case the starting torque at the pitch angle desired at operating angular velocities is insufficient and provided the overall pitch angle can be changed. In that case a high overall pitch angle is chosen to start the rotor from $\Omega = 0$, where the internal resistance (friction in bearings, gearbox, etc.) is large. When an angular speed Ω of a few degrees per second is reached, the pitch angle is diminished to a value close to the optimum one (otherwise the torque will pass a maximum and soon start to decrease again, e.g. as seen from the $\theta_0 = 60°$curve in Fig. 4.1-74). Usually, the internal resistance diminishes as soon as Ω is non-zero. The internal resistance may be represented by an internal torque, $Q_0(\Omega)$, so that the torque available for some load, "external" to the wind power to shaft power converter (e.g. an electric generator), may be written

$$Q^{available} = Q(\Omega) - Q_0(\Omega)$$

If the "load" is characterised by demanding a fixed rotational speed Ω (such as the asynchronous electric generator), it may be represented by a vertical line in the

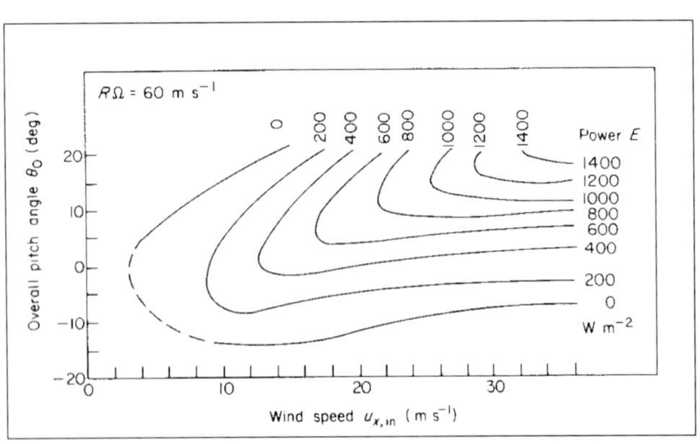

Figure 4.1-73 Same as Fig. 4.1-72, but for a tip speed of 60 m s^{-1}.

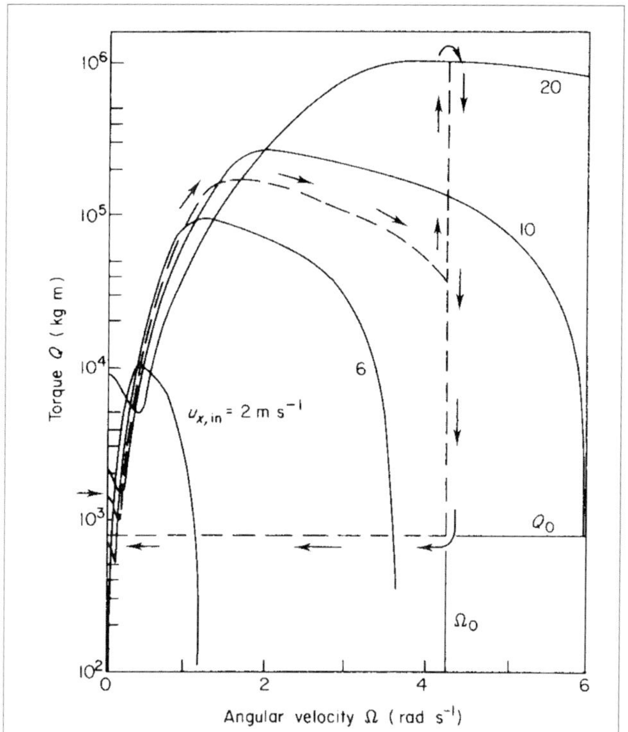

Figure 4.1-75 Torque variation with angular velocity for a wind energy converter of the type described in the previous figures, with radius R = 27 m and overall pitch angle 0°. The dashed, vertical line represents operation at fixed angular velocity (e.g. with a synchronous generator). Q_0 illustrates the internal torque.

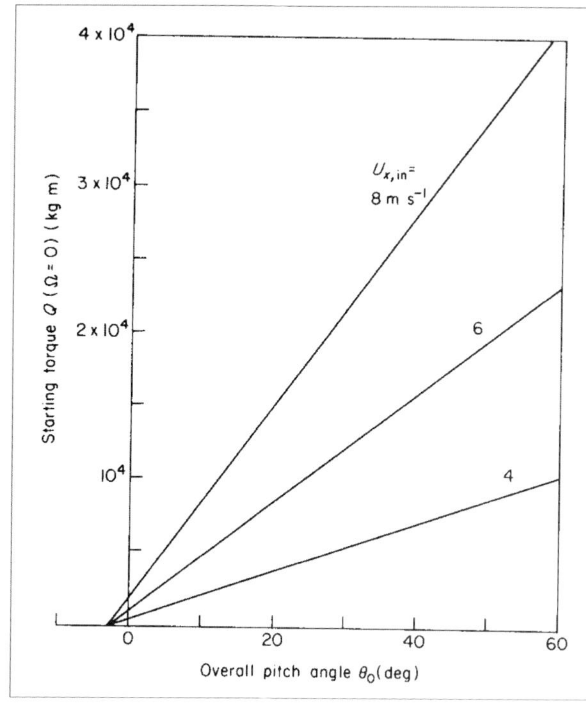

Figure 4.1-76 Starting torque as a function of overall pitch angle for the wind energy converter considered in the preceding figures.

diagram shown in Fig. 4.1-75. As an example of the operation of a wind energy generator of this type, the dashed lines (with accompanying arrows) in Fig. 4.1-75 describe a situation with initial wind speed 8 m s^{-1}, assuming that this provides enough torque to start the rotor at the fixed overall pitch angle $\theta_0 = 0°$ (i.e. the torque produced by the wind at 8 m s^{-1} and $\Omega = 0$ is above the internal torque Q_0, here assumed to be constant). The excess torque makes the angular velocity of the rotor, Ω, increase along the specific curve for $u_{x,in} = 8$ m s^{-1}, until the value Ω_0 characterising the electric generator is reached. At this point the load is connected and the wind energy converter begins to deliver power to the load area. If the wind speed later increases, the torque will increase along the vertical line at $\Omega = \Omega_0$ in Fig. 4.1-75, and if the wind speed decreases, the torque will decline along the same vertical line until it reaches the internal torque value Q_0. Then the angular velocity diminishes and the rotor is brought to a halt.

An alternative type of load may not require a constant angular velocity. Synchronous DC electric generators are of this type, providing an increasing power output with increasing angular velocity (assuming a fixed exchange ratio between rotor angular velocity Ω and generator angular velocity Ωg). Instead of staying on a vertical line in the torque-versus-Ω diagram, for varying wind speed, the torque now varies along some fixed, monotonically increasing curve characterising the generator. An optimum synchronous generator would be characterised by a $Q(\Omega)$ curve, which for each wind speed $u_{x,in}$ corresponds to the value of $\Omega = \lambda u_{x,in}/R$ which provides the maximum power coefficient C_p (Fig. 4.1-70). This situation is illustrated in Fig. 4.1-77, again indicating by a set of dashed curves the torque variation for an initial wind speed of 8 m s^{-1}, followed by an increasing and later again decreasing wind speed. In this case the $Q = Q_0$ limit is finally reached at a very low angular velocity, indicating that power is still delivered during the major part of the slowing-down process.

Non-uniform wind velocity

The velocity field of the wind may be non-uniform in time as well as in spatial distribution. The influence of time variations in wind speed on power output of a propeller-type wind energy converter has been touched upon in the previous subsection, although a detailed investigation involving the actual time dependence of the angular velocity Ω was not included. In general, the direction of the wind velocity is also time dependent, and the conversion device should be able to successively align its rotor axis with the long-range trends in wind direction, or suffer a power

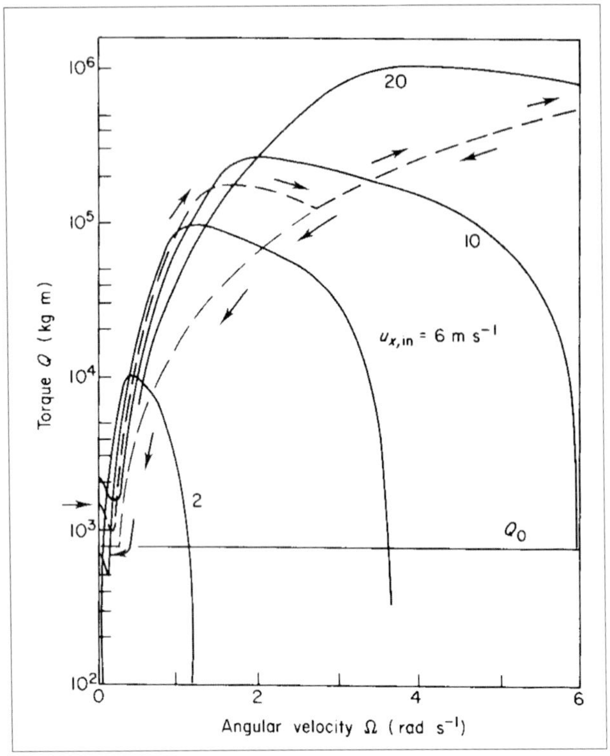

Figure 4.1-77 Same as Fig. 4.1-75, but here operation is illustrated for a load (e.g. asynchronous generator) which is optimised, i.e. corresponds to maximum C_p, at each wind speed.

reduction which is not just the cosine to the angle β between the rotor axis and the wind direction ("yaw angle"), but involves calculating the performance of each blade segment for an apparent wind speed W and angle of attack α different from the ones previously used, and no longer axially symmetric. This means that the quantities W and α (or ϕ) are no longer the same inside a given annular stream-tube, but they depend on the directional position of the segment considered (e.g. characterised by a rotational angle ψ in a plane perpendicular to the rotor axis).

Assuming that both the wind direction and the wind speed are functions of time and of height h (measured from ground level or from the lower boundary of the velocity profile z_0), but that the direction remains horizontal, then the situation will be as depicted in Fig. 4.1-78. The coordinate system (x_0, y_0, z_0) is fixed and has its origin in hub height h_0, where the rotor blades are fastened. Consider now a blade element at a distance r from the origin, on the ith blade. The projection of this position on to the (y_0, z_0)-plane is turned the angle ψi from vertical, where

$$\psi_i = 2\pi/i + \Omega t, \tag{4.1.128}$$

at the time t. The coning angle δ is the angle between the blade and its projection onto the (y_0, z_0)-plane, and the height of the blade element above the ground is

$$h = h_0 + r\cos\delta\cos\psi_i, \tag{4.1.129}$$

where h_0 is the hub height. The height h enters as a parameter in the wind speed $u_{in} = |\,\mathbf{u}_{in}(h, t)\,|$ and the yaw angle $\beta = \beta(h, t)$, in addition to time.

An attempt can now be made to copy the procedure used in the case of a uniform wind velocity along the rotor axis, i.e. to evaluate the force components for an individual stream-tube both by the momentum consideration of section 4.1.1.4 and by the lift and drag approach of section 4.1.3.1. The individual stream-tubes can no longer be taken as annuli, but must be of an area A_s (perpendicular to the wind direction) small enough to

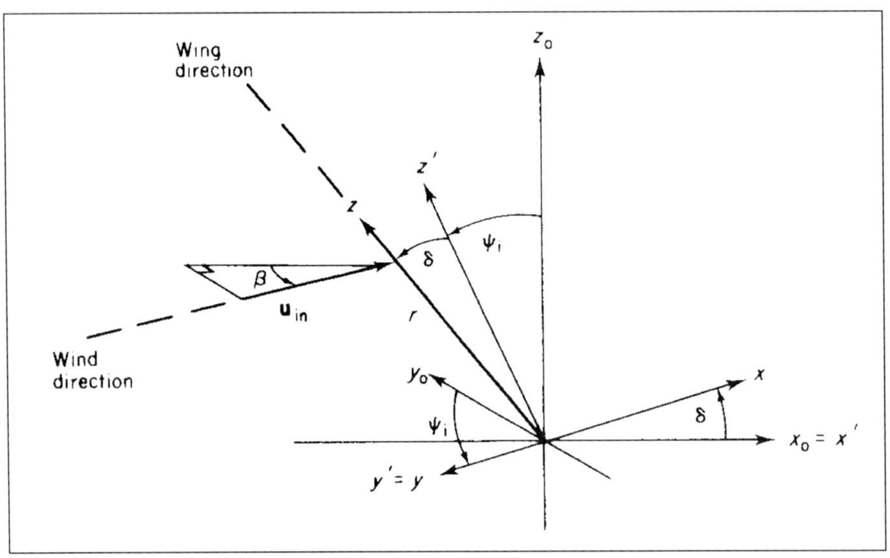

Figure 4.1-78 Definition of coordinate systems for a coning rotor in a non-uniform wind field.

permit the neglect of variations in wind velocity over the area at a given time. It will still be assumed that the flows inside different stream-tubes do not interact, and also, for simplicity, the stream-tubes will be treated as "straight lines", i.e. not expanding or bending as they pass the region of the conversion device (as before, this situation arises when induced radial velocities are left out).

Consider first the "momentum equation" (4.1.43), with the suffix s denoting "in the stream-wise direction" or "along the stream-tube",

$$F_s = -J_m u_s^{ind}.$$

This force has components along the x-, y-, and z-directions of the local coordinate system of a blade turned the angle ψ from vertical (cf. Fig. 4.1-69). The y-direction is tangential to the rotation of the blade element, and in addition to the component of F_s there may be an induced tangential velocity u_t^{ind} (and force) of the same type as the one considered in the absence of yaw (in which case F_s is perpendicular to the y-axis). The total force components in the local coordinate system are thus

$$F_x = -J_m u_s^{ind}(\cos\beta\cos\delta + \sin\beta\sin\delta\sin\psi),$$
$$F_y = -J_m u_s^{ind}\sin\beta\cos\psi + J_m u_t^{ind}. \tag{4.1.130}$$

From the discussion of Fig. 4.1-5, (4.1.44) and (4.1.47),

$$J_m = A_s\rho\, u_s = A_s\rho\, u_{in}(1-a),$$
$$u_s^{ind} = -2au_{in},$$

and in analogy with (4.1.118), a tangential interference factor a' may be defined by

$$u_t^{ind} = 2a' r\Omega\cos\delta.$$

However, the other relation contained in (4.1.118), from which the induced tangential velocity in the rotor plane is half the one in the wake, cannot be derived in the same way without the assumption of axial symmetry. Instead, the variation in the induced tangential velocity as a function of rotational angle ψ is bound to lead to crossing of the helical wake strains and it will be difficult to maintain the assumption of non-interacting stream-tubes. Here, the induced tangential velocity in the rotor plane will still be taken as $\frac{1}{2}u_t^{ind}$, an assumption which at least gives reasonable results in the limiting case of a nearly uniform incident wind field and zero or very small yaw angle.

Secondly, the force components may be evaluated from (4.1.120)–(4.1.122) for each blade segment, defining the local coordinate system (x, y, z) as in Fig. 4.1-78 with the z-axis along the blade and the y-axis in the direction of the blade's rotational motion. The total force, averaged over a rotational period, is obtained by multiplying by the number of blades, B, and by the fraction of time each blade spends in the stream-tube. Defining the stream-tube dimensions by an increment dr in the z-direction and an increment $d\psi$ in the rotational angle ψ, each blade spends the time fraction $d\psi/(2\pi)$ in the stream-tube, at constant angular velocity Ω. The force components are then

$$F_x = B\,(d\psi/2\pi)\tfrac{1}{2}\rho\, cW^2\,(C_D(\alpha)\sin\phi + C_L(\alpha)\cos\phi)dr,$$
$$F_y = B\,(d\psi/2\pi)\tfrac{1}{2}\rho\, cW^2\,(-C_D(\alpha)\cos\phi + C_L(\alpha)\sin\phi)dr, \tag{4.1.131}$$

in the notation of (4.1.122). The angles ϕ and α are given by (4.1.120) and (4.1.121), but the apparent velocity W is the vector difference between the stream-wise velocity $u_{in} + \frac{1}{2}u_s^{ind}$ and the tangential velocity $r\Omega\cos\delta + \frac{1}{2}u_t^{ind}$, both taken in the rotor plane. From Fig. 4.1-78,

$$W_x = u_{in}(1-a)(\cos\beta\cos\delta + \sin\beta\sin\delta\sin\psi),$$
$$W_y = u_{in}(1-a)(\sin\beta\cos\psi + r\Omega\cos\delta\,(1+a'),$$
$$W_z = u_{in}(1-a)(-\cos\beta\sin\delta + \sin\beta\cos\delta\sin\psi). \tag{4.1.132}$$

The appropriate W^2 to insert into (4.1.131) is $W_x^2 + W_y^2$. Finally, the relation between the stream-tube area A_s and the increments dr and $d\psi$ must be established. As indicated in Fig. 4.1-79, the stream-tube is approximately a rectangle with sides dL and dL', given by

$$dL = r\cos\delta\,(\cos 2\beta\cos 2\psi + \sin 2\psi)^{1/2} d\psi, \tag{4.1.133}$$

$$dL' = (\sin 2\delta\,(1+\cos 2\beta) + \cos 2\delta\,(\cos 2\beta\sin 2\psi + \cos 2\psi))^{1/2} dr,$$

and

$$A_s = dL\, dL'.$$

Now, for each stream-tube, a and a' are obtained by equating the x- and y-components of (4.1.130) and (4.1.131) and using the auxiliary equations for (W, ϕ) or (W_x, W_y). The total thrust and torque are obtained by integrating F_x and $r\cos\delta\, F_y$, over dr and $d\psi$ (i.e. over all stream-tubes).

Restoration of wind profile in wake, and implications for turbine arrays

For a wind energy converter placed in the planetary boundary layer (i.e. in the lowest part of the Earth's atmosphere), the reduced wake wind speed in the stream-wise direction, $u_{s,out}$, will not remain below the wind speed $u_{s,in}$ of the initial wind field, provided that this is not diminishing with time. The processes responsible for maintaining the general kinetic motion in the atmosphere making up for the kinetic energy lost by surface friction and other dissipative processes, will also act in the

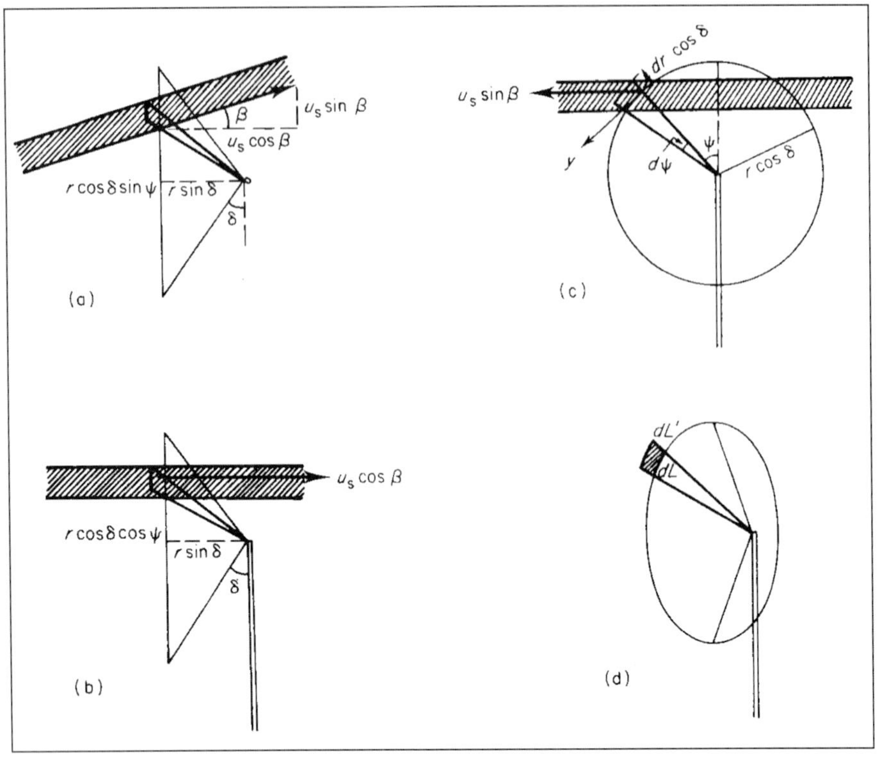

Figure 4.1-79 Stream-tube definition (hatched areas) for a coning rotor in a non-uniform wind field: (a) view from top; (b) side view; (c) front view; (d) view along stream-tube.

direction of restoring the initial wind profile (speed as function of height) in the wake of a power extracting device by transferring energy from higher air layers (or from the "sides") to the partially depleted region (Sørensen, 1996d). The large amounts of energy available at greater altitude (cf. Fig. 3.3-31) make such processes possible almost everywhere at the Earth's surface and not just at those locations where new kinetic energy is predominantly being created.

In the near wake, the wind field is non-laminar, owing to the induced tangential velocity component, u_t^{ind}, and owing to vorticity shedded from the wing tips and the hub region (as discussed in section 4.1.3.2). It is then expected that these turbulent components gradually disappear further down-stream in the wake, as a result of interactions with the random eddy motion of different scales present in the "unperturbed" wind field. "Disappear" here means "get distributed on a large number of individual degrees of freedom", so that no contributions to the time-averaged quantities considered [cf. (2.13)] remain. For typical operation of a propeller-type wind conversion device, such as the situations illustrated in Figs 4.1-72 and 4.1-73, the tangential interference factor a' is small compared to the axial interference factor a, implying that the most visible effect of the passage of the wind field through the rotor region is the change in stream-wise wind speed. This is a function of r, the distance of the blade segment from the hub centre, as illustrated in Fig. 4.1-80, based on wind tunnel measurements. The induced r-dependence of the axial velocity is seen to gradually smear out, although it is clearly visible at a distance of two rotor radii in the set-up studied.

Figure 4.1-81 suggests that under average atmospheric conditions the axial velocity, $u_{s,in}$, will be restored to

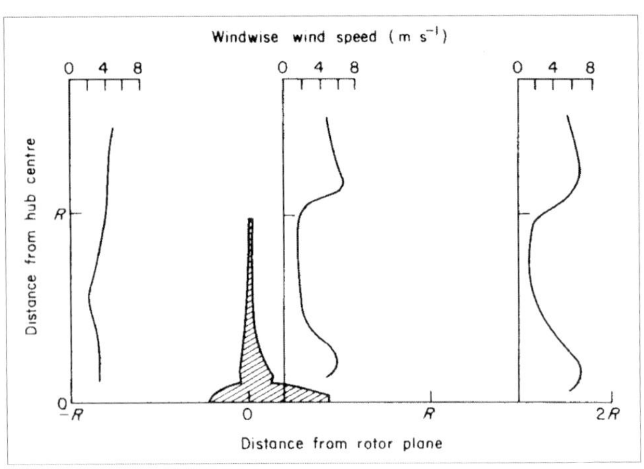

Figure 4.1-80 Wind tunnel results illustrating the distortion of the wind field along the radial direction (R is the rotor radius) at various distances from the rotor plane (based on Hütter, 1976).

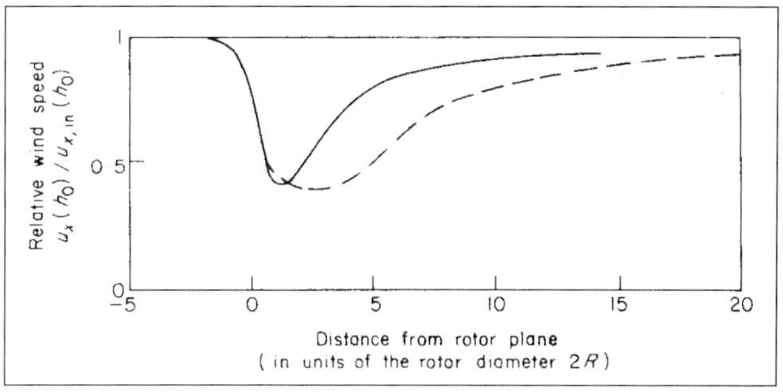

Figure 4.1-81 Wind tunnel results indicating the restoration at stream-wise wind speed behind a rotor (placed at distance 0, the rotor being simulated by a gauze disc). The approaching wind has a logarithmic profile (solid line) or is uniform with approximately laminar flow (dashed line) (based on Pelser, 1975).

better than 90% at about 10 rotor diameters behind the rotor plane and better than 80% at a distance of 5–6 rotor diameters behind the rotor plane, but rather strongly dependent on the amount of turbulence in the "undisturbed" wind field.

A second wind energy converter may be placed behind the first one, in the wind direction, at a distance such that the wind profile and magnitude are reasonably well restored. According to the simplified investigations in wind tunnels (Fig. 4.1-81), supported by field measurements behind buildings, forests and fences, a suitable distance would seem to be 5–10 rotor diameters (increasing to over 20 rotor diameters if "complete restoration" is required). If there is a prevailing wind direction, the distance between conversion units perpendicular to this direction may be smaller (essentially determined by the induced radial velocities, which were neglected in the preceding subsections, but appear qualitatively in Fig. 4.1-5). If, on the other hand, several wind directions are important, and the converters are designed to be able to "yaw against the wind", then the distance required by wake considerations should be kept in all directions.

More severe limitations may possibly be encountered with a larger array of converters, say, distributed over an extended area with the average spacing X between units. Even if X is chosen so that the relative loss in stream-wise wind speed is small from one unit to the next, the accumulated effect may be substantial, and the entire boundary layer circulation may become altered in such a way that the power extracted decreases more sharply than expected from the simple wake considerations. Thus, large-scale conversion of wind energy may even be capable of inducing local climatic changes.

A detailed investigation of the mutual effects of an extended array of wind energy converters and the general circulation on each other requires a combination of a model of the atmospheric motion with a suitable model of the disturbances induced in the wake of individual converters. In one of the first discussions of this problem, Templin (1976) considered the influence of an infinite two-dimensional array of wind energy converters with fixed average spacing on the boundary layer motion to be restricted to a change of the roughness length z_0 in the logarithmic expression for the wind profile, assumed to describe the wind approaching any converter in the array. The change in z_0 can be calculated from the stress τ^{ind} exerted by the converters on the wind, due to the axial force F_x in (4.1.122) or (4.1.124), which according to (4.1.46) can be written

$$\tau^{ind} = F_x/S = \tfrac{1}{2}\rho(u_{s,in})^2 2(1-a)^2 A/S,$$

with S being the average ground surface area available for each converter and A/S being the "density parameter", equal to the ratio of rotor-swept area to ground area. For a quadratic array with regular spacing, S may be taken as X^2; $u_{x,in}$ taken at hub height h_0 can, in the case of a neutral atmosphere, be written

$$u_{x,in}(h_0) = \frac{1}{\kappa}\left(\frac{\tau^0 + \tau^{ind}}{\rho}\right)^{1/2} \log\left(\frac{h_0}{z'_0}\right),$$

where τ^0 is the stress in the absence of wind energy converters, and z'_0 is the roughness length in the presence of the converters, which can now be determined from this equation and τ^{ind} from the previous one.

Figure 4.1-82 shows the results of a calculation for a finite array of converters (Taylor *et al.*, 1993) using a simple model with fixed loss fractions (Jensen, 1994). It is seen that this model is incapable of reproducing the

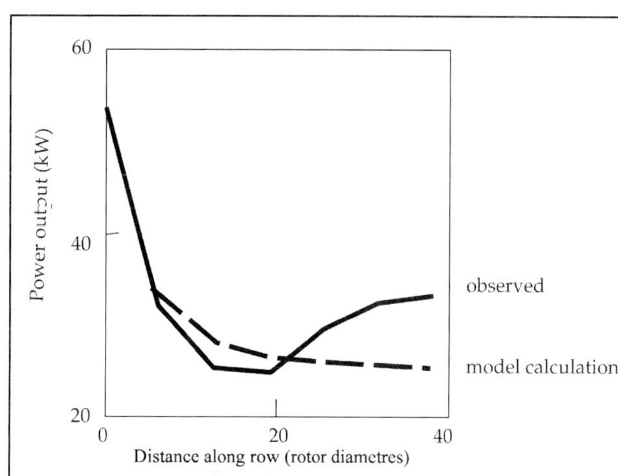

Figure 4.1-82 Calculated (see text) and measured power outputs from one row of turbines perpendicular to the wind direction, in a wind park located at Nørrekær Enge in Denmark. It consists in total of five rows with a turbine spacing of 6 rotor diameters (Taylor *et al.*, 1993).

fast restoration of winds through the turbine array, presumably associated with the propagation of the enhanced wind regions created just outside the swept areas (as seen in Fig. 4.1-80). A three-dimensional fluid dynamics model is required for describing the details of array shadowing effects. Such calculations are in principle possible, but so far no convincing implementation has been presented. The problem is the very accurate description needed, of complex three-dimensional flows around the turbines and for volumes comprising the entire wind farm of maybe hundreds of turbines. This is an intermediate regime between the existing three-dimensional models for gross wind flow over complex terrain and the detailed models of flow around a single turbine used in calculations of aerodynamical stability.

4.1.3.3 Cross-wind converters

Wind energy converters of the cross-wind type have the rotor axis perpendicular to the wind direction. The rotor axis may be horizontal as in wheel-type converters (in analogy to waterwheels) or vertical as in the panemones used in Iran and China. The blades (ranging from simple "paddles" to optimised airfoil sections) will be moving with and against the wind direction on alternative sides of the rotor axis, necessitating some way of emphasising the forces acting on the blades on one side. Possible ways are simply to shield half of the swept area, as in the Persian panemones (Wulff, 1966); to curve the "paddles" so that the (drag) forces are smaller on the convex than on the concave side, as in the Savonius rotor (Savonius, 1931); or to use aerodynamically shaped wing blades producing high lift forces for wind incident on the "front edge", but small or inadequate forces for wind incident on the "back edge", as in the Darrieus rotor (cf. Fig. 4.1-83) and related concepts. Another possibility is to allow for changes in the pitch angle of each blade, as, for example, achieved by hinged vertical blades in the Chinese panemone type (Chen Li, 1951). In this case the blades on one side of a vertical axis have favourable pitch angles, while those on the other side have unfavourable settings. Apart for the shielded ones, vertical axis cross-wind converters are omnidirectional, i.e. they accept any horizontal wind direction on equal footing.

Performance of a Darrieus-type converter

The performance of a cross-wind converter, such as the Darrieus rotor shown in Fig. 4.1-83, may be calculated in a way similar to that used in section 4.1.3.2, by derivation of the unknown, induced velocities from expressing the forces both in terms of lift and drag forces on the blade segments and in terms of momentum changes between incident wind and wake flow. Also, the flow may be divided into a number of streamtubes (assumed to be non-interacting), according to assumptions on the symmetries of the flow field. Figure 4.1-74 gives an example

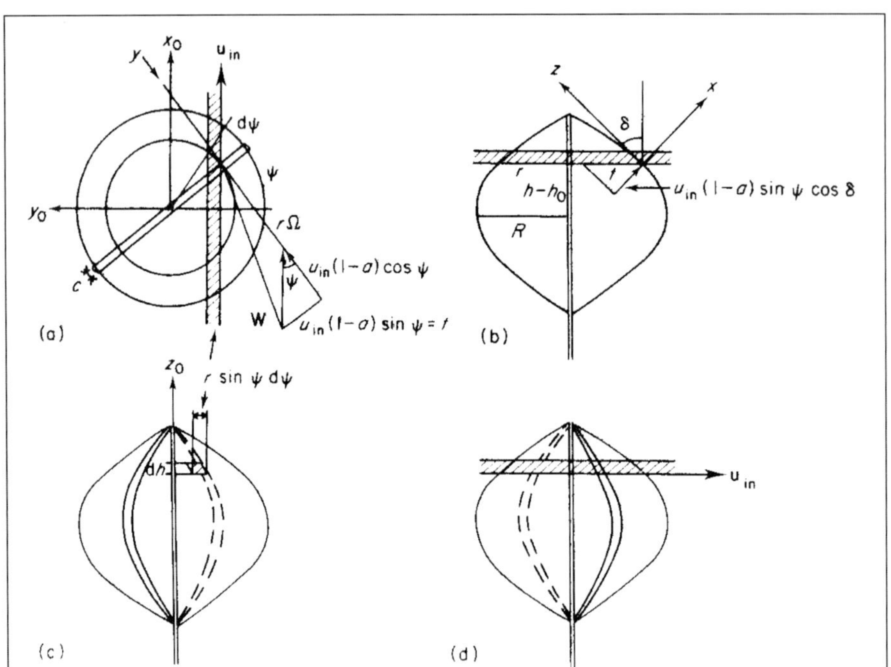

Figure 4.1-83 Streamtube definition (hatched areas) for two-bladed Darrieus rotor, and determination of apparent wind velocity, in the case of negligible cross-wind induced velocity ($u_{y0}^{ind} = 0$): (a) view from top; (b) view along tangent to blade motion; (c) view along streamtube; (4) view perpendicular to streamtube.

of the streamtubes definitions, for a two-bladed Darrieus rotor with angular velocity Ω and a rotational angle ψ describing the position of a given blade, according to (4.1.128). The blade chord, c, has been illustrated as constant, although it may actually be taken as varying to give the optimum performance for any blade segment at the distance r from the rotor axis. As in the propeller rotor case, it is not practical to extend the chord increase to the regions near the axis.

The bending of the blade, characterised by an angle $\delta(h)$ depending on the height h (the height of the rotor centre is denoted h_0), may be taken as a troposkien curve (Blackwell and Reis, 1974), characterised by the absence of bending forces on the blades when they are rotating freely. Since the blade profiles encounter wind directions at both positive and negative forward angles, the profiles are often taken as symmetrical (e.g. NACA 00XX profiles).

Assuming as in section 4.1.3.2 that the induced velocities in the rotor region are half of those in the wake, the stream-tube expressions for momentum and angular momentum conservation analogous to (4.1.124) and (4.1.125) are

$$F_{x0} = -J_m u_s^{ind} = 2\rho A_s (u_{in})^2 (1-a)a, \quad F_{y0} = J_m u_{c.w.}^{ind}, \qquad (4.1.134)$$

where the axial interference factor a is defined as in (4.1.117), and where the streamtube area A_s corresponding to height and angular increments dh and dψ is (cf. Fig. 4.1-83)

$$A_s = r \sin\psi \, d\psi \, dh.$$

The cross-wind induced velocity $u_{c.w.}^{ind}$ is not of the form (4.1.118), since the tangent to the blade rotational motion is not along the y_0-axis. The sign of $u_{c.w.}^{ind}$ will be fluctuating with time, and for low chordal ratio c/R (R being the maximum value of r) it may be permitted to put F_{y0} equal to zero (Strickland, 1975; Lissaman, 1976). This approximation will be made in the following. It will also be assumed that the stream tube area does not change by passage through the rotor region and that individual stream-tubes do not interact (these assumptions being the same as those made for the propeller-type rotor).

The forces along the instantaneous x- and y-axes due to the passage of the rotor blades at a fixed stream-tube location (h, ψ) can be expressed in analogy to (4.1.131), averaged over one rotational period,

$$F_x = B\,(d\psi/2\pi)\tfrac{1}{2}\rho c W^2 (C_D \sin\phi + C_L \cos\phi)\, dh/\cos\delta,$$
$$F_y = B\,(d\psi/2\pi)\tfrac{1}{2}\rho\, c W^2 (-C_D \cos\phi + C_L \sin\phi)\, dh/\cos\delta, \qquad (4.1.135)$$

where

$$\tan\phi = -W_x/W_y; \quad W^2 = W_x^2 + W_y^2$$

and (cf. Fig. 4.1-74)

$$W_x = u_{in}\,(1-a)\sin\psi\cos\delta, \quad W_y = -r\Omega - u_{in}(1-a)\cos\psi, \quad W_z = -u_{in}\,(1-a)\sin\psi\sin\delta. \qquad (4.1.136)$$

The angle of attack is still given by (4.1.121), with the pitch angle θ being the angle between the y-axis and the blade centre chord line (cf. Fig. 4.1-69c). The force components (4.1.135) may be transformed to the fixed (x_0, y_0, z_0) coordinate system, yielding

$$F_{x0} = F_x \cos\delta \sin\psi + F_y \cos\psi, \qquad (4.1.137)$$

with the other components F_{y0} and F_{z0} being neglected due to the assumptions made. Using the auxiliary relations given, a may be determined by equating the two expressions (4.1.134) and (4.1.137) for F_{x0}.

After integration over dh and dψ, the total torque Q and power coefficient C_p can be calculated. Figure 4.1-84 gives an example of the calculated C_p for a low-solidity, two-bladed Darrieus rotor with NACA 0012 blade profiles and a size corresponding to Reynolds number Re $= 3 \times 10^6$. The curve is rather similar to the ones obtained for propeller-type rotors (e.g. Fig. 4.1-70), but the maximum C_p is slightly lower. The reasons why this type of cross-wind converter cannot reach the maximum C_p of 16/27 derived from (4.1.50) (the "Betz limit") are associated with the fact that the blade orientation cannot remain optimal for all rotational angles ψi, as discussed (in terms of a simplified solution to the model presented above) by Lissaman (1976).

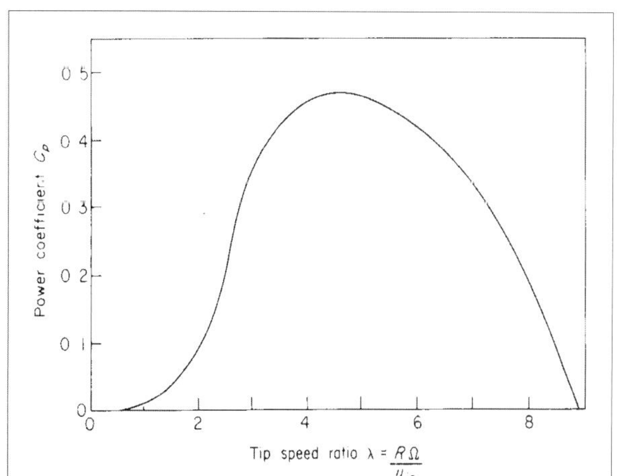

Figure 4.1-84 Power coefficient as a function of tip-speed ratio ("tip" means point furthest away from axis), for a two-bladed Darrieus rotor, with a chord ratio $c/R = 0.1$ and blade data corresponding to a symmetrical profile (NACA 0012 with Reynolds number Re $= 3 \times 10^6$). A stream-tube model of the type described in Fig. 4.1-83 has been used (based on Strickland, 1975).

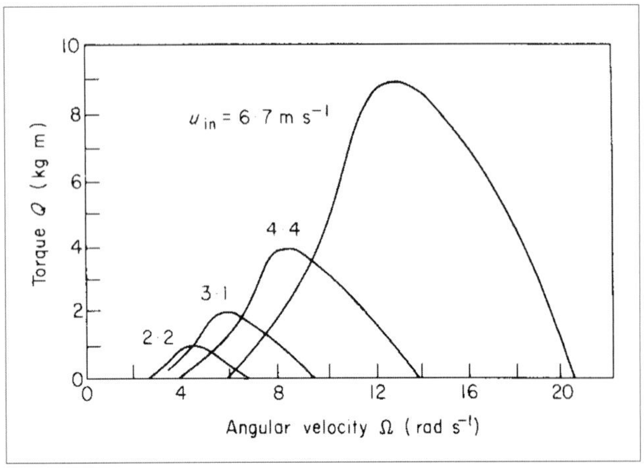

Figure 4.1-85 Torque as a function of angular velocity for a small three-bladed Darrieus rotor and at selected wind speeds. The experimental rotor had a radius $R = 2.25$ m, a chord ratio $c/R = 0.085$ and NACA 0012 blade profiles (based on Banas and Sullivan, 1975).

Figure 4.1-85 gives the torque as a function of angular velocity Ω for a small three-bladed Darrieus rotor. When compared with the corresponding curves for propeller-type rotors shown in Figs 4.1-75 and 4.1-77 (or generally Fig. 4.1-74), it is evident that the torque at $\Omega = 0$ is zero for the Darrieus rotor, implying that it is not self-starting. For application in connection with an electric grid or some other back-up system, this is no problem since the auxiliary power needed to start the Darrieus rotor at the appropriate times is very small compared with the wind converter output, on a yearly average basis. However, for application as an isolated source of power (e.g. in rural areas), it is a disadvantage, and it has been suggested that a small Savonius rotor should be placed on the main rotor axis in order to provide the starting torque (Banas and Sullivan, 1975).

Another feature of the Darrieus converter (as well as of some propeller-type converters), which is evident from Fig. 4.1-85, is that for application with a load of constant Ω (as in Fig. 4.1-75), there will be a self-regulating effect in that the torque will rise with increasing wind speed only up to a certain wind speed. If the wind speed increases further, the torque will begin to decrease. For variable-Ω types of load (as in Fig. 4.1-77), the behaviour of the curves in Fig. 4.1-85 implies that cases of irregular increase and decrease of torque, with increasing wind speed, can be expected.

4.1.3.4 Augmenters and other "advanced" converters

In the preceding sections, it has been assumed that the induced velocities in the converter region were half of those in the wake. This is strictly true for situations where all cross-wind induced velocities (u_t and u_r) can be neglected, as shown in (4.1.45), but if suitable cross-wind velocities can be induced so that the total stream-wise velocity in the converter region, u_x, exceeds the value of $-\frac{1}{2}(u_{x,in} + u_{x,out})$ by a positive amount δu_x^{ind}, then the Betz limit on the power coefficient, $C_p = 16/27$, may be exceeded,

$$u_x = \tfrac{1}{2}(u_{x,in} + u_{x,out}) + \delta u_x^{ind}.$$

A condition for this to occur is that the extra induced stream-wise velocity does not contribute to the induced velocity in the distant wake, u_x^{ind}, which is given implicitly by the above form of u_x, since

$$u_{x,out} = u_{x,in} + u_x^{ind} = u_{x,in}(1 - 2a).$$

The stream-tube flow at the converter, (4.1.44), is then

$$J_m = \rho A_s u_x = \rho A_s u_{x,in}(1 - a + \tilde{a}) \qquad (4.1.138)$$

with $\tilde{a} = \delta u_x^{ind}/u_{x,in}$, and the power (4.1.48) and power coefficient (4.1.50) are replaced by

$$E = \rho A_s (u_{x,in})^3\, 2a(1 - a)(1 - a + \tilde{a}),$$

$$C_p = 4a(1 - a)(1 - a + \tilde{a}), \qquad (4.1.139)$$

where the stream-tube area A_s equals the total converter area A, if the single-stream-tube model is used.

Ducted rotor

In order to create a positive increment $\tilde{a}u_{x,in}$ of axial velocity in the converter region, one may try to take advantage of the possibility of inducing a particular type of cross-wind velocity, which causes the stream-tube area to contract in the converter region. If the stream-tube cross section is circular, this may be achieved by an induced radial outward force acting on the air, which again can be caused by the lift force of a wing section placed at the periphery of the circular converter area, as illustrated in Fig. 4.1-86.

Figure 4.1-86 compares a free propeller-type rotor (top), for which the stream-tube area is nearly constant (as was actually assumed in sections 4.1.3.2 and 4.1.3.3) or expanding as a result of radially induced velocities, with a propeller rotor of the same dimensions, shrouded by a duct-shaped wing-profile. In this case the radial inward lift force F_L on the shroud corresponds (by momentum conservation) to a radial outward force on the air, which causes the stream-tube to expand on both sides of the shrouded propeller; in other words, it causes the streamlines passing through the duct to define a stream-tube, which contracts from an initial cross

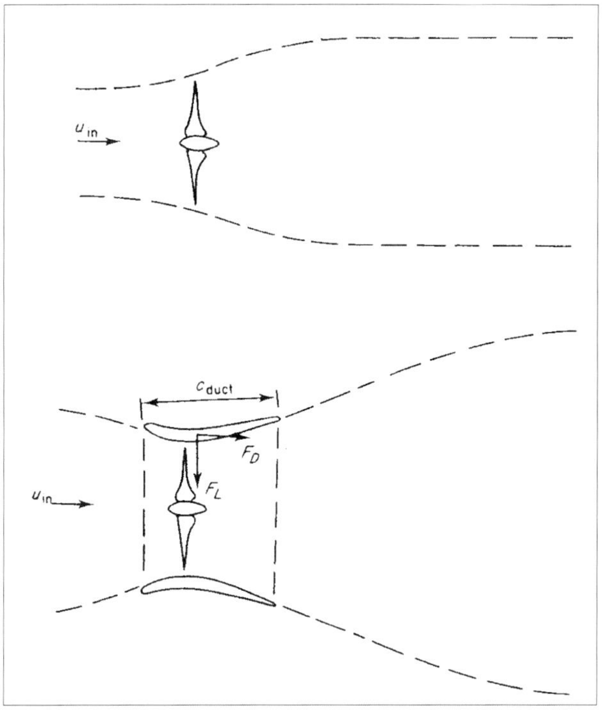

Figure 4.1-86 Peripheral streamlines for free stream (top) and ducted rotor (below).

section to reach a minimum area within the duct and which again expands in the wake of the converter. From (4.1.116), the magnitude of the lift force is

$$F_L^{duct} = ½\, \rho\, W_{duct}^2\, c_{duct}\, C_L^{duct},$$

with the duct chord c_{duct} defined in Fig. 4.1-86 and W_{duct} related to the incoming wind speed $u_{x,in}$ by an axial interference factor a_{duct} for the duct, in analogy with the corresponding one for the rotor itself, (4.1.47),

$$W_{duct} = u_{x,in}(1 - a_{duct}).$$

F_L^{duct} is associated with a circulation Γ around the shroud profile (shown in Fig. 4.1-86), given by (4.1.114). The induced velocity δu_x^{ind} inside the duct appears in the velocity being integrated over in the circulation integral (4.1.114), and it may to a first approximation be assumed that the average induced velocity is simply proportional to the circulation (which in itself does not depend on the actual choice of the closed path around the profile),

$$\delta u_x^{ind} = k_{duct}\Gamma_{duct}/R, \qquad (4.1.140)$$

where the radius of the duct, R (assumed to be similar to that of the inside propeller-type rotor), has been introduced, because the path-length in the integral Γ_{duct} is proportional to R and the factor k_{duct} appearing in (4.1.140) is therefore reasonably independent of R. Writing $\Gamma_{duct} = (\delta\, W_{duct})^{-1}\, F_L^{duct}$ in analogy to (4.1.114), and introducing the relations found above,

$$\tilde{a}_{duct} = k_{duct}c_{duct}C_L^{dust}(1 - a_{duct})/2R. \qquad (4.1.141)$$

If the length of the duct, c_{duct}, is made comparable to or larger than R, and the other factors in (4.1.141) can be kept near to unity, it is seen from (4.1.139) that a power coefficient about unity or larger is possible. This advantage may, however, be outweighed by the much larger amounts of materials needed to build a ducted converter, relative to a free rotor.

Augmenters taking advantage of the lift forces on a suitably situated aerodynamic profile need not provide a fully surrounding duct around the simple rotor device. A vertical cylindrical tower structure (e.g. with a wing profile similar to that of the shroud in Fig. 4.1-86) may suffice to produce a reduction in the widths of the stream-tubes relevant to an adjacently located rotor and thus may produce some enhancement of power extraction.

Rotor with tip-vanes

A formally appealing design, shown in Fig. 4.1-87, places tip-vanes of modest dimensions on the rotor blade tips (van Holten, 1976). The idea is that the tip-vanes act like a duct, without causing much increase in the amount of materials needed in construction. The smaller areas producing lift forces are compensated for by having much larger values of the apparent velocity W_{vane} seen by the vanes than W_{duct} in the shroud case. This possibility occurs because the tip-vanes rotate with the wings of the rotor (in contrast to the duct), and hence see air with an apparent cross-wind velocity given by

$$W_{vane} = R\Omega\,(1 - a_{vane}).$$

The magnitude of the ratio W_{vane}/W_{duct} is thus of the order of the tip-speed ratio) $\lambda = R\Omega/u_{x,in}$, which may be 10 or higher (cf. section 4.1.3.2).

The lift force on a particular vane is given by $F_L^{vane} = ½\, \rho\, W_{vane}^2\, c_{vane}\, C_L^{vane}$, and the average inward force over the periphery is obtained by multiplying this expression by $(2\pi R)^{-1}\, Bb_{vane}$, where B is the number of blades (vanes), and where b_{vane} is the lengths of the vanes (see Fig. 4.1-87).

Using a linearised expression analogous to (4.1.140) for the axial velocity induced by the radial forces,

$$\delta u_x^{ind} = \tilde{a}_{vane}u_{x,in} = k_{vane}\Gamma/R, \qquad (4.1.142)$$

the additional interference factor $\tilde{a}_{vane}$ to use in (4.1.139) may be calculated. Here Γ is the total circulation around the length of the tip-vanes (cf. Fig. 4.1-87),

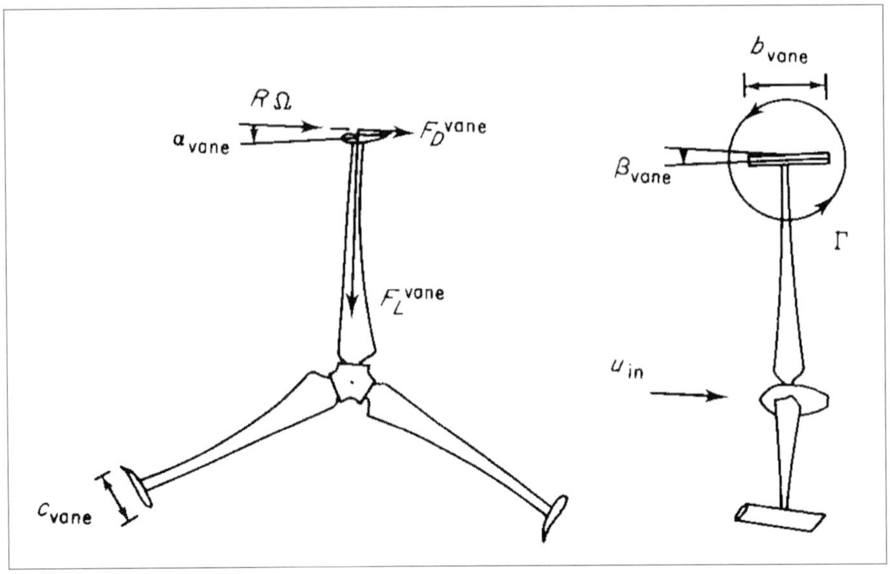

Figure 4.1-87 Geometry of propeller-type rotor with tip-vanes.

and not the circulation $\Gamma_{vane} = (\rho\, W_{vane})^{-1}\, F_L^{vane}$ in the plane of the vane lift and drag forces (left-hand side of Fig. 4.1-87). Therefore, Γ may be written

$$\Gamma = F_L^{vane} B b_{vane}/(2\pi R \rho w''),$$

i.e. equal to the average inward force divided by ρ times the average axial velocity at the tip-vane containing peripheral annulus,

$$w'' = u_{x,in}(1 + a'').$$

Expressions for a" have been derived by van Holten (1976). By inserting the above expressions into (4.1.142), $\tilde{a} = \tilde{a}_{vane}$ is obtained in the form

$$\tilde{a} = k_{vane}(1 - a_{vane})^2 B\, c_{vane}\, b_{vane}\, \lambda^2\, C_L^{vane}/(4\pi R^2(1 + a'')). \quad (4.1.143)$$

The part $(Bc_{vane}b_{vane}/\pi R^2)$ in the above expression is the ratio between the summed tip-vane area and the rotor-swept area. Taking this as 0.05 and [according to van Holten (1976) for $b_{vane}/R = 0.5$] $k_{vane}\,(1 - a_{vane})^2/(1 + a'')$ as 0.7, the tip-speed ratio λ as 10 and $C_L^{vane} = 1.5$, the resulting $\tilde{a}$ is 1.31 and the power coefficient according to the condition which maximises (4.1.139) as function of a,

$$a = (2 + \tilde{a} - ((2 + \tilde{a})^2 - 3(1 + \tilde{a}))^{1/2}) = 0.43;$$
$$C_p = 1.85.$$

The drag forces F_D^{vane} induced by the tip-vanes (cf. Fig. 4.1-87) represent a power loss from the converter, which has not been incorporated in the above treatment. According to van Holten (1976), this loss may in the numerical example studied above reduce C_p to about 1.2. This is still twice the Betz limit, for a tip-vane area which would equal the area of the propeller blades, for a three-bladed rotor with average chord ratio $c/R = 0.05$. It is conceivable that the use of such tip-vanes to increase the power output would in some cases be preferable to achieving the same power increase by increasing the rotor dimensions.

Other concepts

A large number of alternative devices for utilisation of wind energy have been studied, in addition to those treated above and in the preceding sections. The lift-producing profile exposed to the wind may be hollow and contain holes through which inside air is driven out by the lift forces. Through a hollow tower structure, replacement air is then drawn to the wing profiles from sets of openings placed in such a way that the air must pass one or more turbine propellers in order to reach the wings. The turbine propellers provide the power output, but the overall efficiency based on the total dimensions of the device is low (Hewson, 1975). Improved efficiency of devices which try to concentrate the wind energy before letting it reach a modest-size propeller may be obtained by first converting the mainly axial flow of the wind into a flow with a large vorticity of a simple structure, such as a circular air motion. Such vortices are, for example, formed over the sharp edges of highly swept delta wings, as illustrated in Fig. 4.1-88a (Sforza, 1976). There are two positions along the baseline of the delta wing where

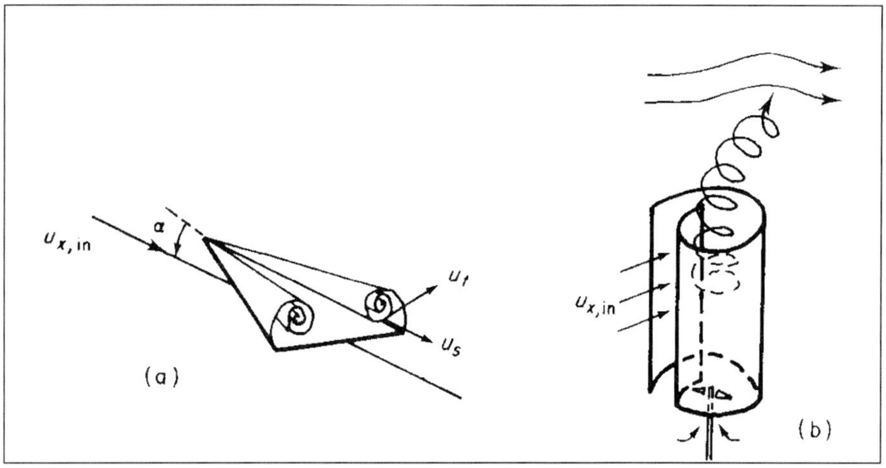

Figure 4.1-88 Augmenter concepts: (a) delta wing forming trailing vortices; (b) ducted vortex system (based on Sforza, 1976; Yen, 1976).

propeller rotors could be placed in an environment with stream-wise velocities u_s/u_{in} of 2–3 and tangential velocities u_t of the same order of magnitude as u_{in} (see Sforza, 1976).

If the wake streamlines can be diffused over a large region, such that interaction with the wind field not contributing to those stream-tubes passing through the converter can transfer energy to the slipstream motion and thereby increase the mass flow J_m through the converter, then further increase in power extraction can be expected. This is the case for the ducted system shown in Fig. 4.1-86 (lower part), but it may be combined with the vorticity concept described above to form a device of the general layout shown in Fig. 4.1-88b (Yen, 1976). Wind enters the vertical cylinder through a vertical slit and is forced to rotate by the inside cylinder walls. The vortex system created in this way (an "artificial tornado") is pushed up through the cylinder by pressure forces and leaves the open cylinder top. Owing to the strong vorticity, the rotating air may retain its identity high up in the atmosphere, where its diffusion extracts energy from the strong winds expected at that height. This energy is supposed to be transferred down the "tornado", strengthening its vorticity and thus providing more power for the turbines placed between the "tornado" bottom and a number of air inlets at the tower foot, which replace air that has left the top of the cylinder (cf. Fig. 4.1-88b). Neither of these two constructions have found practical applications.

Conventional (or unconventional) wind energy converters may be placed on floating structures at sea (where the wind speeds are often higher than on land) or may be mounted on balloons (e.g. a pair of counter-rotating propellers beside one other) in order to utilise the increasing wind speed usually found at elevations not accessible to ordinary building structures on the ground. In order to serve as a power source, the balloons must be guyed to a fixed point on the ground, with wires of sufficient strength.

4.1.3.5 Heat, electrical or mechanical power, and fuel generation

The wind energy converters described in the preceding sections have primarily been converting the power in the wind into rotating shaft power. The conversion system generally includes a further conversion step if the desired energy form is different from that of the rotating shaft.

Examples of this are the electric generators with fixed or variable rotational velocity which were mentioned in connection with Figs 4.1.75 and 4.1.77. The other types of energy can, in most cases, be obtained by secondary conversion of electric power. In some such cases the "quality" of electricity need not be as high as that usually maintained by utility grid systems, in respect to voltage fluctuations and variations in frequency in the (most widespread) case of alternating current. For wind energy converters aimed at constant working angular velocity Ω, it is customary to use a gearbox and an induction-type generator. This maintains an a.c. frequency equal to that of the grid and constant to within about 1%. Alternatively, the correct frequency can be prescribed electronically. In both cases, reactive power is created (i.e. power which, like that of a condenser or coil, is phase shifted), which may be an advantage or disadvantage, depending on the loads on the grid.

For variable-frequency wind energy converters, the electric output would be from a synchronous generator and in the form of variable-frequency alternating current. This would have to be subjected to a time-dependent frequency conversion, and for arrays of wind turbines, phase mismatch would have to be avoided. Several schemes exist for achieving this, for example, semiconductor rectifying devices (thyristors) which first convert the variable frequency AC to DC and then in a second step the DC (direct current) to AC (alternating current) of the required fixed frequency.

If the desired energy form is heat, "low quality" electricity may first be produced, and the heat may then be generated by leading the current through a high ohmic resistance. Better efficiency can be achieved if the electricity can be used to drive the compressor of a heat pump (see section 4.1.6.1), taking the required heat from a reservoir of temperature lower than the desired one. It is also possible to convert the shaft power more directly into heat. For example, the shaft power may drive a pump, pumping a viscous fluid through a nozzle, such that the pressure energy is converted into heat. Alternatively, the shaft rotation may be used to drive a "paddle" through a fluid, in such a way that large drag forces arise and that the fluid is put into turbulent motion, gradually dissipating the kinetic energy into heat. If water is used as the fluid medium, this arrangement is called a "water-brake".

Windmill shaft power has traditionally been used to perform mechanical work of various kinds, including flour milling, threshing, lifting and pumping. Pumping of water, e.g. for irrigation purposes, with a pump connected to the rotating shaft, may be particularly suitable as an application of wind energy, since variable and intermittent power would, in most cases, be acceptable, as long as the average power supply in the form of lifted water over an extended period of time is sufficient.

In other cases an auxiliary source of power may be needed, so that the demand can be met at any time. This can for grid-based systems be achieved by trade of energy (cf. scenarios described in Chapter 6.2). Demand matching can also be ensured if an energy storage facility of sufficient capacity is attached to the wind energy conversion system. A number of such storage facilities will be mentioned in Chapter 5, and among them will be the storage of energy in the form of fuels, such as hydrogen. Hydrogen may be produced, along with oxygen, by electrolysis of water, using electricity from the wind energy converter. The detailed working of these mechanisms over time and space is simulated in the scenarios outlined in Chapter 6 and related studies (Sørensen *et al.*, 2003; Sørensen, 2004a).

The primary interest may also be oxygen, for example, to be dissolved into the water of lakes which are deficient in oxygen (say, as a result of pollution), or to be used in connection with "ocean farming", where oxygen may be a limiting factor in cases where nutrients are supplied in large quantities, e.g. by artificial upwelling. Such oxygen may be supplied by wind energy converters, while the co-produced hydrogen may be used to produce or move the nutrients. Again, in this type of application large power fluctuations may be acceptable.

4.1.3.6 Commercial wind power development

Wind turbines have been in substantial use for more than 1000 years, with early text evidence in China and India and the earliest pictures being from Persia around 1300. These were all vertical axis machines, spreading to the Mediterranean region of Europe, while horizontal axis machines were developed in England, Holland, Denmark, the Baltic region and Russia (cf. overview of wind history in Sørensen, 1995a). Cities like Copenhagen had in the 16th century about 100 turbines with a hub height of over 10 m grinding flour just outside the city walls, and in Holland, some wind turbines were used to pump water as part of the drainage system on low land. Around 1900, wind turbines played a significant role in powering new settlements in the North American plains and Australian outposts, and the first hydrogen-and electricity-producing wind turbines were tested in Denmark (la Cour, 1900; cf. Hansen, 1985). The DC electricity-producing wind turbines kept competing with coal-based power during the first decades of the 20th century, gradually being pushed to island sites and eventually losing them too as undersea cables became extended to most islands. By the end of the 1920s, Denmark had about 30,000 wind turbines or three times as many as today. Experiments with AC generating wind turbines were made before and after World War II, with a comeback in several European countries following the 1957 closure of the Suez Canal (blocking transport of the Middle East oil then used in European power stations). Technically successful was the 200-kW Gedser turbine (Juul, 1961), serving as a model for the subsequent resurrection of wind turbine manufacture following the oil supply cuts in 1973/1974. The development since 1974 has been one of decreasing cost and rapidly increasing installed capacity.

Figure 4.1-89 shows the development in turbine size (height and swept diameter) of commercial wind turbines since 1980, and Fig. 4.1-90 shows corresponding trends in rated capacity and specific power production at the locations of typical installation. During the late 1970s, efforts were directed at several wind technologies: vertical axis machines in the USA and Canada, megawatt-size one- and two-bladed horizontal axis machines in the USA and Germany, and rotors with tip-vanes in the Netherlands. Yet, the winning technology was the Danish three-bladed, horizontal axis machine, starting at a small unit size of 25-50 kW. The reason for

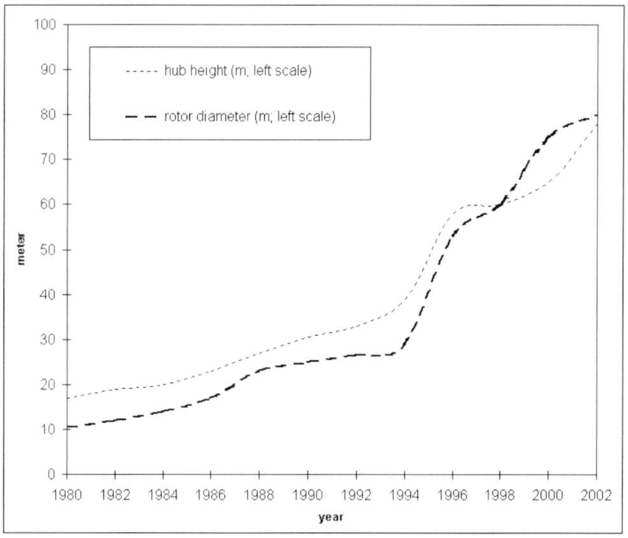

Figure 4.1-89 Development in average height and rotor diameter of Danish wind turbines. For recent years, turbines typically used in new large wind parks are represented, rather than an average of all turbines sold, which would be less indicative of trends, due to shifting policies for replacement of old wind turbines on land (adapted from Danish wind turbine Price Lists and Overviews, 1981–2002).

this was primarily that the initial design was a scaled-down version of the 1957 fixed pitch Gedser turbine with blade stall control, gearbox and asynchronous generator, i.e. an already proven design, and that the further development was for a long time taken in very small, safe enlargement steps gradually reaching megawatt size, rather than facing the problems of a totally new design in one jump (cf. Figs. 4.1-89 and 4.1-90).

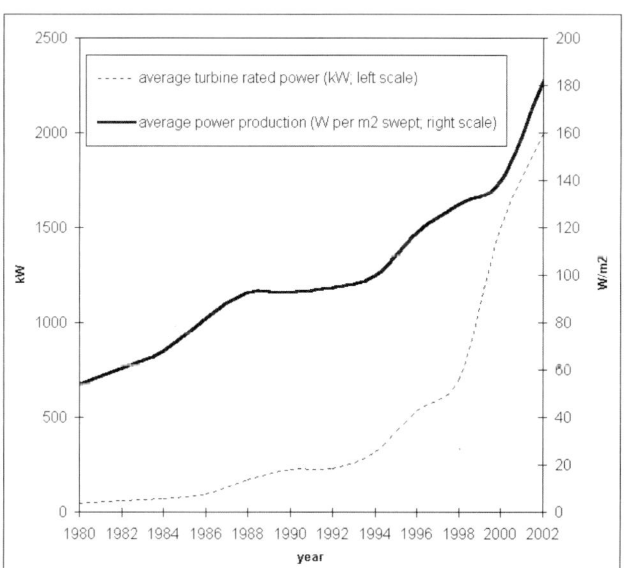

Figure 4.1-90 Development in average rated power and power production of Danish wind turbines. For recent years, turbines typically installed in new large wind parks are represented, rather than an average based on all sales, which would be less indicative of trends, due to shifting overseas markets and Danish policies for replacement of old wind turbines on land (adapted from Danish Windpower Industry Association, 2003; Danish wind turbine Price Lists and Overviews, 1981–2002).

Scaling up means controlling the most critical components, such as the rotor blades that must be able to survive the forces on long pieces of suitable material, experiencing wind loads with gust effects that can cause the blades to bend in different directions and exert large forces on the fastening point of the blades. Use of increasingly sophisticated materials and component dimensioning based on both calculation and experience has allowed the scaling up to proceed smoothly, and it is only in recent years that major design principles have been altered, e.g. doing away with the gearbox, using multipole electric generators operating synchronously and employing variable speed AC to DC to constant frequency a.c. electronic conversions. Also pitch change of blade angles has been introduced and thereby new options for operation or shutdown in high winds. The current production of units of a few megawatt machines is spread over a number of different ones among the design principles mentioned. Interesting questions are being asked about the possibility of operating still longer blades. Presumably, this will at some point mean a transition from the current blades made of glass or plant fibre to advanced coal fibres. It should be considered that the operating life of the blades aimed for is equal to the lifetime of the entire turbine, in contrast, e.g., to fast rotating helicopter rotor blades that have to be replaced at short intervals.

Figure 4.1-89 shows that a major technology shift in terms of increased blade length took place around 1994. In Denmark, this reflects the transition from mainly individually placed wind turbines to utility owned wind "parks" or "farms". Individual owners, mainly farmers, had carried the development of wind power technology to maturity with essentially no help from the established power utility companies. This may be the chief explanation for the early high penetration of wind energy in Denmark, expected to pass 20% in 2003. Fig. 4.1-90 indicates that a significant increase in turbine rated power has appeared since 1998, but that the average production did not increase nearly as much as the rated power. This is the period of starting to go offshore and may reflect that the distribution of wind speeds over water was not sufficiently known when the machines rated at 1–2 MW were designed. Earlier machines were pretty well optimised to give the maximum energy production over the year, when placed at favourable land sites.

The design optimisation is not a trivial question, as it involves considerations of average energy produced, number of hours operating, and maximum loads exerted upon the structure. Figures 4.1-91 and 4.1-92 show the calculated power output at on-shore and off-shore

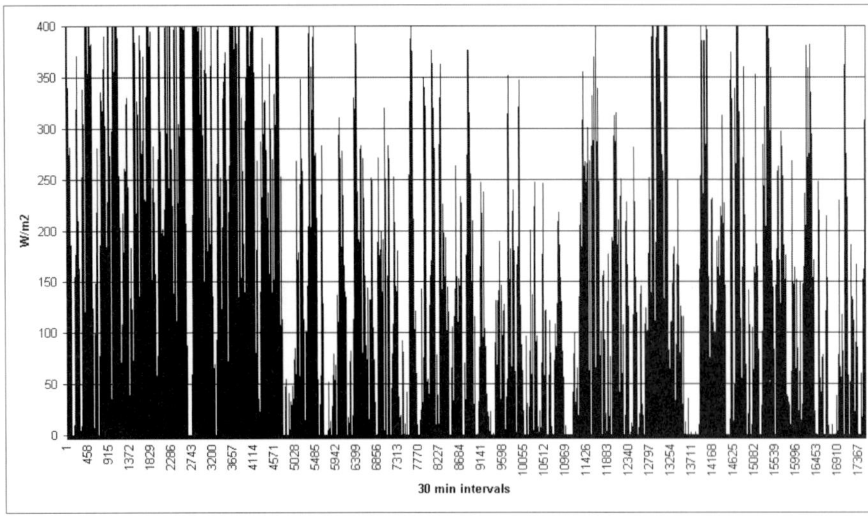

Figure 4.1-91 Power output calculated for a wind turbine placed at Stigsnæs (near Næstved on inner-Baltic coast of Denmark), based on wind speeds measured during 1995 (Risø, 1999) and a turbine power curve rising from 0 to 400 W m^{-2} between wind speeds of 5–12.27 m s^{-1} and then staying at 400 W m^{-2} until 25 m s^{-1} (typical of the behaviour of actual 2 MW Danish turbines; cf. Petersen, 2001).

locations in the inner Danish region, indicating that the imposed power ceiling is reached more often offshore. This situation is expected to occur more frequently at the off-shore locations in the North Sea, such as that of the 160-MW Hornsrev wind farm completed in 2002. It suggests that either the choice of blade profiles, the sizing of the power generator, or the regulation strategies used in current megawatt-size machines could be improved to capture more of the potential power in the wind for turbines placed at the reserved Danish off-shore sites, at least during periods of high winds. Deficiencies in aerodynamic behaviour can be determined by repeating the calculations with different power curves, but it is already clear that at most off-shore sites, just a larger generator would increase the annual energy output.

The model calculation assumes a constant C_p at high wind speeds, up to a maximum wind speed of 25 m s^{-1} where the turbine is brought to a stop. It seems that the manufacturers may have let their designs be influenced by complaints from the power companies, that large surpluses were hard to accept, because the auction-based selling schemes used in the newly established liberalised power markets have a tendency to value such surpluses very lowly. On the other hand, the turbine owner can

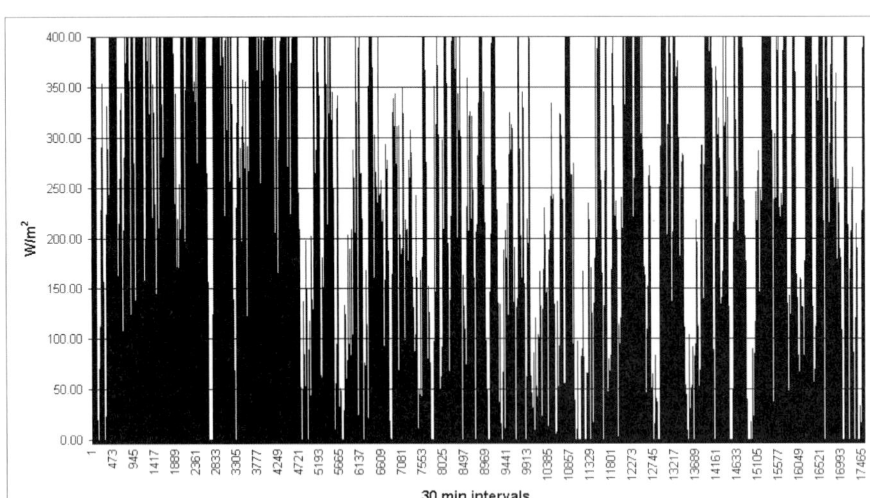

Figure 4.1-92 Power output from wind turbine placed at Vindeby (off-shore North of Lolland, in the inner-Baltic coast of Denmark slightly South of Stigsnæs), based on wind speeds measured during 1995 and the same turbine power curve as used in Fig. 4.1-91. Some missing data sequences (of length from few hours to about a week) have been replaced by measured sequences for the same hours of 1994.

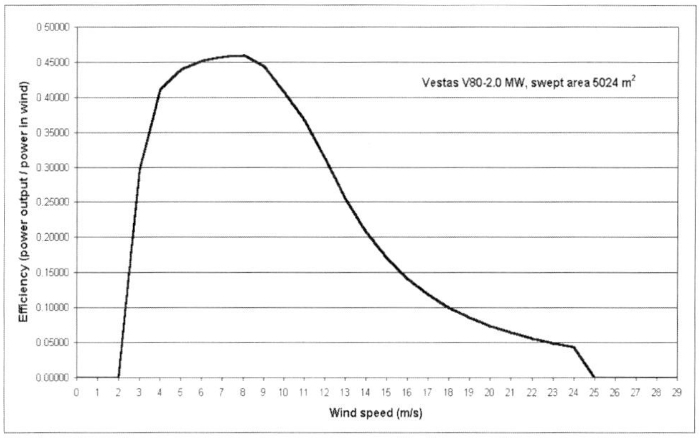

Figure 4.1-93 Efficiency as a function of wind speed for commercial 2-MW wind turbine (cf. Petersen, 2001; Bonefeld *et al.*, 2002).

boast of a high power factor, i.e. the ratio of average power production to constant production at full turbine rated power. For the 2-MW Vestas turbines used at Hornsrev, the power factor is 0.54. The actual efficiency of this turbine (power produced over power in the wind) is shown in Fig. 4.1-93, and again in Fig. 4.1-94, folded with the frequency of wind speeds. It is seen that the efficiency is high at frequently occurring wind speeds below 9 m s^{-1}, and the weighted efficiency (the integral under the curve) is a decent 0.34. However, the discontinuity in the curve and rapid fall-off for wind speeds over 9 m s^{-1} indicate that improved design could greatly increase annual production by better exploiting wind at speeds of 9-15 m s^{-1}, which might be done without serious rotor load problems for blades anyway allowed to operate with up to 25 m s^{-1} winds. It is clear that if the electricity system of a given region can be made to accept larger variations in power production, then the turbine design and regulation may thus be changed in directions that immediately would increase energy production and significantly lower the cost of each kilowatt-hour of energy produced. At present, wind turbine manufacturers insist on offering the same design for use in all wind speed regimes, except for a possible choice of tower height and, for manufacturers interested in arctic markets, some reinforcements for operation in high peak-wind speed, icing-prone environments.

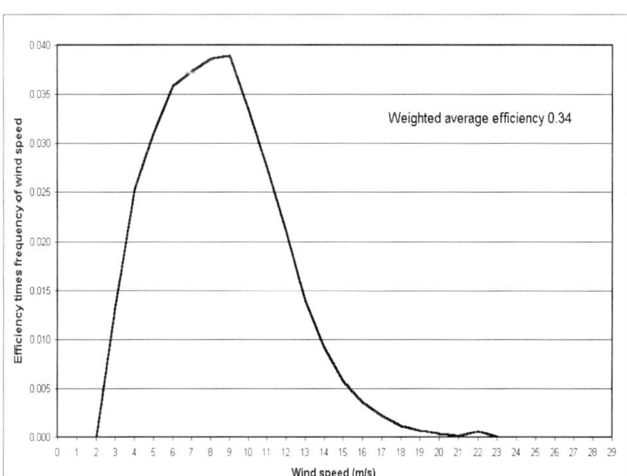

Figure 4.1-94 The efficiency curve of the commercial wind turbine shown in Fig. 4.1-93 has here been folded with the wind speed frequency distribution at Hornsrev, site of a 160-MW Danish offshore wind farm (based on data from Bonefeld *et al.*, 2002).

Current wind turbine unit size is around 2 MW, but prototypes of 4–5 MW have been developed. The economic breakdown of current costs are estimated in Table 4.1-2. There are, of course, variations of installation and grid connection costs depending on location, as there are variations in the cost of the electric energy produced, depending on the actual average power production at the location of the turbines and on financing conditions. Details of the treatment of payments occurring at different times and of interest rates are discussed further elsewhere.

Off-shore foundation and transmission

The possibility of placing wind turbines off-shore, typically in shallow waters of up to 20 m depth, relies on use of low-cost foundation methods developed earlier for harbour and oil-well uses. The best design depends on the material constituting the local water floor, as well as local hydrological conditions, including strength of currents and icing problems. Breakers are currently used to prevent ice from damaging the structure. The presently most common structure in place is a concrete caisson as shown in Fig. 4.1-95a, but in the interest of cost minimizing, the steel solutions shown in Fig. 4.1-95b,c have received increasing attention and the monopile (Fig. 4.1-95c) was selected for the recent Samsø wind farm in the middle of Denmark (Birch and Gormsen, 1999; Offshore Windenergy Europe, 2003). Employing the suction effect (Fig. 4.1-95b) may help cope with short-term gusting forces, while the general stability must be ensured by the properties of the overall structure itself.

The power from an off-shore wind farm is transmitted to an on-shore distribution hub by means of one or more undersea cables, the latter providing redundancy that in the case of large farms adds security against cable disruption or similar failures. Current off-shore wind farms use AC cables of up to 150 kV (Eltra, 2003). New installations use cables carrying all three leads plus control wiring. In the interest of loss minimisation for larger installations, it is expected that future systems may accept the higher cost of DC–AC conversion (on shore, the need for AC–DC conversion at sea depends

Table 4.1-2 Installation price and cost of energy produced (in year 2000 fixed prices), for wind parks on land or off-shore. Estimates have been made on the basis of current price lists for 2-MW turbines, tender data and trend analysis, with use of Danish wind turbine Price Lists and Overviews (2002), Danish Energy Agency (1999), European Commission (2001). A depreciation time of 20 years and a real interest rate of 3% p.a. (roughly the developed world average taken over the entire 20th century) have been assumed

Installed cost	On-shore	Off-shore	Unit
Turbines	0.77	0.77	2000-euro/W_{rated}
Foundations	0.06	0.35	2000-euro/W_{rated}
Grid connection[a]	0.15	0.27	2000-euro/W_{rated}
Engineering & administration	0.04	0.04	2000-euro/W_{rated}
Land rights, access roads, facilities	0.09	0.07	2000-euro/W_{rated}
Operation and maintenance (present value of 20 years' cost)	0.25	0.36	2000-euro/W_{rated}
Total capital cost including O&M	**1.36**	**1.86**	**2000-euro/W_{rated}**
Assumed power production	2.6[b]	3.8	kWh/y/W_{rated}
Cost of power produced	**0.035**	**0.033**	**2000-euro/kWh**

[a]The source data on on-shore grid connection costs are relatively low because of fairly short distances to grids of sufficient strength for Danish wind parks used in the studies underlying the value given. Thus, the additional cost of off-shore cables appears more dominant than would be the case in areas less well served with existing grids.
[b]On the very best Danish inland locations, power production may approach 3.5 kWh/y/W_{rated}.

on the generator type used), similar to the technology presently in use for many undersea cable connections between national grids (e.g. between Denmark and Norway or Sweden). Recent development of voltage source-based high voltage direct current control systems to replace the earlier thyristor-based technology promises better means of regulation of the interface of the DC link to the on-shore AC system (Ackermann, 2002).

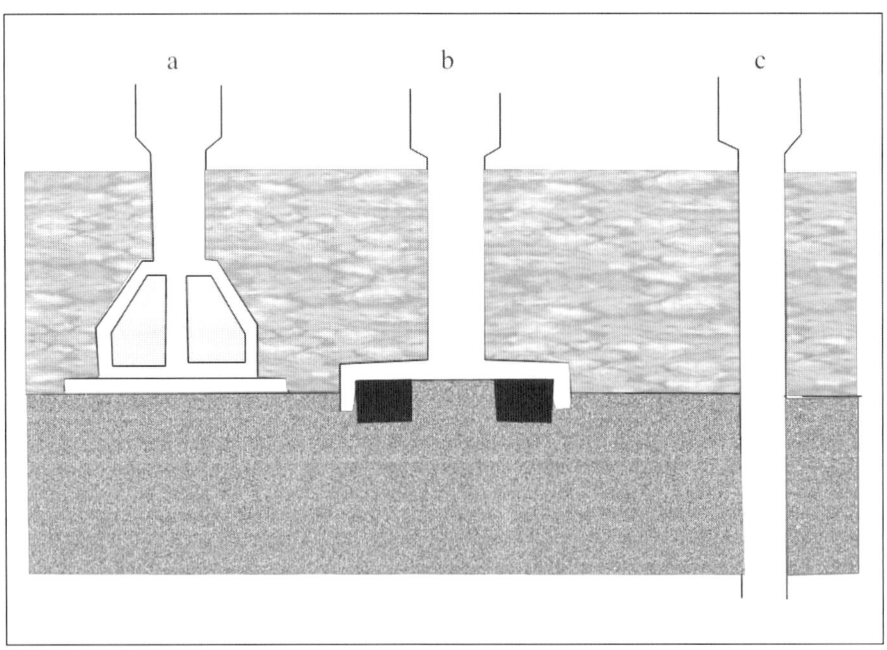

Figure 4.1-95 Some foundation types for off-shore wind turbine towers placed in shallow waters; a: sand-filled caisson standing on the bottom, b: suction buckets, c: steel monopile. These foundation types are in use at locations with clay-till water floor material.

4.1.4 Conversion of wave energy

A large number of devices for converting wave energy to shaft power or compression energy have been suggested, and a few of them have been tested on a modest scale. Reviews of wave devices may be found, for example, in Leichman and Scobie (1975), Isaacs *et al.* (1976), Slotta (1976), Clarke (1981), Sørensen (1999a), Thorpe (2001), CRES (2002), and DEA Wave Program (2002). Below, the technical details will be given for two typical examples of actual devices: the oscillating water column device that has been in successful small-scale operation for many years for powering mid-sea buoys, and the Salter duck which theoretically has a very high efficiency, but has not been successful in actual prototyping experiments. However, first some general remarks will be offered.

The resource evaluation in Chapter 3 indicated that the most promising locations for a wave utilisation apparatus would be in the open ocean rather than in coastal or shallow water regions. Yet all three device types have been researched: (a) shore-fixated devices for concentrating waves into a modestly elevated reservoir, from which the water may drive conventional hydro-turbines; (b) near-shore devices making use of the oscillations of a water column or a float on top of the waves, relative to a structure standing at the sea floor; and (c) floating devices capturing energy by differential movements of different parts of the device.

As a first orientation towards the prospects of developing economically viable wave power devices, a comparison with (e.g. off-shore) wind power may be instructive. One may first look at the weight of the construction relative to its rated power. For an on-shore wind turbine, this number is around 0.1 kg/W_{rated}, while adding the extra foundation weight for off-shore turbines (except caisson in-fill) increases the number to just below 0.2 kg/W_{rated}. For 15 wave power devices studied by the DEA Wave Program (2002), the range of weight to rated power ratios is from 0.4 to 15 kg/W_{rated}. The two numbers below 1.0 are for device concepts not tested and for use on shore or at low water depth, where the power resources are small anyway. So the conclusion is that the weight/power ratio is likely to be at least 2 but likely more than 5 times that of off-shore wind turbines, which to a first approximation is also a statement on the relative cost of the two concepts.

Using the same data, one may instead look at the ratio of actually produced power at a particular location and the weight of the device. For offshore wind in the North Sea, this is around 20 kWh y^{-1} kg^{-1}. For the 15 wave devices, values of 0.1 to 10 are found, by using for all devices the same wave data estimated for a location some 150 km west of the city Esbjerg in Denmark*. Clearly, it is not reasonable to use data for a location 150 km out into the ocean for wave power devices that must stand on the shore or at very shallow water. Omitting these cases, the resulting range reduces to 0.1-1.5 kWh y^{-1} kg^{-1} or over 13 times less that for off-shore wind in the same region. Again, the simplistic translation from weight to cost indicates that wave energy is economically unattractive, because at the same location wind power can be extracted at much lower cost. Also, there are no particular reasons to expect the distribution of weight on less expensive materials (concrete or steel) and more expensive materials (special mechanical and electric equipment) to be substantially different. It is part of this argument, that, where wave power would be feasible, off-shore wind power is also available, because it is the wind that creates the waves. Only at large water depths, say, over 20 m depth, where foundation would be problematic, might the wave devices floating on the surface be more attractive than wind. Yet there are not from the above any indications that the cost of such mid-ocean wave power extraction and cable transmission to land will be economically viable, unless all near-shore wind options have already been exploited.

Finally, there is the argument of the time-distribution of the power from wave devices. As seen from Figs. 3.3.46 and 3.3.49, wave power exhibits large variations with seasons. Also in the North Sea, variations are large: Rambøll (1999) finds 6 times more average wave power in January than in June, where the corresponding factor for wind power is 2 (Sørensen, 2000a). Acceptance of wave power into grids serving electricity demands is thus going to be considerably more difficult than acceptance of wind, which roughly has the same seasonal variation as demand, at least in the Northern Hemisphere. Thus, in addition to an initial cost likely to be substantially higher than that of wind power, additional costs for energy storage or other supply-demand mismatch management must be considered.

4.1.4.1 Pneumatic converter

The only wave energy conversion device that has been in practical use, although on a fairly small scale, is the buoy of Masuda (1971), shown in schematic form in

* The wave data used are calculated from the 10-m wind data shown elsewhere using a fetch model (Rambøll, 1999). The 10-m data are not very reliable, and the grid is coarse, but comparison with the actual measurements at Vyl a few kilometres to the west (shown in Fig. 3.3.45) indicates that the error is well under 50% (however, the calculated data does not, for example, exhibit the two peaks shown in Fig. 3.3.45, but only one).

Fig. 4.1-96. Several similar wave power devices exist, based on an oscillating water column driving an air turbine, in some cases shore based and with only one chamber (see e.g. CRES, 2002). The buoy in Fig. 4.1-96 contains a tube extending downwards, into which water can enter from the bottom, and a "double-action" air turbine, i.e. a turbine that turns in the same way under the influence of pressure and suction (as illustrated by the non-return valves).

The wave motion will cause the whole buoy to move up and down and thereby create an up-and-down motion of the water level in the centre tube, which in turn produces the pressure increases and decreases that make the turbine in the upper air chamber rotate. Special nozzles may be added in order to increase the speed of the air impinging on the turbine blades.

For a simple sinusoidal wave motion but omitting the viscosity-dependent exponential factor, the variables σ_1 and Z describing the water level in the centre tube and the vertical displacement of the entire buoy, respectively, may also be expected to vary sinusoidally, but with phase delays and different amplitudes,

$$\sigma = a \cos(kx - \omega t),$$
$$\sigma_1 = \sigma_{10} \cos(kx - \omega t - \delta_\sigma),$$
$$Z = Z_0 \cos(kx - \omega t - \delta_z),$$

with $\omega = kUw$ in terms of the wave number k and phase velocity Uw. The relative air displacement ρ_1 in the upper part of the centre tube (the area of which is $A_1 = \pi R_1^2$) is now (see Fig. 4.1-96)

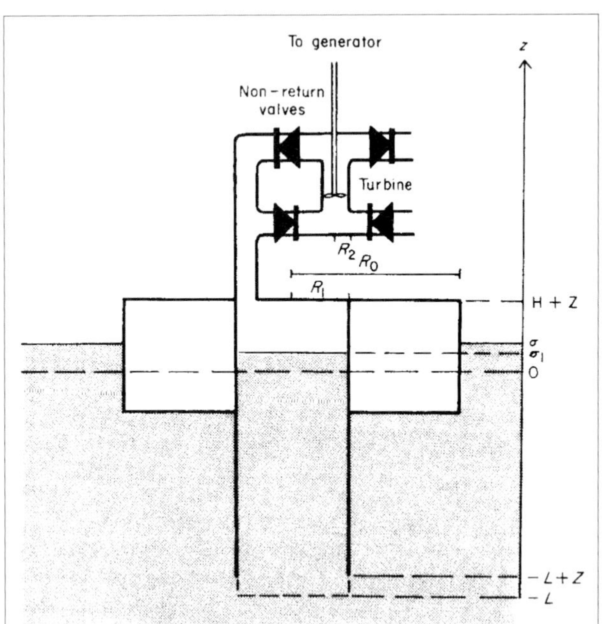

Figure 4.1-96 Masuda's pneumatic wave energy conversion device.

$$\rho_1 = \sigma_1 - Z.$$

Assuming the air flow to be incompressible, continuity requires that the relative air velocity $d\rho_2/dt$ in the upper tubes forming the air chamber (of area $A_2 = \pi R_2^2$) be

$$d\rho_2/dt = (A_1/A_2)d\rho_1/dt.$$

It is now possible to set up the equations of motion for σ_1 and Z, equating the accelerations $d^2\sigma_1/dt^2$ and d^2Z/dt^2 multiplied by the appropriate masses (of the water column in the centre tube and of the entire device) to the sum of forces acting on the mass in question. These are buoyancy and pressure forces, as well as friction forces. The air displacement variable ρ_2 satisfies a similar equation involving the turbine blade reaction forces. The equations are coupled, but some of the mutual interaction can be incorporated as damping terms in each equation of motion. McCormick (1976) uses linear damping terms $-b(d\sigma_1/dt)$, etc., with empirically chosen damping constants, and drops other coupling terms, so that the determination of σ_1 and Z becomes independent. Yet the $d\rho_2/dt$ values determined from such solutions are in good agreement with measured ones, for small wave amplitudes a (or significant heights $H_S = 2 \times 2^{1/2}a$; cf. Chapter 3) and wave periods $T = 2\pi/\omega$ which are rather close to the resonant values of the entire buoy, T_0, or to that of the centre tube water column, T_1. Between these resonant periods the agreement is less good, probably indicating that couplings are stronger between the resonances than near them, as might be expected.

The power transferred to the turbine shaft cannot be calculated from the expression (4.1.12) because the flow is not steady. The mass flows into and out of the converter vary with time, and at a given time they are not equal. However, owing to the assumption that the air is incompressible, there will be no build-up of a compressed air energy storage inside the air chamber, and therefore, the power can still be obtained as a product of a mass flux, i.e. that to the turbine,

$$J_m = \rho_a A_1 d\rho_1/dt$$

(ρ_a being the density of air), and a specific energy change $\Delta\omega$ (neglecting heat dissipation). In addition to the terms (4.1.34)–(4.1.36) describing potential energy changes (unimportant for the air chamber), kinetic energy changes and pressure/enthalpy changes, $\Delta\omega$ will contain a term depending on the time variation of the air velocity $d\rho_2/dt$ at the turbine entrance,

$$\Delta\omega \approx \tfrac{1}{2}\left(\left(\frac{d\rho_1}{dt}\right)^2 - \left(\frac{d\rho_2}{dt}\right)^2\right) + \frac{\Delta P}{\rho_a} - \int\left(\frac{d^2\rho_2}{dt^2}\right)d\rho,$$

where ΔP is the difference between the air pressure in the air chamber (and upper centre tube, since these are assumed equal) and in the free outside air, and ρ is the co-ordinate along a streamline through the air chamber (McCormick, 1976). Internal losses in the turbine have not been taken into account.

Figure 4.1-97 shows an example of the simple efficiency of conversion, η, given by the ratio of $J_m \Delta\omega$ and the power in the incident waves, which for sinusoidal waves is given by (3.3.28). Since the expression (3.3.28) is the power per unit length (perpendicular to the direction of propagation, i.e. power integrated over depth z), it should be multiplied by a dimension of the device which characterises the width of wave acceptance. As the entire buoy is set into vertical motion by the action of the waves, it is natural to take the overall diameter $2R_0$ (cf. Fig. 4.1-96) as this parameter, such that

$$\eta = 8\pi J_m \Delta\omega/(2R_0\rho_\omega g^2 Ta^2)$$
$$= 32\pi J_m \Delta\omega/(\rho_\omega g^2 T H_s^2 R_0). \quad (4.1.144)$$

The efficiency curves as functions of the period T, for selected significant wave heights H_S, which are shown in Fig. 4.1-97, are based on a calculation by McCormick (1976) for a 3650-kg buoy with overall diameter $2R_0 = 2.44$ m and centre tube diameter $2R_1 = 0.61$ m. For fixed H_S, η has two maxima, one just below the resonant period of the entire buoy in the absence of damping terms, $T_0 \approx 3$ s, and one near the air chamber resonant period, $T_1 \approx 5$ s. This implies that the device may have to be "tuned" to the periods in the wave spectra giving the largest power contributions (cf. Fig. 3.3-46), and that the efficiency will be poor in certain other intervals of T. According to the calculation based on solving the equations outlined above and with the assumptions made, the efficiency η becomes directly proportional to H_S. This can only be true for small amplitudes, and Fig. 4.1-97 shows that unit efficiency would be obtained for $H_S \approx 1.2$ m from this model, implying that non-linear terms have to be kept in the model if the power obtained for waves of such heights is to be correctly calculated. McCormick (1976) also refers to experiments which suggest that the power actually obtained is smaller than that calculated.

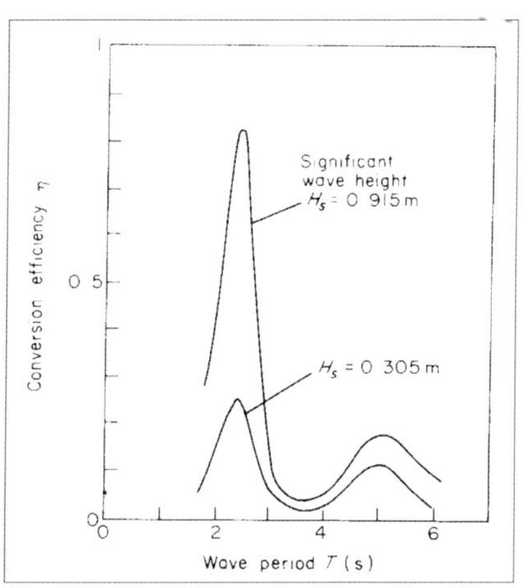

Figure 4.1-97 Efficiency of a pneumatic wave energy converter as function of wave period, calculated for two wave amplitudes (based on McCormick, 1976).

Masuda has proposed placing a ring of pneumatic converters (floating on water of modest depth) such that the diameter of the ring is larger than the wavelength of the waves of interest. He claims that the transmission of wave power through such a ring barrier is small, indicating that most of the wave energy will be trapped (e.g. multiply reflected on the interior barriers of the ring) and thus eventually be absorbed by the buoy and internal fluid motion in the centre tubes. In other developments of this concept, the converters are moored to the sea floor or leaning towards a cliff on the shore (Clarke, 1981; Vindeløv, 1994; Thorpe, 2001).

4.1.4.2 Oscillating vane converter

Since water particles in a sinusoidal wave are moving in circular orbits, it may be expected that complete absorption of the wave energy is only possible with a device possessing two degrees of freedom, i.e. one allowing displacement in both vertical and horizontal directions. Indeed, it has been shown by Ogilvie (1963) that an immersed cylinder of a suitable radius, moving in a circular orbit around a fixed axis (parallel to its own axis) with a certain frequency, is capable of completely absorbing the energy of a wave incident perpendicular to the cylinder axis, i.e. with zero transmission and no wave motion "behind" the cylinder.

Complete absorption is equivalent to energy conversion at 100% efficiency, and Evans (1976) has shown that this is also possible for a half-immersed cylinder performing forced oscillations in two dimensions, around an equilibrium position, whereas the maximum efficiency is 50% for a device which is only capable of performing oscillations in one dimension.

The system may be described in terms of coupled equations of motion for the displacement co-ordinates X_i ($i = 1, 2$),

$$m\mathrm{d}^2X_i/dt^2 = -d_i\mathrm{d}X_i/\mathrm{d}t - k_iX_i + \Sigma_j F_{ij},$$

where m is the mass of the device, d_i is the damping coefficient in the ith direction, k_i correspondingly is the restoring force or "spring constant" (the vertical component of which may include the buoyancy forces), and F_{ij}, is a matrix representing the complete hydrodynamic force on the body. It depends on the incident wave field and on the time derivatives of the coordinates X_i, the non-diagonal terms representing couplings of the motion in vertical and horizontal directions.

The average power absorbed over a wave period T is

$$E = \sum_i T^{-1} \int_0^T \frac{\mathrm{d}X_i}{\mathrm{d}t} \sum_j F_{ij}\, \mathrm{d}t, \quad (4.1.145)$$

and the simple efficiency η is obtained by inserting each of the force components in (4.1.145) as forces per unit length of cylinder, and dividing by the power in the waves (3.3.28). One may regard the following as design parameters: the damping parameters d_i, which depend on the power extraction system, the spring constants k_i and the mass distribution inside the cylinder, which determine the moments and hence influence the forces F_{ij}. There is one value of each d_i for a given radius R of the cylinder and a given wave period, which must be chosen as a necessary condition for obtaining maximum efficiency. A second condition involves k_i, and if both can be fulfilled the device is said to be "tuned" and it will be able to reach the maximum efficiency (0.5 or 1, for oscillations in one or two dimensions). Generally, the second condition can be fulfilled only in a certain interval of the ratio R/λ between the radius of the device and the wavelength $\lambda = gT^2/(2\pi)$.

Figure 4.1-98 shows the efficiency as a function of $2\pi R/\lambda$ for a half-immersed cylinder tuned at $2\pi R/\lambda_0 = 1$ (full line) and for a partially tuned case, in which the choice of damping coefficient corresponds to tuning at $2\pi \cdot R/\lambda_0 = 0.3$, but where the second requirement is not fulfilled. It is seen that such "partial tuning" offers a possibility of widening the range of wave periods accepted by the device.

A system whose performance may approach the theoretical maximum for the cylinder with two degrees of freedom is the oscillating vane or cam of Salter (1974), illustrated in Fig. 4.1-99. Several structures of this cross section indicated are supposed to be mounted on a common backbone providing the fixed axis around which the vanes can oscillate, with different vanes not necessarily moving in phase. The backbone has to be long in order to provide an approximate inertial frame, relative to which the oscillations can be utilised for power extraction, for example, by having the rocking motion create pressure pulses in a suitable fluid (contained in compression chambers between the backbone and oscillating structure). The necessity of such a backbone is also the weakness of the system, owing to the large bending forces along the structure, which must be accepted during storms.

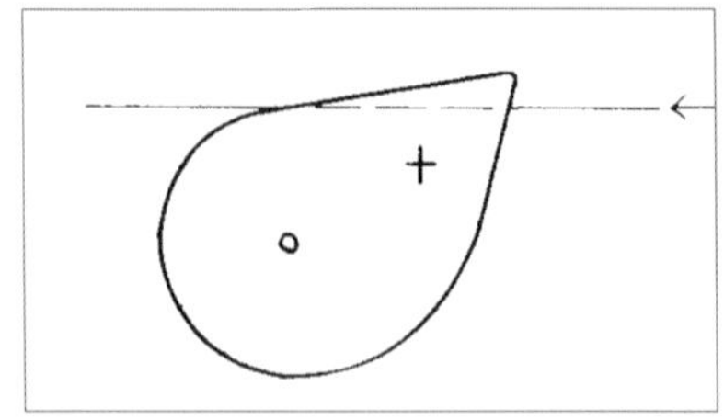

Figure 4.1-99 Salter's design of an oscillating vane wave energy converter. The cross section shows the backbone and axis of rotation as a small circle and the location of the centre of gravity as a cross. The waves are incident from the right-hand side.

The efficiency of wave power absorption for a Salter "cam" or "duck" of the type shown in Fig. 4.1-99 is indicated in Fig. 4.1-100, based on measurements on a model of modest scale ($R = 0.05$ m). Actual devices for use in, say, the good wave energy location in the North Atlantic (cf. Fig. 3.3.46) would need a radius of 8 m or more if tuned at the same value of $\lambda/(2\pi R)$ as the device used to construct Fig. 4.1-100 (Mollison *et al.*, 1976). Electricity generators of high efficiency are available for converting the oscillating vane motion (the angular amplitude of which may be of the order of half a radian)

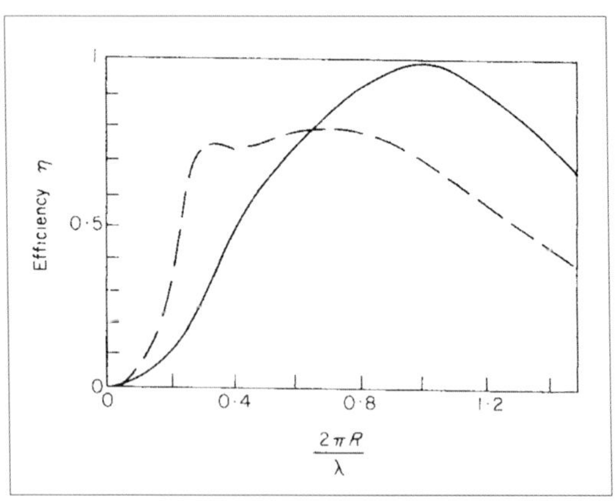

Figure 4.1-98 Efficiency of wave energy absorption by oscillating motion of a half-immersed cylinder of radius *R*. The calculation assumes parameters describing damping and coupling terms, such that the device becomes fully tuned at $2\pi\ R/\lambda = 1$ (solid line) or partially tuned at $2\pi\ R/\lambda = 0.3$ (dashed line) (based on Evans, 1976).

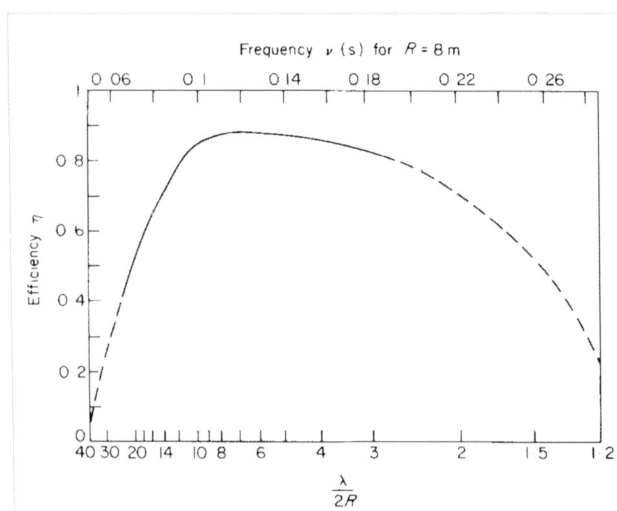

Figure 4.1-100 Efficiency of Salter's device based on model experiments (solid line) and extrapolations (dashed line) (based on Mollison *et al.*, 1976). See also the discussion topics 6.5.2 and 6.5.3 and Figs. 6.98 and 6.99.

to electric power, either directly from the mechanical energy in the rocking motion or via the compression chambers mentioned above, from which the fluid may be let to an electricity-producing turbine. The pulsed compression is not incompatible with continuous power generation (Korn, 1972). Later experiments have shown that the structural strength needed for the backbone is a problem. For all wave devices contemplated placed in favourable locations at mid-ocean, transmission to shore is a problem for the economy of the scheme. (cf. discussion topics 6.5.2 and 6.5.3).

4.1.5 Conversion of water flows or elevated water

Electricity generation from water possessing potential, kinetic or pressure energy (4.1.34)–(4.1.36) can be achieved by means of a turbine, the general theory of which was outlined in section 4.1.1.4. The particular design of turbine to be used depends on whether there is a flow J_m through the device, which must be kept constant for continuity reasons, or whether it is possible to obtain zero fluid velocity after the passage through the turbine.

The form of energy at the entrance of the turbine may be kinetic or pressure energy, causing the forces on the turbine blades to be a combination of "impulse" and "reaction" forces, which can be modified at ease. The potential energy of elevated water may be allowed to "fall", i.e. forming kinetic energy, or it may act as a pressure source through a water-filled tube connecting the elevated water source with the turbine placed below. Conversely, pressure energy may be transformed into kinetic energy by passage through a nozzle.

Typical classical turbine designs are illustrated in Fig. 4.1-101. For high specific energy differences $w_{in}-w_{out}$ (large "heads"), the Pelton turbine, which is a high-speed variant of the simple undershot waterwheel, may be used. It has the inflow through a nozzle, providing purely kinetic energy, and negligible w_{out} (if the reference point for potential energy is taken to correspond to the water level after passing the turbine). Also, the Francis turbine (Fig. 4.1-101b) is used with large water heads. Here the water is allowed to approach the entire rotor, guided to obtain optimum angles of attack, and the rotor moves owing to the reaction forces resulting from both the excess pressure at entrance and the suction at exit.

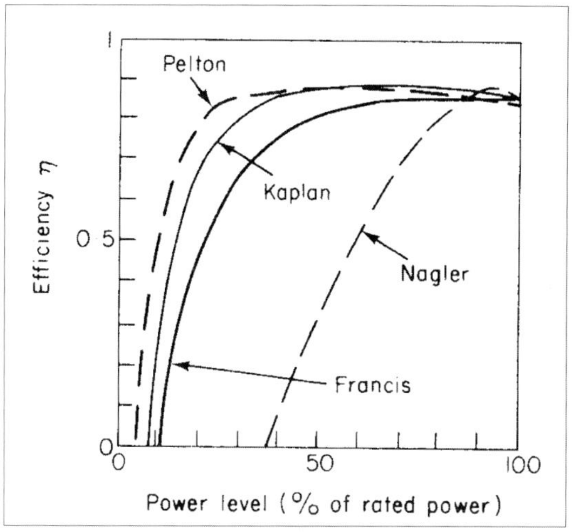

Figure 4.1-102 Efficiency of water turbines as a function of power level (based on Fabritz, 1954).

The third type of turbine, illustrated in Fig. 4.1-101c, can be used for low water heads. Here the rotor is a propeller, designed to obtain high angular speeds. Again, the angle of attack may be optimised by installation of guiding blades at the entrance to the rotor region. If the blade pitch angle is fixed, it is called a Nagler turbine. If it can be varied, it is called a Kaplan turbine.

Figure 4.1-102 gives examples of actual efficiencies for the types of turbine described above (Fabritz, 1954) as functions of the power level. The design point for these turbines is about 90% of the rated power (which is set to 100% on the figure), so the power levels below this point correspond to situations in which the water head is insufficient to provide the design power level.

Pelton and Francis turbines have been used in connection with river flows with rapid descent, including waterfalls, and in many cases it has been possible by building dams to provide a steady energy source

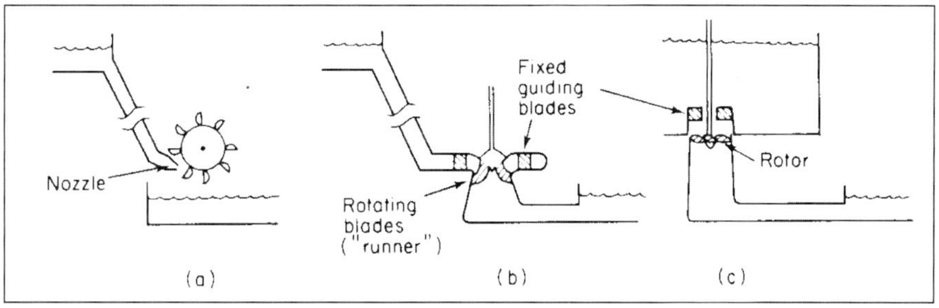

Figure 4.1-101 Pelton (a), Francis (b) and Kaplan (c) water turbines.

throughout most of the year. This implies storing the water at an elevated level in natural or artificial reservoirs and letting the water descend to the turbines only when needed. An example of natural variations inviting regulation was shown in Fig. 3.3.59. The operation of systems that include several sources of flow (springs, glacier or snow melt, rainy regions), several reservoirs of given capacity, a given number of turbine power stations and a load of electricity usage which also varies with time, has been studied and optimised by simulation techniques (see e.g. Jamshidi and Mohseni, 1976).

For many years, hydropower has been the most widely used renewable source of electricity and also – among all types of power plants including fossil and nuclear – the technology inviting the largest schemes of power plants rated at several gigawatts and involving often artificial water reservoirs of gigantic size. The development of such schemes with disregard to social and environmental problems has given hydropower a negative reputation. In developing countries thousands of people have been forcefully removed from their homes with no compensation to give way for flooded reservoirs, causing monumental environmental disruption (this not being restricted to developing countries, as the examples of Tasmania or Norway show) and in some cases destroying priceless sites of archaeological value (e.g. Turkey). In recent decades, it has become clear that in many cases there exist ways of minimising the environmental damage, albeit at a higher cost. The discussion in section 3.3.4.2 mentioned Swiss efforts to use cascading systems to do away with any large reservoirs, accepting that the smaller reservoirs located along the flow of water and designed to minimise local impacts would provide somewhat less regulation latitude. Full consideration of these concerns is today in most societies a prerequisite for considering hydropower as a benign energy source.

Kaplan (or Nagler) turbines are used in connection with low water heads (e.g. the local community power plants in China, cf. *China Reconstructs*, 1975) and tidal plants (André, 1976), and they may be used if ocean currents are to be exploited. The 240-MW tidal plant at la Rance in France has turbines placed in a dam structure across an inlet, which serves as a reservoir for two-way operation of the turbines (filling and emptying of the reservoir). According to André (1976), the turbine efficiency is over 90%, but the turbines are only generating electricity during part of the day, according to the scheme outlined in Fig. 4.1-103. The times at which generation is possible are determined by the tidal cycle, according to the simple scheme of operation, but modifications to suit the load variations better are possible, e.g. by using the turbines to pump water into the reservoir at times when this would not occur as a result of the tidal cycle itself. The installation has had several problems due to siltation, causing it to be operated most of the time only for water flows in one direction, despite its design to accept water inflow from both sides.

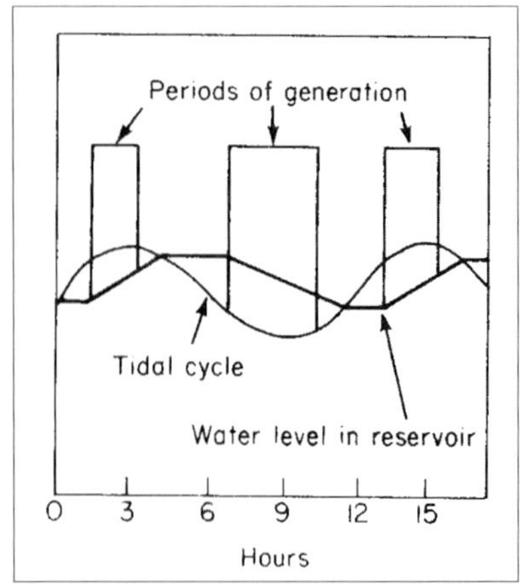

Figure 4.1-103 Operation of the tidal power plant at la Rance (based on André, 1976).

4.1.6 Conversion of heat

4.1.6.1 Application to heating

In some cases, it is possible to produce heat at precisely the temperature needed by primary conversion. However, often the initial temperature is higher than required, even considering losses in transmission and heat drop across heat exchangers of the type shown in Fig. 4.1-20. In such cases, appropriate temperatures are commonly obtained by mixing (if the heat is stored as sensible heat in a fluid such as water, this water may be mixed with colder water, a procedure often used in connection with fossil fuel burners). This procedure is wasteful in the sense of the second law of thermodynamics, since the energy is, in the first place, produced with a higher quality than subsequently needed. In other words, the second law efficiency of conversion (4.1.20) is low, because there will be other schemes of conversion by which the primary energy can be made to produce a larger quantity of heat at the temperature needed at load. An extreme case of a "detour" is the conversion of heat to heat by first generating electricity by thermal conversion (cf. section 4.1.6.2) and then degrading the electricity to heat of low temperature by passing a current through an ohmic resistance ("electric heaters").

Heat pumps

If heat of modest temperature is required, and a high-quality form of energy is available, some device is needed which can derive additional benefits from the high quality of the primary energy source. This can be achieved by using one of the thermodynamic cycles described in section 4.1.1.2, provided that a large reservoir of approximately constant temperature is available. The cycles (cf. Fig. 4.1-3) must be traversed "anti-clockwise", such that high-quality energy (electricity, mechanical shaft power, fuel combustion at high temperature, etc.) is added, and heat energy thereby delivered at a temperature T higher than the temperature T_{ref} of the reference reservoir from which it is drawn. Most commonly the Rankine cycle, also described in section 4.1.1.2 and with a maximum efficiency bounded by (4.1.22), is used (e.g. in an arrangement of the type shown in Fig. 4.1-104). The fluid of the closed cycle, which should have a liquid and a gaseous phase in the temperature interval of interest, may be a fluorochloromethane compound (which needs to be recycled owing to climatic effects caused if it is released to the atmosphere). The external circuits may contain an inexpensive fluid (e.g. water), and they may be omitted if it is practical to circulate the primary working fluid directly to the load area or to the reference reservoir.

The heat pump contains a compressor, which performs step 7-5 in the Rankine cycle depicted in Fig. 4.1-3, and a nozzle, which performs step 2-9. The intermediate steps are performed in the two heat exchangers, giving the working fluid the temperatures T_{up} and T_{low}, respectively. The equations for determining these temperatures are of the form (4.1.99). There are four such equations, which must be supplemented by equations for the compressor and nozzle performance, in order to allow a determination of all the unknown temperatures indicated in Fig. 4.1-104, for given T_{ref}, given load and a certain energy expenditure to the compressor Losses in the compressor are in the form of heat, which in some cases can be credited to the load area.

An indication of the departures from the Carnot limit of the "coefficients of performance", $\varepsilon^{heat\ pump}$, encountered in practice, is given in Fig. 4.1-105, as a function of the temperature difference $T_{up}-T_{low}$ at the heat pump and for selected values of T_{up}. In the interval of temperature differences covered, the $\varepsilon^{heat\ pump}$ is about 50% of the Carnot limit (4.1.22), but it falls more and more below the Carnot value as the temperature difference decreases, although the absolute value of the coefficient of performance increases.

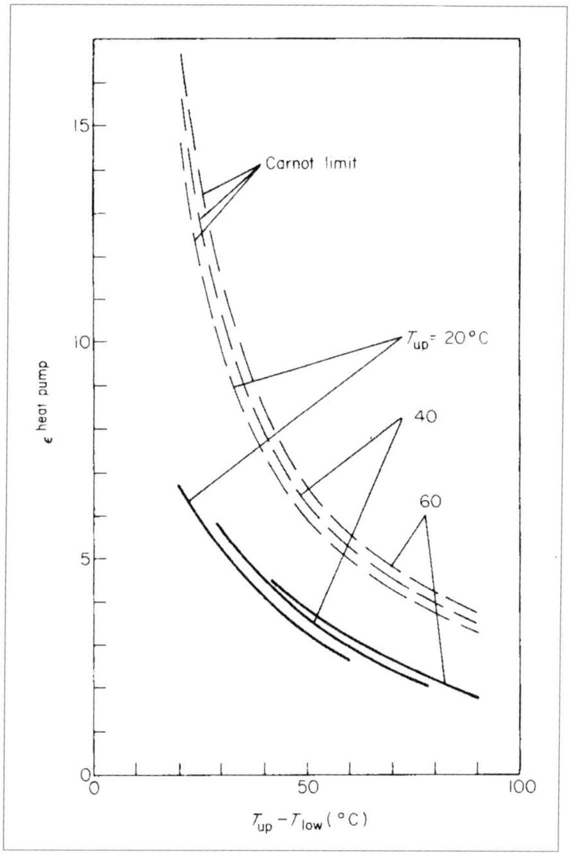

Figure 4.1-105 Measured coefficient of performance, $\varepsilon^{heat\ pump}$, for a heat pump with a semi-hermetic piston-type compressor (solid lines, based on Trenkowitz, 1969), and corresponding curves for ideal Carnot cycles.

Several possibilities exist for the choice of the reference reservoir. Systems in use for space heating or space cooling (achieved by reversing the flow in the compressor and expansion-nozzle circuit, cf. section 4.1.1.2) have utilised river, lake and sea water, and air, as well as soil as reservoirs. The temperatures of such reservoirs are not entirely constant, and it must therefore be acceptable that the performance of the heat pump systems will vary with time. Such variations are damped if water or soil reservoirs at sufficient depth are used, as seen from

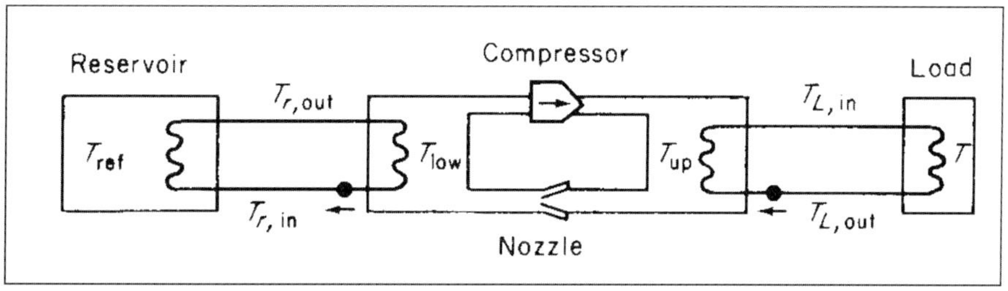

Figure 4.1-104 Schematic picture of a heat pump.

Fig. 3.3.66. Alternative types of reservoirs for use with heat pumps are city waste sites, livestock manure, ventilation air from households or from livestock barns (where the rate of air exchange has to be particularly high), and heat storage tanks connected to solar collectors, etc.

In connection with solar heating systems, the heat pump may be connected between the heat store and the load area (whenever the storage temperature is too low for direct circulation), or it may be connected between the heat store and the collector, such that the fluid let into the solar collector is cooled in order to improve the collector performance. Of course, a heat pump operating on its own reservoir (soil, air, etc.) may also provide the auxiliary heat for a solar heating system of capacity below the demand.

The high-quality energy input to the compressor of a heat pump may also come from a renewable resource, e.g. by wind or solar energy conversion, either directly or via a utility grid carrying electricity generated by renewable energy resources. As for insulation materials, concern has been expressed over the use of CFC gas in the processing or as a working fluid, and substitutes believed to have less negative impacts have been developed.

4.1.6.2 Conversion of heat into work or electricity

The conversion of heat to shaft power or electricity is achieved by one of the thermodynamic cycles, examples of which are shown in Fig. 4.1-3. The cycles may be closed as in Fig. 4.1-3, or they may be "open", in that the working fluid is not recycled through the cooling step (4-1 in most of the cycles shown in Fig. 4.1-3). Instead, new fluid is added for the heating or compression stage, and "used" fluid is rejected after the expansion stage.

It should be kept in mind that the thermodynamic cycles convert heat of temperature T into work plus some residual heat of temperature above the reference temperature T_{ref} (in the form of heated cooling fluid or rejected working fluid). Emphasis should therefore be placed on utilising both the work and the "waste heat". This is done, for example, by co-generation of electricity and water for district heating.

Conversion of heat from solar collectors

Examples of the use of thermodynamic cycles in the conversion of heat derived from solar collectors into work have been given in Figs. 4.1-28–4.1-30 and 4.1-32. The dependence of the limiting Carnot efficiency on temperature is shown in Fig. 4.1-34 for selected values of a parameter describing the concentrating ability of the collector and its short-wavelength absorption to long-wavelength emission ratio. The devices shown in Figs. 4.1-29 and 4.1-30 aim at converting solar heat into mechanical work for water pumping, while the device in Fig. 4.1-32 converts heat from a solar concentrator into electricity.

Ericsson hot-air engine

The engines in these examples were based on the Rankine or the Stirling cycle. It is also possible that the Ericsson cycle (which was actually invented for the purpose of solar energy conversion) will prove advantageous in some solar energy applications. It is based on a gas (usually air) as a working fluid and may have the layout shown in Fig. 4.1-106. In order to describe the cycle depicted in Fig. 4.1-3, the valves must be closed at definite times and the pistons must be detached from the rotating shaft (in contrast to the situation shown in Fig. 4.1-106), such that the heat may be supplied at constant volume. In a different mode of operation, the valves open and close as soon as a pressure difference between the air on the two sides begins to develop. In this case, the heat is added at constant pressure, as in the Brayton cycle, and the piston movement is approximately at constant temperature, as in the Sterling cycle (this variant is not shown in Fig. 4.1-3).

The efficiency can easily be calculated for the latter version of the Ericsson cycle, for which the temperatures T_{up} and T_{low} in the compression and expansion piston cylinders are constant, in a steady situation. This implies that the air enters the heating chamber with temperature T_{low} and leaves it with temperature T_{up}. The heat exchanger equations [corresponding to (4.1.98), but without the assumption that T_3 is constant] take the form

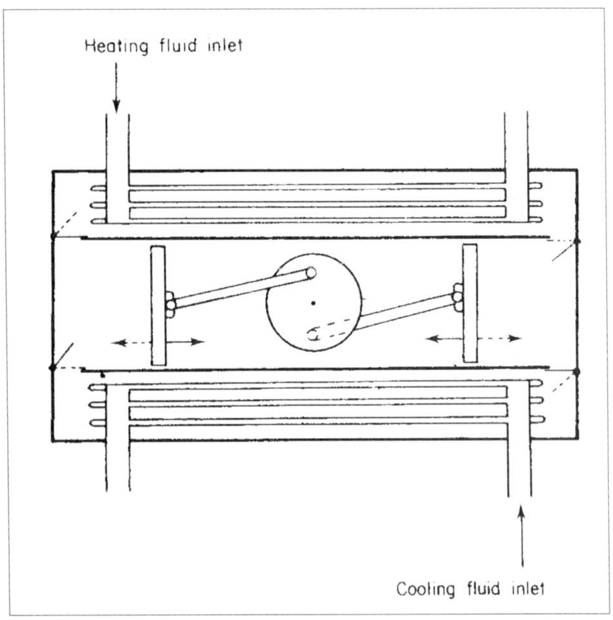

Figure 4.1-106 Example of an Ericsson hot-air engine.

$$-J_m^f C_p^f \, dT^f(x)/dx = h'(T^f(x) - T^g(x)),$$

$$J_m^g C_p^g \, dT^g(x)/dx = h'(T^f(x) - T^g(x)),$$

where the superscript g stands for the gas performing the thermodynamic cycle, f stands for the fluid leading heat to the heating chamber heat exchanger, and x increases from zero at the entrance to the heating chamber to a maximum value at the exit. C_p is a constant-pressure heat capacity per unit mass, and J_m is a mass flow rate. Both these and h', the heat exchange rate per unit length dx, are assumed constant, in which case the equations may be explicitly integrated to give

$$T_{c,in} = T_{c,out} - \frac{J_m^g C_p^g}{J^f C_p^f}(T_{up} - T_{low}) = \frac{J_m^g C_p^g (1-H)}{J_m^g C_p^g + J_m^f C_p^f} T_{low} + \left(H + \frac{J_m^f C_p^f (1-H)}{J_m^g C_p^g + J_m^f C_p^f}\right) T_{c,out}. \tag{4.1.146}$$

Here $T_{c,out} = T$ is the temperature provided by the solar collector or other heat source, and $T_{c,in}$ is the temperature of the collector fluid when it leaves the heat exchanger of the heating chamber, to be re-circulated to the collector or to a heat storage connected to it. H is given by

$$H = \exp(-h((J_m^f C_p^f)^{-1} + (J_m^g C_p^g)^{-1})),$$

where $h = \int h' \, dx$. Two equations analogous to (4.1.146) may be written for the heat exchange in the cooling chamber of Fig. 4.1-106, relating the reject temperature $T_{r,in}$ and the temperature of the coolant at inlet, $T_{r,out} = T_{ref}$, to T_{low} and T_{up}. $T_{c,in}$ may then be eliminated from (4.1.146) and $T_{r,in}$ from the analogous equation, leaving two equations for determination of T_{up} and T_{low} as functions of known quantities, notably the temperature levels T and T_{ref}. The reason for not having to consider equations for the piston motion in order to determine all relevant temperatures is, of course, that the processes associated with the piston motion have been assumed to be isothermal.

The amounts of heat added, Q_{add}, and rejected, Q_{rej}, per cycle are

$$\begin{aligned} Q_{add} &= mC_p^g(T_{up} - T_{low}) + n\mathscr{R}T_{up}\log(V_{max}/V_{min}), \\ Q_{rej} &= mC_p^g(T_{up} - T_{low}) + n\mathscr{R}T_{low}\log(V_{max}/V_{min}) \\ &\quad + Q'_{rej}, \end{aligned} \tag{4.1.147}$$

where m is the mass of air involved in the cycle and n is the number of moles of air involved. $\mathscr{R}$ is the gas constant, and V_{min} and V_{max} are the minimum and maximum volumes occupied by the gas during the compression or expansion stages (for simplicity the "compression ratio" V_{max}/V_{min} has been assumed to be the same for the two processes, although they take place in different cylinders). The ideal gas law has been assumed in calculating the relation between heat amount and work in (4.1.147), and Q'_{rej} represents heat losses not contributing to transfer of heat from working gas to coolant flow (piston friction, etc.). The efficiency is

$$\eta = (Q_{add} - Q_{rej})/Q_{add},$$

and the maximum efficiency which can be obtained with this version of the Ericsson engine is obtained for negligible Q'_{rej} and ideal heat exchangers providing $T_{up} = T$ and $T_{low} = T_{ref}$,

$$\max(\eta) = \left(1 - T_{ref}/T\right) \Big/ \left(1 + \frac{mC_p^g}{n\mathscr{R}\log(V_{max}/V_{min})} \times \left(1 - T_{ref}/T\right)\right) \tag{4.1.148}$$

The ideal Carnot efficiency may even be approached, if the second term in the denominator can be made small (however, to make the compression ratio very large implies an increase in the length of time required per cycle, such that the rate of power production may actually go down, as discussed in section 4.1.1.1.) The power may be calculated by evaluating (4.1.147) per unit time instead of per cycle.

Conversion of geothermal heat

Geothermal heat sources have been utilised by means of thermodynamic engines (e.g. Brayton cycles), in cases where the geothermal heat has been in the form of steam (water vapour). In some regions, geothermal sources exist which provide a mixture of water and steam, including suspended soil and rock particles, such that conventional turbines cannot be used. Work has been done on a special "brine screw" that can operate under such conditions (McKay and Sprankle, 1974).

However, in most regions the geothermal resources are in the form of heat-containing rock or sediments, with little possibility of direct use. If an aquifer passes through the region, it may collect heat from the surrounding layers and allow a substantial rate of heat extraction, for example, by drilling two holes from the surface to the aquifer, separated from each other, as indicated in Fig. 4.107a. Hot water (not developing much steam unless the aquifer lies very deep or its temperature is exceptionally high, cf. Fig. 3.3.68) is pumped or rises by its own pressure to the surface at one hole and is re-injected through a second hole, in a closed cycle, in order to avoid pollution from various undesired chemical substances often contained in the aquifer water. The heat extracted from a heat exchanger may be used directly (e.g. as district heating; cf. Clot, 1977) or may generate electricity through one of the "low-temperature" thermodynamic cycles considered above in connection with solar collectors (Mock *et al.*, 1997).

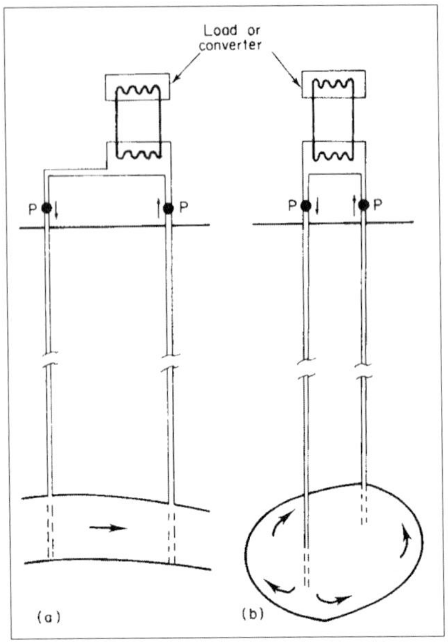

Figure 4.1-107 Examples of the utilisation of geothermal heat: (a) based on the presence of an aquifer; (b) based on a region of fractured rock.

If no aquifer is present to establish a "heat exchange surface" in the heat-containing rock, it may be feasible to create suitable fractures artificially (by explosives or induced pressure). An arrangement of this type is illustrated in Fig. 4.1-107b, counting on the fluid which is pumped down through one drilling hole to make its way through the fractured region of rock to the return drilling hole in such a way that continued heat extraction can be sustained. The heat transfer can only be predicted in highly idealised cases (see e.g. Gringarten *et al.*, 1975), which may not be realised as a result of the fairly uncontrolled methods of rock fractionation available.

One important result of the model calculations is that the heat extraction rate deemed necessary for practical applications is often higher than the geothermal flux into the region of extraction, so that the temperature of the extracted heat will be dropping (e.g. by about 1°C per year for one case considered, the number being highly dependent on fracture distribution, rock structure, etc.). This non-sustainable use of geothermal energy is apparent in actual installations in New Zealand and Italy.

Conversion of ocean thermal energy

As discussed in section 3.3.5.1, downward gradients of temperature exist in most oceans, and they are particularly stable (i.e. without variations with time) in the tropical oceans. The utilisation of such temperature gradients for electricity generation (e.g. by use of a Rankine cycle) has been considered several times since the first suggestions by d'Arsonval (1881).

The temperature differences available over the first 500–1000 m of water depth are only about 25°C, as shown in Fig. 3.3.63. Considering a closed Rankine cycle, with a working fluid (e.g. ammonia) which evaporates and condenses at convenient temperatures, placed near the ocean surface, it will be required to pump colder water through a pipe from the depth to a heat exchanger for condensation of the working fluid. Further, a warm water heat exchanger is required for evaporating the working fluid. If the heat exchange surface is such that, say, 5°C is "lost" at each heat exchanger, the temperature difference available to the thermodynamic cycle is only 15°C, corresponding to a limiting Carnot efficiency of roughly 0.05. For an actual engine, the efficiency is still lower, and from the power generated should be subtracted the power needed to pump hot and cold water through the heat exchangers and to pump cold water from its original depth to the converter level. It is expected that overall efficiencies around 0.02 may be achieved (cf. e.g. McGowan, 1976).

In order to save energy to pump the hot water through the heat exchanger, it has been suggested that these converters be placed in strong currents such as the Gulf Stream (Heronemus, 1975). The possibility of adverse environmental effects from the power extraction from ocean thermal gradients was touched upon in section 3.3.5.1. Such dangers may be increased if ocean currents are incorporated into the scheme, because of the possible sensitivity of the itinerary of such currents to small perturbations, and because the dependence of climatic zones on the course of currents such as the Gulf Stream and the Kuro Shio.

Open thermodynamic cycles have also been suggested for conversion of ocean thermal energy (Claude, 1930; Beck, 1975; Zener and Fetkovich, 1975), for example, based on creating a rising mixture of water and steam bubbles or "foam", which is separated at a height above sea-level, such that the water can be used to drive a turbine rotor.

If viable systems could be developed for conversion of ocean thermal energy, then there would be a number of other applications of such conversion devices in connection with other heat sources of a temperature little higher than that of the surroundings, especially when such heat sources can be regarded as "free". Examples are the reject or "waste" heat flows from the range of other conversion devices operating at higher initial temperature differences, including fuel-based power plants.

4.1.7 Conversion of fuels

A number of technologies are available to convert fuels among themselves and into electricity and heat or motive power. Many of them can also be used for fuels derived from renewable energy sources. Those derived from biomass will be dealt with separately in section 4.1.8. For renewable energy sources such as wind and solar power, one way of dealing with fluctuating production would be to convert surplus electricity to a storable fuel. This could be hydrogen obtained by electrolysis (or by reversible fuel cells), in which case the further conversion of hydrogen to the energy forms in demand can be accomplished by methods already available, for example, for natural gas, such as boilers, engines, gas turbines and fuel cells.

4.1.7.1 Fuel cell technologies

The idea of converting fuel into electricity by an electrode–electrolyte system originated in the 19th century (Grove, 1839). The basic principle behind a hydrogen–oxygen fuel cell was described in section 4.1.1.6. The first practical applications were in powering space vehicles, starting during the 1970s.

Developed for stationary power applications, the phosphoric acid cells use porous carbon electrodes with a platinum catalyst and phosphoric acid as electrolyte and feed hydrogen to the negative electrode, with electrode reactions given by (4.1.72) and (4.1.73). The operating temperature is in the range 175–200°C, and water is continuously removed.

Alkaline cells use KOH as electrolyte and have electrode reactions of the form

$$H_2 + 2OH^- \rightarrow 2H_2O + 2e^-,$$
$$\tfrac{1}{2}O_2 + H_2O + 2e^- \rightarrow 2OH^-.$$

These cells operate in the temperature range 70–100°C, but specific catalysts require maintenance of fairly narrow temperature regimes. Also, the hydrogen fuel must have a high purity and notably not contain any CO_2. Alkaline fuel cells have been used extensively on spacecraft and recently for road vehicles (Hoffmann, 1998a). Their relative complexity and use of corrosive compounds requiring special care in handling make it unlikely that the cost can be reduced to levels acceptable for general-purpose use.

The third fuel cell type in commercial use is the proton exchange membrane (PEM) cell. It has been developed over a fairly short period of time and is considered to hold the greatest promise for economic application in the transportation sector. It contains a solid polymer membrane sandwiched between two gas diffusion layers and electrodes. The membrane material may be polyperfluorosulphonic acid. A platinum or Pt–Ru alloy catalyst is used to break hydrogen molecules into atoms at the negative electrode, and the hydrogen atoms are then capable of penetrating the membrane and reaching the positive electrode, where they combine with oxygen to form water, again with the help of a platinum catalyst. The electrode reactions are again (4.1.72) and (4.1.73), and the operating temperature is 50–100°C (Wurster, 1997). Figure 4.1-108 shows a typical layout of an individual cell. Several of these are then stacked on top of each other. This modularity implies that PEM fuel cells can be used for applications requiring little power (1 kW or less). PEM cell stacks are dominating the current wealth of demonstration projects in road transportation, portable power and special applications. The efficiency of conversion for the small systems is between 40% and 50%, but a 50 kW system has recently shown an efficiency near 60%. As indicated in Fig. 4.1-109, an advantage of particular importance for automotive applications is the high efficiency at part loads, which alone gives a factor of two improvement over current internal combustion engines.

For use in automobiles, compressed and liquefied hydrogen are limited by low energy density and safety precautions for containers and, first of all, by the requirement of a new infrastructure for fuelling. The hydrogen storage problem, which until recently limited fuel cell projects to large vehicles such as buses, may be solved by use of metal hydride or carbon nanofibre stores (see Chapter 5.1). In order to avoid having to make large changes to the current gasoline and diesel fuel filling stations, several current schemes use methanol as the fuel distributed to the vehicle fuel tank. The energy density of methanol is 4.4 kWh litre^{-1}, which is half that of gasoline. This is quite acceptable owing to the higher efficiency of conversion. Hydrogen is then formed onboard by a methanol reformer, before being fed to the fuel cell to produce the electric power for an electric motor. The set-up is illustrated in Fig. 4.1-110. Prototype

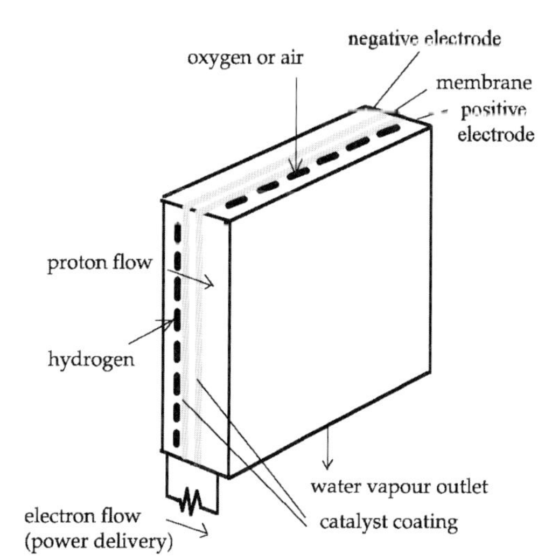

Figure 4.1-108 Layout of a PEM fuel cell layer, several of which may be stacked.

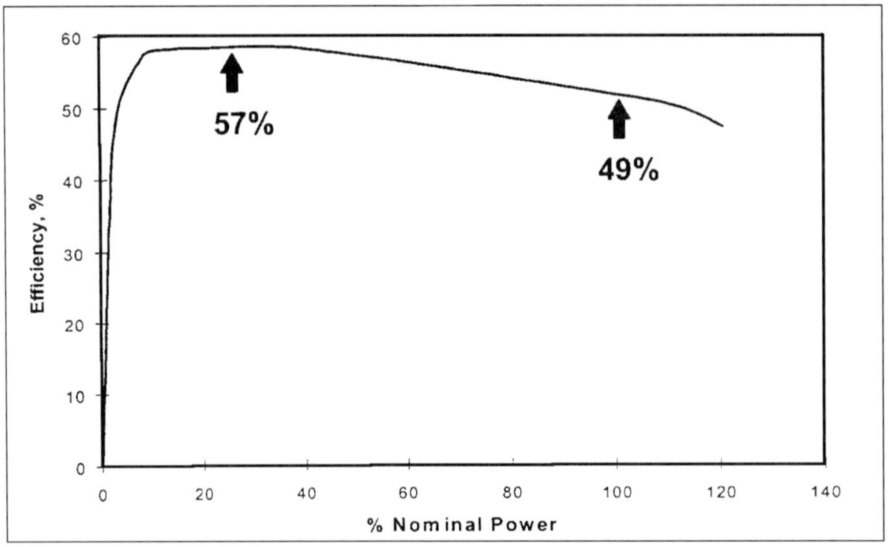

Figure 4.1-109 Expected part-load efficiencies for a 50-kW PEM fuel cell, projected from measurements involving 10–20 cell test stacks (Patil, 1998).

vehicles with this set-up have recently been tested (cf. Takahashi, 1998; Brown, 1998).

Methanol, CH_3OH, may even be used directly in a fuel cell, without the extra step of reforming to H_2. PEM fuel cells accepting as feedstock a mixture of methanol and water are under development (Bell, 1998). As methanol can be produced from biomass, hydrogen may in this way be eliminated from the energy system. On the other hand, handling of surplus production of wind or photovoltaic power might still conveniently involve hydrogen as an intermediate energy carrier, as it may have direct uses and thus may improve the system efficiency by avoiding the losses in methanol production. The electric power to hydrogen conversion efficiency is about 65% in high pressure alkaline electrolysis plants (Wagner *et al.*, 1998), and the efficiency of the further hydrogen (plus CO or CO_2 and a catalyst) to methanol conversion is around 70%. Also, the efficiency of producing methanol from biomass is about 45%, whereas higher efficiencies are obtained if the feedstock is methane (natural gas) (Jensen and Sørensen, 1984; Nielsen and Sørensen, 1998).

For stationary applications, fuel cells of higher efficiency may be achieved by processes operating at

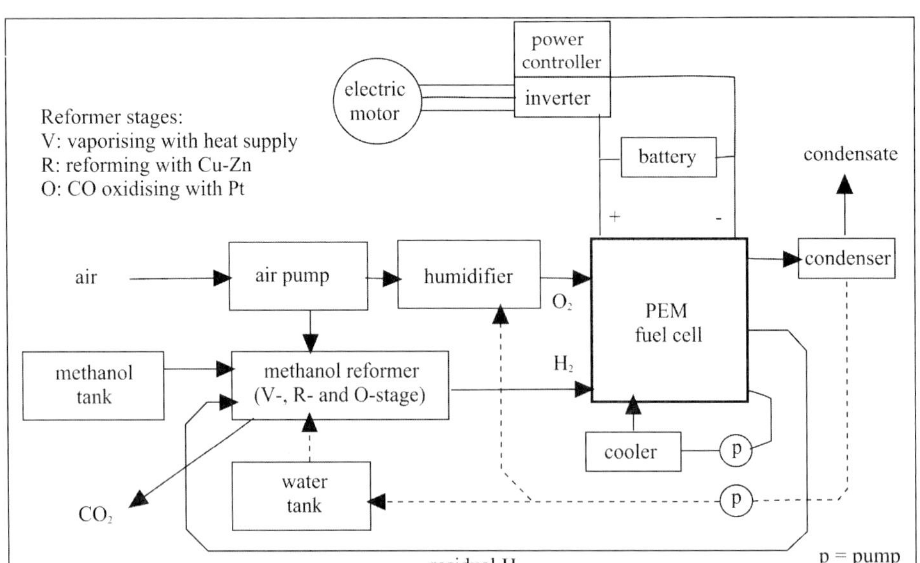

Figure 4.1-110 Layout of the power system for a methanol-to-hydrogen powered vehicle with fuel cell conversion and an electric motor. The power controller allows shift from direct drive to battery charging.

higher temperatures. One line of research has been molten carbonate fuel cells, with electrode reactions

$$H_2 + CO_3^{2-} \rightarrow CO_2 + H_2O + 2e^-,$$
$$\tfrac{1}{2}O_2 + CO_2 + 2e^- \rightarrow CO_3^{2-}.$$

The electrolyte is a molten alkaline carbonate mixture retained within a porous aluminate matrix. The carbonate ions formed at the positive electrode travel through the matrix and combine with hydrogen at the negative electrode at an operating temperature of about 650°C. The process was originally aimed at hydrogen supplied from coal gasification or natural gas conversion. The first full-scale test (250 kW) took place at a power utility company in Bielefeld, Germany (Hoffmann, 1998b). The expected conversion efficiency is about 55%, but additional high-temperature heat may be utilised.

Considerable efforts are dedicated to the solid electrolyte cells. Solid oxide fuel cells (SOFC) use zirconia as the electrolyte layer to conduct oxygen ions formed at the positive electrode. Electrode reactions are

$$H_2 + O^{2-} \rightarrow H_2O + 2e^-,$$
$$\tfrac{1}{2}\,O_2 + 2e^- \rightarrow O^{2-}.$$

The reaction temperature is 700–1000°C. The lower temperatures are desirable, due to lower corrosion problems, and may be achieved by using a thin electrolyte layer (about 10 μm) of yttrium-stabilised zirconia sprayed onto the negative electrode as a ceramic powder (Kahn, 1996). A number of prototype plants (in the 100-kW size range) are in operation. Conversion efficiency was about 55%, but could reach 70–80% in the future (Hoffmann, 1998b).

Particularly for vehicle applications of hydrogen-based technologies, efforts are needed to ensure a high level of safety in collisions and other accidents. Because of the difference between the physical properties of hydrogen fuel and the hydrocarbon fuels currently in use (higher diffusion speed, flammability and explosivity over wider ranges of mixtures with air), there is a need for new safety-related studies, particularly where hydrogen is stored on board in containers at pressures typically of 20–50 MPa. A few such studies have already been made (Brehm and Mayinger, 1989).

A more detailed discussion of fuel cells can be found in a sequel to this book (Sørensen, 2004a).

4.1.8 Conversion of biological material

Traditional uses of organic fuels as energy carriers comprise the use of fresh biomass, i.e. storage and combustion of wood fuels, straw and other plant residues, and in

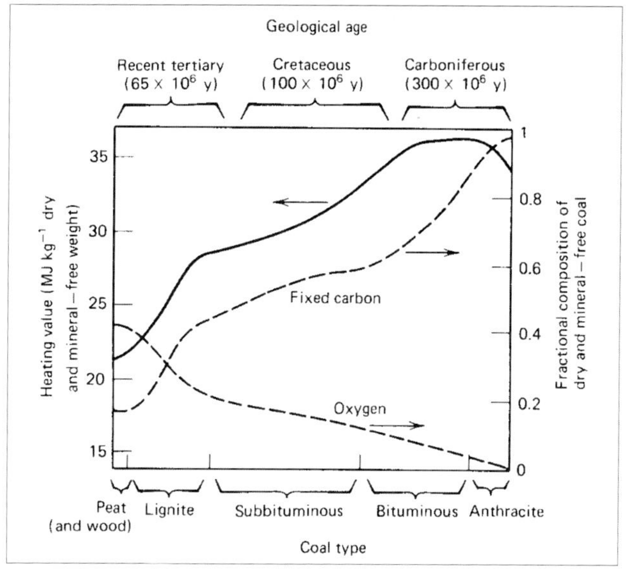

Figure 4.1-111 Coal to peat classification and selected properties. The dashed lines indicate the fractional content of oxygen and of fixed carbon. The remaining fraction consists of volatile matter (based on US DoE, 1979).

recent centuries notably storage and combustion of fossilised biomass, i.e. fuels derived from coal, oil and natural gas. While the energy density of living biomass is in the range of 10–30 MJ per kg of dry weight, the several million years of fossilation processes typically increase the energy density by a factor of two (see Fig. 4.1-111). The highest energy density is found for oil, where the average density of crude oil is 42 MJ kg^{-1}.

The known reserves of fossil fuels that may be economically extracted today, and estimates of further resources exploitable at higher costs, are indicated in Table 4.1-3. The finiteness of such resources is, of course, together with the environmental impact issues, the reason for turning to renewable energy sources, including renewable usage of biomass resources.

However, fresh biomass is a large potential source of renewable energy that in addition to use for combustion may be converted into a number of liquid and gaseous biofuels. This may be achieved by thermochemical or by biochemical methods, as described below.

Biomass is not just stored energy, it is also a store of nutrients and a potential raw material for a number of industries. Therefore, bio-energy is a topic that cannot be separated from food production, timber industries (serving construction purposes, paper and pulp industries, etc.), and organic feedstock-dependent industries (chemical and biochemical industries, notably). Furthermore, biomass is derived from plant growth and animal husbandry, linking the energy aspect to agriculture, livestock, silviculture, aquaculture, and quite generally to the management of the global ecological system. Thus, procuring and utilising organic fuels

Table 4.1-3 Fossil reserves, resources and consumption in EJ (UN, 1981; Jensen and Sørensen, 1984; Nakicenovic et al., 1996)

Source	Reserves (EJ)	Resources (>50% prob.)	Occurrence (speculative)	1990 use (EJ)	Accum. use 1860–1990
Coal				91	5203
hard coal	15 000	100 000	900 000		
brown coal/lignite	4 000	30 000	90 000		
peat	0	4 000	4 000		
Oil				128	3343
conventional	5 000	3 000	13 000		
unconventional	7 000	9 000	20 000		
Natural gas				71	1703
conventional	5 000	5 000	15 000		
unconventional	7 000	20 000	25 000		
in hydrates	0	0	100 000		

constitute a profound interference with the natural biosphere, and the understanding of the range of impacts as well as the elimination of unacceptable ones should be an integral part of any scheme for diversion of organic fuels to human society.

4.1.8.1 Heat production from biomass

Producing heat by burning

Heat may be obtained from biological materials by burning, eventually with the purpose of further conversion. Efficient burning usually requires the reduction of water content, for example, by drying in the Sun. The heat produced by burning cow dung is about 1.5×10^7 J per kg of dry matter, but initially only about 10% is dry matter, so the vaporisation of 9 kg of water implies an energy requirement of 2.2×10^7 J, i.e. the burning process is a net energy producer only if substantial sun-drying is possible. Firewood and other biomass sources constitute stores of energy, since the drying may be performed during summer, such that these biological fuels can be used during winter periods when the heating requirement may be large and the possibility of sun-drying may not exist. The heat produced by burning 1 kg of dry wood or sawmill scrap is about 1.25×10^7 J (1 kg in these cases corresponds to a volume of approximately 1.5×10^{-3} m^3), and the heat from burning 1 kg of straw (assumed water content 15%) is about 1.5×10^7 J. The boilers used for firing with wood or straw have efficiencies that are often considerably lower than those of oil or gas burners. Typical efficiencies are in the range 0.5–0.6 for the best boilers. The rest of the enthalpy is lost, mostly in the form of vapour and heat leaving the chimney (or other smoke exit), and therefore is not available at the load area.

Combustion is the oxidation of carbon-containing material in the presence of sufficient oxygen to complete the process

$$C + O_2 \rightarrow CO_2.$$

Wood and other biomass is burned for cooking, for space heating, and for a number of specialised purposes, such as provision of process steam and electricity generation. In rural areas of many Third World countries a device consisting of three stones for outdoor combustion of twigs is still the most common. In the absence of wind, up to about 5% of the heat energy may reach the contents of the pot resting on top of the stones. In some countries, indoor cooking on simple chulas is common. A chula is a combustion chamber with place for one or more pots or pans on top, resting in such a way that the combustion gases will pass along the outer sides of the cooking pot and leave the room through any opening. The indoor air quality is extremely poor when such chulas are in use, and village women in India using chulas for several hours each day are reported to suffer from severe cases of eye irritation and diseases of the respiratory system.

Earlier, most cooking in Europe and its colonies was done on stoves made of cast iron. These stoves, usually called European stoves, had controlled air intake and both primary and secondary air inlets, chimneys with regulation of gas passage, and several cooking places with ring systems allowing the pots to fit tightly in holes, with a part of the pot indented into the hot gas stream. The efficiency would be up to about 25%, counted as energy delivered to the pots divided by wood energy spent, but such efficiencies would only be reached if all holes were in use and if the different temperatures prevailing at different boiler holes could be made useful, including after-heat. In many cases the average efficiency would hardly have exceeded 10%, but in many of the areas in question the heat lost to the room could also be considered as useful, in which case close to 50% efficiency (useful energy divided by input) could be reached. Today, copies of the European stove are being introduced in several regions of the Third World, with use of local materials such as clay and sand–clay mixtures instead of cast iron.

Wood-burning stoves and furnaces for space heating have conversion efficiencies from below or about 10% (open furnace with vertical chimney) up to 50% (oven with two controlled air inlets and a labyrinth-shaped effluent gas route leading to a tall chimney). Industrial burners and stokers (for burning wood scrap) typically reach efficiencies of about 60%. Higher efficiencies require a very uniform fuel without variations in water content or density.

Most biological material is not uniform, and some pretreatment can often improve both the transportation and storage processes and the combustion (or other final use). Irregular biomass (e.g. twigs) can be chopped or cut to provide unit sizes fitting the containers and burning spaces provided. Furthermore, compressing and pelletising the fuel can make it considerably more versatile. For some time, straw compressors and pelletisers have been available, so that the bulky straw bundles can be transformed into a fuel with volume densities approaching that of coal. Other biomass material can conceivably be pelletised with advantage, including wood scrap, mixed biomass residues, and even aquatic plant material. Portable pelletisers are available (e.g. in Denmark) which allow straw to be compressed in the fields so that longer transport becomes economically feasible and so that even long-term storage (seasonal) of straw residues becomes attractive.

A commonly practised conversion step is from wood to charcoal. Charcoal is easier to store and to transport. Furthermore, charcoal combustion – for example, for cooking – produces less visible smoke than direct wood burning and is typically so much more efficient than wood burning that, despite wood-to-charcoal conversion losses, less primary energy is used to cook a given meal with charcoal than with wood.

Particularly in the rich countries, a considerable source of biomass energy is urban refuse, which contains residues from food preparation and discarded consumer goods from households, as well as organic scrap material from commerce and industry. Large-scale incineration of urban refuse has become an important source of heat, particularly in Western Europe, where it is used mainly for district heating.

For steam generation purposes, combustion is performed in the presence of an abundant water source ("waterwall incineration"). In order to improve pollutant control, fluidised bed combustion techniques may be utilised. The bed consists of fine-grain material, for example, sand, mixed with material to be burned (particularly suited is sawdust, but any other biomass including wood can be accepted if finely chopped). The gaseous effluents from combustion, plus air, fluidise the bed as they pass through it under agitation. The water content of the material in the bed may be high (in which case steam production entails). Combustion temperatures are lower than for flame burning, and this partly makes ash removal easier and partly reduces tar formation and salt vaporisation. As a result, the reactor life is extended and air pollution can better be controlled.

In general, the environmental impacts of biomass utilisation through combustion may be substantial and comparable to, although not entirely of the same nature as, the impacts from coal and oil combustion (see Table 4.1-4). In addition, ashes will have to be disposed of. For boiler combustion, the sulphur dioxide emissions are typically much smaller than for oil and coal combustion, which would give 15–60 kg t^{-1} in the absence of flue gas cleaning. If ash is re-injected into the wood burner, higher sulphur values appear, but these values are still below the fossil fuel emissions in the absence of sulphur removal efforts.

Particulates are not normally regulated in home boilers, but for power plants and industrial boilers, electrostatic filters are employed with particulate removal up to over 99%. Compared to coal burning without particle removal, wood emissions are 5–10 times lower. When starting a wood boiler there is an initial period of very visible smoke emission, consisting of water vapour and high levels of both particulate and gaseous emissions.

Table 4.1-4 Uncontrolled emissions from biomass combustion in boilers (kg per tonne of fuel, based on woody biomass; US EPA, 1980)

Substance emitted	Emissions (kg/10^3kg)
Particulates	12.5–15.0
Organic compounds[a]	1.0
Sulphur dioxide	0–1.5[b]
Nitrogen oxides	5.0
Carbon monoxide	1.0

[a]Hydrocarbons including methane and traces of benzo(a)pyrine.
[b]Upper limit is found for bark combustion. Ten times higher values are reported in cases where combustion ashes are re-injected.

After reaching operating temperatures, wood burns virtually without visible smoke. When stack temperatures are below 60°C, again during start-up and incorrect burning practice, severe soot problems arise.

Nitrogen oxides are typically 2–3 times lower for biomass burning than for coal burning (per kilogram of fuel), often leading to similar emissions if taken per unit of energy delivered.

Particular concern should be directed at the organic compound emissions from biomass burning. In particular, benzo(a)pyrene emissions are found to be up to 50 times higher than for fossil fuel combustion, and the measured concentrations of benzo(a)pyrene in village houses in Northern India ($1.3–9.3 \times 10^{-9}$ kg m^{-3}), where primitive wood-burning chulas are used for several (6–8) hours every day, exceed the German standards of 10^{-11} kg m^{-3} by 2–3 orders of magnitude. However, boilers with higher combustion temperatures largely avoid this problem, as indicated in Table 4.1-4.

The lowest emissions are achieved if batches of biomass are burned at optimal conditions, rather than regulating the boiler up and down according to heating load. Therefore, wood heating systems consisting of a boiler and a heat storage unit (gravel, water) with several hours of load capacity will lead to the smallest environmental problems. This example shows that there can be situations where energy storage would be introduced entirely for environmental reasons.

Finally, it should be mentioned that occupational hazards arise during tree felling and handling. The accident rate is high in many parts of the world, and safer working conditions for forest workers are imperative if wood is to be used sensibly for fuel applications.

Finally, concerning carbon dioxide, which accumulates in the atmosphere as a consequence of rapid combustion of fossil fuels, it should be kept in mind that the carbon dioxide emissions during biomass combustion are balanced in magnitude by the net carbon dioxide assimilation by the plants, so that the atmospheric CO_2 content is not affected, at least by the use of biomass crops in fast rotation. However, the lag time for trees may be decades or centuries, and in such cases, the temporary carbon dioxide imbalance may contribute to climatic alterations.

Composting

Primary organic materials form the basis for a number of energy conversion processes other than burning. Since these produce liquid or gaseous fuels, plus associated heat, they will be dealt with in the following sections on fuel production. However, conversion aiming directly at heat production has also been utilised, with non-combustion processes based on manure from livestock animals and in some cases on primary biomass residues.

Two forms of composting are in use, one based on fluid manure (less than 10% dry matter), and the other based on solid manure (50–80% dry matter). The chemical process is in both cases a bacterial decomposition under aerobic conditions, i.e. the compost heap or container has to be ventilated in order to provide a continuous supply of oxygen for the bacterial processes. The bacteria required for the process (of which lactic acid producers constitute the largest fraction) are normally all present in manure, unless antibiotic residues that kill bacteria are retained after some veterinary treatment. The processes involved in composting are very complex, but it is believed that decomposition of carbohydrates [the inverse of the reaction (3.3.41) is responsible for most of the heat production]. A fraction of the carbon from the organic material is found in new-bred microorganisms.

A device for treating fluid manure may consist of a container with a blower injecting fresh air into the fluid in such a way that it becomes well distributed over the fluid volume. An exit air channel carries the heated airflow to, say, a heat exchanger. Figure 4.1-112 shows an example of the temperature of the liquid manure, along with the temperature outside the container walls, as a function of time. The amount of energy required for the air blower is typically around 50% of the heat energy output, and is in the form of high-quality mechanical energy (e.g. from an electrically driven rotor). Thus, the simple efficiency may be around 50%, but the second law efficiency (4.1.20) may be quite low.

Heat production from solid manure follows a similar pattern. The temperature in the middle of a manure heap ("dunghill") may be higher than that of liquid manure, owing to the low heat capacity of the solid manure (see Fig. 4.1-113). Air may be supplied by blowers placed in the bottom of the heap, and in order to maintain air passage space inside the heap and remove moisture, occasional re-stacking of the heap is required. A certain degree of variation in air supply can be allowed, so that the blower may be driven by a renewable energy

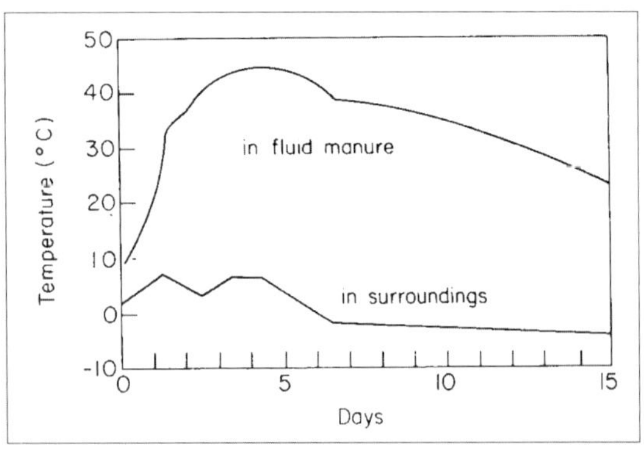

Figure 4.1-112 Temperature development in composting device based on liquid manure from pigs and poultry and with a blower providing constant air supply. Temperature of surroundings is also indicated.

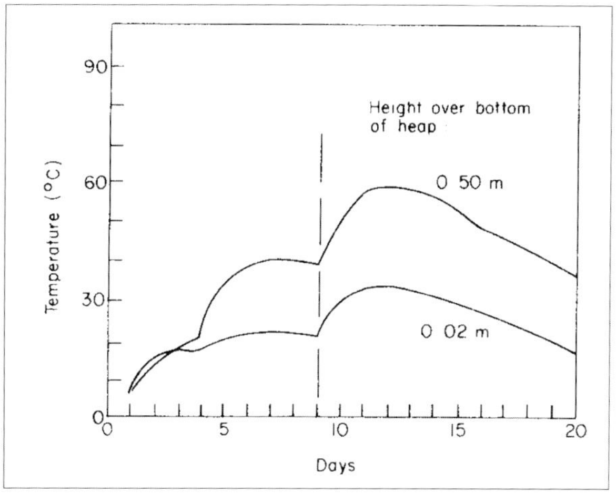

Figure 4.1-113 Temperature development at different locations within a soild manure composing heap (total height 0.8 m). After nine days, air is added from below.

converter, for example, a windmill, without storage or back-up power. With frequent re-stacking, air supply by blowing is not required, but the required amount of mechanical energy input per unit of heat extraction is probably as high as for liquid manure. In addition, the heat extraction process is more difficult, demanding, for example, that heat exchangers be built into the heap itself (water pipes are one possibility). If an insufficient amount of air is provided, the composting process will stop before the maximum heat has been obtained.

Owing to the bacteriological decomposition of most of the large organic molecules present, the final composting product has considerable value as fertiliser.

Metabolic heat

Metabolic heat from the life processes of animals can also be used by man, in addition to the heating of human habitats by man's own metabolic heat. A livestock shed or barn produces enough heat, under most conditions of winter occupancy, to cover the heating requirements of adjacent farm buildings, in addition to providing adequate temperature levels for the animals. One reason for this is that livestock barns must have a high rate of ventilation in order to remove dust (e.g. from the animal's skin, fur, hair or feathers) and water vapour. For this reason, utilisation of barn heat may not require extra feeding of the animals, but may simply consist of heat recovery from air that for other reasons has to be removed. Utilisation for heating a nearby residence building often requires a heat pump (see section 4.1.6.1), because the temperature of the ventilation air is usually lower than that required at the load area, so a simple heat exchanger would not work.

In temperate climates, the average temperature in a livestock shed or barn may be about 15°C during winter. If young animals are present, the required temperature is higher. With an outside temperature of 0°C and no particular insulation of walls, the net heat production of such barns or sheds is positive when the occupants are fully grown animals, but negative if the occupants are young individuals and very much so if newborn animals are present. In chicken or pig farms, the need for heat input may be avoided by having populations of mixed age or heat exchange between compartments for young and adult animals. The best candidates for heat extraction to other applications might then be dairy farms.

A dairy cow transfers about 25% of the chemical energy in the fodder to the milk and a similar amount to the manure. If the weight of the cow remains constant, the rest is converted to heat and is rejected as heat radiation, convection or latent heat of water vaporisation. The distribution of the heat production in sensible and latent forms of heat is indicated in Fig. 4.1-114. It is seen to be strongly dependent on the air temperature in the barn. At 15°C, about two-thirds of the heat production is in the form of sensible heat. Heat transfer to a heat pump circuit may take place from the ventilation air exit. Water condensation on the heat exchanger surface involved may help to prevent dust particles from accumulating on the heat exchanger surface.

4.1.8.2 Fuel production from biomass: overview and generation of gaseous fuels

Fuels are biological material (including fossilised forms), hydrocarbons or just hydrogen, which can be stored and used at convenient times. "Usage" traditionally means

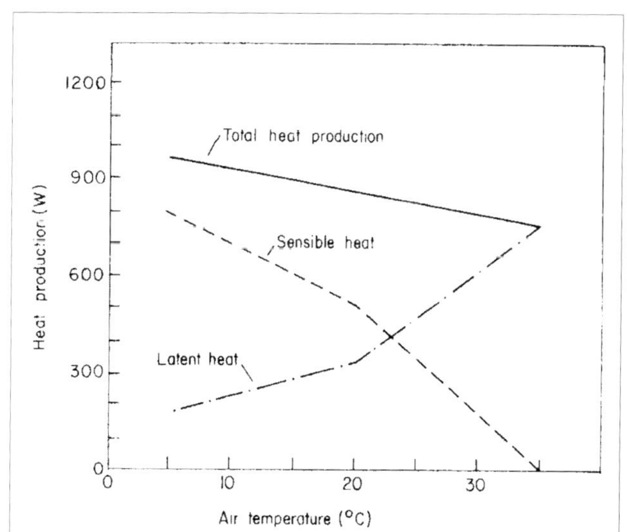

Figure 4.1-114 Average heat production and form of heat for a "standard" cow (the heat production of typical cows of "red" or "black-spotted" races is about 20% higher, while that of typical "Jersey" cows is about 30% lower).

burning in air or pure oxygen, but other means of conversion exist, for example, fuel cells (see sections 4.1.1.6 and 4.1.7).

A number of conversion processes aim at converting one fuel into another which is considered more versatile. Examples are charcoal production from wood (efficiency of conversion about 50%), liquefaction and gasification. Figure 4.1-115 gives an overview of all the non-food uses of biomass. Before discussing the conversion of fresh biomass, the gasification of coal is briefly discussed because of its importance for possible continued use of fossil biomass (coal being the largest such source) and also because of its similarity to processes relevant for other solid biomass.

Inefficient conversion of coal to oil has historically been used by isolated coal-rich but oil-deficient nations (Germany during World War II, South Africa). Coal is gasified to carbon monoxide and hydrogen, which is then, by the Fischer–Tropsch process (passage through a reactor, e.g. a fluidised bed, with a nickel, cobalt, or iron catalyst), partially converted into hydrocarbons. Sulphur compounds have to be removed as they would impede the function of the catalyst. The reactions involved are of the form

$$(2n+1)H_2 + nCO \rightarrow C_nH_{2n+2} + nH_2O,$$
$$(n+1)H_2 + 2nCO \rightarrow C_nH_{2n+2} + nCO_2$$

and conversion efficiencies range from 21 to 55%. Further separation of the hydrocarbons generated may then be performed, for instance gasoline corresponding to the range $4 \leq n \leq 10$ in the above formulae.

Alternative coal liquefaction processes involve direct hydrogenation of coal under suitable pressures and temperatures. Pilot plants have been operating in the United States, producing up to 600 t a day (slightly different processes are named "solvent refining", "H-coal process", and "donor solvent process"). From an energy storage point of view, either coal or the converted product may be stockpiled.

For use in the transportation sector, production of liquid hydrocarbons such as methanol from natural gas could be advantageous. In the long term, methanol is likely to be produced from renewable biomass sources as described below.

Biogas

Conversion of (fresh) biological material into simple hydrocarbons or hydrogen can be achieved by a number of anaerobic fermentation processes, i.e. processes performed by suitable microorganisms and taking place in the absence of oxygen. Such anaerobic "digestion" processes work on the basis of most fresh biological material, wood excepted, provided that the proper conditions are maintained (temperature, population of microorganisms, stirring, etc.). Thus, biological material in forms inconvenient for storage and use may be converted into liquid or gaseous fuels that can be utilised in a multitude of ways, like oil and natural gas products.

The "raw materials" that may be used are slurry or manure (e.g. from dairy farms or "industrial farming" involving large feedlots), city sewage and refuse, farming crop residues (e.g. straw or parts of cereal or fodder plants not normally harvested), or direct "fuel crops", such as ocean-grown algae or seaweeds, water hyacinths (in tropical climates) or fast-growing bushes or trees. The

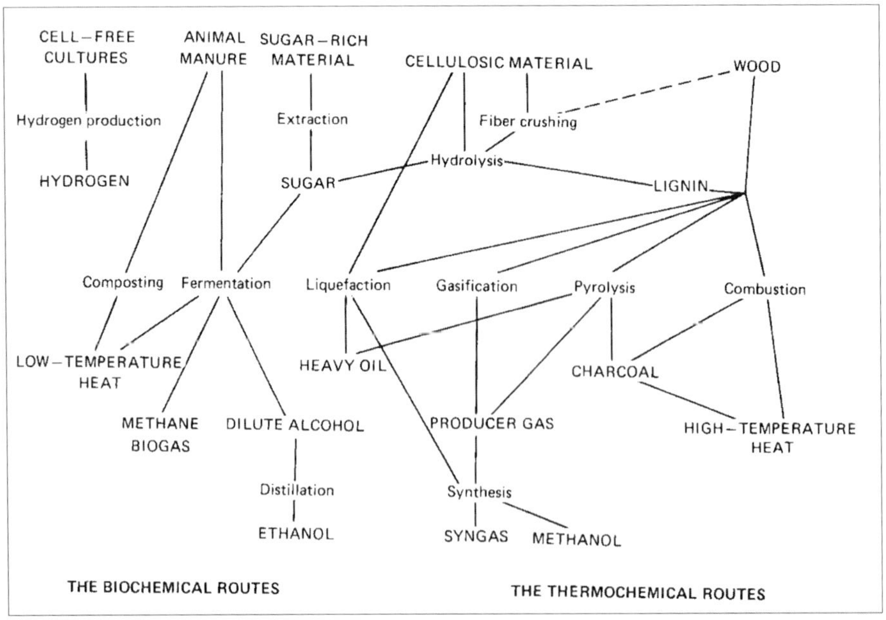

Figure 4.1-115 Non-food energy uses of biomass.

deep harvesting necessary in order to collect crop residues may not be generally advisable, owing to the role of these residues in preventing soil erosion.

Among the fermentation processes, one set is particularly suited for producing gas from biomass in a wet process (cf. Fig. 4.1-115). It is called *anaerobic digestion*. It traditionally used animal manure as biomass feedstock, but other biomass sources can be used within limits, which are briefly discussed in the following. The set of biochemical reactions making up the digestion process (a word indicating the close analogy to energy extraction from biomass by food digestion) is schematically illustrated in Fig. 4.1-116.

There are three discernible stages. In the first, complex biomass material is decomposed by a heterogeneous set of microorganisms, not necessarily confined to anaerobic environments. These decompositions comprise hydrolysis of cellulosic material to simple glucose, using enzymes provided by the microorganisms as catalysts. Similarly, proteins are decomposed to amino acids and lipids to long-chain acids. The significant result of this first phase is that most of the biomass is now water soluble and in a simpler chemical form, suited for the next process step.

The second stage involves dehydrogenation (removing hydrogen atoms from the biomass material), such as changing glucose into acetic acid, carboxylation (removing carboxyl groups) of the amino acids, and breaking down the long-chain fatty acids into short-chain acids, again obtaining acetic acid as the final product. These reactions are fermentation reactions accomplished by a range of acidophilic (acid-forming) bacteria. Their optimum performance requires a pH environment in the range of 6–7 (slightly acid), but because the acids already formed will lower the pH of the solution, it is sometimes necessary to adjust the pH, for example, by adding lime.

Finally, the third phase is the production of biogas (a mixture of methane and carbon dioxide) from acetic acid by a second set of fermentation reactions performed by methanogenic bacteria. These bacteria require a strictly anaerobic (oxygen-free) environment. Often, all processes are made to take place in a single container, but separation of the processes into stages will allow greater efficiencies to be reached. The third phase takes of the order of weeks, the preceding phases of the order of hours or days, depending on the nature of the feedstock.

Starting from cellulose, the overall process may be summarised as

$$(C_6H_{10}O_5)_n + nH_2O \rightarrow 3nCO_2 + 3nCH_4 + 19n\,J\,mol^{-1}. \quad (4.1.149)$$

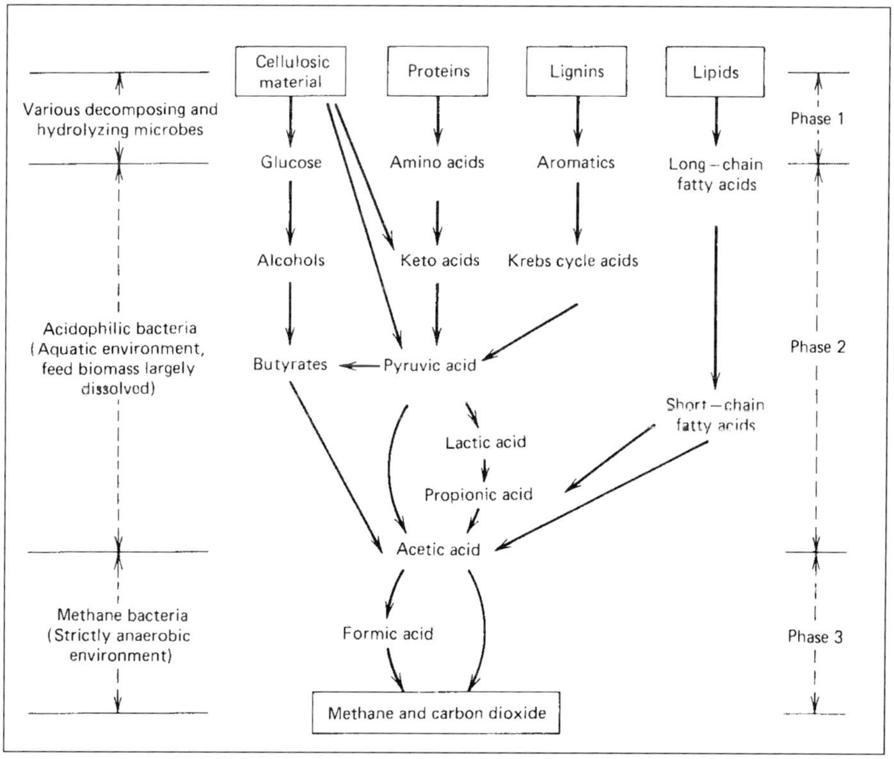

Figure 4.1-116 Simplified description of biochemical pathways in anaerobic digestion of biomass. Used with permission: from D. Stafford, D. Hawkes, and R. Horton, *Methane Production from Waste Organic Matter*. Copyright 1981 by The Chemical Rubber Co., CRC Press, Inc., Boca Raton, FL.

The phase one reactions add up to

$$(C_6H_{10}O_5)_n + nH_2O \rightarrow nC_6H_{12}O_6. \quad (4.1.150)$$

The net result of the phase two reactions is

$$nC_6H_{12}O_6 \rightarrow 3nCH_3COOH, \quad (4.1.151)$$

with intermediate steps such as

$$C_2H_{12}O_6 \rightarrow 2C_2H_5OH + 2CO_2 \quad (4.1.152)$$

followed by dehydrogenation:

$$2C_2H_5OH + CO_2 \rightarrow 2CH_3COOH + CH_4. \quad (4.1.153)$$

The third phase reactions then combine to

$$3nCH_3COOH \rightarrow 3nCO_2 + 3nCH_4. \quad (4.1.154)$$

In order for the digestion to proceed, a number of conditions must be fulfilled. The bacterial action is inhibited by the presence of metal salts, penicillin, soluble sulphides, or ammonia in high concentrations. Some source of nitrogen is essential for the growth of the microorganisms. If there is too little nitrogen relative to the amount of carbon-containing material to be transformed, then bacterial growth will be insufficient and biogas production low. With too much nitrogen (a ratio of carbon to nitrogen atoms below around 15), "ammonia poisoning" of the bacterial cultures may occur. When the carbon–nitrogen ratio exceeds about 30, the gas production starts diminishing, but in some systems carbon–nitrogen values as high as 70 have prevailed without problems. Table 4.1.5 gives carbon–nitrogen values for a number of biomass feedstocks. It is seen that mixing feedstocks can often be advantageous. For instance, straw and sawdust would have to be mixed with some low C:N material, such as livestock urine or clover/lucerne (typical secondary crops that may be grown in temperate climates after the main harvest).

Table 4.1-5 Carbon–nitrogen ratios for various materials (based on Baader *et al.*, 1978; Rubins and Bear, 1942)

Material	Ratio	Material	Ratio
Sewage sludge	13:1	Bagasse	150:1
Cow dung	25:1	Seaweed	80:1
Cow urine	0.8:1	Alfalfa hay	18:1
Pig droppings	20:1	Grass clippings	12:1
Pig urine	6:1	Potato tops	25:1
Chicken manure	25:1	Silage liquor	11:1
Kitchen refuse	6–10:1	Slaughterhouse wastes	3–4:1
Sawdust	200–500:1	Clover	2.7:1
Straw	60–200:1	Lucerne	2:1

If digestion time is not a problem, almost any cellulosic material can be converted to biogas, even pure straw. One initially may have to wait for several months, until the optimum bacterial composition has been reached, but then continued production can take place, and despite low reaction rates an energy recovery similar to that of manure can be achieved with properly prolonged reaction times.

Average manure production for fully bred cows and pigs (in Europe, Australia, and the Americas) is 40 and 2.3 kg wet weight d^{-1}, corresponding to 62 and 6.2 MJ d^{-1}, respectively. The equivalent biogas production may reach 1.2 and 0.18 $m^3\,d^{-1}$. This amounts to 26 and 3.8 MJ d^{-1}, or 42 and 61% conversion efficiency, respectively. A discussion of the overall efficiency including transportation of biomass to the plant is given in Berglund and Börjesson, finding maximum acceptable transport distances of 100–150 km.

The residue from the anaerobic digestion process has a higher value as a fertiliser than the feedstock. Insoluble organics in the original material are, to a large extent, made soluble, and nitrogen is fixed in the microorganisms.

Pathogen populations in the sludge are reduced. Stafford *et al.* (1981) found a 65–90% removal of *Salmonella* during anaerobic fermentation, and there is a significant reduction in streptococci, coliforms, and viruses, as well as an almost total elimination of disease-transmitting parasites such as *Ascaris*, hookworm, *Entamoeba*, and *Schistosoma*.

For this reason, anaerobic fermentation has been used fairly extensively as a cleaning operation in city sewage treatment plants, either directly on the sludge or after growing algae on the sludge to increase fermentation potential. Most newer plants make use of the biogas produced to fuel other parts of the treatment process, but with proper management, sewage plants may well be net energy producers.

The other long-time experience with biogas and associated fertiliser production is in rural areas of a number of Asian countries, notably China and India. The raw materials here are mostly cow dung, pig slurry, and what in India is referred to as human night soil, plus in some cases grass and straw. The biogas is used for cooking, and the fertiliser residue is returned to the fields. The sanitary aspect of pathogen reduction lends strong support to the economic viability of these schemes.

The rural systems are usually based on simple one-compartment digesters with human labour for filling and emptying of material. Operation is either in batches or with continuous new feed and removal of some 3–7% of the reactor content every day. Semi-industrialised plants have also been built during the last decade, for example, in connection with large pig-raising farms, where mechanised and highly automated collection of manure has been feasible. In some cases, these installations have utilised separate acid and methanation tanks.

Figure 4.1-117 shows an example of a town-scale digester plant, where the biogas is used in a combined electric power and district heat generating plant. Expected energy flows are indicated. Storage of biogas for the rural cooking systems is accomplished by variable-volume domes collecting the gas as it is produced (e.g., an inverted, water-locked can placed over the digester core). Biogas contains approximately 23 MJ m^{-3} and is therefore a medium-quality gas. CO_2 removal is necessary in order to achieve pipeline quality. An inexpensive way of achieving over 90% CO_2 removal is by water spraying. This way of producing compressed methane gas from biogas allows for automotive applications, such as the farmer producing tractor fuel on site. In New Zealand such uses have been developed since 1980 (see Fig. 4.1-118). Several demonstration experiences have recently been obtained in Sweden.

Storage of a certain amount of methane at ambient pressure requires over a thousand times more volume than the equivalent storage of oil. However, actual methane storage at industrial facilities uses pressures of about 140 times ambient, so the volume penalty relative to oil storage would then be a factor of 9. Storage of methane in zeolitic material for later use in vehicles has been considered.

If residues are recycled, little environmental impact can be expected from anaerobic digestion. The net impact on agriculture may be positive, owing to nutrients being made more accessible and due to parasite depression. Undesirable contaminants, such as heavy metals, are returned to the soil in approximately the same concentrations as they existed before collection, unless urban pollution has access to the feedstock. The very fact that digestion depends on biological organisms may imply that warning signals in terms of poor digester performance may direct early attention to pollution of crop land or urban sewage systems. In any case, pollutant-containing waste, for example, from industry, should never be mixed with the energy-valuable biological material in urban refuse and sewage. The methane-forming bacteria are more sensitive to changes in environment, such as temperature and acidity, than the acid-forming ones.

The digestion process itself does not emit pollutants if it operates correctly, but gas cleaning, such as H_2S removal, may lead to emissions. The methane gas itself shares many of the accident hazards of other gaseous fuels, being asphyxiating and explosive at certain concentrations in air (roughly 7–14% by volume). For rural cooking applications, the impacts may be compared with those of the fuels being replaced by biogas, notably wood burned in simple stoves. In these cases, from the discussion in section 4.1.8.1, the environment is dramatically improved by introducing biogas digesters.

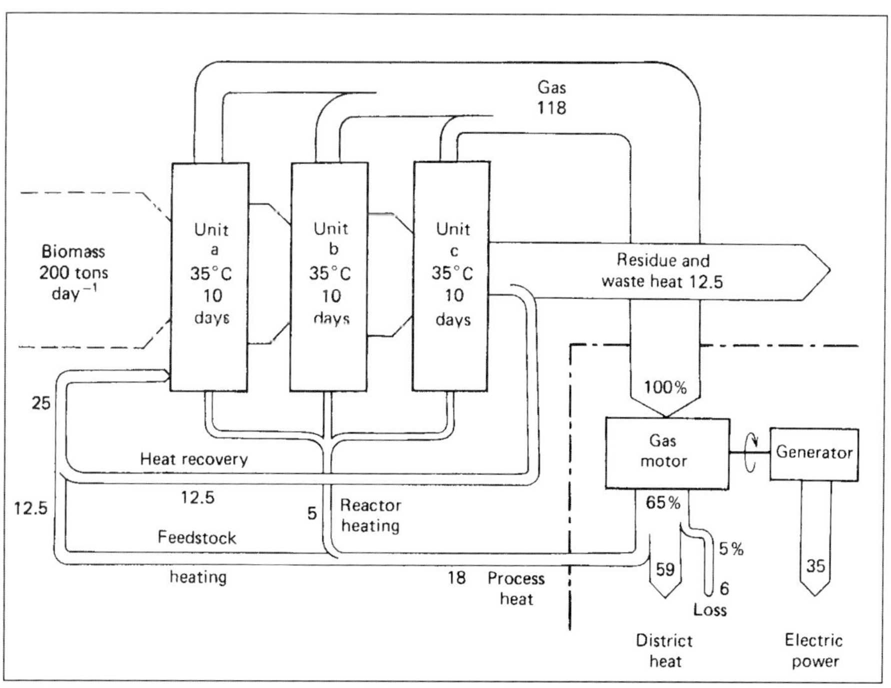

Figure 4.1-117 Calculated energy flows for a town biogas reactor plant, consisting of three successive units with 10 days' residence time in each. A biogas motor drives an electric generator, and the associated heat is, in part, recycled to the digestion process, while the rest is fed into the town district heating lines. Flows (all numbers without indicated unit) are in GJ d^{-1}.

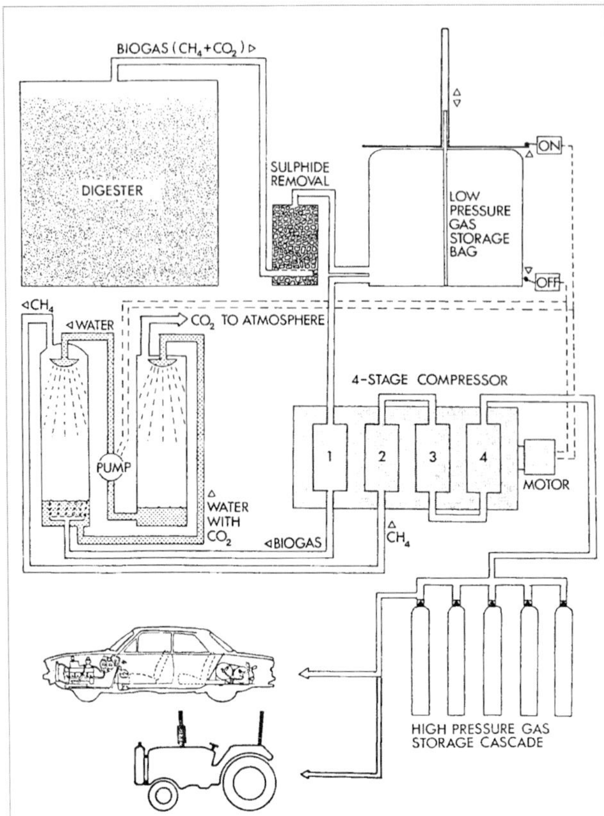

Figure 4.1-118 Schematic view of New Zealand scheme for methane production and vehicle use (from Stewart, D. and McLeod, R., New Zealand Journal of Agriculture, Sept. 1980, 924. With permission).

An example of early biogas plants for use on a village scale in China, India and Pakistan is shown in Fig. 4.1-119. All the reactions take place in one compartment, which does not necessarily lead to optimum conversion efficiency. The time required for the acid-forming step is less than 24 h, whereas the duration of the second step should be 10–30 days. The heating of the fermentation tank required for application in many climatic regions may be derived from solar collectors, which could form the top cover of the tank. Alternatively, the top cover may be an inflatable dome serving as a store of gas, which can smooth cut a certain degree of load variation. Some installations obtain the highest efficiency by batch operation, i.e. by leaving one batch of biological material in the tank for the entire fermentation period. The one shown in Fig. 4.1-119 allows continuous operation, i.e. a fraction of the slurry is removed every day and replaced by fresh biological material.

Examples of predicted biogas production rates, for simple plants of the type shown in Fig. 4.1-119 and based on fluid manure from dairy cows, pigs or poultry, are shown in Table 4.1-6 and Fig. 4.1-120. Table 4.1-6 gives typical biogas production rates, per day and per animal, while Fig. 4.1-120 gives the conversion efficiencies measured, as functions of fermentation time (tank residence time), in a controlled experiment. The conversion efficiency is the ratio of the energy in the biogas (approximately 23 MJ m^{-3} of gas) and the energy in the manure which would be released as heat by complete burning of the dry matter (some typical absolute values are given in Table 4.1.6). The highest efficiency is obtained with pigs' slurry, but the high bacteriological activity in this case occasionally has the side-effect of favouring bacteria other than those helping to produce biogas, e.g. ammonia-producing bacteria, the activity of which may destroy the possibility of further biogas production.

As mentioned, the manure residue from biogas plants has a high value as fertiliser because the decomposition of organic material followed by hydrocarbon conversion leaves plant nutrients (e.g. nitrogen that was originally bound in proteins) in a form suitable for uptake. Malignant bacteria and parasites are not removed to as high a degree as by composting, owing to the lower process temperature.

Some city sewage plants produce biogas (by anaerobic fermentation of the sewage) as one step in the cleaning procedure, using the biogas as fuel for driving mechanical cleaning devices, etc. In this way it is in many cases possible to avoid any need for other energy inputs and in

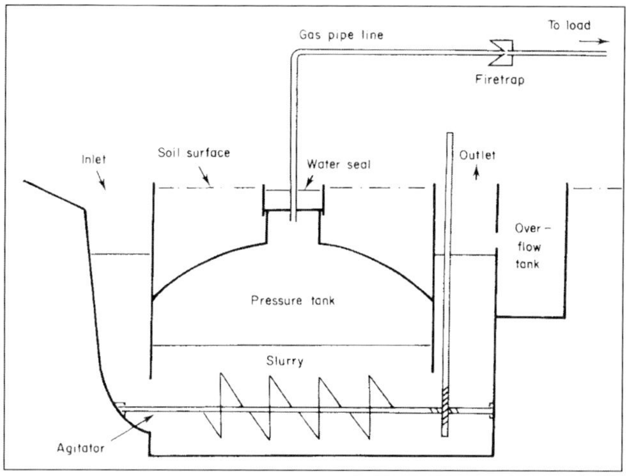

Figure 4.1-119 Single-chamber biogas plant.

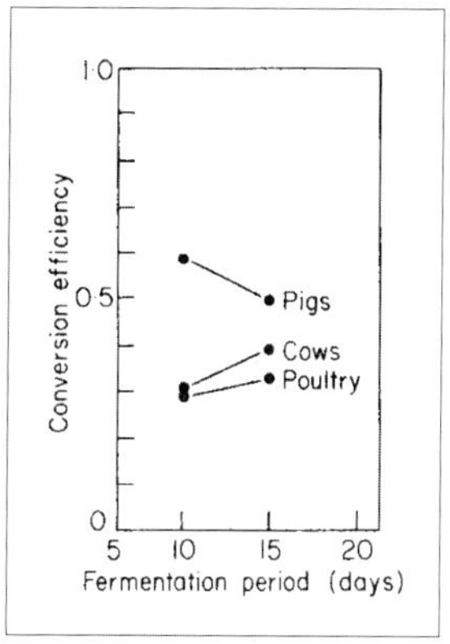

Figure 4.1-120 Measured conversion efficiencies (ratio between energy in biogas gas produced and energy in the manure) for simple biogas plants (like Fig. 4.1-119). Some 10–13 kg of fresh manure was added per day and per m^3 of tank volume.

some cases to become a net energy producer. Figure 4.1-121 shows the system layout for a 300 t of biomass per day biogas plant accepting multiple types of feedstock and capable of delivering both power, process and district heat, and fertiliser. Figure 4.1-122 gives the measured performance data for nine large prototype biogas plants in Denmark.

Energy balance

The 1992 average production of the large-size Danish biogas plants was 35.1 m^3 per m^3 of biomass (at standard pressure, methane content in the biogas being on average 64%), or 806 MJ m^{-3}. In-plant energy use amounted to 90 MJ m^{-3} distributed on 28 MJ electricity and 50 MJ heat, all produced by the biogas plant itself. Fuel used in transporting manure to the plant totalled 35 MJ, and the fertiliser value of the returned residue was estimated at 30 MJ. Thus, the net outside energy requirement is 5 MJ for a production of 716 MJ, or 0.7%, corresponding to an energy payback time of 3 days. If the in-plant biogas use is added, the energy consumption in the process is 13%. To this should be added the energy for construction of the plant, which has not been estimated. However, the best plants roughly break even economically, indicating that the overall energy balance is acceptable.

Greenhouse gas emissions

Using, as in the energy balance section above, the average of large Danish plants as an example, the avoided CO_2 emission from having the combined power and heat production use biomass instead of coal as a fuel is 68 kg per m^3 of biomass converted. Added should be emissions from transportation of biomass, estimated at 3 kg, and the avoided emissions from conventional production of fertiliser replaced by biogas residue,

Table 4.1-6 Manure and potential biogas production for a typical animal per day (based on Taiganides, 1974)

Source	Manure per day	Biogas per day		
	kg wet weight	MJ	m^3	MJ
Cows	40	62	1.2	26
Pigs	2.3	6.2	0.18	3.8
Hens	0.19	0.9	0.011	0.26

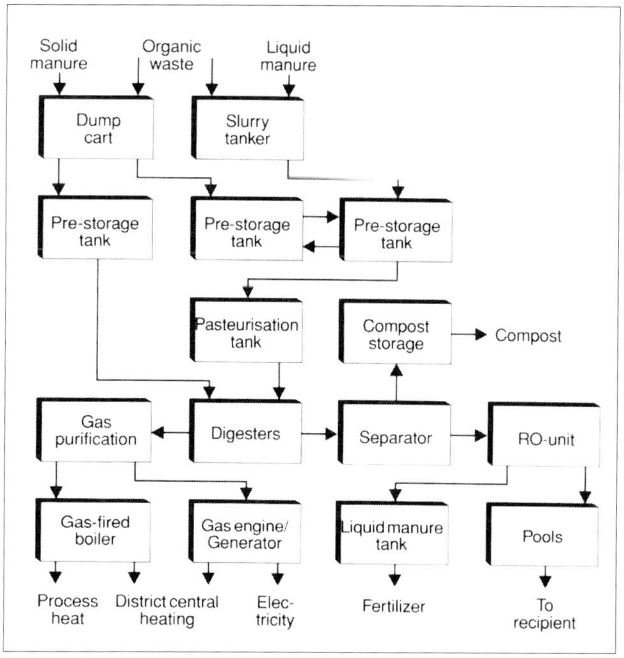

Figure 4.1-121 Layout of Lintrup biogas plant (Danish Energy Agency, 1992).

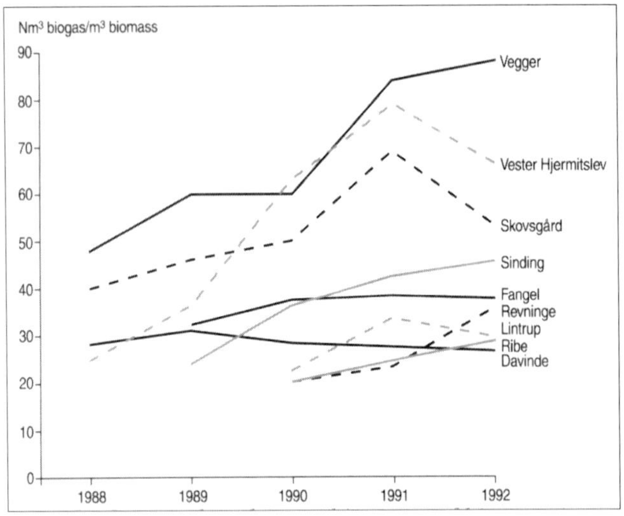

Figure 4.1-122 Annual average production efficiency (m^3 of biogas at standard pressure, denoted Nm^3, per m^3 of biomass feedstock) for nine community-size biogas plants in Denmark.

estimated at 3 kg. Reduced methane emissions, relative to the case of spreading manure directly on the fields, is of the order of 61 kg CO_2 equivalent. As regards nitrous oxide, there is a possible gain by avoiding denitrification in the soil, but high uncertainty has made an actual estimate fortuitous at the present. The overall CO_2 reduction obtained by summing up the estimates given above is then 129 kg for each m^3 of biomass converted to biogas.

Other environmental effects

Compared with the current mix of coal, oil or natural gas plants, biogas plants have a 2–3 times lower SO_2 emission but a correspondingly higher NO_x emission. Higher ammonia content in the digested residue calls for greater care in using fertiliser from biogas plants, in order to avoid loss of ammonia. This is also true as regards avoiding loss of nutrients from fertiliser to waterways. Compared to spreading manure not refined by the biogas production process, a marked gain in fertiliser quality has been noted, including a much better-defined composition, which will contribute to assisting correct dosage and avoiding losses to the environment. The dissemination of biogas plants removes the need for landfills, which is seen as an environmental improvement. Odour is moved from the fields (where manure and slurry would otherwise be spread) to the biogas plant, where it can be controlled by suitable measures (filters, etc.).

Hydrogen-producing cultures

Biochemical routes to fuel production include a number of schemes not presently developed to a technical or economic level of commercial interest. Hydrogen is a fuel that may be produced directly by biological systems. Hydrogen enters in the process of photosynthesis, as it proceeds in green plants, where the net result of the process is

$$2H_2O + \text{solar radiation} \rightarrow 4e^- + 4H^+ + O_2.$$

However, the electrons and protons do not combine directly to form hydrogen,

$$4e^- + 4H^+ \rightarrow 2H_2,$$

but instead transfer their energy to a more complex molecule ($NADPH_2$; cf. Chapter 3.3), which is capable of driving the CO_2 assimilation process. By this mechanism, the plants avoid recombination of oxygen and hydrogen from the two processes mentioned above. Membrane systems keep the would-be reactants apart, and thus the energy-rich compound may be transported to other areas of the plant, where it takes part in plant growth and respiration.

Much thought has been given to modifications of plant material (e.g. by genetic engineering), in such a way that free hydrogen is produced on one side of a membrane and free oxygen on the other side.

While dissociation of water (by light) into hydrogen and oxygen (photolysis) does not require a biological system, it is possible that utilisation of the process on a realistic scale can be more easily achieved if the critical components of the system, notably the membrane and the electron transport system, are of biological origin. Still, a breakthrough is required before any thought can be given to practical application of direct biological hydrogen production cultures.

Thermochemical gasification of biomass

Conversion of fossil biomass such as coal into a gas is considered a way of reducing the negative environmental impacts of coal utilisation. However, in some cases the impacts have only been moved but not eliminated. Consider, for example, a coal-fired power plant with 99% particle removal from flue gases. If it were to be replaced by a synthetic gas-fired power plant with gas produced from coal, then sulphur could be removed at the gasification plant using dolomite-based scrubbers. This would practically eliminate the sulphur oxide emissions, but on the other hand, dust emissions from the dolomite processing would represent particle emissions twice as large as those avoided at the power plant by using gas instead of coal. Of course, the dust is emitted at a different location.

This example, as well as the health impacts associated with coal mining, whether on the surface or in mines (although not identical), has sparked interest in methods of gasifying coal *in situ*. Two or more holes are drilled. Oxygen (or air) is injected through one, and a mixture of gases, including hydrogen and carbon oxides, emerges at the other hole. The establishment of proper communication between holes, and suitable underground contact surfaces, has proved difficult, and recovery rates are modest.

The processes involved would include

$$2C + O_2 \rightarrow 2CO, \quad (4.1.155)$$

$$CO + H_2O \leftrightarrow CO_2 + H_2. \quad (4.1.156)$$

The stoichiometric relation between CO and H_2 can then be adjusted using the shift reaction (4.1.156), which may proceed in both directions, depending on steam temperature and catalysts. This opens the way for methane synthesis through the reaction

$$CO + 3H_2 \rightarrow CH_4 + H_2O. \quad (4.1.157)$$

At present, the emphasis is on improving gasifiers using coal already extracted. Traditional methods include the Lurgi fixed-bed gasifier (providing gas under pressure from non-caking coal at a conversion efficiency as low as 55%) and the Koppers–Totzek gasifier (oxygen input, the produced gas un-pressurised, also of low efficiency).

Improved process schemes include the hy-gas process, requiring a hydrogen input; the bi-gas concept of brute force gasification at extremely high temperatures; and the slagging Lurgi process, capable of handling powdered coal.

Promising, but still at an early stage of development, is catalytic gasification (e.g. potassium catalyst), where all processes take place at a common, relatively low temperature, so that they can be combined in a single reactor (Fig. 4.1-123). The primary reaction here is

$$C + H_2O \rightarrow H_2 + CO \quad (4.1.158)$$

(H_2O being in the form of steam above 550°C), to be followed by (4.1.156) and (4.1.157). The catalyst allows all processes to take place at 700°C. Without catalyst, the gasification would have to take place at 925°C and the shift reaction and methanation at 425°C, that is, in a separate reactor where excess hydrogen or carbon monoxide would be lost.

In a coal gasification scheme storage would be (of coal) before conversion. Peat can be gasified in much the same way as coal, as can wood (with optional subsequent methanol conversion as described in section 4.1.8.3).

Fresh biomass gasification

Gasification of biomass, and particularly wood and other lignin-containing cellulosic material, has a long history. The processes may be viewed as "combustion-like" conversion, but with less oxygen available than needed for burning. The ratio of oxygen available and the amount of oxygen that would allow complete burning is called the "equivalence ratio". For equivalence ratios below 0.1 the process is called "pyrolysis", and only a modest fraction of the biomass energy is found in the gaseous product – the rest being in char and oily residues. If the equivalence ratio is between 0.2 and 0.4, the process is called a proper

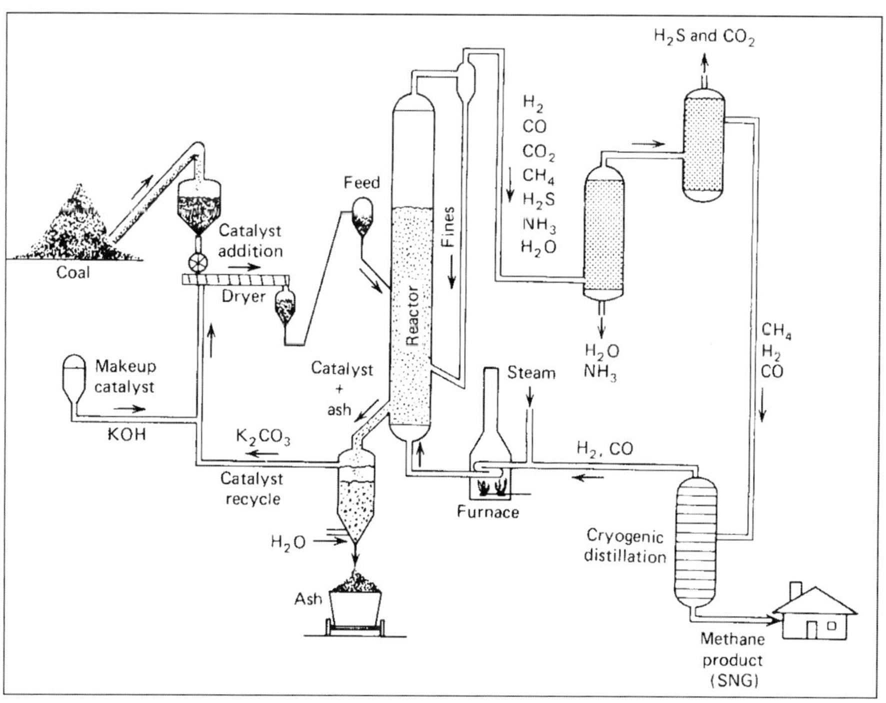

Figure 4.1-123 Schematic diagram of catalytic gasification process (SNG is synthetic natural gas). (From Hirsch *et al.*, 1982. Reprinted from *Science*, **215**, 121–127, 8 January 1982, with permission. Copyright 1982 American Association for the Advancement of Science.)

"gasification". This is the region of maximum energy transfer to the gas.

The chemical processes involved in biomass gasification are similar to the reactions (4.1.155)–(4.1.158) for coal gasification. Table 4.1-7 lists a number of reactions involving polysaccharidic material, including pyrolysis and gasification. In addition to the chemical reaction formulae, the table gives enthalpy changes for idealised reactions (i.e. neglecting the heat required to bring the reactants to the appropriate reaction temperature). Figure 4.1-124 gives the energy of the final products, gas and char, as a function of the equivalence ratio, still based on an idealised thermodynamical calculation. The specific heat of the material is 3 kJ g^{-1} of wood at the peak of energy in the gas, increasing to 21 kJ g^{-1} of wood for combustion at equivalence ratio equal to unity. Much of this sensible heat can be recovered from the gas, so that process heat inputs for gasification can be kept low.

Figure 4.1-125 gives the equilibrium composition calculated as a function of the equivalence ratio. By equilibrium composition is understood the composition of reaction products occurring after the reaction rates and reaction temperature have stabilised adiabatically. The actual processes are not necessarily adiabatic; in particular the low-temperature pyrolysis reactions are not. Still, the theoretical evaluations assuming equilibrium conditions serve as a useful guideline for evaluating the performance of actual gasifiers.

The idealised energy change calculation of Table 4.1-7 assumes a cellulosic composition such as the one given in (4.1.150). For wood, the average ratios of carbon, hydrogen and oxygen are 1:1.4:0.6.

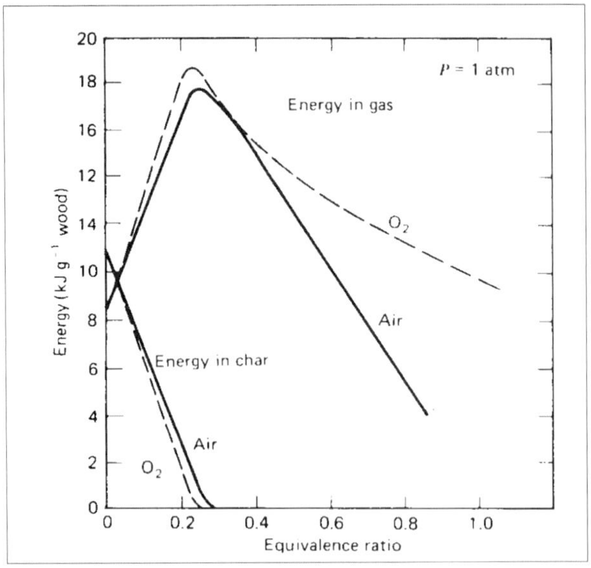

Figure 4.1-124 Calculated energy content in gas and char produced by equilibrium processes between air (or oxygen) and biomass, as a function of equivalence ratio. (From Reed, 1981. Reprinted from *Biomass Gasification* (T. Reed, ed.), with permission. Copyright 1981, Noyes Data Corporation, Park Ridge, NJ).

Figure 4.1-126 shows three examples of wood gasifiers: the updraft, the downdraft, and the fluidised bed types. The drawback of the updraft type is a high rate of oil, tar, and corrosive chemical formation in the pyrolysis zone. This problem is solved by the downdraft version, where such oils and other matter pass through a hot charcoal bed in the lower zone of the reactor and become cracked to simpler gases or char. The fluidised bed reactor may prove superior for large-scale operations,

Table 4.1-7 Energy change for idealised cellulose thermal conversion reactions. (Source: T. Reed (1981), in *Biomass Gasification* (T. Reed, ed.), reproduced with permission. Copyright 1981, Noyes Data Corporation, Park Ridge, NJ)

Chemical reaction	Energy consumed $(kJ\ g^{-1})$[a]	Products/process
$C_6H_{10}O_5 \rightarrow 6C + 5H_2 + 2.5\ O_2$	5.94[b]	Elements, dissociation
$C_6H_{10}O_5 \rightarrow 6C + 5H_2O(g)$	−2.86	Charcoal, charring
$C_6H_{10}O_5 \rightarrow 0.8\ C_6H_8O + 1.8\ H_2O(g) + 1.2\ CO_2$	−2.07[c]	Oily residues, pyrolysis
$C_6H_{10}O_5 \rightarrow 2C_2H_4 + 2CO_2 + H_2O(g)$	0.16	Ethylene, fast pyrolysis
$C_6H_{10}O_5 + \frac{1}{2}O_2 \rightarrow 6CO + 5H_2$	1.85	Synthesis gas, gasification
$C_6H_{10}O_5 + 6H_2 \rightarrow 6\text{“}CH_2\text{”} + 5H_2O(g)$	−4.86[d]	Hydrocarbons – generation
$C_6H_{10}O_5 + 6O_2 \rightarrow 6CO_2 + 5H_2O(g)$	−17.48	Heat, combustion

[a]Specific reaction heat.
[b]The negative of the conventional heat of formation calculated for cellulose from the heat of combustion of starch.
[c]Calculated from the data for the idealised pyrolysis oil C_6H_8O ($\Delta Hc = -745.9$ kcal mol^{-1}, $\Delta H_f = 149.6$ kcal g^{-1}, where H_c = heat of combustion and H_f = heat of fusion).
[d]Calculated for an idealised hydrocarbon with ΔH_c as above. H_2 is consumed.

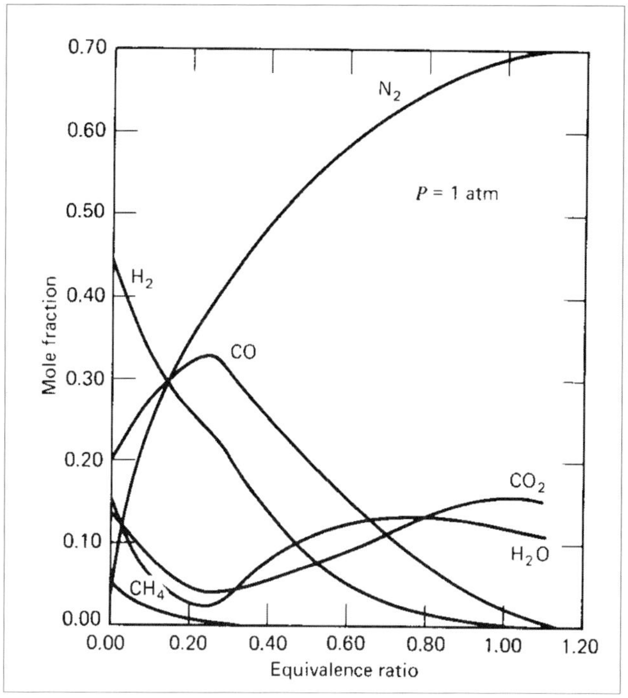

Figure 4.1-125 Calculated gas composition resulting from equilibrium processes between air and biomass, as a function of equivalence ratio. (From Reed, 1981. Reprinted from *Biomass Gasification* (T. Reed, ed.), with permission. Copyright 1981, Noyes Data Corporation, Park Ridge, NJ).

because passage time is smaller. The drawback of this is that ash and tars are carried along with the gas and have to be removed later in cyclones and scrubbers. Several variations on these gasifier types have been suggested.

The gas produced by gasification of biomass is a "medium-quality gas", meaning a gas with burning value in the range 10–18 MJ m^{-3}. This gas may be used directly in Otto or diesel engines, it may be used to drive heat pump compressors, or alternatively, it may by upgraded to pipeline-quality gas (about 30 MJ m^{-3}) or converted to methanol, as discussed in section 4.1.8.3.

Environmental impacts derive from biomass production, collection (e.g. forestry work) and transport to gasification site, from the gasification and related processes, and finally from the use made of the gas. The gasification residues – ash, char, liquid waste water, and tar– have to be disposed of. Char may be recycled to the gasifier, while ash and tars could conceivably be utilised in the road or building construction industry. The alternative of landfill disposal would represent a recognised impact. Investigations of emissions from combustion of producer gas indicate low emissions of nitrous oxides and hydrocarbons, as compared with emissions from combustion of natural gas. In one case, carbon monoxide emissions were found to be higher than for natural gas burning, but it is believed that this can be rectified as more experience in adjusting air-to-fuel ratios is gained.

4.1.8.3 Fuel production from biomass: generation of liquid biofuels

Anaerobic fermentation processes may also be used to produce liquid fuels from biological raw materials. An example is the ethanol production (4.1.152) from glucose, known as standard yeast fermentation in the beer, wine and liquor industries. It has to take place in steps, such that the ethanol is removed (by distillation or dehydrator application) whenever its concentration approaches a value (around 12%) which would impede reproduction of the yeast culture.

In order to reduce the primary biological material (e.g. molasses, cellulose pulp or citrus fruit wastes) to glucose, the hydrolysis process (4.1.154) may be used. Some decomposition takes place in any acid solution, but in order to obtain complete hydrolysis, specific enzymes must usually be provided, either directly or by adding

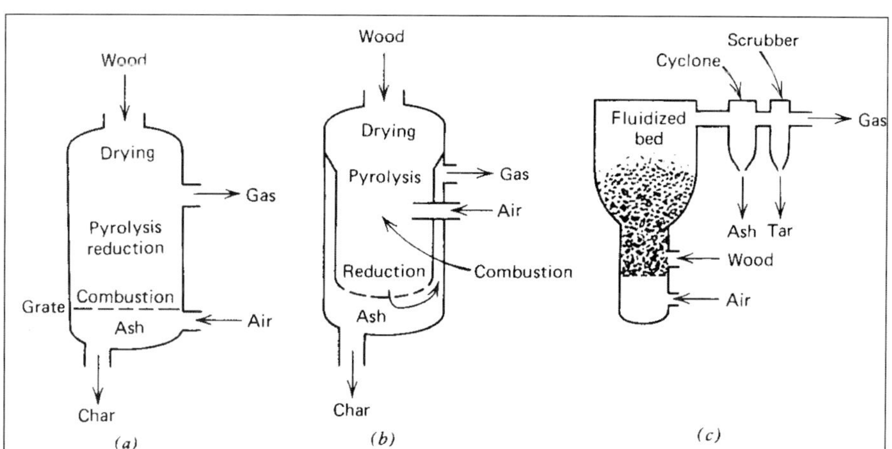

Figure 4.1-126 Gasifier types: (a) updraft, (b) downdraft, and (c) fluidised bed.

microorganisms capable of forming such enzymes. The yeast fungi themselves contain enzymes capable of decomposing polysaccharides into glucose.

The theoretical maximum efficiency of glucose-to-ethanol conversion (described in more detail below) is 97%, and, according to the Brazilian alcohol industry in 1974 obtained 14% of the energy in the raw sugar input, in the form of ethanol produced by fermentation of just the molasses residues from sugar refining, i.e. in addition to the crystallised sugar produced. A current figure is 25% (see Fig. 4.1-128) for an optimised plant design.

Mechanical energy input, e.g. for stirring, could be covered by the fermentation wastes if they were burned in a steam power plant. In the European example, these inputs amount to about a third of the energy inputs through the sugar itself.

Alternative fermentation processes based on molasses or other sugar-containing materials produce acetone–butanol, acetone–ethanol or butanol–isopropanol mixtures, when the proper bacteria are added. In addition, carbon dioxide and small amounts of hydrogen are formed.

Conversion of fossilised biomass into liquid fuels was briefly mentioned at the beginning of section 4.1.8.2, in conjunction with the overview Fig. 4.1-99 showing the conversion routes open for biofuel generation. Among the nonfood energy uses of biomass, there are several options leading to liquid fuels which may serve as a substitute for oil products. The survey of conversion processes given in Fig. 4.1-115 indicates liquid end products as the result of either biochemical conversion using fermentation bacteria or a thermochemical conversion process involving gasification and, for example, further methanol synthesis. The processes, which convert biomass into liquid fuels that are easy to store, are discussed below, but first the possibility of direct production of liquid fuels by photosynthesis is presented.

Direct photosynthetic production of hydrocarbons

Oil from the seeds of many plants, such as rape, olive, groundnut, corn, palm, soy bean, and sunflower, is used as food or in the preparation of food. Many of these oils will burn readily in diesel engines and can be used directly or mixed with diesel oil of fossil origin, as they are indeed in several pilot projects around the world.

However, in most of these cases the oil does not constitute a major fraction of the total harvest yield, and any expansion of these crops to provide an excess of oil for fuel use would interfere with food production. A possible exception is palm oil, because inter-cropping of palm trees with other food crops may provide advantages such as retaining moisture and reducing wind erosion.

Much interest is therefore concentrated on plants that yield hydrocarbons and that, at the same time, are capable of growing on land unsuited for food crops. Calvin (1977) first identified the *Euphorbia* family as an interesting possibility. The rubber tree, *Hevea brasiliensis*, is of this family, and its rubber is a hydrocarbon–water emulsion, the hydrocarbon of which (polyisoprenes) has a large molecular weight, about a million, making it an elastomer. However, other plants of the genus *Euphorbia* yield latex of much lower molecular weight which could be refined in much the same way as crude oil. In the tropical rainforests of Brazil, Calvin found a tree, *Cobaifera langsdorfii*, a specimen of which annually yields some 30 litres of virtually pure diesel fuel. Still, the interest centres on species that are likely to grow in arid areas such as the deserts of the southern United States, Mexico, Australia, and so on.

Possibilities include *Euphorbia lathyris* (gopher plant), *Simmondsia chinensis* (jojoba), *Cucurdia foetidissima* (buffalo gourd) and *Parthenium argentatum* (guayule). The gopher plant has about 50% sterols (on a dry weight basis) in its latex, 3% polyisoprene (rubber), and a number of terpenes. The sterols are suited as feedstocks for replacing petroleum in chemical applications. Yields of first-generation plantation experiments in California are 15–25 barrels of crude oil equivalent or some 144 GJ ha^{-1} (i.e. per 10^4 m^2). In the case of *Hevea*, genetic and agronomic optimisation has increased yields by a factor of 2000 relative to those of wild plants, so quite high hydrocarbon production rates should be achievable after proper development.

Alcohol fermentation

The ability of yeast and bacteria such as *Zymomonas mobilis* to ferment sugar-containing material to form alcohol is well known from beer, wine, and liquor manufacture. If the initial material is cane sugar, the fermentation reaction may be summarised as

$$C_6H_{12}O_6 \rightarrow 2C_2H_5OH + 2CO_2. \quad (4.1.159)$$

The energy content of ethanol is 30 MJ kg^{-1}, and its octane rating is 89–100. With alternative fermentation bacteria, the sugar may be converted into butanol, $C_2H_5(CH_2)_2OH$. In Brazil, the cost of ethanol is finally coming down to that of gasoline.

In most sugar-containing plant material, the glucose molecules exist in polymerised form such as starch or cellulose, of the general structure $(C_6H_{10}O_5)_n$. Starch or hemicellulose is degraded to glucose by hydrolysis (cf. Fig. 4.1-99), while lignocellulose resists degradation owing to its lignin content. Lignin glues the cellulosic material together to keep its structure rigid, whether it be crystalline or amorphous. Wood has a high lignin content (about 25%), and straw also has considerable amounts of lignin (13%), while potato or beet starch contain very little lignin.

Some of the lignin seals may be broken by pre-treatment, ranging from mechanical crushing to the introduction of swelling agents causing rupture.

The hydrolysis process is given by (4.1.150). In earlier times, hydrolysis was always achieved by adding an acid to the cellulosic material. During both world wars, Germany produced ethanol from cellulosic material by acid hydrolysis, but at very high cost. Acid recycling is incomplete; with low acid concentration the lignocelluloses is not degraded, and with high acid concentration the sugar already formed from hemicellulose is destroyed.

Consequently, alternative methods of hydrolysis have been developed, based on enzymatic intervention. Bacterial (e.g. of the genus *Trichoderma*) or fungal (such as *Sporotrichum pulverulentum*) enzymes have proved capable of converting cellulosic material, at near ambient temperatures, to some 80% glucose and a remainder of cellodextrins (which could eventually be fermented, but in a separate step with fermentation microorganisms other than those responsible for the glucose fermentation).

The residue left behind after the fermentation process (4.1.159) can be washed and dried to give a solid product suitable as fertiliser or as animal feed. The composition depends on the original material, in particular with respect to lignin content (small for residues of molasses, beets, etc., high for straws and woody material, but with fibre structure broken as a result of the processes described above). If the lignin content is high, direct combustion of the residue is feasible, and it is often used to furnish process heat to the final distillation.

The outcome of the fermentation process is a water–ethanol mixture. When the alcohol fraction exceeds about 10%, the fermentation process slows down and finally halts. Therefore, an essential step in obtaining fuel alcohol is to separate the ethanol from the water. Usually, this is done by distillation, a step that may make the overall energy balance of the ethanol production negative. The sum of agricultural energy inputs (fertiliser, vehicles, machinery) and all process inputs (cutting, crushing, pre-treatment, enzyme recycling, heating for different process steps from hydrolysis to distillation), as well as energy for transport, is, in existing operations such as those of the Brazilian alcohol programme, around 1.5 times the energy outputs (alcohol and fertiliser if it is utilised). However, if the inputs are domestic fuels, for example, combustion of residues from agriculture, and if the alcohol produced is used to displace imported oil products, the balance might still be quite acceptable from a national economic point of view.

If, further, the lignin-containing materials of the process are recovered and used for process heat generation (e.g. for distillation), then such energy should be counted not only as input but also as output, making the total input and output energy roughly balance. Furthermore, more sophisticated process design, with cascading heat usage and parallel distillation columns operating with a time displacement such that heat can be reused from column to column, could reduce the overall energy inputs to 55–65% of the outputs.

Radically improved energy balances would emerge if distillation could be replaced by a less energy intensive separation method. Several such methods for separating water and ethanol have been demonstrated on a laboratory scale, including: drying with desiccants such as calcium hydroxide, cellulose, or starch; gas chromatography using rayon to retard water, while organic vapours pass through; solvent extraction using di-butyl phthalate, a water-immiscible solvent of alcohols; and passage through semi-permeable membranes or selective zeolite absorbers and phase separation. The use of dry cellulose or starch appears particularly attractive, because over 99% pure alcohol can be obtained with less than 10% energy input, relative to the combustion energy of the alcohol. Furthermore, the cellulosic material may be cost-free, if it can be taken from the input stream to the fermentation unit and returned to it after having absorbed water (the fermentation reaction being "wet" anyway). The energy input of this scheme is for an initial distillation, bringing the ethanol fraction of the aqueous mixture from the initial 5–12% up to about 75%, at which point the desiccation process is started. As can be seen from Fig. 4.1-127, the distillation energy is modest up to an alcohol content of 90% and then starts to rise rapidly. The drying process thus substitutes for the most energy-expensive part of the distillation process.

The ethanol fuel can be stored and used in the transportation sector in much the same way as gasoline. It can be mixed with gasoline or can fully replace gasoline in spark ignition engines with high compression ratios (around 11). The knock resistance and high octane number of ethanol make this possible, and with pre-heating of the alcohol (using combustion heat that is recycled), the conversion efficiency can be improved. Several countries presently use alcohol–gasoline blends with up to 10% ethanol. This does not require any engine modification. Altering the gasoline Otto engines may be inconvenient in a transition period, but if alcohol distribution networks are implemented and existing gas stations modified, then the car engines could be optimised for alcohol fuels without regard to present requirements. A possible alternative to spark ignition engines is compression ignition engines, where auto-ignition of the fuel under high compression (a ratio of 25) replaces spark or glow plug ignition. With additives or chemical transformation into acetal, alcohol fuels could be used in this way. Ethanol does not blend with diesel oil, so mixtures

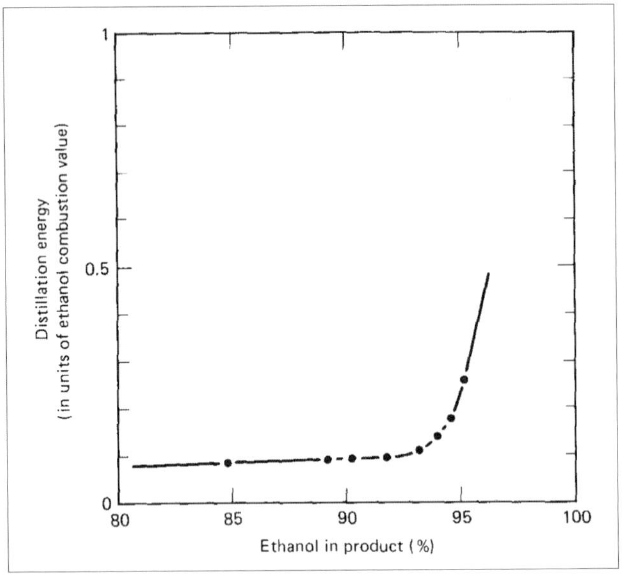

Figure 4.1-127 Distillation energy for ethanol–water mixture, as a function of ethanol content (assumed initial ethanol fraction 12%).

do require the use of special emulsifiers. However, diesel oil can be mixed with other biofuels without problems, e.g. the plant oils (rapeseed oil, etc.) presently in use in Germany.

A number of concerns with regard to the environmental impacts of the ethanol fermentation energy conversion chain must be considered. First of all, the biomass being used may have direct uses as food or may be grown in competition with production of food. The reason is, of course, that the easiest ethanol fermentation is obtained by starting with a raw material with as high a content of elementary sugar as possible, that is, starting with sugar cane or cereal grain. Since sugar cane is likely to occupy prime agricultural land, and cereal production must increase with increasing world population, neither of these biomass resources should be used as fermentation inputs. However, residues from cereal production and from necessary sugar production (present sugar consumption is in many regions of the world too high from a health and nutrition point of view) could be used for ethanol fermentation, together with urban refuse, extra crops on otherwise committed land, perhaps aquatic crops and forest renewable resources. The remarks made in Chapter 3.3 about proper soil management, recycling nutrients, and covering topsoil to prevent erosion are very appropriate in connection with the enhanced tillage utilisation that characterises combined food and ethanol production schemes.

The hydrolysis process involves several potential environmental impacts. If acids are used, corrosion and accidents may occur, and substantial amounts of water would be required to clean the residues for re-use. Most acid would be recovered, but some would follow the sewage stream. Enzymatic hydrolysis would seem less cumbersome. Most of the enzymes would be recycled, but some might escape with waste water or residues. Efforts should be made to ensure that they are made inactive before any release. This is particularly important when, as envisaged, the fermentation residues are to be brought back to the fields or used as animal feed. A positive impact is the reduction of pathogenic organisms in residues after fermentation.

Transport of biomass could involve dust emissions, and transport of ethanol might lead to spills (in insignificant amounts, as far as energy is concerned, but

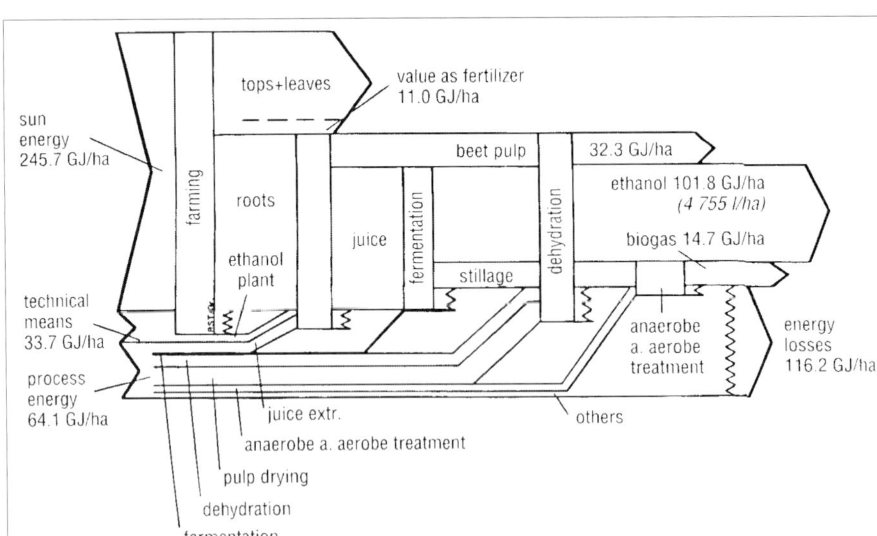

Figure 4.1-128 Energy flows in the production of ethanol from sugar beets. Energy inputs to biomass growth, harvesting and transport are not indicated.

with possible local environmental effects), but overall the impacts from transport would be very small.

Finally, the combustion of ethanol in engines or elsewhere leads to pollutant emissions. Compared with gasoline combustion, emissions of carbon monoxide and hydrocarbons diminish, while those of nitrous oxides, aromatics, and aldehydes increase, assuming that modified ignition engines are used. With special ethanol engines and exhaust controls, critical emissions may be controlled. In any case, the lead pollution still associated with gasoline engines in some countries would be eliminated.

The energy balance of current ethanol production from biomass is not very favourable. A European study has estimated the energy flows for a number of feedstocks. The highest yield of about 100 GJ ha^{-1} is found for sugar beets, shown in Fig. 4.1-128, but the process energy inputs and allotted life-cycle inputs into technical equipment are as large as the energy of the ethanol produced. A part of this may be supplied from biogas coproduced with the ethanol, but the overall energy efficiency remains low.

In a life-cycle analysis of ethanol production, the fact that such production is currently based upon energy crops rather than on residues (sugar cane or beets rather than straw and husk) means that all energy inputs and environmental impacts from the entire agricultural feedstock production should be included along with the impacts pertaining to the ethanol plants and downstream impacts. Clearly, it is very difficult in this mode to balance energy outputs and inputs and to reach acceptable impact levels. The interest should therefore be limited to bio-energy processes involving only residues from an otherwise sensible production (food or wood).

Methanol from biomass

There are various ways of producing methanol from biomass sources, as indicated in Fig. 4.1-115. Starting from wood or isolated lignin, the most direct routes are by liquefaction or by gasification. The pyrolysis alternative gives only a fraction of the energy in the form of a producer gas.

By high-pressure hydrogenation, biomass may be transformed into a mixture of liquid hydrocarbons suitable for further refining or synthesis of methanol, but all methanol production schemes so far have used a synthesis gas, which may be derived from wood gasification or coal gasification. The low-quality "producer gas" resulting directly from the wood gasification (used in cars throughout Europe during World War II) is a mixture of carbon monoxide, hydrogen gas, carbon dioxide, and nitrogen gas (see section 4.1.8.2). If air is used for gasification, the energy conversion efficiency is about 50%, and if pure oxygen is used instead, some 60% efficiency is possible, and the gas produced has less nitrogen content. Gasification or pyrolysis could conceivably be performed with heat from (concentrating) solar collectors, for example, in a fluidised bed gasifier maintained at 500°C.

The producer gas is cleaned, CO_2 and N_2 as well as impurities are removed (the nitrogen by cryogenic separation), and methanol is generated at elevated pressure by the reaction

$$2H_2 + CO \rightarrow CH_3OH. \quad (4.1.160)$$

The carbon monoxide and hydrogen gas (possibly with additional CO_2) is called the "synthesis gas", and it is usually necessary to use a catalyst in order to maintain the proper stoichiometric ratio between the reactants of (4.1.160). A schematic process diagram is shown in Fig. 4.1-129.

An alternative is biogas production from the biomass (section 4.1.8.2) followed by the methane to methanol reaction,

$$2CH_4 + O_2 \rightarrow 2CH_3OH, \quad (4.1.161)$$

also used in current methanol production from natural gas. Change of the H_2/CO stoichiometric ratio for (4.1.160) is obtained by the "shift reaction" (4.1.156) discussed in section 4.1.8.2. Steam is added or removed in the presence of a catalyst (iron oxide, chromium oxide).

The conversion efficiency of the synthesis gas to methanol step is about 85%, implying an overall wood to methanol energy efficiency of 40–45%. Improved catalytic gasification techniques raise the overall conversion efficiency to some 55%. The currently achieved efficiency is about 50%, but all life-cycle estimates of energy inputs have not been included or performed.

The octane number of methanol is similar to that of ethanol, but the heat of combustion is less, amounting to 18 MJ kg^{-1}. However, the engine efficiency of methanol is higher than that of gasoline, by at least 20% for current motor car engines, so an "effective energy content" of 22.5 MJ kg^{-1} is sometimes quoted. Methanol can be mixed with gasoline in standard engines, or used in specially designed Otto or diesel engines, such as a spark ignition engine run on vaporised methanol, with the vaporisation energy being recovered from the coolant flow. Uses are similar to those of ethanol, but several differences exist in the assessment of environmental impacts, from production to use (e.g. toxicity of fumes at filling stations).

The future cost of methanol fuel is expected to reach US$ 8/GJ. The gasification can be made in closed environments, where all emissions are collected, as well as ash and slurry. Cleaning processes in the methanol formation steps will recover most catalysts in re-usable form, but other impurities would have to be disposed of along with the gasification products. Precise schemes for waste disposal have not been formulated, but it seems unlikely that all nutrients could be recycled to agri- or

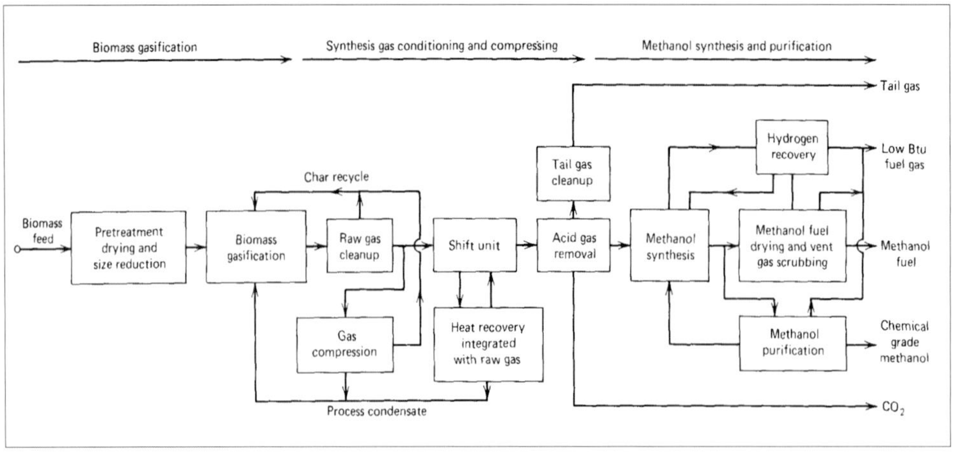

Figure 4.1-129 Schematic flow diagram for biomass to methanol conversion process (From Wan, *et al.*, 1981. Reprinted from *Biomass Gasification* (T. Reed, ed.), with permission. Copyright 1981, Noyes Data Corp., Park Ridge, NJ).

silviculture as in the case of ethanol fermentation. However, the production of ammonia by a process similar to the one yielding methanol is an alternative use of the synthesis gas. Production of methanol from *eucalyptus* rather than from woody biomass has been studied in Brazil. More fundamental studies aiming to better understand the way in which methanol production relies on degradation of lignin are ongoing.

4.1.9 Other conversion processes

This chapter has described a number of general principles, which have been or may become of use in designing actual conversion devices, primarily those aimed at utilising the renewable flows of energy. Not all the existing or proposed devices have been covered.

An attempt has been made to describe at least one example of conversion into a useful form of energy, for each of the renewable resources described in Chapter 3.3. For this reason, the order in which conversion devices have been described in Chapter 4.1 roughly follows the order in which the distribution and magnitude of individual renewable flows or energy stores were discussed in Chapter 3.3. A few digressions from this scheme have been necessary, because some conversion devices (e.g. turbines) are common to several types of resource. This section contains examples of converting more speculative sources of energy, such as the salinity differences identified in section 3.3.7.2 as a renewable energy resource of possible interest. Conversion of nuclear energy is briefly mentioned, without details of actual devices. It is a non-renewable but potentially important energy source, if the long-held visions of safe breeder fission reactors or fusion devices should ever bear fruit.

4.1.9.1 Conversion of salinity gradient resources

As discussed in section 3.3.7.2 (Fig. 3.3.81), a salinity difference, such as the one existing between fresh (river) and saline (ocean) water, may be used to drive an osmotic pump, and the elevated water may in turn be used to generate electricity by an ordinary turbine (cf. section 4.1.5).

An alternative method, aiming directly at electricity production, takes advantage of the fact that the salts of saline water are ionised to a certain degree and thus may be used to derive an electrochemical cell of the type discussed in section 4.1.1.6.

This electrochemical cell, which may be called a "dialytic battery" since it is formally the same as the dialysis apparatus producing saline water with electric power input, is shown schematically in Fig. 4.1-130. In this case, the membrane allows one kind of ion to pass and thereby reach the corresponding electrode (in contrast to the osmotic pump in Fig. 3.3-81, where water could penetrate the membrane but ions could not). The free energy difference between the state with free Na^+ and Cl^- ions (assuming complete ionisation), and the state with no such ions is precisely the difference between saline and fresh water, calculated in section 3.3.7.2 and approximately given by (3.3.51). Assuming further that each Na^+ ion reaching the electrode A neutralises one electron, and that correspondingly each Cl^- ion at the electrode B gives rise to one electron (which may travel in the external circuit), then the electromotoric force, i.e. the electric potential ϕ of the cell, is given in analogy to (4.1.74), with the number of positive ions per mole equal to n_{Na^+}, and ϕ is related to the change in free energy (3.3.51) just as (4.1.74) was related to (4.1.71),

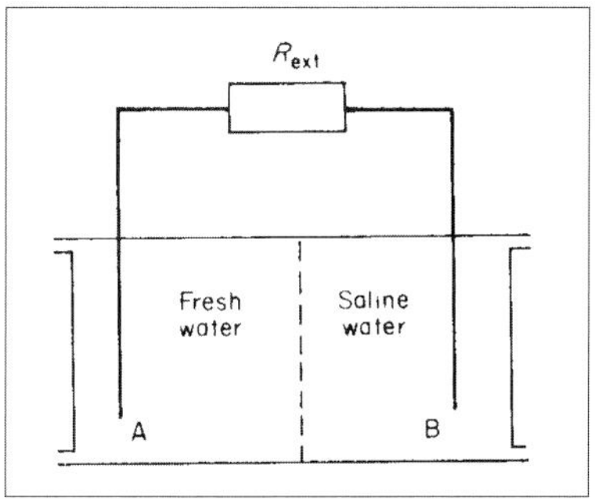

Figure 4.1-130 Schematic picture of dialytic battery.

$$n_e \mathscr{F} \phi = n_{Na^+} \mathscr{F} \phi = \delta G \approx 2 \mathscr{R} T n_{Na^+},$$

or

$$\max(\Delta \phi_{ext}) = \phi \approx 2 \mathscr{R} T / \mathscr{F}. \quad (4.1.162)$$

Inserting the numerical values given in section 3.3.7.2, and $T = 300$ K, the cell electromotive force becomes $\phi \approx 0.05$ V. The actual value of the external potential, $\Delta\phi_{ext}$, may be reduced as a result of various losses, as described in section 4.1.1.6 (e.g. Fig. 4.1-11). ϕ is also altered if the "fresh" water has a finite amount of ions or if ions other than Na^+ and Cl^- are present.

If the load resistance is R_{ext}, the power delivered by the cell is given by the current $I = \Delta\phi_{ext} R_{ext}^{-1}$,

$$E = I \Delta\phi_{ext},$$

and as usual the maximum power corresponds to an external potential difference $\Delta\phi_{ext}$, which is smaller than the open circuit value (4.1.162). The internal losses may also be represented by an internal resistance R_{int} defined by

$$\Delta\phi_{ext} = \phi - IR_{int} = \phi - \Delta\phi_{ext}(R_{int}/R_{ext}).$$

Thus, the power output may also be written

$$E = \phi^2 R_{ext}/(R_{int} + R_{ext})^2.$$

R_{int} depends on electrode properties, as well as on n_{Na^+} and n_{Cl^-} and their variations across the cell.

Several anode-membrane plus cathode-membrane units may be stacked beside each other in order to obtain a sufficiently large current (necessary because the total internal resistance has a tendency to increase strongly, if the current is below a certain value). Figure 4.1-131 shows the results of a small-scale experiment for a stack of about 30 membrane-electrode pairs and an external load, R_{ext}, close to the one giving maximum power output. The power is shown, however, as a function of the salinity of the fresh water, and it is seen that the largest power output is obtained for a fresh water salinity which is not zero but 3–4% of the sea water salinity. The reason is that although the electromotive force (4.1.162) diminishes with increasing fresh water salinity, this is at

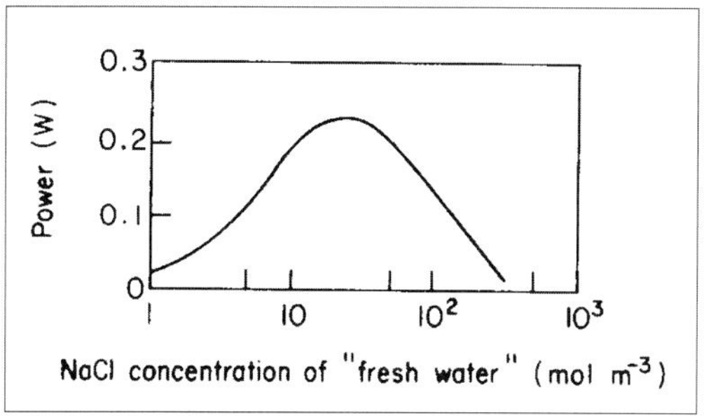

Figure 4.1-131 Measured performance of dialytic battery, containing about 30 membrane pairs in a stack. The external load was $R_{ext} = 10\ \Omega$ (close to power optimisation), and electrode losses had been compensated for by insertion of a balancing power source.

first more than compensated for by the improved conductivity of the solution (decrease in R_{int}), when ions are also initially present in the fresh water compartment.

Needless to say, small-scale experiments of the kind described above are a long way from a demonstration of viability of the salinity gradient conversion devices on a large scale, and it is by no means clear whether the dialysis battery concept will be more or less viable than power turbines based on osmotic pumps. The former seems likely to demand a larger membrane area than the latter, but no complete evaluation of design criteria has been made in either case.

4.1.10 Energy bands in semiconductors

The electrons in a solid move in a potential, which for crystalline forms is periodic, corresponding to the lattice structure of the material. The forces on a given electron are electromagnetic, comprising the attractive interaction with the positively charged nuclei (the effective charge of which may be diminished by the screening effect of more tightly bound electrons) as well as repulsive interaction with other electrons. Owing to the small mass ratio between electrons and nuclei, the positions of the nuclei ($\mathbf{R}i$, $i = 1, 2, 3...$) may to a first approximation be regarded as fixed. This is the basis for saying that the electrons are moving in a "potential". When an electron is close to a definite nucleus, its wavefunction may be approximated by one of the atomic wavefunctions $\psi_{i,n(i)}$ $(r - \mathbf{R}_i)$ for this isolated nucleus. One may therefore attempt to describe an arbitrary electron wavefunction as a linear combination of such atomic wavefunctions ("orbitals"),

$$\psi(\mathbf{r}) = \sum_{i=1}^{N} \sum_{n(i)} c_{i,n(i)}\ \psi_{i,n(i)}(\mathbf{r} - \mathbf{R}_i). \quad (4.1.163)$$

Here N is the total number of atomic nuclei, and $\boldsymbol{r}$ is the position vector of the electron.

A simple example is that of a diatomic molecule ($N = 2$) with only one important level in each atom,

$$\psi(\boldsymbol{r}) = c_{1,n(1)}\psi_{1,n(1)}(\boldsymbol{r} - \boldsymbol{R}_1) + c_{2,n(2)}\psi_{2,n(2)}(\boldsymbol{r} - \boldsymbol{R}_2),$$

for which two normalised solutions exist, with opposite relative sign between the two terms. If the overlap

$$S = \int \psi_{1,n(1)}{}^{*}(\boldsymbol{r} - \boldsymbol{R}_1)\ \psi_{2,n(2)}(\boldsymbol{r} - \boldsymbol{R}_2)\ \mathrm{d}\boldsymbol{r}$$

(the asterisk denotes complex conjugation) is non-zero, the energies of the two superposition states will be different from those ($W_{1,n(1)}$ and $W_{2,n(2)}$) of the atomic wavefunctions.$\int \psi_{1,n(1)}$ and $\psi_{2,n(2)}$. The most tightly bound solution, $\psi_b(\boldsymbol{r})$, will correspond to an energy lower than the lowest of the original ones $W_{1,n(1)}$ and $W_{2,n(2)}$, while the other solution, $\psi_a(\boldsymbol{r})$, will correspond to an energy higher than the highest of the original ones.

If the energies W_b and W_a of the bonding solution $\psi_b(\boldsymbol{r})$ and the anti-bonding solution $\psi_a(\boldsymbol{r})$ are calculated for various relative positions $R = |\ \boldsymbol{R}_1 - \boldsymbol{R}_2\ |$, one may obtain a picture as shown in Fig. 4.1-116 (approximately describing an ionised hydrogen molecule $H_2{}^{+}$).

If the number of (say, identical) nuclei, N, is larger than two, but still placed in a regular geometrical relationship characterised by a single distance R (e.g. the distance between atomic nuclei in a cubic lattice or a diamond type lattice), the superposition type wavefunction (4.1.163) still gives an energy picture similar to Fig. 4.1-132, but now with N curves, the highest and lowest of which may look like the two curves for $N = 2$.

For a solid crystal, N may be of the order of 10^{24}, so the region between the lowest energy state and the highest one will be filled out with energy levels that for all practical purposes can be regarded as continuous. Such a region is called an *energy band*. For each pair of overlapping atomic orbits, $n(1)$ and $n(2)$, there will be a set of linear combinations of wavefunctions in the $N = 2$ case, with bonding and anti-bonding characteristics. In the large-N case, each combination of overlapping atomic orbits will lead to an energy band, such that the overall picture of energy levels as a function of lattice characteristics (in the simple case just R) will be one of a number energy bands inside which all electron energies are "allowed", plus the spaces between energy bands in which no allowed energy is present. The energy band structure may, for example, resemble that shown in Fig. 4.1-133.

If the distance between the nuclei is large, an electron "feels" only one nucleus (plus a number of tightly bound "inner" electrons, if the nucleus has $Z > 2$) at a time, and the energy spectrum is identical to that of an individual atom (right-hand side of Fig. 4.1-133). As the assumed distance between nuclei becomes smaller, energy bands develop and become wider. When the energy bands begin to overlap and cross (left-hand side of Fig. 4.1-133), the basis for the linear assumption (4.1.163) breaks down, at least in the form discussed so far. If the restriction that the bands are formed from one definite atomic orbital [so that there is no sum over $n(i)$ in (4.1.163)], is relaxed, it may be possible still to use an expression of the form (4.1.163) to describe some features of the region of overlapping bands.

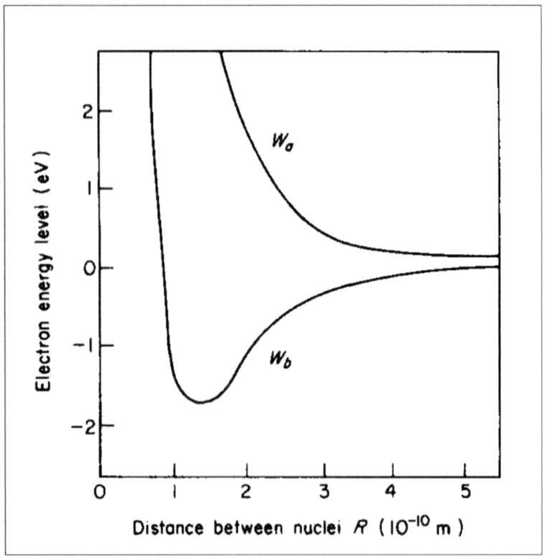

Figure 4.1-132 Electron energy levels in a diatomic molecule as a function of inter-atomic distance (e.g. levels formed from 1s atomic orbitals in the H_2^{+} molecule).

As an example, it is possible to obtain the two separate bands in silicon (see Fig. 4.1-133) by first defining two suitable superpositions of $3s$ and $3p$ atomic orbits [corresponding to the summation over $n(i)$, this procedure being called hybridisation in molecular theory], and then by applying the band theory [corresponding to the summation over i in (4.1.163)]. It is evident that in this region near the band crossing point, it is possible to find energy gaps between adjacent bands, which are much smaller than in the non-overlapping region.

The electron potentials initially represent a situation with one electron per atom being considered. The simplest approximation of many-electron states in one atom is to regard the electrons as independent of each other, except that they have to obey the Pauli exclusion principle according to which there can be at most one electron in each quantum state. Because the electron has an intrinsic spin of $S = ½$, there are two spin states ("spin-up" and "spin-down") for each spatial orbit. Thus, the periodic system would be derived by first placing two electrons in the $1s$-orbit (cf. Fig. 4.1-133), then two in the $2s$-orbit and then six in the $2p$-orbit (comprising three degenerate spatial orbits, because the orbital angular momentum is $L = 1$), etc.

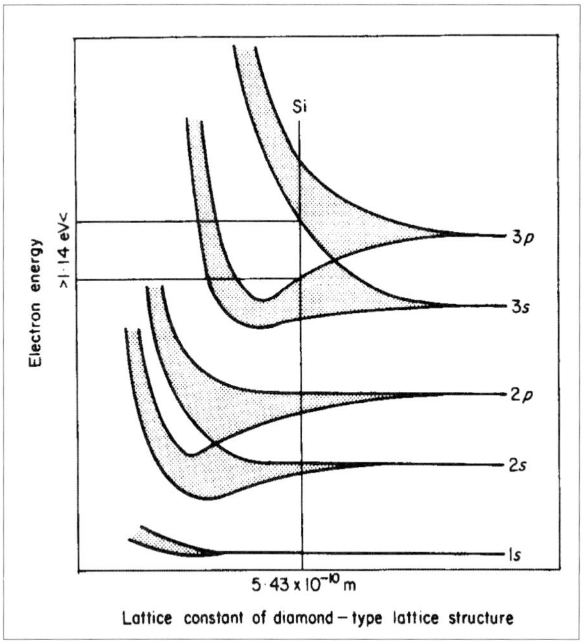

Figure 4.1-133 Energy band structure of a solid as a function of lattice constant (assumed to completely specify inter-atomic distances, as it does for crystals of diamond structure). Atomic physics labelling of orbits are indicated to the right. The band structure in the left-hand region of overlapping bands should be regarded as highly schematic. It may take different forms for different solids. The figure indicates how the overlapping 3s and 3p bands in silicon form the 1.14-eV energy gap further illustrated in Fig. 4.1-134.

A more accurate solution of the Schrödinger equation of quantum mechanics is necessary in order to include the effects of electron–electron interactions, which make the entire energy spectrum depend on electron number in addition to the smooth dependence on the charge of the nucleus ("atomic number"). In a number of cases, these interactions even influence the succession of orbits being filled in the periodic system.

Similar statements can be made for the energy bands in many-atom lattices. The spectrum of energy bands can be used to predict the filling systematics, as a function of the number of electrons per atom, but again the spectrum is not independent of the degree of filling (for this reason no energy scale is provided in Fig. 4.1-133). Thus, a diagram such as Fig. 4.1-133 cannot be used directly to predict the atomic distance (e.g. represented by the lattice constant) that will give the lowest total energy and hence the ground state of the system. This is in contrast to the H_2^+ molecule (Fig. 4.1-132), for which the minimum of the W_b curve determines the equilibrium distance between the nuclei.

Knowing the lattice structure and the lattice constant (the dimension of an elementary cube in the lattice), a vertical cut in a diagram of the type in Fig. 4.1-133 will give the allowed and forbidden energy values, with the reservations made above. Filling the proper number of electrons per atom into the energy bands, with reference to the Pauli principle, the energy of the latest added electron may be determined, as well as the lowest available energy level into which an additional electron added to the system may go.

If the electrons available exactly fill a band, and if there is a substantial distance to the next higher level, the material is an electrical insulator (no electrons can change state, i.e. they cannot "move"). If a band is partially filled, the material is a good conductor (e.g. a metal). The continuum of levels in the partially filled band allows the electrons to move throughout the lattice. If the distance between the highest filled band and an empty band is small, the material is called a semiconductor. At zero absolute temperature a semiconductor is an insulator, but because of the thermal energy spread at higher temperatures, given by (4.1.23) because electrons are Fermi particles, some electrons will be excited into the higher band (the "conduction band"). The conductance of silicon increases by a factor 10^6 between 250 and 450 K.

The bands corresponding to 1*s*, 2*s* and 2*p* atomic orbits are completely filled in Si, but then four electrons remain per atom. According to Fig. 4.1-133, the 3*s* and 3*p* bands are overlapping and allow two mixed bands to be constructed, with an intermediate gap of about 1 eV (1.6×10^{-5} J). The lower band (the "valence band") can hold four electrons per atom, so that this band will be completely full and the other one empty at zero temperature.

If a few of the atoms in a silicon lattice are replaced by an atom with higher Z (atomic number), e.g. phosphorus, the additional electrons associated with the impurity cannot be accommodated in the valence band, but will occupy a discrete level (named the "donor level") just below the conduction band (the energy depression being due to the larger attractive force from the atom of higher Z). A semiconductor material with this type of impurity is called *n*-type. The electrons in the donor-level are very easily excited into the conduction band. Adding *n*-type impurities makes the Fermi level [μ_i in (4.1.23)] move upwards from the initial position approximately half way between the valence and conduction bands.

Impurities with lower Z (e.g. Al in a Si-lattice) lead to electron vacancies or "holes", which are associated with an energy slightly higher than the top of the valence band, again due to the Z-dependence of the attractive force. These "acceptor levels" easily transfer holes to the valence band, a process which may, of course, alternatively be described as the excitation of electrons from the valence band into the acceptor level. Semiconductor material with this type of impurity is called *p*-type. The holes can formally be treated as particles like the electrons, but with positive charges

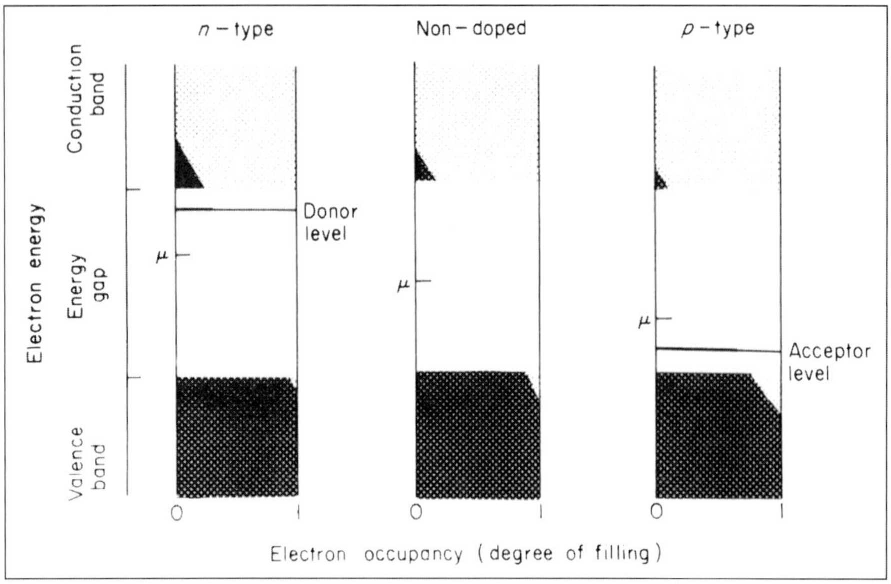

Figure 4.1-134 Energy band structure near the Fermi energy (μ) for a semiconductor material without impurities (middle column) or with n- or p-type doping (cf. text). The dark shading indicates the electron occupancy as a function of energy for a finite temperature (occupancy equal to unity corresponds to the maximum number of electrons at a given energy or energy interval which is consistent with the exclusion principle for Fermi-type particles).

and provided that the state of the pure semiconductor at zero temperature is taken as reference ("vacuum state" in quantum theory).

The energy diagrams of "doped" semiconductors of n- and p-types are sketched in Fig. 4.1-134.

4.1.11 Suggested topics for discussion

4.1.11.1

Discuss the production of heat of the temperature T_L from an amount of heat available, of temperature T above T_L and access to a large reservoir of temperature T_{ref} below T_L.

Quantities of heat with temperatures T and T_{ref} may be simply mixed in order to obtain heat of the desired temperature.

Alternatively, a thermodynamic cycle may be used between the temperatures T and T_{ref} to produce a certain amount of mechanical or electrical energy, which then is used to power the compressor of a heat pump, lifting heat from T_{ref} to the desired temperature T_L.

As a variant, use a thermodynamic cycle between the temperatures T and T_L, producing less work but on the other hand providing reject heat at the desired temperature. Again, a heat pump is used to provide more heat of temperature T_L on the basis of the work output from the thermodynamic cycle.

What are the simple and the second law efficiencies in the individual cases?

4.1.11.2

Discuss design aspects of a propeller-type wind energy converter for which no possibility of changing the overall pitch angle θ_0 is to be provided.

Consider, for instance, a design similar to the one underlying Figs. 4.1.70–4.1.75, but which is required to be self-starting at a wind speed of 2 m s^{-1} and to have its maximum power coefficient at the wind speed of 4 m s^{-1}. Find the tip-speed ratio corresponding to this situation for a suitable overall pitch angle θ_0 and for given assumptions on the internal torque $Q_0(\Omega)$, e.g. that Q_0 is independent of the angular velocity Ω, but proportional to the third power of the blade radius R, with the proportionality factor being given by

$$Q_0 = 10^{-3}\pi\rho R^3 u_{cut\text{-}in}^2,$$

with $u_{cut\text{-}in} = 2$ m s^{-1}.

Compare the power produced per square metre swept at a wind speed of 4 and 6 m s^{-1} to that which could be derived from a solar cell array of the same area.

For an actual design of wind energy converters, the blade shape and twist would not be given in advance, and one would, for a device with fixed angular velocity, first decide on this angular velocity (e.g. from generator and gearbox considerations) and then choose a design wind speed and optimise the blade shape $c(r)$ and blade twist $\theta(r)$, in order to obtain a large power coefficient at the design speed and possibly to avoid a rapid decrease in the power coefficient away from the design point.

The blade shape $c(r)$, of course, depends on blade number B, but to lowest order, the performance is independent of changes, which leave the product of $c(r)$ for the individual blades and B unchanged. Name some other considerations which could be of importance in deciding on the blade number.

4.1.11.3

In section 4.1.3.1, a maximum power coefficient C_p far above unity is suggested for certain cases of sail-driven propulsion on friction-free surfaces. Is this not a contradiction, like taking more power out than there is in the wind?

(Hint: C_p is defined relative to the wind power per unit area, which is of immediate relevance for rotors sweeping a fixed area. But what is the area seen by a sail-ship moving with velocity $\boldsymbol{U}$?).

4.1.11.4

What is, for example based on local data, the ratio between the daily amount of solar radiation, which on a clear day at different seasons intercepts a fully tracking collector system, and that reaching a flat plate of fixed, south-facing orientation and a certain tilt angle s?

What is the relative importance of east–west tracking alone and solar height tracking alone?

4.1.11.5

Discuss shore-based wave energy converters, e.g. based on letting the wave trains ascend an inclined (and maybe narrowing) ramp, such that the kinetic energy is converted into potential energy of elevation, which may be used to drive a turbine.

The simple solutions for gravity waves in a deep ocean cannot be used directly for waves approaching the shallow coastal waters. However, it may be assumed that the wave period remains unchanged, but that the amplitude a of the wave surface increases in such a way that the total power remains constant as the velocity potential becomes altered due to the boundary condition $d\phi/dt = 0$ at the sea bottom. This assumption implies the neglect of frictional dissipation at the sea floor.

There is a maximum ramp height that will permit a wave to reach the top and cross it into an elevated reservoir. If a larger height is required by the turbine, a narrowing ramp construction may be used to give some fraction of the water mass the desired elevation.

4.1.11.6

Combine productivity data such as those reported in section 3.3.6.2 with estimated efficiencies of fuel production by anaerobic fermentation processes (section 4.1.8.2) to give overall conversion efficiencies of bioconversion of solar energy into fuels.

Compare this to other methods of converting solar radiation into mechanical or electrical work, and discuss relative virtues other than conversion efficiency.

4.1.11.7

Estimate the magnitude of and seasonal variation in stored food energy, separately for standing crops on the field, for livestock to be used for food, and for actually stored food (in grain stores, freezers, etc.). Compare food energy stored to the energy value of emergency stores of oil and natural gas, for example, for your own country.

Hint: Some data can be found in Chapter 6.2.

4.1.11.8

Estimate the potential world production of equivalent crude oil from *Euphorbia* plants, assuming only arid land to be utilised and annual harvest yields of 40 MJ m^{-2}.

4.1.11.9

Consider a continuous operation biogas digester for a one-family farm. The digester feedstock is pig slurry (collected daily) and straw (stored). The biogas is used for cooking, for hot water, and for space heating. Estimate the load curves for different seasons and calculate the volume of uncompressed gas storage required, if load is to be met at any time during the year.

4.1.11.10

Consider pure methanol- and pure ethanol-driven cars, and for comparison a gasoline-driven car, weighing 800 kg and going on average 18 km per litre of gasoline and 500 km on a full tank. Calculate the mass penalties for methanol and ethanol fuel tanks, if the same operation range is required. Assume that the fuel usage varies linearly with total mass of the car.

Section Five

Fuel cells

Chapter 5.1

5.1

Fuel cells

5.1.1 Introduction

It is said that the 19th was the century of mechanical engineering, the 20th, that of electronics, and the 21st, that of biology. In fact, the 20th Century could just as well be known as the century of the mechanical heat engine. Counting cars, trucks, and buses, the USA alone, built, from 1900 to 1999, slightly more than 600 million vehicles. If one adds the rest of the world's automotive production, and includes lawn mowers, motorcycles, motorboats, railroad locomotives, airplanes, heavy construction machinery, the production of internal combustion engines in the 20th century probably reached the 2 billion mark!

Mechanical heat engines use the heat released by the reaction of a chemical substance (fuel) with oxygen (usually from from air). The heat is then upgraded to mechanical energy by means of rather complicated machinery. This scheme is inherently inefficient and cumbersome. It is the final outcome of our millenarian struggle to control and use fire. Converting chemical energy directly into electricity is more straightforward, especially in view of the electric nature of the chemical bond that holds atoms in a molecule. Devices that convert chemical energy directly into electricity are called **electrochemical** cells.

Flashlight batteries, automobile batteries and fuel cells are examples of electrochemical cells.

Because electrochemical cells transform chemical energy directly into electricity without requiring an intermediate degradation into heat, they are not limited by the Carnot efficiency.

The words "cell" and "battery" are, in modern parlance, interchangeable. Cell suggest one single unit (although "fuel cell" most frequently consist of a number of series-connected units). Battery suggests a number of units, but a single 1.5 V flashlight cell is commonly called a battery.

If the battery is not worth preserving after its first discharge, it is an **expendable** (also called **primary**) battery. If the device is reusable after discharge, it may fall into one of two categories:

Rechargeable (also called **secondary**) devices, in which the activity is restored by means of an electric **charging** current, as is the case of automobile batteries.

Refuelable devices (**fuel cells**), which deliver a sustained output because their consumables are replenished. To facilitate such replenishment, these consumables are usually fluids, although some fuel cells use solid consumables as is the case of **zinc-air cells**, described later in this chapter.

$$\text{Electrochemical Cells}\begin{cases}\text{Expendable}\\ \text{Nonexpedable}\begin{cases}\text{Rechargeable}\\ \text{Refuelable}\end{cases}\end{cases}$$

Although fuel cells date back to 1839 when Sir William Groves demonstrated his "gaseous voltaic battery," until recently they still remained in their technological infancy.

NASA revived fuel cell research: both *Gemini* and *Apollo* used fuel cells, and so does the space shuttle. The most important applications of fuel cells in the near future are as power sources for buses and automobiles, as central utility power plants, as dispersed (including residential) power suppliers, and as power sources for cell phones and other small electronic devices.

Renewable Energy Focus Handbook 2009; ISBN: 9780123747051

5.1.2 Electrochemical cells

The purpose of electrochemical cells is to provide a flow of electrons in an external circuit where useful work can be done. To this end, the cells must consist of a source and a sink of electrons.

The reactions used in electrochemical cells are called **reduction-oxidation (redox)** reactions, because the buzz word for releasing electrons is **oxidation** and for capturing electrons is **reduction**.

Numerous old scientific terms are confusing or at least not self-explanatory. The terms **reduction** and **oxidation** require explanation.

The word *oxygen* stems from *oxús* = acid or sharp, and means generator of acids, a name that appears in de Morveau and Lavoisier's "Nomenclature Chimique" in 1787, when chemists were under the wrong impression that oxygen was an essential element in acids. Actually, it is hydrogen that is essential. When an acid is dissolved in water, some of its hydrogen atoms lose their electron—the water becomes *acid;* the hydrogen is *oxidized.* By extension any reaction that involves the loss of electrons is called **oxidation.** The reverse reaction—gaining electrons—is called **reduction**.

In electrochemical cells, the full reaction is broken down into two **half-cell reactions** or **half-reactions** that occur in physically separate regions of the device. These regions are interconnected by an **electrolyte** that conducts ions but not electrons. The latter, having been released by the oxidizing half-reaction, can move to the reduction side only via an external circuit. This establishes the external current that is the purpose of the cell. The *conventional* direction of this external current is from the reduction to the oxidizing side—the current exits the device from the reduction side which thus becomes the **cathode** of the cell, and enters the device at the oxidizing side which becomes the **anode.** As in any source of electricity, the cathode is the positive electrode and the anode the negative one, exactly the opposite of what happens in sinks of electricity (**loads**).

As an example of an electrochemical cell, consider a membrane capable of acting as an electrolyte. Put hydrogen in contact with one side of this membrane. At ambient conditions, most of the gas will be in the form of H_2 molecules; however, a small amount will **dissociate:**

$$H_2 \rightarrow 2H, \tag{5.1.1}$$

and some of the resulting H will **oxidize** (ionize)—that is, lose an electron:

$$H \rightarrow H^+ + e^-. \tag{5.1.2}$$

Since the membrane does not conduct electrons, these will remain on its surface while the ions will **diffuse** across it and arrive at the other side. Because the ions carry a positive charge, the hydrogen side becomes negative owing to the excess electrons that remain on it and the opposite side becomes positive owing to the positive ions that arrived there. The resulting electric field causes some of the ions to drift back to the hydrogen side. A **dynamic equilibrium** is established when the diffusion exactly equals the returning drift. It is easy to calculate the potential developed.

Now sprinkle a conducting powder on both sides of the membrane so as to create two porous electron-conducting layers, i.e., two **electrodes**. Interconnect the electrodes externally through a load resistance, R_L. Ions cannot flow through this external connection, but electrons can and, when they do, they flow from the hydrogen side where they are abundant to the opposite side establishing an electric current as indicated in Figure 5.1-2. The reaction of interest that occurs at the hydrogen electrode is

$$\underline{\text{Anode reaction:}} \quad 2H_2 \rightarrow 4H^+ + 4e^-. \tag{5.1.3}$$

The difficulty with this picture is that it contradicts the first law of thermodynamics in that it causes an I^2R_L amount of heat to be generated in the load, while, at the cathode, the incoming electrons will combined with the H^+ that diffused through the membrane regenerating the hydrogen atom, H, and, eventually recreating the H_2 gas used as "fuel." We would generate heat without using any fuel.

The external circuit creates a *path* for the electrons, but cannot by itself force a current to circulate, just as a pipe with one end dipped into a lake cannot cause water to flow up inside it. For the water to flow, the open end of the pipe must be lower than the level of the lake. Similarly, to have an external current, it is necessary to lower the (thermodynamic) potential on the cathode side. This can conveniently be done by introducing oxygen so that, combined with the electrons and the H^+, water is formed:

$$\underline{\text{Cathode reaction:}} \quad 4e^- + 4H^+ + O_2 \rightarrow 2H_2O. \tag{5.1.4}$$

This reaction is strongly exothermal—that is, it releases energy (although, in this case, not mostly as heat, as in the

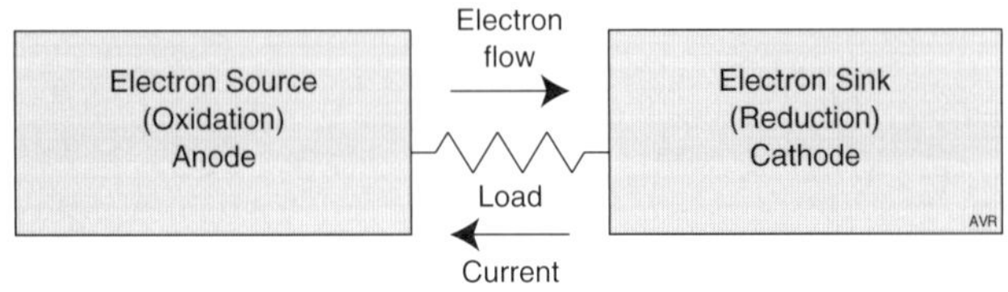

Figure 5.1-1 An electrochemical cell must consist of a source and a sink of electrons.

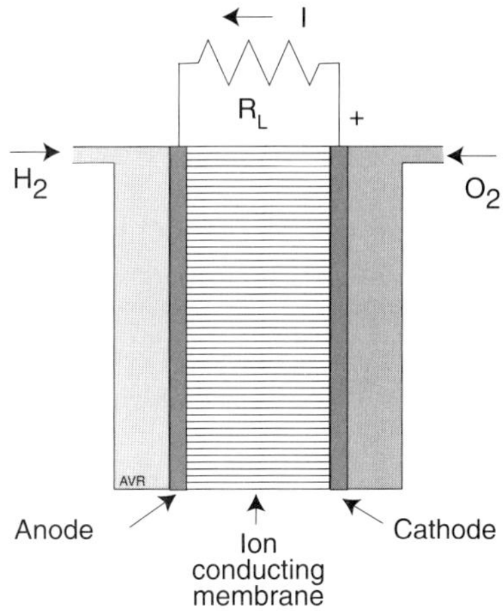

Figure 5.1-2 The simplest electrochemical cell.

case of the combustion of hydrogen, but mainly as electricity.) This is, of course the energy that powers the fuel cell.

The electrochemical cell just described is shown in Figure 5.1-2.

Under STP conditions, the degree of hydrogen dissociation at the anode is small. It can be increased somewhat by altering physical conditions (for example, increasing the temperature,). Remember the Le Chatelier's principle. It also can be increased by the action of catalysts.

The overall cell reaction is:

$$2H_2\ (g) + O_2\ (g) \rightarrow 2H_2O\ (g). \quad (5.1.5)$$

The electrochemical cell invented by Alessandro Volta (1745–1827) in 1800 was the first device capable of generating sustained electrical currents. It consisted of zinc disks separated from silver (or copper) disks by a sheet of paper soaked in salt. A battery or pile was formed by stacking the cells so that there was a direct contact between the copper electrode of one with the zinc electrode of the next—the cells were all in series.

A "Volta" cell can be made by dipping a zinc and a copper electrode into a dilute (say, 10%) sulfuric acid solution. The zinc will oxidize:

$$Zn \rightarrow Zn^{++} + 2e^-, \quad (5.1.6)$$

providing electrons for the external current. Zinc ions are soluble in water. The sulfuric acid, being a strong acid,† will mostly dissociate into ions:

$$H_2SO_4 \rightarrow 2H^+ + SO_4^{--}. \quad (5.1.7)$$

The zinc ions combine with the sulfate ions forming zinc sulfate. The protons, in the form of hydronium, $H^+(H_2O)_x$, migrate through the electrolyte to the copper where they are reduced to hydrogen (by combining with the electrons arriving via the external circuit) and evolve as gas bubbles. This type of cell is of little practical use because soon the copper electrode is covered with adhering hydrogen bubbles that severely limit the current delivered. The so-called "dry cells" use some scheme to avoid the formation of the insulating gas layer at the cathode. The chemical used to absorb the hydrogen is called a **depolarizer**. One of the consumables is in general a metal that can be easily oxidized, zinc being a common choice. Note that the copper in Volta's cell remains unchanged, not undergoing any chemical reaction.

Inexpensive batteries were, until recently, of the **Leclanché** type in which the anode is made of zinc and the cathode of a graphite rod surrounded by pulverized manganese dioxide mixed with carbon (to increase the conductivity). The MnO_2 combines with the liberated hydrogen and keeps it from coating the cathode. The electrolyte is ammonium chloride. More modern batteries use an alkaline electrolyte (**alkaline batteries**).

Perfectly pure zinc is consumed only when a current is drawn. The presence of impurities causes corrosion of the electrode even when the cell is inactive (the impurities form numerous microscopic electrochemical cells within the mass of the metal). To insure a long shelf life the zinc is alloyed with mercury (**amalgamated**). Leclanché cells have been mostly replaced by alkaline ones.

5.1.3 Fuel cell classification

As in many technical areas, there is here a proliferation of acronyms capable of driving the uninitiated to distraction. We will use a few:

AFC	Alkaline fuel cell
DMFC	Direct methanol fuel cell
MCFC	Molten carbonate fuel cell
PAFC	Phosphoric acid fuel cell
SAFC	Solid acid fuel cell
SOFC	Solid oxide fuel cell (ceramic)
SPFC	Solid polymer fuel cell

Just as in the beginning of the 20th century at least three different technologies—steam, electric, and internal combustion—were competing for the automotive market, now, in the beginning of the 21st century, the

† The strength of an acid is a measure of the degree of its dissociation when in aqueous solution. Hydrochloric acid dissociates completely into H^+ and Cl^-; it is a very strong acid. Sulfuric acid is weaker, but is still a strong acid. Surprisingly, hydrofluoric acid, in spite of its corrosiveness, is a weak acid: when in water solution at room temperature, the concentration of H^+ is less than 3% of the concentration of neutral HF molecules.

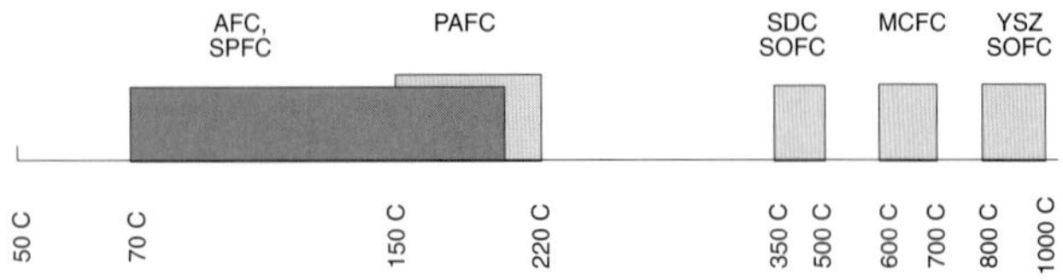

Figure 5.1-3 The operating temperatures of the different fuel cell types fall into relatively narrow ranges. Notice the two different types of SOFC: SDC = samaria-doped ceria, YSZ = yttria-stabilized zirconia.

different fuel cell technologies, listed above, are vying for dominance. Although AFC, MCFC, and PAFC are still in commercial production, it would appear that only SPFC and SOFC have a really good chance of emerging victorious.

Fuel cells can be classified according to several criteria:

5.1.3.1 Temperature of operation

The ideal open-circuit voltage of a fuel cell depends on the nature of the fuel used and, to a small extent, goes down as the temperature increases. However, the maximum deliverable current density rises quickly with increasing temperature because it is related to the rapidity of the chemical reaction, i.e. on the chemical kinetics which are, of course, also a function of the type of fuel and can be improved by catalysts. The higher the temperature, the larger the current the cell can deliver. On the other hand, high temperatures may be incompatible with the electrolyte or with other materials in the cell, and tend to limit life time.

Large power plants that work continuously may profit from high operating temperature: the kinetics are favorable, catalysts are either unnecessary or, when necessary, are immune to CO poisoning,† and exhaust gases can be used for cogeneration increasing the overall efficiency. Plants that operate at somewhat lower temperatures, can produce hot water for general use, an advantage in the case of district or residential generators. For intermittent use, especially in automobiles, it is important to use fuel cells that operate at temperatures low enough to permit short start-up times. However, low temperatures bring the disadvantage of higher vulnerability to carbon monoxide poisoning, of high catalyst loading requirements and of the need for more complicated cooling systems.

At present, the more common SPFC operating at slightly below 100 C are too cool for optimal performance, while SOFC are too hot especially for use in cars. Indeed, at the low temperature of present day SPFC, problems of catalysis and sensitivity to CO poisoning are important, while, in SOFC, the advantages of high temperature mentioned before, are counterbalance by a number of disadvantages listed in subsection 5.1.5.4. For this reason, SPFC research strives for high temperature plastics such as the polybenzimidazole membranes developed by Celanese AG that operate at 200 C, while SOFC research is seeking lower temperature ceramics, such as doped lanthanum gallate (LSGM) and samarium doped ceria (SDC).

5.1.3.2 State of the electrolyte

Most early fuel cells used liquid electrolytes which can leak, require liquid level and concentration management and may be corrosive. Modern cells use either ceramics for high temperatures or plastics for low temperature. Some second generation utility type fuel cells use molten carbonates.

5.1.3.3 Type of fuel

At least in a laboratory, the simplest fuel to use in a fuel cell is hydrogen. However, this gas, is difficult to store (see Chapter 13.1). Especially for automotive use, efforts are being made to use easily storable substances from which hydrogen can be extracted as needed (see Chapter 10.1). The extraction process frequently consists of a **steam reforming** step in which a carbon-bearing feed stock is transformed into a **syngas** or **reformate** consisting mainly of H_2 and CO, followed, when necessary, by a **shift reaction** in which the CO in the reformate is made to react with additional water and is transformed into H_2 and CO_2.

Higher temperature fuel cells such as MCFCs and SOFCs can use CO as a fuel so that the H_2–CO reformate can be fed directly to the cells.

Methanol is used in some fuel fells, especially in small ones for use in portable electronic equipment. High reactivity fuels were used when the technology was quite immature. They included such combinations as hydrazine (NH_2NH_2) with hydrogen peroxide (H_2O_2) as oxidant. These fuels are, however, corrosive and expensive.

5.1.3.4 Chemical nature of the electrolyte

Electrolytes can be alkaline, acid, molten carbonates, or ceramic.

Alkaline electrolytes, such as potassium hydroxide (KOH) are commonly used in **electrolyzers** (the dual of fuel cells—instead of generating electricity from fuels, they produce fuels from electricity). Alkalis are avoided in most fuel cells because the presence of CO_2 in air (used as oxidant) causes the formation of insoluble

† CO poisoning of the catalyst may be a serious problem in low temperature cells operating with fossil fuel-derived hydrogen.

carbonates that spoil the electrolyte. For special applications where pure oxygen is available, KOH fuel cells offer the advantage of high efficiency. This is the case with fuel cells used in space.

Acids tend to be more corrosive than alkalis but relatively weak acids perform well in fuel cells. Phosphoric acid, in particular, was popular.† It tolerates carbon dioxide. For good performance it must operate at temperatures between 150 and 220 C and, to keep the liquid from boiling away, a certain degree of pressurization is needed. At lower temperatures, the conductivity of the solution is too small and at higher temperatures, there are problems with the stability of the materials. Solid acids (subsection 5.1.5.7) have been proposed as electrolytes for fuel cells. They may contribute to the solution of the vexing methanol cross-over problem (subsection 5.1.5.6).

Most ceramic electrolytes, as for instance yttria-stabilized zirconia (YSZ) and samarium-doped ceria (SDC), are anion conductors (conductors of negative ions such as O^-). However, cation conductors have been proposed.

Solid polymers act, in general as proton conductors—that is, as acids although, as in the case of ceramics, cation conductors have been investigated.

5.1.4 Fuel cell reactions

The chemical reaction in a fuel cell depends on both the type of fuel and the nature of the electrolyte. Some of the most common combinations are listed below and are displayed in the illustrations that follow this subsection.

5.1.4.1 Alkaline electrolytes

Hydrogen–oxygen fuel cells with alkaline electrolytes (generally, KOH) use OH^- as the current-carrying ion. Because the ion contains oxygen, water is formed at the anode.

The KOH in the electrolyte dissociates:

$$KOH \rightleftharpoons K^+ + OH^-. \tag{5.1.8}$$

Neutral hydrogen at the anode combines with the hydroxyl ion to form water, releasing the electrons that circulate through the external load:

$$2H_2 + 4OH^- \rightarrow 4H_2O + 4e^-. \tag{5.1.9}$$

At the cathode, the electrons regenerate the hydroxyl ion:

$$4e^- + O_2 + 2H_2O \rightarrow 4OH^-. \tag{5.1.10}$$

The KOH is, of course, not consumed. The overall reaction is:

$$2H_2 + O_2 \rightarrow 2H_2O. \tag{5.1.11}$$

5.1.4.2 Acid electrolytes

When the electrolyte is acid, H^+ ions are available. These can come from the ionization of the hydrogen (as in the SPFC cells) or from the dissociation of the acid in the electrolyte. Take phosphoric acid:

$$H_3PO_4 \rightleftharpoons 3H^+ + PO_4^{---}. \tag{5.1.12}$$

In either case, the H^+ ion is replenished by the anode reaction:

$$2H_2 \rightarrow 4H^+ + 4e^-. \tag{5.1.13}$$

At the cathode, the H^+ is reduced in the presence of O_2 by the electrons that circulate through the load, forming water:

$$4H^+ + 4e^- + O_2 \rightarrow 2H_2O. \tag{5.1.14}$$

The overall reaction is the same as in the previous case. Water is formed at the cathode, and the active ion is hydronium.

5.1.4.3 Molten carbonate electrolytes

Molten carbonate electrolytes are not bothered by carbon oxides. They operate at relatively high temperatures, hence under more favorable kinetics.

When fueled by hydrogen, the reactions are:

$$H_2 + CO_3^{--} \rightarrow H_2O + CO_2 + 2e^-, \tag{5.1.15}$$

$$\frac{1}{2}O_2 + CO_2 + 2e^- \rightarrow CO_3^{--}, \tag{5.1.16}$$

When the fuel is CO, the anode reaction is:

$$CO + CO_3^{--} \rightarrow 2\ CO_2 + 2e^-, \tag{5.1.17}$$

while the cathode reaction is the same as in the hydrogen case.

† Phosphoric acid is a benign acid as attested by its daily consumption by the millions of Coca-Cola drinkers.

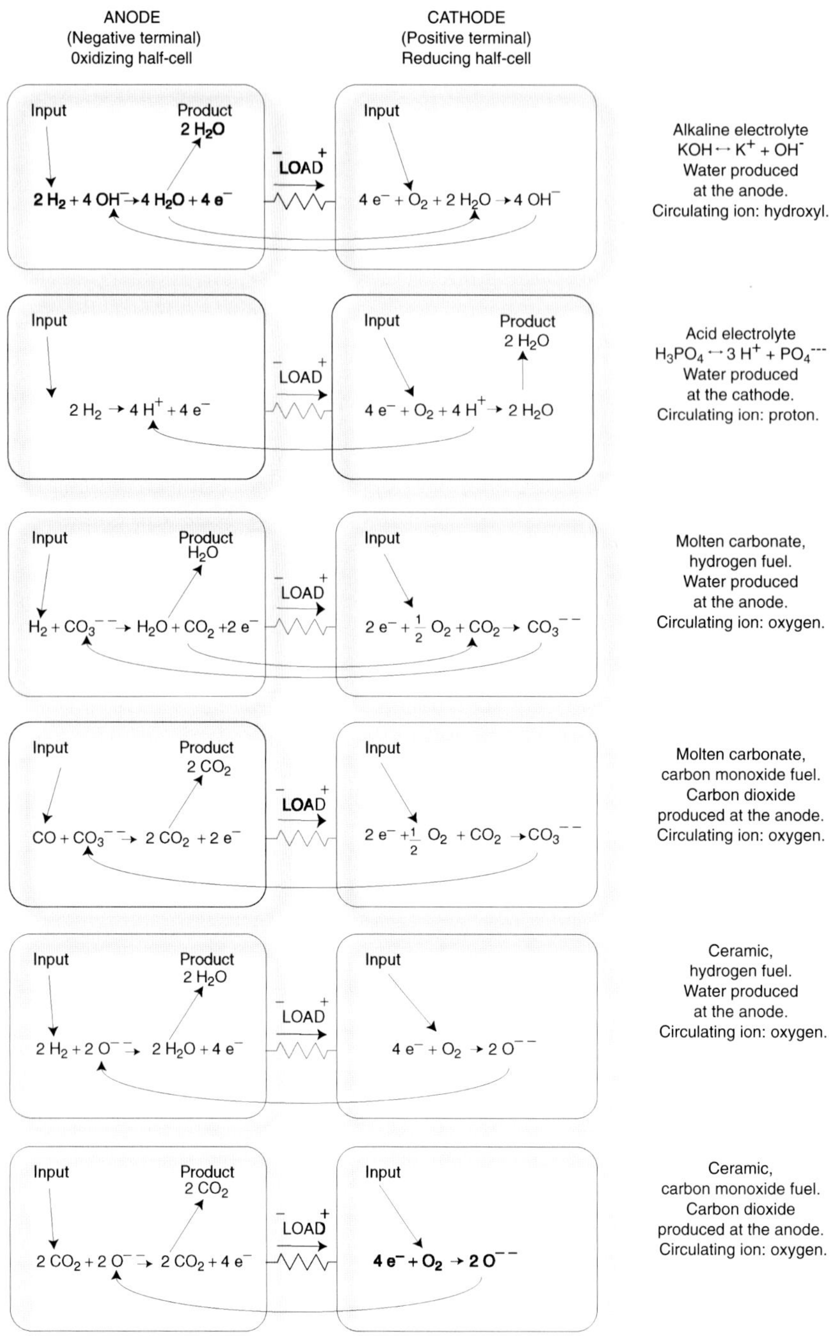

5.1.4.4 Ceramic electrolytes

Ceramic electrolytes usually conduct negative ions. At the cathode, oxygen is ionized by capturing the electrons that constitute the load current:

$$O_2 + 4e^- \rightarrow 2O^{--}. \quad (5.1.18)$$

At the anode, the fuel combines with the O^{--} ions that drifted through the electrolyte, freeing electrons. The fuel may be hydrogen,

$$2H_2 + 2O^{--} \rightarrow 2H_2O + 4e^-, \quad (5.1.19)$$

or carbon monoxide,

$$2CO + 2O^{--} \rightarrow 2CO_2 + 4e^-. \quad (5.1.20)$$

5.1.4.5 Methanol fuel cells

The anode reaction is

$$CH_3OH + H_2O \rightarrow CO_2 + 6H^+ + 6e^-. \quad (5.1.21)$$

The cathode reaction is

$$6H^+ + 6e^- + \frac{3}{2}O_2 \rightarrow 3H_2O. \quad (5.1.22)$$

Thus, the overall reaction is

$$CH_3OH + \frac{3}{2}O_2 \rightarrow CO_2 + 2H_2O. \quad (5.1.23)$$

5.1.5 Typical fuel cell configurations

5.1.5.1 Demonstration fuel cell (KOH)

It is relatively simple to build a small demonstration fuel cell. The design is self-evident from Figure 5.1-4.

The six holes near the rim of the two Lucite covers allow the passage of screws that hold the system together. Diameter of the device is 8 cm.

Fuel sources are two toy balloons, one containing oxygen and the other, hydrogen. Excess gas is vented into a beaker with water.

It can be seen that the cell is symmetrical—there is no structural difference between the anode and the cathode side. The electrodes are made of platinum mesh. Prior to using the cell, platinum black is sprinkled over the mesh to enhance the catalytic action.

Open-circuit voltage is about 1 V and the cell will deliver some 200 mA at 0.7 V, a power of 140 mW.

The electrolyte consists of a disc of filter paper soaked in KOH solution.

The above design is quite primitive and dates from the times when it was difficult to obtain ion exchange membranes. Much better demonstration fuel cells can currently be obtained from a number of vendors. Ask Google for "Fuel cell demonstration kits." Some of the available kits are well designed and easy to operate. They all require distilled water, but that should be no problem. Particularly instructive are kits that incorporate a photovoltaic converter, an electrolyzer and a simple hydrogen storage system, in addition to the fuel cell. They demonstrate that hydrogen can be produced with absolutely no CO_2 emission. The problem is, of course, that it is not yet economically feasible to do so on a large scale.

5.1.5.2 Phosphoric acid fuel cells (PAFC)

A fuel cell battery (Engelhard)

Since fuel cells are low-voltage, high current devices, they must be connected in series. The simplest way of doing so is to use a **bipolar** configuration: a given electrode acts as anode for one cell and (on the other face) as cathode for the next.

The Engelhard phosphoric acid cell (obsolete) is of the air-cooled bipolar type. Its construction can be seen in Figure 5.1-5.

The **bipolar electrode** was made of either aluminum or carbon. Gold was plated on aluminum for protection and flashed on carbon to reduce contact resistance.

The plate was 3 mm thick and had grooves or channels machined on both faces. The oxidant channels ran perpendicular to those for hydrogen on the opposite face. The plate was rectangular with its smaller dimension along the air-flow direction to minimize pressure drop. The channels, in addition to leading in the gases, served also to increase the active surface of the electrodes.

The electrolyte soaked a "cell laminate" held in place by a rubber gasket. It can be seen that this type of construction does not differ fundamentally from that used in the demonstration cell described in the preceding subsection.

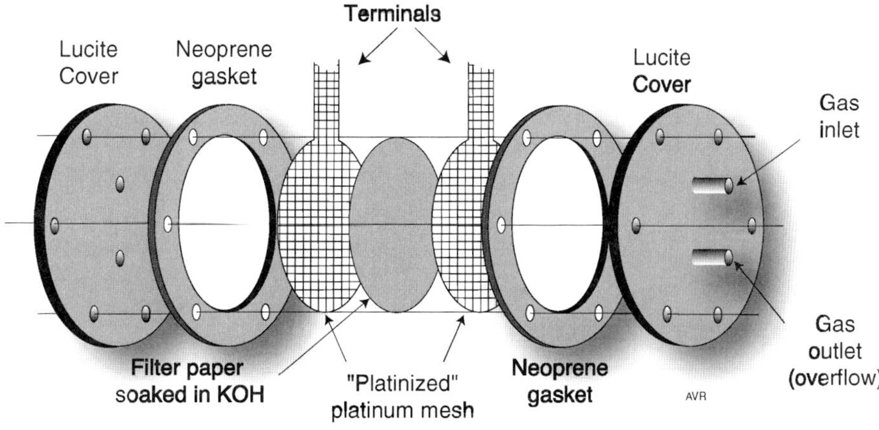

Figure 5.1-4 Exploded view of a demonstration fuel cell.

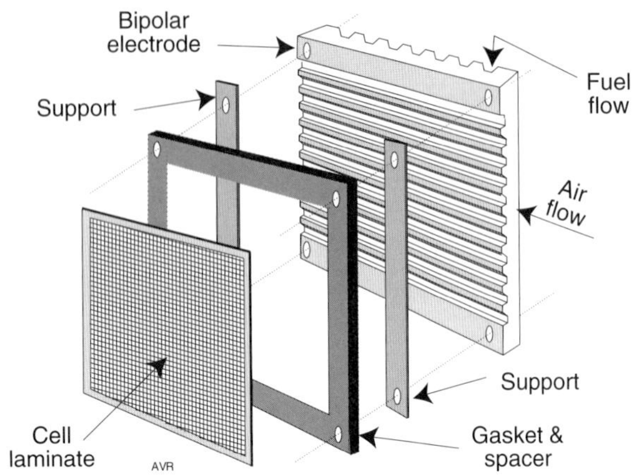

Figure 5.1-5 One element of the Engelhard PAFC.

The cell operated at 125 C. The oxidant was air, which entered at ambient temperature and left hot and moist carrying away excess heat and water (formed by the oxidation of the fuel in the cell). The air flow was adjusted to provide sufficient oxygen and to assure proper heat and water removal.

Apparently, no catalysts were used. This, combined with the relatively low operating temperature (phosphoric acid cells frequently operate at temperatures above 150 C), results in somewhat adverse kinetics and, consequently, in low voltage efficiency.

Engelhard marketed cells of this type in a 750 W battery fed by hydrogen obtained from the cracking of ammonia. Ammonia is a hydride of nitrogen and is a convenient way of storing hydrogen. The system consisted of an ammonia source (a pressure cylinder), a dissociator or **cracker**, a scrubber, the fuel cells and ancillary pumping and control mechanisms.

The fuel cells themselves operated at roughly 45% efficiency (optimistic?) referred to the higher heating value of ammonia, but the system needed to divert fuel to heat the dissociator that worked at 850 C thus reducing the overall efficiency at rated power to about 30% and to much less at lower power levels.

First-generation fuel cell power plant

One early effort to adapt fuel cells for dispersed utility-operated power plant use was made by United Technologies Corporation. It was a 4.8 MW module fueled by natural gas, to be installed in Manhattan.

The UTC fuel cells used phosphoric acid and delivered 0.64 V per cell. At the operating temperature of 150 to 190 C, current densities of 1600 to 3100 A m^{-2} were possible. The life of the cells was expected to be some 40,000 hours (around 5 years).

The UTC project in New York suffered such lengthy delays in the approval process by the city government that when finally it was cleared for operation in 1984, the equipment had deteriorated to the point that it was uninteresting to put it in operation. The first commercial demonstration of this fuel cell application took place in Japan where a 4.5 MW unit started operation in 1983. The same company (Tokyo Electric Power Company) now operates an 11 MW PAFC facility built by Toshiba using US built cells (International Fuel Cell Company a subsidiary of UTC).

ONSI, a subsidiary of the International Fuel Cell Co. manufactured a 200-kW PAFC power plant, known as PC25. Since 1991 until the beginning of 2001, worldwide sales exceeded 220 plants. They delivered both electricity (at 37% efficiency) and heat. The latter could be in the form of 260 kW of hot water (at 60 C) or 130 kW of hot water plus another 130 kW of heat at 120 C. They were fueled by natural gas or, in some cases, by biogas from anaerobic digesters. By 2004, some PC25s could be bought in the surplus market.

Modern PAFCs use platinum catalysts and are vulnerable to CO in the fuel stream (see SPFC, further on). Fortunately, the tolerance of Pt to CO increases with operating temperature: at 175 C, CO concentration can be as high as 1%, and, at 190 C, as high as 2%. This contrasts with the more stringent requirements of the cool SPFCs, which, at 80 C require hydrogen with less than 10 ppm of CO unless special procedures or special catalysts are employed.† On the other hand, whereas the SPFCs have a very stable electrolyte, the phosphoric acid in PAFCs can be degraded by ammonia or sulfur whose concentrations in the fuel stream must be kept low.

5.1.5.3 Molten carbonate fuel cells (MCFCs)

In 1988 the American Public Power Association (APPA) together with the Electric Power Research Institute (EPRI) promoted an international competition to design fuel cells tailored to urban needs. The winning project was a 2-MW MCFC developed by Energy Research Corporation (ERC), now renamed Fuel Cell Energy, Inc. These "second generation" cells were evaluated and in 1996 appeared to have come close to their design performance. The plant achieved the excellent measured efficiency of 43.6% when delivering 1.93 MWAC to the grid.

The cells were assembled in bipolar stacks (see the Engelhard battery) bracketed by nickel-coated stainless steel end plates and separated by bipolar plates made of

† Very roughly, the tolerance, λ (in ppm), of platinum catalysts to CO as a function of the temperature, T (in kelvins), is given by $\lambda = 255 \times 10^{-12} \exp(T/14.5)$.

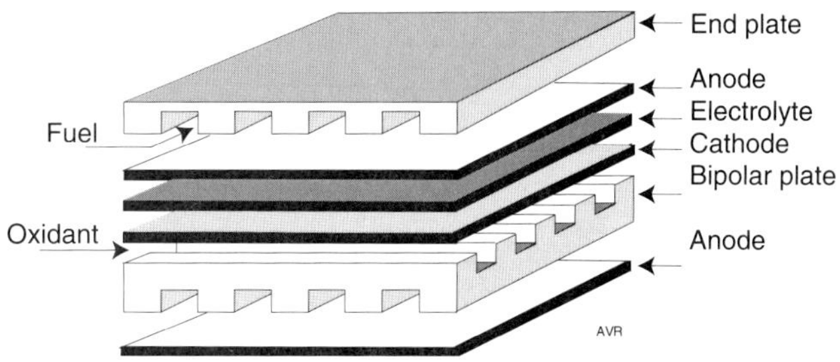

Figure 5.1-6 Exploded view of an MCFC unit.

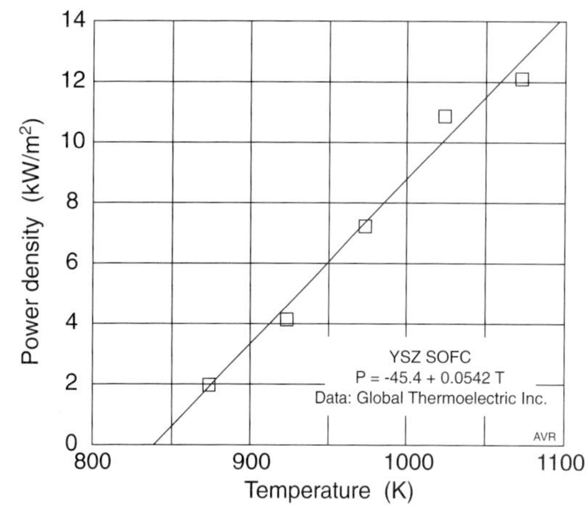

Figure 5.1-7 The power density of a YSZ SOFC rises sharply with operating temperature.

the same material. The cells themselves were a sandwich of an anode, an electrolyte matrix, and a cathode. The nickel-ceramic anode and the nickel oxide cathode were porous. The reactive gases were fed in through the side opposite to the one in contact with the electrolyte, which was a mixture of lithium and potassium carbonates held in lithium aluminate matrix.

Operating temperature was between 600 and 700 C, high enough not to require expensive platinum based catalysts and to permit internal reforming of natural gas. The exhaust gases could efficiently drive a steam turbine in co-generation schemes. Life was limited by the slow dissolution of the cathode and the carbonate poisoning of the reforming catalyst.

Based on the experience gained from operating the Santa Clara plant and from other development work, Fuel Cell Energy, Inc. is offering 2 MW plants for commercial sale.

5.1.5.4 Ceramic fuel cells (SOFCs)

One of the most critical components in any type of fuel cell is the electrolyte. SOFCs are no exception. The ceramics used as electrolytes must have a much higher conductivity for ions than for electrons, a requirement that is not too difficult to achieve; however, high ionic conductivity usually requires operation at elevated temperatures. Figure 5.1.7 illustrates the strong influence of the operating temperature on the achievable power density in a fuel cell using zirconia (YSZ) electrolyte. The performance of the cell is limited by the relatively low conductance of the latter. The data in the figure are from Global Thermoelectric Inc. See Ghosh, D., *et al.* No information on the thickness of the electrolyte was given. There is an obvious temptation to operate SOFC at temperatures lower than the current 1100 K:

1. High temperatures require the use of expensive alloys.
2. Temperature cycling introduces mechanical stresses.
3. The electrodes (but not the electrolyte) must be porous; however, high temperatures promote their sintering causing them to become progressively more impermeable to fuel and air.
4. High temperatures promote diffusion of constituents of the electrodes into the electrolyte.

Items 2 though 4 reduce the life time of the fuel cell.

The conductance of a ceramic electrolyte depends on three factors:

1. The operating temperature discussed above.
2. The thickness of the electrolyte.
3. The nature of the electrolyte, which, in most current cells is yttria- stabilized zirconia (YSZ), typically $(ZrO_2)_{0.9}(Y_2O_3)_{0.1}$.

In order to achieve acceptable conductances, the electrolyte must be quite thin. Global Thermoelectrics has demonstrated 5 μm-thick electrolytes (Gosh *et al.*, 2000) but current commercial cells use thicknesses almost one order of magnitude larger. The reason for this is that the electrolyte must be impermeable to gas, a condition difficult to satisfy when very thin layers are used—porosity will be high and small pin holes are apt to occur. Layer densities of some 98% of the theoretical density are desirable. Additionally, thin electrolytes are exceedingly fragile and must be supported by either the anode or the cathode (i.e. they must be thin compact ceramic layers deposited on one of the two electrodes.)

Lower temperature operation can be achieved by employing ceramic materials other than the popular YSZ. For example, samaria-doped ceria (SDC) is a ceramic electrolyte that, at any given temperature, has much higher ionic conductivity than YSZ. (See subsubsection 9.5.4.2.) Ceramics that conduct protons instead of negative ions have been demonstrated in the laboratory. It is expected that, at 700 C, these ceramics, working with the same power density as zirconia at

1000 C, will deliver some 10% higher efficiency. A promising "low temperature" ceramic proton conductor is $BaCeO_3$. (See Rocky Goldstein (EPRI, Advanced Fossil Power System Business Unit).) Any ceramic used as electrolyte must not react chemically with the materials with which it is in contact.

The most popular cathode material for YSZ fuel cells is strontium-doped lanthanum manganite, LSM ($La_xSr_{1-x}MnO_3$) which is compatible with the electrolyte. The strontium doping increases the electronic coductivity of the material. It is mainly an electron conductor and, consequently, a chemical reaction is required in which the negative ion conducted by the electrolyte must be transformed into electrons conducted by the cathode. This reaction must occur at a triple-point where cathode, electrolyte and oxygen come together. The triple-point area can be increased, as shown by Yoon *et al.* (2000), by coating the pores of the cathode with macroporous YSZ. The cathode material must exhibit high porosity (say, 50%) and must be stable in an oxidizing atmosphere. Porosity is achieved by incorporating a pore-forming substance such as starch in the powder mixture that will form the ceramic.

The fuel electrode must have good catalytic action if internal fuel reforming is used.

The anode, being in a reducing atmosphere, could be made of porous nickel were it not for the high coefficient of thermal expansion of the metal. To correct this, the nickel is dispersed in a matrix of yttria-stabilized zirconia forming a cermet.† YSZ powder is mixed with NiO_2 and sintered together in a environment that reduces the nickel oxide to dispersed metallic nickel leaving behind a porous ceramic. The fuel electrode must have good catalytic action if internal fuel reforming is used. If the temperature is much smaller than some 700 C, internal reformation may become impossible or difficult.

All three layers, anode, electrolyte and cathode, must have matching thermal expansion to avoid delamination and must not react chemically with one another or diffuse into the neighbor.

To form a stack, interconnections between the different cells are required. They must have good mechanical properties and good electric conductivity; they must have appropriate thermal expansion coefficients and good resistance to corrosion and chemical compatibility with the rest of the cell. Above all, they must be of low cost.

Iron alloys, rendered corrosion resistance by the addition of chromium, satisfy most of the above requirements. As in all stainless steels, their rust resistance depends on the formation of a thin scale of chromia on the surface that effectively seals the bulk of the material

Chromium

Chromium forms compounds in which it can assume at least three different oxidation states,† and therefore it has a number of different oxides: chromous oxide, CrO, also called chromium monoxide, oxidation state II, chromic oxide, Cr_2O_3, also called chromium sesquioxide or **chromia,** oxidation state III, and chromium trioxide, CrO_3, oxidation state VI.

The significance, as far as the use of iron–chromium alloys in fuel cells is concerned, is that, whereas chromia is highly refractory (melting point 2435 C), chromium trioxide is quite volatile, having a melting point of only 196 C. The conditions inside a fuel cell lead to the transformation of stable chromium oxides into volatile compounds capable of contaminating the cathode of the cell.

†Oxidation state is the difference between the number of electrons of a neutral atom and that of the corresponding ion in an ionic compound.

from the corroding atmosphere. Unfortunately, chromia, at the high operating temperature of the fuel cells, forms volatile CrO_3 and $CrO_2(OH)_2$, which contaminate the cathode severely degrading the performance. CrO_3 is the dominant chromium vapor in cells using oxygen or very dry air as oxidant, while $CrO_2(OH)_2$ (chromium oxyhydroxide) is dominant when moist air is used. Chromium vaporization can be reduced by operating at low temperature and using dry air. The latter may not be economically attractive.

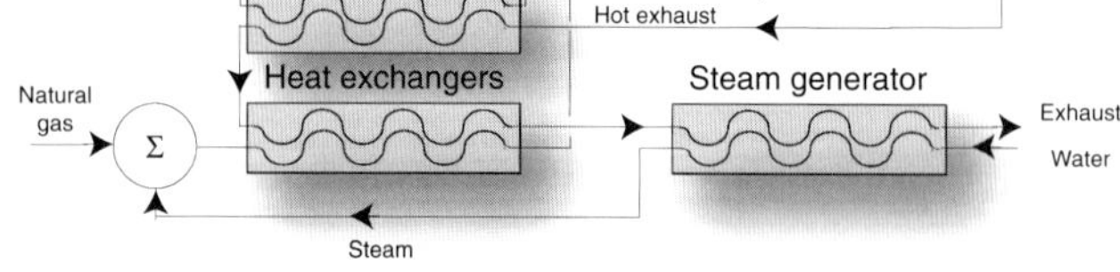

Figure 5.1-8 An arrangement for an SOFC/microturbine combined cycle plant.

† Cermets are combinations of ceramic materials with metals.

Power densities as high as 12 kW/m^2 have been demonstrated in the laboratory (Pham *et al.*, 2000). These levels are below the 20 kW/m^2 reached by modern SPFC (Ballard) and refer to individual cells, not to the whole stack. SOFC in or near production show a more modest power density of some 3 kW/m^2 (Siemens Westinghouse). This may not be a major disadvantage in stationary applications, but may be more of a problem in automotive use where compactness is desirable.

SOFC have high efficiencies (over 50%) especially if used in hybrid (co-generation) arrangements when overall efficiencies approaching 60% have been achieved. Since no corrosives are used (such as phosphoric acid in PAFC or molten carbonates in MCFC), long life is possible.

Continuous operation for over 35,000 hours has been demonstrated. (See Bessette and Pierre, 2000). SOFCs have unlimited shelf life. Many of the ceramics used in fuel cells have a perovskite structure. (See the box opposite.)

In negative ion solid electrolyte fuel cells, the air electrode (cathode) reaction is that of Equation 5.1.21 and the fuel electrode reaction (anode) is either that of Equation 5.1.22 or 5.1.23 or both, depending on the fuel used.

SOFCs are usually either planar or tubular. The latter, as exemplified by the Siemens Westinghouse cells, have the great advantage of doing away with seals that plague planar cells.

High temperature ceramic fuel cells

Much work in the area of tubular cells has been done by Siemens Westinghouse. This company had in 2001 a 100 kWe plant in the Netherlands that operated unattended, generating power into the grid. It also had a 220 kWe hybrid (operating in conjunction with a micro gas turbine) at the Southern California Edison Co., in Los Angeles, CA. A 1 MWe plant with 58% efficiency was being installed in Ft Meade, Maryland, for the Environmental Protection Agency while another plant of equal power was planned for Marbach, Germany.

When starting up and when stopping, SOFCs are subjected to large amplitude thermal cycling, which creates serious problems with seals. Siemens Westinghouse has come up with a clever seal-free configuration using tubular cells. Each individual cell consists of a triple layered ceramic tube, as depicted in Figure 5.1-9. The inner layer is the air electrode (cathode), the middle layer, the electrolyte, and the outer, the fuel electrode (anode). Manufacture of the fuel cell starts with an air electrode made of lanthanum manganite ($La_{0.9}Sr_{0.1}$). The YSZ electrolyte is built up on the cathode tube and the anode is deposited on top of that.

Connections to the anode are made directly to the outer electrode. To reach the cathode, an interconnecting strip is placed longitudinally on the tube and penetrates to the inner layer (see Figure 5.1-9). This interconnecting strip must be stable at both the oxidizing environment of the air electrode and the reducing one at the fuel electrode. It must also be impermeable to gases. These requirements are met by lanthanum chromite. To enhance conductivity, the material is doped with Ca, Mg, Sr or other metals of low valence.

Perovskite

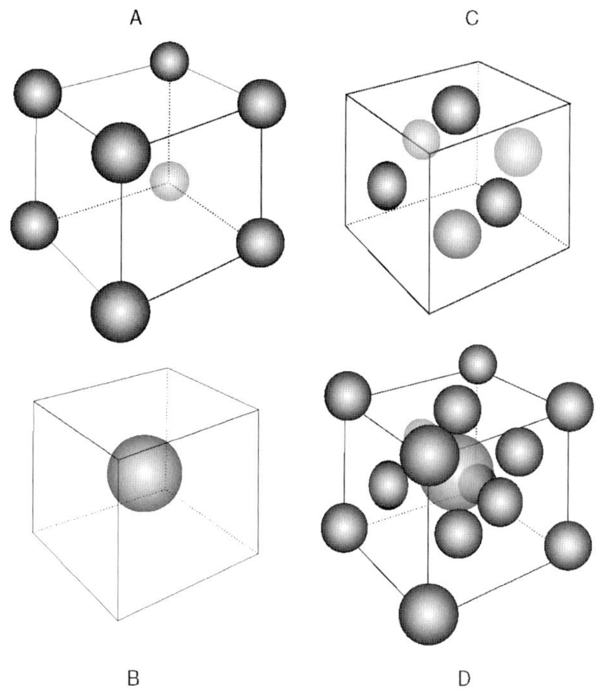

Perovskite is the mineral $CaTiO_3$. By extension, the word designates any substance having the empirical formula ABC_3 (where C usually corresponds to oxygen) which crystallizes in the perovskite arrangement.

Consider a cubic crystalline unit cell having 8 atoms, A, at the eight vertices (see "A" in the figure). When these unit cells are stacked together, each of the atoms is shared among 8 different, adjacent cells. In other words, there is, on average, exactly 1 atom A per cell.

Now refer to "B" in the figure. There is 1 single B atom per unit cell.

Finally, refer to "C." C atoms are centered on the face of the cube. There are 6 faces, but each atom is shared by 2 adjacent unit cells. Thus, on average, there are 3 C atoms per cell. In "D", the three arrangements are put together yielding a unit cell with the empirical formula ABC_3.

Perovskites are important in SOFC electrolytes and in high temperature superconductors.

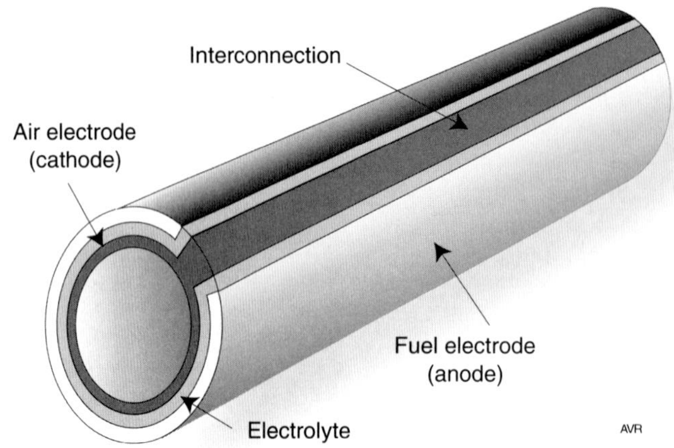

Figure 5.1-9 Tubular concentric electrodes of the Siemens Westinghouse SOFC.

Figure 5.1-10 shows how bundles of tubes can be stacked in series-parallel connections to form generator modules. Nickel felt made of nickel fibers sintered together provide mechanically compliant interconnections. Notice that all these interconnections take place in a chemically reducing environment.

The modules consist of two chambers as indicated in Figure 5.1-11. The lower and longer **fuel reaction chamber** receives the fuel from openings at the bottom. The fuel flows up and most of it is used up by the anode.

Unreacted fuel, some 15% of the total, exits into the **spent fuel combustion chamber** where it is mixed with the excess air that comes out from the top of each tube and burns producing heat used in preheating the input air and increasing the temperature (up to 900 C) of the exhaust gases in order to drive a bottoming cycle generator such as a turbine.

If an adequate amount of water vapor is mixed with fossil fuel gases, automatic reformation will take place thanks to the catalytic action of the nickel in the anode. (See subsection on modern hydrogen production in Chapter 10.1.)

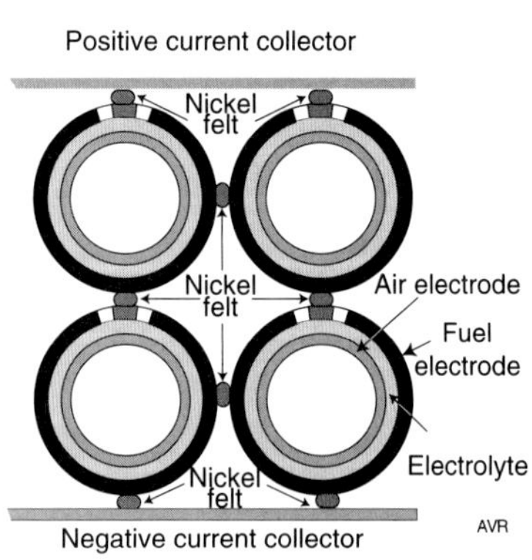

Figure 5.1-10 Tubular cells can easily be stacked in series–parallel connections.

$$CH_4 + H_2O \rightarrow CO + 3H_2.$$

"Standard" Siemens Westinghouse cells are 22 mm diameter tubes of varying lengths, usually 1.5 m. Better performance is obtained by the use of alternate geometries.

Flattening the cells into the "ribbed" configuration does not only improve the stacking factor (so that more units fit into a given volume) but also reduces production cost. These flattened cells have incorporated improvements that substantially increase cell efficiency. See the comparison of the performances of cylindrical and "ribbed" cell shown in Figure 5.1-13.

Low temperature ceramic fuel cells

It would be useful to operate SOFC at temperatures lower than those currently used with YSZ. Among other things the life of the stacks would presumably be longer.

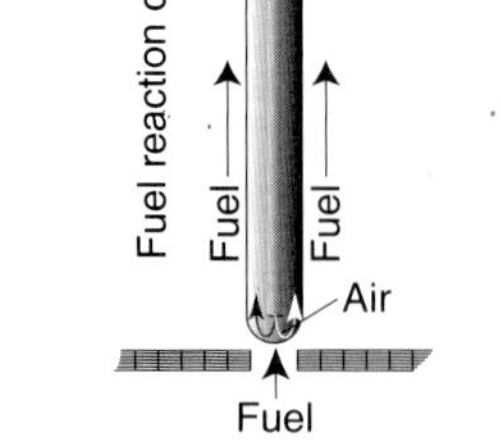

Figure 5.1-11 Tubular cells are stacked in seal-free modules.

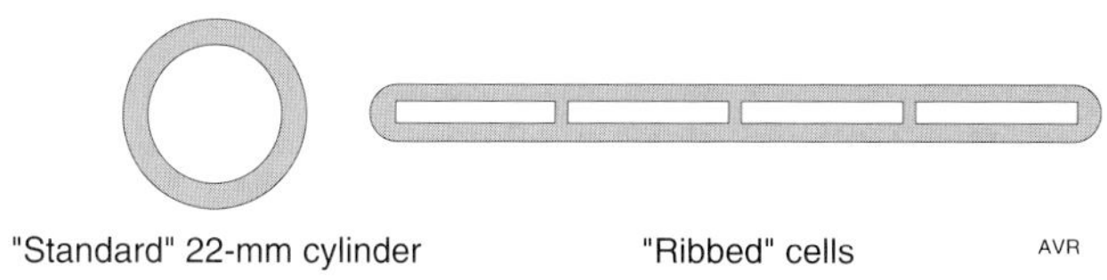

Figure 5.1-12 Cross-section of cylindrical and flattened "ribbed" cells.

It would also be simpler if one could build a **single chamber** fuel cell.

In the usual fuel cell, fuel and oxidizer are fed separately to the device, each gas to its own electrode. In a single chamber cell, the fuel is mixed with air in proportions which are too rich to allow an explosion. The mixture is fed simultaneously to both electrodes, one of which reacts preferentially with the fuel, the other, with the oxygen.

In the cell described by Hibino, the SDC electrolyte was a ceramic disc ground to 0.15 mm thickness. The anode was a layer of nickel-enriched SDC while the cathode consisted of $Sm_{0.5}Sr_{0.5}CoO_3$—samarium strontium cobalt oxide.

The ethane (18%)/air mixture in contact with the nickel-rich anode is reformed into a hydrogen/carbon monoxide mixture according to

$$C_2H_6+O_2 \rightarrow 3H_2 + 2CO \tag{5.1.24}$$

The two gases in the reformate are oxidized:

$$3H_2 + 3O^{--} \rightarrow 3H_2O + 6e^-, \tag{5.1.25}$$

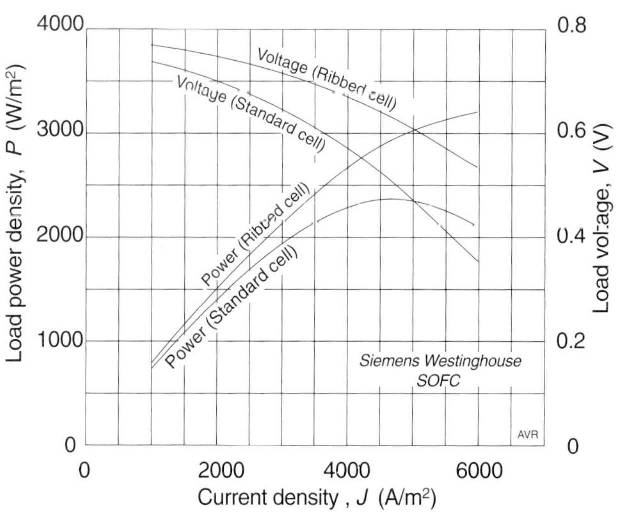

Figure 5.1-13 "Ribbed" cells show substantially better performance than cylindrical ones.

$$2CO + 2O^{--} \rightarrow 2CO_2 + 4e^-, \tag{5.1.26}$$

Thus, each ethane molecule yields 10 electrons that circulate in the load. The 5 oxygen ions come from the cathode (by moving through the electrolyte). They are produced by combining the 10 electrons with 2.5 oxygen molecules from the fuel/air mixture:

$$10e^- + 2.5\,O_2 \rightarrow 5\,O^{--} \tag{5.1.27}$$

The single chamber configuration simplifies the construction and makes the cell more resistant to both thermal and mechanical shocks.

The performance of some of the small experimental cells prepared by Hibino is quite promising: over 4 kW/m^2 at 500 C and over 1 kW/m^2 at 350 C (Hibino et al., 2000). Compare this with the 20 kW/m^2 of the much more developed Ballard SPFC.

The v–i characteristics of one of the experimental SDC cells, with 0.15 mm electrode thickness, is shown in Fig 5.1-15.

5.1.5.5 Solid-polymer electrolyte cells

This is conceptually the simplest type of fuel cell and, potentially, the easiest to manufacture. As the name suggests, the electrolyte is a solid membrane made of an ion conducting polymer. The ion can be positive (a cation), usually a proton, or negative (an anion) usually an OH^-.

SPFCs are safer than liquid or molten electrolyte cells because they use non-corrosive (and of course, nonspillable) electrolytes. They can stand fairly large differential pressures between the fuel and the oxidant sides, thus simplifying the management of these gases.

A short history of the development of SPFC cells can be found in an article by David S. Watkins (Watkins, 1993).

SPFCs were pioneered by GE, starting in 1959. In 1982, the technology was transferred to United Technology Corporation/Hamilton Standard. Little additional progress was made until Ballard Power Systems† took up further development and pushed it, with the collaboration of Daimler Benz, to the production stage. Prior to Ballard, fuel cell cars were predicted to come on the market by 2020. Now, the general expectation is for fuel cell cars to be in general use by 2010!

The progress made by the original GE effort is illustrated by the growth of power densities from the initial 50 W/m^2 in 1959 to over 8 kW/m^2, in 1982, a 160-fold improvement.

Many factors contributed to such progress, including better membranes (early devices used sulfonated

† For Ballard's history, read Tom Koppel (Koppel, 1999).

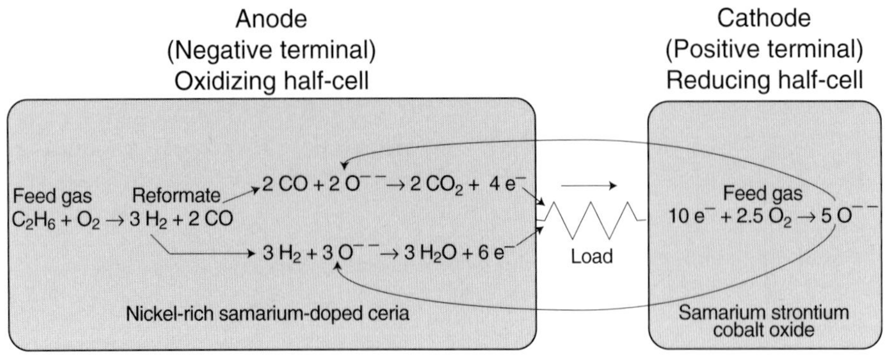

Figure 5.1-14 Cell reactions of the SDC low temperature SOFC.

polystyrene; later ones used Nafion), thinner membranes (from 250 μm to 123 μm), higher operating temperatures (from 25 C to 150 C), and better catalysts.

SPFCs are fast emerging as the preferred solution for automotive use and as a competitor for fixed power plants. The extremely fast advances in this type of cell, a sign of a young and immature technology, is attested by the exponential improvement in power density of Ballard's cells illustrated in Figure 5.1-16. Ballard cells exceed 20 kW per square meter of active surface (compare with 8 kW/m^2 for the best GE cells). The power-to-mass ratio (gravimetric power density) has also improved substantially—it exceeds 0.7 kW/kg, approaching the values for a modern aircraft engine.[††] Thus, power-to-volume and power-to-mass ratios of a modern SPFC are already well into the range of acceptability in automotive applications. What needs to be improved—substantially—is the power-to-cost ratio.

Cost can be reduced in many areas. Most obvious is the need to create a markets large enough to permit mass production. Cheaper membranes and cheaper catalysts are a must. Daimler forecasts prices of \$20 to \$30 per kW for the fuel cell stack in automobiles when the technology is mature.

Cell construction

The construction of Ballard fuel cells illustrates the manner in which SPFCs are put together. The heart of a cell is the **membrane/electrode assembly (MEA)**, a triple-layer structure consisting of two thin porous sheet electrodes (frequently carbon cloth) separated by the ion exchange membrane electrolyte.

A small amount of catalyst is applied to the side of the electrodes in contact with the electrolyte. The three layers are bonded together by heating the sandwich under pressure to a point where the electrolyte becomes soft.

The MEA measures less than a millimeter in thickness.

Contacts to the electrodes are made through gas flow plates (metal or, more commonly, graphite) on whose surface grooves have been formed to distribute the fuel

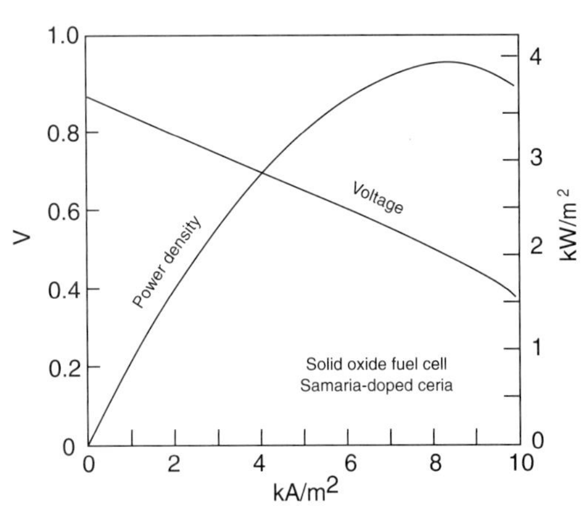

Figure 5.1-15 Characteristics of an experimental SDC fuel cell developed by Hibino.

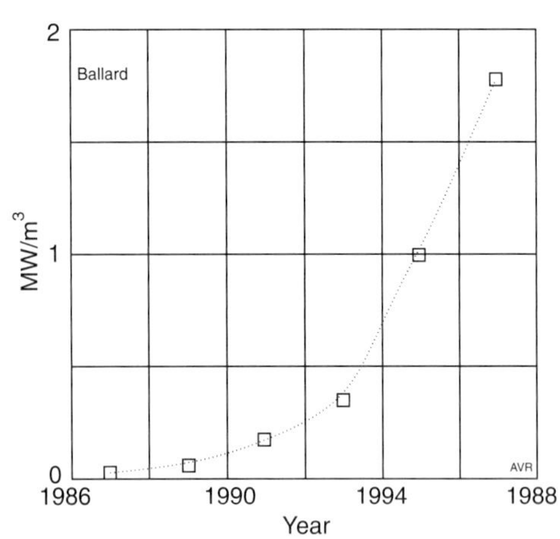

Figure 5.1-16 The power density of Ballard fuel cells has risen exponentially since 1986. Now, more than 2 MW can be generated by 1 m^3 of cells (2 kW/liter).

[††] To be fair, one must compare the aircraft engine weight with the **sum** of the fuel cell stack weight plus that of the electric motor.

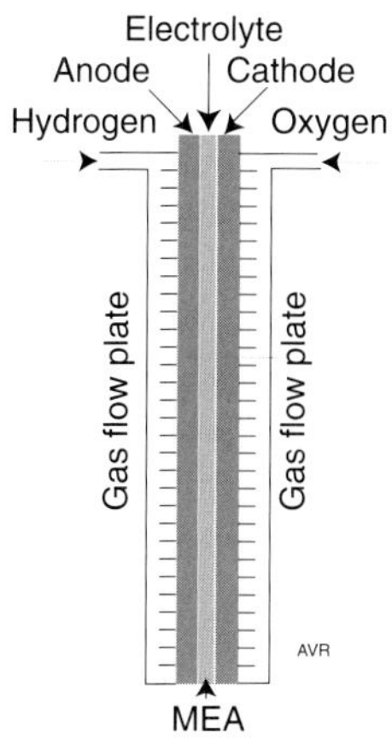

Figure 5.1-17 Structure of an SPFC.

Figure 5.1-18 Ballard fuel cell. Grooves are formed on the surface of the gas flow plates to distribute fuel and oxidant to the cell electrodes. In this cathode face of the plate, oxidant enters and exits through the large, roughly triangular holes at the upper right-hand and lower left-hand corners of the plate. The remaining holes are for hydrogen and water.

and the oxidant uniformly over the whole active area of the cell. An example of the complicated pattern of grooves in the gas flow plates appears in Figure 5.1-18.

The plates have a number of holes near the corners constituting a manifold through which fluids flow. All but the two end plates are bipolar—one face connects to the anode, the other, to the cathode of the cells. The plate depicted shows its cathode side to which oxidant is fed and from which oxygen-depleted oxidant and product water are removed. The anode side of the flow plate is fed through the smaller hydrogen holes. This is because air is delivered in larger quantities than what would be demanded by stoichiometry—the excess air is used to cool the cells and remove the product water.

The design of the groove pattern can greatly influence the performance of the cell. The grooves should not trap any of the reaction water thereby "drowning" areas of the cell and must distribute the reactants uniformly. The bipolar configuration facilitates stacking the cells so that they are electrically in series.

Membrane

Though most membranes are of the proton-exchange type, anion-exchange ones have been proposed claiming a number of advantages. Among these are their immunity to CO_2 fouling, simplified water management requirements, and the possibility of using less expensive non-noble metal catalysts.

Membranes that act as electrolyte must be excellent ionic conductors, but must not conduct electrons. They must be reasonably impermeable to gases and sufficiently strong to withstand substantial pressure differences.

Many such membranes consist of a Teflon-like backbone to which side chains containing a sulfonic (SO_3) group are attached. These acid groups are immobile, and cannot be diluted or leached out.

The thinner the membrane, the smaller its resistance to the flow of protons. Typically, membranes 50 to 180 μm thick are used.† If they are too thin, gas crossover occurs and the performance of the cell is degraded, as witnessed by a reduction in the V_{oc} of the cell.

For a membrane to be a good proton conductor, its SO_3 concentration must be sufficiently high. The ratio between the mass of dry polymer (in kg) and the number of kilomoles of SO_3 sites is called the **equivalent weight (EW)**. Smaller EWs lead to higher ion conductivity. For membranes with the same backbone, the shorter the side chains, the smaller the EW. Thus Nafion having a side chain,

$$-O - CF_2 - CF_3\,CF - O - CF_2 - CF_2 - SO_3H,$$

has a larger EW than the Dow membrane with its shorter chain,

$$O - CF_2 - CF_2 - SO_3H.$$

In fact, Nafion has a typical EW of 1100 kg per kmoles whereas the Dow membrane has an EW of some 800.

The proton conductivity of a membrane also depends on it being thoroughly hydrated—a dry membrane loses most of its conductivity and, for this reason, careful water management is essential (see below).

† The paper commonly used in copying machines is roughly 100 μm thick.

Membranes can be surprisingly costly. In the 80s, Nafion sold at \$800 per m^2. One reason is that this material was created to last 100,000 hours of operation in a chlorine producing plant. On the other hand, an automobile is not expected to survive much more than 300,000 km, which, even at a very modest speed of 30 km/h, represents only some 10,000 hours of operation. Nafion is, as far as automotive fuel cells are concerned, "overdesigned" by one order of magnitude. Ballard set out to develop its own proprietary membrane which contains much less fluorine and is much cheaper in spite of being even more efficient than Nafion and the Dow membrane. Surprisingly, the life of the membrane exceeds 15,000 hours, more than enough for an automotive fuel cell.

Catalysts

Because of their low operating temperature, the chemical kinetics of SPFC are unacceptable unless special catalysts are used. Catalysts typically use expensive noble metals and greatly contribute to the cost of the cells. Great effort has been made to lower the amount of material used. In earlier cells, platinum loading was about 4 mg/cm^2. This platinum loading would add more than \$500 to the cost of each square meter of active fuel cell area.

Fortunately, techniques that permit a reduction of a factor of 10 in the platinum loading are now practical and the Los Alamos National Laboratory has had success with even less catalyst. The British firm, Johnson and Matthey, producers of catalytic converters for conventional (IC) cars, is working with Ballard to reduce the cost of catalysts in SPFCs. The quest for reduced platinum loading is partially based on the fact that in a catalyst it is only the surface that is active. By reducing the size of the grains, the surface to volume ratio is increased—that is, less platinum is needed for a given catalyst area.

Another approach is to use small grains of graphite upon which a thin layer of platinum has been deposited. It is also possible to reduce the amount of platinum by alloying it with ruthenium which is cheaper, acts as a catalyst, and is more CO-tolerant. Indeed, one problem with some catalysts is their vulnerability to the presence of certain contaminants—especially CO—in the hydrogen stream. If the cell is to be fed with hydrogen extracted from a carbon-bearing "carrier" such as methane or methanol, then the reformate will invariably contain some carbon monoxide—perhaps as much as 1% to 2%. Much lower concentrations are required for proper operation of the cells. However, it may be costly to reduce the CO content below 100 ppm.

One possibility is to force the reformate through a palladium filter which will eliminate most of the contaminant. The Los Alamos National Laboratory has built such a filter based on a thin tantalum sheet plated with palladium on both sides. Idatech, of Bend, OR, has developed fuel processors capable of producing very pure hydrogen thanks to the use of palladium filters. Other types of catalysts are being considered and it may even be possible to use enzymes for this purpose.

Hydrogen containing small amounts of CO (say 100 ppm) will drastically impair the performance of low temperature fuel cells with pure Pt catalysts as shown in Figure 5.1-19. However, if a small amount of O_2 is added to the fuel, it will selectively combine with the carbon monoxide and almost complete recovery of the cell performance is achieved.

The CO vulnerability of Pt catalysts diminishes with increasing temperature. At 80 C, the CO concentration must be kept below 10 ppm. At 90 C, where Ballard cells operate, higher concentrations are acceptable.

Alloy catalysts (Pt + Ru, for instance) are less vulnerable to CO. Ballard has shown that a reformate containing as much as 2% CO can be used if it is passed over a platinum-on-alumina catalyst and Pt-Ru is used in the anode.

There is a possibility of avoiding altogether the use of expensive platinum based catalysts. The anode reaction consists of splitting the hydrogen molecule into two protons and two electrons (see Equation 5.1.3), a feat performed by many anaerobic bacteria (Chapter 12.1) that use enzymes called **hydrogenases**. Synthetic hydrogenases have been created by Robert Hembre of the University of Nebraska, Lincoln, using compounds containing ruthenium and iron. Ruthenium, being substantially cheaper than platinum, may have a future as a catalyst in low temperature fuel cells.

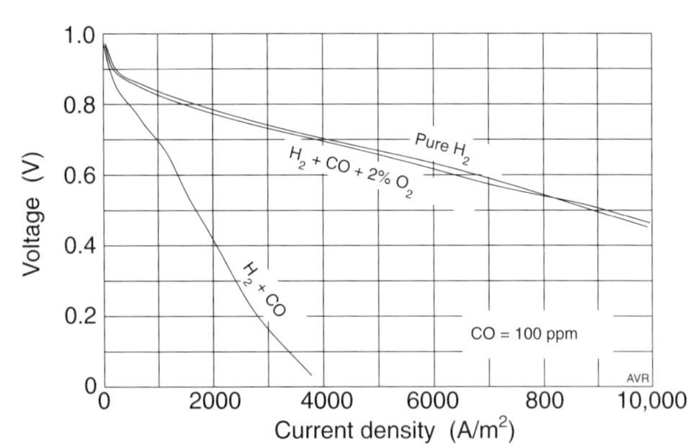

Figure 5.1-19 When an SPFC using pure Pt catalysts is fed hydrogen contaminated by 100 ppm of CO, its performance suffers drastically. However, if a small amount of oxygen is added to the fuel stream, it will selectively combine with the CO and the performance is then essentially indistinguishable from that with pure hydrogen.

If the fuel cell catalysts become poisoned, all is not lost—a short exposure to pure H_2 will completely rejuvenate the cell.

CO is not the only contaminant to avoid. SPFCs have extremely low tolerance for sulfur compounds (H_2S, for instance) and for ammonia (NH_3).

Water management

Present-day ion-exchange membranes are an acid electrolyte. Therefore, water forms at the cathode where it would collect and "drown" the electrode—that is, it would impede the penetration of the oxygen to the active surface. To reduce the tendency of the water to collect in the pores and interstices of the MEA, a hydrophobic material is applied (typically a Teflon emulsion).

Water is removed by the flow of oxidant which is generally supplied well in excess of the stoichiometric requirement. Usually the amount of air circulated is double that needed to furnish the correct amount of oxygen.

Although the anode reaction (Equation 5.1.3) indicates the formation of protons, the ion that actually migrates through the electrolyte is hydronium—a hydrated proton. This means that water is consumed at the anode because it is electrically "pumped" to the cathode, tending to desiccate the membrane. This drying is compensated, in part, by water from the cathode that diffuses back through the membrane driven by the concentration gradient. Nevertheless, there is a tendency to dehydrate the proton-exchange membrane which tends to wrinkle and have its proton conductivity drop catastrophically. To avoid desiccation, a partial pressure of at least 50 kPa of H_2O must be maintained in both the fuel and oxidant streams. If the feed pressure is 100 kPa (1 atmos.), the partial pressure of H_2 would be some 50 kPa and the output voltage would fall correspondingly. For this reason, the fuel cell is usually pressurized to 300 kPa (3 atmos.) leading to a hydrogen partial pressure of 250 kPa, compatible with efficient fuel cell operation.

Careful water management is necessary to avoid either "drowning" or "desiccation" of the membrane.

5.1.5.6 Direct methanol fuel cells

As discussed in the preceding subsection, one attractive alternative for carrying hydrogen on board a fuel-cell vehicle is to use a hydrogen carrier, a substance from which hydrogen can be extracted as needed by some chemical reforming processes. Methanol is a leading candidate as a hydrogen carrier for vehicular applications. Nevertheless, the need for separate fuel processing equipment in the vehicle adds to the cost and complexity of the system. An obvious solution is to develop fuel cells that can use methanol directly without the need of preprocessing.

The Jet Propulsion Laboratory (JPL) has done considerable work on the development of direct methanol fuel cells (DMFC) (*see Halpert et al.*, 1997).

The fuel used is a low concentration (3%) methanol in water solution while air is the oxidant.

The anode reaction is

$$CH_3OH + H_2O \rightarrow CO_2 + 6H^+ + 6e^-. \qquad (5.1.28)$$

The cathode reaction is

$$6H^+ + 6e^- + \frac{3}{2}O_2 \rightarrow 3H_2O. \qquad (5.1.29)$$

Thus, the overall reaction is

$$CH_3OH + \frac{3}{2}O_2 \rightarrow CO_2 + 2H_2O. \qquad (5.1.30)$$

Halpert (1997) reports a cell producing 5.0 MJ of energy per liter of methanol consumed when operated at 90 C and 2.4 atmospheres of air pressure. This can be interpreted as an efficiency of 28% if referred to the higher heat of combustion of the alcohol or 32% if referred to the lower.

The JPL DMFC is a solid-polymer fuel cell using a Nafion 117 membrane with a fairly high loading of noble-metal catalyst (2 to 3 mg of Pt-Ru per cm^2). The combination of Nafion with high catalyst loading leads to high costs. The search for better anode catalysts can be greatly accelerated by the technique described by Reddington *et al.* (1998) that permits massively parallel testing of large numbers of catalyst samples. (See also Service, 1998.)

One advantage of DMFCs is that since the anode is in direct contact with water, problems of membrane dehydration are circumvented and the constantly flowing liquid simplifies heat removal.

Pollution problems are greatly alleviated because only carbon dioxide and water are produced whereas carbon monoxide is generated when methanol is reformed into hydrogen for use in other types of cell.

"Methanol crossover" (i.e. the transport of methanol from anode to cathode through the Nafion membrane) severely reduces the cell efficiency. In the JPL cell mentioned above, crossover consumed 20% of the fuel. Efforts are being made to develop new low cost membranes less subject to this difficulty. It appears that the figure of 10% has already been achieved. Lower crossover rates will permit operation with more concentrated methanol mixtures yielding better efficiencies. JPL feels confident that efficiencies above 40% can be attained.

Crossover reduces the efficiency of the cell, not only because some fuel is diverted, but also because the

Table 5.1-1 Thermodynamic data for methanol fuel cells at RTP

Methanol	Water	$\Delta H°$ MJ/kmole	$\Delta G°$ MJ/kmole
liquid	gas	−638.5	−685.3
gas	gas	−676.5	−689.6
liquid	liquid	−726.5	−702.4
gas	liquid	−764.5	−706.7

Data from Dohl *et al.* (2000)

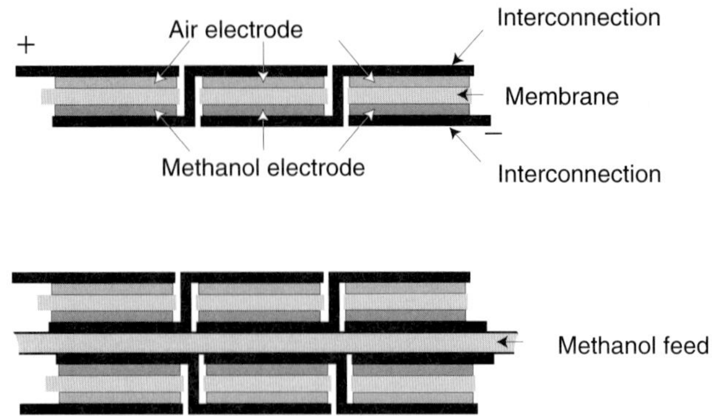

Figure 5.1-20 Bipolar configuration reduces the series resistance between individual cells, an advantage when high currents are involved. Flat-packs can be extremely thin, making it easy to incorporate them into small-sized equipment. In the flat-pack configuration (developed by the Jet Propulsion Laboratory, see Narayanan *et al.*, 2000) cell interconnections pierce the membrane as shown. Two flat-packs deployed back-to-back permit the sharing of the methanol feed.

methanol that reaches the cathode is prone to undergo the same reaction (Equation 5.1.28) that normally takes place at the anode, generating a counter-voltage and, thus, reducing the voltage delivered to the load. International Fuel Cells Corporation has a patent for a catalyst that promotes the reaction of Equation 5.1.29, but not that of Equation 5.1.28.†

If the methanol at the cathode is not consumed, perhaps it can be recovered. This can be done by condensing the methanol–water vapor mixture exhausted from the cathode.

At present, direct methanol fuel cells are being considered for transportation applications and for powering portable electronic equipment. For the latter, "air breathing"—that is, delivery of air to the cathode by diffusion only (no pumps)—is important.

Instead of employing the conventional "bipolar" configuration, direct methanol fuel cells designed for the powering of cell phones, laptops and other small portable equipment may employ the "flat-pack" design shown in Figure 5.1-20 (top) or the "twin-pack" design (bottom). This lends to the batteries (or "stacks") a shape that is more compatible with the equipment with which they are used.

5.1.5.7 Solid acid fuel cells

Phosphoric acid and most solid polymer fuel cells are examples of cells that use hydrated acid electrolytes. Solid acids, on the other hand, can exhibit anhydrous proton conductivity and may have certain advantages over the popular SPFC. See Boysen *et al.* (2000).

H_2SO_4 is a common acid. However, it is liquid at temperatures one would expect fuel cells to operate. On the other hand, replacing one of the hydrogens by a cesium atom leads to $CsHSO_4$, an acid (known as cesium hydrogen sulfate) with a high melting point. Above 414 K, this acid becomes a good proton conductor (commonly referred to as a **superprotonic** or **superionic** conductor) and has been proposed as a fuel cell electrolyte by groups at the California Institute of Technology and the University of Washington, among others.

Since these solid acids are soluble in water and since water is produced by the fuel cell reaction, it is essential to operate the cells at temperatures high enough to assure that any water that comes in contact with the electrolyte is in vapor form.

This means that operating temperatures should be above some 150 C.

A major difficulty with $CsHSO_4$ (and with the corresponding selenate) is that they react with hydrogen forming, in the sulfate case, hydrogen sulfide. Greater stability is expected from phosphate-based compounds. CsH_2PO_4 fuel cells, for instance, were demonstrated at 235 C (see Boysen *et al.*, 2000).

The advantage of SAFC over SPFC is that, operating at higher temperatures, they are less susceptible to CO poisoning and require less catalyst. In addition, since the water is always in vapor form, they do not need the careful water management required by the polymer cells.

Perhaps the most promising aspect of SAFC is that the electrolyte is not permeable to methanol, thus eliminating the serious methanol cross-over problem of

† Chu *et al.* at the U.S. Army Research Laboratory, have developed an iron-cobalt tetraphenyl porphyrin catalyst that can promote the oxygen reduction to water (Eq. 5.1.29) but will not catalyze methanol oxidation (Eq. 5.1.28). Consequently, it has possibilities as a cathode catalyst in DMFC.

polymer cells. (See the subsection on direct methanol fuel cells.) The power densities of solid acid methanol fuel cells (Boysen *et al.*, 2000) are within a factor of 5 of the density of the most advanced DMFC (2004). The electrolyte thickness of these cells was 260 μm. Future cells with substantially thinner electrolytes may exhibit considerable improvement in their performance. However, one important difficulty with this type of cell is the fragility of the electrolyte.

5.1.5.8 Rechargeable fuel cells and the nickel metal hydride battery (NiMH)

The growingly popular NiMH battery is useful in, among other applications, electrical vehicles. Although not strictly a fuel cell, it deserves mention here. It is a secondary battery (i.e., it is rechargeable) that derives its energy from the change in the Ni oxidation state from +3 to +2.

A metal alloy anode is immersed in a 30% (by weight) solution of KOH in water. This strong alkali dissociates almost completely into K^+ and OH^- ions. Hydrogen supplied to this electrode will combine with the hydroxyl ion forming water and liberating an electron which becomes available to circulate through an external load constituting the useful output of the cell:

$$H + OH^- \rightarrow H_2O + e^-. \tag{5.1.31}$$

The cathode is made of nickel oxyhydroxide (NiOOH) which, upon receiving an electron, is reduced to nickel hydroxide liberating a hydroxyl ion:

$$NiOOH + H_2O + e^- \rightarrow Ni(OH)_2 + OH^-. \tag{5.1.32}$$

Notice that the average amount of water and of hydroxyl ions is constant—at the anode water is consumed at exactly the rate at which it is produced at the cathode. This contrasts with NiCd batteries in which, during the discharge, water is consumed at both electrodes. The OH^- concentration remains constant during the discharge because this ion is produced at the anode and consumed at the cathode.

The materials consumed, hydrogen and nickel oxyhydroxide, are regenerated during the charge when an external power supply forces electrons into the metal electrode. This causes the water to be electrolyzed into H^+ and OH^-. At the other electrode, the nickel hydroxide is oxidized into nickel oxyhydroxide consuming

Table 5.1-2 USABC performance goals compared with prototype NiMH battery performance

Property	USABC	NiMH
Mass energy density ($MJ\ kg^{-1}$)	0.29 (0.36 desired)	0.29
Volume energy density ($MJ\ m^{-3}$)	486	775
Mass power density ($kW\ kg^{-1}$)	0.15 (>0.2 desired)	175
Volume power density ($kW\ m^{-3}$)	250	470
Cycle life (cycles)	600	1000
Life (years)	5	10
Environmental operating temp.	−30 to 65 C	−30 to 60 C
Recharge time	<6 hours	15 min (60%) <1 hour (100%)
Self discharge	<15% in 48 hours	<10% in 48 hours

Energy density values were measured with 50-Ah prototype cells, discharged at a rate that exhausted the cells in 3 hours. The mass power density was determined by discharging the battery in 30 seconds to 20% of its capacity. In determining the cycle life, the battery was repeatedly discharged to 20%.

hydroxyl ions. The metallic alloy on the hydrogen side absorbs this gas, storing it in the form of hydride (see Chapter 12.1), so that no gas evolves. The hydrogen becomes available during the discharge. Nominal voltage delivered by each element of an NiMH battery is 1.2 V, some 60% of that of a lead–acid cell.

When first introduced, prototypes of this battery exceeded most of the requirements of the US Advanced Battery Consortium (USABC),[†] as shown in Table 5.1-2 taken from an article by Ovshinsky *et al.* (1993).

The larger the energy density of the battery, the larger the range of the vehicle. The GM EV1 (the first car designed from the ground up as an electric vehicle) used lead–acid batteries storing 49 MJ of energy (roughly the energy of one kg of gasoline) and had a range of 140 km. Equipped with NiMH batteries of the same mass, a more modern version of the vehicle had a range of over 250 km.

Solectria, a manufacturer of electric vehicles, has built a demonstration car[††] with a composite body and equipped with a 32.5 kWh (117 MJ), 210-cell NiMH Ovonic battery and has covered a 601.25 km range (through rural roads and city streets) on a single charge.

[†] USABC, under the Department of Energy, promotes the development of batteries for electrical vehicles (EVs).
[††] The "Sunrise" carries two passengers and has a curb weight of 1000 kg.

This was a new record for EV. Ovonic Battery Co. of Troy, Mich. is the developer of MiMH batteries.

As reported by the Argonne National Laboratories, the NiMH battery compares favorably with other candidates for EV application:

Table 5.1-3 Peak power delivered by different batteries (at 25% depth of discharge)

Battery type	Peak power W/kg
NiMH	235
NiCd	200
NaS	145
NiFe	120
Pb–Acid	100
LiS	100
ZnBr	65

Material requirements of the NiMH battery are complicated. The energy storage capacity of the battery depends on the amount of hydrogen that can be absorbed by the metal alloy electrode. It is necessary that the hydride formation be easily reversible (see Chapter 12.1) and this is determined by the enthalpy of formation of the hydride that must fall in the 25 to 50 MJ/kmole range. If the enthalpy of formation is too small, hydrogen will fail to react with the alloy and gas will evolve. If too large, the electrode will be oxidized. Surface properties of the metal alloy are critical in determining the catalytic activity, electric conductivity, and the porosity and area of the surface.

The crucial question of cost of the battery will depend on many factors, especially on the demand for these devices. Nevertheless, at the present stage of development, the NiMH battery looks like a promising candidate as a storage system for EVs.

Small NiMH batteries are also popular for use in camcorders, cellular phones and other small devices. In these applications, NiMH batteries face a serious competitor—lithium ion batteries.

5.1.5.9 Metallic fuel cells and zinc–air fuel cells

From the practical point of view, one of the advantages of **refuelable** cells, over **rechargeable** ones, is that refueling tends to be much faster than recharging (minutes versus hours). However, hydrogen fuel cells suffer from the difficulty of transporting and storing the gas. The use of certain metals as fuel may lead to simple and safe transportability and large volumetric energy density. Metallic cells are an exception to the rule of using fluids as fuel. Two of the metals considered for this purpose are aluminum and zinc. Aluminum, however, is corroded by the caustic electrolyte even when no current is being generated. Aluminum-Power, Inc. has developed a type of cell in which the electrolyte is pumped away when the cell is not in use. This apparently leads to a cumbersome system. It appears that it is easier to use zinc. Metallic Power, Inc. has created a promising zinc–air fuel cell. Small zinc pellets (about 1 mm in diameter) are one of the reactants, the other being oxygen, in general just plain air. The electrolyte in the cells is a concentrated KOH solution.

At the anode (the negative electrode) the reaction is

$$Zn + 4OH^{--} \rightarrow Zn(OH)_4^{--} + 2e^-. \quad (5.1.33)$$

The ion on the right-hand side of the equation is a zincate ion which consists of a positively charged Zn^{++} surrounded by four hydroxyls (OH^-). The electrons, after circulating through the load, recombine with the oxygen at the cathode and with water regenerating half of the hydroxyl ions used at the anode,

$$2\,e^- + \frac{1}{2}O_2 + H_2O \rightarrow 2OH^-. \quad (5.1.34)$$

The electrolyte containing the zincate ion is pumped to an **Electrolyte Managing unit**—an integral part of the fuel cell—where the reaction,

$$Zn(OH)_4^{--} \rightarrow ZnO + H_2O + 2OH^-, \quad (5.1.35)$$

not only supplies the water consumed at the cathode and the other half of the needed hydroxyls, but also generates zinc oxide which precipitates out and can be removed from the cell for recovery of the zinc.

The overall fuel cell reaction is

$$Zn + \frac{1}{2}O_2 \rightarrow ZnO. \quad (5.1.36)$$

The cell is refueled by loading its hopper with zinc pellets. As the zinc is consumed, the hopper automatically drops additional pellets into the reaction area. The zinc oxide is transferred to a stand-alone **zinc regeneration/refueling system** in which the the oxide is electrochemically reduced to metal which is then pelletized for delivery to the fuel cell.

5.1.6 Fuel cell applications

Fuel cells are in a stage of rapid development and are on the verge of achieving maturity. As time goes on and their economic potential becomes well documented, this technology will occupy a growing number of niches.

At present, much of the effort is being concentrated in two distinct areas of application: stationary power plants and automotive power plants.

5.1.6.1 Stationary power plants

Stationary power plants of various types include central utility-operated power plants of large capacity (say, up to 1 GW), dispersed utility-operated power plants (perhaps in the tens of MW sizes) and on-site electrical generators (some 10 to 100 kW).

For these applications, fuel cells present the following advantages over conventional heat engines:

1. Absence of noise.
2. Little pollution.
3. Ease of expansion (owing to modular construction).
4. Susceptibility to mass production (again, owing to modularity).
5. Possibility of dispersion of power plants. Owing to the low pollution and low noise, plants can be operated even in residential areas, thus economizing transmission lines.
6. Possibility of using reject heat for ambient heating because fuel cells can be near populated areas where there is a demand for hot water.
7. Possibility of cogeneration, using the high temperature exhaust gases in some types of plant.
8. Fast response to demand changes.
9. Good efficiency at a fraction of rated power.
10. Extremely good overload characteristics.
11. Small mass/power ratio, in some types of plant.
12. Small volume/power ratio, in some types of plant.
13. Great reliability (potentially).
14. Low maintenance cost (potentially).

Owing to the modular construction of many fuel cells, plant capacity can be easily expanded as demand grows. Capital investment can be progressive, considerably lessening the financial burden.

Clearly, not all of the above advantages can be realized simultaneously. Cogeneration can only be achieved with high temperature fuel cells such as MCFCs and SOFCs, especially the latter. When cogeneration is used, the low noise advantage may be lost.

5.1.6.2 Automotive power plants

Imagine a hypothetical country having a modern industrial base but, for some unexplained reason, lacking completely any automotive industry. Imagine also that someone wants to build a small number of automobile engines of the type that drives present-day cars and that in the United States, Germany, or Japan, can be bought for roughly $4000 a piece. Since specialized mass production machines do not exist, the engines have to be built one by one and their cost would be certainly more than an order of magnitude higher than that of the imported model.

Exactly the same situation prevails, all over the world, regarding fuel-cell power plants for cars. Current technology is already quite adequate as far as efficiency, life time, power-to-mass ratio, power-to-volume ratio, and so forth are concerned. Further improvements will be made, but the existing technology is acceptable. However, the cost of manufacturing is not. It will only come down when a sufficient number of units can be sold; a sufficient number will only be sold when the cost comes down. This vicious circle is hard to break. One must remember that the low cost of automobiles results from millions upon millions being sold. The retooling for a new model runs to 1 or 2 billion dollars and can only be justified by large sales. Once the vicious circle above is broken, a second, very attractive, area of application of fuel cells will be in transportation.

For small and medium sized (and perhaps even for large) vehicles, the compact SPFCs are nearly ideal. Among the many advantages that can be claimed for them, one must count the extremely low pollution, the high efficiency (guaranteeing fuel economy), and the high power density (permitting compact designs). In addition, their low operating temperature permits rapid startups.

The expected long life time of SPFC stacks may lead to an unusual situation. A cell with a plausible 50,000 hour life would, even at only 40 km/h average speed, drive a car some 2 million kilometers. The fuel cell would outlive by far the automobile body. It would, then, make sense to re-use the cell in a new car. Different vehicle models could be designed to operate with one of a small number of standardized fuel cell types. This would have the adverse effect of reducing the market for fuel cells.

Hydrogen is the usual fuel for SPFCs. The gas can be used as such (compressed or liquefied) or in the form of a hydrogen "carrier" such as methanol or a metallic hydride. Hydrogen carriers can be derived from fossil fuels or from biomass.

Fuel-cell cars may need energy storage devices (batteries, flywheels or ultracapacitors) to provide startup power, and to accumulate the energy recovered from dynamic braking. This stored energy can be used to supplement the fuel cell during fast accelerations. Hence, the needed fuel-cell power might be closer to the average power required by the vehicle than to the peak power. This would cut down the size and cost of the cell stack. Under all circumstances, the total electrically stored energy would be only a small fraction of that of a purely electric car.

Ballard Power Systems of Vancouver, BC, is one of the manufacturers closest to bringing fuel cells for vehicular use into regular production. In 1999, their fuel cell buses were already in revenue service in Vancouver and in Chicago. These vehicles, fueled by compressed hydrogen, have a range of 560 km, with excellent acceleration and a top speed of over 95 km/h. They are driven by a 275-hp motor and are totally non-polluting.

Daimler-Benz, which has a major financial interest in Ballard, produces a fuel-cell van, the "Necar II," that is being tested by various users in Germany. The company plans to soon have production fuel cell cars for sale. Following this example, Ford, Toyota, Mazda and General motors have made similar promises. Honda delivered the first few fuel cell FCX vehicles in 2003.

Manufacturers have been accumulating experience with cars that may help the transition away from purely internal combustion engines. Extremely efficient cars like the General Motors EV1 (whose production has stopped in 1999) can travel over 180 km on 14 kWh of electricity. This energy can be obtained from less than 1 kg of hydrogen stored in less than 100 kg of hydride. Honda also produced a limited number of electric cars—the EV Plus. Toyota has sold thousands of hybrid (electric-IC) cars—the Prius.

5.1.6.3 Other applications

The first practical applications of fuel cells was in space, where considerations of reliability far outweigh cost. The cells work with the hydrogen and the oxygen already available for other uses in the spacecraft and provide valuable drinking water as an output. GE SPFCs started the trend; at the moment AFCs are more popular—they were used in the Apollo program and currently supply energy for the space shuttle and for the International Space Station.

Small submersibles as well as full scale submarines benefit from the clean operation of fuel cells. The German navy developed 400-kW fuel cells for their submarines. This is half the power of their standard diesel engines. The fuel cells will probably be used in a hybrid combination with the diesels. Some military applications benefit from the low heat signature and absence of noise. Once fuel cells become even more compact and light, they may allow the operation of "cold" aircraft invisible to heat seeking missiles. Fuel cells are important in the propulsion of other naval vessels. Very compact power plants can be built in combination with superconducting motors. The U.S. Department of Defense is actively pursuing this development. Presumably a number of other nations are doing the same.

Micro-fuel cells appear to be a promising solution for trickle charging cellular phones and other portable electronics. They will probably be direct methanol devices. In this particular application, micro-fuel cells may have to face stiff competition from **nuclear batteries** using β^- emitters such as tritium or nickel-63. Such batteries take advantage of the enormous energy density of nuclear materials, (thousands of times more than chemical batteries). One implementation consists of a metallic target that, collecting the emitted electrons, becomes negatively charged. The resulting electrostatic attraction causes a thin silicon cantilever to bend until the target touches the source and discharges. The cantilever snaps back and the process repeats. The mechanical energy of the vibrating cantilever is transformed into electrical energy by means of a piezoelectric converter. Efficiencies of 4% have been demonstrated, but Lal and Blanchard (2004) expect 20% with more advanced configurations. It should be pointed out that beta-emitters are relatively harmless.

5.1.7 The thermodynamics of fuel cells

Symbology

G, free energy,

H, enthalpy

Q, heat,

S, entropy, and

U, internal energy:

1 Capital letters will indicate the quantity associated with an arbitrary amount of matter or energy.

2 Lowercase letters indicate the quantity per unit. A subscript may be used to indicate the species being considered. For example, the free energy per kilomole of H_2 will be represented by $\bar{g}_{H_2}$.

g = free energy per kilogram.

$\bar{g}$ = free energy per kilomole.

g^* = free energy per kilogram, at 1 atmosphere pressure.

$\bar{g}^*$ = free energy per kilomole, at 1 atmosphere pressure.

$\bar{g}_f$ = free energy of formation per kilomole.

$\bar{g}_f^o$ = free energy of formation per kilomole, at 298 K, 1 atmosphere, i.e., at RTP, (Standard Free Energy of Formation).

5.1.7.1 Heat of combustion

Let us return to reaction 5.1.5

$$2H_2 + O_2 \rightarrow 2H_2O$$

(in section 5.1.2) in which hydrogen and oxygen combine to form water. In rearranging 4 hydrogen atoms and 2 oxygen atoms to form 2 molecules of water, some energy is left over. The force that binds atoms into molecules is electric in nature. Nevertheless, when H_2 reacts directly with O_2, only heat results because electrons and ions collide and their energy is randomized.

Assume that a measured amount, μ, of hydrogen is introduced into a constant pressure calorimeter containing sufficient oxygen and is made to react. A certain amount of heat, Q, is released. The ratio, $Q/\mu = \bar{h}_{comb}$, is the **heat of combustion**, of hydrogen in an atmosphere of oxygen (joules/kmole). The exact amount of heat released depends on the temperature of both the reactants and the product and on the state of the latter. If the water produced is liquid, the heat of combustion is larger than when water is in the form of gas because the heat released includes the heat of condensation of water. As explained in Section 3, all fuels containing hydrogen have two different heats of combustion: the **higher heat of combustion** in the case of liquid water and the **lower heat of combustion** in the case of water vapor.

Since the reaction occurs at constant pressure, the heat released equals the change in enthalpy, ΔH, of the system.

$$Q = \Delta H0. \tag{5.1.37}$$

Q is taken as positive if it is added to the system as happens in a heat engine. In **exothermic** reactions, $Q < 0$ and so is ΔH.

Notice the sign convention used here. Energies introduced into a system, $\sum W_{in}$, are taken as positive; energies rejected by the system, $\sum W_{out}$, are negative. Consequently the energy balance reads

$$\sum W_{in} + \sum W_{out} = 0. \tag{5.1.38}$$

Enthalpy, as any other form of energy, has no absolute value; only changes can be measured. We can, therefore, arbitrarily assume any value for the enthalpy of the reactants. By convention, *at 298.15 K, the enthalpy of all elements in their natural state in this planet is taken as zero*. This means that, at 298.15 K, the enthalpies of H_2 and O_2 are zero, but those of H and O are not.

The difference between the enthalpy of a product and those of its elements is the **enthalpy of formation**, ΔHf, of the product. If both reactants and products are at the **reference temperature and pressure (RTP)**,† then the enthalpy of formation is called the **standard** enthalpy of formation, $\Delta Hf°$.

From the calorimeter experiment, for water (expressing the enthalpies in a per kilomole basis),

$$H_2O\ (g): \bar{h}^{\circ}_{f_g} = -241.8\ \text{MJ/kmol}, \tag{5.1.39}$$

$$H_2O\ (l): \bar{h}^{\circ}_{f_l} = -285.9\ \text{MJ/kmol}. \tag{5.1.40}$$

The subscripts "g" and "l" indicate the state of the product water. The difference of −44.1 MJ/kmol between the enthalpies of formation of liquid and gaseous water is the latent **heat of condensation**, $\bar{h}_{con}$, of water. Clearly, $\bar{h}_{con} = -\bar{h}_{vap}$, where $\bar{h}_{vap}$ is the **latent heat of vaporization**.

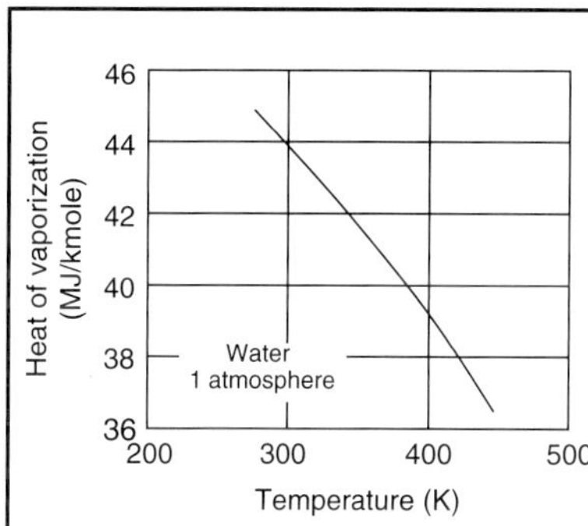

For coarse estimates, the latent heat of vaporization can be taken as constant. However, it is, in fact, temperature dependent, falling somewhat with rising temperature as can be seen in the figure.

5.1.7.2 Free energy

If the reaction under consideration occurs not in a calorimeter but in an ideal fuel cell, part (but not all) of the energy will be released in the form of electricity. It is important to investigate how much of the ΔH of the reaction will be converted into electric energy and to understand why some of the energy must take the form of heat even in the ideal case.

Let V_{rev} be the voltage produced by the cell. Each kilomole of water (or of any other substance) contains N_0 molecules, where $N_0 = 6.022 \times 10^{26}$ is **Avogadro's number.**†† From Equation 5.1.4, Section 5.1.2, describing the cathode reaction of the cell

† **Standard temperature and pressure (STP)**, frequently used by chemists, corresponds to 1 atmosphere and 0 C (273.15 K); the *CRC Handbook of Chemistry and Physics*, however, lists the standard thermodynamic properties at 1 atmosphere and 298.15 K, which we will call RTP.
†† Observe that, owing to our use of kilomoles instead of moles, Avogadro's number is three orders of magnitude larger than the value usually listed.

$$4e^- + 4H^+ + O_2 \rightarrow 2H_2O$$

we see that for each molecule of water, 2 electrons (charge q) circulate in the load. The energy delivered to the load is the product of the charge, $2qN_0$, times the voltage, V_{rev}. More generally, the electric energy produced per kilomole of product by a reversible fuel cell is

$$W_e = n_e q N_0 V_{rev} \equiv n_e F V_{rev}, \quad (5.1.41)$$

where

n_e = number of kmoles of electrons released per kilomole of products;
q = charge of the electron in coulombs;
N_0 = Avogadro's number in MKS;

$$F \equiv qN_0 \equiv \textbf{faraday} = 1.602 \times 10^{-19} \times 6.022 \times 10^{26}$$
$$= 96.47 \times 10^6 \text{ coulombs/kmole}$$
$$= \text{charge in one kmole of electrons;}$$

V_{rev} = voltage.

Consider a hypothetical experiment in which a reversible fuel cell is constructed and its open-circuit voltage is accurately measured. If such experiment were carried out, the voltage measured at **RTP** would be 1.185 V. The voltage delivered by a reversible fuel cell is called the **reversible voltage** and is, usually, designated by V_{rev}, as was done above.† Thus, the electric energy produced by a reversible fuel cell of this type is

$$|W_e| = 2 \times 96.47 \times 10^6 \times 1.185$$
$$= 228.6 \text{ MJ/kmole.} \quad (5.1.42)$$

The electric energy delivered to a load by a reversible cell is called the **free energy change** owing to the reaction (designated ΔG). Usually, if there is a single product, ΔG is given per kilomole of product, and is represented by $\overline{g}_f^{\circ}$ (if at RTP). Again, since the cell *delivers* electric energy, the free energy change is negative, conforming to the convention for the sign of the enthalpy change of the reaction. As in the case of enthalpies, by convention, at RTP, the *free energy of all elements in their natural state in this planet is taken as zero.*

In most cases (but not always—see Problem 5.1.3), $|\Delta G| < |\Delta H|$. This means that energy from the reaction usually exceeds the electric energy delivered to a load even in an ideal reversible cell. The excess energy, $\Delta H - \Delta G$ must appear as heat. Consider the entropies involved.

Each substance has a certain entropy which depends on its state. Entropies are tabulated in, among others, the *Handbook of Chemistry and Physics*, CRC Press. For the substances involved in the reaction we are examining, the absolute entropies, at RTP, are:

H_2 (g) : $\overline{s}^{\circ} = 130.6$ kJ K^{-1}kmol^{-1}
O_2 (g) : $\overline{s}^{\circ} = 205.0$ kJ K^{-1}kmol^{-1}
H_2O (g) : $\overline{s}^{\circ} = 188.7$ kJ K^{-1}kmol^{-1}

When 1 kilomole of water is formed, 1 kilomole of H_2 and 0.5 kilomoles of O_2 disappear and so do the corresponding entropies: a total of 130.6 + 205.0/2 = 233.1 kJ/K disappear. This is, in part, compensated by the appearance of 188.7 kJ/K corresponding to the entropy of the water formed. Nevertheless, the matter balance leads to an entropy change of 188.7 − 233.1 = −44.4 kJ/K.

In a closed system, the entropy cannot decrease (Second Law of Thermodynamics), and at best—under reversible conditions—its change is zero. Consequently, an amount of entropy, $\Delta S = Q/T$ must appear as heat:

$$Q = T\Delta S = 298 \times (-44.4 \times 10^3)$$
$$= -13.2 \text{ MJ/kmol.} \quad (5.1.43)$$

This amount of heat must come from the enthalpy change of the reaction leaving −241.8 − (−13.2) = −228.6 MJ/kmol as electricity.

Chemical energy can be thought of as consisting of two parts: an entropy free part, called **free energy**, that can entirely be converted to electricity and a part that must appear as heat. The free energy,†† G, is the enthalpy, H, minus the energy, TS, that must appear as heat:

$$G = H - TS \quad (5.1.44)$$

and

$$\Delta G = \Delta H - \Delta(TS). \quad (5.1.45)$$

In isothermic cases (as in the example above),

$$\Delta G = \Delta H - T\Delta S. \quad (5.1.46)$$

The electric energy, W_e, delivered by the reversible fuel cell is ΔG:

† Owing to irreversibilities in practical cells, such a measurement cannot be carried out accurately. However, the reversible voltage can be estimated by connecting a voltage generator to the cell and observing its voltage-versus-current characteristic as the applied voltage is varied. If this voltage is sufficiently high, current will be driven into the cell and if it is low, current will be delivered by the cell to the generator. The characteristic should be symmetrical around the reversible voltage.

†† The idea of free energy was first proposed by Josiah Willard Gibbs (1839–1903), hence the symbol "G."

Table 5.1-4 Some thermodynamic values

	$\Delta \bar{h}f°$ (MJ/kmole)	$\Delta \bar{g}f°$ (MJ/kmole)	$\bar{s}°$ (kJ K^{-1} $kmole^{-1}$)
H_2O (g)	−241.8	−228.6	188.7
H_2O (l)	−285.9	−237.2	70.0
H_2 (g)	0	0	130.6
O_2 (g)	0	0	205.0

$$\Delta G = -n_e q N_0 N |V_{rev}|. \tag{5.1.47}$$

where N is the number of kilomoles of water produced. Note that since ΔG is removed from the cell it must be < 0. Per kilomole of water,

$$\bar{g}_f = -n_e q N_0 |V_{rev}|. \tag{5.1.48}$$

The dual of a fuel cell is an electrolyzer. A fuel cell may use hydrogen and oxygen, generating electricity and producing water and heat. The electrolyzer consumes water and electricity, producing hydrogen and oxygen. In the ideal case, the electrolyzer absorbs heat from the environment, acting as a heat pump. If there is insufficient heat flow from the environment to the electrolyzer, the latter will cool down.

For a given amount of gas handled, the electric energy generated by the reversible fuel cell is precisely the same as that required by the reversible electrolyzer and the heat produced by the reversible fuel cell is precisely the same as that absorbed by the electrolyzer. This amount of heat is reversible.

Clearly the reversibility is destroyed if the system has losses. A lossy (read, practical) fuel cell generates more heat than $T\Delta S$, while a lossy electrolyzer will generate heat which may (and frequently does) exceed the thermodynamically absorbed heat. Nevertheless, some realizable electrolyzers may operate with sufficient efficiency to actually cool down during operation.

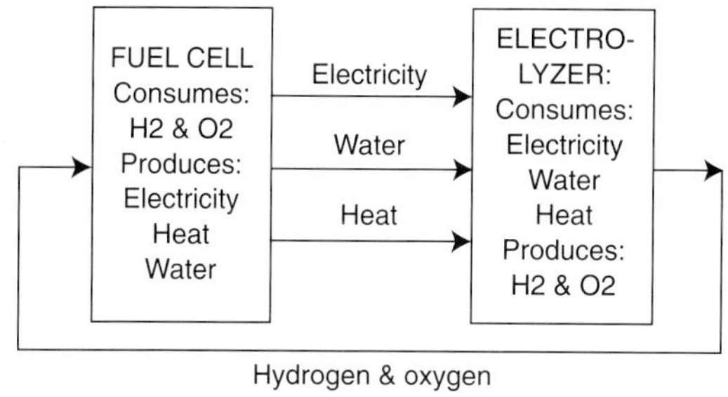

Figure 5.1-21 An ideal fuel cell and its dual, the ideal electrolyzer, act as a reversible system. The inputs of one are precisely the same as the output of the other. For this to be true, the input to the electrolyzer must be water vapor.

5.1.7.3 Efficiency of reversible fuel cells

The efficiency of a fuel cell is the ratio of the electric energy generated to the enthalpy change of the chemical reaction involved. In reversible cells,

$$\eta_{rev} = \frac{\Delta G}{\Delta H}. \tag{5.1.49}$$

It is of interest to examine how this efficiency depends on the temperature and the pressure of reactants and products, or, in other words, how the enthalpy and free energy changes of the reaction depend on such variables.

Figure 5.1-22 (left) depicts the manner in which ΔH and ΔG, vary with temperature, and (right) how

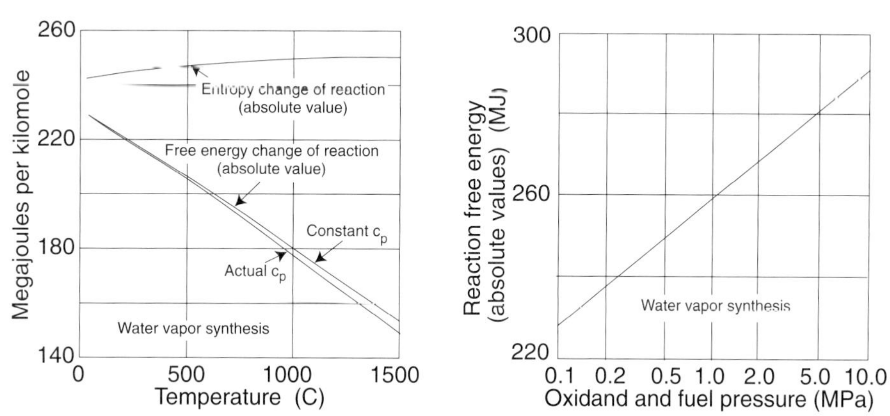

Figure 5.1-22 (Left) At constant pressure (0.1 MPa), increasing equally the temperature of reactants and product increases slightly the absolute value of the enthalpy change of the reaction, but reduces substantially the corresponding free energy. (Right) At constant temperature (298 K), increasing equally the pressure of reactants and product increases substantially the absolute value of the free energy change of the reaction.

ΔG varies with pressure. The data assume that both reactants, H_2 and, O_2 and the product, H_2O, are at identical pressures and identical temperatures.

Increasing the temperature while maintaining the pressure constant has a small effect on ΔH, but reduces appreciably the available free energy (the electric energy generated). Thus, for reversible cells, the higher the temperature the lower the efficiency. It must be noted that exactly the opposite happens with practical cells. The reason is that the improvement in the cell kinetics with temperature more than offsets the free energy loss. This point will be discussed in more detail farther on.

We will now derive the dependence of the enthalpy and the free energy changes of a reaction on temperature and pressure.

5.1.7.4 Effects of pressure and temperature on the enthalpy and free energy changes of a reaction

Enthalpy dependence on temperature

Let ΔH be the enthalpy change owing to a chemical reaction.

By definition, ΔH is equal to the sum of the enthalpies of the products of a reaction minus the sum of the enthalpies of all reactants. Thus, $\Delta H =$ *enthalpy of products − enthalpy of reactants*. When there is a single product, then the enthalpy change of the reaction can be expressed per kilomole of product by dividing ΔH by the number of kilomoles of product created. An equivalent definition holds for the free energies.

In mathematical shorthand,

$$\Delta H = \sum n_{P_i} H_{P_i} - \sum n_{R_i} H_{R_i}. \quad (5.1.50)$$

Here,

n_{P_i} = number of kilomoles of the *ith* product,
n_{R_i} = number of kilomoles of the *ith* reactant,
H_{P_i} = enthalpy of the *ith* product,
H_{R_i} = enthalpy of the *ith* reactant.

For the water synthesis reaction,

$$\Delta H = 1 \times H_{H_2O} - 1 \times H_{H_2} - \frac{1}{2} \times H_{O_2}. \quad (5.1.51)$$

where H_{H_2O} is the enthalpy of water, and so on.

Of course, at RTP, for water vapor, $H_{H_2O} = -241.8$ MJ/kilomole, and H_{H_2} and H_{O_2} are both zero by choice.[†] Hence, $\Delta H = -241.8$ MJ/kilomole.

If the value of the enthalpy, H_0 of a given substance (whether product or reactant) is known at a given temperature, T_0, then at a different temperature, T, the enthalpy is

$$H = H_0 + \int_{T_0}^{T} c_p dT, \quad (5.1.52)$$

where c_p is the specific heat at constant pressure.

c_p is somewhat temperature dependent and its value for each substance can be found either from tables or from mathematical regressions derived from such tables. For rough estimates, c_p, can be taken as constant. In this latter case,

$$H = H_0 + c_p \Delta T, \quad (5.1.52a)$$

where $\Delta T \equiv T - T_0$.

Replacing the various enthalpies in Equation 5.1.50 by their values as expressed in Equation 5.1.52

$$\begin{aligned}\Delta H &= \sum n_{P_i} H_{0_{P_i}} + \sum n_{P_i} \int_{T_{0_{P_i}}}^{T_{P_i}} c_{P_{P_i}} dT \\ &\quad - \sum n_{R_i} H_{0_{R_i}} + \sum n_{R_i} \int_{T_{0_{R_i}}}^{T_{R_i}} c_{P_{R_i}} dT \\ &= \Delta H_0 + \sum n_{P_i} \int_{T_{0_{P_i}}}^{T_{P_i}} c_{P_{P_i}} dT - \sum n_{R_i} \int_{T_{0_{R_i}}}^{T_{R_i}} c_{P_{R_i}} dT.\end{aligned} \quad (5.1.53)$$

For a simple estimate of the changes in ΔH, one can use the constant c_p formula:

$$\begin{aligned}\Delta H &= \Delta H_0 + \sum n_{P_i} c_{P_{P_i}} (T_{P_i} - T_{0_{P_i}}) \\ &\quad - \sum n_{R_i} c_{P_{R_i}} (T_{R_i} - T_{0_{R_i}}).\end{aligned} \quad (5.1.53a)$$

Enthalpy dependence on pressure

Earlier, we defined enthalpy as the sum of the internal energy, U, with the pressure-volume work, pV:

$$H \equiv U + pV. \quad (5.1.54)$$

The internal energy of the gas is the energy stored in its molecules. Such storage can take the form of excitation, ionization, etc. However, in this chapter, we will limit ourselves to the internal energy stored as kinetic energy of the molecules, a quantity measured by the temperature of the gas.

[†] Remember that, by covention, the enthalpy of all elements in their natural state under normal conditions on earth is taken as zero.

Example 1

The standard enthalpy of formation of water vapor is −241.8 MJ/kmoles. What is the enthalpy of formation when both reactants and product are at 500 K?

We need to know the specific heats at constant pressure of H_2, O_2 and $H_2O(g)$. To obtain an accurate answer, one must use Equation 5.1.53 together with tabulated values of the specific heat as a function of temperature. An approximate answer can be obtained from Equation 5.1.53a using constant values of the specific heats. We shall do the latter.

We saw earlier that, if one can guess the number of degrees of freedom, ν, of a molecule, one can estimate the specific heat by using the formula

$$c_p = R\left(1 + \frac{\nu}{2}\right).$$

The advantage of this procedure is that it is easier to remember ν than c_p. For diatomic gases, ν can be taken as 5, yielding a $c_p = 29.1$ kJ kmole^{-1}K^{-1}, and for water vapor, ν can be taken as 7, yielding $c_p = 37.4$ kJ kmole^{-1}K^{-1}.

Using Equation 5.1.53a (in this problem, all ΔT are the same: 500 K −298 K),

$$\begin{aligned}\Delta H &= -241.8 + (500 - 298)\,[0.0374 - (0.0291 \\ &\quad + \frac{1}{2} \times 0.0291)] \\ &= -243.1 \text{ MJ per kmole of water vapor.}\end{aligned}$$

Since the enthalpy changes only little with temperature, these approximate results are close to the correct value of −243.7 MJ/kmole obtained through the use of Equation 5.1.53.

Changing the pressure at constant temperature does not change the average energy of the molecules, and since the mass of gas (the number of molecules) is constant, the total internal energy remains unaltered—that is, U remains constant when p is changed provided T is unchanged.

We have $pV = \mu RT$. At constant temperature, pV must also be constant.

Thus, neither U nor pV changes when the pressure is altered isothermally. Consequently, H does not depend on pressure provided the temperature and the mass of the gas is unaltered.

Free energy dependence on temperature

The free energy is

$$G = H - TS. \qquad (5.1.55)$$

The behavior of H as a function of temperature was discussed previously. We must now investigate the behavior of the entropy, S.

From earlier discussion, we have the relationship (for isobaric processes and per kilomole of gas),

$$dS = c_p\frac{dT}{T}, \qquad (5.1.56)$$

which integrates to

$$S = S_0 + \int_{T_0}^{T} c_p\frac{dT}{T}. \qquad (5.1.57)$$

The change in free energy when the temperature is changed under constant pressure is

$$\begin{aligned}G - G_0 &= H - H_0 - (TS - T_0S_0) \\ &= \int_{T_0}^{T} c_p dT - \left(TS_0 + T\int_{T_0}^{T} c_p\frac{dT}{T} - T_0S_0\right) \\ &= \int_{T_0}^{T} c_p dT - T\int_{T_0}^{T} c_p\frac{dT}{T} - S_0\Delta T.\end{aligned} \qquad (5.1.58)$$

Given a table of values for c_p as a function of T, the change, $G - G_0$, in the free energy can be numerically calculated from Equation 5.1.58 and the ΔG of reaction can be obtained from

$$\Delta G = \sum n_{P_i} G_{P_i} - \sum n_{R_i} G_{R_i}, \qquad (5.1.59)$$

an equation equivalent to Equation 5.1.50 for ΔH.

For the case when c_p is assumed constant, Equation 5.1.58 reduces to

$$G - G_0 = (c_p - S_0)\Delta T - Tc_p \ln\frac{T}{T_0}. \qquad (5.1.58a)$$

More accurate values can be obtained from numerical integration of Equation 5.1.59. The values of c_p for H_2, O_2, and H_2O can be read from experimentally determined tables (reproduced below from Haberman and John, 1989).

The data of Table 5.1-5 are displayed in the plots of Figures 5.1-23 through 5.1-25. They show that the specific heats at constant pressure, c_p, and the γ of the

Example 2

Estimate the free energy of the $H_2(g) + \frac{1}{2}O_2(g) \rightarrow H_2O(g)$ reaction at standard pressure and 500 K, using constant c_p.

The necessary values are

	Entropy kj K^{-1} $kmole^{-1}$ (at RTP)	Specific heat KJ K^{-1} $kmole^{-1}$
H_2 (g)	130.6	29.1
O_2 (g)	205.0	29.1
H_2O (g)	188.7	37.4

Since the product and all reactants are at the same temperature, $\Delta T = 500 - 298 = 202$ for all gases. Let us calculate individually their change in free energy remembering that elements, in their natural state at RTP, have zero free energy.

From Equation 5.1.58a,

$$G_{H_2} = 0 + (29.1 \times 10^3 - 130.6 \times 10^3) \times 202 - 500 \times 29.1 \times 10^3 \times \ln\frac{500}{298} = -28.03 \times 10^6 \text{ J/kmole},$$

$$G_{O_2} = 0 + (29.1 \times 10^3 - 205.0 \times 10^3) \times 202 - 500 \times 29.1 \times 10^3 \times \ln\frac{500}{298} = -43.06 \times 10^6 \text{ J/kmole},$$

$$G_{H_2O} = -228.6 \times 10^6 + (37.4 \times 10^3 - 188.7 \times 10^3) \times 202 - 500 \times 37.4 \times 10^3 \times \ln\frac{500}{298} = -268.8 \times 10^6 \text{ J/kmole}.$$

Thus, the ΔG of the reaction is

$$\Delta G = G_{H_2O} - G_{H_2} - \frac{1}{2}G_{O_2} = -268.8 - (-28.0) - \frac{1}{2}(-43.1) = -219.3 \text{ MJ/kmole}.$$

This estimate came fortuitously close to the correct value of −219.4 MJ/kmole. If we had calculated ΔG at, say, 2000 K, we could be making a substantial error.

Nevertheless, the assumption of constant c_p leads to estimates that indicate the general manner in which the free energy depends on temperature.

three gases of interest, far from being temperature independent as suggested by simple theory, do vary substantially.

The value of c_p derived from a guess of the number of degrees of freedom is indicated by the horizontal dotted line in each figure. For hydrogen and oxygen, two diatomic molecules, a reasonable number of degrees of freedom would be $\nu = 5$ which leads to $c_p = 29.1$ kJ K^{-1} per kilomole. This matches the actual value for hydrogen at temperatures between 350 K and 600 K. At higher temperatures, the c_p of this gas seems to be heading towards the 37.4 kJ K^{-1} per kilomole that correspond to 7 degrees of freedom. A somewhat more detailed discussion of these changing degrees of freedom is referred to earlier.

Water, with its presumed 7 degrees of freedom, should have $c_p = 37.4$ kJ K^{-1} per kilomole. This is actually the correct value at some 700 K. However, the c_p of water, like that of most gases, varies fairly rapidly with temperature.

We fitted a 5th order polynomial to the data in Table 5.1-5, so that the values can be calculated as a function of T with reasonable accuracy:

$$c_p = a + bT + cT^2 + dT^3 + eT^4 + fT^5 \qquad (5.1.60)$$

where the constants a through f are given in Table 5.1-6.

It should be noticed that these regressions must be used only in the 220 K < T < 1800 K interval. Outside this range, the errors become, in some cases, unacceptably large.

Free energy dependence on pressure

The free energy, G, is defined as

$$G = H - TS \qquad (5.1.61)$$

We saw that the enthalpy does not change when the pressure is altered isothermally. Thus, isothermally, pressure can only alter the free energy through its effect on the entropy, S:

$$\Delta G = -T\Delta S \qquad (5.1.62)$$

When an amount, Q, of heat is added to a system at constant temperature, the entropy increases by Q/T:

$$\Delta S = \frac{Q}{T}. \qquad (5.1.63)$$

From the first law of thermodynamics,

$$\Delta U = Q - W \qquad (5.1.64)$$

Table 5.1-5 Specific heats at constant pressure and gammas

T (C)	T (K)	H_2 c_p (kJ/K)/kmole	H_2 γ	O_2 c_p (kJ/K)/kmole	O_2 γ	H_2O c_p (kJ/K)/kmole	H_2O γ
−50	223.18	27.620	1.426	29.152	1.399	33.318	1.333
0	273.18	28.380	1.410	29.280	1.397	33.336	1.332
25	298.18	28.560	1.406	29.392	1.406	33.489	1.330
50	323.18	28.740	1.402	29.504	1.403	33.642	1.328
100	373.18	28.920	1.399	29.888	1.386	34.020	1.323
150	423.18	28.980	1.398	30.336	1.378	34.434	1.318
200	473.18	29.020	1.397	30.816	1.369	34.902	1.312
226.8	500	29.031	1.397	31.090	1.365	35.172	1.309
250	523.18	29.040	1.397	31.328	1.361	35.406	1.307
300	573.18	29.080	1.396	31.840	1.354	35.946	1.301
350	623.18	29.120	1.395	32.320	1.346	36.522	1.295
400	673.18	29.180	1.394	32.768	1.340	37.098	1.288
450	723.18	29.240	1.393	33.184	1.334	37.710	1.283
500	773.18	29.340	1.391	33.536	1.329	38.322	1.277
550	823.18	29.440	1.389	33.888	1.325	38.952	1.271
600	873.18	29.560	1.387	34.208	1.321	39.564	1.266
650	923.18	29.720	1.384	34.496	1.318	40.194	1.261
700	973.18	29.880	1.381	34.752	1.315	40.788	1.256
750	1023.2	30.040	1.378	34.976	1.312	41.382	1.251
800	1073.2	30.240	1.375	35.200	1.309	41.958	1.247
850	1123.2	30.420	1.372	35.392	1.307	42.642	1.242
900	1173.2	30.640	1.369	35.584	1.305	43.326	1.237
950	1223.2	30.840	1.365	35.744	1.303	43.920	1.233
1000	1273.2	31.060	1.362	35.904	1.301	44.514	1.229
1050	1323.2	31.280	1.358	36.064	1.300	45.072	1.226
1100	1373.2	31.500	1.355	36.224	1.298	45.630	1.223
1150	1423.2	31.720	1.351	36.352	1.296	46.170	1.219
1200	1473.2	31.940	1.348	36.480	1.295	46.674	1.216
1250	1523.2	32.140	1.345	36.608	1.294	47.178	1.214
1300	1573.2	32.360	1.342	36.736	1.292	47.664	1.211
1350	1623.2	32.560	1.339	36.864	1.291	48.114	1.209
1400	1673.2	32.760	1.337	36.992	1.290	48.564	1.206
1450	1723.2	32.960	1.344	37.120	1.289	48.978	1.204
1500	1773.2	33.160	1.331	37.248	1.287	49.392	1.202

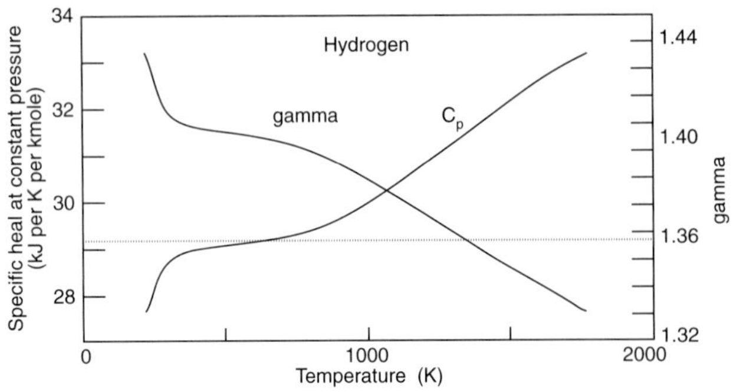

Figure 5.1-23 Specific heat and gamma of hydrogen.

But we saw previously that in an isothermal compression, $\Delta U = 0$, hence, $Q = W$. However, the work done per kilomole of gas isothermally compressed is

$$W = RT \ln \frac{p_1}{p_0}. \quad (5.1.65)$$

Thus,

$$-\Delta G = T\Delta S = Q = RT \ln \frac{p_1}{p_0}. \quad (5.1.66)$$

Consequently, the energy of isothermal compression of a gas is entirely free energy. This is an important effect. It is possible to change the efficiency (and the voltage) of a fuel cell by changing the pressure of products and reactants.

Voltage dependence on temperature

We have derived expressions that show how the free energy depends on both pressure and temperature. This was done under the assumption that the specific heats of the different substances involved in the reactions are independent of temperature. We will now derive the temperature dependence of the voltage of an ideal fuel cell in a rigorous manner. We will, however, limit ourselves to the constant pressure case.

By definition,

$$G = H - TS \quad (5.1.67)$$

and

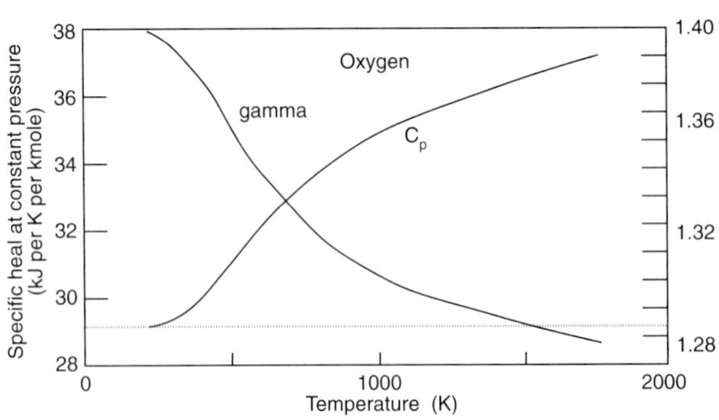

Figure 5.1-24 Specific heat and gamma of oxygen.

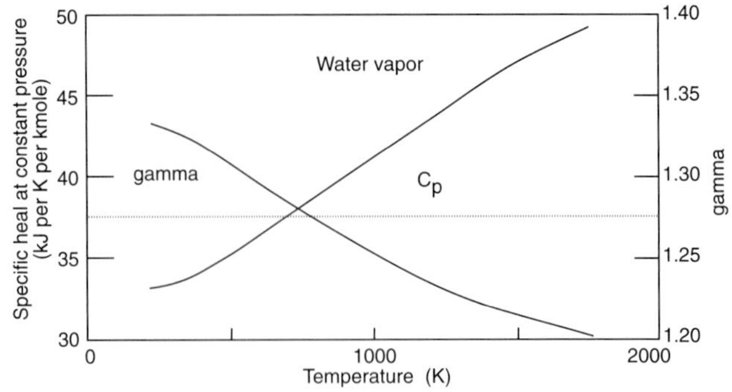

Figure 5.1-25 Specific heat and gamma of water vapor.

$$H = U + pV, \quad (5.1.68)$$

hence

$$G = U + pV - TS, \quad (5.1.69)$$

from which

$$dG = dU + pdV + Vdp - TdS - SdT. \quad (5.1.70)$$

From the combined laws of thermodynamics,

$$dU = TdS - pdV, \quad (5.1.71)$$

hence

$$dG = Vdp - SdT. \quad (5.1.72)$$

G is a function of the independent variables p and T. Thus, formally,

$$dG = \left(\frac{\partial G}{\partial p}\right)_T dp + \left(\frac{\partial G}{\partial T}\right)_p dT. \quad (5.1.73)$$

Comparing the last two equations, one can see that

$$\left(\frac{\partial G}{\partial T}\right)_p = -S. \quad (5.1.74)$$

From here on, V represents voltage, not volume, as before.

$$\sum n_{P_i} n_e q N_0 V = -\Delta G = -\left(\sum n_{P_i} G_{P_i} - \sum n_{R_i} G_{R_i}\right), \quad (5.1.75)$$

$$\sum n_{P_i} n_e q N_0 \left(\frac{\partial V}{\partial T}\right)_p = -\left(\sum n_{P_i} \frac{\partial G_{P_i}}{\partial T} - \sum n_{R_i} \frac{\partial G_{R_i}}{\partial T}\right)$$
$$= \sum n_{P_i} S_{P_i} - \sum n_{R_i} S_{R_i} = \frac{\sum n_{P_i} T S_{P_i} - \sum n_{R_i} T S_{R_i}}{T}. \quad (5.1.76)$$

Table 5.1-6 Coefficients of the polynomial used to calculate the specific heats at constant pressure 100,000 PA (1 atmosphere)

Gas	a	b	c	d	f	
H_2	22.737	37.693E-3	−85.085E-6	89.807E-9	−42.908E-12	7.6821E-15
O_2	30.737	−19.954E-3	72.554E-6	−80.005E-9	38.443E-12	−6.8611E-15
H_2O	32.262	1.2532E-3	11.285E-6	−3.7103E-9	—	—

But $TS = H - G$, hence,

$$\sum n_{P_i} n_e q N_0 \left(\frac{\partial V}{\partial T}\right)_p = \frac{\sum n_{P_i} H_{P_i} - \sum n_{R_i} H_{R_i} - \sum n_{P_i} G_{P_i} + \sum n_{R_i} G_{R_i}}{T}. \qquad (5.1.77)$$

Therefore,

$$\left(\frac{\partial V}{\partial T}\right)_p = \frac{V + \Delta H / (\sum n_{P_i} n_e N_0 q)}{T}, \qquad (5.1.78)$$

where $\Delta H/(\sum n_{Pi} n_e N_0 q)$ is the voltage the cell would have if all the enthalpy-change of the reaction were transformed into electrical energy. Let us call this the **enthalpy voltage**.

For an H_2/O_2 fuel cell producing water vapor,

$$\frac{\Delta H}{\sum n P_i n_e N_0 q} = \frac{-241.8 \times 10^6}{2 \times 6.022 \times 10^{26} \times 1.6 \times 10^{-19}} = -1.225\ \text{V},$$

where we set $\sum n P_i = 1$ because the value of ΔH used is the one for a single kilomole of water.

$$\left(\frac{\partial V}{\partial T}\right)_p = \frac{1.185 - 1.255}{298} = -2.3 \times 10^{-4}\ \text{V/K}.$$

Example 3

A reversible fuel cell, when fed hydrogen and oxygen at RTP, delivers a voltage of 1.185 V. Calculate the voltage delivered by the same cell if air (at RTP) replaces the oxygen.

Air contains roughly 20% of oxygen. Thus the partial pressure of this gas is 0.2 atmosphere, a 5:1 decompression relative to the pure oxygen case.

The energy of isothormal decompression is

$$W_{decompr.} = \frac{1}{2} RT \ln \frac{1}{5} = \frac{1}{2}\ 8314 \times 298 \times \ln 0.2 = -2 \times 10^6\ \text{J/kmole}.$$

The factor, 1/2, results from the stoichiometric proportion of one half kilomole of oxygen per kilomole of water. This energy must be *subtracted* from the ΔG of the reaction.

$$\Delta G = -228.6 - (-2) = -226.6\ \text{MJ/kmole}.$$

The voltage is now

$$V = \frac{|\Delta G|}{n_e q N_0} = \frac{226.6}{2 \times 1.60 \times 10^{-19} \times 6.02 \times 10^{26}} = 1.174\ \text{V}.$$

5.1.8 Performance of real fuel cells

In examining the performance of real fuel cells, we must inquire

1. What current can the cell deliver?
2. What is the efficiency of the cell?
3. What are the current/voltage characteristics?
4. What is the heat balance?
5. How can the excess heat be removed?

5.1.8.1 Current delivered by a fuel cell

If $\dot{N}$ is the rate (in kilomoles/sec) at which the product is generated (water, in case of hydrogen/oxygen cells) and n_e is the number of electrons per molecule of product (2, for hydrogen/oxygen cells), then the rate at which electrons are delivered by the cell to the load is $n_e \dot{N}$ or, expressing this in kilomoles of electrons per second, $n_e N_0 \dot{N}$. Consequently, the current is

$$I = q n_e N_0 \dot{N}. \qquad (5.1.79)$$

One defines a **current efficiency** as the ratio of the actual load current, I_L, to the theoretical current calculated above. In many cases, one can safely assume 100% current efficiency.

5.1.8.2 Efficiency of practical fuel cells

It was shown that the theoretical efficiency of a reversible fuel cell is

$$\eta_{rev} = \frac{\Delta G}{\Delta H}.$$

One has the choice of using for ΔH either the higher or the lower heat of combustion of the reactants, and, in stating a given efficiency, reference should be made to which was chosen. As to ΔG, one should use the one appropriate for the formation of water vapor (if, indeed, water is involved).

The efficiency of practical fuel cells is defined as the ratio of the electric power, P_L, delivered to the load to the heat power that would be generated by combining the reactants in a calorimeter, under the same temperature and pressure used in the cell,

$$\eta_{practical} = \frac{P_L}{P_{in}} = \frac{I_L V_L}{\Delta \bar{h} \dot{N}} = \frac{q n_e N_0 V_L}{\Delta \bar{h}} \quad (5.1.80)$$

Again, one has a choice of which $\Delta \bar{h}$ to use. Assuming 100% current efficiency, for hydrogen/oxygen fuel cell at **RTP** referred to the lower heat of combustion, the efficiency is, at **RPP**,

$$\eta_{pract_{RTP}} = 0.798 V_L. \quad (5.1.81)$$

Practical fuel cells have lower efficiency than ideal ones owing to the following:

1. Not all reactants used up take part in the desired reaction— some may simply escape, others may take part in undesired side-reactions. Sometimes, part of the fuel is consumed to operate ancillary devices such as heating catalytic crackers, and so on, or it may be "after burned" to raise the exhaust gas temperature in co-generation arrangements.
2. Not all the current produced will go through the load—some may leak through parallel paths (a minor loss of current) or be used to drive ancillary equipment such as compressors.
3. The voltage, V_L, the cell delivers to a load is smaller than V_{rev}, the **reversible voltage** (the theoretical voltage associated with the change in the free energy).

   A number of factors contribute to such voltage loss:

   3.1 Unavoidably, fuel cells have internal resistance to the flow of electrons and, in the electrolyte, to the flow of ions.

   3.2 The rate at which the chemical reactions take place—the **kinetics** of the chemical reaction—limits the rate at which electrons are liberated—that is, limits the current produced.

   3.3 Unwanted reactions may generate voltages that oppose the normal potential of the cell. For instance, some fuel may leak to the oxidizer side (**fuel crossover**), a worrisome problem with direct methanol fuel cells.
4. The effective electrode area may become reduced because

   4.1 excess water may "drown" the electrodes of SPFC disturbing the "triple point" contact between reactants, electrolyte and electrodes;

   4.2 insufficient moisture may dry out the solid electrolyte membrane of SPFCs further increasing the resistance to ion flow.

Incomplete use of fuel (Effect **1**), diversion of some of the generated current (Effect **2**), and loss of effective area (Effect **4**) are problems dealt mostly by system design whereas voltage loss (Effect **3**) is an inherent property of the individual cell itself. We will discuss this voltage loss by examining the voltage–current (or voltage–current density) characteristics of the cell.

5.1.8.3 Characteristics of fuel cells

Reversible fuel cells will deliver to a load a voltage, $V_L = V_{rev}$, which is independent of the current generated. Their *V–I* characteristic is a horizontal line. Such ideal cells require reaction kinetics fast enough to supply electrons at the rate demanded by the current drawn. Clearly, reversible cells cannot be realized.

In practical fuel cells, two major deviation from the ideal are observed:

1. The open-circuit voltage, V_{oc}, is smaller than V_{rev}.
2. The load voltage, V_L, decreases as the load current, I_L, increases.

Frequently, the *V–I* characteristics approaches a straight line with some curvature at low currents as, for example, those of the Figure 5.1.26 (left), while other cells exhibit characteristics in which the linearity can be seen only in a limited region (right).

The representative *V–I* characteristic of Figure 5.1.26 (left) can be mathematically described by

$$V_L = V_{oc} - I_L R_{int} - V_{act}, \quad (5.1.82)$$

where $I_L R_{int}$ is the voltage drop owing to the internal resistance, R_{int}, and V_{act}— which we shall call the **activation voltage**—is a voltage that has exactly the correct

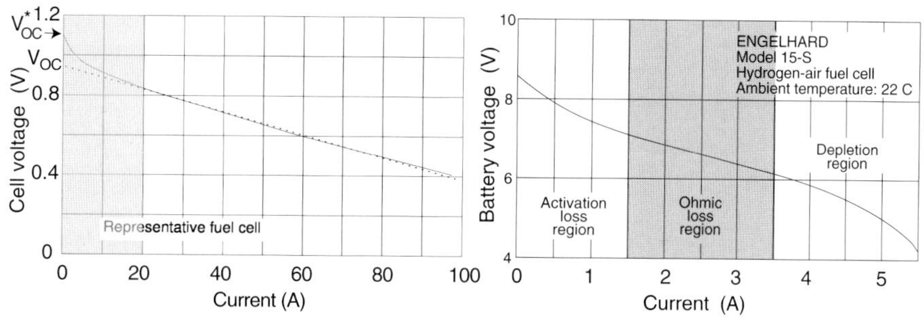

Figure 5.1-26 The typical modern fuel cell tends to have *V–I* characteristic consisting of a long stretch of apparently linear relationship between current and voltage with a small curvature at the low current end (left, above). The small Engelhard liquid-electrolyte demonstration cell had a limited region in which voltage decreases linearly with current. At both lower and higher currents the characteristic exhibited marked curvature (right, above).

dependence on I_L needed to reproduce the observed values of V_L. Having good data of V_L as a function of I_L should permit the determination of the value of R_{int} and of the functional dependence of V_{act} on I_L. However, some additional theoretical considerations should help in this task.

As discussed later on, there are reasons to believe that V_{act} should have a logarithmic dependence on I_L. At reasonably large currents,

$$V_{act} = V_2 \ln \frac{I_L}{I_0}, \tag{5.1.83}$$

where V_2 and I_0 are parameters derived from the measured characteristics.

The slope of the *V–I* characteristic is not an exact measure of the internal resistance. The actual slope is (except at very low currents)

$$-\frac{dV_L}{dI_L} = R_{int} + \frac{V_2}{I_L} \equiv R_{app}. \tag{5.1.84}$$

As defined above, R_{app} is a function of I_L. However, in many cases we expect to operate entirely in a region of the characteristics in which there is an apparent straight-line relationship between V and I, i.e. $V_2/I_L << R_{int}$. This allows the modeling of the cell as a voltage generator in series with an internal resistance, R_{int}, as suggested by the diagram in Figure 5.1.27.

The two circuits of Figure 5.1-27 are entirely equivalent. The one on the left is obvious; in the one on the right the open-circuit voltage, V_{oc}, is represented by two opposing voltage generators, V_{rev} and $V_{rev} - V_{oc}$. Such representation facilitates the calculation of internal heat generation later in this chapter.

V_{oc} is, of course, the intercept of the *V–I* (or *V–J*) line with the ordinate axis.

From the circuit model

$$V_L = V_{oc} - I_L R_{int} = V_{oc} - J A R_{int}, \tag{5.1.85}$$

Not uncommonly, voltages are plotted versus current densities, $J = I/A$, A being the effective surface area of the electrodes:

$$V_L = V_{oc} - JAR_{int} = V_{oc} - \mathcal{R}J \tag{5.1.86}$$

where $\mathcal{R}$, the **specific resistance** of the cell, has dimensions of ohms × m^2.

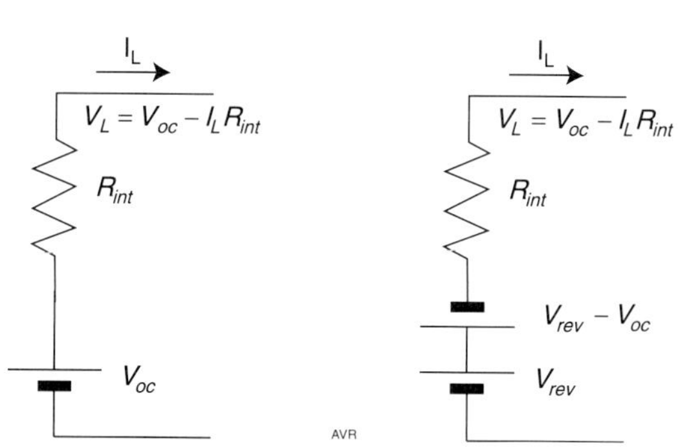

Figure 5.1-27 Circuit model for a fuel cell with straight line *V–I* characteristics.

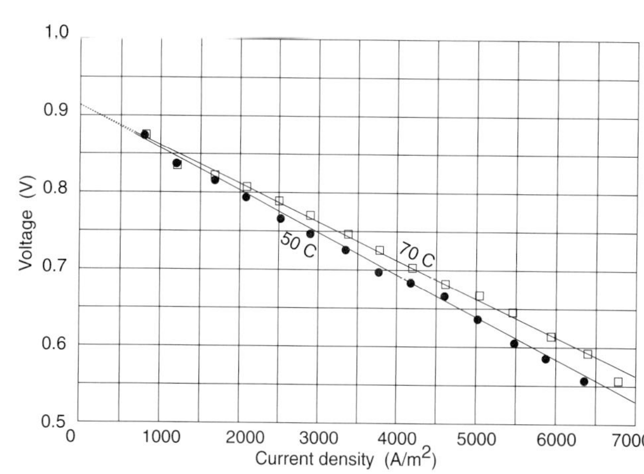

Figure 5.1-28 Characteristics of a Ballard SPFC cell.

For the particular Ballard cells whose characteristics appear in Figure 5.1-28, the linear regressions derived from the measured data lead to:

$$\textit{at } 50\text{ C},\qquad V_L = 0.912 - 54.4 \times 10^{-6} J,$$
$$\textit{and at } 70\text{ C},\qquad V_L = 0.913 - 49.3 \times 10^{-6} J,$$

where J is the current density in amperes per m^2.

Since the active area of this particular cell is $A = 0.0232\ m^2$, the equations above can be written in terms of the current, I, where $I = J \times A$,

$$\textit{at } 50\text{ C},\qquad V_L = 0.912 - 2.34 \times 10^{-3} I,$$
$$\textit{and at } 70\text{ C},\qquad V_L = 0.913 - 2.12 \times 10^{-3} I.$$

The open-circuit voltages are

$$\textit{at } 50\text{ C},\ 0.912\text{ V or } 77.4\%\text{ of } V_{rev} \text{ which is } 1.178\text{ V},$$
$$\textit{and at } 70\text{ C},\ 0.913\text{ V or } 78.0\%\ \textit{of } V_{rev} \text{ which is } 1.171\text{ V}.$$

The open-circuit voltage is only slightly influenced by the temperature. As explained before, V_{rev} became a bit smaller with the increase in temperature, while V_{oc} actually became marginally larger owing to improved kinetics.

The internal resistance was affected by the temperature in a more substantial way. It fell from 2.34 mΩ to 2.12 mΩ (more than 9%) with the 20 K increase in temperature.

Scaling fuel cells

Some of the data on fuel cell performance are presented in the form of V–J rather than V–I characteristics. Such practice permits the scaling of the cells— estimating the performance of a larger cell based on the data from a smaller one of the same type.

Consider a fuel cell with an active area A_0 and having a V–I characteristic

$$V_L = V_{oc} - R_{int_0} I. \tag{5.1.87}$$

Since

$$J = \frac{I}{A_0}, \tag{5.1.88}$$

the equation can be written

$$V_L = V_{oc} - R_{int_0} A_0 J = V_{oc} - \mathcal{R} J, \tag{5.1.89}$$

where, remember, $\mathcal{R}$ is the **specific resistance** of the fuel cell.

If now another fuel cell is built with exactly the same configuration and the same materials but with a different active area, A, then plausibly its internal resistance will be

$$R_{int} = R_{int_0} \frac{A_0}{A}, \tag{5.1.90}$$

because $R = \rho A/L$, (assuming that the thickness of the cell does not change).

V_{oc} does not depend on the area and will, therefore, remain the same. Thus, the load voltage will be

$$V_L = V_{oc} - R_{int_0} \frac{A_0}{A} I = V_{oc} - \mathcal{R} J. \tag{5.1.91}$$

In other words, the specific resistance of the larger cell is the same as that of the smaller and both have the same V–J characteristics although they may have quite different V–I characteristics.

In practice, scaling up a particular fuel cell is not necessarily simple. Problems with heat removal and water management arise when dimensions are changed. The same occurs when cells are operated together in "stacks" to raise the overall voltage, as it is almost invariably the case.

5.1.8.4 More complete V–I characteristics of fuel cells

In the previous subsection, it was suggested that the activation voltage, V_{act}, is logarithmically dependent on the current. Here, we will first infer this dependence from empirical observations and then present a relatively simple argument as theoretical support. We will use real data from an experimental KHO fuel cell. These data were published in a book titled *Fuel Cells* by Will Mitchell, Jr (1963). We simply scaled a graph that appeared in page 153 of the book. This introduced additional noise in the data but still permits the carrying out of the necessary computer experimentation. The device is designated "New Cell" and refers to an an old (1960) alkaline (KOH), high pressure hydrogen–oxygen fuel cell described (in the Mitchell book) by Adams *et al.* It operates at 200 C and, to keep the electrolyte from boiling away, had to be pressurized to 42 atmospheres. Owing to its high operating temperature this experimental cell is reasonably efficient as witnessed by the large open-circuit voltage.

We re-scaled the published data obtaining a tabulation of V_L as a function of I. The values are re-plotted and these tabulated values in Figure 5.1-29.

Our goal is to develop a mathematical formula that, given a certain load current, yields the correct load voltage. This can be accomplished by writing,

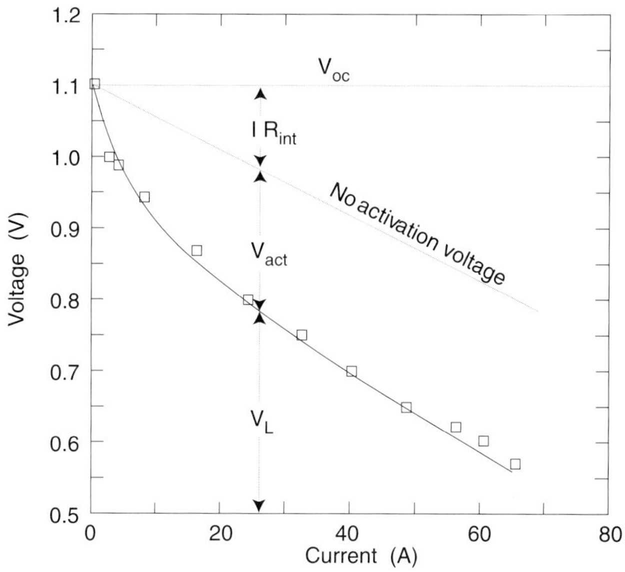

Figure 5.1-29 *V–I* characteristics of a high pressure hydrogen–oxygen KOH fuel cell of 1960.

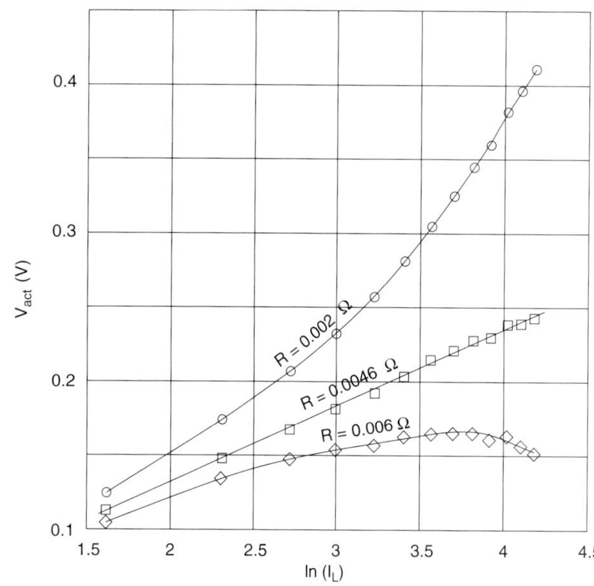

Figure 5.1-30 Only a specific value of R_{int} will yield a linear relationship between V_{act} and ln *I*.

$$V_L = V_{OC} - R_{int}I - V_{act}, \qquad (5.1.92)$$

where V_{act} is a function of I constructed in such a way as to yield the correct V_L. This definition of V_{act} is not unique: it depends on our the choice of R_{int}. To resolve this uncertainty, we must impose the additional requirement that V_{act} be a relatively straightforward function of I and, if possible, find a physical mechanism that justifies such a behavior. A mechanism does indeed exist and it requires, as we are going to show farther on, that V_{act} has a logarithmic dependence on I. This would lead to

$$V_L = V_{OC} - R_{int}I - V_2 \ln \frac{I}{I_0}, \qquad (5.1.93)$$

where V_2 and I_0 are appropriate constants.

The procedure to find the correct expression for V_{act} and incidentally, the correct value of R_{int} is as follows.

Choose an arbitrary (but not implausible) value for R_{int}. If the only loss mechanism in the cell were this internal resistance, then the *V–I* characteristic would simply be the straight line marked "No activation voltage" in the figure. Owing, however, to the existence of an activation voltage drop, the true load voltage is given by Equation 5.1.93. Solving for V_{act}

$$V_{act} = V_{OC} - R_{int}I - V_L. \qquad (5.1.94)$$

For each selected value of I, our data provide a value for V_L, and, since we know V_{OC} and assumed a value for R_{int}, we can calculate the corresponding V_{act}. Thus a table of V_{act} versus ln I can be constructed and these data can be plotted as seen, for example, in the curve marked $R = 0.002\ \Omega$ in Figure 5.1-30.

Observe that the curve in question is far from straight—that is, the relationship between V_{act} and ln I is not linear, as hoped. Repeating the procedure for $R = 0.006\ \Omega$, still results in a non-straight graph but with a reversed concavity. A little experimentation reveals that using $R_{int} = 0.0046\ \Omega$ yields an acceptably linear relationship between V_{act} and ln I. One can then write a first degree regression between these two variables,

$$V_{act} = V_1 + V_2 \ln I \qquad (5.1.95)$$

This empirical expression is called the **Tafel Equation**. For the current example, the Tafel Equation is

$$V_{act} = 0.0277 + 0.0521 \ln I. \qquad (5.1.96)$$

This can also be written as

$$V_{act} = V_2 \ln \frac{I}{I_0} = 0.0521 \ln \frac{I}{0.588}, \qquad (5.1.97)$$

because

$$I_0 = \exp\left(-\frac{V_1}{V_2}\right) = \exp\left(-\frac{0.0277}{0.0521}\right) = 0.588. \qquad (5.1.98)$$

The characteristics of the fuel cell are

$$V_L = 1.111 - 0.0046I - 0.0521 \ln \frac{I}{0.588}. \quad (5.1.99)$$

The open-circuit voltage is 1.111.

The equation above fits well the measured data, with one major exception. At low currents, the predicted V_L exceeds the observed value. At 0.1 A, the predicted value is 1.20 V, which is larger than the observed V_{OC}, and when $I = 0$, the equation predicts a value of $V_L = \infty$ which is patently absurd.

It is obvious that our regression for calculating V_{act} is not valid when small load currents are considered. In fact, we discarded the value, $I_L = 0$, when we made our calculations. Can we modify Equation 5.1.96 or 5.1.97 so as to extend our model into the low current range?

Equation 5.1.97 can be inverted

$$I = I_0 \exp\left(\frac{V_{act}}{V_2}\right). \quad (5.1.100)$$

The preceding expression reminds us of Boltzmann's law which states: "The probability of finding molecules in a given spatial arrangement varies exponentially with the negative of the potential energy of the arrangement, divided by kT."

The potential energy of the electron in the presence of a voltage, V_{act}, is qV_{act}. Equation 5.1.100 can then be written as

$$I = I_0 \exp\left(\alpha \frac{qV_{act}}{kT}\right) \quad (5.1.101)$$

Here, we introduced the arbitrary factor, α, to adjust the magnitude of the argument of the exponential. Clearly, if Equation 5.1.101 is the same as Equation 5.1.100, then

$$\frac{V_{act}}{V_2} = \alpha \frac{qV_{act}}{kT}, \quad (5.1.102)$$

and, for the present example in which $T = 473$ K

$$\alpha = \frac{kT}{qV_2} = 0.783. \quad (5.1.103)$$

Observe that after all the manipulations, Equation 5.1.101 is simply a mathematical representation of most of the experimental data. However, it fails badly in representing the obvious condition that V_{act} must be zero when $I = 0$. On the other hand, Equation 5.1.104 will also fit the data provided β is sufficiently large (because then the second term is essentially zero). However, this same equation does also fit the condition, $V_{act} = 0$ for $I = 0$.

$$I = I_0 \exp\left(\alpha \frac{qV_{act}}{kT}\right) - I_0 \exp\left(-\beta \frac{qV_{act}}{kT}\right). \quad (5.1.104)$$

Later on we are going to show that there are theoretical reasons for assuming that $\alpha + \beta = 1$, or $\beta = 1 - \alpha$. If so, Equation 5.1.104 becomes

$$I = I_0 \exp\left(\alpha \frac{qV_{act}}{kT}\right) - I_0 \exp\left((\alpha - 1) \frac{qV_{act}}{kT}\right), \quad (5.1.105)$$

and, for the current example,

$$\begin{aligned} I &= 0.588 \exp(0.783 \times 24.5V_{act}) \\ &\quad - 0.588 \exp(-0.217 \times 24.5V_{act}) \\ &= 0.588 \left[\exp(19.2V_{act}) - \exp(-5.32V_{act})\right], \end{aligned} \quad (5.1.106)$$

where 24.5 is the value of q/kT when the temperature is 473 K.

The first term of the preceding expression is, of course, Equation 5.1.101. The second term is equal to the first when $V_{act} = 0$ forcing $I = 0$ under such conditions. As V_{act} grows, the second term quickly decreases in magnitude becoming negligible for even small values of the activation voltage so that Equations 5.1.101 and 5.1.106 yield then the same numerical result.

Thus, Equation 5.1.105 can be made to represent with acceptable accuracy the characteristics of a fuel cell. It is, nevertheless totally empirical. There is a simple, albeit somewhat handwaving, model that justifies its use.

When two dissimilar materials, at uniform temperature, are placed in contact with one another, a contact potential develops. The most familiar case (at least for the electrical engineer) is the potential that appears across a *p–n* junction.

In a single semiconductor crystal consisting of an *n*-region and a *p*-region in close juxtaposition, free electrons, more abundant in the *n*-side, diffuse toward the *p*-side, whereas holes from the *p*-side migrate to the *n*-side.

Were these particles uncharged, the diffusion process would only stop when the concentration became uniform across the crystal. This does not occur because a compensating drift current causes carriers to flow back in a direction opposite to that of the diffusion current. The drift current is driven by a **contact potential** created by the following mechanism: being charged, the migrating electrons not only transport negative charges to the *p*-side but, also, leave uncompensated positively charged donors in the *n*-side. The holes also contribute to the accumulation of positive charges in the *n*-side and, the uncompensated acceptors, to the accumulation of

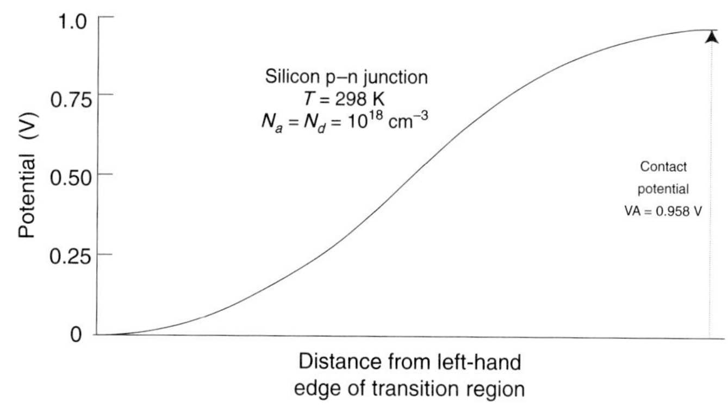

Figure 5.1-31 The potential varies with distance from the edge of a p-n junction in a simple and predictable manner.

negative charges in the *p*-side. Thus, the *n*-side becomes positive and the *p*-side negative.

In an open-circuited junction in equilibrium, there is no net current. This is the result of the precise cancellation of the diffusion current by the drift current. Although these **exchange currents** add to exactly zero,† they are not individually zero; their magnitude is surprisingly large. In silicon, under normal conditions, they may be of the order of 1 million A cm^{-2}.

The above is a good example of the dynamic equilibria that occur frequently in nature—a zero net effect is the result of the precise cancellation of two large effects.

One peculiarity of the contact potential is that it cannot be measured directly because around any closed loop the potentials cancel each other.

The behavior of the electric potential across a *p–n* junction is easy to predict (see Figure 5.1-31).

In the metal-electrolyte junction, the situation is more complicated because, at this interface, there is a change of carriers—in the metal electrode, the current is transported by electrons, while in the electrolyte, it is transported by ions, either positive or negative (**cations** or **anions**, respectively). Thus, the flow of current from electrode to electrolyte or vice versa always involves a chemical reaction.

Consider an inert metal electrode in simultaneous contact with an electrolyte and with adsorbed hydrogen atoms, some of which will ionize spontaneously.

The H^+ goes into solution and the electrons stay in the metal. As a consequence, the solution becomes more positive than the metal and some of the dissolved ions are attracted back to the electrode. Just as in the *p–n* case, two exchange currents are established:

1. a current carried by the ions that leave the metal and diffuse into the electrolyte under the influence of the gradient of ion concentration near the metal;
2. a current carried by the ions from the electrolyte that drift back to the metal under the influence of the electric field.

When there is no external electric connection between metal and electrolyte, the two currents must, under steady state conditions, be equal and opposite so that their sum is zero. If the diffusion current were larger than the return drift current, the ion concentration in the electrolyte would grow, causing the latter to become even more positive. This would reduce the potential barrier for the returning ions, increasing the drift current until it matched the diffusion current.

However, the hydrogen ionization reaction is not straightforward. Intermediate high-energy species are formed near the metal and the potential distribution acquires a hump as seen in Figure 5.1-32. This is the usual situation in chemical reactions. Catalysts are used to reduce the height of the hump. Notice that the difference in potential between the bulk of the electrolyte and the metal is independent of the size or the presence of the hump.

To diffuse from metal to electrolyte, ions must have the energy, qV_A, of the hump, or more. Owing to their Maxwellian energy distribution, the fraction of ions with more than an energy, qV_A, is proportional to $\exp(-qV_a/kT)$.††

The diffusion current, i_{f0}, must be of the form

$$i_{f_0} = I_f \exp\left(-\frac{qV_A}{kT}\right). \tag{5.1.107}$$

The return current consists of ions from the electrolyte that have energy larger than qV_B. Consequently, the return current must be of the form

$$i_{r_0} = I_r \exp\left(-\frac{qV_B}{kT}\right). \tag{5.1.108}$$

† There are minute statistical fluctuations that give rise to the radio noise.
†† This is a situation like that of thermionic emission.

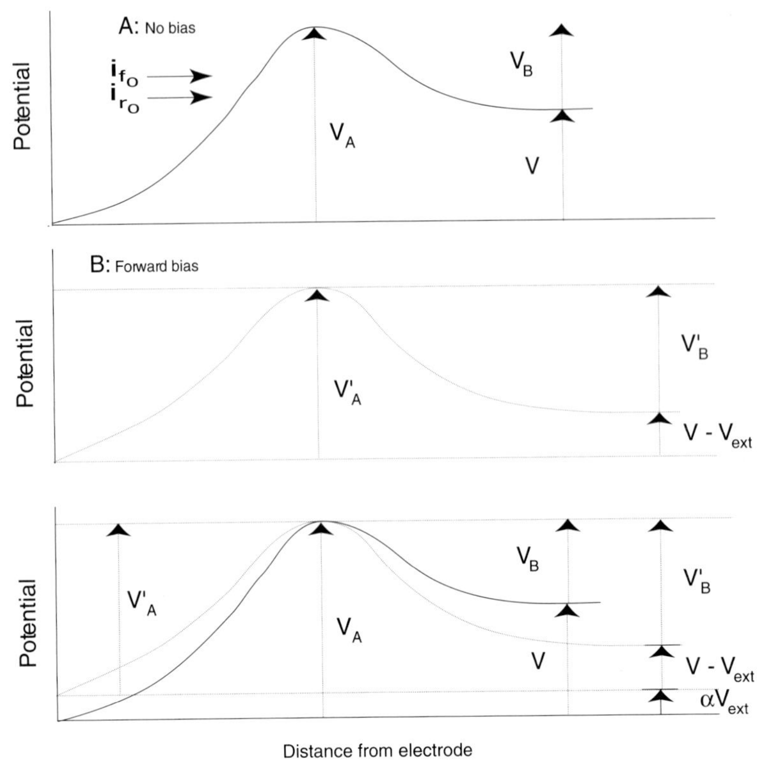

Figure 5.1-32 Potential versus distance from electrode at a metal–electrolyte interface. A — Unbiased, B — Biased, C — Biased and unbiased superposed.

The total current, i_0, is of course

$$i_0 = i_{f_0} + i_{r_0} = 0. \tag{5.1.109}$$

The subscript "0" indicates that these currents are in absence of bias. Notice the positive direction of i_{r_0} as defined in Figure 5.1-32A. This means that $i_{r_0} < 0$ because it represents H^+ ions moving *towards* the electrode.

Now, assume that a voltage is applied to the system so that the potential between the electrode and the solution is reduced by an amount, V_{ext}. In other words, V_{ext} forward-biases the metal-electrolyte junction (see Figure 5.1-32B). The magnitude of the potential barrier for ions diffusing from the electrode is now

$$V'_A = V_A - \alpha V_{ext}, \tag{5.1.110}$$

where $\alpha < 1$ is a fraction of the applied potential. Its value depends on the particular circumstances.

The barrier for ions from the electrolyte returning to the electrode is now

$$V'_B = V_B + (1 - \alpha)V_{ext}. \tag{5.1.111}$$

This can be shown by observing (Figure 5.1-32B) that

$$V'_A = V_A - \alpha V_{ext} = V - V_{ext} + V'_B, \tag{5.1.112}$$

and (Figure 5.1-32A) that

$$V_A = V + V_B. \tag{5.1.113}$$

Thus, the forward bias reduces the barrier to the diffusion current and enhances that of the return drift current. These currents no longer cancel one another and a net current circulates: The two currents are now

$$i_f = I_f \exp\left[-\frac{q(V_A - \alpha V_{ext})}{kT}\right] = i_{f_0} \exp\left[\alpha \frac{qV_{ext}}{kT}\right], \tag{5.1.114}$$

$$i_r = I_r \exp\left[-q\frac{V_B + (1-\alpha)V_{ext}}{kT}\right] = i_{r_0}\exp\left[-(1-\alpha)\frac{qV_{ext}}{kT}\right]. \tag{5.1.115}$$

The total current that circulates in the external circuit is

$$i = i_f + i_r = i_{f_0}\exp\left[\alpha\frac{qV_{ext}}{kT}\right] + i_{r_0}\exp\left[-(1-\alpha)\frac{qV_{ext}}{kT}\right]. \tag{5.1.116}$$

If we let $i_{f_0} = I_0$ and $i_{r0} = -I_0$ (remember that $i_{f0} = -i_{r0}$), then the expression in Equation 5.1.116 becomes the same as that in Equation 5.1.109, derived empirically. The lower the potential barrier, V_A, the larger the exchange current, I_0. This current is, thus, a measure of the catalytic activity that controls the barrier.

In the derivation of Equation 5.1.109, V_{act} is the total activation voltage drop of the cell—that is, the sum of the voltage drop between anode and electrolyte plus the drop between electrolyte and cathode. In Equation 5.1.116, V_{ext} is the drop between a single electrode and the electrolyte.

5.1.8.5 Heat dissipation by fuel cells

Departures from reversibility constitute losses that appear as heat. Under thermal equilibrium conditions, the rate of heat rejection is equal to the difference between the total power available from the reaction and the electric power delivered to the load. The power available from the reaction is $|\Delta\bar{h}|\dot{N}$ where $\Delta\bar{h}$ is the enthalpy change owing to the reaction per kilomole of product and $\dot{N}$ is the rate at which the product is being generated. If P_{heat} is the heat rejected, and P_L is the electric power in the load,

$$P_{heat} = |\Delta\bar{h}|\dot{N} - P_L. \quad (5.1.117)$$

In practical cells there are several mechanisms for heat generation:

1. The thermodynamic heat power, $P_{thermodynamic} = T|\Delta\bar{s}|\dot{N}$, discussed previously. This heat is rejected (rarely, absorbed) even by reversible cells.

Example 4

Consider the Ballard fuel cell of Figure 5.1-28. What is the maximum power that can transferred to a load and what heat is generated? What is the efficiency of the cell? Use the *V–I* characteristics for 70 C, but, to simplify the problem, assume that the operating conditions are at RTP. The product water is removed from the cell in vapor form.

The *V–J* characteristic of the cells is

$$V_L = 0.913 - 49.3 \times 10^{-6} J, \quad (5.1.118)$$

consequently the power output is,

$$P_L = V_L J = 0.913J - 49.3 \times 10^{-6} J^2 \quad \text{W m}^{-2}. \quad (5.1.119)$$

This is the power the cell delivers to a load per square meter of active electrode surface. The cell delivers maximum power when

$$\frac{dP}{dJ} - 0.913 - 98.6 \times 10^{-6} J - 0, \quad (5.1.120)$$

or

$$J = 9260 \quad \text{A/m}^2. \quad (5.1.121)$$

At this current, the cell would deliver 4230 W/m^2.

With 100% current efficiency the rate of water synthesis is

$$\dot{N} = \frac{J}{qn_eN_0} = \frac{9260}{1.60 \times 10^{-19} \times 2 \times 6.02 \times 10^{26}}$$
$$= 48 \times 10^{-6} \text{ kilomoles } (H_2O)\text{s}^{-1}\text{m}^{-2}. \quad (5.1.122)$$

Hence, the energy input to the cell is

$$P_{in} = \Delta\bar{h}\dot{N} = 242 \times 10^6 \times 48 \times 10^{-6} = 11,600 \quad \text{W/m}^2. \quad (5.1.123)$$

Of these, 4230 W/m^2 appear as electric energy in the load, so that 11,600 − 4230 = 7370 W/m^2 of heat are generated.

Notice that the Joule losses inside the cell amount to

$$P_{Joule} = R_{int}J^2 = 49.3 \times 10^{-6} \times 9260^2 = 4220 \quad \text{W/m}^2. \quad (5.1.124)$$

This is, of course, equal to the power delivered to the load because, for maximum power transfer the load resistance must equal the internal resistance of the generator.

The thermodynamic heat is

$$P_{therm} = T|\Delta\bar{s}|\dot{N} = 298 \times 44.4 \times 10^3 \times 48 \times 10^{-6}$$
$$= 635 \quad \text{W/m}^2. \quad (5.1.125)$$

The losses owing to V_{oc} being different from V_{rev} amount to

$$P_{OC} = (V_{rev} - V_{OC})I = (1.185 - 0.913) \times 9260$$
$$= 2519 \quad \text{W/m}^2. \quad (5.1.126)$$

Necessarily,

$$P_{in} = P_{therm} + P_{OC} + P_{Joule} + P_L \quad (5.1.127)$$

The efficiency of this cell is

$$\eta = \frac{P_L}{P_{in}} = \frac{4230}{11,600} = 0.365. \quad (5.1.128)$$

2. The heat resulting from the difference between the reversible, voltage, V_{rev}, and the open-circuit voltage, V_{oc}. It is $P_{oc} = (V_{rev} - V_{oc})I$ watts.
3. The heat dissipated in the internal resistance, R_{int}, of the cell. This amounts to $P_{Joule} = I^2 R_{int}$ watts.
4. The heat owing to other departures, V_{extra}, of the cell voltage from the simple $V_L = V_{oc} - IR_{int}$ behavior. These departure may be due to the activation voltage drop, V_{act}, or due to the voltage drop resulting from electrolyte depletion, as mentioned. This amounts to $P_{extra} = V_{extra} I$.
5. The heat of condensation, $P_{cond} = |\Delta \bar{h}_{cond}| \dot{N}$, of the product water. If $\Delta \bar{h}$ of the reaction (Equation 117) is the value for formation of water vapor and the product water is removed as vapor, or if $\Delta \bar{h}$ is for the formation of liquid water, and the water is removed as liquid, then P_{con} must be taken as zero.

In the above example, the current delivered by the cell (9260 A/m^2) exceeds the maximum current (7000 A/m^2) in the Ballard data. This probably means that the cell cannot deliver all this power—that is, it cannot dissipate the 7,400 W/m^2 of heat it generates.

The efficiency of the cell, if operated at lower output levels, will be substantially large than that in the extreme example above. In fact, at $J = 7000$, the efficiency would be about 45%.

Heat removal from fuel cells

The operating temperature of a fuel cell depends on the type of cell In any cell, if the temperature is to remain unchanged, an amount of heat power, P_{remov}, equal to the generated heat power, P_{heat} must be removed from the device.

This heat removal can be accomplished passively or by using especial heat exchange schemes. Heat can, for example, be removed by using a rate of air flow in excess of that which is required to satisfy the oxygen demand of the cell. This same stream of air may also be useful in the removal of the reaction water.

References

Adams, A. M., F. T. Bacon, and R. G. H. Watson, *The high pressure hydrogen-oxygen cell*, Chapter 4 of Fuel Cells, ed. Will Mitchell, Jr. Academic Press, **1963**.

Bessette, N. F. and J. F. Pierre, Status of SiemensWestinghouse tubular solid oxide fuel cell technology and development program, *Fuel Cells—Powering the 21st Century*, Oct. 30 to Nov. 2, **2000**, Portland, Oregon.

Boysen, Dane A., Tetsuya Uda, Calum R.I. Chisholm, and Sossina M. Haile, *High-performance solid acid fuel cells through humidity stabilization*, *Science* **303**, 68 2 January 2004. Chu, Deryn, R. Jiang and C. Walker, Methanol tolerant catalyst for direct methanol fuel cell applications, *Fuel Cells—Powering the 21st Century*, Oct. 30 to Nov. 2, **2000**, Portland, Oregon.

Dodelet, J. P., M. C. Denis, P. Gouérec, D. Guay, and R. Scholz, CO tolerant anode catalysts for fuel cells made by high energy ball-milling, *Fuel Cells—Powering the 21st Century, Fuel Cell Seminar*, October 30–November 2,**2000**, Portland, OR.

Dohl, H., S. von Adrian, J. Divisek, B. Höhlein, J. Meusinger, Development and process analysis of direct methanol fuel cell systems, Fuel Cells—Powering the 21st Century, October 30–November 2, **2000**, Portland, Oregon.

Forbes, C. A. and J. F. Pierre, The solid fuel-cell future, *IEEE Spectrum*, Oct. **1993**, p 40.

Goldstein, Rocky, *Proton conductors for solid electrolyte fuel cells*, Exploratory Research Letter, Electric Power Research Institute (EPRI), Palo Alto, CA.

Ghosh, D., M.E. Pastula, R. Boersma, D. Prediger, M. Perry, A. Horvath, and J. Devitt, *Development of low temperature SOSF systems for remote power and home cogen applications*, page 511, *Fuel Cells—Powering the 21st Century, Fuel Cell Seminar*, October 30–November 2, **2000**, Portland, OR.

Haberman,William L., and James E. A. John, *Engineering thermodynamics with heat transfer*, Allyn and Bacon, **1989**.

Halpert, Gerald, Sekharipuram R. Narayanan, Thomas Valdez, William Chun, Harvey Frank, Andrew Kindler, and Subbarao Surampudi (Jet Propulsion Laboratory), and Jack Kosek, Cecelia Cropley, and Anthony LaConti (Giner Inc), Progress with the direct methanol liquid-feed fuel cell system, *IECEC*, **1997**.

Hibino, Takashi, Atsuko Hashimoto, Takao Inoue, Jun-ichi Tokuno, Shin-ichiro Yoshida and Mitsuru Sano, *A low-operating-temperature solid oxide fuel cell in hydrocarbon-air mixtures*, *Science* **288**, 2031 16 June **2000**.

Koppel, Tom, Powering the Future (The Ballard fuel cell and the race to change the world), JohnWiley, **1999**.

Lal, Amit, and James Blanchard, the Daintiest dynamos, *IEEE Spectrum* September **2004**.

Mitchell, Jr., Will, Fuel Cells, Academic Press, **1963**.

Narayanan, S,T., T.I Valdez and F. Clara, Design and development of miniature direct methanol fuel cell sources for cellular phone applications. *Fuel Cells—Powering the 21st Century*, Oct 30 to Nov 2 **2000**, Portland, Oregon.

Ovshinsky, S. R., M. A Fetcenko and J. Ross, A nickel metal hydride battery for electrical vehicles, *Science* **260**, 176, April 9 **1993**.

Pham, A. Q., B. Chung, J. Haslam, J. DiCarlo, and R. S. Glass, Solid oxide fuel cell development at Lawrence Livermore National Laboratory, *Fuel Cells—Powering the 21st Century, p. 787*, Oct. 30 to Nov. 2 **2000**, Portland, Oregon.

Reddington, E. et.al., Combinatorial Electrochemistry: a highly parallel, optical screening method for discovery of better electrocatalysts, *Science*, 280, p. 1735, June 12 **1998**.

Service, R. F., The fast way to a better fuel cell, Science, 280, p. 1690, June 12 **1998**.

Watkins, David S., Research, development and demonstration of solid polymer fuel cell systems, *Fuel cell systems*, Leo J. M. J Blomen and Michael N. Nugerwa, editors, Plenum Press, New York, **1993**.

Yoon, S. P., S. W. Nam, T.-H. Lim, I.-H Oh, H. Y. Ha, and S.-A Hong, Enhancement of cathode performance by sol-gel coating of yttria-stabilized zirconia, *Fuel Cells–Powering the 21st Century, Fuel Cell Organizing Committee*, page 611–614, Oct 30–Nov 2 **2000**, Portland Oregon.

Further reading

European Fuel Cell R&D Review, Sep. **1994**, *Argonne National Laboratory.*

Fuel Cells, A Handbook (Revision 3), Jan. **1994**, *USDE, Office of Fossil Energy.*

Prater, Keith B., Polymer electrolyte fuel cells: a review of recent developments, *J. Power Sources*, 51, p. 129, **1994.**

Reynolds,W. C., Thermodynamic properties in SI, *Department of Mechanical Engineering*, Stanford University, **1979.**

Technology Development Goals for Automotive Fuel Cell Power Systems, Jul. **1995**, *Argonne National Laboratory.*

Williams, Robert H., The clean machine, *Technology Review*, Apr. **1994.**

Problems

5.1.1 Every substance is endowed with a certain amount of energy and a certain amount of entropy. While the latter is well defined, the former—the energy— has no absolute value; only changes in energy can be measured. For this reason (entirely by convention), the enthalpy of formation of elements in their natural state is taken as zero.

Consider aluminum and oxygen. In their natural states, their standard enthalpy of formation (i.e., the energy of formation at RTP) is, as we said, zero. Every kilogram of aluminum has (at RTP) an entropy of 1.05 kJ/K, whereas every kilogram of oxygen has an entropy of 6.41 kJ/K.

Aluminum burns fiercely forming an oxide (Al_2O_3) and releasing energy. The standard enthalpy of formation of the oxide is −1.67 GJ/kmole. The entropy of the oxide is 51.0 kJ/K per kilomole.

According to the second law of thermodynamics, the entropy of a closed system suffering any transformation can not diminish. It can, at best, remain unchanged as the result of the transformation or else, it must increase. If you add up the entropies of the aluminum and of the oxygen, you will discover that the sum exceeds the entropy of the oxide formed. This means that when aluminum combines with oxygen, only part of the chemical energy can be transformed into electricity, while the rest must appear as the heat associated with the missing entropy. That part that can (ideally) be converted into electricity is called the **free energy**.

Calculate the free energy of the aluminum/oxygen reaction.

5.1.2 Daniel cells used to be popular last century as a source of electricity, especially in telegraph systems. These cells consisted of a container divided into two compartments by a membrane permeable to ions. In one compartment, a zinc electrode was dipped in a zinc sulfate solution and, in the other, a copper electrode in a copper sulfate solution.

The zinc oxidizes (i.e., loses electrons):

$$Zn \rightarrow Zn^{++} + 2e^-$$

The zinc is eroded, going into the solution in the form of ions.

At the other electrode, the copper is reduced (i.e., the copper ions in the sulfate accept electrons and the copper from the sulfate plates out onto the copper electrode):

$$Cu^{++} + 2e^- \rightarrow Cu$$

The cell will deliver current until it runs out of either zinc or sulfate, whichever is less.

Assume the cell has 95 g of zinc and 450 ml of a 0.1 M $CuSO_4$ solution. M stands for molarity: moles of solute per liter of solution. How long can this cell deliver a current of 2 A to a load?

5.1.3 A fuel cell has the following reactions:

$$\text{ANODE}: \quad C + 2O^{--} \rightarrow CO_2 + 4e^-, \qquad (1)$$

$$\text{CATHODE}: \quad 4e^- + O_2 \rightarrow 2O^{--}. \qquad (2)$$

At RTP, the changes of enthalpy and of free energy, per kilomole of CO_2, are:

$$\Delta \bar{h}_f^\circ = -393.5 \text{ MJ},$$

$$\Delta \bar{g}_f^\circ = -394.5 \text{ MJ}.$$

What is the overall reaction? What is the ideal emf (i.e., V_{rev})? What is the difference in entropy between reactants and products?

Assume that the internal resistance of the cell is 1 milliohm. Otherwise, the cell behaves as an ideal voltage source. How much carbon is needed to deliver 1 MWh of electricity to the load in minimum possible time? What is the load resistance under such conditions?

5.1.4 The enthalpies and the free energies of formation (at RTP) of each species of interest in this problem are:

	$\Delta \bar{h}_f^{\circ}$ (MJ/kmole)	$\Delta \bar{g}_f^{\circ}$
CH_3OH (g)	−201.2	−161.9
CH3OH (l)	−238.6	−166.2
O_2 (g)	0	0
CO_2 (g)	−393.5	−394.4
H_2O (g)	−241.8	−228.6
H_2O (l)	−285.9	−237.2

Owing to the reaction, the changes in enthalpy and in free energy are:

Methanol	Water	$\Delta \bar{h}^{\circ}$ MJ/kmole	$\Delta \bar{g}^{\circ}$ MJ/kmole
liquid	gas	−638.5	−685.3
gas	gas	−676.5	−689.6
liquid	liquid	−726.5	−702.4
gas	liquid	−764.5	−706.7

Data from Dohle *et al.*

Consider methanol, a fuel that has been proposed for both internal combustion (IC) engines and for fuel cells. Methanol can be derived from fossil fuels and also from biomass. Being liquid at RTP conditions, it is a convenient fuel for automobiles. It has reasonable reactivity in fuel cells.

In IC engines, methanol is first evaporated and then burned. The engine exhaust water vapor. In fuel cells, the methanol reacts in liquid form but the product water is in vapor form.

1. How much heat do you get by burning 1 kg of methanol in an IC engine?
2. How much electric energy will an ideal fuel cell (using methanol and air) produce per kg of fuel?
3. How much heat does the cell reject?
4. A practical OTTO cycle engine has an efficiency of, say, 20%, while a practical methanol fuel cell may have an efficiency of 60% (this is the efficiency of the practical cell compared with that of the ideal cell). If a methanol fueled IC car has a highway performance of 10 km per liter, what is the performance of the fuel cell car assuming that all the other characteristics of the cars are identical?
5. If you drive 2000 km per month and a gallon of methanol costs $1.20, how much do you save in fuel per year when you use the fuel cell version compared with the IC version? Can you think of other savings besides that in fuel?
6. Suppose you get a ten year loan such that the yearly repayments of principal plus interest are 18% of the initial amount borrowed. By how much can the initial cost of the fuel-cell car exceed that of the IC car for you to break even? Assume that after 10 years the car is totally depreciated.
7. What is the open circuit voltage of an ideal methanol fuel cell at RTP? To answer this question, you need to make an intelligent guess about the number of electrons freed per molecule of methanol.

In the above questions, assume 100% current efficiency and 100% efficiency of the electric motor.

5.1.5 An inventor wants to build a hydrogen manometer based on the dependence of the output voltage of a fuel cell on the pressure of the reactants. Take a H_2/O_2 fuel cell operating at 298 K. Assume that it produces water vapor and that the open circuit voltage is that of an ideal cell. The oxygen pressure is maintained at a constant 0.1 MPa while the hydrogen pressure, p_{H_2}, is the quantity to be measured.

1. What is the output voltage when p_{H_2} is 0.1 MPa?
2. What is the output voltage when p_{H_2} is 1 MPa?
3. Develop an expression showing the rate of change of voltage with p_{H_2}. What is this rate of change when p_{H_2} is 0.1 MPa?
4. The output voltage of the cell is sensitive to temperature. Assume that a ±10% uncertainty in pressure measurement can be tolerated (when the pressure is around 1 MPa). In other words, assume that when a voltage corresponding to 1 MPa and 298 K is read, the actual pressure is 0.9 MPa because the temperature of the gases is no longer 298 K. What is the maximum tolerable temperature variation?

5.1.6 A certain gas, at 10^5 Pa, has a specific heat given by the expression

$$c_p = a + bT + cT^2 \quad (1)$$

for 298 K $<$ T $<$ 2000 K.

$$a = 27.7\ \text{kJK}^{-1}\ \text{kmole}^{-1},$$

$$b = 0.8 \times 10^{-3}\ \text{kJK}^{-2}\ \text{kmole}^{-1},$$

$$c = 10^{-6}\ \text{JK}^{-3}\ \text{kmole}^{-1}.$$

The enthalpy of the gas is 0 at 298.0 K and its entropy, at this temperature, is 130.0 kJ K^{-1} $kmole^{-1}$.

What are H, G and S of the gas (per kilomole) at $T = 1000$ K and $p = 10^5$ Pa? Please calculate with 4 significant figures.

5.1.7 A fuel cell has the reactions:

$$\text{ANODE}: 2\,A_2 = 4\,A^{++} + 8\,e^{-}. \quad (1)$$

$$\text{CATHODE}: 4\,A^{++} + B_2 + 8\,e^{-} = 2\,A_2B \quad (2)$$

All data are at RTP.

The overall reaction, $2\,A_2 + B_2 = 2\,A_2B$, releases 300 MJ per kmole of A_2B in a calorimeter. The entropies of the different substances are:

A_2 : 200 KJ K^{-1} $kmole^{-1}$,

B_2 : 400 KJ K^{-1} $kmole^{-1}$,

A_2B : 150 KJ K^{-1} $kmole^{-1}$,

A_2 and B_2 are gases, whereas A_2B is liquid.

What is the voltage of an ideal fuel cell using this reaction at RTP?

Estimate the voltage at standard pressure and 50 C.

How much heat does the ideal fuel cell produce per kilomole of A_2B, at RTP? What is the voltage of the cell if the gases are delivered to it at 100 MPa? The operating temperature is 25 C.

If the internal resistance of the cell (operating at RTP) is 0.001 Ω, what is the maximum power the cell can deliver to a load? What is the fuel consumption rate of the cell under these circumstances? What is the efficiency of the cell?

5.1.8 Owing to its ceramic electrolyte, a fuel cell can operate at 827 C. Pure oxygen is used as oxidizer. The gases are fed to the cell at a pressure of 1 atmosphere.

Use the data below:

	$\Delta\bar{h}_f^{\circ}$ MJ /kmol	$\Delta\bar{g}_f^{\circ}$ MJ /kmol	γ	$\bar{s}^{\circ}$ kJ K^{-1} /kmol
CO(g)	−110.54	−137.28	1.363	197.5
CO_2(g)	−393.51	−394.38	1.207	213.7
O_2(g)	0	0	1.341	205.0

The values of γ are those appropriate for the the 25 C to 827 C interval.

We want to examine the influence of temperature on the performance of an ideal fuel cell.

1. Calculate the reversible voltage and the efficiency of the above (ideal) cell at both 25 C and 827 C.
2. As expected (if you did the calculations correctly), you will have found that both V_{rev} and η are larger at the lower temperature. Yet, the cell is operated at 827 C where its ideal performance is not as good. Why? Explain in some detail.

5.1.9 A fuel cell was tested in the laboratory and yielded the following:

Open-circuit voltage	0.600 V
Internal resistance	0.01 Ω
Voltage (I = 1 A)	0.490 V
Voltage (I = 10 A)	0.331 V

Thermodynamic data indicate that the reversible voltage of the fuel cell is 0.952 V and that the enthalpy change of the reaction is 1.26 times the reaction free energy. Two electrons circulate in the load per molecule of product.

1. What is the power the cell delivers to a load when the current is 5 A?
2. What is the heat power dissipated internally when 5 A are being delivered?

Assume the Tafel Equation is valid. The internal resistance given above is the slope of the straight line portion of the v–i characteristic of the cell.

5.1.10 A hydrogen–oxygen fuel cell, operating at RTP, has the following v–i characteristics:

$$V_L = 0.8 - 0.0001 I_L.$$

Assume 100% current efficiency.

1. What is the hydrogen consumption rate (in mg/s) when the cell delivers 1 kW to a load?
2. What is the heat generated by the cell? Liquid water is produced.

5.1.11 A fuel cell is prismatic in shape and measures $d \times 2d \times 2d$ (where d = 33 cm). It is fed with H_2 and O_2 which are admitted at 25 C and 1 atmosphere. Product water is exhausted at 1 atmosphere and 110 C.

The inside of the cell is evenly at 110 C. The outside is maintained at 50 C by immersing it totally in running water admitted at 20 C and exhausted at 45 C. Liquid water has a heat capacity of 4 MJ per K per m^3. Assume that the temperature gradient across the walls is uniform. The walls are made of 10-mm thick stainless steel with a heat conductivity of 70 W per m per K. The only energy input to the cell comes from the fuel gases admitted. Heat is removed by both the coolant water and the product water that is exhausted from the cell at a rate of $\dot{N}$ kmoles/s.

The load voltage is $V_L = 0.9 - R_{int} I$ volts. $R_{int} = 10^{-7}$ Ω.

1. What heat energy is removed every second by the coolant water?
2. What is the flow rate of the coolant water?

3. Express the heat removal rate by the product water in terms of $\dot{N}$.
4. What is the input power in terms of $\dot{N}$?
   What is the power delivered to the load in terms of $\dot{N}$?
5. Write an equation for thermal equilibrium of the cell using your results from the above.
6. What is the value of I that satisfies the above equation?
   To simplify the solution, assume that the fuel cell reaction proceeds at RTP and liquid water is produced at 25 C. This water is then heated up so that the exhaust is at the 110 C prescribed.

5.1.12 A fuel cell has a cooling system that allows accurate measurement of the heat dissipated and precisely controls the operating temperature which is kept, under all circumstances, at 298 K.

When a current of 500 A is generated, the cooling system removes 350 W of heat, while when the current is raised to 2000 A, the amount of heat removed is 2000 W

Estimate the heat removed when the current is 1000 A. The input gases are at RTP and liquid water is created in the process.

Assume a linear dependence of the load voltage on the load current.

5.1.13 An ideal fuel cell operates at 1000 K. Two reactant gases (not necessarily H_2 and O_2) are fed in at 1 atmosphere. The reversible voltage is 1.00 V. What will the voltage be if gas A is fed in at 100 atmospheres and gas B at 200 atmospheres, both still at 1000 K? The reaction is

$$2A + 3B \rightarrow A_2B_3.$$

The product is liquid (in spite of the high temperature). A total of 6 electrons circulate in the load for each molecule of product.†

5.1.14 A fuel cell employs the reaction below,

$$A + B \rightarrow AB.$$

At STP, the relevant thermodynamic data are

	$\Delta \bar{h}_f^\circ$ MJ/kmole MJ/kmole	$\bar{s}$ kJ/(K kmole)
A(g)	0	100
B(g)	0	150
AB(g)	−200	200

What is the reversible voltage of the above fuel cell? For each molecule of AB, 2 electrons circulate in the load.

5.1.15 A fuel cell battery is to be used in a satellite power supply. It must deliver a steady 2 kW at 24 V for 1 week. The mass of the cell must be the minimum possible.

The fuel cell manufacturer has a design with the following characteristics:

Open-circuit voltage: 1.10 V,
Internal resistivity: 92×10^{-6} ohm m^2,
Cell mass: 15 kg per m^2 of active electrode area.
There is a linear relationship between V_L and I_L.

How many cells must be connected in series?
What is the total mass of all fuel cells in the battery?

5.1.16 The open circuit voltage of a hydrogen-oxygen fuel cell operating at RTP is 0.96 V and its internal resistance is 1.2 mΩ. The activation voltage drop is given by

$$V_{act} = 0.031 + 0.015 \ln I,$$

where I is in amperes.

From thermodynamic data, the reversible voltage, V_{rev}, is known to be 1.20 V.

200 of the above cells are connected in series forming a battery that feeds a resistive load rated at 2.5 kW at 100 V.

What is the actual load voltage?
How much heat does the battery generate internally?

5.1.17 A fuel cell battery is fed by H_2 and O_2 (both at 300 K) and produces water vapor that promptly condenses inside the cell.

ΔT is the difference, in kelvins, between the temperature of the active area of the cell and 300 K, which is supposed to be the temperature of the cooling liquid and of the environment.

Two main mechanisms remove heat from the cells:

1. Some heat is conducted away at a rate of 40 W per m^2 of active electrode surface for each kelvin of ΔT. This, of course, implies some cooling system whose exact nature is irrelevant as far as this problem is concerned.
2. To simplify the solution, assume that water vapor is synthesized in the cell at the temperature of the incoming gases (300 K) and then immediately condenses at this temperature and then heats up by an amount ΔT to reach the operating temperature of the cell. The water is then removed, carrying with it a certain amount of heat and, thus, cooling the cell. If the temperature of the product water exceeds 100 C, assume that the cell is pressurized so that water does not boil. However, assume that all the

† The very large difference in pressure between the two reactant gases would require a strong diaphragm or electrolyte. This suggests a ceramic electrolyte and hence the high operating temperature. Clearly, the realizability of this fuel cell is highly doubtful.

reactions actually occur at RTP, i.e., use thermodynamic data for RTP.

The V–I characteristic of the cell is

$$V = 1.05 - 95.8 \times 10^{-6} J,$$

where J is the current density in A/m^2.

The current efficiency is 100%.

Although the cell will operate at conditions that differ from RTP, use RTP thermodynamic data to simplify the problem.

1. The battery is not going to be operated at full power because it probably will exceed the maximum allowable temperature. Nevertheless, calculate what the equilibrium temperature would be if full power operation were attempted.
2. In fact, the battery will operate at a much lower power. It must deliver 20 kW to a load at 12 V. It must consist of several cells connected in series. The mass of each cell is 15 kg for each m^2 of active electrode area.

   The battery must deliver this power for a week. The total mass (fuel plus battery) must be minimized. Ignore the mass of the fuel tanks.

   How many cells must be employed?
   What is the total mass?
   How many kg of H_2 and how many of O_2 are needed?
   What is the operating temperature of the cell?

5.1.18 Fill in the answers as follows:

If output voltage rises, mark "R," if it falls, mark "F," if there is no effect, mark "N."

	IDEAL FUEL CELL	PRACTICAL FUEL CELL
Higher temperature		
Higher reactant pressure		
Higher product pressure		

5.1.19 A fuel cell, generating water vapor, has a straight line v–i characteristic:

$$V_L = V_0 - R_{int} I$$

Both V_0 and R_{int} are temperature dependent and are given by the expressions below, over the temperature range of interest.

$$V_0 = \beta_0(1 + \alpha_v T)V_{rev},$$

$$R_{int} = (1 + \alpha_R T)R_{int_0}.$$

The coefficients are:

1. $\beta_0 = 0.677$,

2. $\alpha v = 443.5 \times 10^{-6}$ per K,

3. $R_{int_0} = 0.00145\ \Omega$,

4. $\alpha_R = -1.867 \times 10^{-3}$ per K.

What are the efficiencies of the fuel cell at 298 K and at 500 K when feeding a 1 milliohm load?

5.1.20 A fuel cell battery is to be used aboard the space station. The bus voltage (the voltage the battery has to deliver under full load) is 24 V when delivering 30 kW. Since the space craft has a hydrogen and oxygen supply, the battery will use these gases which are delivered to it at 1 atmosphere and 298 K.

A manufacturer-submitted sample cell was tested in the laboratory with the following results:

When no current is drawn from the cell, its voltage is 1.085 V. When delivering 2000 A, the voltage is 0.752 V. A straight line relationship was found to exist between V and I.

The cell masses 75 kg and, when taken apart, it was found that the active electrode area is 1.5 m^2.

It is clear that if the battery is to deliver 30 kW under 24 V, it must generate a current of 1250 A. Since all the cells are in series, this is also the current through each cell.

The sample cell operates with a current density of 1250/1.5= 833.3 A/m^2. If the manufacturer constructs a cell, in all aspects identical to the sample, except with different active electrode area, S, the new cell must still deliver the 1250 A but under different current density and, consequently, under different cell load voltage. Since the battery load voltage must still be 24 V, the battery will contain a different number, N, of cells.

Assume that the mass of the new cell is proportional to the active area of the electrodes.

The total mass (mass, M_B, of the battery plus mass, M_F, of the fuel, H_2 and O_2) is to be minimized for a 30-day long mission during which the battery delivers a steady 30 kW at 24 V.

Ignore the mass of the fuel tanks. The current efficiency is 100%.

Calculate this minimum total mass. How many cells are needed in series?

In the cell above, assume that water is synthesized as vapor at the temperature of the incoming gases (298 K) and promptly condenses into a liquid and then heats up to T_{op}, the operating temperature of the device. The product water is continuously removed from the cell at this latter temperature.

In addition, a cooling system also removes heat. It does this at a rate of 6 W per degree of temperature difference (T_{op} – 298 K) for each square meter of active electrode surface.

What is T_{op} when the battery delivers 30 kW as in the first part of this problem?

5.1.21 A hydrogen–oxygen fuel cell has the following characteristics when both reactants are supplied at the pressure of 1 atm:

$$V_{oc} = 0.75 + 0.005T \quad \text{V},$$

$$R_{\text{int}} = 0.007 - 0.000015T \quad \Omega.$$

Estimate, roughly, the power this fuel cell delivers to a 5 milliohm load. In any fuel cell, heat may be removed by

1. circulation of a coolant,
2. excess reactants that leave the cell at a temperature higher than the input temperature,
3. products that leave the cell at a temperature higher than that at which they were synthesized.

To simplify this problem assume that the contribution of mechanism a is always 30 times that of mechanism c and that of mechanism b is negligible.

Assume that the reactants are fed in at 298.2 K and the product water is created as a vapor at this temperature and then heated to T_{op} by the heat rejected by the cell.

Pure hydrogen and pure oxygen are supplied at 1 atm.

1. What is the temperature of the cell when temperature equilibrium has been reached?
2. What is the load current and the power delivered to the load under the above conditions?

5.1.22

R_L	I
Ohms	Amps
0.05	14.98
0.10	8.23
0.15	5.71
0.20	4.37
0.25	3.54

The reactions in a fuel cell are:

$$\text{Anode}: A \rightarrow A^{++} + 2e^-$$

$$\text{Cathode}: A^{++} + 2e^- + B \rightarrow AB$$

The gases, A and B, are fictitious (and so are their properties). The atomic mass of A is 16 daltons and that of B is 18 daltons. Both A and B behave, over the temperature range of interest, as if they had 5 degrees of freedom, while the product, AB, as if it had 7.

The fuel cell was tested in a laboratory by observing the current delivered as the load resistance was altered.

The results are displayed in the table.

1. What are the open circuit voltage and the internal resistance of the cell?
2. Careful calorimetric observations show that when the fuel cell is delivering 10.0 A to a load, the heat dissipated internally is 3.40 W.

   From this information, determine the ΔH of the reaction.
3. Notice that P_{heat} depends on $(V_{rev} - V_{oc})I$. This would suggest that it is possible to determine V_{rec} from the knowledge of P_{heat}. Demonstrate that it is not possible to do so, i.e., that for a fixed V_{oc} and R_{int}, P_{heat} is not sensitive to the value of V_{rev}.
4. In order to estimate the ΔG of the reaction, an external voltage was applied to the fuel cell so as to cause it to act as an electrolyzer. When this external voltage was 1.271 V, the electrolyzed produced A at a rate of 2.985 g/hour.

   Making plausible assumptions, estimate the ΔG of the reaction.

5.1.23 A hydrogen/oxygen fuel cell has a *V–J* characteristic (at **RTP**) given by

$$V_L = 0.98 - 10^{-3} J.$$

The active area of its electrodes is 0.444 m². The water is exhausted from the cell in gaseous form.

1. What is the rate of heat production when the cell is connected to a load of

   1.1 Open circuit?

   1.2 Short circuit?

   1.3 A load that maximizes the power output?
2. What are the efficiencies of the cell under the 3 conditions above?
3. What is the efficiency of the cell if it delivers half of its maximum power? Use the more efficient solution.
4. Assume that V_{oc} is a constant fraction of V_{rev}. Thus, under all circumstances

   $$V_{oc} = (0.98/1.185) \times V_{rev} = 0.827 V_{rev}.$$

   What is the *V–I* characteristic of the cell when fed air at 1 atmos. and 25 C instead of oxygen?
5. To simplify this problem assume that the only way to remove heat from the cell is via the exhaust stream

which consists of water vapor and excess input gases. The input gases (hydrogen and air) are at 298 K. Assume that the water is produced at 298 K and then heated by the fuel cell to the exhaust temperature, *T.*

What is the value of *T* when the cell, fed by the minimum amount of air that satisfies the oxygen requirement of the device, produces the electric output of Item 3 (half its maximum power). Although the oxidizer is air, not pure oxygen, use, for simplicity, the *V–I* characteristic for pure oxygen as given in Item 1.

6. If you made no mistake, you have found that the temperature calculated in the preceding item is too high. Much more vigorous cooling will be necessary. This an be accomplished by injecting much more air that is required by the stoichiometry. Assume that the temperature raise should not exceed 80 K. How much must the flow of air be compared with that required in Item 5?

5.1.24 The EV1 was an exceedingly well designed automobile, having excellent aerodynamics and, all over, low losses. With an energy supply of 14 kWh, it had range of over 100 km. Its 100 kW motor allowed very good acceleration making it a "sexy" machine. The problem was that, no matter how good a battery it used, it took a long time to recharge it. If instead of a battery, it had used fuel cells, then *refueling* would take only minutes versus hours for *recharging.*

Imagine that you want to replace the NiMH batteries by a fuel cell battery which, of course, must supply 100 kW of power. The *V–I* characteristic of the available Hydrogen/Oxygen fuel cell operating at RTP is

$$V_l = 1.1 - 550 \times 10^{-6} I.$$

The maximum internal heat dissipation capability is 300 W. Product water exits the cell in vapor form.

The fuels cells deliver their energy to a power conditioning unit (inverter) that changes its dc input into ac power The efficiency of this unit can be taken as 100%.

1. What is the input voltage of the power conditioning unit, in other words, what is the voltage that the fuel cell battery (at 100 kW), must deliver assuming the the smallest possible number of individual cells are used?

2. The 100 kW are needed only for acceleration. For cruising at 110 km/h, only 20 kW are required.† How many kg of hydrogen are needed to provide a range of 800 km to the car (using 20 kW)?

3. If the hydrogen is stored at 500 atmospheres, how much volume does it occupy at 298 K?

5.1.25 A single chamber low-operating-temperature solid oxide fuel cell somewhat similar to the one described by Hibino *et al.*, when operated at a current density of 6000 A/m^2, delivers a load voltage that depends on the thickness of the electrolyte in the manner indicated in the table below:

Electrolyte thickness mm	Load voltage V
0.15	0.616
0.35	0.328
0.50	0.112

This type of cell has essentially straight *V–I* characteristics. The specific resistance, $\mathscr{R}$, of the cell can be considered as consisting of two components, $\mathscr{R}_1 + \mathscr{R}_2$ where $\mathscr{R}_1$ is the resistance of the electrolyte and $\mathscr{R}_2$ represents all other resistances of the cell. The open-circuit voltage is 0.892 V.

1. If it were possible to use a vanishingly thin electrolyte, what maximum power would the cell be able to deliver?

2. What would be the corresponding load resistance if the cell has an effective electrode area of 10 by 10 cm?

3. Compare the power output of the cell with that for the cell with 0.15 mm-thick electrolyte.

5.1.26 Solid Oxide fuel cells manufactured by Siemens Westinghouse have a very pronounced curvature in their *V–J* characteristics. One class of cells using "ribbed" units behave according to

$$V_L = 0.781 - 1.607 \times 10^{-6} J - 6.607 \times 10^{-9} J^2,$$

where J is the current density in A/m^2 and V_L is the load voltage in V.

1. What is the open-circuit voltage of the cell?

2. What is the voltage of the cell when delivering maximum power to a load?

5.1.27 The *V–I* characteristics of a given fuel cell (measured with incredible precision) are tabulated as shown. See plot. The measurements were made at RTP. Water leaves the cell as a gas.

1. Calculate the efficiency of the cell when 10 A are being delivered.

2. Calculate the rate of heat generated by the cell when $I_L = 10$ A.

3. Visually, the characteristics appear as a straight line for sufficiently large current. This suggests that, in the relatively large current region, one can use the equation

† Just a wild guess!

$$V_L = V_{oc} - R_{app}I,$$

where R_{app} is the apparent internal resistance of the cell as inferred from the straight line. Estimate the value of R_{app} using

3.1 using the region $30 \leq I \leq 41$ A.

3.2 using the region $10 \leq I \leq 41$ A.

4. For each of the values of R_{app} above, determine the magnitude of the various sources of heat (Joule effect, etc.) when the cell delivers 10 A to the load. Clearly, because you used a straight-line approximation, the activation voltage does not contribute to the heat calculation.

5. Now, determine accurately the value of the internal resistance, R_{int}, i.e., include the activation voltage in the V–I characteristics.

6. Write a set of equations describing the manner in which the load voltage depends on the load current. Check the values obtained from your equations against the tabulated data. Do this for $I_L = 40$ A and for $I_L = 0.5$ A.

7. Explain why your equation overestimates V_L at small currents.

5.1.28 Although low voltage automotive batteries have been standardized at 12 V, no such standards have been agreed for automotive traction batteries. Some hybrid cars use 275 V motors and 275 V batteries (some use 550 V motors powered by 275 V batteries.)

Consider a fuel cell battery rated 100 kW at 275 V. It uses pure hydrogen and pure oxygen, both at 1 atmosphere pressure. The battery, consisting of 350 cells, operates at 390 K. To simplify the problem, assume 298 K thermodynamics.

Assume a linear V vs I characteristic for the fuel cells.

1. What is the hydrogen consumption (in kg of H_2 per hour) when the battery delivers 100 kW?

2. The retarding force on a car can be represented by a power series in U (the velocity of the car):

$$F = a_0 + a_1U + a_2U^2. \qquad (1)$$

a_1U represents mostly the force associated with the deformation of the tires. a_2U^2 is the aerodynamic retarding force and is

$$a_2 = \frac{1}{2}\rho C_D A U^2 \qquad (2)$$

Load Current (A)	Load Voltage (V)	Load Current (A)	Load Voltage (V)
0.00	0.90000		
0.5	0.84919		
1.00	0.83113	21.00	0.72622
2.00	0.81144	22.00	0.72390
3.00	0.79922	23.00	0.72163
4.00	0.79018	24.00	0.71942
5.00	0.78291	25.00	0.71725
6.00	0.77677	26.00	0.71514
7.00	0.77142	27.00	0.71306
8.00	0.76664	28.00	0.71103
9.00	0.76230	29.00	0.70902
10.00	0.75831	30.00	0.70706
11.00	0.75461	31.00	0.70512
12.00	0.75114	32.00	0.70322
13.00	0.74786	33.00	0.70134
14.00	0.74476	34.00	0.69949
15.00	0.74180	35.00	0.69766
16.00	0.73896	36.00	0.69586
17.00	0.73624	37.00	0.69408
18.00	0.73361	38.00	0.69232
19.00	0.73107	39.00	0.69058
20.00	0.72861	40.00	0.68886
		41.00	0.68715

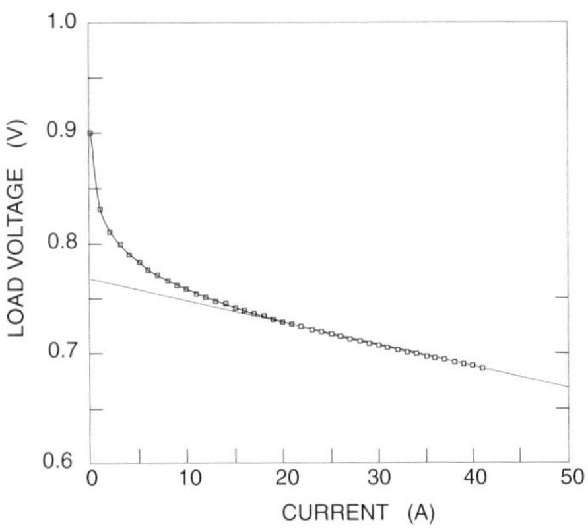

where $\rho = 1.29\ kg/m^3$ is the density of air, $C_D = 0.2$ is the drag coefficient, and $A = 2\ m^2$, is the frontal area of the vehicle. Make $a_0 = 0$.

3. When delivering 50 kW, the battery voltage is 295 V. When cruising at a constant, moderate speed of 80 km/hr, the car uses only 15 kW. What is the range of the car under such conditions if the hydrogen tank can store 4 kg of the gas? This assumes flat, horizontal roads and no wind.
4. How slow must the car drive to do 1000 km on 4 kg of H_2?

5.1.29 To test a fuel cell in a laboratory, an ac voltage generator (peak-to-peak voltage $v_{pp} = 0.001$ V) was connected in series with the load and an ac ammeter (peak-to-peak current i_{pp}) was used to measure the load current fluctuations caused by the varying V_L. The frequency used was low enough to cause any reactive component in the measurement to be negligible. The following measurements were obtained:

I_L A	V_L V	i_{pp} A
5.34	0.956	0.186
10.67	0.936	0.366

It was observed that there was a 180° phase relationship between v_{pp} and i_{pp}—that is, that increasing the voltage, actually reduced the current.

Calculate the true internal resistance of the cell.

5.1.30 Hydrogen–oxygen fuel cell. Although the temperature of the cell will vary throughout its operation, use thermodynamic data for RTP so as not to complicate the computation.

Each cell is 3 mm thick and has a total area of 10 by 10 cm.

The density of each cell is equal to twice the density of water, and the specific heat capacity of the cell is 10% of the specific heat of water. This means that it takes 24 J of heat to raise the cell temperature by 1 kelvin.

Under all circumstances, the product water is removed in vapor form.

The highest allowable operating temperature of the cell is 450 K.

Although heat is removed from the cell by several different mechanisms, the net effect is that the rate of heat removal is proportional to $T - 300$: in fact the heat removal rate, $\dot{Q}_{rem} = 0.3(T - 300)$W.

Laboratory tests reveal that when the load current is 2 A, the load voltage is 0.950 V, and when the load current is 20 A, the load voltage falls to 0.850 V.

1. Write an equation relating V_L to I_L, assuming a linear relationship between these variables.
2. What is the maximum power the cell can deliver?
3. Show that this maximum power cannot be delivered continuously because it would cause the cell temperature to exceed the maximum allowable operating temperature.
4. What is the maximum power that the cell can deliver continuously to a load?
5. Although the cell cannot deliver maximum power continuously, it can do so for a short time if it starts out cold—that is, if its initial temperature is 300 K. It will generate more heat than it can shed and its temperature will rise. How long can the cell (initially at 300 K) deliver maximum power to the load without exceeding the 450 K temperature limit?

Section Six

Solar power

Chapter 6.1

6.1

Solar power

Solar energy is the most important source of energy available to the earth and its inhabitants. Without it there would be no life. It is the energy source that drives the photosynthesis reaction. As such, it is responsible for all the biomass on the surface of the earth and is the origin of fossil fuels, the products of photosynthesis millions of years ago and now buried beneath the earth's surface. Solar energy creates the world's winds, it evaporates the water which is responsible for rain; waves and ocean thermal power are both a result of insolation. In fact, apart from nuclear energy, geothermal energy and tidal power, the sun is responsible for all the forms of energy which are exploited by man.

All these different sources of energy, each derived from the sun, can be used to generate electricity. However solar energy can also be used directly to generate electricity. This can be achieved most simply by exploiting the heat contained in the sun's radiation, but electricity can also be generated directly from light using an electronic device called a *solar cell*. Both methods are valuable renewable sources of electricity.

The solar energy resource

The energy radiated by the sun is around 7% ultraviolet light, 47% visible light and 46% infrared light. Its energy content at the distance of the earth from the sun is around 1.4 kW/m^2. Each year around 1500 million TWh of solar energy reaches the earth.

Not all this energy reaches the surface of the earth. Much of the shorter wavelength ultraviolet radiation is absorbed in the atmosphere. Water vapour and carbon dioxide absorb longer wavelength energy while dust particles scatter more radiation, dispersing some of it back into space. Clouds also reflect light back into space.

When all these factors are taken into account, around 47% of the energy, 700 million TWh actually reaches the surface. This is 14,000 times the amount of energy, 50,000 TWh, used by mankind each year. Much of this solar energy strikes the world's oceans and is inaccessible. Even so, with reasonably efficient energy conversion systems, less than 1% of the world's land area would provide sufficient energy to meet global electricity demand, around 15,000 TWh.

Let us put this into a more practical perspective. A group of solar thermal power plants were built in California in the late 1980s and early 1990s. The design of these plants was based on an estimated solar energy input of 2725 kWh/m^2/year. This is equivalent to 22.75 GWh for each hectare each year. On this basis, assuming a conversion efficiency of 10%,[1] 10 million ha, or 100,000 km^2 (316 km $\times$ 316 km), could generate enough energy to supply the entire USA.

This may appear to be a large area but the demand is not onerous. Such an area could be found quite easily, particularly if desert areas were exploited. In fact solar electricity generation takes up less land than most hydropower projects where these include reservoirs. Indeed the land requirements of some hydro schemes can be as much as 50 times a typical solar project yielding the same output.[2]

But in spite of its enormous potential, global solar electricity generating capacity is tiny. According to European Union estimates, there was probably less than 800 MW of installed capacity in 1995 (including all types of solar generation technologies). Between 1995 and the end of 2003, gross world production of solar cells was

Renewable Energy Focus Handbook 2009; ISBN: 9780123747051

around 2600 MW. With little other additional solar capacity, total global solar generating capacity may have been 3400 MW at the beginning of 2004.

Sites for solar power generation

In principle solar power can be generated anywhere on the earth but some regions are better than others. Places where the sun shines frequently and regularly are preferable to regions where cloud cover is common. The brighter the sunlight, the greater the output and the more advantageous the economics of the generating plant. Many of the world's developing countries, where demand for electricity is growing rapidly, offer good conditions for solar electricity generation.

Solar generating stations do not take up enormous amounts of land but they do require many times the space of a similarly sized fossil fuel power plant. But solar power does not necessarily require large contiguous areas of land in order to generate electricity. Solar panels can be made in small modular units which can be incorporated into buildings so that power generation can share space used for other purposes.

Distributed generation of this type has many advantages. In California, and elsewhere, there is a major daytime grid demand peak resulting from the use of air-conditioning systems. As the air-conditioning systems are used to combat heat generated by the sun, distributed solar electricity generation matches this demand perfectly. Recent experience has shown that domestic solar panels virtually eliminate this additional demand from the houses to which they are fitted.[3]

Solar technology

There are two ways of turning the energy contained in sunlight into electricity. The first, called *solar thermal generation,* involves using the sun simply as a source of heat. This heat is captured, concentrated and used to drive a heat engine. The heat engine may be a conventional steam turbine, in which case the heat will be used to generate steam, but it could also be a gas turbine or a sterling engine.

The second way of capturing solar energy and converting it into electricity involves use of the *photovoltaic* or *solar cell.* The solar cell is a solid-state device like a transistor or microchip. It uses the physical characteristics of a semiconductor such as silicon to turn the sunlight directly into electricity.

The simplicity of the solar cell makes it an extremely attractive method of generating electricity. However the manufacture of the silicon required for solar cells is energy intensive. The solar thermal plant, although more complex is currently cheaper and uses more conventional power station technology.

Whatever its type, a solar power plant has a major weakness. It can only generate electricity when the sun is shining. During the night there is no sunlight and so no electricity. In order to circumvent this problem, a solar power station must either have some form of conventional fuel back-up, or it must incorporate energy storage. Solar cells are frequently coupled with rechargeable batteries in order to provide continuous power in remote locations. Solar thermal power plants can also be designed with heat storage systems which allow them to supply power in the absence of the sun.

Solar thermal power generation

The sun is a source of high-quality heat which can easily be exploited for power generation. This was recognised as early as 1907 when the first patent for a solar collector was granted in Germany to Dr W. Maier. The development of modern solar thermal power technology began in the 1970s and was finally proved in the late 1980s with a series of commercial solar thermal power plants in California.

In spite of the success of these plants no further commercial plants have been built anywhere in the world. Research has continued, however, and interest accelerated at the beginning of the twenty-first century following renewed support from government and international agencies. It seems likely that several new projects will be built before the end of the first decade of the new century.

Modern solar thermal research has concentrated on three different approaches to converting solar energy into electricity. All require sunlight to be collected and concentrated to provide a high-energy source. The first uses a parabolic trough-shaped mirror to focus the energy contained in sunlight onto an energy collector at the focus of the parabola. These parabolic trough solar units can be deployed in massive arrays to provide a large generating capacity.

The second approach, called a *solar tower,* employs a solar energy collector mounted atop a large tower. A field of mirrors is used to direct sunlight onto the collector where the concentrated heat is used in a power generation system. Both this and the parabolic trough system can be used to build utility-sized power plants. The third system, usually called the *solar dish,* comprises a parabolic dish with a solar heat engine mounted at its focus. Dishes are usually only 10–50 kW in capacity but can achieve high-energy conversion efficiency. Fields of dishes are needed to produce a high-capacity power plant.

There is also a novel technique being explored in Australia called a *solar chimney.* This involves building a massive greenhouse, in the centre of which is an extremely tall chimney. The chimney sucks hot air from the greenhouse, creating a massive updraft. Fans or turbines placed inside the chimney can capture energy from this updraft to generate electricity. It is estimated that 40 km^2 of greenhouse and a chimney 1000 m high will be needed to generate 200 kW.

Parabolic troughs

The sunlight which reaches the earth, while it can feel extremely hot, does not contain sufficient energy in the diffuse form in which it arrives to constitute the basis for a thermal power generation system. In order to make it useful, the sunlight from a large area must be concentrated. This can be achieved with a magnifying lens but lenses form a relatively expensive way of concentrating sunlight. Much better is a concentrating reflector.

The parabola is the ideal shape for a solar reflector because it concentrates all the light incident on it from the sun at a single point called the *focus*. A complete parabola is circular; this forms the basis for solar dish system (see below). However there is a limit to the size of dish which can be built. For large-scale solar concentration, a trough-shaped reflector has proved more effective. If the trough is built with a parabolic cross-section, the reflector will bring the incident sunlight to focus at a line rather than at a single point, a line running along the length of the trough. This is the basis for the solar trough.

The reflecting panels in a solar trough are made from mirrored glass, although cheaper options are also being developed. These reflecting panels are mounted on a substantial substructure capable of supporting their weight. Along the length of each trough, mounted exactly at the focus of the parabolic cross-section, is a heat absorbing tube. This is where the solar heat is collected.

An individual solar trough reflector may be up to 150 m in length. Arrays of parallel troughs provide the required collection and generating capacity. A large number or troughs will be required to build a utility-sized power station. For example, a single 30 MW power plant

Figure 6.1-1 A solar trough system.

in California employs 980 parabolic trough collectors, each 47 m long.

In order to achieve the highest efficiency, the solar troughs should track the sun across the sky. If the solar troughs are aligned north-south, a system that tilts the troughs about their long axes can be used to follow the sun from east to west. This is the arrangement which has been used in existing solar trough power plants.

Once sunlight has been concentrated it must be captured and converted into a form of energy that can be used to generate electricity. The simplest way of achieving this is to place a tube containing a heat absorbing liquid at the focus of the parabolic trough. The liquid is pumped through the tube, absorbing heat as it passes and this heat is used to provide energy to drive a heat engine.

The small number of commercial solar trough power plants in operation all use a heat absorbing oil as the heat collection and transfer medium. This is pumped through the absorber tubes in the solar troughs, where it eventually reaches close to 400°C. The hot oil is then pumped through a heat exchanger where it is used to heat water and raise steam in a secondary system. The steam drives a steam turbine which generates power.

Commercial experience with parabolic trough solar power is based on nine plants in California built from the end of the 1980s with capacities ranging from 3.8 to 80 MW. These plants employed a secondary firing system utilising natural gas so that output could be maintained when solar input was low. In total the fossil fuel input could provide up to 25% of the plant output. In 2000, these plants achieved a peak solar-to-electric energy conversion efficiency of 23% and an annual efficiency of 15%.

The nine Californian plants have proved a success, commercially, but they were expensive to construct and would be unlikely to attract support in today's deregulated electricity market. However new designs are expected to become competitive. Research suggests that costs can be reduced significantly if new reflector fabrication and mounting methods are developed. Another significant cost reduction can be achieved by dispensing with the intermediate heat transfer oil used in existing plants and heating water directly to generate steam. Some form of heat storage system will improve flexibility and economics by allowing solar plants to generate power at night.

Another approach is to build on the idea of a combination of solar and fossil fuel energy by developing sophisticated hybrid power plants. The most attractive scheme involves building a parabolic trough collector array adjacent to a combined cycle power plant and using the heat collected by the solar array to supplement the heat from the gas turbine exhaust of the combined cycle plant. Both the gas turbine exhaust heat and the solar

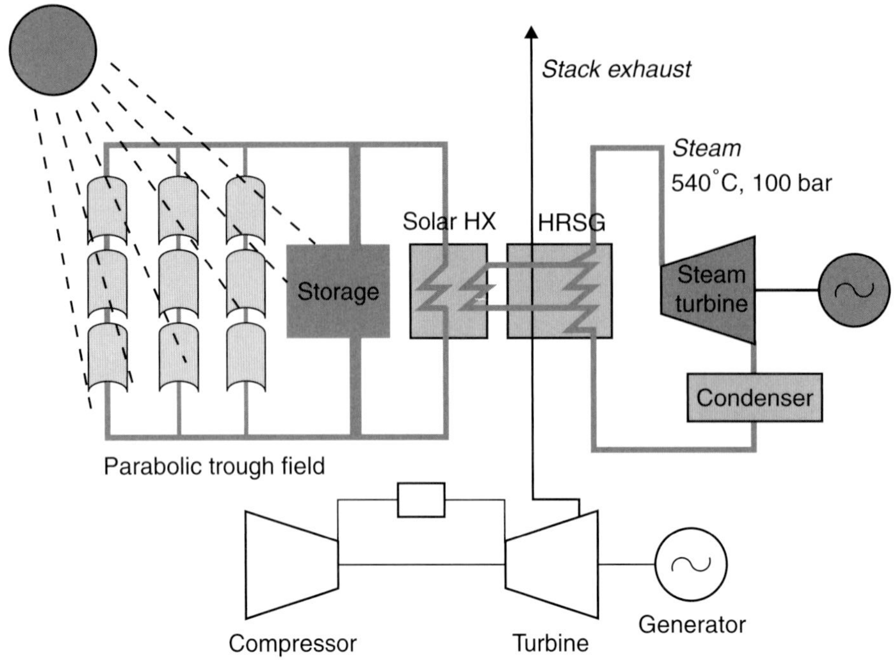

Figure 6.1-2 An integrated-solar-thermal/combined cycle power plant utilising solar troughs.

heat are then used to raise steam and drive a steam turbine.

This arrangement, called an *integrated-solar combined cycle* (ISCC) plant, makes good use of solar energy when it is available but is not reliant on the sun. Incorporating solar collection into a conventional fossil fuel power plant reduces the cost of the solar energy system significantly and this approach has attracted the support of the World Bank's Global Environment Facility. Such plants might involve a solar generating capacity of 30–40 MW out of a total of 140–300 MW. While the ISCC plant-still relies primarily on fossil fuel, it does represent an economical way of introducing significant solar capacity.

Solar towers

The solar tower takes a slightly different approach to solar thermal power generation. Whereas the parabolic trough array uses a heat collection system spread throughout the array, the solar tower concentrates heat collection and utilisation at a single central facility.

The central facility includes a large solar energy receiver and heat collector which is fitted to the top of a tower. The tower is positioned in the centre of a field of special mirrors called *heliostats*, each of which is controlled to focus the sunlight that reaches it onto the tower-mounted solar receiver.

The mirrors used as heliostats must be parabolic in section, just like the trough mirrors, but because they have a very long focal length they appear almost flat. Each mirror has to be able to track the sun independently so that the incident light remains directed at the solar receiver. The heliostat field can be very large, large enough to supply energy to generate several hundred megawatts of electricity.

The most technically demanding component of a solar tower system is the heat capture and transfer system. At the top of the solar tower is a solar receiver containing tubes through which a heat transfer fluid flows. This has to be capable of absorbing the heat from the whole heliostat field. Once heated, the fluid is pumped to a heat exchanger where the heat is used to generate steam for a steam turbine.

This arrangement is much like the parabolic trough power plant, but modern designs include a crucial difference. The heat transfer fluid in a solar tower is not pumped directly from the solar receiver to the heat exchanger. Instead it is taken to a high-temperature storage tank, a heavily insulated tank where the hot fluid can be stored for up to 24 h. From here the fluid is taken as needed and pumped through the heat exchanger to generate steam, and then it is stored in a low-temperature reservoir. Fluid from this low-temperature reservoir is supplied to the solar receiver to be reheated.

By careful sizing of the storage system and power generation system, a solar tower can be constructed so that it can supply power continuously, not just during daylight hours. This means that the plant can be employed like a normal base-load fossil-fuel-fired power

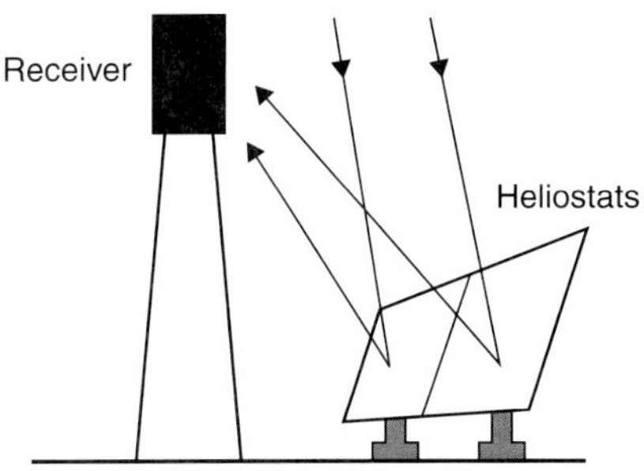

Figure 6.1-3 A solar tower system.

station, making it much more flexible and therefore much more valuable on a grid system.

A key component of the solar tower is the heat transfer fluid. The most successful has proved to be a molten salt comprising a mixture of sodium and potassium nitrates. This will melt at 220°C. It is normally kept at around 300°C in the low-temperature reservoir and is heated to 550°C in the solar receiver for storage in the high-temperature reservoir.

No commercial solar tower power plants have been built but a number of pilot-scale projects have been operated. Most important of these have been two projects at Barstow in California, called Solar One and Solar Two. Solar One operated from 1982 until 1988. It was later cannibalised to build Solar Two which started up in 1996 and operated until 1999. Both had power generating capacities of 10 MW.

Solar One used water as its heat transfer medium but it was converted to a molten salt system with two storage tanks for Solar Two. The latter comprised a 91-m high tower surrounded by 2000 heliostats with tracking systems which were computer controlled. At the top of the tower was the solar receiver, a system of vertical pipes which carried a molten salt. This molten salt reached a temperature of 565°C when heated by the sun. Storage capacity was 30,000 kWh.

The solar tower is a source of very high-grade heat. While pilot plants have operated with temperatures of around 550°C it is quite plausible to raise the temperature to 1000°C. Such a high temperature could be used to heat air to drive a gas turbine, instead of for raising steam. A gas-turbine-based system could prove more efficient than the current steam turbine solar tower design, but the scheme has yet to be proved. There may also be ways of combining a solar tower with a fossil fuel power plant in a hybrid arrangement similar to that being considered for the ISCC power plant described above.

The solar tower concept has never been proved commercially but it is considered to be perhaps the most cost effective of the solar thermal technologies. However it will be the second or third decades of the twenty-first century before it reaches commercial maturity.

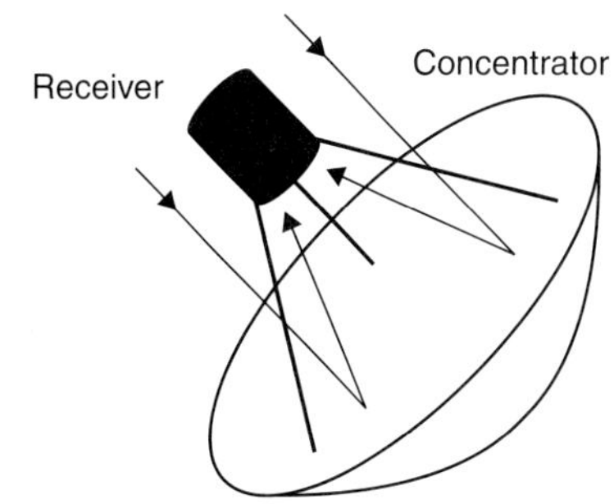

Figure 6.1-4 A solar dish.

Solar dish collectors

The third type of solar thermal power unit is the solar dish. A solar dish is more accurately a parabolic mirror, at the centre of which is placed a small heat collector and electricity generator. The reflector tracks the sun and focuses its energy onto the collector.

Unlike the two preceding technologies which are being developed for utility scale generation, the solar disk will always be a relatively small-scale electricity plant. Those currently being tested have diameters of between 5 and 15 m and outputs of 5–50 kW. Larger dishes seem possible and there is a plan to build one with a capacity of greater than 1 MW, but even so utility multi-megawatt capacities can only be achieved by installing large numbers of individual units.

The two key components of a dish system are the parabolic reflector and the heat engine. Since the reflector must track the sun, a tracking system must also be included. Reflectors can be made using traditional glass-based techniques but these are very heavy and new, lighter fabrication methods are needed to bring down costs.

The most popular type of engine for use with a solar dish is a sterling engine. This is a piston engine (see Chapter 6) but a piston engine in which the pistons are part of a completely closed system. The energy source, heat, is applied externally. Consequently this is perfectly suited to solar dish applications.

The solar dish is the most efficient of all the solar thermal technologies. The best recorded solar-to-electrical conversion efficiency is 30%, but the Stirling engine is theoretically capable of 40% efficiency. This is of importance because of the area needed for a solar power plant. While parabolic trough systems require 2.2–3.4 ha for each megawatt of generating capacity, solar dishes need 1.2–1.6 ha.[4]

Solar dishes are currently expensive but costs can be reduced significantly. However they are unlikely to be a cost effective as the solar tower. Their main use is likely to be for stand-alone remote generation where their high efficiency and reliability could eventually challenge that of the solar cell, the solar device currently used most widely for such applications.

Photovoltaic devices

The solar photovoltaic device, more commonly known as the *solar cell*, exploits a completely different means of converting sunlight into electricity. This depends on the physical characteristics of materials called *semiconductors*.

The solar cell is a solid-state device which shares a heritage with the diode, the transistor and the microchip. It was developed in the Bell Laboratories in the early 1950s and soon found action in the US space programme. Today it remains the most widely used means of providing electric power to satellites and space vehicles.

Solar cells began to be used for terrestrial applications during the 1980s, mainly in remote locations where reliable power was needed without regular human intervention. As costs began to fall (although they remained extremely expensive) their use was extended to a wider range of applications. From 1990 onwards, grid-connected solar cells began to appear in domestic and some commercial applications. This usage continued to expand during the first years of the twenty-first century.

Solar photovoltaic technology

The solar cell is made from a thin layer of semiconducting material. The key feature of this semiconductor layer is that it will absorb photons of radiation in the visible region of the electromagnetic spectrum. Each photon of light energy is absorbed by an electron within the solid material. In absorbing the energy, the electron acquires an electrical potential. This potential can be made available as electrical energy, as an electric current. The current is produced at a specific fixed voltage called the *cell voltage*. The cell voltage is a property of the semiconducting material. For silicon it is around 0.6 V.

The energy contained in light increases as the frequency increases from infrared through red to blue and ultraviolet light. However a solar cell must throw away some of these frequencies. It can only absorb light which is above a certain energy threshold, called the *cell threshold*. Light with energy content below that threshold will simply pass through the cell or be reflected.

Ideally, then, one might want to choose the lowest-feasible threshold in order to utilise as much of the solar spectrum as possible. There is, however, another difficulty. The cell threshold determines the cell voltage. If the threshold frequency is very low, the solar cell will provide a low-output voltage. When light is absorbed with an energy higher than the threshold, all the excess energy is simply thrown away, wasted. So setting the threshold too low wastes energy too. Thus the threshold must be set at an optimum level.

Most commercial solar cells use silicon as the semiconducting layer. Silicon does not represent the optimum solar absorber but it is relatively easy to work with, it is extremely abundant and therefore cheap, and it benefits from an accumulation of experience with the material as a result of its use in the manufacture of transistors and microchips.

Other materials are being introduced. Among the most promising are cadmium telluride and copper indium selenide. Gallium arsenide has also been used in space. Silicon, however, appears likely to form the backbone of the solar cell industry for the immediate future.

Types of solar cell

Microchips and transistors are universally fabricated using slices cut from perfect crystals of silicon. These crystals are carefully grown under controlled conditions and are expensive to produce.

Solar cells can be made with single crystal silicon too. Indeed, cells using this material have provided the best performance of any silicon solar cells, with solar-to-electrical conversion efficiencies of up to 24%. Long-term performance of single crystal silicon cells is good too, but the cost of the single crystal material remains relatively high.

In an attempt to bring fabrication costs down, many manufacturers have experimented with alternatives to single crystal silicon. One of the most widely used is a form called *polycrystalline silicon*; this is silicon made up of lots of tiny individual crystals instead of one large crystal. Such material is much cheaper to produce but has proved less efficient than the single crystal material. However it does produce reliable and stable cells, at a significantly lower cost. Efficiencies of over 18% have been achieved.

Cheaper still is a completely non-crystalline form of silicon called *amorphous silicon*. This was initially found to suffer from a serious degradation problem when exposed to light, an effect which reduced efficiencies by 20–40%. With extensive redesign, it has now proved possible to circumvent the most serious aspect of this problem. The amorphous cell does still suffer degradation of around 20% but its output eventually stabilises. Cell efficiencies of around 13% can now be achieved (after degradation has taken place).

The cheaper and simpler amorphous silicon fabrication process has allowed some more complex solar cell designs to be developed. For example, some amorphous cells have been fabricated as three cells one on top of the other, designed to absorb first blue, then green and finally red light. This three cell design offers the potential for higher efficiency than a single cell absorbing the whole spectrum.

All silicon solar cells require extremely pure silicon. The manufacture of pure silicon is both expensive and energy intensive. The traditional method of production requires 90 kWh of electricity for each kilogram of silicon. Newer methods have been able to reduce this to 15 kWh/kg. This still means that a silicon solar cell takes 2 years to generate the energy needed to make it.[5] This compares with around 5 months for a solar thermal power plant.[6]

Silicon-based solar cells dominate the market today. Single crystal and polycrystalline silicon cells remain the most popular, accounting together for 89% of production in 2003. Of that, polycrystalline cells accounted for 62%.[7] Amorphous silicon adds a further 3%. These materials are likely to remain dominant for several years. There are alternatives to silicon. Most promising of these are cells fabricated from cadmium telluride or from copper indium selenide. Manufacture of solar cells based on these materials has begun, with a total production of 7 MW in 2003.[8]

Fifty years of experience with silicon solar cells means that long-term performance can now be assessed with accuracy. Modern silicon solar cells sold in panel form for installation on roofs come with a 25-year warranty. Longer lifetimes still should eventually be plausible. The long-term performance of newer materials has yet to be established.

Solar cell manufacture

Solar cell production is a highly technical process and this has severely limited the number of companies involved in the industry. Virtually all global solar cell production in 2003 was carried out by just ten companies. The largest, by far, of these was the Japanese company Sharp.

Regionally, Japan has come to dominate production, followed by Europe, the USA and the rest of the world. Production figures for 2003 are broken down regionally in Table 6.1-1. Total global production was 744 MW. That was 32% higher than in 2002. This rate of growth in production has been typical over the past decade.

Table 6.1-1 Solar cell production, 2003, by region

	Production (MW)
Japan	364
Europe	193
USA	103
Rest of the World	84
Total	744

Source: Renewable Energy World.[15]

Solar panels and inverters

A single modern silicon solar cell will produce between 2 and 3 W of power depending on its size. This will equate to between 3 and 5 A at 0.6 V. In order to provide a usable current and voltage, groups of cells are connected both in series and in parallel. For example, 36 solar cells connected in series will provide an output of about 20 V, suitable for a battery charger designed to recharge a 12 V battery.

For grid-connected applications, more cells are necessary. Typical units are designed to provide up to 200 W. The individual cells are normally mounted behind a glass protective barrier similar to a vehicle windscreen. The whole assembly is then encapsulated to protect it from the weather and framed with aluminium extrusions. Such assemblies are called *solar panels*.

A solar panel provides a stable direct current output. If this is to form a part of a grid-connected solar power system, perhaps on the roof of a household, it must be converted to AC at the grid voltage. This is carried out by an inverter. A typical household system will require a 2 kW inverter.

Solar cell deployment

There are three key ways of deploying solar cells. These are referred to as residential photovoltaics, utility array photovoltaics and solar concentrators. In the early years of the twenty-first century residential photovoltaics have become the most important but all three can, and may have a part to play in the future of solar electricity generation.

Utility photovoltaic arrays

A utility photovoltaic array is solar cell-based power plant with a generating capacity similar to that of a fossil fuel power plant. Construction of such a plant would involve an enormous number of individual solar cells, mounted in solar panels, and the solar panels themselves mounted in groups, each group having its own support structure.

The groups of cells would probably be fitted with a system to track the sun across the sky. Both single- and double-axis tracking are possible, the latter providing the best solar input, but at the expense of greater complexity.

Costs remain too high for this to be a competitive means of generating electricity but demonstration projects in the USA have shown that it is theoretically feasible. Initial commercial plants might have a capacity of 20 MW.

The potential for this type of technology is vast. In the USA, price permitting, it could provide up to 10% of grid-connected utility generation, or up to 200 GW.[9] Although this represents a massive market, it is in the developing world that they may find their greatest application. The small, modular nature of the utility array makes it ideally suited to remote regions where grid connection is either too expensive or geographically impossible.

Solar concentrators

A solar concentrator uses a lens or reflector to capture sunlight from a wide area and focus it onto a small area where a photovoltaic convertor device is located. Sunlight concentration may be as little as 2× or as high as 2400×. Like utility-scale photovoltaic arrays, solar concentrators are essentially a large-capacity deployment technique.

Concentrators require much smaller quantities of semiconductor material than conventional photovoltaic arrays. As a consequence it becomes cost effective to use the most efficient material available, even if this is much more expensive than the material used in large photovoltaic arrays. With only a small area of semiconductor, most of the concentrator is made from relatively cheap and readily available materials. This means that scaling up to larger sizes is easy and economical to achieve.

There are various ways of building concentrators. These range from arrays of small cells, each with a lens focusing sunlight from a small area onto a tiny photovoltaic device, to a 10–20 m parabolic reflector collecting sunlight and concentrating it at a small central receiver where the photovoltaic convertor is situated. Common to them all is the need to track the sun in order to achieve good performance, because concentrators generally rely on the incident sunlight being perpendicular to the actual solar cell.

While concentrators have yet to gain much of a foothold terrestrially, they are attracting the attention of the space industry. Their advantage in space over conventional photovoltaic arrays is a reduced exposure to radiation damage because the sensitive photovoltaic convertor is shielded inside the device. This gives them, potentially, a much longer life.

Residential photovoltaic arrays

The third category of photovoltaic applications, residential photovoltaic arrays, forms the most significant area of photovoltaic expansion as a result of major national programmes, particularly in Japan, Germany and the USA, to install solar cells on the roofs of homes.

The residential array is a type of distributed generation. As such it offers some important benefits. The power from the rooftop array is delivered directly to the place where it is needed. This is the most efficient way of generating power because there are no transmission and distribution losses involved. Further, extensive local deployment of this type reduces the need to increase transmission system capacities as demand grows.

Rooftop deployment avoids the problem of finding space to install solar arrays. The space, the roof, is already available. In addition the roof structure provides the support for the array so the expense of a dedicated support structure is avoided.

Placing solar arrays on the roofs of existing buildings will generally be a compromise because the buildings will rarely offer the optimum orientation and inclination for solar collection. Even so efficient production is possible. Far better is to design modern buildings with solar generation in mind. The solar arrays can then become the roofs, saving in building materials. They may exploit appropriately oriented wall surfaces too.

As the solar panel will only generate power during daylight hours, grid connection is essential to provide supply at nighttime. For this reason building arrays, although designated residential arrays, are actually most effective on commercial buildings where the daily cycle matches closely that of the sun.

Single household arrays offer the simplest application but the most cost effective may be a group deployment where tens or hundreds of units are fitted in a compact area. Such projects could be underwritten or owned by a local utility or distribution company. Equally they could be owned and operated by a residents' co-operative. The larger deployment attracts economies of scale that are not available for individual householders.

Environmental considerations

Solar power is considered to be one of the most environmentally benign methods of generating electricity. Neither solar thermal nor solar photovoltaic power plants generate any atmospheric emissions during operation. A photovoltaic installation makes no noise either, and a solar thermal plant very little. Nevertheless both types of plant do have an environmental impact.

On a utility scale, both types of solar power plant require a significant amount of space, more than that required by a fossil fuel power plant. However the best sites for such plants are likely to be in arid areas where this should not pose a problem. Construction of a large plant is likely to involve some local environmental disruption. Once in operation there may be some benefits locally from the shade created by the arrays of solar collectors.

When solar panels are installed on rooftops or incorporated into new buildings they share space used for other purposes. Retrofit of solar panels can be unsightly but where a building has been designed to incorporate solar panels there is no excuse for any negative visual impact.

This type of deployment has environmental benefits because it reduces the need to central power station capacity, it reduces the need to reinforce transmission and distribution systems, and it provides electricity at the point of use so energy losses should be much lower than when power is transmitted many kilometres.

Solar thermal power plants rely on conventional mechanical and electrical components. There may be spillages of heat transfer fluid but these should be easy to control. Otherwise their construction, operation and decommissioning should be easily managed without major affects on the local environment.

Solar photovoltaic devices use less commonplace materials. The predominant material for solar cells today is silicon. This is very energy intensive to produce in its pure form. Lifetime analysis of photovoltaic systems show a relatively high level of emissions of carbon dioxide and other atmospheric emissions as a result of the emissions from the predominantly fossil-fuel-fired power plants generating the electricity used in the production of the silicon.

Lifetime analysis of photovoltaic generation suggests that such a plant will release between 100 and 170 g of carbon dioxide for every kilowatt-hour of electricity it generates. This is much higher than from a solar thermal power plant for which the equivalent figures are 30–40 g/kWh. It is, nevertheless, much lower than a gas-fired power station (430 g/kWh) or a coal-fired power station (960 g/kWh).[10] In the future this impact should be reduced as global renewable capacity grows and with it a wider availability of cleaner electricity.

The large-scale deployment of solar cells will involve much larger quantities of semiconducting material than has been manufactured for microprocessors. Some newer semiconductor materials contain toxic elements; the cadmium in cadmium telluride is a good example. This semiconductor is a stable material but it will be important to ensure that conditions cannot occur which would permit cadmium to enter the environment. This will be particularly important when a plant is decommissioned. The processes involved in the manufacture of both silicon and other solar cells involves toxic organic chemicals and these, too, have to be strictly contained.

Financial risks

The risks associated with the deployment of solar photovoltaic and solar thermal power generation technologies are primarily the risks always associated with new technologies. These relate to reliability, plant lifetime, and long-term operation and maintenance costs.

Solar photovoltaic devices are now well understood and the reliability of the predominant silicon technology has been broadly established. Solar thermal plant experience is limited to nine plants in California and while the data from these plants is encouraging it cannot be considered exhaustive, particularly since any new plants are likely to use modified plant designs to reduce costs.

The Californian plants all use solar trough technology. The two other solar thermal technologies, solar towers and solar dishes, have yet to be tested commercially so any gauge of performance must rely on data from demonstration projects. The perceived risk of these technologies is likely to be higher, but all three solar thermal technologies can expect to be viewed with a degree of scepticism by financial institutions.

From a resource perspective, solar energy is well understood. Solar insolation records exist just about everywhere so there should be no problem establishing the expected solar input at any site on the earth.

Even then, unforeseen effects can occur. The solar plants at Kramer Junction in California recorded drops in gross sol output during 1991 and 1992 that were attributed to the global ramifications resulting from the eruption of Mount Pinatubo in the Philippines.

There could also be a risk associated with the diurnal nature of solar power generation. Solar power can replace conventional sources of power during daylight hours. But without some form of energy storage it cannot supply power when it is dark.

It can, therefore, be argued that solar power is best deployed for peak power generation (demand peaks in hot countries often coincide with highest temperatures where there is widespread use of air conditioning). There is a danger, if it is deployed in this way, that future demand management programmes may reduce the peak demand level. Then the marginal value of the solar output may fall. The significance of this will depend on the mode of operation of the solar project and the details of any power purchase agreement.

The cost of solar power

Solar thermal and solar photovoltaic power plants share a number of features such as short deployment times and additional benefits from dispersed deployment that affect the cost and value of both technologies. However the technologies themselves have different roots and the costs associated with them have to be considered separately.

Solar thermal costs

Table 6.1-2 lists costs for solar thermal power plants estimated by the Sandia National Laboratory and the National Renewable Energy Laboratory, both run under the auspices of the US Department of Energy.

The only one of the three technologies listed that is operating in a commercial environment is the solar trough technology, exemplified by the nine plants built during the late 1980s and early 1990s in California. Operational costs have fallen at these plants in recent years and performance has increased so they may well be generating electricity at around $0.11–$0.12/kWh, in line with predictions in Table 6.1-2. This is too expensive for the technology to compete in the bulk power market in the USA but is low enough to enable it to compete in niche markets. Perhaps more importantly, it can also compete in the peak power market, which is where power from the Californian plants is sold.

The other two solar thermal technologies are currently at an earlier stage of development than solar trough technology. Consequently they are not operating under commercial conditions.

Costs for all three technologies are expected to fall. By 2010, the Department of Energy (DOE) predicts that solar towers will be capable of generating power at a levelised cost of $0.05/kWh, solar dishes at $0.06/kWh and solar troughs at $0.09/kWh. These levelised prices have been estimated on the basis of projects being built by an independent power producer with private financing. Other estimates have predicted an even lower cost for solar trough power plants, perhaps as low as $0.06/kWh by the middle of the next decade.

Costs are likely to be lower still for an ISCC power plant. A World Bank assessment put the cost of a near-term ISCC plant based on solar trough technology at 1080/kW, and the generating cost at less than 0.07/kWh[11]. This could fall to 0.05/kWh over the longer term.

Table 6.1-2 Solar thermal costs

	Capital cost ($/kW)	O & M costs ($/kW)	Levelised energy cost ($/kW)	
			2000	2010
Solar trough	2900	1.0	0.11	0.09
Solar tower	2400–2900	0.7	0.09	0.05
Solar dish	2900	2.0	0.13	0.06

Note: The levelised energy cost is for private financing.
Source: US Department of Energy.

Plants such as solar trough facilities with large solar arrays could be cheaper to build in the developing world where labour costs are lower than in the developed. A 100-MW solar trough plant could cost 19% less in Brazil than in the USA, for example.[12]

Solar photovoltaic costs

The main market for solar photovoltaic technology in 2003 was grid-connected residential and domestic installations. These accounted for 365 MW of total annual production of 744 MW, or roughly 50%.[13]

The cost of a grid-connected solar photovoltaic system based on silicon can be divided roughly into thirds. One-third is for the actual silicon to make the cell (the module), a further one-third for the manufacture of the solar cell and panel or module, and one-third for installation and ancillary equipment.[14]

In the USA in 2003, the cost of an installed rooftop system of this type was $6500–$8000/kW (see Table 6.1-3). This compares with $7000–$9000/kW in 2001 and $12000/kW in 1993. Even so, this makes solar photovoltaic technology one of the most expensive available today for generating electricity.

The cost of the solar cell accounts for a major part of the overall cost. Table 6.1-3 shows, this is between one-third and one-half of the total cost. Newer technologies may offer hope of reduced costs. Amorphous silicon and cadmium telluride modules were selling for $2000/kW and $3000/kW in 2003. The manufacture of silicon designed specially for solar cell applications may also reduce costs of silicon further.

Apart from the introduction of new technologies, the main hope for a reduction in the cost of solar cells comes from economies of scale. This effect is already bringing prices down slowly and global capacity rises. Government sponsored schemes to encourage the use of grid-connected photovoltaic arrays in commercial and

Table 6.1-3 Solar photovoltaic costs

	Photovoltaic module ($/kW)	Installed AC system ($/kW)
1993	4250	12,000
1995	3750	11,000–12,000
1997	4150	10,000–12,000
1999	3500	9000–11,000
2001	3500	7000–9000
2003	3000	6000–8000

Source: Renewable Energy World.[16]

domestic situations in countries like Japan, Germany and the USA are helping to increase demand.

The cost of electricity from solar photovoltaic power plants remains high. At an installed cost of $5000/kW, electricity probably costs around $0.25/kWh. This can be competitive with the peak power costs in somewhere like California but is way above the cost of base-load power, $0.025–0.050/kWh, in markets with well-developed infrastructures. Nevertheless the cost has reduced to a point where widespread installation is feasible.

End notes

1. In fact the Californian plants have achieved 20% conversion efficiency.
2. This is a World Bank estimate.
3. Photovoltaics come down to earth, Bill Yerkes, Modern Power Systems (July 2004) p. 30.
4. Renewable Energy Technology Characterization, The US Department of Energy and the Electric Power Research Institute (Topical Report 109496) 1997, available at. http://www.eere.energy.gov/power/pdfs/techchar.pdf
5. Refer *supra* note 3.
6. Solar Thermal Power 2020, Greenpeace, 2004.
7. PV Market Update, Paul Maycock, Renewable Energy World (July–August 2004).
8. Refer *supra* note 7.
9. Refer *supra* note 7.
10. US Energy Information Administration estimate.
11. Lifecycle figures are taken from Benign Energy? The Environmental Impact of Renewables, published by the International Energy Agency.
12. Cost Reduction Study for Solar Thermal Power Plants, Final Report, World Bank (May 1999).
13. Refer *supra* note 6.
14. Refer *supra* note 7.
15. Refer *supra* note 3.
16. Refer *supra* note 7.

Chapter 6.2

6.2

Solar thermal collectors and applications

Soteris A. Kalogirou

Nomenclature

A_a absorber area (m^2)
A_c total collector aperture area (m^2)
A_f collector geometric factor
A_r receiver area (m^2)
b bond width (m)
b_0 incidence angle modifier constant
b_1 incidence angle modifier constant
c_p specific heat at constant pressure (J/kg K)
c_0 intercept efficiency [$=F_R\tau\alpha$]
c_1 first-order coefficient of the collector efficiency (W/m^2 °C)
c_2 second-order coefficient of the collector efficiency (W/m^2 °C^2)
C collector concentration ratio [$= A_a/A_r$], factor given by Eq. (6.2.26), investment cost ($)
C_b bond conductance (W/m °C)
C_{FA} cost rate for auxiliary energy ($/kJ)
C_{FL} cost rate for conventional fuel ($/kJ)
d market discount rate (%), interest rate (%)
d_r displacement of receiver from focus (m)
d^* universal non-random error parameter due to receiver mislocation and reflector profile errors ($d^* = d_r/D$)
D riser tube outside diameter (m), monthly total heating load for space heating and hot water or demand (J)
D_i tube inside diameter (m)
D_o tube outside diameter (m)
$E_{x,in}$ exergy in (W)
$E_{x,out}$ exergy out (W)
f focal distance (m), solar contribution, factorgiven by Eq. (6.2.25)
F' collector efficiency factor
F fin efficiency, cash flow
F_R heat removal factor
F'_R collector heat exchanger efficiency factor
G_b beam (or direct) irradiation (W/m^2)
G_t total (direct plus diffuse) solar energy incident on the collector aperture (W/m^2)
h hour angle (degrees)
$\overline{H}_T$ monthly average daily radiation incident on the collector surface per unit area (J/m^2)
h_{fi} heat transfer coefficient inside absorber tube (W/m^2 °C)
h_w wind heat transfer coefficient (W/m^2 °C)
h_p height of the parabola (m)
i inflation rate (%)
I total horizontal radiation per unit area (W/m^2)
I_{bT} incident beam radiation per unit area (W/m^2)
I_d horizontal diffuse radiation per unit area (W/m^2)
k absorber thermal conductivity (W/m °C)
$k_{\alpha\tau}$ incidence angle modifier
k_b bond thermal conductivity (W/m °C)
k_0 intercept efficiency [$=F_R n_0$]
k_1 first-order coefficient of the collector efficiency (W/m^2 °C) [$=c_1/C$]
k_2 second-order coefficient of the collector efficiency (W/m^2 °C^2) [$=c_2/C$]
L half distance between two consecutive riser pipes [$=(W-D)/2$], collector length (m)
m mass flow rate of fluid (kg/s), factor given in Eq. (6.2.6)
M mass flow number
n_c collector efficiency
n_o collector optical efficiency
N days in month, number of years
N_g number of glass covers
N_s entropy generation number
PW_N present worth after N years
q^* irradiation per unit of collector area (W/m^2)
q_u rate of useful energy delivered by the collector (W)
q'_u useful energy gain per unit length (J/m)
q'_{fin} useful energy conducted per unit fin length (J/m)
q'_{tube} useful energy conducted per unit tube length (J/m)
q^*_o radiation falling on the receiver (W/m^2)
Q rate of heat transfer output (W)

Renewable Energy Focus Handbook 2009; ISBN: 9780123747051

Q_{aux} auxiliary energy (J)
Q_{load} load or demand energy (J)
Q^* solar radiation incident on collector (W)
Q_o rate of heat loss to ambient (W)
R receiver radius (m)
s specific entropy (J/kg K)
S absorbed solar energy (kJ/m^2)
S_{gen} generated entropy (J/K)
t time
T absolute temperature (K)
T_a ambient temperature (°C)
T_b local base temperature (°C)
$\overline{T}_a$ monthly average ambient temperature (°C)
T_{av} average collector fluid temperature (°C)
T_f local fluid temperature (°C)
T_{fi} temperatures of the fluid entering the collector (°C)
T_i temperatures of the fluid entering the collector (°C)
T_o ambient temperature (K), temperature of the fluid leaving the collector (°C)
T_{oi} collector outlet initial water temperature (°C)
T_{ot} collector outlet water temperature after time t (°C)
T_p average temperature of the absorbing surface (°C), stagnation temperature (°C)
T_r temperature of the absorber (°C), receiver temperature (K)
T_{ref} an empirically derived reference temperature [=100 °C]
T_s apparent black body temperature of the sun (~6000 K)
T^* apparent sun temperature as an exergy source (~4500 K)
U_b bottom heat loss coefficient (W/m^2 °C)
U_e edges heat loss coefficient (W/m^2 °C)
U_L solar collector heat transfer loss coefficient (W/m^2 °C)
U_o heat transfer coefficient from fluid to ambient air (W/m^2 °C)
U_r receiver-ambient heat transfer coefficient based on A_r (W/m^2 K)
U_t top heat loss coefficient (W/m^2 °C)
W distance between riser tubes (m), wind velocity (m/s)
W_a collector aperture (m)
x factor used in Eq. (6.2.72) [$=(T_i-T_a)/G_t$]
X dimensionless parameter given by Eq. (6.2.83)
X_c corrected value of X
y factor used in Eq. (6.2.74) [$=(T_i-T_a)/G_b$]
Y dimensionless parameter given by Eq. (6.2.84)

Greek symbols

α_a absorber absorptance
α fraction of solar energy reaching surface that is absorbed, absorptivity
β incidence angle (degrees), collector slope (degrees), misalignment angle error (degrees)
β^* universal non-random error parameter due to angular errors ($\beta^*=\beta C$)
δ absorber (fin) thickness (m), declination angle (degrees)
ΔT temperature difference [$=T_i - T_a$]
Δx elemental fin or riser tube distance (m)
ϵ_g emissivity of glass covers
ϵ_p absorber plate emittance
γ collector intercept factor, average bond thickness (m)
ρ density (kg/m^3), mirror reflectance
ρ_m mirror reflectance
$\rho_{(\theta)}$ distance, ρ; along a tangent from the receiver to the curve given by Eq. (6.2.101)
θ dimensionless temperature [$=T/T_o$], angle of incidence (degrees)
θ_A acceptance half angle for CPC collectors (degrees)
θ_m collector half acceptance angle (degrees)
θ_{sky} effective incidence angle for evaluating the incidence angle modifier of flat-plate collector for sky diffuse radiation
θ_{gnd} effective incidence angle for evaluating the incidence angle modifier of flat-plate collector for ground reflected radiation
Φ zenith angle (degrees)
Ψ factor given in Eq. (6.2.7)
σ Stefan-Boltzmann constant [$= 5.67 \times 10^{-8}$ W/m^2 K^4]
σ^* universal random error parameter ($\sigma^*=\sigma C$)
σ_{sun} standard deviation of the energy distribution of the sun's rays at normal incidence
σ_{slope} standard deviation of the distribution of local slope errors at normal incidence
σ_{mirror} standard deviation of the variation in diffusivity of the reflective material at normal incidence
τ_α absorber transmittance
$\tau\alpha$ transmittance–absorptance product
$\overline{\tau\alpha}$ monthly average transmittance–absorptance product
$(\tau\alpha)_b$ transmittance–absorptance product for estimating incidence angle modifier for beam radiation
$(\tau\alpha)_s$ transmittance–absorptance product for estimating incidence angle modifier for sky radiation
$(\tau\alpha)_g$ transmittance–absorptance product for estimating incidence angle modifier for ground reflected radiation
φ parabolic angle (degrees): the angle between the axis and the reflected beam at focus of the parabola
φ_r collector rim angle (degrees)

Abbreviations

AFP advanced flat-plate
CLFR compact linear Fresnel reflector
CPC compound parabolic collector
CTC cylindrical trough collector
ED electrodialysis
ER energy recovery
E–W east–west
ETC evacuated tube collector
FPC flat-plate collector
HFC heliostat field collector
ICPC integrated compound parabolic collector
LCR local concentration ratio
LCS life cycle savings
LFR linear Fresnel reflector
MEB multiple effect boiling
MSF multistage flash
N-S north–south
PDR parabolic dish reflector
PTC parabolic trough collector
PWF present worth factor
RO reverse osmosis
TI transparent insulation
VC vapor compression

6.2.1 Introduction

The sun is a sphere of intensely hot gaseous matter with a diameter of 1.39×10^9 m. The solar energy strikes our planet a mere 8 min and 20 s after leaving the giant furnace, the sun which is 1.5×10^{11} m away. The sun has an effective blackbody temperature of 5762 K [1]. The temperature in the central region is much higher and it is estimated at 8×10^6 to 40×10^6 K. In effect the sun is a continuous fusion reactor in which hydrogen is turned into helium. The sun's total energy output is 3.8×10^{20} MW which is equal to 63 MW/m^2 of the sun's surface. This energy radiates outwards in all directions. Only a tiny fraction, 1.7×10^{14} kW, of the total radiation emitted is intercepted by the earth [1]. However, even with this small fraction it is estimated that 30 min of solar radiation falling on earth is equal to the world energy demand for one year.

Man realised that a good use of solar energy is in his benefit, from prehistoric times. The Greek historian Xenophon in his "memorabilia" records some of the teachings of the Greek philosopher Socrates (470–399 BC) regarding the correct orientation of dwellings in order to have houses which were cool in summer and warm in winter.

Since prehistory, the sun has dried and preserved man's food. It has also evaporated sea water to yield salt. Since man began to reason, he has recognised the sun as a motive power behind every natural phenomenon. This is why many of the prehistoric tribes considered Sun as "God". Many scripts of ancient Egypt say that the Great Pyramid, one of the man's greatest engineering achievements, was built as a stairway to the sun [2].

Basically, all the forms of energy in the world as we know it are solar in origin. Oil, coal, natural gas and woods were originally produced by photosynthetic processes, followed by complex chemical reactions in which decaying vegetation was subjected to very high temperatures and pressures over a long period of time [1]. Even the wind and tide energy have a solar origin since they are caused by differences in temperature in various regions of the earth.

The greatest advantage of solar energy as compared with other forms of energy is that it is clean and can be supplied without any environmental pollution. Over the past century fossil fuels have provided most of our energy because these are much cheaper and more convenient than energy from alternative energy sources, and until recently environmental pollution has been of little concern.

Twelve winter days of 1973 changed the economic relation of fuel and energy when the Egyptian army stormed across the Suez Canal on October the 12th provoking an international crisis and for the first time, involved as part of Arab strategy, the threat of the "oil weapon". Both the price and the political weapon issues quickly came to a head when the six Gulf members of the Organisations of Petroleum Exporting Countries (OPEC), met in Kuwait and quickly abandoned the idea of holding any more price consultations with the oil companies, announcing that they were raising the price of their crude oil by 70%.

The reason for the rapid increase in oil demand occurred mainly because increasing quantities of oil, produced at very low cost, became available during the 50s and 60s from the Middle East and North Africa. For the consuming countries imported oil was cheap compared with indigenously produced energy from solid fuels.

But the main problem is that proved reserves of oil and gas, at current rates of consumption, would be adequate to meet demand for another 40 and 60 years, respectively. The reserves for coal are in better situation as they would be adequate for at least the next 250 years.

If we try to see the implications of these limited reserves we will be faced with a situation in which the price of fuels will be accelerating as the reserves are decreased. Considering that the price of oil has become firmly established as the price leader for all fuel prices then the conclusion is that energy prices will increase over the next decades at something greater than the rate of inflation or even more. In addition to this is also the concern about the environmental pollution caused by the burning of the fossil fuels. This issue is examined in Section 1.1.

In addition to the thousands of ways in which the sun's energy has been used by both nature and man through time, to grow food or dry clothes, it has also been deliberately harnessed to perform a number of other jobs. Solar energy is used to heat and cool buildings (both active and passive), to heat water for domestic and industrial uses, to heat swimming pools, to power refrigerators, to operate engines and pumps, to desalinate water for drinking purposes, to generate electricity, for chemistry applications, and many more. The objective of this paper is to present the various types of collectors used to harness solar energy, their thermal analysis and performance, and a review of applications.

There are many alternative energy sources which can be used instead of fossil fuels. The decision as to what type of energy source should be utilised must, in each case, be made on the basis of economic, environmental and safety considerations. Because of the desirable environmental and safety aspects it is widely believed that solar energy should be utilised instead of other alternative energy forms, even when the costs involved are slightly higher.

6.2.1.1 Energy related environmental problems

Energy is considered a prime agent in the generation of wealth and a significant factor in economic development.

The importance of energy in economic development is recognised universally and historical data verify that there is a strong relationship between the availability of energy and economic activity. Although at the early 70s, after the oil crisis, the concern was on the cost of energy, during the past two decades, the risk and reality of environmental degradation have become more apparent. The growing evidence of environmental problems is due to a combination of several factors since the environmental impact of human activities has grown dramatically. This is due to the increase of the world population, energy consumption and industrial activities. Achieving solutions to environmental problems that humanity faces today requires long-term potential actions for sustainable development. In this respect, renewable energy resources appear to be one of the most efficient and effective solutions.

A few years ago, most environmental analysis and legal control instruments concentrated on conventional pollutants such as sulphur dioxide (SO_2), nitrogen oxides (NO_x), particulates, and carbon monoxide (CO). Recently however, environmental concern has extended to the control of hazardous air pollutants, which are usually toxic chemical substances which are harmful even in small doses, as well as to other globally significant pollutants such as carbon dioxide (CO_2). Additionally, developments in industrial processes and structures have led to new environmental problems. A detailed description of these gaseous and particulate pollutants and their impacts on the environment and human life is presented by Dincer [3,4].

One of the most widely accepted definitions of sustainable development is: "development that meets the needs of the present without compromising the ability of future generations to meet their own needs". There are many factors that can help to achieve sustainable development. Today, one of the main factors that must be considered is energy and one of the most important issues is the requirement for a supply of energy that is fully sustainable [5,6]. A secure supply of energy is generally agreed to be a necessary, but not a sufficient requirement for development within a society. Furthermore, for a sustainable development within a society it is required that a sustainable supply of energy and effective and efficient utilization of energy resources are secured. Such a supply in the long-term should be readily available at reasonable cost, be sustainable and be able to be utilized for all the required tasks without causing negative societal impacts. This is why there is a close connection between renewable sources of energy and sustainable development.

Pollution depends on energy consumption. Today the world daily oil consumption is 76 million barrels. Despite the well-known consequences of fossil fuel combustion on the environment, this is expected to increase to 123 million barrels per day by the year 2025 [7]. There are a large number of factors which are significant in the determination of the future level of energy consumption and production. Such factors include population growth, economic performance, consumer tastes and technological developments. Furthermore, governmental policies concerning energy and developments in the world energy markets will certainly play a key role in the future level and pattern of energy production and consumption [8].

Another parameter to be considered is the world population. This is expected to double by the middle of this century and as economic development will certainly continue to grow, the global demand for energy is expected to increase. Today much evidence exists which suggests that the future of our planet and of the generations to come will be negatively impacted if humans keep degrading the environment. Currently, three environmental problems are internationally known; these are the acid precipitation, the stratospheric ozone depletion, and the global climate change. These are analysed in more detail below.

Acid rain

This is a form of pollution depletion in which SO_2 and NO_x produced by the combustion of fossil fuels are transported over great distances through the atmosphere and deposited via precipitation on the earth, causing damage to ecosystems that are exceedingly vulnerable to excessive acidity. Therefore, it is obvious that the solution to the issue of acid rain deposition requires an appropriate control of SO_2 and NO_x pollutants. These pollutants cause both regional and transboundary problems of acid precipitation.

Recently, attention also has been given to other substances such as volatile organic compounds (VOCs), chlorides, ozone and trace metals that may participate in a complex set of chemical transformations in the atmosphere resulting in acid precipitation and the formation of other regional air pollutants. A number of evidences that show the damages of acid precipitation are reported by Dincer and Rosen [6].

It is well known that some energy-related activities are the major sources of acid precipitation. Additionally, VOCs are generated by a variety of sources and comprise a large number of diverse compounds. Obviously, the more energy we spend the more we contribute to acid precipitation; therefore, the easiest way to reduce acid precipitation is by reducing energy consumption.

Ozone layer depletion

The ozone present in the stratosphere, at altitudes between 12 and 25 km, plays a natural equilibrium-maintaining role for the earth, through absorption of ultraviolet (UV) radiation (240–320 nm) and absorption of infrared radiation [3]. A global environmental problem is the depletion of the stratospheric ozone layer which is caused by the emissions of CFCs, halons (chlorinated and brominated organic compounds) and NO_x. Ozone depletion can lead to increased levels of damaging UV radiation reaching the ground, causing increased rates of skin cancer and eye damage to humans and is harmful to many biological species. It should be noted that energy related activities are only partially (directly or indirectly) responsible for the emissions which lead to stratospheric ozone depletion. The most significant role in ozone depletion is played by the CFCs, which are mainly used in air conditioning and refrigerating equipment as refrigerants, and NO_x emissions which are produced by the fossil fuel and biomass combustion processes, the natural denitrification and nitrogen fertilizers.

In 1998 the size of the ozone hole over Antarctica was 25 million km^2. It was about 3 million km^2 in 1993 [7]. Researchers expect the Antarctic ozone hole to remain severe in the next 10–20 years, followed by a period of slow healing. Full recovery is predicted to occur in 2050; however, the rate of recovery is affected by the climate change [8].

Global climate change

The term greenhouse effect has generally been used for the role of the whole atmosphere (mainly water vapour and clouds) in keeping the surface of the earth warm. Recently however, it has been increasingly associated with the contribution of CO_2 which is estimated contributes about 50% to the anthropogenic greenhouse effect. Additionally, several other gasses such as CH_4, CFCs, halons, N_2O, ozone and peroxyacetylnitrate (also called greenhouse gases) produced by the industrial and domestic activities can also contribute to this effect, resulting in a rise of the earth's temperature. Increasing atmospheric concentrations of greenhouse gases increase the amount of heat trapped (or decrease the heat radiated from the earth's surface), thereby raising the surface temperature of the earth. According to Colonbo [9] the earth's surface temperature has increased by about 0.6°C over the last century, and as a consequence the sea level is estimated to have risen by perhaps 20 cm. These changes can have a wide range of effects on human activities all over the world. The role of various greenhouse gasses is summarized in Ref. [6].

Humans contribute through many of their economic and other activities to the increase of the atmospheric concentrations of various greenhouse gases. For example, CO_2 releases from fossil fuel combustion, methane emissions from increased human activity and CFC releases all contribute to the greenhouse effect. Predictions show that if atmospheric concentrations of greenhouse gases, mainly due to fossil fuel combustion, continue to increase at the present rate, the earth's temperature may increase by another 2–4°C in the next century. If this prediction is realized, the sea level could rise by between 30 and 60 cm before the end of this century [9]. The impacts of such sea level increase could easily be understood and include flooding of coastal settlements, displacement of fertile zones for agriculture toward higher latitudes, and decrease the availability of fresh water for irrigation and other essential uses. Thus, such consequences could put in danger the survival of entire populations.

Renewable energy technologies

Renewable energy technologies produce marketable energy by converting natural phenomena into useful forms of energy. These technologies use the sun's energy and its direct and indirect effects on the earth (solar radiation, wind, falling water and various plants, i.e. biomass), gravitational forces (tides), and the heat of the earth's core (geothermal) as the resources from which energy is produced. These resources have massive energy potential; however, they are generally diffused and not fully accessible, most of them are intermittent, and they have distinct regional variabilities. These characteristics give rise to difficult, but solvable, technical and economical challenges. Nowadays, significant progress is made by improving the collection and conversion efficiencies, lowering the initial and maintenance costs, and increasing the reliability and applicability.

Worldwide research and development in the field of renewable energy resources and systems has been carried out during the last two decades. Energy conversion systems that are based on renewable energy technologies appeared to be cost effective compared to the projected high cost of oil. Furthermore, renewable energy systems can have a beneficial impact on the environmental, economic, and political issues of the world. At the end of 2001 the total installed capacity of renewable energy systems was equivalent to 9% of the total electricity generation [10]. By applying a renewable energy intensive scenario the global consumption of renewable sources by 2050 would reach 318 exajoules [11].

The benefits arising from the installation and operation of renewable energy systems can be placed into three categories: energy saving, generation of new working posts and the decrease of environmental pollution.

The energy-saving benefit derives from the reduction in consumption of the electricity and/or diesel which are used conventionally to provide energy. This benefit can be directly translated into monetary units according to the corresponding production or avoiding capital expenditure for the purchase of imported fossil fuels.

Another factor which is of considerable importance in many countries is the ability of renewable energy technologies to generate jobs. The penetration of a new technology leads to the development of new production activities contributing to the production, market distribution and operation of the pertinent equipment. Specifically in the case of solar energy collectors, job creation mainly relates to the construction and installation of the collectors. The latter is a decentralised process since it requires the installation of equipment in every building involving every individual consumer.

The most important benefit of renewable energy systems is the decrease of environmental pollution. This is achieved by the reduction of air emissions due to the substitution of electricity and conventional fuels. The most important effects of air pollutants on the human and natural environment are their impact on the public health, agriculture and on ecosystems. It is relatively simple to measure the financial impact of these effects when they relate to tradable goods such as the agricultural crops; however, when it comes to non-tradable goods, like human health and ecosystems, things becomes more complicated. It should be noted that the level of the environmental impact and therefore the social pollution cost largely depends on the geographical location of the emission sources. Contrary to the conventional air pollutants, the social cost of CO_2 does not vary with the geographical characteristics of the source as each unit of CO_2 contributes equally to the climate change threat and the resulting cost.

In this paper emphasis is given to solar thermal systems. Solar thermal systems are non-polluting and offer significant protection of the environment. The reduction of greenhouse gas pollution is the main advantage of utilising solar energy. Therefore, solar thermal systems should be employed whenever possible in order to achieve a sustainable future.

6.2.1.2 History of solar energy

The idea of using solar energy collectors to harness the sun's power is recorded from prehistoric times when at 212 BC the Greek scientist/physician Archimedes devised a method to burn the Roman fleet. Archimedes reputedly set the attacking Roman fleet afire by means of concave metallic mirror in the form of hundreds of polished shields, all reflecting on the same ship [2].

The Greek historian Plutarch (AD 46–120) referred to the incident saying that the Romans, seeing that indefinite mischief overwhelmed them from no visible means, began to think they were fighting with the gods. The basic question was whether or not Archimedes knew enough about the science of optics to device a simple way to concentrate sunlight to a point where ships could be burned from a distance. Archimedes had written a book "On burning Mirrors" but no copy has survived to give evidence [12].

Eighteen hundred years after Archimedes, Athanasius Kircher (1601–1680) carried out some experiments to set fire to a woodpile at a distance in order to see whether the story of Archimedes had any scientific validity but no report of his findings survived [12].

Amazingly, the very first applications of solar energy refer to the use of concentrating collectors, which are by their nature (of accurate shape and construction) and the requirement to follow the sun, more "difficult" to apply. During the 18th century, solar furnaces capable of melting iron, copper and other metals were being constructed of polished-iron, glass lenses and mirrors. The furnaces were in use throughout Europe and the Middle East. One furnace designed by the French scientist Antoine Lavoisier, attained the remarkable temperature of 1750°C. The furnace used a 1.32 m lens plus a secondary 0.2 m lens to obtain such temperature which turned out to be the maximum achieved by man for one hundred years.

During the 19th century the attempts to convert solar energy into other forms were based upon the generation of low-pressure steam to operate steam engines. August Monchot pioneered this field by constructing and operating several solar-powered steam engines between the years 1864 and 1878 [12]. Evaluation of one built at Tours by the French government showed that it was too expensive to be considered feasible. Another one was set up in Algeria. In 1875, Mouchot made a notable advance in solar collector design by making one in the form of a truncated cone reflector. Mouchot's collector consisted of silver-plated metal plates and had a diameter of 5.4 m and a collecting area of 18.6 m^2. The moving parts weighed 1400 kg.

Abel Pifre was a contemporary of Mouchot who also made solar engines [12,13]. Pifre's solar collectors were parabolic reflectors made of very small mirrors. In shape they looked rather similar to Mouchot's truncated cones.

In 1901 A.G. Eneas installed a 10 m diameter focusing collector which powered a water pumping apparatus at a California farm. The device consisted of a large umbrella-like structure open and inverted at an angle to receive the full effect of sun's rays on the 1788 mirrors which lined the inside surface. The sun's rays were concentrated at a focal point where the boiler was located. Water within the boiler was heated to produce steam which in turn powered a conventional compound engine and centrifugal pump [1,12].

In 1904 a Portuguese priest, Father Himalaya, constructed a large solar furnace. This was exhibited at the St Louis World's fair. This furnace appeared quite modern in structure, being a large, off-axis, parabolic horn collector [12].

In 1912 Shuman, in collaboration with C.V. Boys, undertook to build the world's largest pumping plant in Meadi, Egypt. The system was placed in operation in 1913 and it was using long parabolic cylinders to focus sunlight onto a long absorbing tube. Each cylinder was 62 m long, and the total area of the several banks of cylinders was 1200 m^2. The solar engine developed as much as 37–45 kW continuously for a 5 h period [1,12,13]. Despite the plant's success, it was completely shut down in 1915 due to the onset of World War I and cheaper fuel prices.

During the last 50 years many variations were designed and constructed using focusing collectors as a means of heating the transfer or working fluid which powered mechanical equipment. The two primary solar technologies used are the central receivers and the distributed receivers employing various point and line-focus optics to concentrate sunlight. Central receiver systems use fields of heliostats (two-axis tracking mirrors) to focus the sun's radiant energy onto a single tower-mounted receiver [14]. Distributed receiver technology includes parabolic dishes, Fresnel lenses, parabolic troughs, and special bowls. Parabolic dishes track the sun in two axes and use mirrors to focus radiant energy onto a point-focus receiver. Troughs and bowls are line-focus tracking reflectors that concentrate sunlight onto receiver tubes along their focal lines. Receiver temperatures range from 100°C in low-temperature troughs to close to 1500°C in dish and central receiver systems [14]. More details of the basic types of collectors are given in Section 2.

Another area of interest, hot water and house heating appeared in the mid 1930s, but gained interest in the last half of the 40s. Until then millions of houses were heated by coal burn boilers. The idea was to heat water and feed it to the radiator system that was already installed.

The manufacture of solar water heaters (SWH) began in the early 60s. The industry of SWH expanded very quickly in many countries of the world. Typical SWH in many cases are of the thermosyphon type and consist of two flat-plate solar collectors having an absorber area between 3 and 4 m^2, a storage tank with capacity between 150 and 180 l and a cold water storage tank, all installed on a suitable frame. An auxiliary electric immersion heater and/or a heat exchanger, for central heating assisted hot water production, are used in winter during periods of low solar insolation. Another important type of SWH is the force circulation type. In this system only the solar panels are visible on the roof, the hot water storage tank is located indoors in a plantroom and the system is completed with piping, pump and a differential thermostat. Obviously, this latter type is more appealing mainly due to architectural and aesthetic reasons, but also more expensive especially for small-size installations [15]. These together with a variety of other systems are described in Section 5.

Becquerel had discovered the photovoltaic effect in selenium in 1839. The conversion efficiency of the "new" silicon cells developed in 1958 was 11% although the cost was prohibitively high ($1000/W) [12]. The first practical application of solar cells was in space where cost was not a barrier and no other source of power is available. Research in the 1960s resulted in the discovery of other photovoltaic materials such as gallium arsenide (GaAS). These could operate at higher temperatures than silicon but were much more expensive. The global installed capacity of photovoltaics at the end of 2002 was near 2 GWp [16]. Photovoltaic (PV) cells are made of various semiconductors, which are materials that are only moderately good conductors of electricity. The materials most commonly used are silicon (Si) and compounds of cadmium sulphide (Cds), cuprous sulphide (Cu_2S), and GaAs.

Amorphous silicon cells are composed of silicon atoms in a thin homogenous layer rather than a crystal structure. Amorphous silicon absorbs light more effectively than crystalline silicon, so the cells can be thinner. For this reason, amorphous silicon is also known as a "thin film" PV technology. Amorphous silicon can be deposited on a wide range of substrates, both rigid and flexible, which makes it ideal for curved surfaces and "fold-away" modules. Amorphous cells are, however, less efficient than crystalline based cells, with typical efficiencies of around 6%, but they are easier and therefore cheaper to produce. Their low cost makes them ideally suited for many applications where high efficiency is not required and low cost is important.

Amorphous silicon (a-Si) is a glassy alloy of silicon and hydrogen (about 10%). Several properties make it an attractive material for thin-film solar cells:

1. Silicon is abundant and environmentally safe.
2. Amorphous silicon absorbs sunlight extremely well, so that only a very thin active solar cell layer is required (about 1 μm as compared to 100 μm or so for crystalline solar cells), thus greatly reducing solar-cell material requirements.
3. Thin films of a-Si can be deposited directly on inexpensive support materials such as glass, sheet steel, or plastic foil.

A number of other promising materials such as cadmium telluride and copper indium diselenide are now being used for PV modules. The attraction of these technologies is that they can be manufactured by relatively inexpensive industrial processes, in comparison to crystalline silicon technologies, yet they typically offer higher module efficiencies than amorphous silicon.

The PV cells are packed into modules which produce a specific voltage and current when illuminated. PV modules can be connected in series or in parallel to produce larger voltages or currents. Photovoltaic systems can be used independently or in conjunction with other electrical power sources. Applications powered by PV systems include communications (both on earth and in space), remote power, remote monitoring, lighting, water pumping and battery charging.

The two basic types of PV applications are the stand alone and the grid connected. Stand-alone PV systems are used in areas that are not easily accessible or have no access to mains electricity. A stand-alone system is independent of the electricity grid, with the energy produced normally being stored in batteries. A typical stand-alone system would consist of PV module or modules, batteries and charge controller. An inverter may also be included in the system to convert the direct current generated by the PV modules to the alternating current form (AC) required by normal appliances.

In grid connected applications the PV system is connect to the local electricity network. This means that during the day, the electricity generated by the PV system either can be used immediately (which is normal for systems installed in offices and other commercial buildings), or can be sold to one of the electricity supply companies (which is more common for domestic systems where the occupier may be out during the day). In the evening, when the solar system is unable to provide the electricity required, power can be bought back from the network. In effect, the grid is acting as an energy storage system, which means the PV system does not need to include battery storage.

When PVs started to be used for large-scale commercial applications, about 20 years ago, their efficiency was well below 10%. Nowadays, their efficiency increased to about 15%. Laboratory or experimental units can give efficiencies of more than 30%, but these have not been commercialized yet. Although 20 years ago PVs were considered as a very expensive solar system the present cost is around 5000$ per kW_e and there are good prospects for further reduction in the coming years. More details on photovoltaics are beyond the scope of this paper.

The lack of water was always a problem to humanity. Therefore among the first attempts to harness solar energy were the development of equipment suitable for the desalination of sea-water. Solar distillation has been in practice for a long time. According to Malik et al. [17], the earliest documented work is that of an Arab alchemist in the 15th century reported by Mouchot in 1869. Mouchot reported that the Arab alchemist had used polished Damascus mirrors for solar distillation.

The great French chemist Lavoisier (1862) used large glass lenses, mounted on elaborate supporting structures, to concentrate solar energy on the contents of distillation flasks [17]. The use of silver or aluminium coated glass reflectors to concentrate solar energy for distillation has also been described by Mouchot.

The use of solar concentrators in solar distillation has been reported by Pasteur (1928) [17] who used a concentrator to focus solar rays onto a copper boiler containing water. The steam generated from the boiler was piped to a conventional water cooled condenser in which distilled water was accumulated.

Solar stills are one of the simplest types of desalination equipment which uses the greenhouse effect to evaporate salty water. Solar stills were the first to be used on large-scale distilled water production. The first water distillation plant constructed was a system built at Las Salinas, Chile, in 1874 [12,17]. The still covered 4700 m^2 and produced up to 23 000 l of fresh water per day (4.9 l/m^2), in clear sun. The still was operated for 40 years and was abandoned only after a fresh-water pipe was installed supplying water to the area from the mountains.

The renewal of interest on solar distillation occurred after the First World War at which time several new devices had been developed such as: roof type, tilted wick, inclined tray and inflated stills. Some more details on solar stills are given in section 6.2.5.5. In this section it is also indicated how solar collectors can be used to power conventional desalination equipment. More information on solar desalination is given in Ref. [18].

Another application of solar energy is solar drying. Solar dryers have been used primarily by the agricultural industry. The objective in drying an agricultural product is to reduce its moisture contents to that level which prevents deterioration within a period of time regarded as the safe storage period. Drying is a dual process of heat transfer to the product from the heating source, and mass transfer of moisture from the interior of the product to its surface and from the surface to the surrounding air.

The objective of a dryer is to supply the product with more heat than is available under ambient conditions, increasing sufficiently the vapour pressure of the moisture held within the crop, thus enhancing moisture migration from within the crop and decreasing significantly the relative humidity of the drying air, thus increasing its moisture carrying capability and ensuring a sufficiently low equilibrium moisture content.

In solar drying, solar energy is used as either the sole source of the required heat or as a supplemental source, and the air flow can be generated by either forced or natural convection. The heating procedure could involve the passage of the pre-heated air through the product, by directly exposing the product to solar radiation or a combination of both. The major requirement is the transfer of heat to the moist product by convection and conduction from surrounding air mass at temperatures above that of the product, or by radiation mainly from the sun and to a small extent from surrounding hot surfaces, or conduction from heated surfaces in contact with the product. Details of solar dryers are beyond the scope of this paper. More information on solar dryers can be found in Ref. [19].

Section 6.2.2 gives a brief description of several of the most common collectors available in the market.

6.2.2 Solar collectors

Solar energy collectors are a special kind of heat exchangers that transform solar radiation energy to internal

Table 6.2-1 Solar energy collectors

Motion	Collector type	Absorber type	Concentration ratio	Indicative temperature range (°C)
Stationary	Flat plate collector (FPC)	Flat	1	30–80
	Evacuated tube collector (ETC)	Flat	1	50–200
	Compound parabolic collector (CPC)	Tubular	1–5	60–240
Single-axis tracking			5–15	60–300
	Linear Fresnel reflector (LFR)	Tubular	10–40	60–250
	Parabolic trough collector (PTC)	Tubular	15–45	60–300
	Cylindrical trough collector (CTC)	Tubular	10–50	60–300
Two-axes tracking	Parabolic dish reflector (PDR)	Point	100–1000	100–500
	Heliostat field collector (HFC)	Point	100–1500	150–2000

Note: Concentration ratio is defined as the aperture area divided by the receiver/absorber area of the collector.

energy of the transport medium. The major component of any solar system is the solar collector. This is a device which absorbs the incoming solar radiation, converts it into heat, and transfers this heat to a fluid (usually air, water, or oil) flowing through the collector. The solar energy thus collected is carried from the circulating fluid either directly to the hot water or space conditioning equipment, or to a thermal energy storage tank from which can be drawn for use at night and/or cloudy days.

There are basically two types of solar collectors: non-concentrating or stationary and concentrating. A non-concentrating collector has the same area for intercepting and for absorbing solar radiation, whereas a sun-tracking concentrating solar collector usually has concave reflecting surfaces to intercept and focus the sun's beam radiation to a smaller receiving area, thereby increasing the radiation flux. A large number of solar collectors are available in the market. A comprehensive list is shown in Table 6.2-1 [20].

In this section a review of the various types of collectors currently available will be presented. This includes FPC, ETC, and concentrating collectors.

6.2.2.1 Stationary collectors

Solar energy collectors are basically distinguished by their motion, i.e. stationary, single axis tracking and two-axes tracking, and the operating temperature. Initially, the stationary solar collectors are examined. These collectors are permanently fixed in position and do not track the sun. Three types of collectors fall in this category:

1. Flat plate collectors (FPC);
2. Stationary compound parabolic collectors (CPC);
3. Evacuated tube collectors (ETC).

6.2.2.1.1 Flat-plate collectors

A typical flat-plate solar collector is shown in Fig. 6.2-1. When solar radiation passes through a transparent cover and impinges on the blackened absorber surface of high absorptivity, a large portion of this energy is absorbed by the plate and then transferred to the transport medium in the fluid tubes to be carried away for storage or use. The underside of the absorber plate and the side of casing are well insulated to reduce conduction losses. The liquid tubes can be welded to the absorbing plate, or they can be an integral part of the plate. The liquid tubes are connected at both ends by large diameter header tubes.

The transparent cover is used to reduce convection losses from the absorber plate through the restraint of the stagnant air layer between the absorber plate and the glass. It also reduces radiation losses from the collector as the glass is transparent to the short wave radiation received by the sun but it is nearly opaque to long-wave thermal radiation emitted by the absorber plate (greenhouse effect).

FPC are usually permanently fixed in position and require no tracking of the sun. The collectors should be oriented directly towards the equator, facing south in the northern hemisphere and north in the southern. The optimum tilt angle of the collector is equal to the latitude of the location with angle variations of 10–15° more or less depending on the application [20].

A FPC generally consists of the following components as shown in Fig. 6.2-2:

Glazing. One or more sheets of glass or other diathermanous (radiation-transmitting) material.

Tubes, fins, or passages. To conduct or direct the heat transfer fluid from the inlet to the outlet.

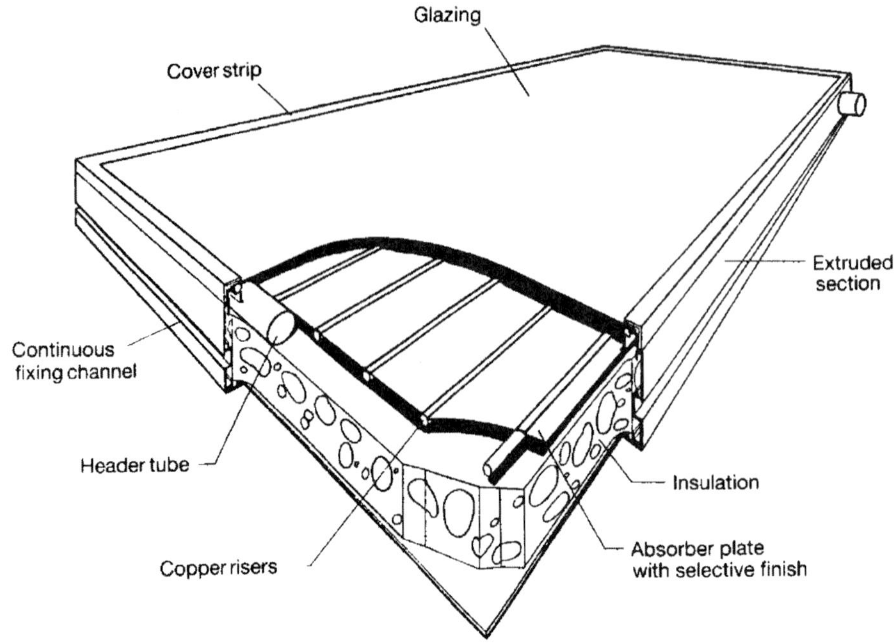

Fig. 6.2-1 Pictorial view of a flat-plate collector.

Absorber plates. Flat, corrugated, or grooved plates, to which the tubes, fins, or passages are attached. The plate may be integral with the tubes.

Headers or manifolds. To admit and discharge the fluid.

Insulation. To minimise the heat loss from the back and sides of the collector.

Container or casing. To surround the aforementioned components and keep them free from dust, moisture, etc.

FPC have been built in a wide variety of designs and from many different materials. They have been used to heat fluids such as water, water plus antifreeze additive, or air. Their major purpose is to collect as much solar energy as possible at the lower possible total cost. The collector should also have a long effective life, despite the adverse effects of the sun's ultraviolet radiation, corrosion and clogging because of acidity, alkalinity or hardness of the heat transfer fluid, freezing of water, or deposition of dust or moisture on the glazing, and breakage of the glazing because of thermal expansion, hail, vandalism or other causes. These causes can be minimised by the use of tempered glass.

More details are given below about the glazing and absorber plate materials. Most of these details apply also to other types of collector.

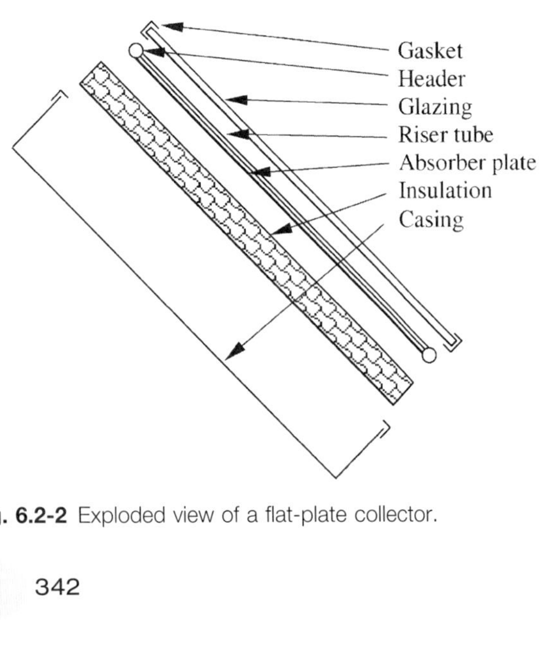

Fig. 6.2-2 Exploded view of a flat-plate collector.

Glazing materials

Glass has been widely used to glaze solar collectors because it can transmit as much as 90% of the incoming shortwave solar irradiation while transmitting virtually none of the longwave radiation emitted outward by the absorber plate. Glass with low iron content has a relatively high transmittance for solar radiation (approximately 0.85–0.90 at normal incidence), but its transmittance is essentially zero for the longwave thermal radiation (5.0–50 μm) emitted by sun-heated surfaces.

Plastic films and sheets also possess high shortwave transmittance, but because most usable varieties also have transmission bands in the middle of the thermal radiation spectrum, they may have longwave transmittances as high as 0.40. Plastics are also generally limited in the temperatures they can sustain without deteriorating or undergoing dimensional changes. Only a few types of plastic can withstand the sun's ultraviolet radiation for long periods. However, they are not broken by hail or stones, and, in the form of thin films, they are completely flexible and have low mass.

The commercially available grades of window and greenhouse glass have normal incidence transmittances of about 0.87 and 0.85, respectively. For direct radiation, the transmittance varies considerably with the angle of incidence [21].

Antireflective coatings and surface texture can also improve transmission significantly. The effect of dirt and dust on collector glazing may be quite small, and the cleansing effect of an occasional rainfall is usually adequate to maintain the transmittance within 2–4% of its maximum value.

The glazing should admit as much solar irradiation as possible and reduce the upward loss of heat as much as possible. Although glass is virtually opaque to the longwave radiation emitted by collector plates, absorption of that radiation causes an increase in the glass temperature and a loss of heat to the surrounding atmosphere by radiation and convection. These are analysed in more details in section 6.2.3.

Various prototypes of transparently insulated FPC and CPC have been built and tested in the last decade [22,23]. Low cost and high temperature resistant transparent insulating (TI) materials have been developed so that the commercialisation of these collectors becomes feasible. A prototype FPC covered by TI was developed by Benz *et al.* [24]. It was experimentally proved that the efficiency of the collector was comparable with that of ETC. However, no commercial collectors of this type are available in the market.

Collector absorbing plates

The collector plate absorbs as much of the irradiation as possible through the glazing, while losing as little heat as possible upward to the atmosphere and downward through the back of the casing. The collector plates transfer the retained heat to the transport fluid. The absorptance of the collector surface for shortwave solar radiation depends on the nature and colour of the coating and on the incident angle. Usually black colour is used, however various colour coatings have been proposed in Refs [25–27] mainly for aesthetic reasons.

By suitable electrolytic or chemical treatments, surfaces can be produced with high values of solar radiation absorptance (α) and low values of longwave emittance (ε). Essentially, typical selective surfaces consist of a thin upper layer, which is highly absorbent to shortwave solar radiation but relatively transparent to longwave thermal radiation, deposited on a surface that has a high reflectance and a low emittance for longwave radiation. Selective surfaces are particularly important when the collector surface temperature is much higher than the ambient air temperature. Lately, a low-cost mechanically manufactured selective solar absorber surface method has been proposed [28].

An energy efficient solar collector should absorb incident solar radiation, convert it to thermal energy and deliver the thermal energy to a heat transfer medium with minimum losses at each step. It is possible to use several different design principles and physical mechanisms in order to create a selective solar absorbing surface. Solar absorbers are based on two layers with different optical properties, which are referred as tandem absorbers. A semiconducting or dielectric coating with high solar absorptance and high infrared transmittance on top of a non-selective highly reflecting material such as metal constitutes one type of tandem absorber. Another alternative is to coat a non-selective highly absorbing material with a heat mirror having a high solar transmittance and high infrared reflectance [29].

Today, commercial solar absorbers are made by electroplating, anodization, evaporation, sputtering and by applying solar selective paints. Much of the progress during recent years has been based on the implementation of vacuum techniques for the production of fin type absorbers used in low temperature applications. The chemical and electrochemical processes used for their commercialization were readily taken over from the metal finishing industry. The requirements of solar absorbers used in high temperature applications, however, namely extremely low thermal emittance and high temperature stability, were difficult to fulfil with conventional wet processes. Therefore, large-scale sputter deposition was developed in the late 70s. The vacuum techniques are nowadays mature, characterized by low cost and have the advantage of being less environmentally polluting than the wet processes.

For fluid-heating collectors, passages must be integral with or firmly bonded to the absorber plate. A major problem is obtaining a good thermal bond between tubes and absorber plates without incurring excessive costs for labour or materials. Material most frequently used for collector plates are copper, aluminium, and stainless steel. UV-resistant plastic extrusions are used for low temperature applications. If the entire collector area is in contact with the heat transfer fluid, the thermal conductance of the material is not important.

Fig. 6.2-3 shows a number of absorber plate designs for solar water and air heaters that have been used with varying degrees of success [30]. Fig. 6.2-3A shows a bonded sheet design, in which the fluid passages are integral with the plate to ensure good thermal conduct between the metal and the fluid. Fig. 6.2-3B and C shows fluid heaters with tubes soldered, brazed, or otherwise fastened to upper or lower surfaces of sheets or strips of

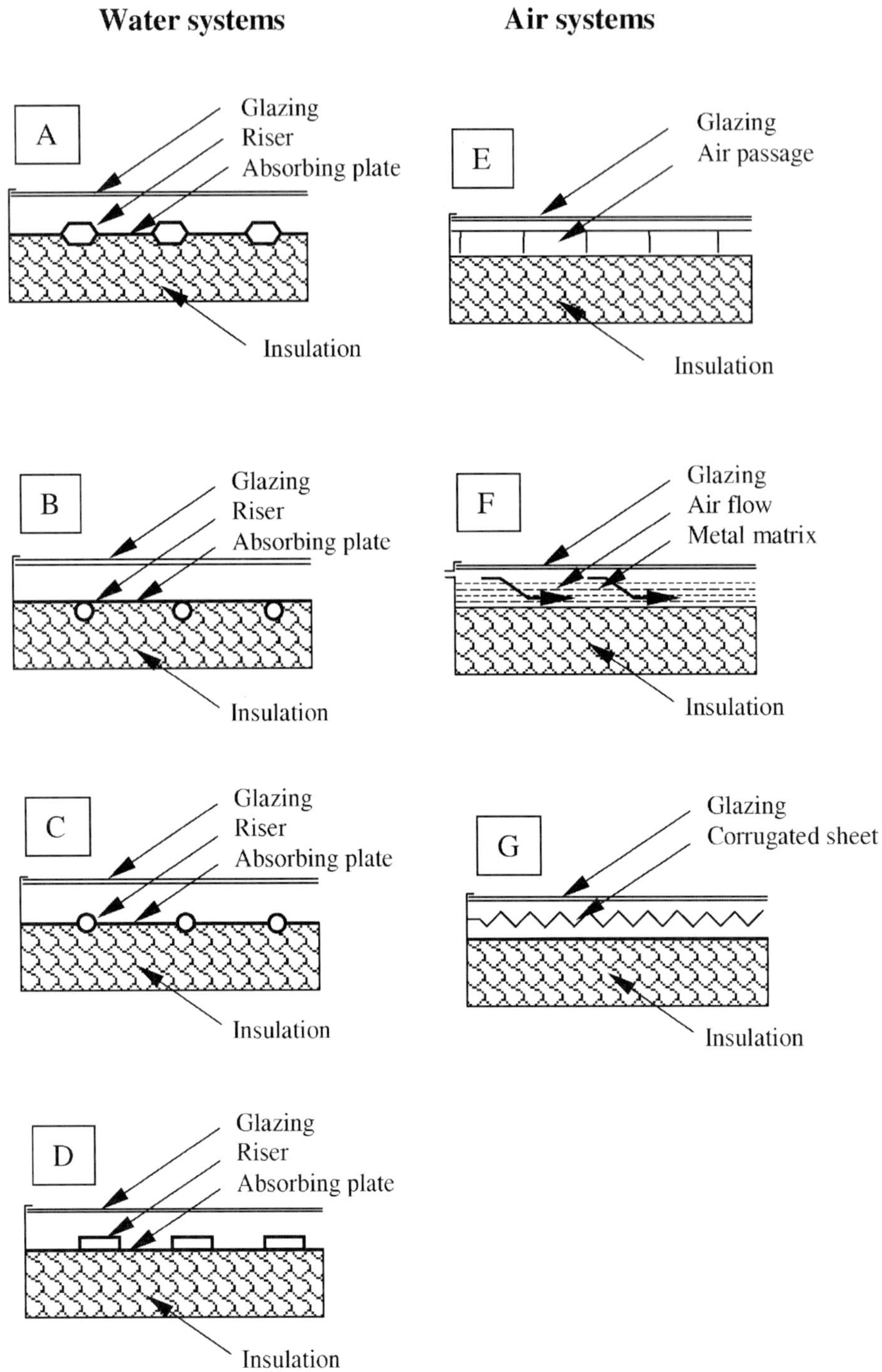

Fig. 6.2-3 Various types of flat-plate solar collectors.

copper. Copper tubes are used most often because of their superior resistance to corrosion.

Thermal cement, clips, clamps, or twisted wires have been tried in the search for low-cost bonding methods. Fig. 6.2-3D shows the use of extruded rectangular tubing to obtain a larger heat transfer area between tube and plate. Mechanical pressure, thermal cement, or brazing may be used to make the assembly. Soft solder must be avoided because of the high plate temperature encountered at stagnation conditions.

Air or other gases can be heated with FPC, particularly if some type of extended surface (Fig. 6.2-3E) is used to counteract the low heat transfer coefficients between metal and air [30]. Metal or fabric matrices (Fig. 6.2-3F) [13,30], or thin corrugated metal sheets (Fig. 6.2-3G) may be used, with selective surfaces applied to the latter when a high level of performance is required. The principal requirement is a large contact area between the absorbing surface and the air. Various applications of solar air collectors are reported in Refs [31–37]. A design procedure for solar air heating systems is presented in Ref. [38] whereas the optimisation of the flow passage geometry is presented in Ref. [39].

Reduction of heat loss from the absorber can be accomplished either by a selective surface to reduce radiative heat transfer or by suppressing convection. Francia [40] showed that a honeycomb made of transparent material, placed in the airspace between the glazing and the absorber, was beneficial.

Another category of collectors which is not shown in Fig. 6.2-3 is the uncovered or unglazed solar collector [41]. These are usually low-cost units which can offer cost-effective solar thermal energy in applications such as water preheating for domestic or industrial use, heating of swimming pools [42,43], space heating and air heating for industrial or agricultural applications.

FPC are by far the most used type of collector. FPC are usually employed for low temperature applications up to 100°C, although some new types of collector employing vacuum insulation and/or TI can achieve slightly higher values [24]. Due to the introduction of highly selective coatings actual standard FPC can reach stagnation temperatures of more than 200°C. With these collectors good efficiencies can be obtained up to temperatures of about 100°C.

The characteristics of a typical water FPC are shown in Table 6.2-2.

Lately some modern manufacturing techniques have been introduced by the industry like the use of ultrasonic welding machines, which improve both the speed and the quality of welds. This is used for the welding of fins on risers in order to improve heat conduction. The greatest advantage of this method is that the welding is performed at room temperature therefore deformation of the welded parts is avoided. These collectors with selective coating are called advance FPC and the characteristics of a typical type are also shown in Table 6.2-2.

Compound parabolic collectors

CPC are non-imaging concentrators. These have the capability of reflecting to the absorber all of the incident radiation within wide limits. Their potential as collectors of solar energy was pointed out by Winston [44]. The necessity of moving the concentrator to accommodate the changing solar orientation can be reduced by using a trough with two sections of a parabola facing each other, as shown in Fig. 6.2-4.

Compound parabolic concentrators can accept incoming radiation over a relatively wide range of angles. By using multiple internal reflections, any radiation that is entering the aperture, within the collector acceptance angle, finds its way to the absorber surface located at the bottom of the collector. The absorber can take a variety of configurations. It can be cylindrical as shown in Fig. 6.2-4 or flat. In the CPC shown in Fig. 6.2-4 the lower portion of the reflector (AB and AC) is circular, while the upper portions (BD and CE) are parabolic. As the upper part of a CPC contribute little to the radiation reaching the absorber, they are usually truncated thus forming a shorter version of the CPC, which is also cheaper. CPCs are usually covered with glass to avoid dust and other materials from entering the collector and thus reducing the reflectivity of its walls.

These collectors are more useful as linear or trough-type concentrators. The acceptance angle is defined as

Table 6.2-2 Characteristics of a typical water FPC system

Paramcter	Simple flat plate collector	Advanced flat plate collector
Fixing of risers on the absorber plate	Embedded	Ultrasonically welded
Absorber coating	Black mat paint	Chromium selective coating
Glazing	Low-iron glass	Low-iron glass
Efficiency mode	$nv_s(T_i - T_a)/G$	$nv_s(T_i - T_a)/G$
G_{test}-flow rate per unit area at test conditions (kg/s m^2)	0.015	0.015
c_0-intercept efficiency	0.79	0.80
c_1-negative of the first-order coefficient of the efficiency (W/m^2 °C)	6.67	4.78
b_0-incidence angle modifier constant	0.1	0.1
Collector slope angle	Latitude +5 to 10°	Latitude +5 to 10°

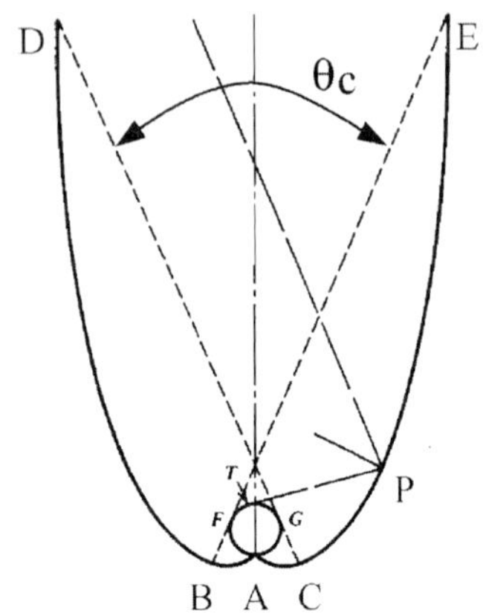

Fig. 6.2-4 Schematic diagram of a compound parabolic collector.

the angle through which a source of light can be moved and still converge at the absorber. The orientation of a CPC collector is related to its acceptance angle (θ_c, in Fig. 6.2-4). Also depending on the collector acceptance angle, the collector can be stationary or tracking. A CPC concentrator can be orientated with its long axis along either the north–south or the east–west direction and its aperture is tilted directly towards the equator at an angle equal to the local latitude. When orientated along the north–south direction the collector must track the sun by turning its axis so as to face the sun continuously. As the acceptance angle of the concentrator along its long axis is wide, seasonal tilt adjustment is not necessary. It can also be stationary but radiation will only be received the hours when the sun is within the collector acceptance angle. When the concentrator is orientated with its long axis along the east–west direction, with a little seasonal adjustment in tilt angle the collector is able to catch the sun's rays effectively through its wide acceptance angle along its long axis. The minimum acceptance angle in this case should be equal to the maximum incidence angle projected in a north–south vertical plane during the times when output is needed from the collector. For stationary CPC collectors mounted in this mode the minimum acceptance angle is equal to 47°. This angle covers the declination of the sun from summer to winter solstices (2 × 23.5°). In practice bigger angles are used to enable the collector to collect diffuse radiation at the expense of a lower concentration ratio. Smaller (less than 3) concentration ratio CPCs are of greatest practical interest. These according to Pereira [45] are able to accept a large proportion of diffuse radiation incident on their apertures and concentrate it without the need of tracking the sun.

A method to estimate the optical and thermal properties of CPCs is presented in Ref. [46]. In particular, a simple analytic technique was developed for the calculation of the average number of reflections for radiation passing through a CPC, which is useful for computing optical loses. Many numerical examples are presented which are helpful in designing a CPC.

Table 6.2-3 Characteristics of a typical CPC system

Parameter	Value
F' : collector fin efficiency factor	0.9
U_L: overall loss coefficient of collector per unit aperture area (W/m^2 °C)	1.5
ρ_R : reflectivity of walls of CPC	0.85
θ_c : half-acceptance angle of CPC (degrees)	45
Ratio of truncated to full height of CPC	0.67
Axis orientation	Receiver axis is horizontal and in a plane with a slope of 35° (transverse)
a: absorbtance of absorber plate	0.95
N_G : number of cover plates	1
η_R : index of refraction of cover material	1.526
K_L : product of extinction coefficient and the thickness of each cover plate	0.0375
Collector slope angle	(local latitude)

Two basic types of CPC collectors have been designed: the symmetric and the asymmetric. These usually employ two main types of absorbers: fin type with pipe and tubular absorbers [47–50].

Practical design considerations such as the choice of the receiver type, the optimum method for introducing a gap between receiver and reflector to minimise optical and thermal loses and the effect of a glass envelope around the receiver are given in Ref. [51]. Other practical design considerations for CPCs with multichannel and bifacial absorbers are given in Refs. [52] and [53], respectively, whereas design considerations and performance evaluation of cost-effective asymmetric CPCs are given in Ref. [54].

The characteristics of a typical CPC are shown in Table 6.2–3.

Evacuated tube collectors

Conventional simple flat-plate solar collectors were developed for use in sunny and warm climates. Their

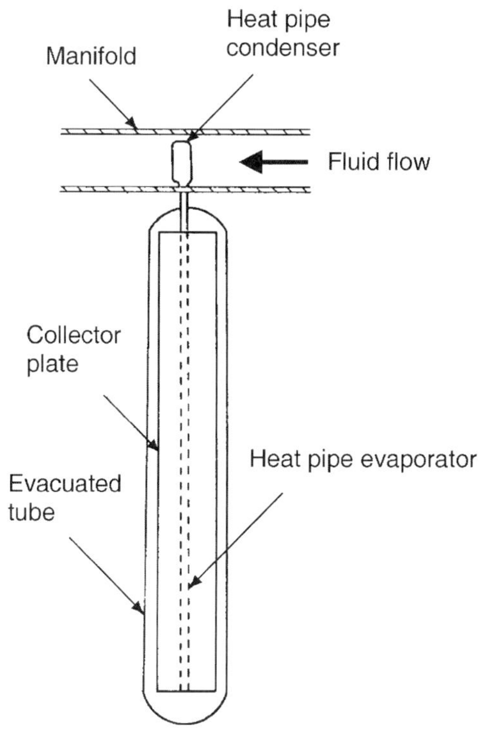

Fig. 6.2-5 Schematic diagram of an evacuated tube collector.

benefits however are greatly reduced when conditions become unfavourable during cold, cloudy and windy days. Furthermore, weathering influences such as condensation and moisture will cause early deterioration of internal materials resulting in reduced performance and system failure. Evacuated heat pipe solar collectors (tubes) operate differently than the other collectors available on the market. These solar collectors consist of a heat pipe inside a vacuum-sealed tube, as shown in Fig. 6.2–5.

ETC have demonstrated that the combination of a selective surface and an effective convection suppressor can result in good performance at high temperatures [21]. The vacuum envelope reduces convection and conduction losses, so the collectors can operate at higher temperatures than FPC. Like FPC, they collect both direct and diffuse radiation. However, their efficiency is higher at low incidence angles. This effect tends to give ETC an advantage over FPC in day-long performance.

ETC use liquid–vapour phase change materials to transfer heat at high efficiency. These collectors feature a heat pipe (a highly efficient thermal conductor) placed inside a vacuum-sealed tube. The pipe, which is a sealed copper pipe, is then attached to a black copper fin that fills the tube (absorber plate). Protruding from the top of each tube is a metal tip attached to the sealed pipe (condenser). The heat pipe contains a small amount of fluid (e.g. methanol) that undergoes an evaporating-condensing cycle. In this cycle, solar heat evaporates the liquid, and the vapour travels to the heat sink region where it condenses and releases its latent heat. The condensed fluid return back to the solar collector and the process is repeated. When these tubes are mounted, the metal tips up, into a heat exchanger (manifold) as shown in Fig. 6.2–5. Water, or glycol, flows through the manifold and picks up the heat from the tubes. The heated liquid circulates through another heat exchanger and gives off its heat to a process or to water that is stored in a solar storage tank.

Because no evaporation or condensation above the phase-change temperature is possible, the heat pipe offers inherent protection from freezing and overheating. This self-limiting temperature control is a unique feature of the evacuated heat pipe collector.

ETC basically consist of a heat pipe inside a vacuum-sealed tube. A large number of variations of the absorber shape of ETC are on the market [55]. Evacuated tubes with CPC-reflectors are also commercialised by several manufacturers. One manufacturer recently presented an all-glass ETC, which may be an important step to cost reduction and increase of lifetime. Another variation of this type of collector is what is called Dewar tubes. In this two concentric glass tubes are used and the space in between the tubes is evacuated (vacuum jacket). The advantage of this design is that it is made entirely of glass and it is not necessary to penetrate the glass envelope in order to extract heat from the tube thus leakage losses are not present and it is also less expensive than the single envelope system [56]. The characteristics of a typical ETC are shown in Table 6.2–4.

Another type of collector developed recently is the integrated compound parabolic collector (ICPC). This is an ETC in which at the bottom part of the glass tube a reflective material is fixed [57]. The collector combines the vacuum insulation and non-imaging stationary concentration into a single unit. In another design a tracking ICPC is developed which is suitable for high temperature applications [58].

6.2.2.2 Sun tracking concentrating collectors

Energy delivery temperatures can be increased by decreasing the area from which the heat losses occur. Temperatures far above those attainable by FPC can be reached if a large amount of solar radiation is concentrated on a relatively small collection area. This is done by interposing an optical device between the source of radiation and the energy absorbing surface. Concentrating collectors exhibit certain advantages as compared with the conventional flat-plate type [59]. The main ones are:

1. The working fluid can achieve higher temperatures in a concentrator system when compared to a flat-plate system of the same solar energy collecting surface.

Table 6.2-4 Characteristics of a typical ETC system

Parameter	Value
Glass tube diameter	65 mm
Glass thickness	1.6 mm
Collector length	1965 mm
Absorber plate	Copper
Coating	Selective
Absorber area for each collector	0.1 m^2
Efficiency mode	$nv_s(T_i - T_a)/G$
G_{test} : flow rate per unit area at test conditions (kg/s m^2)	0.014
c_0 : intercept efficiency	0.82
c_1 : negative of the first-order coefficient of the efficiency (W/m^2 °C)	2.19
b_0 : incidence angle modifier constant	0.2
Collector slope angle	Latitude +5 to 10°

This means that a higher thermodynamic efficiency can be achieved.

2. It is possible with a concentrator system, to achieve a thermodynamic match between temperature level and task. The task may be to operate thermionic, thermodynamic, or other higher temperature devices.
3. The thermal efficiency is greater because of the small heat loss area relative to the receiver area.
4. Reflecting surfaces require less material and are structurally simpler than FPC. For a concentrating collector the cost per unit area of the solar collecting surface is therefore less than that of a FPC.
5. Owing to the relatively small area of receiver per unit of collected solar energy, selective surface treatment and vacuum insulation to reduce heat losses and improve the collector efficiency are economically viable.

Their disadvantages are:

1. Concentrator systems collect little diffuse radiation depending on the concentration ratio.
2. Some form of tracking system is required so as to enable the collector to follow the sun.
3. Solar reflecting surfaces may lose their reflectance with time and may require periodic cleaning and refurbishing.

Many designs have been considered for concentrating collectors. Concentrators can be reflectors or refractors, can be cylindrical or parabolic and can be continuous or segmented. Receivers can be convex, flat, cylindrical or concave and can be covered with glazing or uncovered. Concentration ratios, i.e. the ratio of aperture to absorber areas, can vary over several orders of magnitude, from as low as unity to high values of the order of 10,000. Increased ratios mean increased temperatures at which energy can be delivered but consequently these collectors have increased requirements for precision in optical quality and positioning of the optical system.

Because of the apparent movement of the sun across the sky, conventional concentrating collectors must follow the sun's daily motion. There are two methods by which the sun's motion can be readily tracked. The first is the altazimuth method which requires the tracking device to turn in both altitude and azimuth; i.e. when performed properly, this method enables the concentrator to follow the sun exactly. Paraboloidal solar collectors generally use this system. The second one is the one-axis tracking in which the collector tracks the sun in only one direction either from east to west or from north to south. Parabolic trough collectors (PTC) generally use this system. These systems require continuous and accurate adjustment to compensate for the changes in the sun's orientation. Relations on how to estimate the angle of incidence of solar radiation for these tracking modes are given in section 6.2.3.2.

The first type of solar concentrator, shown in Fig. 6.2–6, is effectively a FPC fitted with simple flat reflectors which can markedly increase the amount of direct radiation reaching the collector. This is a concentrator because the aperture is bigger than the absorber but the system is stationary. A comprehensive analysis of such a system is presented in Ref. [60]. The model facilitates the prediction of the total energy absorbed by the collector at any hour of the day for any latitude for random tilt angles and azimuth angles of the collector and reflectors. This simple enhancement of FPC was initially suggested by Tabor in 1966 [61]. Other important studies on this area were presented by Seitel [62] and Perers *et al.* [63].

Another type of collector, already covered under the stationary collectors, the CPC is also classified as a concentrator. This, depending on the acceptance angle, can

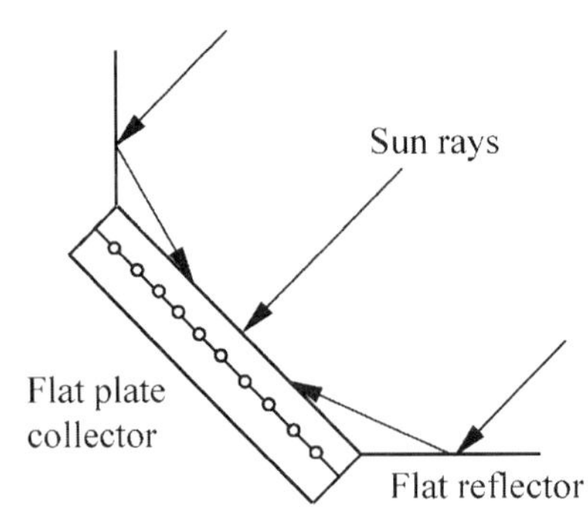

Fig. 6.2-6 Flat plate collector with flat reflectors.

be stationary or tracking. When tracking is used this is very rough or intermittent as concentration ratio is usually small and radiation can be collected and concentrated by one or more reflections on the parabolic surfaces.

As was seen above one disadvantage of concentrating collectors is that, except at low concentration ratios, they can use only the direct component of solar radiation, because the diffuse component cannot be concentrated by most types. However, an additional advantage of concentrating collectors is that, in summer, when the sun rises well to the north of the east–west line, the sun-follower, with its axis oriented north–south, can begin to accept radiation directly from the sun long before a fixed, south-facing flat-plate can receive anything other than diffuse radiation from the portion of the sky that it faces. Thus, in relatively cloudless areas, the concentrating collector may capture more radiation per unit of aperture area than a FPC.

In concentrating collectors solar energy is optically concentrated before being transferred into heat. Concentration can be obtained by reflection or refraction of solar radiation by the use of mirrors or lens. The reflected or refracted light is concentrated in a focal zone, thus increasing the energy flux in the receiving target. Concentrating collectors can also be classified into non-imaging and imaging depending on whether the image of the sun is focused at the receiver or not. The concentrator belonging in the first category is the CPC whereas all the other types of concentrator belong to the imaging type.

The collectors falling in this category are:

1. Parabolic trough collector;
2. Linear Fresnel reflector (LFR);
3. Parabolic dish;
4. Central receiver.

Parabolic trough collectors

In order to deliver high temperatures with good efficiency a high performance solar collector is required. Systems with light structures and low cost technology for process heat applications up to 400°C could be obtained with parabolic through collectors (PTCs). PTCs can effectively produce heat at temperatures between 50 and 400°C.

PTCs are made by bending a sheet of reflective material into a parabolic shape. A metal black tube, covered with a glass tube to reduce heat losses, is placed along the focal line of the receiver (Fig. 6.2-7). When the parabola is pointed towards the sun, parallel rays incident on the reflector are reflected onto the receiver tube. It is sufficient to use a single axis tracking of the sun and thus long collector modules are produced. The collector can be orientated in an east–west direction, tracking the sun from north to south, or orientated in a north–south direction and tracking the sun from east to west. The

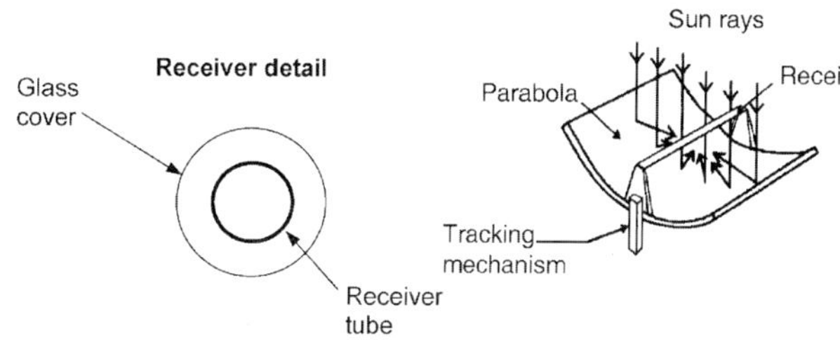

Fig. 6.2-7 Schematic of a parabolic trough collector.

advantages of the former tracking mode is that very little collector adjustment is required during the day and the full aperture always faces the sun at noon time but the collector performance during the early and late hours of the day is greatly reduced due to large incidence angles (cosine loss). North–south orientated troughs have their highest cosine loss at noon and the lowest in the mornings and evenings when the sun is due east or due west.

Over the period of one year, a horizontal north–south trough field usually collects slightly more energy than a horizontal east–west one. However, the north–south field collects a lot of energy in summer and much less in winter. The east–west field collects more energy in the winter than a north–south field and less in summer, providing a more constant annual output. Therefore, the choice of orientation usually depends on the application and whether more energy is needed during summer or during winter [64].

Parabolic trough technology is the most advanced of the solar thermal technologies because of considerable experience with the systems and the development of a small commercial industry to produce and market these systems. PTCs are built in modules that are supported from the ground by simple pedestals at either end.

PTCs are the most mature solar technology to generate heat at temperatures up to 400°C for solar thermal electricity generation or process heat applications. The biggest application of this type of system is the Southern California power plants, known as solar electric generating systems (SEGS), which have a total installed capacity of 354 MW_e [65]. More details on this system are given in Section 6.2.5.6. Another important application of this type of collector is installed at Plataforma Solar de Almeria (PSA) in Southern Spain mainly for experimental purposes. The total installed capacity of the PTCs is equal to 1.2 MW [66].

The receiver of a parabolic trough is linear. Usually, a tube is placed along the focal line to form an external surface receiver (Fig. 6.2-7). The size of the tube, and therefore the concentration ratio, is determined by the size of the reflected sun image and the manufacturing tolerances of the trough. The surface of the receiver is typically plated with selective coating that has a high absorptance for solar radiation, but a low emittance for thermal radiation loss.

A glass cover tube is usually placed around the receiver tube to reduce the convective heat loss from the receiver, thereby further reducing the heat loss coefficient. A disadvantage of the glass cover tube is that the reflected light from the concentrator must pass through the glass to reach the absorber, adding a transmittance loss of about 0.9, when the glass is clean. The glass envelope usually has an antireflective coating to improve transmissivity. One way to further reduce convective heat loss from the receiver tube and thereby increase the performance of the collector, particularly for high temperature applications, is to evacuate the space between the glass cover tube and the receiver.

In order to achieve cost effectiveness in mass production, not only the collector structure must feature a high stiffness to weight ratio so as to keep the material content to a minimum, but also the collector structure must be amenable to low-labour manufacturing processes. A number of structural concepts have been proposed such as steel framework structures with central torque tubes or double V-trusses, or fibreglass [67]. A recent development in this type of collector is the design and manufacture of EuroTrough, a new PTC, in which an advance lightweight structure is used to achieve cost efficient solar power generation [68,69]. Based on environmental test data to date, mirrored glass appears to be the preferred mirror material although self-adhesive reflective materials with 5–7 years' life exist in the market.

The design of this type of collector is given in a number of publications. The optimization of the collector aperture and rim angle is given in Ref. [59]. Design of other aspects of the collector is given in Refs [70,71].

A tracking mechanism must be reliable and able to follow the sun with a certain degree of accuracy, return the collector to its original position at the end of the day or during the night, and also track during periods of intermittent cloud cover. Additionally, tracking mechanisms are used for the protection of collectors, i.e. they turn the collector out of focus to protect it from the hazardous environmental and working conditions, like wind gust, overheating and failure of the thermal fluid flow mechanism. The required accuracy of the tracking mechanism depends on the collector acceptance angle. This is described in detail in section 6.2.4.3.

Various forms of tracking mechanism, varying from complex to very simple, have been proposed. They can be divided into two broad categories, namely mechanical [72–74] and electrical/electronic systems. The electronic systems generally exhibit improved reliability and tracking accuracy. These can be further subdivided into the following:

1. Mechanisms employing motors controlled electronically through sensors, which detect the magnitude of the solar illumination [75–77].
2. Mechanisms using computer controlled motors with feedback control provided from sensors measuring the solar flux on the receiver [78–80].

A tracking mechanism developed by the author uses three light dependent resistors which detect the focus, sun/cloud, and day or night conditions and give instruction to a DC motor through a control system to focus the collector, to follow approximately the sun path when cloudy conditions exist and to return the collector to the east during the night. More details are given in Ref. [81].

New developments in the field of PTC aim at cost reduction and improvements of the technology. In one system the collector can be washed automatically thus reducing drastically the maintenance cost.

After a period of research and commercial development of the PTC in the 80s a number of companies entered into the field producing this type of collector, for the temperature range between 50 and 300°C, all of them with one-axis tracking. One such example is the solar collector produced by the Industrial Solar Technology (IST) Corporation. IST erected several process heat installations in the United States with up to 2700 m^2 of collector aperture area [82].

The IST parabolic trough has thoroughly been tested and evaluated by SANDIA [83] and the German Aerospace Centre (DLR) [82] for efficiency and durability. Improvements of the optical performance, which recently have been discussed [84], would lead to a better incident angle modifier and a higher optical efficiency.

The characteristics of the IST collector system are shown in Table 6.2-5.

Linear Fresnel reflector

LFR technology relies on an array of linear mirror strips which concentrate light on to a fixed receiver mounted on a linear tower. The LFR field can be imagined as a broken-up parabolic trough reflector (Fig. 6.2-8); but unlike parabolic troughs, it does not have to be of parabolic shape, large absorbers can be constructed and the absorber does not have to move. A representation of an element of an LFR collector field is shown in Fig. 6.2-9. The greatest advantage of this type of system is that it uses flat or elastically curved reflectors which are cheaper compared to parabolic glass reflectors. Additionally, these are mounted close to the ground, thus minimizing structural requirements.

The first to apply this principle was the great solar pioneer Giorgio Francia [85] who developed both linear and two-axis tracking Fresnel reflector systems at Genoa, Italy, in the 60s. These systems showed that elevated temperatures could be reached using such systems but he moved on to two-axis tracking, possibly because advanced selective coatings and secondary optics were not

Table 6.2-5 Characteristics of the IST PTC system

Parameter	Value/type
Collector rim angle	70°
Reflective surface	Silvered acrylic
Receiver material	Steel
Collector aperture	2.3 m
Receiver surface treatment	Highly selective blackened nickel
Absorptance	0.97
Emittance (80°C)	0.18
Glass envelope transmittance	0.96
Absorber outside diameter	50.8 mm
G_{test} : flow rate per unit area at test conditions (kg/s m^2)	0.015
k_0 : intercept efficiency	0.762
k_1 : negative of the first-order coefficient of the efficiency (W/m^2 °C)	0.2125
k_2 : negative of the second-order coefficient of the efficiency (W/m^2 °C^2)	0.001672
b_0 : incidence angle modifier constant	0.958
b_1 : incidence angle modifier constant	−0.298
Tracking mechanism accuracy	0.05°
Collector orientation	Axis in N–S direction
Mode of tracking	E W horizontal

available [86]. Two of the early published works on this area are given in Refs. [87,88], whereas some later papers are given in Refs. [89,90].

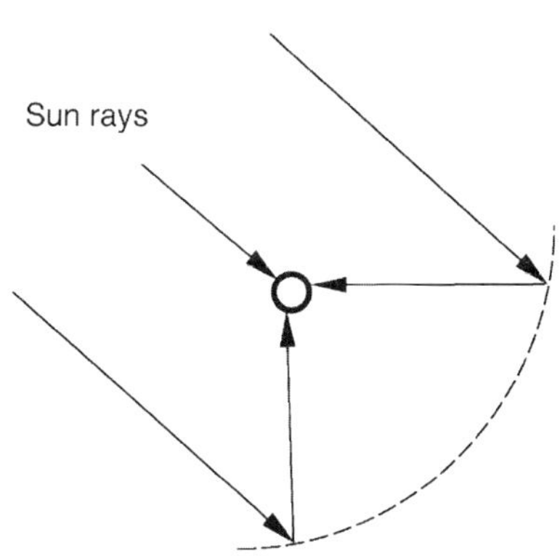

Fig. 6.2-8 Fresnel type parabolic trough collector.

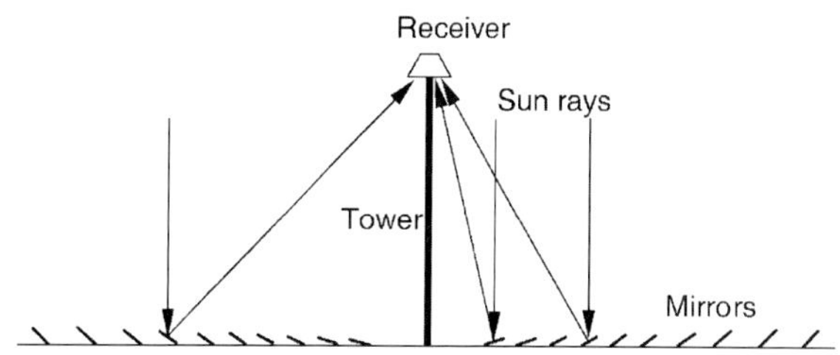

Fig. 6.2-9 Schematic diagram of a downward facing receiver illuminated from an LFR field.

In 1979, the FMC Corporation produced a detailed project design study for 10 and 100 MW$_e$ LFR power plants for the Department of Energy (DOE) of the US. The larger plant would have used a 1.68 km linear cavity absorber mounted on 61 m towers. The project however was never put into practice as it ran out of DOE funding [86].

A latter effort to produce a tracking LFR was made by the Israeli Paz company in the early 90s by Feuermann and Gordon [91]. This used an efficient secondary CPC-like optics and an evacuated tube absorber.

One difficulty with the LFR technology is that avoidance of shading and blocking between adjacent reflectors leads to increased spacing between reflectors. Blocking can be reduced by increasing the height of the absorber towers, but this increases cost. Compact linear Fresnel reflector (CLFR) technology has been recently developed at Sydney University in Australia. This is in effect a second type of solution for the Fresnel reflector field problem which has been overlooked until recently. In this design adjacent linear elements can be interleaved to avoid shading. The classical LFR system has only one receiver, and there is no choice about the direction and orientation of a given reflector. However, if it is assumed that the size of the field will be large, as it must be in technology supplying electricity in the MW class, it is reasonable to assume that there will be many towers in the system. If they are close enough then individual reflectors have the option of directing reflected solar radiation to at least two towers. This additional variable in the reflector orientation provides the means for much more densely packed arrays, because patterns of alternating reflector orientation can be such that closely packed reflectors can be positioned without shading and blocking [86]. The interleaving of mirrors between two receiving towers is shown in Fig. 6.2-10. The arrangement minimizes beam blocking by adjacent reflectors and allows high reflector densities and low tower heights to be used. Close spacing of reflectors reduces land usage but this is in many cases not a serious issue as in deserts. The avoidance of large reflector spacing and tower heights is an important cost issue when the cost of ground preparation, array substructure cost, tower structure cost,

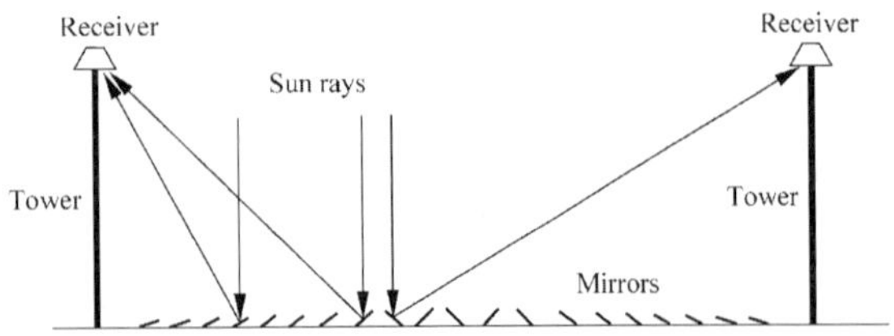

Fig. 6.2-10 Schematic diagram showing interleaving of mirrors in a CLFR with reduced shading between mirrors.

steam line thermal losses and steam line cost are considered. If the technology is to be located in an area with limited land availability such as in urban areas or next to existing power plants, high array ground coverage can lead to maximum system output for a given ground area [86].

Parabolic dish reflector (PDR)

A parabolic dish reflector, shown schematically in Fig. 6.2-11, is a point-focus collector that tracks the sun in two axes, concentrating solar energy onto a receiver located at the focal point of the dish. The dish structure must track fully the sun to reflect the beam into the thermal receiver. For this purpose tracking mechanisms similar to the ones described in the previous section are employed in double so that the collector is tracked in two axes.

The receiver absorbs the radiant solar energy, converting it into thermal energy in a circulating fluid. The thermal energy can then either be converted into electricity using an engine-generator coupled directly to the receiver, or it can be transported through pipes to a central power-conversion system. Parabolic-dish systems can achieve temperatures in excess of 1500°C. Because the receivers are distributed throughout a collector field, like parabolic troughs, parabolic dishes are often called

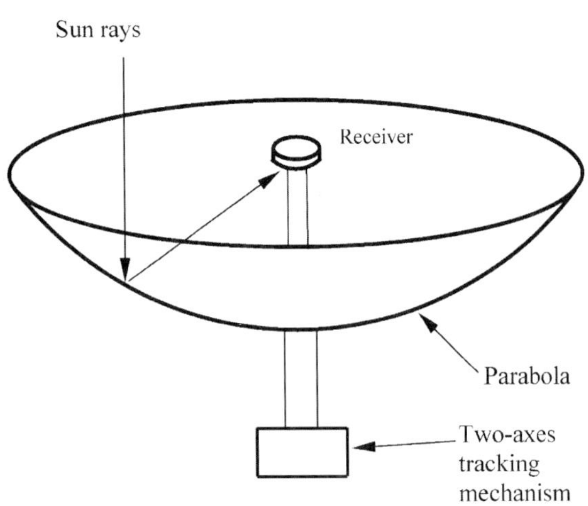

Fig. 6.2-11 Schematic of a parabolic dish collector.

distributed-receiver systems. Parabolic dishes have several important advantages:

1. Because they are always pointing at the sun, they are the most efficient of all collector systems.
2. They typically have concentration ratio in the range of 600–2000, and thus are highly efficient as thermal-energy absorption and power conversion systems.
3. They have modular collector and receiver units that can either function independently or as part of a larger system of dishes.

The main use of this type of concentrator is for parabolic dish engines. A parabolic dish-engine system is an electric generator that uses sunlight instead of crude oil or coal to produce electricity. The major parts of a system are the solar dish concentrator and the power conversion unit. More details on this system are given in section 6.2.5.6.

Parabolic-dish systems that generate electricity from a central power converter collect the absorbed sunlight from individual receivers and deliver it via a heat-transfer fluid to the power-conversion systems. The need to circulate heat-transfer fluid throughout the collector field raises design issues such as piping layout, pumping requirements, and thermal losses.

Systems that employ small generators at the focal point of each dish provide energy in the form of electricity rather than as heated fluid. The power conversion unit includes the thermal receiver and the heat engine. The thermal receiver absorbs the concentrated beam of solar energy, converts it to heat, and transfers the heat to the heat engine. A thermal receiver can be a bank of tubes with a cooling fluid circulating through it. The heat transfer medium usually employed as the working fluid for an engine is hydrogen or helium. Alternate thermal receivers are heat pipes wherein the boiling and condensing of an intermediate fluid is used to transfer the heat to the engine. The heat engine system takes the heat from the thermal receiver and uses it to produce electricity. The engine-generators have several components: a receiver to absorb the concentrated sunlight to heat the working fluid of the engine, which then converts the thermal energy into mechanical work; an alternator attached to the engine to convert the work into electricity; a waste-heat exhaust system to vent excess heat to the atmosphere; and a control system to match the engine's operation to the available solar energy. This distributed parabolic dish system lacks thermal storage capabilities, but can be hybridised to run on fossil fuel during periods without sunshine. The Sterling engine is the most common type of heat engine used in dish-engine systems. Other possible power conversion unit technologies that are evaluated for future applications are microturbines and concentrating photovoltaics [92].

Heliostat field collector

For extremely high inputs of radiant energy, a multiplicity of flat mirrors, or heliostats, using altazimuth mounts, can be used to reflect their incident direct solar radiation onto a common target as shown in Fig. 6.2-12. This is called the heliostat field or central receiver collector. By using slightly concave mirror segments on the heliostats, large amounts of thermal energy can be directed into the cavity of a steam generator to produce steam at high temperature and pressure.

The concentrated heat energy absorbed by the receiver is transferred to a circulating fluid that can be stored and later used to produce power. Central receivers have several advantages:

1. They collect solar energy optically and transfer it to a single receiver, thus minimizing thermal-energy transport requirements.
2. They typically achieve concentration ratios of 300–1500 and so are highly efficient both in collecting energy and in converting it to electricity.
3. They can conveniently store thermal energy.
4. They are quite large (generally more than 10 MW) and thus benefit from economies of scale.

Each heliostat at a central-receiver facility has from 50 to 150 m^2 of reflective surface. The heliostats collect and concentrate sunlight onto the receiver, which absorbs the concentrated sunlight, transferring its energy to a heat-transfer fluid. The heat-transport system, which consists primarily of pipes, pumps, and valves, directs the transfer fluid in a closed loop between the receiver, storage, and power-conversion systems. A thermal-storage system typically stores the collected energy as sensible heat for later delivery to the power-conversion system. The storage system also decouples the collection of solar energy from its conversion to electricity. The power-conversion system consists of a steam generator, turbine generator, and support equipment, which convert the thermal energy into electricity and supply it to the utility grid.

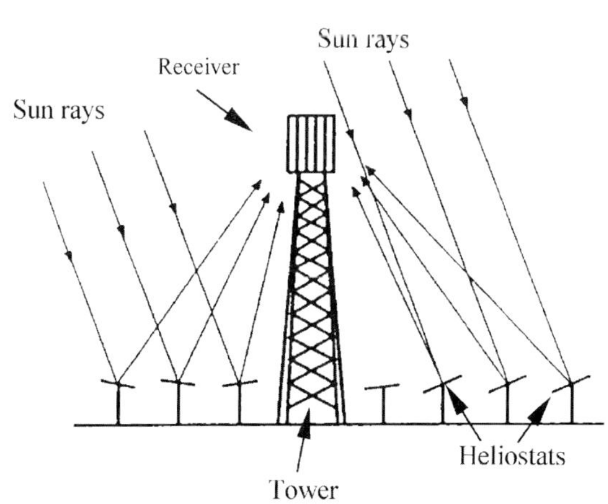

Fig. 6.2-12 Schematic of central receiver system.

In this case incident sunrays are reflected by large tracking mirrored collectors, which concentrate the energy flux towards radiative/convective heat exchangers, where energy is transferred to a working thermal fluid. After energy collection by the solar system, the conversion of thermal energy to electricity has many similarities with the conventional fossil-fuelled thermal power plants [93].

The average solar flux impinging on the receiver has values between 200 and 1000 kW/m^2. This high flux allows working at relatively high temperatures of more than 1500°C and integration of thermal energy in more efficient cycles. Central receiver systems can easily be integrated into fossil-fuelled plants for hybrid operation in a wide variety of options and have the potential to operate more than half the hours of each year at nominal power using thermal energy storage.

Central receiver systems are considered to have a large potential for mid-term cost reduction of electricity compared to parabolic trough technology since they allow many intermediate steps between the integration in a conventional Rankine cycle up to the higher energy cycles using gas turbines at temperatures above 1000°C, and this subsequently leads to higher efficiencies and larger throughputs [94,95]. Another alternative is to use Brayton cycle turbines, which require higher temperature than the ones employed in the Rankine cycle.

There are three general configurations for the collector and receiver systems. In the first, heliostats completely surround the receiver tower, and the receiver, which is cylindrical, has an exterior heat-transfer surface. In the second, the heliostats are located north of the receiver tower (in the northern hemisphere), and the receiver has an enclosed heat-transfer surface. In the third, the heliostats are located north of the receiver tower, and the receiver, which is in a vertical plane, has a north-facing heat-transfer surface.

In the final analysis, however, it is the selection of the heat-transfer fluid, thermal-storage medium, and power-conversion cycle that defines a central-receiver plant. The heat-transfer fluid may be water/steam, liquid sodium, or molten nitrate salt (sodium nitrate/potassium nitrate), whereas the thermal-storage medium may be oil mixed with crushed rock, molten nitrate salt, or liquid sodium. All rely on steam-Rankine power-conversion systems, although a more advanced system has been proposed that would use air as the heat-transfer fluid, ceramic bricks for thermal storage, and either a steam-Rankine or open-cycle Brayton power-conversion system.

6.2.3 Thermal analysis of collectors

In this section the thermal analysis of the collectors is presented. The two major types of collectors, i.e. flat-plate and concentrating are examined separately. The basic parameter to consider is the collector thermal efficiency. This is defined as the ratio of the useful energy delivered to the energy incident on the collector aperture. The incident solar flux consists of direct and diffuse radiation. While FPC can collect both, concentrating collectors can only utilise direct radiation if the concentration ratio is greater than 10 [96].

6.2.3.1 Flat-plate collectors performance

In this section various relations that are required in order to determine the useful energy collected and the interaction of the various constructional parameters on the performance of a collector are presented.

Under steady-state conditions, the useful heat delivered by a solar collector is equal to the energy absorbed by the heat transfer fluid minus the direct or indirect heat losses from the surface to the surroundings. The useful energy collected from a collector can be obtained from the following formula:

$$q_u = A_c[G_t\tau\alpha - U_L(T_p - T_a)] = mc_p[T_o - T_i] \tag{6.2.1}$$

Eq. (6.2.1) can be modified by substituting inlet fluid temperature (T_i) for the average plate temperature (T_p), if a suitable correction factor is included. The resulting equation is

$$q_u = A_c F_R[G_t(\tau\alpha) - U_L(T_i - T_a)] \tag{6.2.2}$$

where F_R is the correction factor, or collector heat removal factor.

Heat removal factor can be considered as the ratio of the heat actually delivered to that delivered if the collector plate were at uniform temperature equal to that of the entering fluid. In Eq. (6.2.2) the temperature T_i of the inlet fluid depends on the characteristics of the complete solar heating system and the hot water demand or heat demand of the building. However, F_R is affected only by the solar collector characteristics, the fluid type, and the fluid flow rate through the collector. F_R may be obtained from Ref. [97]

$$F_R = \frac{mc_p}{A_c U_L}\left(1 - \exp\left[\frac{U_L F' A_c}{mc_p}\right]\right) \tag{6.2.3}$$

where F' is the collector efficiency factor. It represents the ratio of the actual useful energy gain that would result if the collector-absorbing surface had been at the local fluid temperature.

The collector efficiency factor can be calculated by considering the temperature distribution between two pipes of the collector absorber and by assuming that the temperature gradient in the flow direction is negligible [97]. This analysis can be performed by considering the sheet tube configuration shown in Fig. 6.2-13, where the distance between the tubes is W, the tube diameter is D, and the sheet thickness is δ. As the sheet metal is usually made from copper or aluminum which are good conductors of heat, the temperature gradient through the sheet is negligible, therefore the region between the centerline separating the tubes and the tube base can be considered as a classical fin problem.

The fin, shown in Fig. 6.2-13(a) is of length $L = (W - D)/2$. An elemental region of width Δx and unit length in the flow direction is shown in Fig. 6.2-13(b). An energy balance on this element gives

$$S\,\Delta x - U_L\,\Delta x(T - T_a) + \left(-k\delta\frac{dT}{dx}\right)\Big|_x - \left(-k\delta\frac{dT}{dx}\right)\Big|_{x+\Delta x} = 0 \tag{6.2.4}$$

where S is the absorbed solar energy. By dividing through with Δx and finding the limit as Δx approaches zero, gives:

$$\frac{d^2T}{dx^2} = \frac{U_L}{k\delta}\left(T - T_a - \frac{S}{U_L}\right) \tag{6.2.5}$$

The two boundary conditions necessary to solve this second-order differential equation are:

$$\frac{dT}{dx}\Big|_{x=0} = 0, \quad \text{and} \quad T\big|_{x=L} = T_b$$

For convenience the following two variables are defined:

$$m = \sqrt{\frac{U_L}{k\delta}} \tag{6.2.6}$$

$$\Psi = T - T_a - \frac{S}{U_L} \tag{6.2.7}$$

Therefore, Eq. (6.2.5) becomes

$$\frac{d^2\Psi}{dx} - m^2\Psi = 0 \tag{6.2.8}$$

which has the boundary conditions:

$$\frac{d\Psi}{dx}\Big|_{x=0} = 0 \quad \text{and} \quad \Psi\big|_{x=L} = T_b - T_a - \frac{S}{U_L}$$

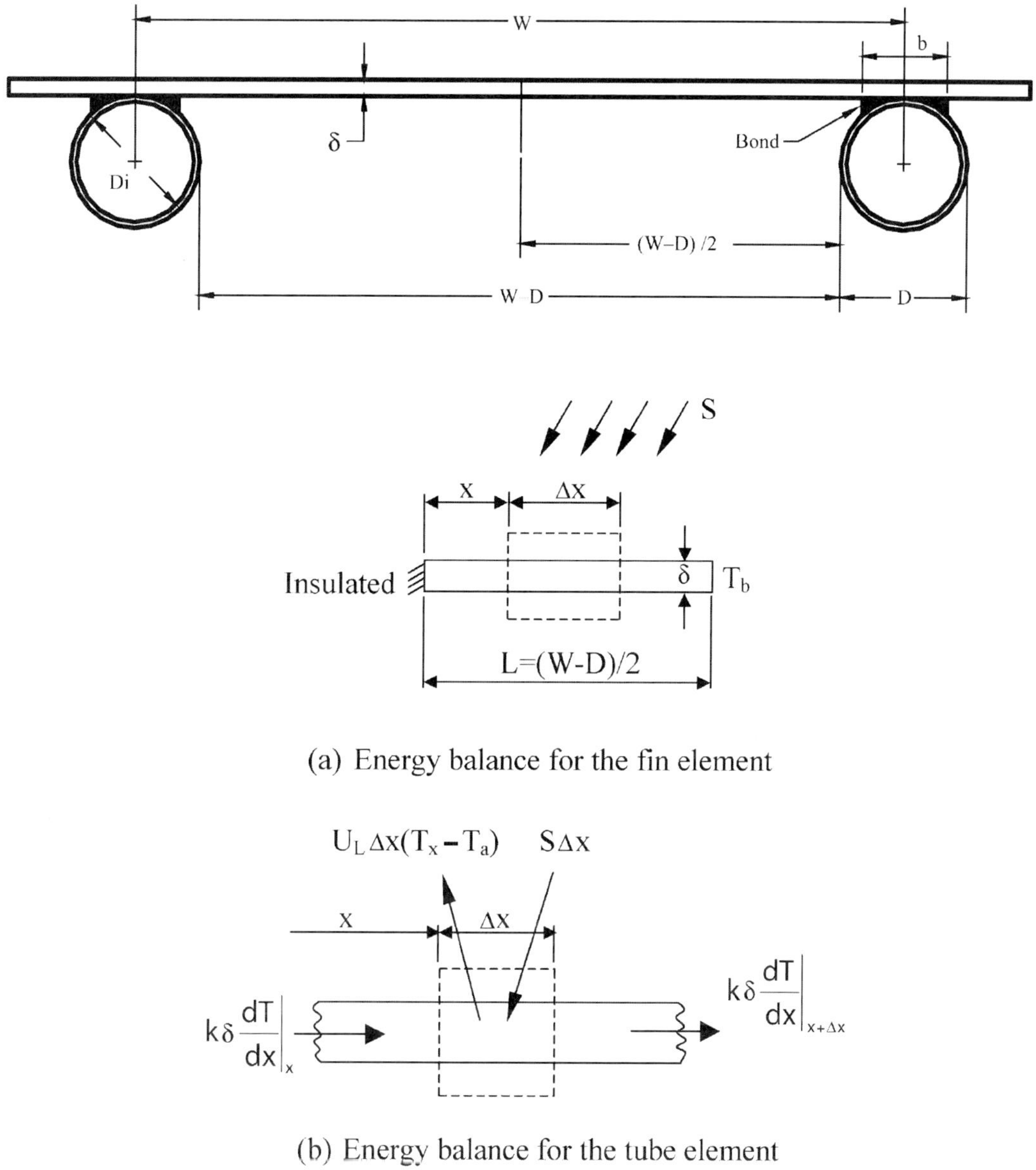

Fig. 6.2-13 Flat-plate sheet and tube configuration.

Eq. (6.2.8) is a second-order homogeneous linear differential equation whose general solution is:

$$\Psi = C_1' e^{mx} + C_2' e^{-mx} = C_1 \sinh(mx) + C_2 \cosh(mx) \quad (6.2.9)$$

The first boundary yields $C_1 = 0$, and the second boundary condition yields:

$$\Psi = T_b - T_a - \frac{S}{U_L} = C_2 \cosh(mL) \quad \text{or}$$

$$C_2 = \frac{T_b - T_a - S/U_L}{\cosh(mL)}$$

With C_1 and C_2 known, Eq. (6.2.9) becomes:

$$\frac{T - T_a - S/U_L}{T_b - T_a - S/U_L} = \frac{\cosh(mx)}{\cosh(mL)} \quad (6.2.10)$$

This equation gives the temperature distribution in the x-direction at any given y.

The energy conducted to the region of the tube per unit length in the flow direction can be found by evaluating Fourier's law at the fin base:

$$q'_{fin} = -k\delta \frac{dT}{dx}\Big|_{x=L} = \frac{k\delta m}{U_L}\left[S - U_L(T_b - T_a)\right]\tanh(mL) \quad (6.2.11)$$

but $k\delta m/U_L$ is just 1/m. Eq. (6.2.11) accounts for the energy collected on only one side of the tube; for both sides, the energy collection is

$$q'_{fin} = (W-D)[S-U_L(T_b-T_a)]\frac{\tanh[m(W-D)/2]}{m(W-D)/2} \quad (6.2.12)$$

or with the help of fin efficiency

$$q'_{fin} = (W-D)F[S-U_L(T_b-T_a)] \quad (6.2.13)$$

where factor F in Eq. (6.2.13) is the standard fin efficiency for straight fins with rectangular profile, obtained from:

$$F = \frac{\tanh[m(W-D)/2]}{m(W-D)/2} \quad (6.2.14)$$

The useful gain of the collector also includes the energy collected above the tube region. This is given by:

$$q'_{tube} = D[S-U_L(T_b-T_a)] \quad (6.2.15)$$

Accordingly, the useful energy gain per unit length in the direction of the fluid flow is:

$$\begin{aligned} q'_u &= q'_{fin} + q'_{tube} \\ &= [(W-D)F+D][S-U_L(T_b-T_a)] \end{aligned} \quad (6.2.16)$$

This energy must be ultimately transferred to the fluid, which can be expressed in terms of two resistances as:

$$q'_u = \frac{T_b-T_f}{\dfrac{1}{h_{fi}\pi D_i}+\dfrac{1}{C_b}} \quad (6.2.17)$$

In Eq. (6.2.17), C_b is the bond conductance which can be estimated from knowledge of the bond thermal conductivity k_b, the average bond thickness γ, and the bond width b. The bond conductance on a per unit length basis is given by:

$$C_b = \frac{k_b b}{\gamma} \quad (6.2.18)$$

The bond conductance can be very important in accurately describing the collector performance and generally it is necessary to have good metal-to-metal contact so that the bond conductance is greater than 30 W/m K and preferably the tube should be welded to the fin.

Solving Eq. (6.2.17) for T_b, substituting it into Eq. (6.2.16) and solving the result for the useful gain, we get

$$q'_u = WF'[S-U_L(T_f-T_a)] \quad (6.2.19)$$

where the collector efficiency factor F' is given by:

$$F' = \frac{\dfrac{1}{U_L}}{W\left[\dfrac{1}{U_L[D+(W-D)F]}+\dfrac{1}{C_b}+\dfrac{1}{\pi D_i h_{fi}}\right]} \quad (6.2.20)$$

A physical interpretation of F' is that it represents the ratio of the actual useful energy gain to the useful energy gain that would result if the collector absorbing surface had been at the local fluid temperature. It should be noted that the denominator of Eq. (6.2.20) is the heat transfer resistance from the fluid to the ambient air. This resistance can be represented as $1/U_o$. Therefore, another interpretation of F' is:

$$F' = \frac{U_o}{U_L} \quad (6.2.21)$$

The collector efficiency factor is essentially a constant factor for any collector design and fluid flow rate. The ratio of U_L to C_b, the ratio of U_L to h_{fi}, and the fin efficiency F are the only variables appearing in Eq. (6.2.20) that may be functions of temperature. For most collector designs F is the most important of these variables in determining F'. The factor F' is a function of U_L and h_{fi}, each of which has some temperature dependence, but it is not a strong function of temperature. Additionally, the collector efficiency factor decreases with increased tube center-to-center distances and increases with increases in both material thicknesses and thermal conductivity. Increasing the overall loss coefficient decreases F' while increasing the fluid-tube heat transfer coefficient increases F'.

The overall heat loss coefficient is a complicated function of the collector construction and its operating conditions and it is given by the following expression

$$U_L = U_t + U_b + U_c \quad (6.2.22)$$

i.e. it is the heat transfer resistance from the absorber plate to the ambient air.

In addition to serving as a heat trap by admitting shortwave solar radiation and retaining longwave thermal radiation, the glazing also reduces heat loss by convection. The insulating effect of the glazing is enhanced by the use of several sheets of glass, or glass plus plastic. The top loss coefficient in Eq. (6.2.22) is given by [98]:

$$U_t = \frac{\dfrac{1}{N_g}}{\dfrac{C}{T_p}\left[\dfrac{T_{av}-T_a}{N_g+f}\right]^{0.33}+\dfrac{1}{h_w}} + \frac{\sigma(T_{av}^2+T_a^2)(T_{av}-T_a)}{\dfrac{1}{\varepsilon_p+0.05N_g(1-\varepsilon_p)}+\dfrac{2N_g+f-1}{\varepsilon_g}-N_g} \quad (6.2.23)$$

where

$$h_w = 5.7 + 3.8W \quad (6.2.24)$$

$$f = (1 - 0.04h_w + 0.0005h_w^2)(1 + 0.091N_g) \quad (6.2.25)$$

$$C = 365.9(1 - 0.00883\beta + 0.0001298\beta^2) \quad (6.2.26)$$

and T_p is the collector stagnation temperature, i.e. the temperature of the absorbing plate when the flow rate is equal to zero, and is obtained from:

$$T_p = \frac{G_t(\tau\alpha)}{U_L} + T_a \quad (6.2.27)$$

As good insulation is usually used in the collector construction, the loss coefficient for the bottom and edges of the collector, U_b and U_e in Eq. (6.2.22) is constant, and its estimation is straightforward. The heat loss from the back of the plate rarely exceeds 10% of the upward loss.

The overall transmittance–absorptance product $(\tau\alpha)$ is determined as:

$$(\tau\alpha) = \frac{I_{bT}(\tau\alpha)_b + I_d\left(\frac{1+\cos\beta}{2}\right)(\tau\alpha)_s + \rho I\left(\frac{1-\cos\beta}{2}\right)(\tau\alpha)_g}{I} \quad (6.2.28)$$

Finally, the collector efficiency can be obtained by dividing q_u by (G_tA_c). Therefore,

$$n = F_R\left[\tau\alpha - \frac{U_L(T_i - T_a)}{G_t}\right] \quad (6.2.29)$$

For incident angles below about 35°, the product τ times α is essentially constant and Eqs (6.2.2) and (6.2.29) are linear with respect to the parameter $(T_i - T_a)/G_t$, as long as U_L remains constant.

6.2.3.2 Concentrating collector performance

For concentrating collectors both optical and thermal analyses are required.

Optical analysis

The concentration ratio (C) is defined as the ratio of the aperture area to the receiver/absorber area, i.e.

$$C = \frac{A_a}{A_r} \quad (6.2.30)$$

For FPC with no reflectors, C = 1. For concentrators C is always greater than 1. For a single axis tracking collector the maximum possible concentration is given by [1,97]:

$$C_{max} = \frac{1}{\sin(\theta_m)} \quad (6.2.31)$$

and for a two-axes tracking collector [1,97]

$$C_{max} = \frac{1}{\sin^2(\theta_m)} \quad (6.2.32)$$

where θ_m is the half acceptance angle. The half acceptance angle denotes coverage of one-half of the angular zone within which radiation is accepted by the concentrator's receiver. Radiation is accepted over an angle of $2\theta_m$ because radiation incident within this angle reaches the receiver after passing through the aperture. This angle describes the angular field within which radiation can be collected by the receiver without having to track the concentrator.

Eqs (6.2.31) and (6.2.32) define the upper limit of concentration that may be obtained for a given collector viewing angle. For a stationary CPC the angle θ_m depends on the motion of the sun in the sky. For example, for a CPC having its axis in a N–S direction and tilted from the horizontal such that the plane of the sun's motion is normal to the aperture, the acceptance angle is related to the range of hours over which sunshine collection is required, e.g. for 6 h of useful sunshine collection $2\theta_m = 90°$ (sun travels 15°/h). In this case $C_{max} = 1/\sin(45°) = 1.41$.

For a tracking collector θ_m is limited by the size of the sun's disk, small scale errors and irregularities of the reflector surface and tracking errors. For a perfect collector and tracking system C_{max} depends only on the sun's disk, which has a width of 0.53° (32′) [97]. Therefore,

$$\text{For single axis tracking: } C_{max} = 1/\sin(16') = 216$$

$$\text{For full tracking: } C_{max} = 1/\sin^2(16') = 46,747$$

It can, therefore, be concluded that the concentration ratio for moving collectors is much higher. However, high accuracy of the tracking mechanism and careful construction of the collector is required with increased concentration ratio as θ_m is very small. In practice, due to various errors, much lower values than the above maximum ones are employed.

Another factor that needs to be determined is the incidence angle for the various modes of tracking. This can be about a single axis or about two axes. In the case of single axis mode the motion can be in various ways, i.e. east–west, north–south or parallel to the earth's axis.

The mode of tracking affects the amount of incident radiation falling on the collector surface in proportion to the cosine of the incidence angle. The amount of energy falling on a surface of 1 m^2 for four modes of tracking for the summer and winter solstices and the equinoxes is

Table 6.2-6 Comparison of energy absorbed for various modes of tracking

Tracking mode	Solar energy (kW h/m^2)			Percent to full tracking		
	E	SS	WS	E	SS	WS
Full tracking	8.43	10.60	5.70	100.0	100.0	100.0
E–W polar	8.43	9.73	5.23	100.0	91.7	91.7
N–S horizontal	6.22	7.85	4.91	73.8	74.0	86.2
E–W horizontal	7.51	10.36	4.47	89.1	97.7	60.9

Note: E: equinoxes, SS: summer solstice, WS: winter solstice.

shown in Table 6.2-6 [64]. The amount of energy shown in Table 6.2-6 is obtained by applying a radiation model [12]. This is affected by the incidence angle which is different for each mode.

The performance of the various modes of tracking can be compared to the full tracking mode, which collects the maximum amount of solar radiation, shown as 100% in Table 6.2-6. Relations for the estimation of the angle of incidence for the various modes of tracking are given in Table 6.2-7.

The optical efficiency is defined as the ratio of the energy absorbed by the receiver to the energy incident on the collector's aperture. The optical efficiency depends on the optical properties of the materials involved, the geometry of the collector, and the various imperfections arising from the construction of the collector. In equation form [99]:

$$n_o = \rho\tau\alpha\gamma[(1 - A_f \tan(\theta))\cos(\theta)] \quad (6.2.33)$$

The geometry of the collector dictates the geometric factor A_f, which is a measure of the effective reduction of the aperture area due to abnormal incidence effects. For a PTC, its value can be obtained by the following relation [100]:

$$A_f = \frac{2}{3}W_a h_p + fW_a\left[1 + \frac{W_a^2}{48f^2}\right] \quad (6.2.34)$$

The most complex parameter involved in determining the optical efficiency of a PTC is the intercept factor. This is defined as the ratio of the energy intercepted by the receiver to the energy reflected by the focusing device, i.e. parabola [99]. Its value depends on the size of

Table 6.2-7 Relations for the estimation of the angle of incidence (θ) for the various modes of tracking

Mode of tracking	Incidence angle	Remarks
Full tracking	$\cos(\theta) = 1$	This depends on the accuracy of the tracking mechanism. This mode collects the maximum possible sunshine
Collector axis in N–S axis polar E–W tracking	$\cos(\theta) = \cos(\delta)$	For this mode the sun is normal to the collector at equinoxes ($\delta=0°$) and the cosine effect is maximum at the solstices. When more than one collector is used, front collectors cast shadows on adjacent ones
Collector axis in N–S axis horizontal E–W tracking	$\cos(\theta) = \sqrt{\sin^2(\alpha) + \cos^2(\delta)\sin^2(h)}$ or $\cos(\theta) = \cos(\phi)\cos(h) + \cos(\delta)\sin^2(h)$	The greatest advantage of this arrangement is that very small shadowing effects are encountered when more than one collector is used. These are present in the first and last hours of the day
Collector axis in E–W axis horizontal N–S tracking	$\cos(\theta) = \sqrt{1 - \cos^2(\delta)\sin^2(h)}$ or $\cos(\theta) = \sqrt{\sin^2(\delta) + \cos^2(\delta)\cos^2(h)}$	The shadowing effects of this arrangement are minimal. The principal shadowing is caused when the collector is tipped to a maximum degree south ($\delta = 23.5°$) at winter solstice. In this case the sun casts a shadow toward the collector at the north

Notes: δ: declination angle, h: hour angle, ϕ : zenith angle. Relations to determine these angles can be found in many solar energy books [1,97].

the receiver, the surface angle errors of the parabolic mirror, and solar beam spread.

The errors associated with the parabolic surface are of two types, random and non-random [101]. Random errors are defined as those errors which are truly random in nature and, therefore, can be represented by normal probability distributions. Random errors are identified as apparent changes in the sun's width, scattering effects caused by random slope errors (i.e. distortion of the parabola due to wind loading) and scattering effects associated with the reflective surface. Non-random errors arise in manufacture/assembly and/or in the operation of the collector. These can be identified as reflector profile imperfections, misalignment errors and receiver location errors. Random errors are modeled statistically, by determining the standard deviation of the total reflected energy distribution, at normal incidence [102] and are given by:

$$\sigma = \sqrt{\sigma^2_{sun} + 4\sigma^2_{slope} + \sigma^2_{mirror}} \tag{6.2.35}$$

Non-random errors are determined from a knowledge of the misalignment angle error β (i.e. the angle between the reflected ray from the centre of sun and the normal to the reflector's aperture plane) and the displacement of the receiver from the focus of the parabola (d_r). As reflector profile errors and receiver mislocation along the Y axis essentially have the same effect a single parameter is used to account for both. According to Guven and Bannerot [102] random and non-random errors can be combined with the collector geometric parameters, concentration ratio (C) and receiver diameter (D) to yield error parameters universal to all collector geometries. These are called 'universal error parameters' and an asterisk is used to distinguish them from the already defined parameters. Using the universal error parameters the formulation of the intercept factor γ is possible [101]:

$$\gamma = \frac{1-\cos\phi_r}{2\sin\phi_r} \times \int_0^{\phi_r} \mathrm{Erf}\left(\frac{\sin\phi_r(1+\cos\phi)(1-2d^*\sin\phi)-\pi\beta^*(1+\cos\phi_r)}{\sqrt{2}\pi\sigma^*(1+\cos\phi_r)}\right) - \mathrm{Erf}\left(-\frac{\sin\phi_r(1+\cos\phi)(1+2d^*\sin\phi)+\pi\beta^*(1+\cos\phi_r)}{\sqrt{2}\pi\sigma^*(1+\cos\phi_r)}\right) \times \frac{d\phi}{(1+\cos\phi)} \tag{6.2.36}$$

where

- d^* universal non-random error parameter due to receiver mislocation and reflector profile errors ($d^* = d_r/D$)
- β^* universal non-random error parameter due to angular errors ($\beta^* = \beta C$)
- σ^* universal random error parameter ($\sigma^* = \sigma C$)
- C collector concentration ratio [$= A_a/A_r$]
- D riser tube outside diameter (m)
- d_r displacement of receiver from focus (m)
- β misalignment angle error (degrees)

Another parameter that needs to be determined is the radiation concentration distribution on the receiver of the collector, called local concentration ratio (LCR). For the PTC this distribution is as shown in Fig. 6.2-14. The shape of the curves depends on the same types or error mentioned above and on the angle of incidence. Analysis of these effects is presented in Ref. [103] and may not be repeated here. It should be noted that the distribution for half the receiver is shown in Fig. 6.2-14. Another more representative way to show this distribution for the

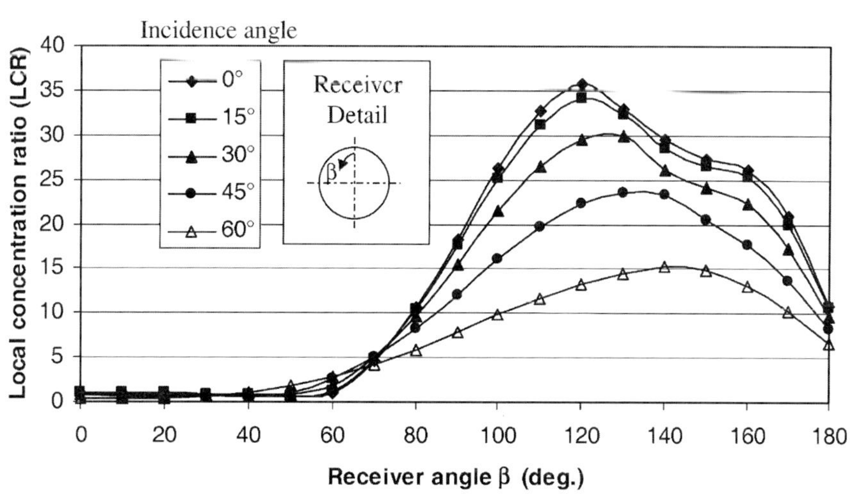

Fig. 6.2-14 Local concentration ratio on the receiver of a parabolic trough collector.

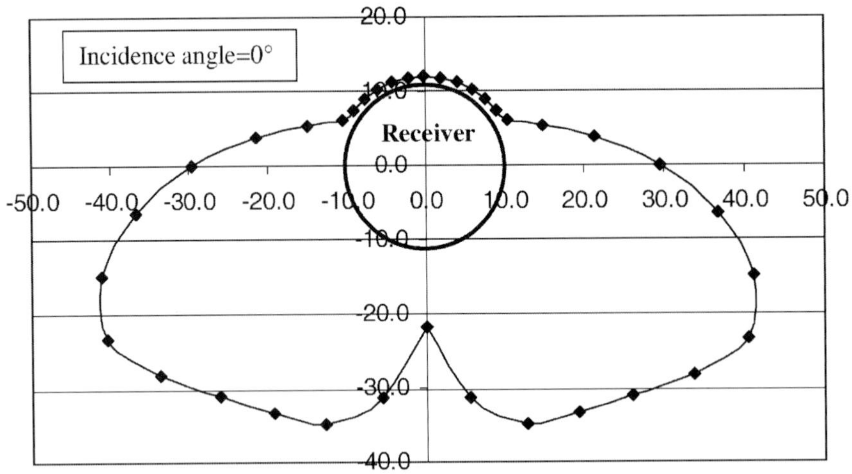

Fig. 6.2-15 More representative view of LCR for a collector with receiver diameter of 22 mm and rim angle of 90°.

whole receiver is given in Fig. 6.2-15. As can be seen from these figures, the top part of the receiver receives essentially only direct sunshine from the sun and the maximum concentration, about 36 suns, occurs at zero incidence angle and at an angle β, shown in Fig. 6.2-14, of 120°.

Thermal analysis

The generalised thermal analysis of a concentrating solar collector is similar to that of a FPC. It is necessary to derive appropriate expressions for the collector efficiency factor F', the loss coefficient U_L and the collector heat removal factor F_R. For the loss coefficient, standard heat transfer relations for glazed tubes can be used.

The instantaneous efficiency of a concentrating collector may be calculated from an energy balance of its receiver. Eq. (6.2.1) also may be adapted for use with concentrating collectors. Therefore, the useful energy delivered from a concentrator is:

$$q_u = G_b n_o A_a - A_r U_L (T_r - T_a) \quad (6.2.37)$$

The useful energy gain per unit of collector length can be expressed in terms of the local receiver temperature T_r as:

$$q'_u = \frac{q_u}{L} = \frac{A_a n_o G_b}{L} - \frac{A_r U_L}{L}(T_r - T_a) \quad (6.2.38)$$

In terms of the energy transfer to the fluid at local fluid temperature T_f

$$q'_u = \frac{\left(\frac{A_r}{L}\right)(T_r - T_f)}{\frac{D_o}{h_{fi} D_i} + \left(\frac{D_o}{2k}\ln\frac{D_o}{D_i}\right)} \quad (6.2.39)$$

If T_r is eliminated from Eqs (6.2.38) and (6.2.39) we have:

$$q'_u = F'\frac{A_a}{L}\left[n_o G_b - \frac{U_L}{C}(T_f - T_a)\right] \quad (6.2.40)$$

where F' is the collector efficiency factor given by:

$$F' = \frac{1/U_L}{\frac{1}{U_L} + \frac{D_o}{h_{fi} D_i} + \left(\frac{D_o}{2k} + \ln\frac{D_o}{D_i}\right)} \quad (6.2.41)$$

Similarly as for the FPC the heat removal factor can be used and Eq. (6.2.37) can be written as:

$$q_u = F_R[G_b n_o A_a - A_r U_L (T_i - T_a)] \quad (6.2.42)$$

And the collector efficiency can be obtained by dividing q_u by $(G_b A_a)$. Therefore,

$$n = F_R\left[n_o - U_L\left(\frac{T_i - T_a}{G_b C}\right)\right] \quad (6.2.43)$$

where C is the concentration ratio $[C = A_a/A_r]$. For F_R a relation similar to Eq. (6.2.3) is used by replacing A_c to A_r.

Another analysis usually performed for PTCs is by applying a piecewise two-dimensional model of the receiver by considering the circumferential variation of solar flux shown in Figs 6.2-14 and 6.2-15. Such an analysis can be performed by dividing the receiver into longitudinal and isothermal nodal sections as shown in Fig. 6.2-16 and applying the principle of energy balance to the glazing and receiver nodes [104]. This analysis can give the temperature distribution along the circumference and length of the receiver, and thus any points of high temperature, which might reach a temperature above the degradation temperature of the receiver selective coating, can be determined.

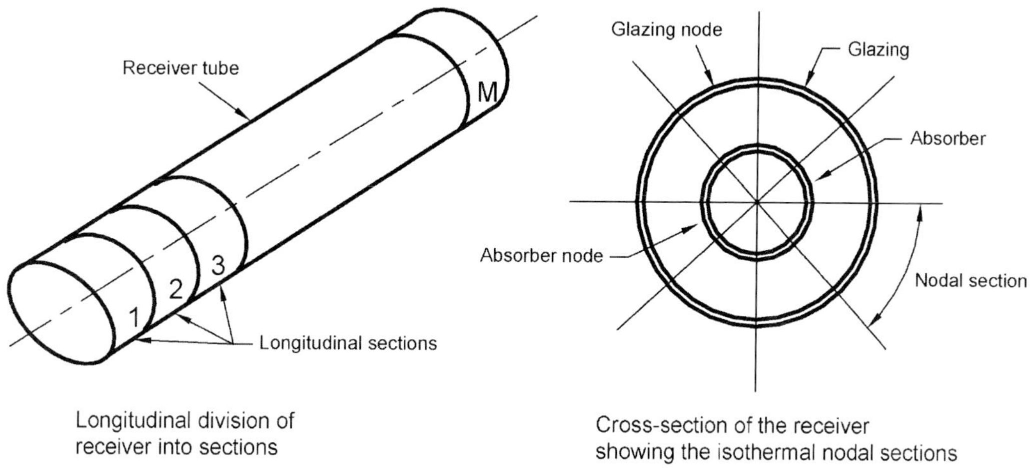

Fig. 6.2-16 Piecewise two-dimensional model of the receiver assembly.

6.2.3.3 Second law analysis

The analysis presented here is based on Bejan's work [105,106]. The analysis however is adapted to imaging collectors because entropy generation minimisation is more important to high temperature systems. Consider that the collector has an aperture area (or total heliostat area) A_a and receives solar radiation at the rate Q^* from the sun as shown in Fig. 6.2-17. The net solar heat transfer Q^* is proportional to the collector area A_a and the proportionality factor q^* (W/ m^2) which varies with geographical position on the earth, the orientation of the collector, meteorological conditions and the time of day. In the present analysis q^* is assumed to be constant and the system is in steady state, i.e.

$$Q^* = q^* A_a \tag{6.2.44}$$

For concentrating systems q^* is the solar energy falling on the reflector. In order to obtain the energy falling on the collector receiver the tracking mechanism accuracy, the optical errors of the mirror including its reflectance and the optical properties of the receiver glazing must be considered.

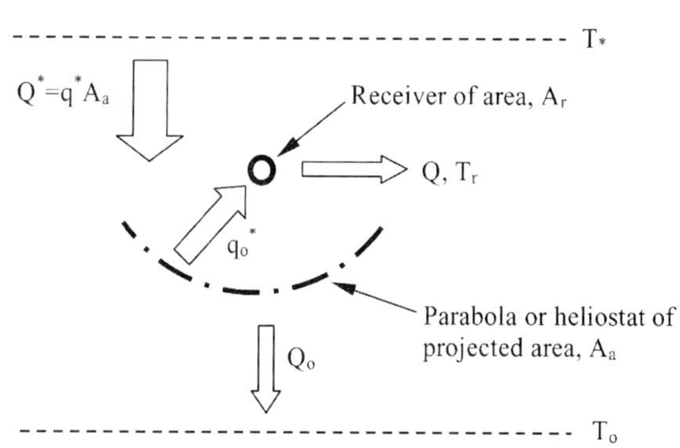

Fig. 6.2-17 Imaging concentrating collector model.

Therefore, the radiation falling on the receiver q_o^* is a function of the optical efficiency, which accounts for all the above errors. For the concentrating collectors, Eq. (6.2.33) can be used. The radiation falling on the receiver is:

$$q_o^* = n_o q^* = \frac{n_o Q^*}{A_a} \tag{6.2.45}$$

The incident solar radiation is partly delivered to a power cycle (or user) as heat transfer Q at the receiver temperature T_r. The remaining fraction Q_o represents the collector-ambient heat loss:

$$Q_o = Q^* - Q \tag{6.2.46}$$

For imaging concentrating collectors Q_o is proportional to the receiver-ambient temperature difference and to the receiver area as:

$$Q_o = U_r A_r (T_r - T_o) \tag{6.2.47}$$

where U_r is the overall heat transfer coefficient based on A_r. It should be noted that U_r is a characteristic constant of the collector.

Combining Eqs. (6.2.46) and (6.2.47) it is apparent that the maximum receiver temperature occurs when $Q = 0$, i.e. when the entire solar heat transfer Q^* is lost to the ambient. The maximum collector temperature is given in dimensionless form by:

$$\theta_{max} = \frac{T_{r,max}}{T_o} = 1 + \frac{Q^*}{U_r A_r T_o} \tag{6.2.48}$$

Combining Eqs. (6.2.45) and (6.2.48):

$$\theta_{max} = 1 + \frac{q_o^* A_a}{n_o U_r A_r T_o} \tag{6.2.49}$$

Considering that $C = A_a/A_r$, then:

$$\theta_{max} = 1 + \frac{q_o^* C}{n_o U_r T_o} \tag{6.2.50}$$

As can be seen from Eq. (6.2.50), θ_{max} is proportional to C, i.e. the higher the concentration ratio of the collector the higher is θ_{max} and $T_{r,max}$. The term $T_{r,max}$ in Eq. (6.2.48) is also known as the stagnation temperature of the collector, i.e. the temperature that can be obtained at no flow condition. In dimensionless form the collector temperature $\theta = T_r/T_o$ will vary between 1 and θ_{max}, depending on the heat delivery rate Q. The stagnation temperature θ_{max} is the parameter that describes the performance of the collector with regard to collector-ambient heat loss as there is no flow through the collector and all the energy collected is used to raise the temperature of the working fluid to stagnation temperature which is fixed at a value corresponding to the energy collected equal to energy loss to ambient. Thus the collector efficiency is given by:

$$\eta_c = \frac{Q}{Q^*} = 1 - \frac{\theta - 1}{\theta_{max} - 1} \tag{6.2.51}$$

Therefore, η_c is a linear function of collector temperature. At stagnation point the heat transfer Q carries zero exergy or zero potential for producing useful work.

Minimum entropy generation rate

The minimization of the entropy generation rate is the same as the maximization of the power output. The process of solar energy collection is accompanied by the generation of entropy upstream of the collector, downstream of the collector and inside the collector as shown in Fig. 6.2-18.

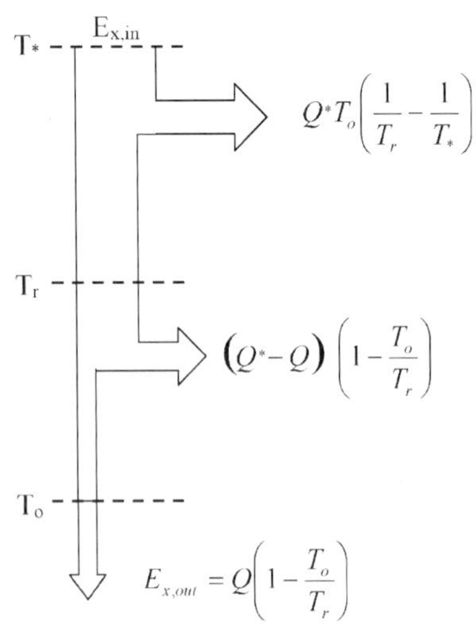

Fig. 6.2-18 Exergy flow diagram.

The exergy inflow coming from the solar radiation falling on the collector surface is:

$$E_{x,in} = Q^*\left(1 - \frac{T_o}{T_*}\right) \tag{6.2.52}$$

where T_* is the apparent sun temperature as an exergy source. In this analysis the value suggested by Petela [107] is adopted, i.e. T_* is approximately equal to $3/4T_s$, where T_s is the apparent black body temperature of the sun, which is about 6000 K. Therefore, T_* considered here is 4500 K. It should be noted that in this analysis T_* is also considered constant and as its value is much greater than T_o, $E_{x,in}$ is very near Q^*. The output exergy from the collector is given by:

$$E_{x,out} = Q\left(1 - \frac{T_o}{T_r}\right) \tag{6.2.53}$$

whereas the difference between the $E_{x,in} - E_{x,out}$ represents the destroyed exergy. From Fig. 6.2-18, the entropy generation rate can be written as:

$$S_{gen} = \frac{Q_o}{T_o} + \frac{Q}{T_r} - \frac{Q^*}{T^*} \tag{6.2.54}$$

This equation can be written with the help of Eq. (6.2.46) as:

$$S_{gen} = \frac{1}{T_o}\left[Q^*\left(1 - \frac{T_o}{T_*}\right) - Q\left(1 - \frac{T_o}{T_r}\right)\right] \tag{6.2.55}$$

By using Eqs. (6.2.52) and (6.2.53), Eq. (6.2.55) can be written as:

$$S_{gen} = \frac{1}{T_o}(E_{x,in} - E_{x,out}) \tag{6.2.56}$$

or

$$E_{x,out} = E_{x,in} - T_o S_{gen} \tag{6.2.57}$$

Therefore, if we consider $E_{x,in}$ constant, the maximisation of the exergy output ($E_{x,out}$) is the same as the minimisation of the total entropy generation S_{gen}.

Optimum collector temperature

By substituting Eqs (6.2.46) and (6.2.47) into Eq. (6.2.55) the rate of entropy generation can be written as:

$$S_{gen} = \frac{U_r A_r (T_r - T_o)}{T_o} - \frac{Q^*}{T_*} + \frac{Q^* - U_r A_r (T_r - T_o)}{T_r} \tag{6.2.58}$$

By applying Eq. (6.2.50) in Eq. (6.2.58) and by performing various manipulations:

$$\frac{S_{gen}}{U_r A_r} = \theta - 2 - \frac{q_o^* C}{n_o U_r T_*} + \frac{\theta_{max}}{\theta} \tag{6.2.59}$$

The dimensionless term S_{gen}/U_rA_r accounts for the fact that the entropy generation rate scales with the finite size of the system which is described by $A_r = A_a/C$.

By differentiating Eq. (6.2.59) with respect to θ and setting to zero, the optimum collector temperature (θ_{opt}) for minimum entropy generation is obtained:

$$\theta_{opt} = \sqrt{\theta_{max}} = \left(1 + \frac{q_o^*C}{n_oU_rT_o}\right)^{1/2} \quad (6.2.60)$$

By substituting θ_{max} by $T_{r,max}/T_o$ and θ_{opt} by $T_{r,opt}/T_o$, Eq. (6.2.60) can be written as:

$$T_{r,opt} = \sqrt{T_{r,max}T_o} \quad (6.2.61)$$

This equation states that the optimal collector temperature is the geometric average of the maximum collector (stagnation) temperature and the ambient temperature. Typical stagnation temperatures and the resulting optimum operating temperatures for various types of concentrating collector are shown in Table 6.2-8. The stagnation temperatures shown in Table 6.2-8 are estimated by considering mainly the collector radiation losses.

As can be seen from the data presented in Table 6.2-8 for high performance collectors, like the central receiver, it is better to operate the system at high flow rates in order to lower the temperature around the value shown instead of operating at very high temperature, in order to obtain higher thermodynamic efficiency from the collector system.

By applying Eq. (6.2.60) to Eq. (6.2.59), the corresponding minimum entropy generation rate is:

$$\frac{S_{gen,min}}{U_rA_r} = 2(\sqrt{\theta_{max}} - 1) - \frac{\theta_{max} - 1}{\theta_*} \quad (6.2.62)$$

where $\theta_* = T_*/T_o$. It should be noted that for flat-plate and low concentration ratio collectors, the last term of Eq. (6.2.62) is negligible as θ_* is much bigger than $\theta_{max} - 1$, but it is not for higher concentration collectors, like the central receiver and the parabolic dish ones, which have stagnation temperatures of several thousands of degrees.

By applying the stagnation temperatures shown in Table 6.2-8 to Eq. (6.2.62), the dimensionless entropy generated against the collector concentration ratios considered here as shown in Fig. 6.2-19 is obtained.

Non-isothermal collector

So far the analysis has been carried out by considering an isothermal collector. For a non-isothermal one, which is a more realistic model particularly for the long PTC, and by applying the principle of energy conservation:

$$q^* = U_r(T - T_o) + mc_p\frac{dT}{dx} \quad (6.2.63)$$

where x is from 0 to L (the collector length). The generated entropy can be obtained from:

$$S_{gen} = mc_p \ln\frac{T_{out}}{T_{in}} - \frac{Q^*}{T_*} + \frac{Q_o}{T_o} \quad (6.2.64)$$

From an overall energy balance, the total heat loss is:

$$Q_o = Q^* - mc_p(T_{out} - T_{in}) \quad (6.2.65)$$

Substituting Eq. (6.2.65) into Eq. (6.2.64) and performing the necessary manipulations the following relation is obtained:

$$N_s = M\left(\ln\frac{\theta_{out}}{\theta_{in}} - \theta_{out} + \theta_{in}\right) - \frac{1}{\theta_*} + 1 \quad (6.2.66)$$

where $\theta_{out} = T_{out}/T_o$, $\theta_{in} = T_{in}/T_o$, N_s is the entropy generation number and M is the mass flow number given by:

$$N_s = \frac{S_{gen}T_o}{Q^*}, \quad \text{and} \quad M = \frac{mc_pT_o}{Q^*} \quad (6.2.67)$$

If the inlet temperature is fixed $\theta_{in} = 1$, then the entropy generation rate is a function of only M and θ_{out}. These parameters are interdependent because the collector outlet temperature depends on the mass flow rate.

Table 6.2-8 Optimum collector temperatures for various types of concentrating collector

Collector type	Concentration ratio	Stagnation temperature (°C)	Optimal temperature (°C)
Parabolic trough	50	565	227
Parabolic dish	500	1285	408
Central receiver	1500	1750	503

Notes: Ambient temperature considered = 25°C.

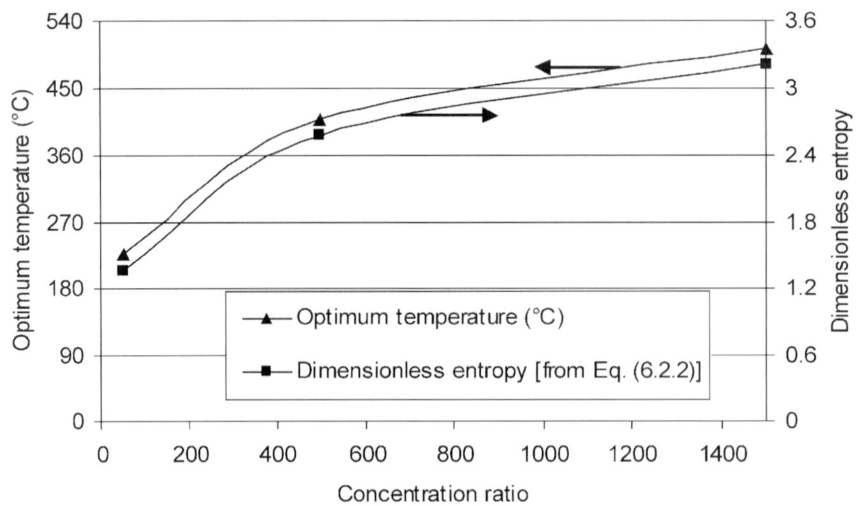

Fig. 6.2-19 Entropy generated and optimum temperatures against collector concentration ratio.

6.2.4 Performance of solar collectors

ASHRAE Standard 93:1986 [108] for testing the thermal performance of collectors is undoubtedly the one most often used to evaluate the performance of flat-plate and concentrating solar collectors. The thermal performance of the solar collector is determined partly by obtaining values of instantaneous efficiency for different combinations of incident radiation, ambient temperature, and inlet fluid temperature. This requires experimental measurement of the rate of incident solar radiation falling onto the solar collector as well as the rate of energy addition to the transfer fluid as it passes through the collector, all under steady-state or quasi-steady-state conditions. In addition, tests must be performed to determine the transient thermal response characteristics of the collector. The variation of steady-state thermal efficiency with incident angles between the direct beam and the normal to collector aperture area at various sun and collector positions is also required [108].

ASHRAE Standard 93:1986 [108] gives information on testing solar energy collectors using single-phase fluids and no significant internal storage. The data can be used to predict performance in any location and under any weather conditions where load, weather, and insolation are known.

6.2.4.1 Collector thermal efficiency

In reality the heat loss coefficient U_L in Eqs (6.2.2) and (6.2.42) is not constant but is a function of collector inlet and ambient temperatures. Therefore:

$$F_R U_L = c_1 + c_2(T_i - T_a) \tag{6.2.68}$$

Applying Eq. (6.2.68) in Eqs (6.2.2) and (6.2.42) we have for FPC:

$$q_u = A_a F_R[\tau\alpha G_t - c_1(T_i - T_a) - c_2(T_i - T_a)^2] \tag{6.2.69}$$

and for concentrating collectors:

$$q_u = F_R[G_b n_o A_a - A_r c_1(T_i - T_a) - A_r c_2(T_i - T_a)^2] \tag{6.2.70}$$

Therefore for FPC, the efficiency can be written as:

$$n = F_R\tau\alpha - c_1\frac{(T_i - T_a)}{G_t} - c_2\frac{(T_i - T_a)^2}{G_t} \tag{6.2.71}$$

and if we denote $c_0 = F_R\tau\alpha$ and $x = (T_i - T_a)/G_t$ then:

$$n = c_o - c_1 x - c_2 G_t x^2 \tag{6.2.72}$$

And for concentrating collectors the efficiency can be written as:

$$n = F_R n_o - \frac{c_1(T_i - T_a)}{CG_b} - \frac{c_2(T_i - T_a)^2}{CG_b} \tag{6.2.73}$$

and if we denote $k_0 = F_R n_o, k_1 = c_1/C, k_2 = c_2/C$ and $y = (T_i - T_a)/G_b$ then:

$$n = k_0 - k_1 y - k_2 G_b y^2 \tag{6.2.74}$$

Usually, the second-order terms are neglected in which case $c_2 = 0$ and $k_2 = 0$ (or the third-term in the above equations is neglected). Therefore, Eqs (6.2.71) and (6.2.73) plot as a straight line on a graph of efficiency versus the heat loss parameter $(T_i - T_a)/G_t$ for the case of FPCs and $(T_i - T_a)/G_b$ for the case of concentrating collectors. The intercept (intersection of the line with the vertical efficiency axis) equates to $F_R\tau\alpha$ for the FPCs and $F_R n_o$ for the concentrating ones. The slope of the line, i.e. the efficiency difference divided by the corresponding horizontal scale difference, equates to $-F_R U_L$ and $-F_R U_L/C$, respectively. If experimental data on collector heat delivery at various temperatures and solar conditions are plotted, with efficiency as the vertical axis and $\Delta T/G$ (G_t or G_b is used according to the type of collector) as the horizontal axis, the best straight line through the data points correlates collector performance with solar and temperature conditions. The intersection of the line with the vertical axis is where the temperature of the fluid entering the collector equals the ambient temperature, and collector efficiency is at its maximum. At the intersection of the line with the horizontal axis, collector efficiency is zero. This condition corresponds to such a low radiation level, or to such a high temperature of the fluid into the collector, that heat losses equal solar absorption, and the collector delivers no useful heat. This condition, normally called stagnation, usually occurs when no fluid flows in the collector.

A comparison of the efficiency of various collectors at irradiation levels of 500 and 1000 W/m^2 is shown in Fig. 6.2-20. Five representative collector types are considered:

- Flat-plate collector.
- Advanced flat-plate collector (AFP). In this collector the risers are ultrasonically welded to the absorbing plate, which is also electroplated with chromium selective coating.
- Stationary CPC orientated with its long axis in the east–west direction.
- Evacuated tube collector.
- Parabolic trough collector with E–W tracking.

As seen in Fig. 6.2-20 the higher the irradiation level the better the efficiency, and the higher performance collectors like the CPC, ETC and PTC retain high efficiency even at higher collector inlet temperatures. It should be noted that the radiation levels examined are considered as global radiation for all collector types except the PTC for which the same radiation values are used but considered as beam radiation.

As it can be seen from Fig. 6.2-20 the advantage of concentrating collectors is that the heat losses are inversely proportional to the concentration ratio C. This leads to the small slope of the collector performance curve. Thus the efficiency of concentrating collectors remains high at high inlet-water temperatures.

The difference in performance can also be seen from the performance equations. For example, the performance of a good FPC is given by

$$n = 0.792 - 6.65\left(\frac{\Delta T}{G_t}\right) - 0.06\left(\frac{\Delta T^2}{G_t}\right) \tag{6.2.75}$$

whereas the performance equation of the IST collector (obtained by the Sandia tests [83]) as given by the manufacturer is:

$$n = 0.762 - 0.2125\left(\frac{\Delta T}{G_b}\right) - 0.001672\left(\frac{\Delta T^2}{G_b}\right) \tag{6.2.76}$$

Eqs. (6.2.71) – (6.2.74) include all important design and operational factors affecting steady-state performance except collector flow rate and solar incidence angle. Flow rate inherently affects performance through the average absorber temperature. If the heat removal rate is reduced, the average absorber temperature increases, and more heat is lost. If the flow is increased, collector absorber temperature and heat loss decreases. The effect of solar incidence angle is accounted for by the incidence angle modifier.

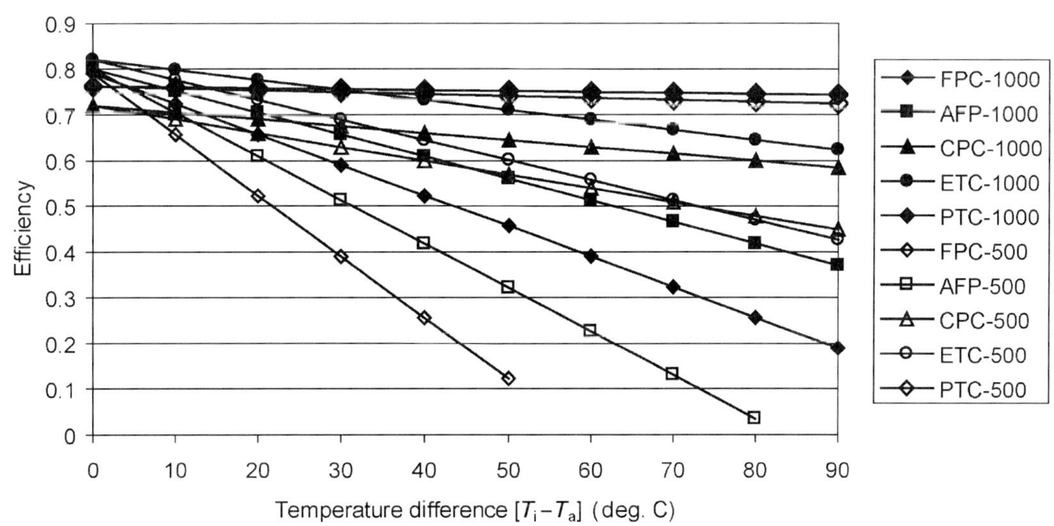

Fig. 6.2-20 Comparison of the efficiency of various collectors at two irradiation levels, 500 and 1000 W/m^2.

6.2.4.2 Collector incidence angle modifier

Flat-plate collectors

The above performance equations (6.2.69) and (6.2.71) assume that the sun is perpendicular to the plane of the collector, which rarely occurs. For the glass cover plates of a FPC, specular reflection of radiation occurs thereby reducing the $(\tau\alpha)$ product. The incidence angle modifier $k_{\alpha\tau}$ defined as the ratio of $\tau\alpha$ at some incidence angle θ to $\tau\alpha$ at normal radiation $(\tau\alpha)_{n,}$ is described by the following expression:

$$k_{\alpha\tau} = 1 - b_0\left(\frac{1}{\cos(\theta)} - 1\right) - b_1\left(\frac{1}{\cos(\theta)} - 1\right)^2 \quad (6.2.77)$$

For a single glass cover, a single-order equation can be used with $b_0 = -0.1$ and $b_1 = 0$.

With the incidence angle modifier, the collector efficiency equation (6.2.71) can be modified as:

$$n = F_R(\tau\alpha)_n k_{\alpha\tau} - c_1\frac{(T_i - T_a)}{G_t} - c_2\frac{(T_i - T_a)^2}{G_t} \quad (6.2.78)$$

Concentrating collectors

Similarly, for concentrating collectors the performance equations (6.2.70) and (6.2.73) described previously are reasonably well defined as long as the direct beam of solar irradiation is normal to the collector aperture. However, for off-normal incidence angles, the optical efficiency term (n_o) is often difficult to be described analytically because it depends on the actual concentrator geometry, concentrator optics, receiver geometry and receiver optics, which may differ significantly. As the incidence angle of the beam radiation increases these terms become more complex. Fortunately, the combined effect of these three parameters at different incidence angles can be accounted for with the incidence angle modifier. This is simply a correlation factor to be applied to the efficiency curve and is only a function of the incidence angle between the direct solar beam and the outward drawn normal to the aperture plane of the collector. It describes how the optical efficiency of the collector changes as the incidence angle changes. With the incidence angle modifier Eq. (6.2.73) becomes:

$$n = F_R K_{\alpha\tau} n_o - \frac{c_1(T_i - T_a)}{CG_b} - \frac{c_2(T_i - T_a)^2}{CG_b} \quad (6.2.79)$$

If the inlet fluid temperature is maintained equal to ambient temperature, the incidence angle modifier can be determined from:

$$K_{\alpha\tau} = \frac{n(T_{fi} = T_a)}{F_R[n_o]_n} \quad (6.2.80)$$

where $n(T_{fi} = T_a)$ is the measured efficiency at the desired incident angle and for an inlet fluid temperature equal to the ambient temperature. The denominator in Eq. (6.2.80) is the test intercept taken from the collector efficiency test with Eq. (6.2.73) with $[n_o]_n$ being the normal optical efficiency, i.e. at normal angle of incidence.

As an example the results obtained from such a test (Fig. 6.2-21) are denoted by the small squares. By using

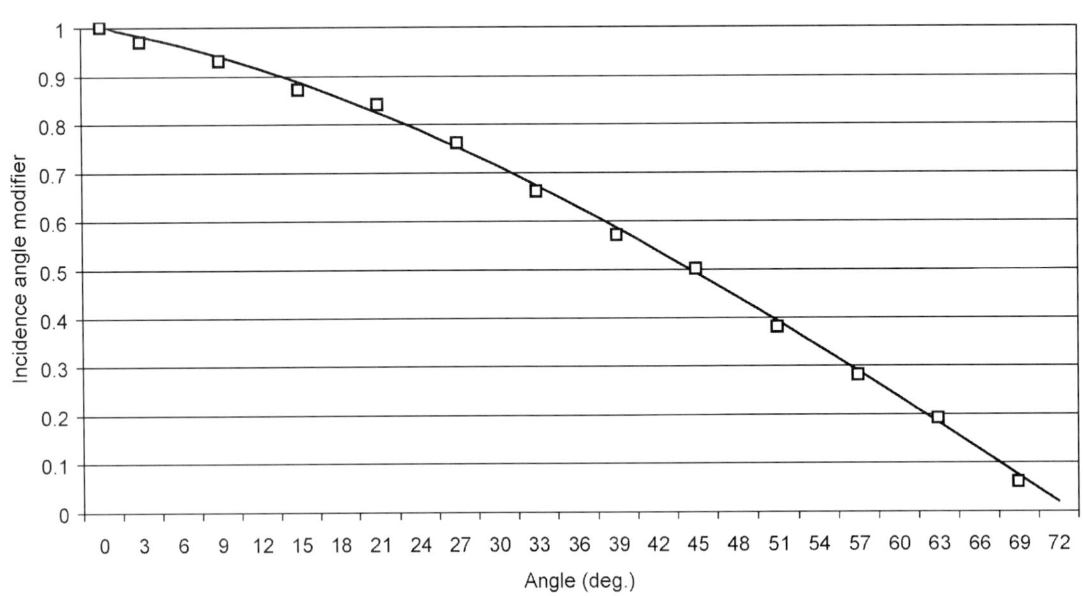

Fig. 6.2-21 Parabolic trough collector incidence angle modifier test results.

a curve fitting method (second-order polynomial fit), the curve that best fits the points can be obtained [59]:

$$K_{\alpha\tau} = 1 - 0.00384(\theta) - 0.000143(\theta)^2 \quad (6.2.81)$$

For the IST collector, the incidence angle modifier $k_{\alpha\tau}$ of the collector, given by the manufacturer is:

$$k_{\alpha\tau} = \cos(\theta) + 0.0003178(\theta) - 0.00003985(\theta)^2 \quad (6.2.82)$$

6.2.4.3 Concentrating collector acceptance angle

Another test required for the concentrating collectors is the determination of the collector acceptance angle, which characterises the effect of errors in the tracking mechanism angular orientation.

This can be found with the tracking mechanism disengaged and measuring the efficiency at various out of focus angles as the sun is travelling over the collector plane. An example is shown in Fig. 6.2-22 where the angle of incidence measured from the normal to the tracking axis (i.e. out of focus angle) is plotted against the efficiency factor, i.e. the ratio of the maximum efficiency at normal incidence to the efficiency at a particular out of focus angle.

A definition of the collector acceptance angle is the range of incidence angles (as measured from the normal to the tracking axis) in which the efficiency factor varies by no more than 2% from the value of normal incidence [108]. Therefore from Fig. 6.2-22, the collector half-acceptance angle, $\theta_{m,}$ is 0.5°. This angle determines the maximum error of the tracking mechanism.

6.2.4.4 Collector time constant

A last aspect of collector testing is the determination of the heat capacity of a collector in terms of a time constant. It is also necessary to determine the time response of the solar collector in order to be able to evaluate the transient behaviour of the collector, and to select the correct time intervals for the quasi-steady state or steady-state efficiency tests. Whenever transient conditions exist, Eqs. (6.2.69)–(6.2.74) do not govern the thermal performance of the collector since part of the absorbed solar energy is used for heating up the collector and its components.

The time constant of a collector is the time required for the fluid leaving the collector to reach 63% of its ultimate steady value after a step change in incident radiation. The collector time constant is a measure of the time required for the following relationship to apply [108]:

$$\frac{T_{ot} - T_i}{T_{oi} - T_i} = \frac{1}{e} = 0.368 \quad (6.2.83)$$

where T_{ot} is the collector outlet water temperature after time t (°C); T_{oi} is the collector outlet initial water temperature (°C); T_i is the collector inlet water temperature (°C).

The procedure for performing this test is to operate the collector with the fluid inlet temperature maintained at the ambient temperature. The incident solar energy is then abruptly reduced to zero by either shielding a FPC, or

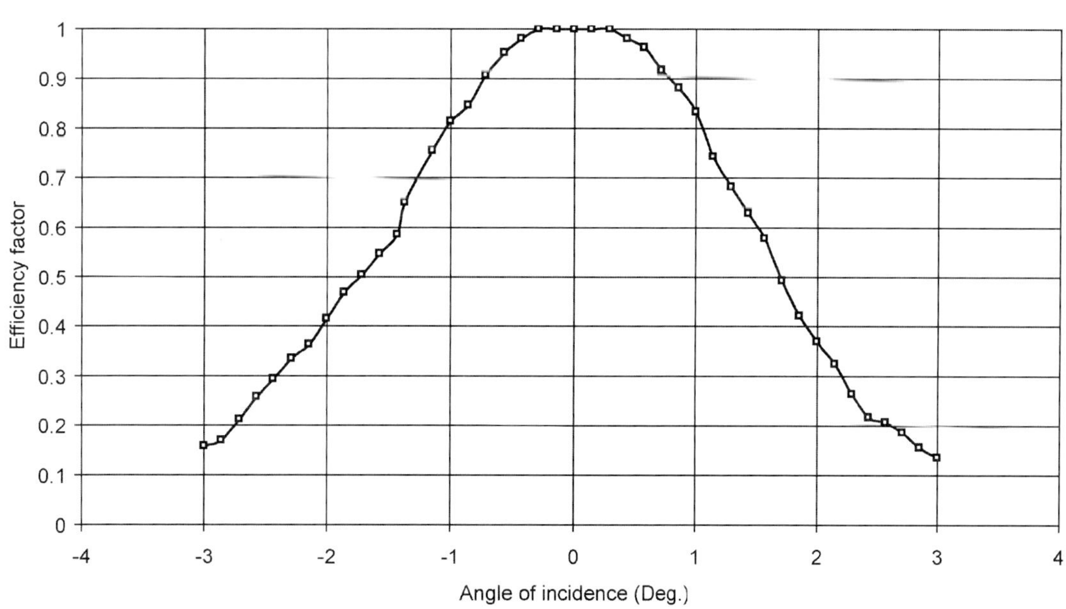

Fig. 6.2-22 Parabolic trough collector acceptance angle test results.

Table 6.2-9 Sequence of quality tests for solar collectors [109]

Sequence	Test
1	Internal pressure
2	High temperature resistance
3	Exposure
4	External thermal shock
5	Internal thermal shock
6	Rain penetration
7	Freeze resistance
8	Internal pressure (re-test)
9	Thermal performance
10	Impact resistance
11	Final inspection

defocusing a concentrating one. The temperatures of the transfer fluid are continuously monitored as a function of time until Eq. (6.2.83) is satisfied. Results of tests carried out on a PTC constructed by the author are given in Ref. [71].

6.2.4.5 Collector test results and preliminary collector selection

Collector testing is required in order to evaluate the performance of solar collectors and compare different collectors to select the most appropriate one for a specific application. As can be seen from sections 6.2.4.1–6.2.4.4 the tests show how a collector absorbs solar energy, how it loses heat, the effects of angle of incidence of solar radiation and the significant heat capacity effects which are determined from the collector time constant.

Finally, tests are performed on the solar collectors in order to determine their quality. In particular the ability of a collector to resist extreme operating conditions is examined as specified in International Standard ISO 9806-2 (1995) [109]. The tests are required to be applied in the sequence specified in Table 6.2-9 so that possible degradation in one test will be exposed in a later test.

Final selection of a collector should be made only after energy analyses of the complete system, including realistic weather conditions and loads, have been conducted for one year. Also, a preliminary screening of collectors with various performance parameters should be conducted in order to identify those that best match the load. The best way to accomplish this is to identify the expected range of the parameter $\Delta T/G$ for the load and climate on a plot of efficiency n as a function of the heat loss parameter, as indicated in Fig. 6.2-23.

Collector efficiency curves may be used for preliminary collector selection. However, efficiency curves illustrate only the instantaneous performance of a collector. They do not include incidence angle effects, which vary throughout the year, heat exchanger effects, probabilities of occurrence of T_i, T_a, solar irradiation, system heat loss, or control strategies. Final selection requires determining the long-term energy output of a collector as well as performance cost-effectiveness studies. Estimating the annual performance of a particular collector and system requires the aid of appropriate analysis tools such as *F*-Chart, Watsun, or TRNSYS. These are analysed briefly in section 6.2.4.6.

6.2.4.6 Modelling of solar systems

The proper sizing of the components of a solar system is a complex problem which includes both predictable

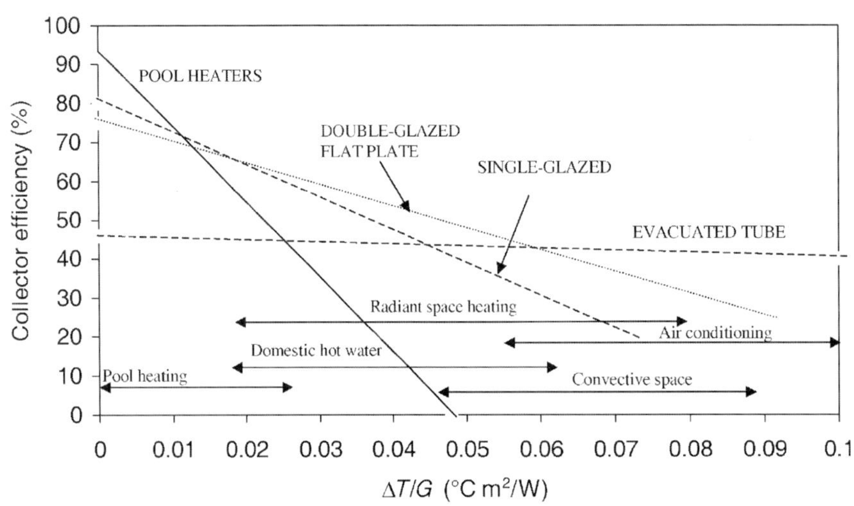

Fig. 6.2-23 Collector efficiencies of various liquid collectors.

(collector and other components performance characteristics) and unpredictable (weather data) components. In this section an overview of the simulation techniques and programs suitable for solar heating and cooling systems is presented.

Computer modelling of thermal systems presents many advantages, the most important of which are the following [110]:

1. Eliminate the expense of building prototypes.
2. Complex systems are organised in an understandable format.
3. Provide thorough understanding of system operation and component interactions.
4. It is possible to optimise the system components.
5. Estimate the amount of energy delivery from the system.
6. Provide temperature variations of the system.
7. Estimate the design variable changes on system performance by using the same weather conditions.

The initial step in modelling a system is the derivation of a structure to be used to represent the system. It will become apparent that there is no unique way of representing a given system. Since the way the system is represented often strongly suggests specific modelling approaches, the possibility of using alternative system structures should be left open while the modelling approach selection is being made. The structure that represents the system should not be confused with the real system. The structure will always be an imperfect copy of reality. However, the act of developing a system structure and the structure itself will foster an understanding of the real system. In developing a structure to represent a system, system boundaries consistent with the problem being analysed are first established. This is accomplished by specifying what items, processes, and effects are internal to the system and what items, processes, and effects are external.

Simplified analysis methods have advantages of computational speed, low cost, rapid turnaround, which is especially important during iterative design phases, and easy of use by persons with little technical experience. Disadvantages include limited flexibility for design optimisation, lack of control over assumptions, and a limited selection of systems that can be analysed. Thus, if the system application, configuration, or load characteristics under consideration are significantly non-standard, a detailed computer simulation may be required to achieve accurate results. The following sections describe briefly four software programs TRNSYS, WATSUN, Polysun and *F*-Chart as well as artificial neural networks applied in solar energy systems modelling and prediction.

TRNSYS simulation program

TRNSYS is an acronym for a "transient simulation" which is a quasi-steady simulation model. This program [111] was developed at the University of Wisconsin by the members of the Solar Energy Laboratory. The program consists of many subroutines that model subsystem components. The mathematical models for the subsystem components are given in terms of their ordinary differential or algebraic equations. With a program such as TRNSYS which has the capability of interconnecting system components in any desired manner, solving differential equations and facilitating information output, the entire problem of system simulation reduces to a problem of identifying all the components that comprise the particular system and formulating a general mathematical description of each.

Once all the components of the system have been identified and a mathematical description of each component is available, it is necessary to construct an information flow diagram for the system. The purpose of the information flow diagram is to facilitate identification of the components and the flow of information between them. Each component is represented as a box, which requires a number of constant PARAMETERS and time dependent INPUTS and produces time dependent OUTPUTS. An information flow diagram shows the manner in which all system components are interconnected. A given OUTPUT may be used as an INPUT to any number of other components. From the flow diagram a deck file has to be constructed containing information on all the components of the system, weather data file, and the output format.

Subsystem components in the TRNSYS include solar collectors, differential controllers, pumps, auxiliary heaters, heating and cooling loads, thermostats, pebble-bed storage, relief valves, hot water cylinders, heat pumps and many more. There are also subroutines for processing radiation data, performing integrations, and handling input and output. Time steps down to 1/1000 h (3.6 s) can be used for reading weather data, which makes the program very flexible with respect to using measured data in simulations. Simulation time steps at a fraction of an hour is also possible.

Model validation studies have been conducted in order to determine the degree to which the TRNSYS program serves as a valid simulation program for a physical system. It has been shown by analysing the results of these validation studies that the TRNSYS program provides results with a mean error between the simulation results and the measured results on actual operating systems under 10% [112]. The use of TRNSYS for the modelling of a thermosyphon SWH was also validated by the author and found to be accurate within 4.7% [110]. TRNSYS is not a user-friendly program, although some graphical interfacing has been developed recently, like IISiBat, but is the most versatile with respect to the detail that systems are modelled on.

More details about the TRNSYS program can be found in the program manual [111] and in Ref. [113]. There are numerous applications of the program in the literature. Some typical examples are for the modelling of a thermosyphon system [110], modelling and performance evaluation of solar domestic hot water systems [114], investigation of the effect of load profile [115], modelling of industrial process heat applications [20,116,117] and modelling and simulation of a lithium bromide absorption system [118].

WATSUN simulation program

WATSUN simulates active solar systems and has been developed by the Watsun Simulation Laboratory of the University of Waterloo in Canada [119]. It is a ready-made program that the user can learn and operate easily. It combines collection, storage, and load information provided by the user with hourly weather data for a specific location, and calculates the system state every hour. For convenience, a monthly summary is also provided. Both hourly and monthly reports include data about incident solar radiation, energy collected, load and auxiliary energy. WATSUN provides information necessary for long-term performance calculations. Also included with WATSUN is an economic analysis option that can be used to assess the costs and profits generated by the use of the solar energy systems.

WATSUN uses weather data consisting of hourly values for global radiation on a horizontal surface, dry bulb temperature, wind speed and relative humidity. For those locations where hourly data is not available, synthetic hourly data can be generated using the WATGEN synthetic weather generator, which needs only monthly average values as input.

The WATSUN simulation interacts with the outside world through a series of files. A file is a collection of information, labelled and placed in a specific location. Files are used by the program to input and output information. There is one input file defined by the user, called the simulation data file. The simulation program then produces three output files, a listing file, an hourly data file, and a monthly data file.

The system is an assembly of collection devices, storage devices, and load devices that the user wants to access. The system is defined in the simulation data file. The file is made up of data blocks that contain groups of related parameters.

The simulation data file controls the simulation. The parameters in this file specify the simulation period, weather data and output options. There are many systems that can be modelled, including domestic hot water, pool systems, and industrial process heating systems. Different data must be entered for each type of system.

The simulation data file also contains information about the physical characteristics of the collector device, the storage device(s), the heat exchangers, and the load. When the simulation data file has been fully delineated, the simulation requires one more file, the weather file, before it can run.

Polysun simulation program

The Polysun program provides dynamic annual simulations of solar thermal systems and helps to optimise them [120]. It operates with dynamic time steps from 1 s to 1 h, and thus simulation can be more stable and exact. The program is user friendly and the graphic–user interface permits a comfortable and clear input of all system parameters. All aspects of the simulation are based on physical models that work without empirical correlation terms. In addition the program performs economic viability analysis and ecological balance, which includes emissions from the eight most significant greenhouse gases; thus the emissions of systems working only with conventional fuel and systems employing solar energy can be compared. Program Polysun was validated by Gantner [121] and was found to be accurate to within 5–10%.

F-Chart method and program

The method was developed by Beckman et al. [122]. The method provides a means for estimating the fraction of a total heating and cooling that will be supplied by solar energy for a given solar heating system. The primary design variable is the collector area whereas secondary variables are collector type, storage capacity, fluid flow rates, and load and collector heat exchanger sizes. The method is a correlation of the results of many hundreds of thermal performance simulations of solar heating systems performed with TRNSYS. The conditions of simulations were varied over appropriate ranges of parameters of practical system designs. The resulting correlations give f, the fraction of the monthly load supplied by solar energy as a function of two dimensionless parameters. One is related to the ratio of collector losses to heating loads, and the other is related to the ratio of absorbed solar radiation to heating loads. The f-charts have been developed for three standard system configurations, liquid and air systems for space (and hot water) heating and systems for service hot water only. Detailed simulations of these systems have been used to develop correlations between dimensionless variables and f. The two dimensionless groups are:

$$X = \frac{A_c F'_R U_L (T_{ref} - \overline{T}_a)\Delta t}{D} \qquad (6.2.84)$$

$$Y = \frac{A_c F'_R(\overline{\tau\alpha})\overline{H}_T N}{D} \tag{6.2.85}$$

For the purpose of calculating the values of the dimensionless parameters X and Y, Eqs (6.2.84) and (6.2.85) are usually rearranged to read:

$$X = F_R U_L \frac{F'_R}{F_R}(T_{ref} - \overline{T}_a)\Delta t \frac{A_c}{D} \tag{6.2.86}$$

$$Y = F_R(\tau\alpha)_n \frac{F'_R}{F_R}\left[\frac{(\overline{\tau\alpha})}{(\tau\alpha)_n}\right]\overline{H}_T N \frac{A_c}{D} \tag{6.2.87}$$

The reason for the rearrangement is that the factors $F_R U_L$ and $F_R(\tau\alpha)_n$ are readily available form standard collector tests (section 6.2.4.1). The dimensionless parameters X and Y have some physical significance. The parameter X represents the ratio of the reference collector total energy loss to total heating load or demand (D) during the period Δt whereas the parameter Y represents the ratio of the total absorbed solar energy to the total heating load or demand (D) during the same period.

The method can be used to simulate standard water and air systems configurations. The fraction f of the monthly total load supplied by the solar space system and air or water heating system is given as a function of the two parameters, X and Y, which can be obtained from charts [122] or from the following equations:

For air heating systems:

$$f = 1.040Y - 0.065X - 0.159Y^2 + 0.00187X^2 - 0.0095Y^3 \tag{6.2.88}$$

For liquid-based systems:

$$f = 1.029Y - 0.065X - 0.245Y^2 + 0.0018X^2 + 0.0215Y^3 \tag{6.2.89}$$

The F-Chart was developed for a storage capacity of 0.25 m^3 of pebbles per square metre of collector area for air systems and 75 l of stored water per square meter of collector area for water systems. Other storage capacities can be used by modifying X by a storage size correction factor X_c/X as given by Duffie and Beckman [97].

For air heating systems for $0.50 \leq$ (actual/standard storage capacity) ≤ 4.0:

$$X_c/X = (\text{Actual/Standard storage capacity})^{-0.30} \tag{6.2.90}$$

For liquid-based systems for $0.50 \leq$ (actual/standard storage capacity) ≤ 4.0:

$$X_c/X = (\text{Actual/Standard storage capacity})^{-0.25} \tag{6.2.91}$$

Also air heating systems must be corrected for the flow rate. The standard collector flow rate is 10 l/s of air per square meter of collector area. The performance of systems having other collector flow rates can be estimated by using appropriate values of F_R and Y and then modifying the value of X by a collector air flow rate correction factor X_c/X to account for the degree of stratification in the pebble bed.

For $0.50 \leq$ (actual/standard air flow rate) ≤ 2.0:

$$X_c/X = (\text{Actual/Standard air flow rate})^{0.28} \tag{6.2.92}$$

Although the F-Chart method is simple in concept, the required calculations are tedious, particularly the manipulation of radiation data. The use of computers greatly reduces the effort required. Program F-Chart [123] as developed by the originators of TRNSYS is very easy to use and gives predictions very quickly. The model is accurate only for solar heating systems of a type comparable to that which was assumed in the development of the F-Chart. However, the model does not provide the flexibility of detail simulations and performance investigations as TRNSYS does.

F-Chart method was used by Datta *et al.* [124] for the optimisation of the collector inclination angle. It was also used by the author for a feasibility study on the use of PTC for hot water production [125].

Artificial neural networks in solar energy systems modelling and prediction

Artificial neural networks mimic somewhat the learning process of a human brain. They are widely accepted as a technology offering an alternative way to tackle complex and ill specified problems. They can learn from examples, are fault tolerant in the sense that they are able to handle noisy and incomplete data, are able to deal with non-linear problems, and once trained can perform prediction and generalisation at high speed. They have been used in diverse applications in control, robotics, pattern recognition, forecasting, medicine, power systems, manufacturing, optimisation, signal processing, and social/psychological sciences. They are particularly useful in system modelling such as in implementing complex mappings and system identification. Artificial neural networks have been used by the author in the field of solar energy, for modelling the heat-up response of a solar steam generating plant, for the estimation of a PTC intercept factor, for the estimation of PTC local concentration ratios and for the design of a solar steam generation system. A review of these models together with other applications in the field of renewable energy is given in Ref. [126]. In all those models a multiple hidden layer architecture has been used. Errors reported are well within acceptable limits, which clearly suggest that artificial neural networks can be used for modelling and prediction in other fields of solar energy production and use. What is required is to have a set of data (preferably experimental) representing the past history of a system so that a suitable neural network can be trained to learn the dependence of expected output on the input parameters.

Limitations of simulations

Simulations are powerful tools for process design offering a number of advantages as outlined in the previous sections. However, there are limits to their use. For example, it is easy to make mistakes, such as assuming erroneous constants and neglecting factors, which may be important. Like other engineering calculations, a high level of skill and scientific judgement is required in order to produce correct and useful results.

It is possible to model a system to a high degree of accuracy in order to extract the required information. In practice, however, it may be difficult to represent in detail some of the phenomena occurring in real systems. Additionally, physical world problems such as leaks, plugged or restricted pipes, scale on heat exchangers, failure of controllers, poor installation of collectors and other equipment, and poor insulation, cannot be easily modelled or accounted for. Simulation programs are dealing only with thermal processes but mechanical and other considerations can also affect the thermal performance of solar systems.

There is no substitute to carefully executed experiments. A combination of simulation and physical experiments can lead to better systems and better understanding of how processes work. These can reveal whether or not theory is adequate and where difficulties are present in the design and/or operation of the systems. As a conclusion, simulations are powerful tools for the modelling, design, prediction of performance and research and development. They must, however, be used with care and skill.

No study of solar systems is complete unless an economic analysis is carried out. For this purpose a life cycle analysis is usually performed as explained briefly in the following section.

6.2.4.7 Economic analysis

The economic analysis of solar energy systems is carried out in order to determine the least cost of meeting the energy needs, considering both solar and non-solar alternatives. The method employed for the economic analysis is called the life savings analysis. This method takes into account the time value of money and allows detailed consideration of the complete range of costs. Solar processes are generally characterised by high initial cost and low operating costs. Thus, the basic economic problem is of comparing an initial known investment with estimated future operating costs.

Life cycle cost (LCC) is the sum of all the costs associated with an energy delivery system over its lifetime in today's money, and takes into account the time value of money. The life cycle savings (LCS), for a solar plus auxiliary system, is defined as the difference between the LCC of a conventional fuel-only system and the LCC of the solar plus auxiliary system. This is equivalent to the total present worth (PW) of the gains from the solar system compared to the fuel-only system.

All software programs described in previous section have routines for the economic analysis of the modelled systems. The economic analysis of solar systems can also be performed with a spreadsheet program. Spreadsheet programs are especially suitable for economic analyses as their general format is a table with cells which can contain values or formulae and they incorporate many built-in functions. The economic analysis is carried out for every year for which various economic parameters are calculated in different columns. A detailed description of the method of economic analysis of solar systems using spreadsheets is given in Ref. [127].

Time value of money

It must be noted that a sum of money at hand today is worth less than the same sum in the future, because the money at hand can be invested at some compounding interest to generate a bigger sum in the future. Therefore, a sum of money or cash flow in the future must be discounted and worth less at present-day value. A cash flow F occurring N years from now can be reduced to its present value P by:

$$P = \frac{F}{(1+d)^N} \tag{6.2.93}$$

where d is the market discount rate (%).

Similarly, the amount of money needed to purchase an item is increasing because the value of money is decreasing. With an annual inflation rate i, a purchase cost C at the end of year N will become a future cost F according to:

$$F = C(1+i)^{N-1} \tag{6.2.94}$$

Method description

In general, the PW (or discounted cost) of an investment or cost C at the end of year N, at a discount rate of d and interest rate of i is obtained by combining Eqs (6.2.93) and (6.2.94) as:

$$PW_N = \frac{C(1+i)^{N-1}}{(1+d)^N} \tag{6.2.95}$$

Eq. (6.2.95) can easily be incorporated into a spreadsheet with the parameters d and i entered into separate cells and referencing them in the formulae as absolute cells. In this way a change in either d or i will cause automatic recalculation of the spreadsheet.

The fuel savings are obtained by subtracting the annual cost of the conventional fuel used for the auxiliary energy from the fuel needs of a fuel-only system. The integrated cost of the auxiliary energy use for the first year, i.e. solar back up, is given by the formula:

$$C_{aux} = \int_0^t C_{FA} Q_{aux}\, dt \tag{6.2.96}$$

The integrated cost of the total load for the first year, i.e. cost of conventional fuel without solar, is:

$$C_{load} = \int_0^t C_{FL} Q_{load} dt \tag{6.2.97}$$

where C_{FA} and C_{FL} are the cost rates for auxiliary energy and conventional fuel, respectively.

Such analysis is performed annually and the following are evaluated:

- Fuel savings;
- Extra mortgage payment;
- Extra maintenance cost;
- Extra parasitic cost;
- Extra tax savings;
- Solar savings.

In some countries some other costs may be present, i.e. extra property tax to cover the new system. In this case these costs should be considered as well. The word "extra" appearing in some of the above items assumes that the associated cost is also present for a fuel-only system and therefore only the extra part of the cost incurred by the installation of the solar system should be included. The inflation, over the period of economic analysis, of the fuel savings is estimated by using Eq. (6.2.94) with i equal to the fuel inflation rate. The parasitic cost is the energy required to power auxiliary items like the pump, fan and controllers. This cost is also increased at an inflational rate over the period of economic analysis by using Eq. (6.2.94) with i equal to the annual increase of electricity price.

The mortgage payment is the annual value of money required to cover the funds borrowed at the beginning to install the system. This includes interest and principal payment. The estimation of the annual mortgage payment can be found by dividing the amount borrowed by the present worth factor (PWF). The PWF is estimated by using the inflation rate equal to zero (equal payments) and with the market discount rate equal to the mortgage interest rate. The PWF can be obtained from tables or calculated by the following equation [97]:

$$PWF = \frac{1}{d}\left[1 - \left(\frac{1}{1+d}\right)^N\right] \tag{6.2.98}$$

where d is the interest rate, and N is the number of years (equal instalments).

Solar savings for each year are the sums of the items as:

$$\begin{aligned}\text{Solar savings} = {} & \text{Extra mortgage payment}\\ & + \text{Extra maintenance cost}\\ & + \text{Extra parasitic cost} + \text{Fuel savings}\\ & + \text{Extra tax savings}\end{aligned} \tag{6.2.99}$$

Actually, the savings are positive and the costs are negative. Finally, the PW of each year's solar savings is determined by using Eq. (6.2.95). The results are estimated for each year. These annual values are then added up to obtain the LCS according to the equation:

$$PW_{LCS} = \sum_{N=1}^{N} \frac{\text{Solar savings}}{(1+d)^N} \tag{6.2.100}$$

6.2.5 Solar collector applications

Solar collectors have been used in a variety of applications. These are described in this section. In Table 6.2-10 the most important technologies in use are listed together with the type of collector that can be used in each case.

6.2.5.1 Solar water heating systems

The main part of a SWH is the solar collector array that absorbs solar radiation and converts it into heat. This heat is then absorbed by a heat transfer fluid (water, non-freezing liquid, or air) that passes through the collector. This heat can then be stored or used directly. Portions of the solar energy system are exposed to the weather conditions, so they must be protected from freezing and from overheating caused by high insolation levels during periods of low energy demand.

In solar water heating systems, potable water can either be heated directly in the collector (direct systems) or indirectly by a heat transfer fluid that is heated in the collector, passing through a heat exchanger to transfer its heat to the domestic or service water (indirect systems). The heat transfer fluid is transported either naturally (passive systems) or by forced circulation (active systems). Natural circulation occurs by natural convection (thermosyphoning), whereas for the forced circulation systems pumps or fans are used. Except for thermosyphon and integrated collector storage (ICS) systems, which

Table 6.2-10 Solar energy applications and types of collector used

Application	System	Collector
Solar water heating		
Thermosyphon systems	Passive	FPC
Integrated collector storage	Passive	CPC
Direct circulation	Active	FPC, CPC ETC
Indirect water heating systems	Active	FPC, CPC ETC
Air systems	Active	FPC
Space heating and cooling		
Space heating and service hot water	Active	FPC, CPC ETC
Air systems	Active	FPC
Water systems	Active	FPC, CPC ETC
Heat pump systems	Active	FPC, CPC ETC
Absorption systems	Active	FPC, CPC ETC
Adsorption (desiccant) cooling	Active	FPC, CPC ETC
Mechanical systems	Active	PDR
Solar refrigeration		
Adsorption units	Active	FPC, CPC ETC
Absorption units	Active	FPC, CPC ETC
Industrial process heat		
Industrial air and water systems	Active	FPC, CPC ETC
Steam generation systems	Active	PTC, LFR
Solar desalination		
Solar stills	Passive	–
Multistage flash (MSF)	Active	FPC, CPC ETC
Multiple effect boiling (MEB)	Active	FPC, CPC ETC
Vapour compression (VC)	Active	FPC, CPC ETC
Solar thermal power systems		
Parabolic trough collector systems	Active	PTC
Parabolic tower systems	Active	HFC
Parabolic dish systems	Active	PDR
Solar furnaces	Active	HFC, PDR

need no control, solar domestic and service hot water systems are controlled using differential thermostats.

Five types of solar energy systems can be used to heat domestic and service hot water: thermosyphon, ICS, direct circulation, indirect, and air. The first two are called passive systems as no pump is employed, whereas the others are called active systems because a pump or fan is employed in order to circulate the fluid. For freeze protection, recirculation and drain-down are used for direct solar water heating systems and drain-back is used for indirect water heating systems.

All these systems offer significant economic benefits with payback times, depending on the type of fuel they replace, between 4 years (electricity) and 7 years (diesel oil). Of course, these payback times depend on the economic indices, like the inflation rates and fuel price applied in a country. A wide range of collectors have been used for solar water heating systems. A review of the systems manufactured in the last 20 years is given in Ref. [128].

Thermosyphon systems (passive)

Thermosyphon systems, shown schematically in Fig. 6.2-24, heat potable water or heat transfer fluid and use natural convection to transport it from the collector to storage. The water in the collector expands becoming less dense as the sun heats it and rises through the collector into the top of the storage tank. There it is replaced by the cooler water that has sunk to the bottom of the tank, from which it flows down the collector. The circulation is continuous as long as there is sunshine. Since the driving force is only a small density difference, larger than normal pipe sizes must be used to minimise pipe friction. Connecting lines must be well insulated to prevent heat losses and sloped to prevent formation of air pockets which would stop circulation. At night, or whenever the collector is cooler than the water in the tank, the direction of the thermosyphon flow will reverse, thus cooling the stored water. One way to prevent this is to place the top of the collector well below (about 30 cm) the bottom of the storage tank.

The main disadvantage of thermosyphon systems is the fact that they are comparatively tall units, which makes them not very attractive aesthetically. Usually, a cold water storage tank is installed on top of the solar collector, supplying both the hot water cylinder and the cold water needs of the house, thus making the collector unit taller and even less attractive. Additionally, extremely hard or acidic water can cause scale deposits that clog or corrode the absorber fluid passages. For direct systems, pressure-reducing valves are required when the city water is used directly (no cold water storage tank) and pressure is greater than the working pressure of the collectors.

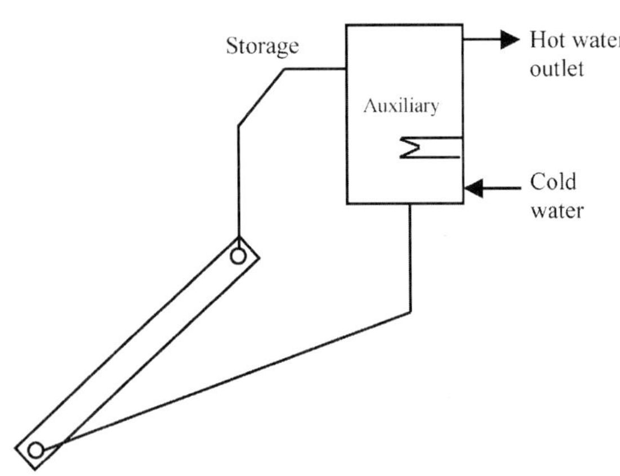

Fig. 6.2-24 Schematic diagram of a thermosyphon solar water heater.

There have been extensive analyses of the performance of thermosyphon SWH, both experimentally and analytically by numerous researchers. Some of the most important are shown here.

Gupta and Garg [129] developed one of the first models for thermal performance of a natural circulation SWH with no load. They represented solar radiation and ambient temperature by Fourier series, and were able to predict a day's performance in a manner that agreed substantially with experiments.

Ong performed two studies [130,131] to evaluate the thermal performance of an SWH. He instrumented a relatively small system with five thermocouples on the bottom surface of the water tubes and six thermocouples on the bottom surface of the collector plate. A total of six thermocouples were inserted into the storage tank and a dye tracer mass flow meter was employed. Ong's studies appear to be the first detailed ones on a thermosyphonic system.

Kudish *et al.* [132] in their study measured the thermosyphon flow rate directly by adapting a simple and well-known laboratory technique, a constant level device, to a solar collector in the thermosyphon mode. The thermosyphon flow data gathered were utilised to construct a standard efficiency test curve, thus showing that this technique can be applied for testing collectors in the thermosyphon mode. Also, they determined the instantaneous collector efficiency as a function of time of day.

Morrison and Braun [133] have studied system modelling and operation characteristics of thermosyphon SWH with vertical or horizontal storage tank. They found that the system performance is maximised when the daily collector volume flow is approximately equal to the daily load flow, and the system with horizontal tank did not perform as well as a vertical one.

Hobson and Norton [134] in their study developed a characteristic curve for an individual directly heated thermosyphon solar energy water heater obtained from data of a 30 days tests. Using such a curve, the calculated annual solar fraction agreed well with the corresponding value computed from the numerical simulation. Furthermore, the analysis was extended, and they produced a simple but relatively accurate design method for direct thermosyphon solar energy water heaters.

Shariah and Shalabi [135] have studied optimisation of design parameters for a thermosyphon SWH for two regions in Jordan represented by two cities, namely Amman and Aqaba through the use of TRNSYS simulation program. Their results indicate that the solar fraction of the system can be improved by 10–25% when each studied parameter is chosen properly. It was also found that the solar fraction of a system installed in Aqaba (hot climate) is less sensitive to some parameters than the solar fraction of a similar system installed in Amman (mild climate).

Integrated collector storage systems (passive)

ICS systems use hot water storage as part of the collector, i.e. the surface of the storage tank is used also as an absorber. As in all other systems, to improve stratification, the hot water is drawn from the top of the tank and cold make-up water enters the bottom of the tank on the opposite side.

The main disadvantage of the ICS systems is the high thermal losses from the storage tank to the surroundings since most of the surface area of the storage tank cannot be thermally insulated as it is intentionally exposed for the absorption of solar radiation. In particular, the thermal losses are greatest during the night and overcast days with low ambient temperature. Due to these losses the water temperature drops substantially during the night, especially during the winter. Various techniques have been used to avoid this from happening. Tripanagnostopoulos *et al.* [136] presented a number of experimental units in which the reduction of thermal losses was achieved by considering single and double cylindrical horizontal tanks properly placed in truncated symmetric and asymmetric CPC reflector troughs.

Details of an ICS unit developed by the author are presented here [137]. The system employs a non-imaging CPC cusp type collector. A fully developed cusp concentrator for a cylindrical receiver is shown in Fig. 6.2-25. The particular curve illustrated has an acceptance half-angle, θ_A, of 60°, or a full acceptance angle, $2\theta_A$, of 120°. Each side of the cusp has two mathematically distinct segments smoothly joined at a point P related to θ_A. The first segment, from the bottom of the receiver to point P, is the involute of the receiver's circular cross-section. The second segment is from point P to the top of the curve, where the curve becomes parallel to the y–axis [138].

With reference to Fig. 6.2-26, for a cylindrical receiver the radius R and acceptance half-angle, θ_A, the distance, ρ, along a tangent from the receiver to the curve, is related to the angle θ, between the radius to the bottom of the receiver and the radius to the point of tangency, T, by the following expressions for the two sections of the curve [138]:

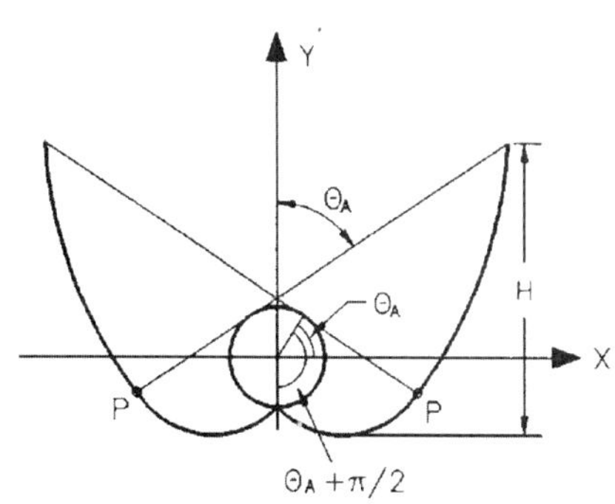

Fig. 6.2-25 Fully developed cusp.

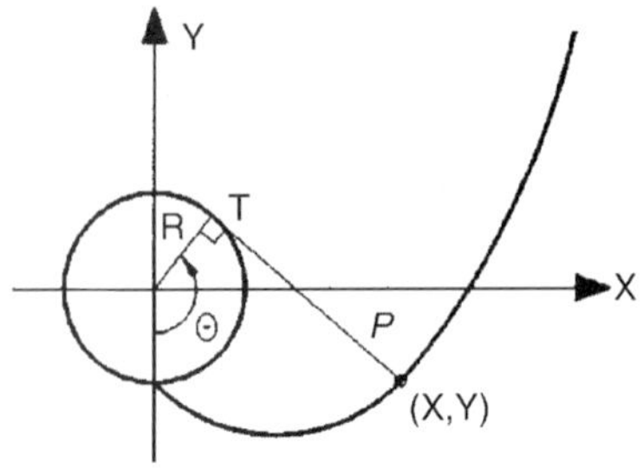

Fig. 6.2-26 Mirror co-ordinates for ideal non-imaging cusp concentrator.

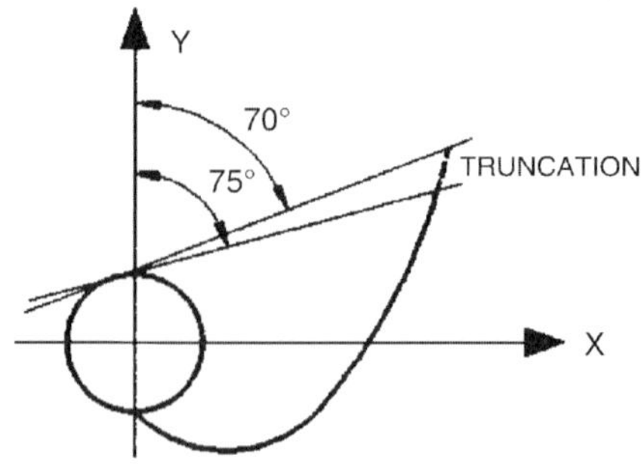

Fig. 6.2-27 Truncation of non-imaging concentrator.

$$\rho(\theta) = R\theta, |\theta| \leq \theta_A + \pi/2 \qquad (6.2.101)$$

(the involute part of the curve)

$$\rho(\theta) = R\left\{\frac{\{\theta + \theta_A + \pi/2 - \cos(\theta - \theta_A)\}}{1 + \sin(\theta - \theta_A)}\right\}$$

$$\theta_A + \pi/2 \leq \theta \leq 3\pi/2 - \theta_A$$

The two expressions for $\rho(\theta)$ are equivalent for the point P in Fig. 6.2-25, where $\theta = \theta_A + \pi/2$. The curve is generated by incrementing θ in radians, calculating ρ, and then calculating the co-ordinates, X and Y, by:

$$X = R\sin\theta - \rho\cos\theta, \quad Y = -R\cos\theta - \rho\sin\theta \qquad (6.2.102)$$

Fig. 6.2-25 shows a full untruncated curve which is the mathematical solution for a reflector shape with the maximum possible concentration ratio. The reflector shape shown in Fig. 6.2-25 is not the most practical design for a cost-effective concentrator, because reflective material is not effectively used in the upper portion of the concentrator. As in the case of the CPC, a theoretical cusp curve should be truncated to a lower height and slightly smaller concentration ratio. Graphically, this is done by drawing a horizontal line across the cusp at a selected height and discarding the part of the curve above the line. Mathematically, the curve is defined to a maximum angle θ value less than $3\pi/2-\theta_A$. The shape of the curve below the cut-off line is not changed by truncation, so the acceptance angle used for the construction of the curve (using Eq. (6.2.101)) of a truncated cusp is equal to the acceptance angle of the fully developed cusp from which it was truncated.

A large acceptance angle of 75° is used in this design so that the collector would be able to collect as much diffuse radiation as possible [137]. The fully developed cusp together with the truncated one is shown in Fig. 6.2-27. The receiver radius considered in the construction of the cusp is 0.24 m. The actual cylinder is 0.20 m. This is done in order to create a gap at the underside of the receiver and the edge of the cusp in order to minimise the optical and conduction losses. The final design is shown in Fig. 6.2-28. The collector aperture is 1.77 m^2, which in combination with the absorber diameter used, gives a concentration ratio of 1.47 [137]. It should be noted that, as shown in Fig. 6.2-28, the system is inclined at the local latitude in order to work effectively.

Direct circulation systems (active)

In direct circulation systems, shown schematically in Fig. 6.2-29, a pump is used to circulate potable water from storage to the collectors when there is enough available solar energy to increase its temperature and then return the heated water to the storage tank until it is needed. As a pump circulates the water, the collectors can be mounted either above or below the storage tank. The optimum flow rate for such units is about 0.015 l/m^2 of collector area. Direct circulation systems can be used in areas where freezing is not frequent. For extreme weather conditions, freeze protection is usually provided by recirculating warm water from the storage tank. Direct circulation systems often use a single storage tank equipped with an auxiliary water heater, but two-tank storage systems can also be used.

Direct circulation systems can be used with water supplied from a cold water storage tank or connected directly to city water mains. Pressure-reducing valves and pressure relief valves are required, however, when

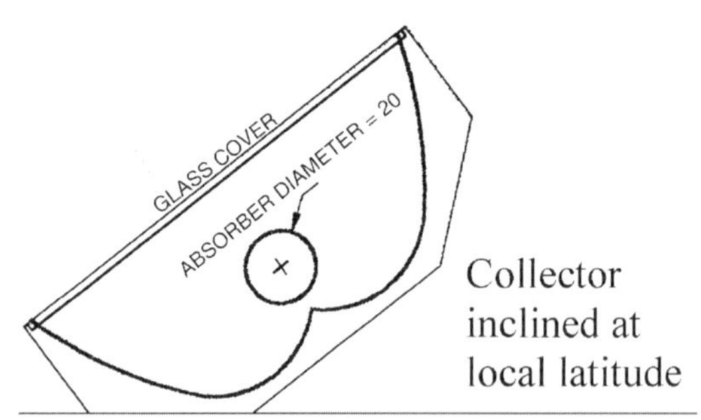

Fig. 6.2-28 The final collector.

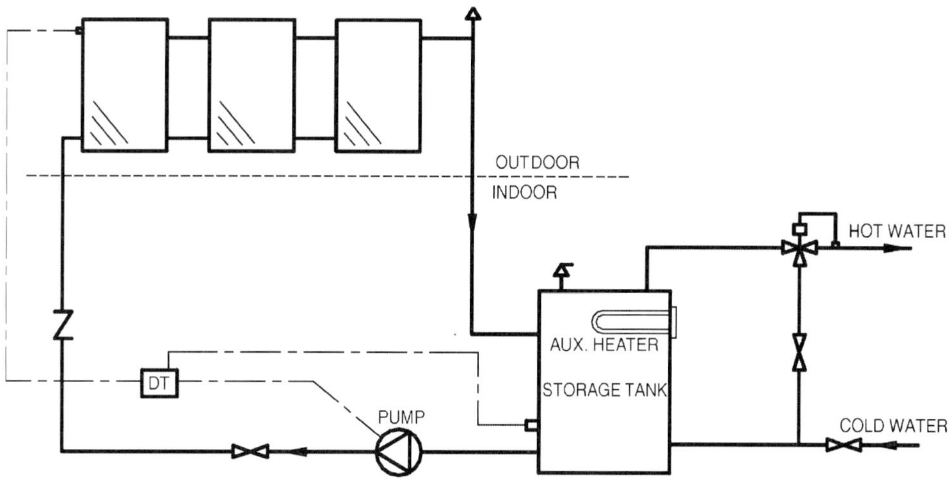

Fig. 6.2-29 Direct circulation system.

the city water pressure is greater than the working pressure of the collectors. Direct water heating systems should not be used in areas where the water is extremely hard or acidic because scale deposits may clog or corrode the collectors.

A variation of the direct circulation system is the drain-down system shown in Fig. 6.2-30. In this case also, potable water is pumped from storage to the collector array where it is heated. When a freezing condition or a power failure occurs, the system drains automatically by isolating the collector array and exterior piping from the make-up water supply and draining it using the two normally open (NO) valves, shown in Fig. 6.2-30. It should be noted that the solar collectors and associated piping must be carefully sloped to drain the collector's exterior piping when circulation stops. The same comments about pressure and scale deposits apply here as for the direct circulation systems.

Indirect water heating systems (active)

Indirect water heating systems, shown schematically in Fig. 6.2-31, circulate a heat transfer fluid through the closed collector loop to a heat exchanger, where its heat is transferred to the potable water. The most commonly used heat transfer fluids are water/ethylene glycol solutions, although other heat transfer fluids such as silicone oils and refrigerants can also be used. When fluids that are non-potable or toxic are used, double-wall heat exchangers should be employed. The heat exchanger can be located inside the storage tank, around the storage tank (tank mantle) or can be external. It should be noted that the collector loop is closed and therefore an expansion tank and a pressure relief valve are required. Additional over-temperature protection may be needed to prevent the collector heat transfer fluid from decomposing or becoming corrosive.

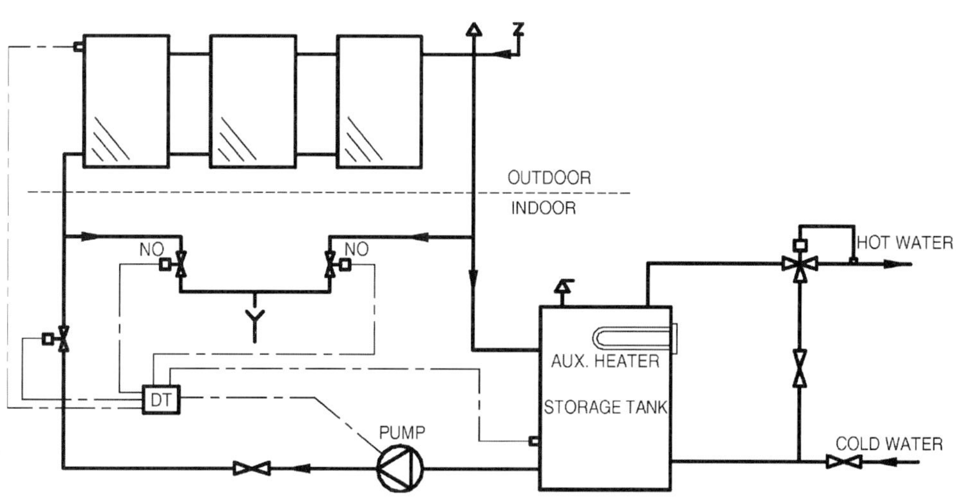

Fig. 6.2-30 Drain-down system.

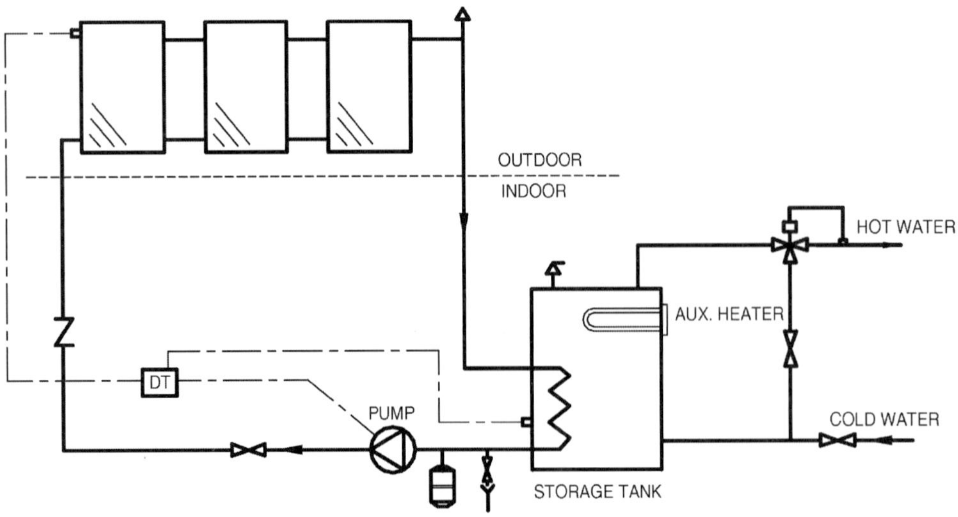

Fig. 6.2-31 Indirect water heating system.

A variation of indirect water heating systems is the drain-back system. Drain-back systems are generally indirect water heating systems that circulate water through the closed collector loop to a heat exchanger, where its heat is transferred to the potable water. Circulation continues as long as usable energy is available. When the circulation pump stops the collector fluid drains by gravity to a drain-back tank. If the system is pressurised the tank serves also as an expansion tank when the system is operating and in this case it must be protected with a temperature and pressure relief valves. In the case of an unpressurised system (Fig. 6.2-32), the tank is open and vented to the atmosphere.

As the collector loop is isolated from the potable water, no valves are needed to actuate draining, and scaling is not a problem, however, the collector array and exterior piping must be adequately sloped to drain completely.

Air systems

Air systems are indirect water heating systems that circulate air via ductwork through the collectors to an air-to-liquid heat exchanger. In the heat exchanger, heat is transferred to the potable water, which is also circulated through the heat exchanger and returned to the storage tank. Fig. 6.2-33 shows a double storage tank system. This type of system is used most often, because air systems are generally used for preheating domestic hot water and thus auxiliary is used only in one tank as shown.

The main advantage of the system is that air does not need to be protected from freezing or boiling, is non-corrosive, and is free. The disadvantages are that air handling equipment (ducts and fans) need more space than piping and pumps, air leaks are difficult to detect, and parasitic power consumption is generally higher than that of liquid systems.

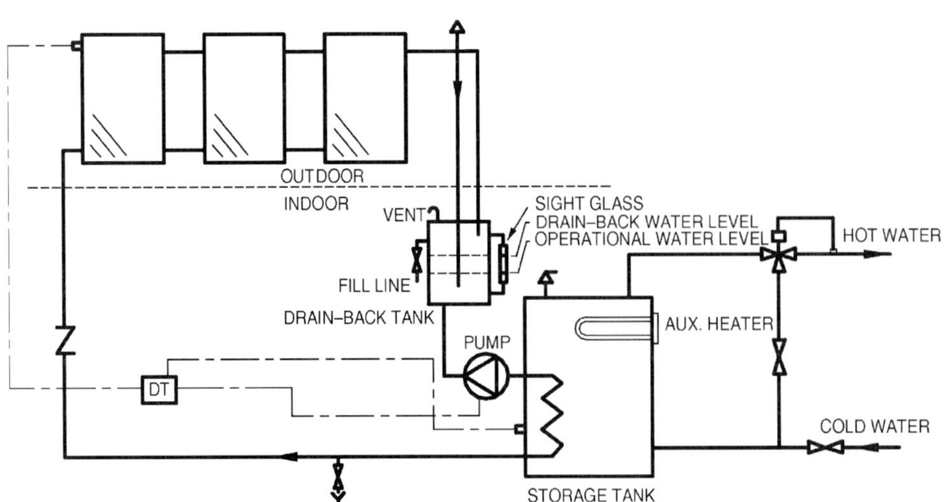

Fig. 6.2-32 Drain-back system.

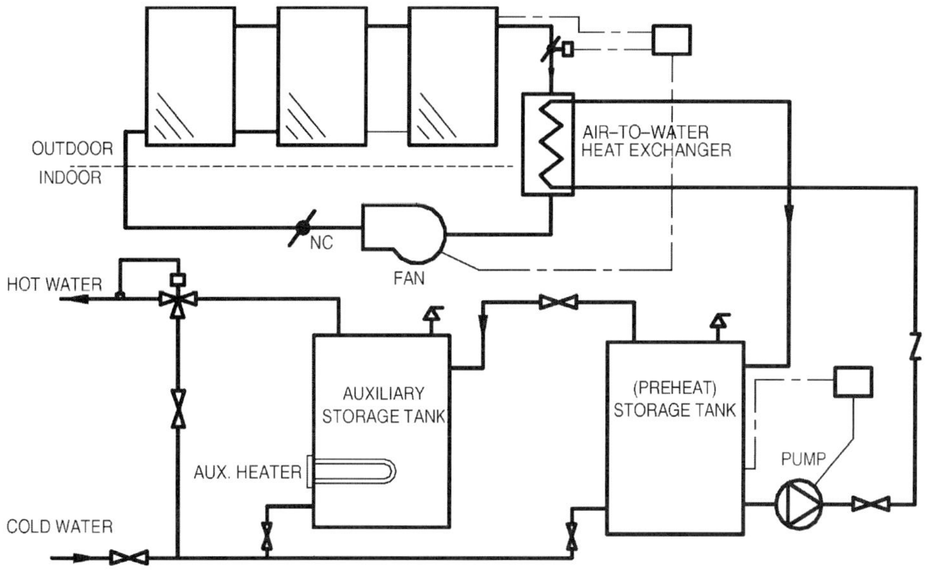

Fig. 6.2-33 Air system.

6.2.5.2 Solar space heating and cooling

The components and subsystems discussed in section 6.2.5.1 may be combined to create a wide variety of building solar heating and cooling systems. There are again two principal categories of such systems, passive and active.

The term passive system is applied to buildings that include as integral parts of the building elements that admit, absorb, store and release solar energy and thus reduce the need for auxiliary energy for comfort heating. As no solar collectors are employed in the passive systems discussed here, only active systems are considered.

Systems for space heating are very similar to those for water heating, and as the same considerations for combination with an auxiliary source, boiling and freezing, controls, etc., apply to both, these will not be repeated again. The most common heat transfer fluids are water, water and antifreeze mixtures, and air. The load is the building to be heated. Although it is technically possible to construct a solar heating or cooling system which can satisfy the design load 100%, such a system would be non-viable since it would be oversized for most of the time. The size of the solar system may be determined by a life-cycle cost analysis described in section 6.2.4.7.

Active solar space systems use collectors to heat a fluid, storage units to store solar energy until needed, and distribution equipment to provide the solar energy to the heated spaces in a controlled manner. A complete system includes additionally pumps or fans for transferring the energy to storage or to the load which require a continuous availability of non-renewable energy, generally in the form of electricity.

The load can be space cooling, heating, or a combination of these two with hot water supply. In combination with conventional heating equipment solar heating provides the same levels of comfort, temperature stability, and reliability as conventional systems.

Active solar energy systems can also be combined with heat pumps for water heating and/or space heating. In residential heating the solar system can be used in parallel with a heat pump, which supplies auxiliary energy when the sun is not available. Additionally, for domestic water systems requiring high water temperatures, a heat pump can be placed in series with the solar storage tank.

During daytime the solar system absorbs solar radiation with collectors and conveys it to storage using a suitable fluid. As the building requires heat it is obtained from storage. Control of the solar system is exercised by differential temperature controllers, i.e. the controller compares the temperature of the collectors and storage and whenever the temperature difference is more than a certain value (7–10°C), the solar pump is switched ON. In locations where freezing conditions are possible to occur, a low-temperature sensor is installed on the collector which controls the solar pump when a pre-set temperature is reached. This process wastes some stored heat, but it prevents costly damages to the solar collectors.

Solar cooling of buildings is an attractive idea as the cooling loads and availability of solar radiation are in phase. Additionally, the combination of solar cooling and heating greatly improves the use factors of collectors compared to heating alone. Solar air conditioning can be accomplished by three types of systems: absorption cycles, adsorption (desiccant) cycles and solar mechanical processes. Some of these cycles are also used in solar refrigeration systems and are described in section 6.2.5.3. The rest of this section deals with solar heating and service hot water production. It should be noted that the same solar collectors are used for both space heating and cooling systems when both are present.

A review of the various solar heating and cooling systems is presented in Ref. [139]. A review of solar and low energy cooling technologies is presented in Ref. [140].

Space heating and service hot water

It is useful to consider solar systems as having five basic modes of operation, depending on the conditions that exist in the system at a particular time [97]:

1. If solar energy is available and heat is not needed in the building, energy gain from the collector is added to storage.
2. If solar energy is available and heat is needed in the building, energy gain from the collector is used to supply the building need.
3. If solar energy is not available, heat is needed in the building, and the storage unit has stored energy in it, the stored energy is used to supply the building need.
4. If solar energy is not available, heat is needed in the building, and the storage unit has been depleted, auxiliary energy is used to supply the building need.
5. The storage unit is fully heated, there are no loads to be met, and the collector is absorbing heat.

When the last mode occurs, there is no way to use or store the collected energy, and this energy must be discarded. This can be achieved through the operation of pressure relief valves or if the stagnant temperature will not be detrimental to the collector materials, the flow of fluids is turned off, and thus the collector temperature will rise until the absorbed energy is dissipated by thermal losses. This is more suitable to solar air collectors.

Additional operational modes can also be employed such as to provide service hot water. It is also possible to combine modes, i.e. to operate in more than one mode at a time. Moreover, many systems do not allow direct heating from solar collector to building, but always transfer heat from collector to storage whenever this is available and from storage to load whenever needed. In Europe solar heating systems for combined space and water heating are known as combisystems. The following sections describe the design of residential-scale installations.

Air systems

A schematic of a basic solar heating system using air as the heat transfer fluid, with pebble bed storage unit and auxiliary heating source is shown in Fig. 6.2-34. The various modes of operations are achieved by appropriate positioning of the dampers. In most air systems it is not practical to combine the modes of adding energy to and removing energy from storage at the same time. Auxiliary energy can be combined with energy supplied from collector or storage to top-up the air temperature in order to cover the building load. As shown in Fig. 6.2-34, it is possible to bypass the collector and storage unit when auxiliary alone is being used to provide heat. Fig. 6.2-35 shows a more detailed schematic of an air system. Blowers, controls, means of obtaining service hot water, and more details of ducting are shown.

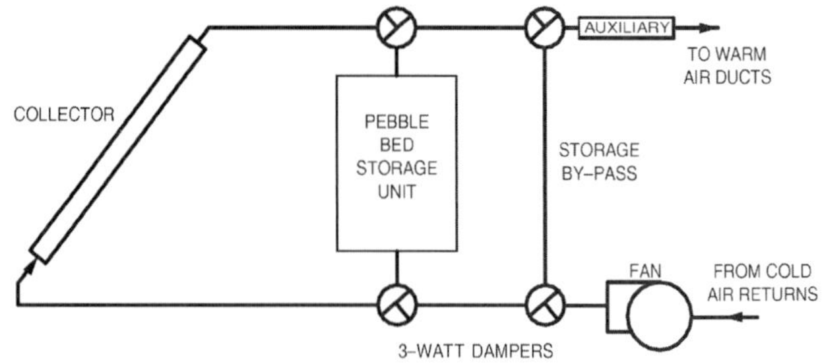

Fig. 6.2-34 Schematic of basic hot air system.

The advantages of using air as a heat transfer fluid have been outlined earlier in Air systems. Additionally, other advantages include the high degree of stratification possible in the pebble bed which leads to lower collector inlet temperatures. The working fluid is air, and warm air heating systems are in common use. Control equipment that can be applied to those systems is also readily available. Additional to the disadvantages of water heating air systems is the difficulty of adding solar air conditioning to the systems. Finally, air collectors are operated at lower fluid capacitance rates and thus with lower values of F_R than the liquid heating collectors.

Usually, air heating collectors in space heating systems are operated at fixed air flow rates; thus the outlet temperature varies through the day. It is also possible to operate them at a fixed outlet temperature by varying the flow rate. This, however, results in reduced F_R and thus reduced collector performance when flow rates are low.

Water systems

There are many variations of systems used for both solar space heating and service hot water production. The basic configuration is similar to that of the solar water heating systems outlined earlier. When used for both space and hot water production these systems allow independent control of the solar collector-storage and storage-auxiliary-load loops as solar-heated water can be added to storage at the same time that hot water is removed from storage to meet building loads. Usually, a bypass is provided around the storage tank to avoid

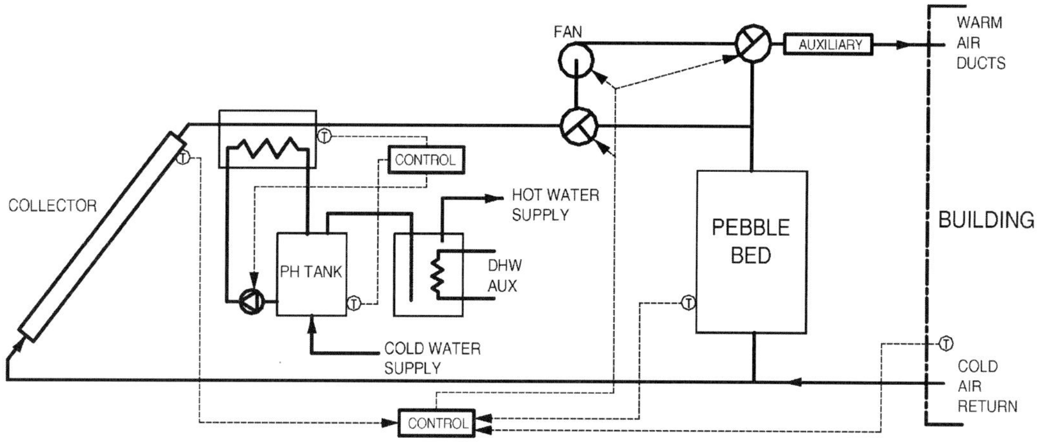

Fig. 6.2-35 Detail schematic of a solar air heating system.

heating the storage tank, which can be of considerable size, with auxiliary energy.

A detailed schematic of a liquid-based system is shown in Fig. 6.2-36 [97]. In this case a collector heat exchanger is shown between the collector and the storage tank, which allows the use of antifreeze solutions with the collector circuit. Relief valves are also required for dumping excess energy if the collector temperature reaches saturation. Means of extracting energy for service hot water are indicated. Auxiliary energy for heating is added so as to "top off" that available from solar energy system.

A load heat exchanger is shown in Fig. 6.2-36 to transfer energy from the tank to the air in the heated spaces. The load heat exchanger must be adequately designed to avoid excessive temperature drop and corresponding increase in the tank and collector temperatures.

Advantages of liquid heating systems include high collector F_R, smaller storage volume, and relatively easy adaptation to supply energy to absorption air conditioners.

Heat pump systems

Heat pumps use mechanical energy to transfer thermal energy from a source at a lower temperature to a sink at a higher temperature. Electrically driven heat pump heating systems have two advantages compared to electric resistance heating or expensive fuels. The heat pump's COP is high enough to yield 11 to 15 MJ of heat for each kW h of energy supplied to the compressor [21], which saves on purchase of energy, and is useful for air conditioning in the summer. Water-to-air heat pumps, which use solar heated water from the storage tank as the

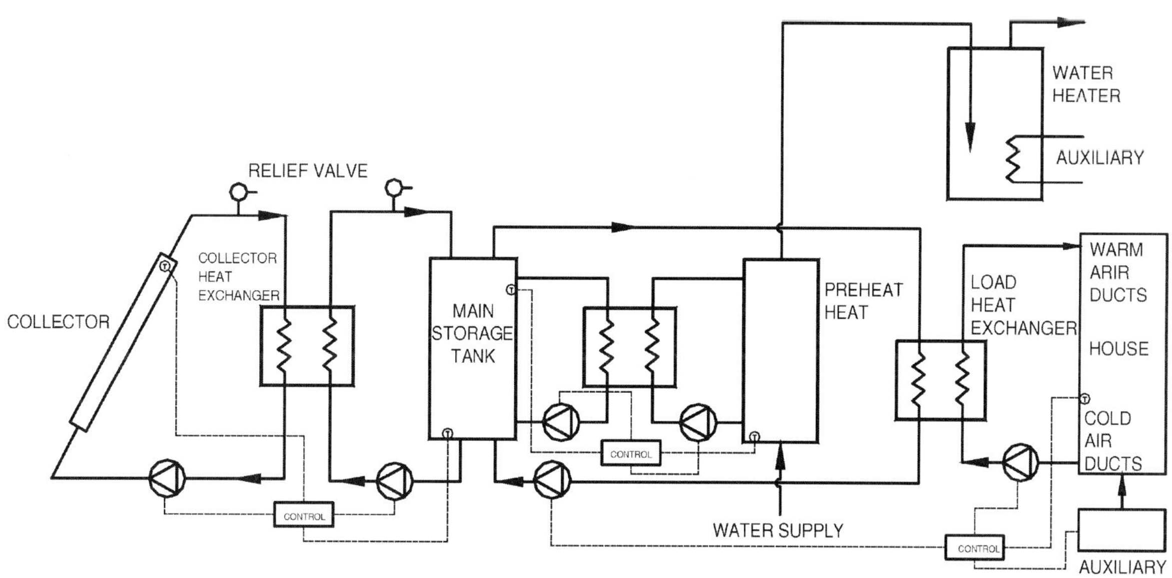

Fig. 6.2-36 Detail schematic of a solar water heating system.

evaporator energy source, are an alternative auxiliary heat source. Use of water involves freezing problems which need to be taken into consideration. Solar heating systems using liquids will operate at lower temperatures than conventional hydronic systems and will require more baseboard heater area to transfer heat into the building.

6.2.5.3 Solar refrigeration

Solar cooling can be considered for two related processes: to provide refrigeration for food and medicine preservation and to provide comfort cooling. Solar refrigeration systems usually operate at intermittent cycles and produce much lower temperatures (ice) than in air conditioning. When the same cycles are used in space cooling they operate on continuous cycles. The cycles employed for solar refrigeration are absorption and adsorption. During the cooling portion of the cycles, the refrigerant is evaporated and reabsorbed. In these systems the absorber and generator are separate vessels. The generator can be an integral part of the collector, with refrigerant absorbent solution in the tubes of the collector circulated by a combination of a thermosyphon and a vapour lift pump.

There are many options available which enable the integration of solar energy into the process of "cold" production. Solar refrigeration can be accomplished by using either a thermal energy source supplied from a solar collector or electricity supplied from photovoltaics. This can be achieved by using either thermal adsorption or absorption units or conventional refrigeration equipment powered from photovoltaics. Solar refrigeration is employed mainly to cool vaccine stores in areas with no mains electricity and for solar space cooling.

Photovoltaic refrigeration, although using standard refrigeration equipment, which is an advantage, has not achieved widespread use because of the low efficiency and high cost of the photovoltaic cells. As photovoltaics are not covered in this paper details are given only on the solar adsorption and absorption units with more emphasis on the latter.

Adsorption units

Porous solids, called adsorbents, can physically and reversibly adsorb large volumes of a vapour, called the adsorbate. Though this phenomenon, called solar adsorption, was recognised in the 19th century its practical application in the field of refrigeration is relatively recent. The concentration of adsorbate vapours in a solid adsorbent is a function of the temperature of the pair, i.e. the mixture of adsorbent and adsorbate, and the vapour pressure of the latter. The dependence of adsorbate concentration on temperature, under constant pressure conditions, makes it possible to adsorb or desorb the adsorbate by varying the temperature of the mixture. This forms the basis of the application of this phenomenon in the solar-powered intermittent vapour sorption refrigeration cycle.

An adsorbent–refrigerant working pair for a solar refrigerator requires the following characteristics:

1. A refrigerant with a large latent heat of evaporation.
2. A working pair with high thermodynamic efficiency.
3. A small heat of desorption under the envisaged operating pressure and temperature conditions.
4. A low thermal capacity.

Water–ammonia has been the most widely used sorption–refrigeration pair and research has been undertaken to utilise the pair for solar-operated refrigerators. The efficiency of such systems is limited by the condensing temperature, which cannot be lowered without introduction of advanced and expensive technology. For example, cooling towers or desiccant beds have to be used to produce cold water to condense ammonia at lower pressure. Amongst the other disadvantages inherent in using water and ammonia as the working pair are the heavy gauge pipe and vessel walls required to withstand the high pressure, the corrosiveness of ammonia, and the problem of rectification, i.e. removing water vapour from ammonia during generation. A number of different solid adsorption working pairs such as zeolite–water, zeolite–methanol, and activated carbon–methanol, have been studied in order to find the one that performs better. The activated carbon–methanol working pair was found to perform the best [19].

Because complete physical property data are available for only a few potential working pairs, the optimum performance remains unknown at the moment. In addition, the operating conditions of a solar-powered refrigerator, i.e. generator and condenser temperature, vary with its geographical location [19].

The development of three solar/biomass adsorption air conditioning refrigeration systems is presented by Critoph [141]. All systems use active carbon–ammonia adsorption cycles and the principle of operation and performance prediction of the systems are given.

Thorpe [142] presented an adsorption heat pump system which uses ammonia with granular active adsorbate. A high COP is achieved and the cycle is suitable for the use of heat from high temperature (150–200°C) solar collectors for air conditioning.

Absorption units

Absorption is the process of attracting and holding moisture by substances called desiccants. Desiccants are sorbents, i.e. materials that have an ability to attract and hold other gases or liquids, which have a particular affinity for water. During absorption the desiccant undergoes a chemical change as it takes on moisture; for example, the table salt, which changes from a solid to a liquid as it absorbs moisture. The characteristic of the binding of desiccants to moisture, makes the desiccants very useful in chemical separation processes [143].

Absorption systems are similar to vapour-compression air conditioning systems but differ in the pressurisation stage. In general an absorbent, on the low-pressure side, absorbs an evaporating refrigerant. The most usual combinations of fluids include lithium bromide-water ($LiBr–H_2O$) where water vapour is the refrigerant and ammonia–water ($NH_3–H_2O$) systems where ammonia is the refrigerant.

The pressurisation is achieved by dissolving the refrigerant in the absorbent, in the absorber section (Fig. 6.2-37). Subsequently, the solution is pumped to a high pressure with an ordinary liquid pump. The addition of heat in the generator is used to separate the low-boiling refrigerant from the solution. In this way the refrigerant vapour is compressed without the need of large amounts of mechanical energy that the vapour-compression air conditioning systems demand.

The remainder of the system consists of a condenser, expansion valve and evaporator, which function in a similar way as in a vapour-compression air conditioning system.

The $NH_3–H_2O$ system is more complicated than the $LiBr–H_2O$ system, since it needs a rectifying column that assures that no water vapour enters the evaporator where it could freeze. The $NH_3–H_2O$ system requires generator temperatures in the range of 125–170°C with air-cooled absorber and condenser and 95–120°C when water-cooling is used. These temperatures cannot be obtained with FPCs. The coefficient of performance (COP), which is defined as the ratio of the cooling effect to the heat input, is between 0.6 and 0.7.

The $LiBr–H_2O$ system operates at a generator temperature in the range of 70–95°C with water used as a coolant in the absorber and condenser and has COP higher than the $NH_3–H_2O$ systems. The COP of this system is between 0.6 and 0.8 [97]. A disadvantage of the $LiBr–H_2O$ systems is that their evaporator cannot operate at temperatures much below 5°C since the refrigerant is water vapour. Commercially available absorption chillers for air conditioning applications usually operate with a solution of lithium bromide in water and use steam or hot water as the heat source. In the market two types of chillers are available, the single and the double effect.

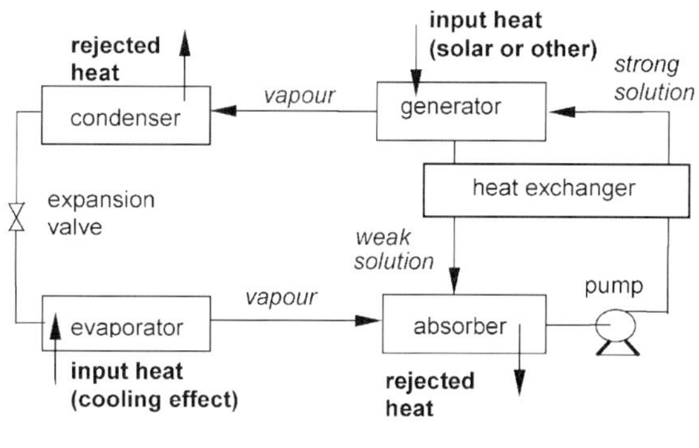

Fig. 6.2-37 Basic principle of the absorption air conditioning system.

The single effect absorption chiller is mainly used for building cooling loads, where chilled water is required at 6–7°C. The COP will vary to a small extent with the heat source and the cooling water temperatures. Single effect chillers can operate with hot water temperature ranging from about 80 to 150°C when water is pressurised [118].

The double effect absorption chiller has two stages of generation to separate the refrigerant from the absorbent. Thus the temperature of the heat source needed to drive the high-stage generator is essentially higher than that needed for the single-effect machine and is in the range of 155–205°C. Double effect chillers have a higher COP of about 0.9–1.2 [144]. Although double effect chillers are more efficient than the single-effect machines they are obviously more expensive to purchase. However, every individual application must be considered on its merits since the resulting savings in capital cost of the single-effect units can largely offset the extra capital cost of the double effect chiller.

The Carrier Corporation pioneered lithium–bromide absorption chiller technology in the United States, with early single-effect machines introduced around 1945. Due to the success of the product soon other companies joined the production. The absorption business thrived until 1975. Then the generally held belief that natural gas supplies were lessening, let to US government regulations prohibiting the use of gas in new constructions and together with the low cost of electricity led to the declination of the absorption refrigeration market [145]. Today the major factor on the decision on the type of system to install for a particular application is the economic trade-off between the different cooling technologies. Absorption chillers typically cost less to operate, but they cost more to purchase than vapour compression units. The payback period depends strongly on the relative cost of fuel and electricity assuming that the operating cost for the needed heat is less than the operating cost for electricity.

The technology was exported to Japan from the US early in the 1960s, and the Japanese manufacturers set a research and development program to improve further the absorption systems. The program led to the introduction of the direct-fired double-effect machines with improved thermal performance.

Today gas-fired absorption chillers deliver 50% of commercial space cooling load worldwide, but less than 5% in the US, where electricity-driven vapour compression machines carry the majority of the load [145].

Many researchers have developed solar assisted absorption refrigeration systems. Most of them have been produced as experimental units and computer codes were written to simulate the systems. Some of these designs are presented here.

Hammad and Audi [146] described the performance of a non-storage, continuous, solar operated absorption refrigeration cycle. The maximum ideal COP of the system was determined to be equal to 1.6, while the peak actual COP was determined to be equal to 0.55.

Haim et al. [147] performed a simulation and analysis of two open-cycle absorption systems. Both systems comprise a closed absorber and evaporator as in conventional single stage chillers. The open part of the cycle is the regenerator, used to reconcentrate the absorber solution by means of solar energy. The analysis was performed with a computer code developed for modular simulation of absorption systems under varying cycle configurations (open- and closed-cycle systems) and with different working fluids. Based on the specified design features, the code calculates the operating parameters in each system. Results indicate a definite performance advantage of the direct-regeneration system over the indirect one.

Hawlader et al. [148] developed a lithium bromide absorption cooling system employing an 11×11 m^2 collector/regenerator unit. They also have developed a computer model, which they validated against real experimental values with good agreement. The experimental results showed a regeneration efficiency varying between 38 and 67% and the corresponding cooling capacities ranged from 31 to72 kW.

Ameel et al. [149] give performance predictions of alternative low-cost absorbents for open cycle absorption using a number of absorbents. The most promising of the absorbents considered, was a mixture of two elements, lithium chloride and zinc chloride. The estimated capacities per unit absorber area were 50–70% less than those of lithium bromide systems.

Ghaddar et al. [150] presented modelling and simulation of a solar absorption system for Beirut. The results showed that, for each ton of refrigeration, it is required to have a minimum collector area of 23.3 m^2 with an optimum water storage capacity ranging from 1000 to 1500 l, for the system to operate solely on solar energy for about 7 h per day. The monthly solar fraction of total energy use in cooling is determined as a function of solar collector area and storage tank capacity. The economic analysis performed showed that the solar cooling system is marginally competitive only when it is combined with domestic water heating.

Erhard and Hahne [151] simulated and tested a solar-powered absorption cooling machine. The main part of the device is an absorber/desorber unit, which is mounted inside a concentrating solar collector. Results obtained from field tests are discussed and compared with the results obtained from a simulation program developed for this purpose.

Hammad and Zurigat [152] described the performance of a 1.5 ton solar cooling unit. The unit comprises a 14 m^2 flat-plate solar collector system and five shell and tube heat exchangers. The unit was tested in April and May in Jordan. The maximum value obtained for actual COP was 0.85.

Zinian and Ning [153] describe a solar absorption air conditioning system which uses an array of 2160 evacuated tubular collectors of total aperture area of 540 m^2 and a LiBr absorption chiller. Thermal efficiencies of the collector array are 40% for space cooling, 35% for space heating and 50% for domestic water heating. It was found that the cooling efficiency of the entire system is around 20%.

A new family of ICPC designs was developed by Winston et al. [154] which allows a simple manufacturing approach to be used and solves many of the operational problems of previous ICPC designs. A low concentration ratio is used that requires no tracking together with an off-the-shelf 20 ton double effect LiBr direct fired absorption chiller, modified to work with hot water. The new ICPC design and double effect chiller was able to produce cooling energy for the building using a collector field that was about half the size of that required for a more conventional collector and chiller.

A method to design, construct and evaluate the performance of a single stage lithium bromide–water absorption machine is presented in Ref. [155]. In this the necessary heat and mass transfer relations and appropriate equations describing the properties of the working fluids are specified. Information on designing the heat exchangers of the LiBr–water absorption unit is also presented. Single-pass vertical-tube heat exchangers have been used for the absorber and for the evaporator. The solution heat exchanger was designed as a single-pass annulus heat exchanger. The condenser and the generator were designed using horizontal tube heat exchangers.

6.2.5.4 Industrial process heat

Beyond the low temperature applications there are several potential fields of application for solar thermal energy at a medium and medium–high temperature level (80–240°C). The most important of them is heat production for industrial processes. The industrial heat demand constitutes about 15% of the overall demand of final energy requirements in the southern European countries. The 2000 demand in the EU for medium and medium-high temperatures was estimated to be about 300 T W h/yr [117].

From a number of studies on industrial heat demand, several industrial sectors have been identified with favourable conditions for the application of solar energy. The most important industrial processes using heat at a mean temperature level are: sterilising, pasteurising, drying, hydrolysing, distillation and evaporation, washing and cleaning, and polymerisation. Some of the most important processes and the range of the temperatures required for each are outlined in Ref. [20].

Large-scale solar applications for process heat benefit from the effect of scale. Therefore, the investment costs should be comparatively low, even if the costs for the collector are higher. One way to cause economically easy terms is to design systems without heat storage, i.e. the solar heat is fed directly into a suitable process (fuel saver). In this case the maximum rate at which the solar energy system delivers energy must not be appreciably larger than the rate at which the process uses energy. This system, however, cannot be cost-effective in cases where heat is needed at the early or late hours of the day or at nighttime when the industry operates on a double shift basis.

The types of industries that spend most energy are the food industry and the manufacture of non-metallic mineral products. Particular types of food industries which can employ solar process heat, are the milk and cooked pork meats (sausage, salami, etc.) industries and breweries. Most of the process heat is used in food and textile industry for such diverse applications as drying, cooking, cleaning, extraction and many others. Favourable conditions exist in the food industry, because food treatment and storage are processes with high energy consumption and high running time. Temperatures for these applications may vary from near ambient to those corresponding to low-pressure steam, and energy can be provided either from flat-plate or low concentration ratio concentrating collectors.

The principles of operation of components and systems outlined in the previous sections apply directly to industrial process heat applications. The unique features of the latter lie in the scale on which they are used, and the integration of the solar energy supply system with the auxiliary energy source and the industrial process.

The two primary problems that need to be considered when designing an industrial process heat application concern the type of energy to be employed and the temperature at which the heat is to be delivered. For example, if a process requires hot air for direct drying, an air heating system is probably the best solar energy system option. If hot water is needed for cleaning in food processing, the solar energy will be a liquid heater. If steam is needed to operate an autoclave or sterilizer, the solar energy system must be designed to produce steam probably with concentrating collectors. Another important factor in determination of the most suitable system for a particular application is the temperature of the fluid to the collector. Other requirements concern the fact that the energy may be needed at particular temperature or over a range of temperatures, and that possible sanitation requirements of the plant must also be met as for example in food processing applications.

The investments in industrial processes are generally large, and the transient and intermittent characteristics of solar energy supply are so unique that the study of options in solar industrial applications can be done by modelling methods at costs that are very small compared to the investments.

Many industrial processes use large amounts of energy in small spaces. If solar is to be considered for these applications, the location of collectors can be a problem. It may be necessary to locate the collector arrays on adjacent buildings or grounds, resulting in long runs of pipes or ducts. Where feasible, collectors can be mounted on the roof of a factory especially when no land area is available. In this case shading between adjacent collector rows should be avoided and considered. However, the collector area may be limited by roof area and orientation. Existing buildings are generally not designed or orientated to accommodate arrays of collectors, and in many cases structures to support collector arrays must be added to the existing structures. New buildings can be readily designed, often at little or no incremental cost, to allow for collector mounting and access.

In a solar process heat system, interfacing of the collectors with conventional energy supplies must be done in a way compatible with the process. The easiest way to accomplish this is by using heat storage, which can also allow the system to work in periods of low irradiation and/or nighttime.

The central system for heat supply in most factories uses hot water or steam at a pressure corresponding to the highest temperature needed in the different processes. Hot water or low pressure steam at medium temperatures (<150°C) can be used for preheating of water (or other fluids) used for processes (washing, dyeing, etc.) or for steam generation or by direct coupling of the solar system to an individual process working at temperatures lower than that of the central steam supply (Fig. 6.2-38). In the case of water preheating, higher efficiencies are obtained due to the low input temperature to the solar system; thus low-technology collectors can work effectively and the required load supply temperature has no or little effect on the performance of the solar system.

A number of research papers on the subject have been presented recently by a number of researchers. Norton

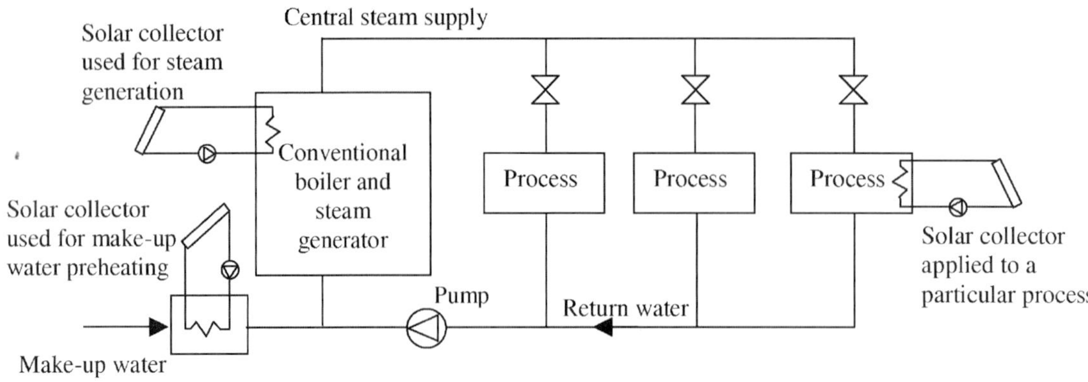

Fig. 6.2-38 Possibilities of combining a solar system with existing heat supply.

[156] presented the most common applications of industrial process heat. In particular the history of solar industrial and agricultural process applications were presented and practical examples were described.

A system for solar process heat for decentralised applications in developing countries was presented by Spate *et al.* [157]. The system is suitable for community kitchen, bakeries and post-harvest treatment. The system employs a fix-focus parabolic collector, a high temperature FPC and a pebble bed oil storage.

Benz *et al.* [158] presented the planning of two solar thermal systems producing process heat for a brewery and a dairy in Germany. In both industrial processes the solar yields were found to be comparable to the yields of solar systems for domestic solar water heating or space heating. In another paper, Benz *et al.* [159] presented a study for the application of non-concentrating collectors for the food industry in Germany. In particular, the planning of four solar thermal systems producing process heat for a large and a small brewery, a malt factory and a dairy, is presented. In the breweries, the washing machines for the returnable bottles were chosen as a suitable process to be fed by solar energy; in the dairy, the spray-dryers for milk and whey powder production; and in the malt factory, the wither and kiln processes. Up to 400 kW h/m^2 per annum were delivered from the solar collectors, depending on the type of collector.

Solar industrial air and water systems

There are two types of applications employing solar air collectors: open circuit, and recirculating applications. In open circuit, heated ambient air is used in industrial applications where, because of contaminants, recirculation of air is not possible. Examples are drying, paint spraying, and supplying fresh air to hospitals. It should be noted that heating of outside air is an ideal operation for the collector, as it operates very close to ambient temperature, thus more efficiently.

In recirculating air systems a mixture of recycled air from the dryer and ambient air is supplied to the solar collectors. Solar-heated air supplied to a drying chamber, can be applied to a variety of materials, including food crops, and lumber. In these applications, adequate control of the rate of drying, which can be obtained by controlling the temperature and humidity of the supply air, can lead to improved product quality.

Similarly, there are also two types of applications employing solar water collectors: once-through systems and recirculating water heating applications. The latter are very similar to domestic water heating systems presented earlier. Once-through systems are employed in cases where large quantities of water are used for cleaning in food industries, and recycling of used water is not practical because of the contaminants picked up by the water in the cleaning process.

Solar steam generation systems

PTC are frequently employed for solar steam generation, because relatively high temperatures can be obtained without any serious degradation in the collector efficiency. Low temperature steam can be used in industrial applications, sterilisation, and for powering desalination evaporators.

Three methods have been employed to generate steam using PTC [160]:

1. The steam-flash concept, in which pressurised water is heated in the collector and then flashed to steam in a separate vessel.
2. The direct or in situ concept, in which two-phase flow is allowed in the collector receiver so that steam is generated directly.
3. The unfired-boiler concept, in which a heat-transfer fluid is circulated through the collector and steam is generated via heat-exchange in an unfired boiler.

All these systems have certain advantages and disadvantages. In a steam-flash system, shown schematically in

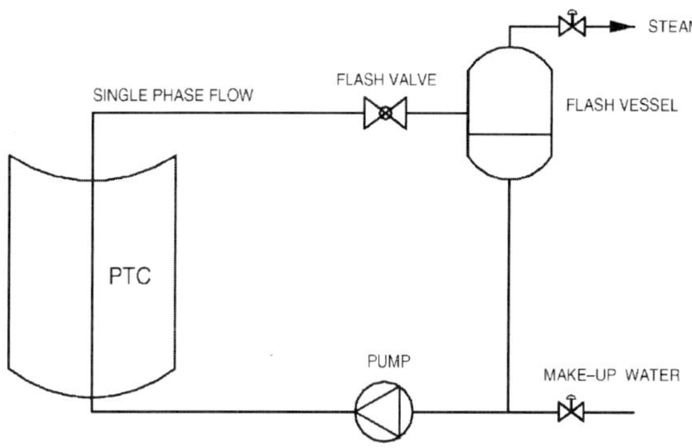

Fig. 6.2-39 The steam-flash steam generation concept.

Fig. 6.2-39, water, pressurised to prevent boiling, is circulated through the collector and then flashed across a throttling valve into a flash vessel. Treated feedwater input maintains the level in the flash vessel and the subcooled liquid is recirculated through the collector. The in situ boiling concept, shown in Fig. 6.2-40, uses a similar system configuration without a flash valve. Subcooled water is heated to boiling and steam forms directly in the receiver tube. Capital costs associated with a direct-steam and a flash-steam system would be approximately the same [161].

Although both systems use water, a superior heat transport fluid, the in situ boiling system is more advantageous. The flash system uses a sensible heat change in the working fluid, which makes the temperature differential across the collector relatively high. The rapid increase in water vapour pressure with temperature requires a corresponding increase in system operating pressure to prevent boiling. Increased operating temperatures reduce the thermal efficiency of the solar collector. Increased pressures within the system require a more robust design of collector components, such as receivers and piping. The differential pressure over the delivered steam pressure required to prevent boiling is supplied by the circulation pump and is irreversibly dissipated across the flash valve. When boiling occurs in the collectors, as in an in situ boiler, the system pressure drop and consequently, electrical power consumption, is greatly reduced. In addition, the latent heat-transfer process minimises the temperature rise across the solar collector. Disadvantages of in situ boiling are the possibility of a number of stability problems [162] and the fact that even with a very good feedwater treatment system, scaling in the receiver is unavoidable. In multiple row collector arrays, the occurrence of flow instabilities could result in loss of flow in the affected row. This in turn could result in tube dryout with consequent damage of the receiver selective coating. No significant instabilities were reported by Hurtado and Kast [161] when experimentally testing a single row 36 m system. Once through systems have been developed on a pilot scale for direct steam generation in which PTC have been used inclined at 2–4° [163].

A diagram of an unfired boiler system is shown in Fig. 6.2-41. In this system, the heat-transfer fluid should be non-freezing and non-corrosive, system pressures are low and control is straightforward. These factors largely overcome the disadvantages of water systems, and are the main reasons for the predominant use of heat-transfer oil systems in current industrial steam-generating solar systems.

The major disadvantages of the system result from the characteristics of the heat-transfer fluid. Such fluid is hard to contain, and most heat-transfer fluids are flammable. Decomposition, when the fluids are exposed to air, can greatly reduce ignition-point temperatures, and leaks into certain types of insulation can cause combustion at temperatures that are considerably lower than measured self-ignition temperatures. Heat-transfer fluids are also relatively expensive and present a potential pollution problem that makes them unsuitable for food industry applications [164]. Heat-transfer fluids have much poorer heat-transfer characteristics than water. They are more viscous at ambient temperatures, are less dense, and have lower specific heats and thermal conductivities than water. These characteristics mean that higher flow rates, higher collector differential temperatures, and greater pumping power are required to obtain the equivalent quantity of energy transport when compared to a system using water.

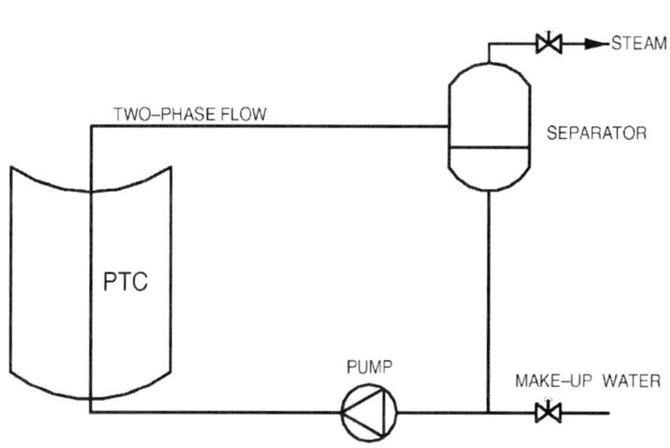

Fig. 6.2-40 The direct steam generation concept.

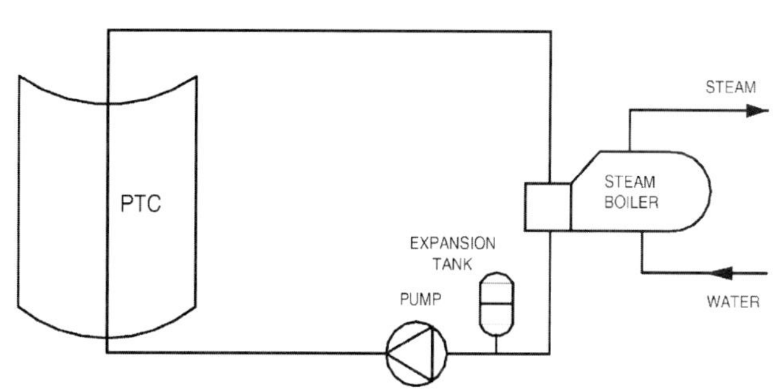

Fig. 6.2-41 The unfired-boiler steam generation concept.

In addition, heat-transfer coefficients are lower, so there is a larger temperature differential between the receiver tube and the collector fluid. Higher temperatures are also necessary to achieve cost effective heat exchange. These effects result in reduced collector efficiency.

For every application the suitable system has to be selected by taking into consideration all the above factors and constrains.

6.2.5.5 Solar desalination systems

Water is one of the most abundant resources on earth, covering three-fourths of the planet's surface. About 97% of the earth's water is salt water in the oceans; 3% of all fresh water is in ground water, lakes and rivers, which supply most of human and animal needs. Water is essential to life. The importance of supplying potable water can hardly be overstressed. Man has been dependent on rivers, lakes and underground water reservoirs for fresh water requirements in domestic life, agriculture and industry. However, rapid industrial growth and the population explosion all over the world have resulted in a large escalation of demand for fresh water. Added to this is the problem of pollution of rivers and lakes by industrial wastes and the large amounts of sewage discharged. On a global scale, manmade pollution of natural sources of water is becoming the single largest cause for fresh water shortage [17]. The only nearly inexhaustible sources of water are the oceans. Their main drawback, however, is their high salinity. It would be attractive to tackle the water-shortage problem with desalination of this water.

Desalination can be achieved by using a number of techniques. These may be classified into the following categories:

(i) phase-change or thermal processes; and

(ii) membrane or single-phase processes.

In Table 6.2-11, the most important technologies in use are listed. In the phase-change or thermal processes, the distillation of sea water is achieved by utilising a thermal energy source. The thermal energy may be obtained from a conventional fossil-fuel source, nuclear energy or from a non-conventional solar energy source. In the membrane processes, electricity is used either for driving high pressure pumps or for ionisation of salts contained in the sea water.

Table 6.2-11 Desalination processes

Phase-change processes	Membrane processes
(1) Multistage flash (MSF)	(1) Reverse osmosis (RO)
(2) Multiple effect boiling (MEB)	RO without energy recovery
(3) Vapour compression (VC)	RO with energy recovery (ER-RO)
(4) Freezing	(2) Electrodialysis (ED)
(5) Humidification/dehumidification	
(6) Solar stills	
Conventional stills	
special stills	
wick-type stills	
multiple-wick-type stills	

Desalination processes require significant quantities of energy to achieve separation. This is highly significant as it is a recurrent cost which few of the water-short areas of the world can afford. Many countries in the Middle East, because of oil income, have enough money to invest and run desalination equipment. People in many other areas of the world have neither the cash nor the oil resources to allow them to develop in a similar manner. It is estimated that the installed capacity of desalinated water systems in year 2000 is about 25 million m^3/day, which is expected to increase drastically in the next decades. The dramatic increase in desalinated water supply will create a series of problems, the most significant of which are those related to energy consumption. It has been estimated that the production of 25 million m^3 /day requires 230 million tons of oil per year. Even if oil were much more widely available, could we afford to burn it on the scale needed to provide everyone with fresh water? Given current understanding of the greenhouse effect and the importance of CO_2 levels, this use of oil is debatable. Thus, apart from satisfying the additional energy demand, environmental pollution would be a major concern. Fortunately, there are many parts of the world that are short of water but have exploitable renewable sources of energy that could be used to drive desalination processes.

Solar energy can be used for sea-water desalination either by producing the thermal energy required to drive the phase-change processes or by producing electricity required to drive the membrane processes. Solar desalination systems are thus classified into two categories, i.e. direct and indirect collection systems. As their name implies, direct collection systems use solar energy to produce distillate directly in the solar collector, whereas in indirect collection systems, two sub-systems are employed (one for solar energy collection and one for desalination). Conventional desalination systems are similar to solar systems since the same type of equipment is applied. The prime difference is that in the former, either a conventional boiler is used to provide the required heat or mains electricity is used to provide the required electric power, whereas in the latter, solar energy is applied.

A representative example of direct collection systems is the conventional solar still, which uses the greenhouse effect to evaporate salty water. It consists of a basin, in which a constant amount of seawater is enclosed in a veeshaped glass envelope. The sun's rays pass through the glass roof and are absorbed by the blackened bottom of the basin. As the water is heated, its vapour pressure is increased. The resultant water vapour is condensed on the underside of the roof and runs down into the troughs, which conduct the distilled water to the reservoir. The still acts as a heat trap because the roof is transparent to the incoming sunlight, but it is opaque to the infrared radiation emitted by the hot water (greenhouse effect). The roof encloses all of the vapour, prevents losses and, at the same time, keeps the wind from reaching the salty water and cooling it. The stills require frequent flushing, which is usually done during the night. Flushing is performed to prevent salt precipitation [165]. Design problems encountered with solar stills are brine depth, vapour tightness of the enclosure, distillate leakage, methods of thermal insulation, and cover slope, shape and material [165,166]. A typical still efficiency, defined as the ratio of the energy utilised in vaporising the water in the still to the solar energy incident on the glass cover, is 35% (maximum) and daily still production is about 34 l/m^2 [167]. The interested readers can find more details and a survey of indirect systems in Ref. [18]. For these systems a number of collectors ranging from stationary to low concentration ratio PTC can be used according to the temperature required by the desalination process. The usual temperature that the thermal desalination evaporators work is around 100°C. The use of PTC for seawater desalination is described in Ref. [168].

6.2.5.6 Solar thermal power systems

Conversion of solar to mechanical and electrical energy has been the objective of experiments for more than a century, starting from 1872 when Mouchot exhibited a steam-powered printing press at the Paris Exhibition. The idea is to use concentrating collectors to produce and supply steam to heat engines. A historical review of this and other experiments is given earlier. Much of the early attention to solar thermal–mechanical systems was for small scale applications (up to 100 kW) and most of them were designed for water pumping. Since 1975 there have been several large-scale power systems constructed and operated. Commercial plants of 30 and 80 MW electric (peak) generating capacity have been in operation for more than a decade.

The process of conversion of solar to mechanical and electrical energy by thermal means is fundamentally similar to the traditional thermal processes. These systems differ from the ones considered so far as these operate at much higher temperatures.

This section is concerned with generation of mechanical and electrical energy from solar energy by processes based mainly on concentrating collectors and heat engines. There are also another three kinds of power systems, which are not covered in this paper. These are the photovoltaic cells for the direct conversion of solar to electrical energy by solid state devices, solar-biological processes that produce fuels for operation of conventional engines or power plants and solar ponds.

The basic process for conversion of solar to mechanical energy is shown schematically in Fig. 6.2-42. Energy is collected by concentrating collectors, stored (if appropriate), and used to operate a heat engine. The main problem of these systems is that the efficiency of the collector is reduced as its operating temperature increases, whereas the efficiency of the heat engine increases as its operating temperature increases. The maximum operating temperature of stationary collectors is low relative to desirable input temperatures of heat engines, therefore concentrating collectors are used exclusively for such applications.

Identifying the best available sites for the erection of solar thermal power plants is a basic issue of project development. Recently the planning tool STEPS was developed by the German Aerospace Centre (DLR) [169], which uses satellite and Geographic Information System (GIS) data in order to select a suitable site. The factors taken into account are the slope of the terrain, land use (forest, desert, etc.), geomorphological features, hydrographical features, the proximity to infrastructure (power lines, roads, etc.) and of course solar irradiation of the area.

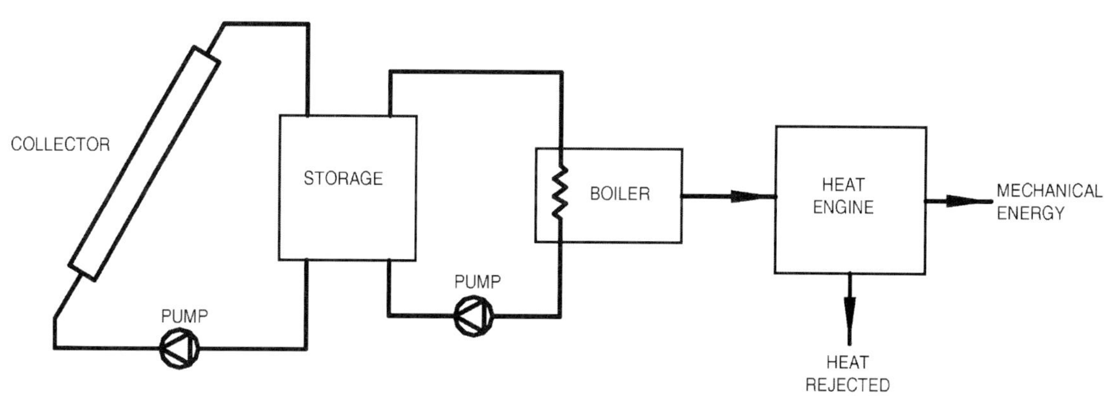

Fig. 6.2-42 Schematic of a solar-thermal conversion system.

Three system architectures have been used for such applications, the PTC system, the power tower system, and the dish system. These are described in this section.

Parabolic trough collector systems

Several parabolic trough solar thermal systems have been build and operated throughout the world. Most of these systems provide process steam to industry. They displace fossil fuels such as oil or natural gas as the energy source for producing steam. These systems incorporate fields of PTC having aperture areas from 500 to 5000 m^2. Most of these systems, however, supply industrial process steam from 150 to 200°C.

The most current example of power production using parabolic trough is the nine commercial solar energy generating systems (SEGS). The total installed capacity of SEGS is 354 MW and they are designed, installed and operated in the Mojave Desert of Southern California. These plants are based on large parabolic trough concentrators providing steam to Rankine power plants. The first of these plants is a 14 MW electric (MW_e) plant, the next six are 30 MW_e plants, and the two latest are 80 MW_e [65].

The plants can supply peaking power, using solely solar energy, solely natural gas, or a combination of the two, regardless of time or weather, within the constraint of the annual limit on gas use. The most critical time for power generation and delivery, and the time in which the selling price of the power per kW h is highest, is between noon and 6 p.m. in the months from June to September. Operating strategy is designed to maximise solar energy use. Natural gas is used to provide power during cloudy periods. The turbine-generator efficiency is best at full load, therefore the use of natural gas supplement to allow full-load operation maximises plant output.

A schematic of a typical plant is shown in Fig. 6.2-43. As it can be seen the solar and natural gas loops are in parallel to allow operation with either or both of the energy resources. The plants do not have energy storage facilities. The major components in the systems are the collectors, the fluid transfer pumps, the power generation system, the natural gas auxiliary subsystem, and the controls.

A synthetic heat transfer fluid is heated in the collectors and is piped to the solar steam generator and superheater where it generates the steam which drives the turbine. Reliable high-temperature circulating pumps are critical to the success of the plants, and significant engineering effort has gone into assuring that pumps will stand the high fluid temperatures and temperature cycling. The normal temperature of the fluid returned to the collector field is 304°C and that leaving the field is 390°C. Experience indicates that availability of the collector fields is about 99% [97].

The power generation system consists of a conventional Rankine cycle reheat steam turbine with feedwater heaters deaerators, etc. The condenser cooling water is cooled in forced draft cooling towers.

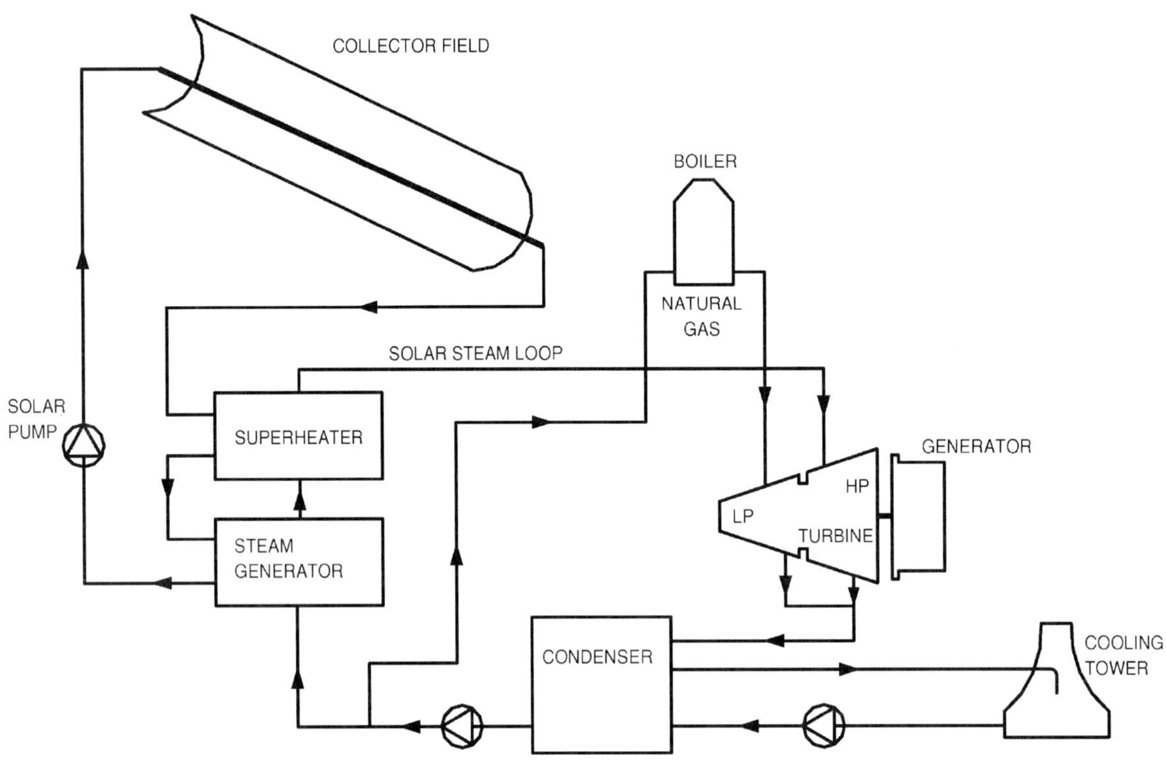

Fig. 6.2-43 Typical schematic of SEGS plant.

The reflectors are made of black-silvered, low-iron float-glass panels which are shaped over parabolic forms. Metallic and lacquer protective coatings are applied to the back of the silvered surface, and no measurable degradation of the reflective material has been observed [97]. The glass is mounted on truss structures, with the position of large arrays of modules adjusted by hydraulic drive motors. The reflectance of the mirrors is 0.94 when clean. Maintenance of high reflectance is critical to plant operation. With a total of 2.32×10^6 m^2 of mirror area, mechanised equipment has been developed for cleaning the reflectors, which is done regularly at intervals of about 2 weeks.

The receivers are 70 mm diameter steel tubes with cement selective surfaces surrounded by a vacuum glass jacket in order to minimise heat losses. The selective surfaces have an absorptance of 0.96 and an emittance of 0.19 at 350°C.

The collectors rotate about horizontal north–south axes, an arrangement which results in slightly less energy incident on them over the year but favours summertime operation when peak power is needed and its sale brings the greatest revenue. Tracking of the collectors is controlled by a system that utilises an optical system to focus radiation on two light-sensitive sensors. Any imbalance of radiation falling on the sensors causes corrections in the positioning of the collectors. There is a sensor and controller on each collector assembly; the resolution of the sensor is 0.5°.

A promising new configuration that combines SEGS parabolic-trough technology with a gas-turbine combined-cycle power plant is conceived to meet utility needs for continuous operation and peaking power with minimal environmental damage. Such a hybrid combined-cycle plant uses the solar field as the evaporation stage of an integrated system, with the gas-turbine exhaust being recycled for superheating and preheating; thus, the solar field serves as the boiler in an otherwise conventional combined-cycle plant. This approach has several advantages:

1. The direct steam generation system can take advantage of the steam turbine, generator, and other facilities of the combined-cycle plant at a modest increase in capital cost.
2. Adding the direct steam generation facility requires no additional operators or electrical interconnection equipment.
3. Thermodynamic efficiencies are maximized because steam is evaporated outside the waste-heat recovery system; only the remaining thermal-heat exchange processes take place in the recovery heat exchanger. Thus, higher working-steam conditions can be achieved for the same degree of heat use which increases overall cycle efficiency.

This new configuration is preferable from the perspective of the second law of thermodynamics because the solar field reduces the production of entropy in the system.

Power tower systems

In power tower systems, heliostats reflect and concentrate sunlight onto a central tower-mounted receiver where the energy is transferred to a heat transfer fluid. This energy is then passed either to storage or to power-conversion systems which convert the thermal energy into electricity and supply it to the grid.

The major components of the system are the heliostat field, the heliostat controls, the receiver, the storage system, and the heat engine which drives the generator. The heliostat design must ensure that radiation is delivered to the receiver at the desired flux density at minimum cost. Various receiver shapes have been considered, including cavity receivers and cylindrical receivers. The optimum shape is a function of the radiation intercepted and absorbed, thermal losses, receiver cost and design of the heliostat field. For a large heliostat field a cylindrical receiver has advantages when used with Rankine cycle engines, particularly for radiation from heliostats at the far edges of the field. Cavity receivers with larger tower height to heliostat field area ratios are used for higher temperatures required for the operation of Brayton cycle turbines.

As the collector represents the largest cost in the system an efficient engine is justified to obtain maximum useful conversion of the collected energy. Several possible thermodynamic cycles can be considered. Brayton or Stirling gas cycle engines operated at inlet temperatures of 800–1000°C provide high engine efficiencies, but are limited by low gas heat transfer coefficients and by practical constrains on collector design (i.e. the need for cavity receivers) imposed by the requirements of very high temperatures. Rankine cycle engines employing turbines driven from steam generated in the receiver at 500–550°C have several advantages over the Brayton cycle. Heat transfer coefficients in the steam generator are high, allowing the use of high energy densities and smaller receivers. Cavity receivers are not needed and cylindrical receivers that are usually employed permit larger heliostat fields to be used. The use of reheat cycles improves steam turbine performance, but entail mechanical design problems. Additionally, it is also possible to use steam turbines with steam generated from an intermediate heat transfer fluid circulated through the collector or boiler. With such systems the fluids could be molten salts or liquid metals, and cylindrical receivers could be operated at around 600°C. In fact, these indirect systems are the only ones that can be combined with thermal storage.

Power tower plants are defined by the options chosen for a heat transfer fluid, for the thermal storage medium and for the power-conversion cycle. The heat transfer fluid may be water/steam, molten nitrate salt, liquid metals or air. Thermal storage may be provided by phase change materials or ceramic bricks. Power tower systems usually achieve concentration ratios of 300–1500, can operate at temperature up to 1500°C, and are quite large, generally 10 MW_e or more.

Power tower systems currently under development use either nitrate salt or air as the heat transfer medium. In the USA, the Solar One plant in Barstow, CA was originally a water/steam plant and is now converted to Solar Two, a nitrate salt system. The use of nitrate salt for storage allows the plant to avoid tripping off line during cloudy periods and also allows the delivery of power after sunset. The heliostat system consists of 1818 individually oriented reflectors, each consisting of 12 concave panels with a total area of 39.13 m^2, for a total array of 71 100 m^2. The reflective material is back-silvered glass. The receiver is a single pass superheated boiler, generally cylindrical in shape, 13.7 m high, 7 m in diameter, with the top 90 m above the ground. It is an assembly of 24 panels, each 0.9 m wide and 13.7 m long. Six of the panels on the south side, which receives the least radiation, are used as feedwater preheaters and the balance is used as boilers. The panels are coated with a non-selective flat black paint which was heat cured in place with solar radiation. The receiver was designed to produce 50 900 kg/h of steam at 516°C with absorbing surface operating at a maximum temperature of 620°C [66].

Meanwhile the PHOEBUS consortium, a European industry group, is leading the way with air-based systems. Gaseous heat transfer media allow for significantly higher receiver outlet temperatures, but require higher operating pressures. Pressure-tolerant gas-cooled ceramic-tube receivers have, however, relatively high heat losses compared to water/steam or advance receivers. The PHOEBUS consortium is developing a novel Technology Solar Air (TSA) receiver, a volumetric air receiver which distributes the heat-exchanging surface over a three-dimensional volume and operates at ambient pressures. Because of its relative simplicity and safety, these plants can be used for applications in developing countries [170].

Future work will concentrate on the scaling up of the nitrate salt and TSA/PHOEBUS systems. The target size for nitrate salt plants in south-west USA is 100–200 MW_e, while a 30 MW_e plant is the aim for the PHOEBUS consortium. In addition to these two systems, a 20 MW Solgas plant, using a combined cycle plant with a solar power tower back-up, is planned for southern Spain [66].

Recent research and development efforts have focused on polymer reflectors and stretched-membrane heliostats. A stretched-membrane heliostat consists of a metal ring, across which two thin metal membranes are stretched. A focus control system adjusts the curvature of the front membrane, which is laminated with a silvered-polymer reflector, usually by adjusting the pressure (a very slight vacuum) in the plenum between the two membranes. Stretched-membrane heliostats are potentially much cheaper than glass/metal heliostats because they weigh less and have fewer parts.

Parabolic dish systems

A parabolic dish concentrates solar energy onto a receiver at its focal point. The receiver absorbs the energy and converts it into thermal energy. This can be used directly as heat or supply for power generation. The thermal energy can either be transported to a central generator for conversion, or it can be converted directly into electricity at a local generator coupled to the receiver.

Dishes track the sun on two axes, and thus they are the most efficient collector systems because they are always focused. Concentration ratios usually range from 600 to 2000, and they can achieve temperatures in excess of 1500°C. Rankine-cycle engines, Brayton-cycle engines, and sodium-heat engines have been considered for systems using dish-mounted engines; the greatest attention, though, was given to Stirling-engine systems.

Current developments in the USA and Europe are focused on 7.5 kW_e systems for remote applications. In Europe, three dish/Stirling systems are demonstrated at PSA in Spain, whereas in the USA a program has been set to demonstrate water pumping and village power applications [171]. Stretched-membrane concentrators are currently the focus of considerable attention because they are most likely to achieve the goals of low production cost and adequate performance. Both multifaceted and single-facet designs are being pursued. Recently, a 7-meter single-facet dish was developed, which demonstrated excellent performance in tests.

The greatest challenge facing distributed-dish systems is developing a power-conversion unit, which would have low capital and maintenance costs, long life, high conversion efficiency, and the ability to operate automatically. Several different engines, such as gas turbines, reciprocating steam engines, and organic Rankine engines, have been explored, but in recent years, most attention has been focused on Stirling-cycle engines. These are externally heated piston engines in which heat is continuously added to a gas (normally hydrogen or helium at high pressure) that is contained in a closed system. The gas cycles between hot and cold spaces in the engine store and release the heat that is added during expansion and rejected during compression.

6.2.5.7 Solar furnaces

Solar furnaces are made of high concentration and thus high temperature collectors of the parabolic dish and heliostat type. They are primarily used for material processing. Solar material processing involves affecting the chemical conversion of materials by their direct exposure to concentrated solar energy. A diverse range of approaches are being researched for applications related to high added-value products such as fullerenes, large carbon molecules with major potential commercial applications in semiconductors and superconductors, to commodity products such as cement [172]. None of these processes however, have achieved large-scale commercial adoption. Some pilot systems are shortly described here.

A solar thermochemical process has been developed by Steinfeld et al. [173] which combines the reduction of zinc oxide with reforming of natural gas leading to the co-production of zinc, hydrogen and carbon monoxide. At the equilibrium chemical composition in a black-body solar reactor operated at a temperature of 1250 K at atmospheric pressure with solar concentration of 2000, efficiencies between 0.4 and 0.65 have been found, depending on product heat recovery. A 5 kW solar chemical reactor has been employed to demonstrate this technology in a high-flux solar furnace. Particles of zinc oxide were introduced continuously in a vortex flow natural gas contained within a solar cavity receiver exposed to concentrated insolation from a heliostat field. The zinc oxide particles are exposed directly to the high radiative flux avoiding the inefficiencies and cost of heat exchangers.

A 2 kW concentrating solar furnace has been used to study the thermal decomposition of titanium dioxide at temperatures of 2300–2800 K in an argon atmosphere [174]. The decomposition rate was limited by the rate at which oxygen diffuses from the liquid–gas interface. It was shown that this rate is accurately predicted by a numerical model which couples the equations of chemical equilibrium and steady-state mass transfer [174].

6.2.5.8 Solar chemistry applications

Solar energy is essentially unlimited and its utilization is ecologically benign. However, solar radiation reaching the earth is intermittent and not distributed evenly. There is thus a need to store solar energy and transport it from the sunny uninhabited regions to the industrialized populated regions where energy is needed. The way to achieve this is by the thermochemical conversion of solar energy into chemical fuels. This method provides a thermochemically efficient path for storage and transportation. For this purpose high concentration ratio collectors similar to the ones used for power generation are employed. Thus by concentrating solar radiation in receivers and reactors, energy can be supplied to high-temperature processes to drive endothermic reactions. Solar energy can also assist in the processing of energy-intensive and high-temperature materials.

Applications include the solar reforming of low hydrocarbon fuels such as LPG and natural gas and upgrade it into a synthesis gas that can be used in gas turbines. Thus weak gas resources diluted with carbon dioxide can be used directly as feed components for the conversion process. Therefore, natural gas fields currently not exploited due to high CO_2 content might be opened to the market. Furthermore, gasification products of non-conventional fuels like biomass, oil shale and waste asphaltenes can also be fed into the solar upgrade process [175].

Other applications include the solar gasification of biomass and the production of solar aluminium the manufacture of which is one of the most energy intensive processes. Another interesting application is the solar zinc and syngas production which are both very valuable commodities. Zinc finds application in Zn/air fuel cells and batteries. Zinc can also react with water to form hydrogen which can be further processed for heat and electricity generation. Syngas can be used to fuel highly efficient combined cycles or can be used as the building block of a wide variety of synthetic fuels, including methanol, which is a very promising substitute of gasoline for fuelling cars [175].

A model for solar volumetric reactors for hydrocarbons reforming operation at high temperature and pressure is presented by Yehesket et al. [176]. The system is based on two achievements: the development of a volumetric receiver tested at 5000–10 000 suns, gas outlet temperature of 1200°C and pressure at 20 atm and a laboratory scale chemical kinetics study of hydrocarbons reforming. Other related applications are a solar driven ammonia based thermochemical energy storage system [177] and an ammonia synthesis reactor for a solar thermochemical energy storage system [178].

Another field of solar chemistry applications is the solar photochemistry. Solar photochemical processes make use of the spectral characteristics of the incoming solar radiation to effect selective catalytic transformations which find application in the detoxification of air and water and in the processing of fine chemical commodities.

In solar detoxification photocatalytic treatment of non-biodegradable persistent chlorinated water contaminants typically found in chemical production processes is achieved. For this purpose PTC with glass absorbers are employed and the high intensity of solar radiation is used for the photocatalytic decomposition of organic contaminants. The process uses ultraviolet (UV) energy, available in sunlight, in conjunction with the photocatalyst, titanium dioxide, to decompose organic chemicals into non-toxic compounds [179]. Another application concerns the development of a prototype employing lower concentration CPC [175]. Recent developments in photocatalytic detoxification and disinfection of water and air are presented by Goswami [180].

The development of a compound parabolic concentrator technology for commercial solar detoxification applications is given in Ref. [181]. The objective is to develop a simple, efficient and commercially competitive water treatment technology. A demonstration facility is planned to be erected by the project partners at PSA in Southern Spain.

6.2.6 Conclusions

Several of the most common types of solar collectors are presented in this paper. The various types of collectors described include flat-plate, compound parabolic, evacuated tube, parabolic trough, Fresnel lens, parabolic dish and Heliostat field collector (HFC). The optical, thermal and thermodynamic analysis of collectors is also presented as well as methods to evaluate their performance. Additionally, typical applications are described in order to show to the reader the extent of their applicability. These include water heating, space heating and cooling, refrigeration, industrial process heat, desalination, thermal power systems, solar furnaces and chemistry applications. It should be noted that the applications of solar energy collectors are not limited to the above areas. There are many other applications which are not described here either because they are not fully developed or are not matured yet. The application areas described in this paper show that solar energy collectors can be used in a wide variety of systems, could provide significant environmental and financial benefits, and should be used whenever possible.

References

[1] Kreith F, Kreider JF. Principles of solar engineering. New York: McGraw-Hill; 1978.

[2] Anderson B. Solar energy: fundamentals in building design. New York: McGraw-Hill; 1977.

[3] Dincer I. Energy and environmental impacts: present and future perspectives. Energy Sources 1998; 20(4/5):427–53.

[4] Dincer I. Renewable energy, environment and sustainable development. Proceedings of the World Renewable Energy Congress V, Florence, Italy; 1998. p. 2559–62.

[5] Rosen MA. The role of energy efficiency in sustainable development. Technol Soc 1996; 15(4):21–6.

[6] Dincer I, Rosen MA. A worldwide perspective on energy, environment and sustainable development. Int J Energy Res 1998;22(15):1305–21.

[7] www.worldwatch.org.

[8] Dincer I. Environmental impacts of energy. Energy Policy 1999;27(14): 845–54.

[9] Colonbo U. Development and the global environment. In: Hollander JM, editor. The energy-environment connection. Washington: Island Press; 1992. p. 3–14.

[10] Sayigh AAW. Renewable energy: global progress and examples. Renewable Energy 2001, WREN 2001;15–17.

[11] Johanson TB, Kelly H, Reddy AKN, Williams RH. Renewable fuels and electricity for a growing world economy. In: Johanson TB, Kelly H, Reddy AKN, Williams RH, editors. Renewable energy-sources for fuels and electricity. Washington, DC: Island Press; 1993. p. 1–71.

[12] Meinel AB, Meinel MP. Applied solar energy: an introduction. Reading, MA: Addison-Wesley; 1976.

[13] Kreider JF, Kreith F. Solar heating and cooling. New York: McGraw-Hill; 1977.

[14] SERI. Power from the Sun: principles of high temperature solar thermal technology; 1987.

[15] Kalogirou S. Solar water heating in Cyprus. Current status of technology and problems. Renewable Energy 1997;10: 107–12.

[16] Lysen E. Photovoltaics: an outlook for the 21st century. Renewable Energy World 2003;6(1):43–53.

[17] Malik MAS, Tiwari GN, Kumar A, Sodha MS. Solar distillation. New York: Pergamon Press; 1985.

[18] Kalogirou S. Survey of solar desalination systems and system selection. Energy: Int J 1997;22: 69–81.

[19] Norton B. Solar energy thermal technology. London: Springer; 1992.

[20] Kalogirou S. The potential of solar industrial process heat applications. Appl Energy 2003;76:337–61.

[21] ASHRAE. Handbook of HVAC Applications, Atlanta; 1995 [chapter 30].

[22] Spate F, Hafner B, Schwarzer K. A system for solar process heat for decentralised applications in developing countries. Proceedings of ISES Solar World Congress, Jerusalem, Israel on CD-ROM; 1999.

[23] Schweiger H. Optimisation of solar thermal absorber elements with transparent insulation. Thesis, Universitat Politecnica de Catalunya, Terrassa, Barcelona, Spain; 1997.

[24] Benz N, Hasler W, Hetfleish J, Tratzky S, Klein B. Flat-plate solar collector with glass TI. Proceedings of Eurosun'98 Conference on CD-ROM, Portoroz, Slovenia; 1998.

[25] Tripanagnostopoulos Y, Souliotis M, Nousia Th. Solar collectors with colored absorbers. Solar Energy 2000;68: 343–56.

[26] Wazwaz J, Salmi H, Hallak R. Solar thermal performance of a nickel-pigmented aluminium oxide selective absorber. Renewable Energy 2002;27:277–92.

[27] Orel ZC, Gunde MK, Hutchins MG. Spectrally selective solar absorbers in different non-black colours. Proceedings of WREC VII, Cologne on CD-ROM; 2002.

[28] Konttinen P, Lund PD, Kilpi RJ. Mechanically manufactured selective solar absorber surfaces. Solar Energy Mater Solar Cells 2003;79(3):273–83.

[29] Wackelgard E, Niklasson GA, Granqvist CG. Selective solar-absorbing coatings. In: Gordon J, editor. Solar energy: the state of the art. Germany: ISES; 2001. p. 109–44.

[30] Kreider JF. The solar heating design process. New York: McGraw-Hill; 1982.

[31] Close DJ. Solar air heaters. For low and moderate temperature applications. Solar Energy 1963; 7(3):117–24.

[32] Gupta CL, Garg HP. Performance studies on solar air heaters. Solar Energy 1967;11(1):25–31.

[33] Wijeydundera NE, Lee LAh, Tjioe LEk. Thermal performance study of two-pass solar air heaters. Solar Energy 1982; 28(5):363–70.

[34] Samuel TD. Heat withdrawal from multi-layer thermal trap collectors. Solar Energy 1983;30(3): 261–70.

[35] Biondi P, Cicala L, Farina G. Performance analysis of solar air heaters of conventional design. Solar Energy 1988;41(1): 101–7.

[36] Parker BF, Lindley MR, Colliver DG, Murphy WE. Thermal performance of three solar air heaters. Solar Energy 1993; 51(6): 467–79.

[37] Kolb A, Winter ERF, Viskanta R. Experimental studies on a solar air collector with metal matrix absorber. Solar Energy 1999;65(2):91–8

[38] Klein SA, Beckman WA, Duffie JA. A design procedure for solar air heating systems. Solar Energy 1977; 19:509–12.

[39] Hollands KGT, Shewen EC. Optimization of flow passage geometry for air-heating, plate-type solar collectors. J Solar Energy Engng 1981;103:323–30.

[40] Francia G. A new collector of solar radiant energy. UN Conf New Sources Energy, Rome 1961;4:572.

[41] Soltau H. Testing the thermal performance of uncovered solar collectors. Solar Energy 1992;49(4): 263–72.

[42] Molineaux B, Lachal B, Gusian O. Thermal analysis of five outdoor swimming pools heated by unglazed solar collectors. Solar Energy 1994; 53(1):21–6.

[43] Winter F. Twenty-year progress report on the copper development association do-it-yourself solar swimming pool heating manual and on the associated prototype heater. Solar Energy 1994;53(1): 33–6.

[44] Winston R. Solar concentrators of novel design. Solar Energy 1974;16: 89–95.

[45] Pereira M. Design and performance of a novel non-evacuated 1.2x CPC type concentrator. Proceedings of Intersol Biennial Congress of ISES, Montreal, Canada, vol. 2.; 1985. p. 1199- 204.

[46] Rabl A. Optical and thermal properties of compound parabolic collectors. Solar Energy 1976;18: 497-511.

[47] Mills DR, Giutronich JE. Asymmetrical non-imaging cylindrical solar concentrators. Solar Energy 1978;20:45–55.

[48] McIntire WR. Optimization of stationary nonimaging reflectors for tubular evacuated receivers aligned north south. Solar Energy 1980;24: 169–75.

[49] O'Gallagher JJ, Snail K, Winston R, Peek C, Garrison JD. A new evacuated CPC collector tube. Solar Energy 1982;29(6): 575–7.

[50] Ronnelid M, Perers B, Karlsson B. Construction and testing of a large-area CPC-collector and comparison with a flat plate collector. Solar Energy 1996;57(3):177-84.

[51] Rabl A, Goodman NB, Winston R. Practical design considerations for CPC solar collectors. Solar Energy 1979;22:373–81.

[52] Tripanagnostopoulos Y, Yianoulis P. CPC solar collectors with multichannel absorber. Solar Energy 1996;58(1–3): 49–61.

[53] Tripanagnostopoulos Y, Yianoulis P, Papaefthimiou S, Zafeiratos S. CPC collectors with flat bifacial absorbers. Solar Energy 2000;69(3): 191–203.

[54] Tripanagnostopoulos Y, Yianoulis P, Papaefthimiou S, Souliotis M, Nousia Th. Cost effective asymmetric CPC solar collectors. Renewable Energy 1999;16:628–31.

[55] Lin Q, Furbo S. Solar heating systems with evacuated tubular solar collector. Proceedings of the Eurosun'98 Conference on CD-ROM, Portoroz, Slovenia; 1998.

[56] Morrison GL. Solar collectors. In: Gordon J, editor. Solar energy: the state of the art. Germany: ISES; 2001. p. 145–221.

[57] Winston R, O'Gallagher J, Muschaweck J, Mahoney A, Dudley V. Comparison of predicted and measured performance of an integrated compound parabolic concentrator (ICPC). Proceedings of ISES Solar World Congress on CDROM, Jerusalem, Israel; 1999.

[58] Grass C, Benz N, Hacker Z, Timinger A. Tube collector with integrated tracking parabolic concentrator. Proceedings of the Eurosun'2000 Conference on CD-ROM, Copenhagen, Denmark; 2000.

[59] Kalogirou S, Eleftheriou P, Lloyd S, Ward J. Design and performance characteristics of a parabolic-trough solarcollector system. Appl Energy 1994;47:341–54.

[60] Garg HP, Hrishikesan DS. Enhancement of solar energy on flat-plate collector by plane booster mirrors. Solar Energy 1998;40(4): 295–307.

[61] Tabor H. Mirror boosters for solar collectors. Solar Energy 1966;10: 111–8.

[62] Seitel SC. Collector performance enhancement with flat reflectors. Solar Energy 1975;17:291 5.

[63] Perers B, Karlsson B, Bergkvist M. Intensity distribution in the collector plane from structured booster reflectors with rolling grooves and corrugations. Solar Energy 1994;53(2): 215–26.

[64] Kalogirou S. Solar energy utilisation using parabolic trough collectors in Cyprus. MPhil Thesis. The Polytechnic of Wales; 1991.

[65] Kearney DW, Price HW. Solar thermal plants-LUZ concept (current status of the SEGS plants). Proceedings of the Second Renewable Energy Congress, Reading UK, vol. 2.; 1992. p. 582–8.

[66] Grasse W. Solar PACES Annual Report, DLR Germany; 1995.

[67] Kalogirou S, Eleftheriou P, Lloyd S, Ward J. Low cost high accuracy parabolic troughs: construction and evaluation. Proceedings of the world renewable energy congress III, Reading, UK, vol. 1.; 1994. p. 384–6.

[68] Lupfert E, Geyer M, Schiel W, Zarza E, Gonzalez-Anguilar RO, Nava P. Eurotrough: a new parabolic trough collector with advanced light weight structure. Proceedings of Solar Thermal 2000 International Conference, on CD-ROM, Sydney, Australia; 2000.

[69] Geyer M, Lupfert E, Osuna R, Esteban A, Schiel W, Schweitzer A, Zarza E, Nava P, Langenkamp J, Mandelberg E. Eurotrough: parabolic trough collector developed for cost efficient solar power generation. Proceedings of 11th Solar PACES International Symposium on Concentrated Solar Power and Chemical Energy Technologies on CD-ROM, Zurich, Switzerland; 2002.

[70] Bakos GC, Adamopoulos D, Tsagas NF., Soursos M. Design and construction of a line-focus parabolic trough solar concentrator for electricity generation. Proceedings of ISES Solar World Congress on CD-ROM, Jerusalem, Israel; 1999.

[71] Kalogirou S. Parabolic trough collector system for low temperature steam generation: design and performance characteristics. Appl Energy 1996; 55(1):1–19.

[72] Kupta KC, Mirakhur PK, Sathe AP. A simple solar tracking system. SUN, Proceedings of the International Solar Energy Society, New Delhi, India: Pergamon Press; 1978.

[73] Cope NA, Ingley HA, Farber EA, Morrison CA. Dynamic response analysis of a solar powered heliotropic fluid mechanical drive system. University of Florida; 1981.

[74] Singh TAK, Dinesh PS. Liquid vapour balance based sun tracking system. Proceedings of the 25th National Renewable Energy Convention 2001 of the Solar Society of India, Warangal, India; 2001. p. 401–6.

[75] Hession PJ, Bonwick WJ. Experience with a Sun tracker system. Solar Energy 1984; 32:311.

[76] Zogbi R, Laplaze D. Design and construction of a Sun tracker. Solar Energy 1984;33:369–72.

[77] Mori Y, Hijikata K, Himero N. Fundamental research on heat transfer performances of solar focusing and tracking collector. Solar Energy 1977;19:595–600.

[78] Boultinghouse KD. Development of a solar flux tracker for parabolic trough collectors. Albuquerque, USA: Sandia National Labs; 1982.

[79] Briggs F. Tracking-refinement modelling for solar-collector control. Albuquerque, USA: Sandia National Labs; 1980.

[80] Nuwayhid RY, Mrad I, Abu-Said R. The realisation of a simple solar tracking concentrator for university research applications. Renewable Energy 2001;24:207–22.

[81] Kalogirou SA. Design and construction of a one-axis Suntracking mechanism. Solar Energy 1996;57(6):465–9.

[82] Kruger D, Heller A, Hennecke K, Duer K. Parabolic trough collectors for district heating systems at high latitudes: a case study. Proceedings of Eurosun'2000 on CD ROM, Copenhagen, Denmark; 2000.

[83] Dudley V. SANDIA Report test results for industrial solar technology parabolic trough solar collector. SAND94-1117, Albuquerque, USA: Sandia National Laboratory; 1995.

[84] Riffelmann KJ, Fend Th, Pitz-Paal R. Parabolic trough collector efficiency improvement activities. In: Kreetz H, Lovegrove K, Meike W, editors. 10th International Symposium—Solar PACES—Solar Thermal Concentrating Technologies, Sydney, Australia. 2000. p. 121–9.

[85] Francia G. Pilot plants of solar steam generation systems. Solar Energy 1968;12:51–64.

[86] Mills DR. Solar thermal electricity. In: Gordon J, editor. Solar energy: the state of the art. Germany: ISES; 2001. p. 577–651.

[87] Nelson DT, Evans DL, Bansal RK. Linear Fresnel lens concentrators. Solar Energy 1975;17:285–9.

[88] Collares-Pereira M. High temperature solar collector with optimal concentration: non-focusing Fresnel lens with secondary concentrator. Solar Energy 1979;23: 409–20.

[89] Kritchman EM, Friesem AA, Yekutieli G. Efficient Fresnel lens for solar concentration. Solar Energy 1989;22:119–23.

[90] Lorenzo E, Minano JC. Design of one-axis tracked linear Fresnel lenses. Solar Energy 1986;36(6): 531–4.

[91] Feuermann D, Gordon JM. Analysis of a two-stage linear Fresnel reflector solar concentrator. ASME J Solar Energy Engng 1991;113: 272–9.

[92] Pitz-Paal R. Concentrating solar technologies: the key to renewable electricity and process heat for a wide range of applications. Proceedings of the World Renewable Energy Congress VII on CD-ROM, Cologne, Germany; 2002.

[93] Romero M, Buck R, Pacheco JE. An update on solar central receiver systems projects and technologies. J Solar Energy Engng 2002;124(2): 98-108.

[94] Schwarzbözl P, Pitz-Paal R, Meinecke W, Buck R. Costoptimized solar gas turbine cycles using volumetric air receiver technology. Proceedings of the Renewable Energy for the New Millennium, Sydney, Australia; 2000. p. 171–7.

[95] Chavez JM, Kolb GJ, Meinecke W. In: Becker M, Klimas PC, editors. Second generation central receiver technologies: a status report. Karlsruhe, Germany: Verlag; 1993.

[96] Prapas DE, Norton B, Probert SD. Optics of parabolic trough solar energy collectors possessing small concentration ratios. Solar Energy 1987;39:541–50.

[97] Duffie JA, Beckman WA. Solar engineering of thermal processes. New York: Wiley; 1991.

[98] Klein SA. Calculation of flat plate collector loss coefficients. Solar Energy 1975;17:79–80.

[99] Sodha MS, Mathur SS, Malik MAS, Reviews of renewable energy resources, vol. 2. New York: Wiley; 1984.

[100] Jeter MS. Geometrical effects on the performance of trough collectors. Solar Energy 1983;30: 109–13.

[101] Guven HM, Bannerot RB. Derivation of universal error parameters for comprehensive optical analysis of parabolic troughs.

Proceedings of the ASME-ISES Solar Energy Conference, Knoxville, USA; 1985. p. 168–74.
[102] Guven HM, Bannerot RB. Determination of error tolerances for the optical design of parabolic troughs for developing countries. Solar Energy 1986;36:535–50.
[103] Jeter MJ. Calculation of the concentrated flux density distribution in parabolic trough collectors by a semi-finite formulation. Solar Energy 1986; 37(5):335–45.
[104] Karimi A, Guven HM, Thomas A. Thermal analysis of direct steam generation in parabolic trough collectors. Proceedings of the ASME Solar Energy Conference; 1986. p. 458–64.
[105] Bejan A, Kearney DW, Kreith F. Second law analysis and synthesis of solar collector systems. J Solar Energy Engng 1981;103:23–8.
[106] Bejan A. Entropy generation minimization. 2nd ed. Boca Raton: CRC Press; 1995. [chapter 9].
[107] Petela R. Exergy of heat radiation. ASME J Heat Transfer 1964;68:187.
[108] ASHRAE Standard 93. Methods of testing to determine the thermal performance of solar collectors, ANSI/ASHRAE 93-1986; 1986.
[109] ISO 9806-2, Solar energy-test methods for solar energy collectors. Part 2. Qualification test procedures. Geneva: International Standards Organisation; 1995.
[110] Kalogirou S, Papamarcou C. Modelling of a thermosyphon solar water heating system and simple model validation. Renewable Energy 2000;21(3/4):471–93.
[111] TRNSYS Program Manual. Solar Energy Laboratory, University of Wisconsin, Madison, USA; 1996.
[112] Kreider JF, Kreith F. Solar energy handbook. New York: McGrawHill; 1981.
[113] Beckman WA. Modern computing methods in solar energy analysis. Proceedings of Eurosun'98 on CD-ROM, Portoroz, Slovenia; 1998.
[114] Oishi M, Noguchi T. The evaluation procedure on performance of SDHW system by TRNSYS simulation for a yearly performance prediction. Proceedings of Eurosun'2000 on CD-ROM, Copenhagen, Denmark; 2000.
[115] Jordan U, Vajen K. Influence of the DHW load profile on the fractional energy savings: a case study of a solar combisystem with TRNSYS simulations. Proceedings of Eurosun' 2000 on CD-ROM, Copenhagen, Denmark; 2000.
[116] Benz N, Gut M, Belkircher Th, Russ W. Solar process heat with non-concentrating collectors for food industry. Proceedings of ISES Solar world Congress on CD-ROM, Jerusalem, Israel; 1999.
[117] Schweiger H, Mendes JF, Benz N, Hennecke K, Prieto G, Gusi M, Goncalves H. The potential of solar heat in industrial processes. a state of the art review for Spain and Portugal. Proceedings of Eurosun'2000 Copenhagen, Denmark on CD-ROM; 2000.
[118] Florides G, Kalogirou S, Tassou S, Wrobel L. Modelling and simulation of an absorption solar cooling system for Cyprus. Solar Energy 2002; 72(1):43–51.
[119] Watsun Users Manual and Program Documentation. Watsun Simulation Laboratory, University of Waterloo, Canada; 1992.
[120] Polysun. User's manual for Polysun 3.3, SPF, Switzerland; 2000.
[121] Gantner M. Dynamische simulation thermischer solaranlagen. Diploma Thesis, Hochschule für Technik Rapperswil (HSR), Switzerland; 2000.
[122] Beckman WA, Klein SA, Duffie JA. Solar heating design by the *f*-chart method. New York: Wiley/ Interscience; 1977.
[123] Klein SA, Beckman WA. *F*-Chart User's Manual, University of Wisconsin; 1981.
[124] Datta G, Broman L, Garg HP. Optimisation of collector tilts using *F*-chart method. Proceedings of the Second World Renewable Energy Congress, Reading, UK; 1992. p. 1051– 5.
[125] Kalogirou S, Lloyd S. Use of solar parabolic trough collectors for hot water production in Cyprus. A feasibility study. Renewable Energy 1992;2(2):117–24.
[126] Kalogirou S. Artificial neural networks in renewable energy systems: a review. Renewable Sustainable Energy Rev 2001; 5(4): 373–401.
[127] Kalogirou S. Economic analysis of solar energy systems using spreadsheets. Proceedings of the World Renewable Energy Congress IV, Denver, Colorado, USA, vol. 2; 1996. p. 1303–7.
[128] Morrison G, Wood B. Packaged solar water heating technology: twenty years of progress. Proceedings of ISES Solar World Congress on CD-ROM, Jerusalem, Israel; 1999.
[129] Gupta GL, Garg HP. System design in solar water heaters with natural circulation. Solar Energy 1968;12: 163–82.
[130] Ong KS. A finite difference method to evaluate the thermal performance of a solar water heater. Solar Energy 1974;16: 137–47.
[131] Ong KS. An improved computer program for the thermal performance of a solar water heater. Solar Energy 1976;18: 183–91.
[132] Kudish AI, Santamaura P, Beaufort P. Direct measurement and analysis of thermosyphon flow. Solar Energy 1985;35: 167–73.
[133] Morrison GL, Braun JE. System modelling and operation characteristics of thermosyphon solar water heaters. Solar Energy 1985;34:389–405.
[134] Hobson PA, Norton B. A design nomogram for direct thermosyphon solar energy water heaters. Solar Energy 1989;43:89–95.
[135] Shariah AM, Shalabi B. Optimal design for a thermosyphon solar water heater. Renewable Energy 1997;11:351–61.
[136] Tripanagnostopoulos Y, Souliotis M, Nousia Th. CPC type integrated collector storage systems. Solar Energy 2002; 72(4):327–50.
[137] Kalogirou S. Design, construction, performance evaluation, and economic analysis of an integrated collector storage system. Renewable Energy 1997;12(2):179–92.
[138] McIntire WR. Truncation of nonimaging cusp concentrators. Solar Energy 1979;23:351–5.
[139] Hahne E. Solar heating and cooling. Proceedings of Eurosun'96, Freiburg, Germany, vol. 1; 1996. p. 3–19.
[140] Florides G, Tassou S, Kalogirou S, Wrobel L. Review of solar and low energy cooling technologies for buildings. Renewable Sustainable Energy Rev 2002;6(6):557–72.

[141] Critoph RE. Development of three solar/biomass adsorption air conditioning refrigeration systems. Proceedings of the World Renewable Energy Congress VII on CD-ROM, Cologne, Germany; 2002.

[142] Thorpe R. Progress towards a highly regenerative adsorption cycle for solar thermal powered air conditioning. Proceedings of the World Renewable Energy Congress VII on CD-ROM, Cologne, Germany; 2002.

[143] ASHRAE. Handbook of fundamentals, Atlanta; 1989.

[144] Dorgan CB, Leight SP, Dorgan CE. Application guide for absorption cooling/refrigeration using recovered heat. American Society of Heating, Refrigerating and Air Conditioning Engineers, Inc.; 1995.

[145] Keith EH. Design challenges in absorption chillers. Mech Engng: CIME 1995;117(10):80–4.

[146] Hammad MA, Audi MS. Performance of a solar LiBr water absorption refrigeration system. Renewable Energy 1992; 2(3):275–82.

[147] Haim I, Grossman G, Shavit A. Simulation and analysis of open cycle absorption systems for solar cooling. Solar Energy 1992;49(6): 515–34.

[148] Hawlader MNA, Noval KS, Wood BD. Unglazed collector/regenerator performance for a solar assisted open cycle absorption cooling system. Solar Energy 1993;50(1):59–73.

[149] Ameel TA, Gee KG, Wood BD. Performance predictions of alternative, Low cost absorbents for open-cycle absorption solar cooling. Solar Energy 1995;54(2):65–73.

[150] Ghaddar NK, Shihab M, Bdeir F. Modelling and simulation of solar absorption system performance in Beirut. Renewable Energy 1997; 10(4):539–58.

[151] Erhard A, Hahne E. Test and simulation of a solar-powered absorption cooling machine. Solar Energy 1997;59(4-6):155–62.

[152] Hammad M, Zurigat Y. Performance of a second generation solar cooling unit. Solar Energy 1998;62(2):79–84.

[153] Zinian HE, Ning Z. A solar absorption air-conditioning plant using heat-pipe evacuated tubular collectors. Proceedings of ISES Solar World Congress on CD-ROM, Jerusalem, Israel; 1999.

[154] Winston R, O'Gallagher J, Duff W, Henkel T, Muschaweck J, Christiansen R, Bergquam J. Demonstration of a new type of icpc in a double-effect absorption cooling system. Proceedings of ISES Solar World Congress on CD-ROM, Jerusalem, Israel; 1999.

[155] Florides G, Kalogirou S, Tassou S, Wrobel L. Design and construction of a lithium bromide water absorption machine. Energy Conversion Mgmt 2003;44(15): 2483–508.

[156] Norton B. Solar process heat: distillation, drying, agricultural and industrial uses. Proceedings of ISES Solar World Congress, Jerusalem, Israel on CD-ROM, Jerusalem, Israel; 1999.

[157] Spate F, Hafner B, Schwarzer K. A system for solar process heat for decentralised applications in developing countries. Proceedings of ISES Solar World Congress on CD-ROM, Jerusalem, Israel; 1999.

[158] Benz N, Gut M, Rub W. Solar process heat in breweries and dairies. Proceedings of EuroSun 98, Portoroz, Slovenia on CD-ROM; 1998.

[159] Benz N, Gut M, Beikircher T. Solar process heat with nonconcentrating collectors for food industry. Proceedings of ISES Solar World Congress on CD ROM, Jerusalem, Israel; 1999.

[160] Kalogirou S, Lloyd S, Ward J. Modelling optimisation and performance evaluation of a parabolic trough collector steam generation system. Solar Energy 1997;60(1):49–59.

[161] Hurtado P, Kast M. Experimental study of direct in-situ generation of steam in a line focus solar collector, SERI; 1984.

[162] Peterson RJ, Keneth E. Flow instability during direct steam generation in line-focus solar collector system, SERI/TR-1354; 1982.

[163] Zarza E, Hennecke K, Coebel O. Project DISS (Direct Solar Steam) update on project status and future planning. Proceedings of ISES Solar World Congress on CD-ROM, Jerusalem, Israel; 1999.

[164] Murphy LM, Keneth E. Steam generation in line-focus solar collectors: a comparative assessment of thermal performance, operating stability, and cost issues. SERI/TR-1311; 1982.

[165] Mustacchi C, Cena V. Solar desalination: design, performances, economics, Sogesta; 1981.

[166] Eibling JA, Talbert SG, Lof GOG. Solar stills for community use: digest of technology. Solar Energy 1971;13:263.

[167] Daniels F. Direct use of the sun's energy. 6th ed. London: Yale University Press; 1974 [chapter 10].

[168] Kalogirou S. Use of parabolic trough solar energy collectors for sea-water desalination. Appl Energy 1998; 60(2):65–88.

[169] Kronshage S, Schillings C, Trieb F. Country analysis for solar thermal power stations using remote sensing methods. Proceedings of the World Renewable Energy Congress VII on CD-ROM, Cologne, Germany; 2002.

[170] Meurer C, Barthels H, Brocke WA, Emonts B, Groehn HG. PHOEBUS: an autonomous supply system with renewable energy-six years of operational experience and advanced concepts. Proceedings of ISES Solar World Congress on CD-ROM, Jerusalem, Israel; 1999.

[171] Solar Thermal Power and Solar Chemical Energy Systems, Solar PACES Program of the International Energy Agency. Birmingham, UK: The Franklin Company Consultants Ltd; 1994.

[172] Norton B. Solar process heat. In: Gordon J, editor. Solar energy: the state of the art. Germany: ISES; 2001. p. 477–96.

[173] Steinfeld A, Larson C, Palumbo R, Foley M. Thermodynamic analysis of the co-production of zinc and synthesis gas using solar process heat. Energy 1996;21:205–22.

[174] Palumbo R, Rouanet A, Pichelin G. Solar thermal decomposition of TiO_2 at temperatures above 2200 K and its use in the production of Zn and ZnO. Energy 1995;20:857–68.

[175] Grasse W. Solar PACES Annual Report. DLR Germany; 1998.

[176] Yehesket J, Rubin R, Berman A, Karni J. Chemical kinetics of high temperature hydrocarbons reforming using a solar reactor. Proceedings of Eurosun'2000 on CD-ROM, Copenhagen, Denmark; 2000.

[177] Lovegrove K, Luzzi A, Kreetz H. A solar driven ammonia based thermochemical energy storage system. Proceedings of ISES Solar

World Congress on CD-ROM, Jerusalem, Israel; 1999.

[178] Kreetz H, Lovegrove K. Theoretical analysis and experimental results of a 1 kW_{chem} ammonia synthesis reactor for a solar thermochemical energy storage system. Proceedings of ISES Solar World Congress on CD-ROM, Jerusalem, Israel; 1999.

[179] Mehos M, Turchi C, Pacheco J, Boegel AJ, Merrill T, Stanley R. Pilot-scale study of the solar detoxification of VOC-contaminated groundwater, NREL/TP-432-4981; 1992.

[180] Goswami DY. Recent developments in photocatalytic detoxification and disinfection of water and air. Proceedings of ISES Solar World Congress on CD-ROM, Jerusalem, Israel; 1999.

[181] Blanco J, Malato S, Fernandez P, Vidal A, Morales A, Trincado P, Oliveira J, Minero C, Musci M, Casalle C, Brunotte M, Tratzky S, Dischinger N, Funken K, Sattler C, Vincent M, Collares-Pereira M, Mendes J, Rangel C. Compound parabolic concentrator technology development to commercial solar detoxification applications. Proceedings of ISES Solar World Congress on CD-ROM, Jerusalem, Israel; 1999.

Section Seven

Ocean, wave and tidal power

Chapter 7.1

7.1

Ocean power

7.1.1 Ocean thermal energy conversion

OTEC relies on the principle exploited in most forms of electricity generation that a source of heat and a source of cold[1] can be used to drive an engine. In the case of OTEC the source of heat is the surface of a tropical or subtropical sea while the source of cold is the deep sea.

The possibility of extracting energy from the sea in this way was recognised in the latter half of the nineteenth century and the first practical system was proposed by the French inventor D'Arsonval in 1881. D'Arsonval's system employed a closed cycle ammonia turbine and it was finally demonstrated in 1979 in a small pilot project of Hawaii. However a different, open cycle system was tested during the 1930s by another Frenchman, Georges Claude. Claude's system proved the theory on which it was based but was not successful commercially. Work was not revived on his system until late in the twentieth century.

Tropical oceans and seas have surface water temperatures of between 24°C and 33°C. Below 500 m, the temperature will drop to between 9°C and 5°C. This provides a maximum exploitable temperature difference of 28°C. In practice the temperature difference is likely to be closer to 20°C, providing a theoretical energy conversion efficiency of 6.7%.[2] When account is taken of the need to pump cold water up from the depths, efficiency falls to 2–3%.

Though cool water may be available at 500 m, in practice the depth of 1000 m is normally considered necessary. If this is to be made accessible from land, a very long cold water pipe will be required to pump the deep water to the plant. This pipe will need to be 2000 m, or more, in length and if care is not taken, the cold water will become warmed before it reaches the plant. The alternative is to build an OTEC plant on a floating platform from which the cold water pipe stretches vertically downwards. Even with this arrangement, the cold water pipe will need to be 1000 m long.

In order to generate 1 MW of electricity, an OTEC plant requires 4 m^3/s of warm seawater and 2 m^3/s of cold seawater. This will require a cold water pipe of around 11 m in diameter to supply a 100 MW plant, the largest size considered practical.[3] The discharge of mixed hot and cold water from a plant of this size would be equivalent to that of the Colorado River in the USA discharging into the Pacific Ocean. Such massive quantities of water could have significant environmental impact.

7.1.1.1 Open and closed cycle ocean thermal energy conversion

The siting of an OTEC plant, either onshore or offshore, represents one of the key decision for any proposed project. The other key decision is the type of cycle to use. There are two principal options, an open cycle plant or

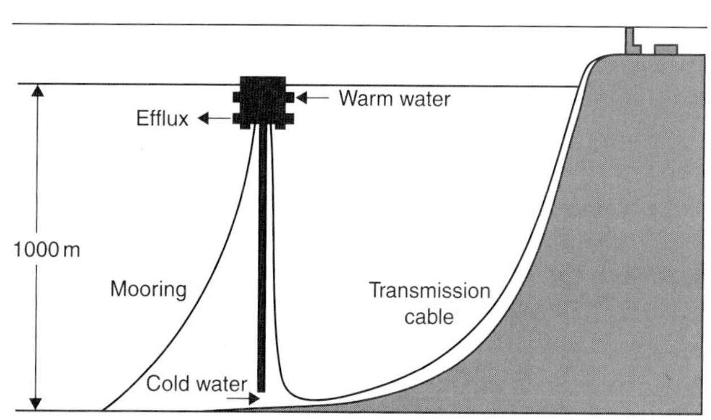

Figure 7.1-1 Schematic diagram of a floating OTEC plant.

Renewable Energy Focus Handbook 2009; ISBN: 9780123747051

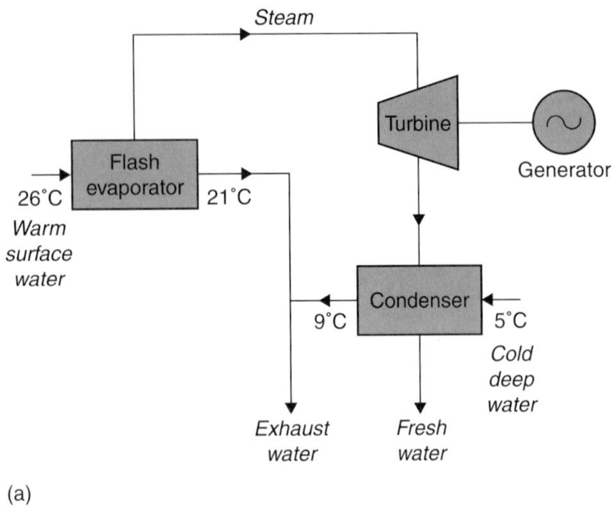

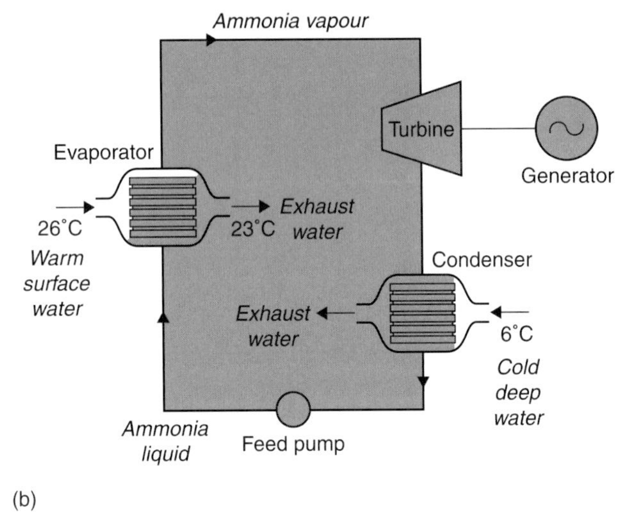

Figure 7.1-2 Block diagrams of (a) an open cycle OTEC plant and (b) a closed cycle OTEC plant.

a closed cycle plant. Hybrids of the two have also been proposed.

A closed cycle OTEC plant employs a thermodynamic fluid such as ammonia or a refrigerant like freon. This is contained in a completely closed system including the plant turbine. Hot surface seawater is used to evaporate the fluid and the vapour is then exploited to drive the turbine. The vapour from the turbine exhaust is condensed using the cold, deep ocean water, and returned to the beginning of the cycle where it can be reheated. A 50 kW closed cycle OTEC plant was built in Hawaii in 1979 and operated for a few months. A consortium of Japanese companies has also operated a 100 kW closed cycle OTEC plant in Nauru. Again this plant operated for only a few months to prove the concept. Neither was large enough to be commercially viable. Indeed, closed cycle OTEC is unlikely to be commercially viable in sizes of less than 40 MW.

In an open cycle OTEC system the seawater itself is used to provide the thermodynamic fluid. Warm seawater is expanded rapidly in a partially evacuated chamber where some of it "flashes" to steam. This steam is then used to drive a steam turbine. From the exhaust of the turbine, the vapour is condensed using cold seawater. The vapour produced by flashing warm seawater is at a relatively low pressure so it requires a very large turbine to operate effectively. Practical limitations mean that the largest open cycle turbine that can be built today is around 2.5 MW, much smaller than for a closed cycle system.

One of the major advantages of the open cycle system is that the water condensed from the turbine exhaust is fresh, not salt water, and so the plant can also serve as a source of drinking water as well as electricity. A 210 kW open cycle OTEC pilot plant operated in Hawaii between 1993 and 1998.

In a hybrid OTEC plant warm seawater is flashed to produce steam and this steam is then employed as the heat source for a closed cycle system. This system is more complex that either of the other cycles but it marries the compact closed cycle system with the ability to produce drinking water.

7.1.1.2 Technical challenges

The major challenge facing OTEC is the development of cold water pipe technology. The cold water pipe has to pump water from a depth of around 1000 m. For a floating OTEC plant the pipe will be at least this long. Land-based plants will require significantly longer cold water pipes. A 40 MW project proposed for Hawaii would involve a cold water pipe of over 3.5 km in length. Designing pumps and piping capable of delivering the volumes of water required over this distance without significant temperature rise represents a considerable engineering challenge.

Heat exchangers are also important components of an OTEC plant. The two heat exchangers are likely to be the largest and most costly components and require careful optimisation. As already noted, the turbine in an open cycle OTEC plant is also large. Special low-speed turbine designs will be needed to achieve high outputs from this type of OTEC plant.

Offshore OTEC plants will require deep-water moorings. Modern offshore oil and gas expertise should provide a good starting point for developing such moorings but further work will be needed to tailor mooring systems to the needs of the OTEC facility.

7.1.1.3 Hybrid applications

The OTEC plant is designed primarily to generate electricity. As such it could prove to be a significant source of power for a small island community where the conditions for OTEC exploitation exist. Such communities will also often benefit from fresh water production from an open cycle or hybrid OTEC facility. This combination will prove important for the future of OTEC.

There is a further resource available from an OTEC plant, a supply of nutrient-rich and bacteria-free deep ocean water. This can be used for forms of aquaculture as well as to provide cooling. When combined with electricity and drinking water production, aquaculture could make OTEC more attractive economically.

7.1.1.4 Browsing ocean thermal energy conversion

The short-term application of OTEC will be for land-based or floating inshore plants providing services to a local community, with power possibly supplied to a grid system. However, most OTEC potential is far offshore. This could be exploited by browsing OTEC plants, which cruise the oceans looking for the hottest surface water temperatures. Such plants would not be able to transfer electricity directly to land. However, they could generate hydrogen and potable water, both of which could be stored for later transportation to land. Current costs would not make this economically practical today.

7.1.1.5 The environmental impact of ocean thermal energy conversion

The main environmental impact of an OTEC plant results from the pumping of water from below 1000 m and then returning it at a much lower depth. The surface water for an OTEC plant is taken at around 20 m depth, so the mixed hot and cold water must be returned at around 60 m depth to prevent it returning directly to the hot water input.

The volumes of water involved are enormous and the movement of water from lower to upper regions of the seas and oceans could have a significant impact on the local marine environment. There is little evidence available today to indicate what the effect would be, but the danger of this impact is likely to limit the exploitation of OTEC, at least until extensive environmental studies have been carried out.

The warm water heat exchanger in a closed cycle OTEC plant is likely to be subject to biofouling. This must be prevented to maintain efficiency. The only way to prevent biofouling is by chemical treatment, probably with chlorine, leading to some release into the sea. Such releases would need to be closely monitored and would have to fall within legal limits.

Construction of an OTEC plant will lead to some seabed disruption but this is likely to be short term and relatively minor. There will also be the danger of small releases of oil and, in the case of a closed cycle OTEC plant, thermodynamic fluid. The impact of such releases should be of a similar nature to those from existing offshore facilities.

7.1.1.6 The cost of ocean thermal energy conversion

OTEC is still under development and any costing must be considered extremely tentative. Like many renewable technologies, OTEC is capital intensive with capital costs starting at around $4000/kW. This is expensive for a power plant but may appear less so if drinking water production is taken into account. However, pre-commercial demonstration of the technology is still required, to prove that it is viable. There is too little experience to provide any realistic estimate of the cost of electricity generated from an OTEC plant.

7.1.2 Wave energy

All seas contain energy in the form of surface waves which can be exploited by wave energy conversion devices. Not all seas provide an economically exploitable resource. A good wave regime will normally be the first consideration. Such regimes are found principally on western coasts facing the world's great oceans. Wave energy exploitation is limited to coastal and near-shore sites, so the opportunity for deployment is restricted. Even so, it is possible that wave energy could supply up to 10% of global electricity demand.[4]

The development of modern wave energy conversion technology started after the oil price rises of the 1970s. Much work was carried out in the UK but national funding was withdrawn in 1989. Work continued in other countries in Europe and in the USA and Japan. The environmental concerns of the 1990s added further impetus to the wave energy industry and the UK resumed funding in 1999. By the beginning of the twenty-first century a wide variety of different wave generation devices were under development.

Unlike virtually all other power generation technologies, wave energy conversion requires completely unique energy conversion devices. These devices have to convert the wave motion at the sea surface into electricity. Engineers who have addressed the problem have devised a range of novel solutions. These can be categorised in a number of ways but perhaps the simplest is to divide them into two groups. The first comprises shore and bottom mounted near-shore devices and the second comprises–offshore devices.

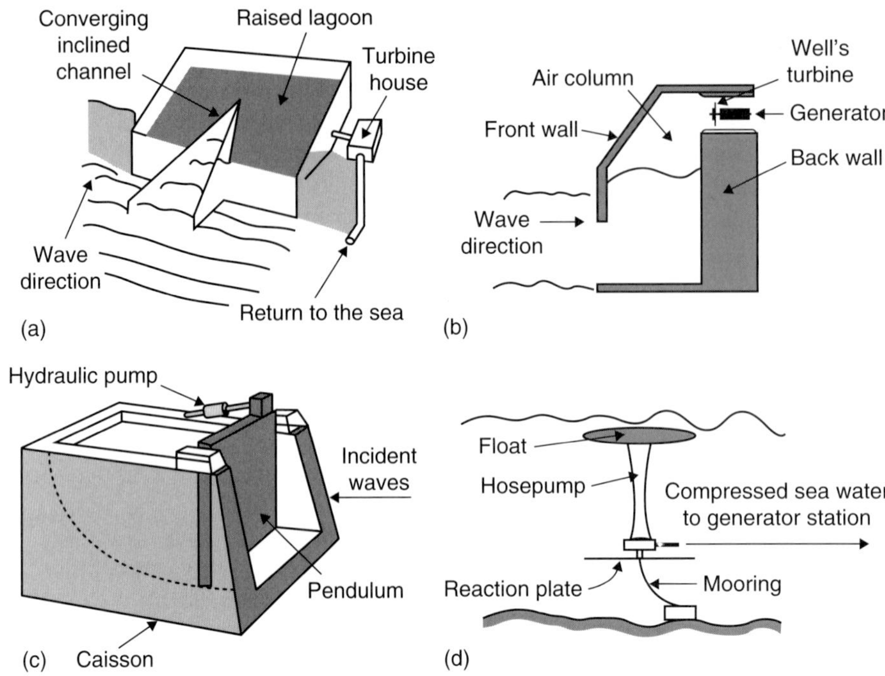

Figure 7.1-3 Wave energy conversion devices. (a) Tapered channel device (Tapchan); (b) Oscillating water column (OWN); (c) Pivoting flap device (the pendulor); (d) Heaving buoy device (the hosepump).

7.1.2.1 Shore and near-shore wave converters

1. Oscillating water columns

Perhaps the most widely tested of shore and near-shore devices is the oscillating water column. If a tube, sealed at one end, is placed so that its open end is just beneath the surface of the sea, as waves pass the tube, the level of water inside the tube will rise and fall, alternately compressing and expanding the air column within the tube.

If, instead of a seal, the upper end of this tube is open and houses a device that acts like a wind turbine, then the moving water will cause the air to move in the tube and this will make the turbine rotor turn backwards and forwards. This air movement forms the basis for an oscillating water column wave energy converter.

Oscillating water columns can either be shore or bottom mounted. They normally comprise some form of concrete structure, which is designed to create an enclosure containing air; this is open to the sea at the bottom. A special type of turbine called a *Well's turbine* is frequently mounted at the top. This can derive continuous power from movement of air both up and down without the need for a complex arrangement of valves.

Oscillating water columns have been tested in many parts of the world including Europe and Japan. The Well's turbine, developed in the 1970s, has most usually been employed in these prototypes but newer bidirectional turbines with greater efficiency are under development. The economics of oscillating water column converters and of other shoreline devices can be improved if they are built into breakwaters.

2. Tapered channels

Another approach to wave energy conversion uses and amplifies the height of a wave in order to create a head of water which can be used to drive a conventional low-head hydro turbine. Devices like this usually employ a tapered channel with its mouth open to the sea. The side walls of the channel rise above the normal sea level and beyond them is a reservoir.

Waves travelling towards the coast are focused into the channel. As these waves flow along the channel they become more and more restricted by the taper and this forces the height of the wave to increase, until water starts to fall over the upper edges of the channel walls. This water is captured, creating a reservoir of water, which is above the sea level. This water can then be run back into the sea through a hydro turbine, generating electricity.

A system of this type, called *Tapchan*, was built on the Norwegian coast in the late 1980s. The technology is relatively simple but construction costs can be high. Deployment is restricted by the need for a relatively low tidal range. Otherwise the operation of the converter is compromised.

3. Oscillating flaps

The energy contained in moving waves is sufficient to cause a pendulum or flap to move backwards and forwards, and this too has been used as the basis for a shore-based wave energy convertor. The best known converter of this type is a Japanese device called the *pendulor*, which comprises a box, open to the sea on one side, but with the open side closed using a flap hinged horizontally from the top. When waves strike the flap they cause it to oscillate to and fro like a pendulum and this motion can be converted into electricity using hydraulic[5] rams. Small devices of this type have been built and tested and a plant of around 200 kW has been designed for a site in Sri Lanka.

7.1.2.2 Offshore devices

The three types of device discussed above can all be exploited offshore provided they can be moored so that they remain stationary relative to the waves. However, most offshore devices try to exploit the wave motion in different ways.

1. Float pumps

The hosepump, developed in Sweden, is based on an elastic tube that changes its internal volume as it is stretched. One end of the tube is sealed and attached to a float while the other end is open, and is connected to a moored plate close to the bottom of the sea. As waves pass the device, the float moves up and down, alternately stretching and relaxing the tube. This pumps water in and out of the lower end of the tube and the pressurised water is used in a hydraulic energy conversion system to generate electricity.

A wave power float pump developed in Denmark takes a slightly different approach. In this case a float at the surface is attached to a rod, which bears in turn on a shaft (like a crankshaft), attached to a piston-pump device. Movement of the float up and down causes the rod to rotate the (crank) shaft, turning the vertical motion into rotary motion from which electricity can be extracted, exactly as in a piston engine.

The Archimedes Wave Swing, developed in the Netherlands, adopts a similar principle but the movement up and down of a buoyant floater is converted into electricity by means of a linear generator.

2. Ducks, wave pumps and other water snakes

There are a number of wave energy devices designed in hinged sections which all float. As waves pass these devices, the different sections move relative to one another and this differential motion is used to derive hydraulic energy, which is then converted into electricity.

The first of these is called *Salter's Duck* after British designer Stephen Salter. The prototype appeared in the 1970s but the concept is still under development. The duck has a beak-shaped float, which is fixed by a hinge to a second anchored section. The beak moves in the waves relative to the anchored spine and this relative motion is used to extract energy.

The McCabe wave pump comprises three rectangular pontoons connected through hinges. The central pontoon has a damper plate attached to it, which slows its vertical motion relative to the two outer pontoons. This generates relative motion between the three sections, which again can be exploited to generate electricity.

The Pelamis looks like, and is named after, a sea snake. It comprises a series of buoyant cylindrical sections joined end to end. The device is tethered at one end. As waves pass along it, the sections move relative to one another and hydraulic cylinders extract energy from this motion. A 750 kW prototype is under construction.

3. Piezoelectric devices

A US programme is developing a wave energy converter, which is based on a piezoelectric material in sheet form. The device, called the *eel*, can produce an electric current when bent by waves. It is still in a very early stage of development.

7.1.2.3 The environmental implications of wave energy converters

Wave energy converters remove energy from waves. This means that both near-shore bottom mounted devices and offshore floating devices will calm the sea. This will be broadly beneficial since it will protect the coast from waves.

Land and near-shore bottom mounted wave converters will cause some disruption to the marine environment during construction but this should be short lived. Once in place they should have little impact. There may be a visual impact and oscillating water column converters may generate noise from their air turbines.

Floating offshore devices should be less disruptive since they will normally be built onshore and then towed to the site where they are deployed. However, floating devices are likely to be a hazard to shipping and sites will need to be selected that cause minimum disruption. It may well make sense to deploy such devices at offshore wind sites; they need to be away from shipping lanes and well marked to prevent the danger of collision. A floating device could become a serious hazard if it slipped its mooring, so strict monitoring would be essential.

7.1.2.4 The cost of wave energy conversion

Wave energy conversion is still in an early development stage and it is impossible to gain a realistic idea of costs. However, UK estimates have suggested that wave power may be able to produce electricity for between €0.06/kWh and €0.12/kWh. This may make wave power more economical for remote coastal and island communities that currently rely on diesel generation. Capital costs range from €800/kW to €40,000/kW though the former must be considered optimistic at this stage while the latter refers to a technology in an early stage of development. Commercial wave power plants are unlikely to be deployed before 2010.

7.1.3 Ocean current generation

The movement of water within oceans and seas is the basis for ocean current energy conversion. This movement may be caused by tidal ebb and flow. In this case the water movement will follow a sinusoidal variation in speed and direction, the latter normally reversing twice every 24 h.

Other currents are caused by thermal gradients. The most prominent of these is the Gulf stream which moves around 80 million m^3 of water each second[5] but there are other, lesser currents in many parts of the world. These currents usually flow in one direction only, and are relatively constant in strength.

The conversion of ocean currents into electricity involves similar considerations and technology to that employed by wind power plants. The main difference is that the energy density contained in a current of water is much higher than that of air. As a result turbines can be much smaller. For a tidal current which varies regularly a current of around 1.5 m/s is considered sufficient to exploit. Where the current is continuous in a single direction, a flow speed of 1 m/s is exploitable. The latter also offers a higher-capacity factor, around 80%, whereas a tidal current will provide a capacity factor of 40–50%.

Sea currents hold one further advantage over wind: they are predictable. Thus whereas a wind farm cannot guarantee its output, the output from an ocean current power plant should be entirely predictable. This has significant implications for network operation and dispatching since a reliable source of electricity is much more valuable than an unpredictable source.

As with wind turbines, there are two different configurations of tidal stream turbine, a horizontal axis and a vertical axis turbine. Both types are under development.

7.1.3.1 Horizontal axis turbines

The horizontal axis turbine, or propeller turbine, comprises a propeller with two or more blades. The turbine can either be mounted on a tower fixed to the seabed or it can be deployed below a floating support. The former method is most suitable for shallow waters whereas the floating support can be deployed in deeper water. In order to increase the efficiency of a horizontal axis system, water flow around the turbine can be controlled using a shroud.

As a result of the high-energy density, water turbines are much smaller than wind turbines. A unit with a diameter of 10–15 m can produce between 200 kW and 700 kW. Prototypes include a 15 kW unit tested in a Scottish Loch and a 300 kW unit deployed off the coast of southern England in 2003.[6]

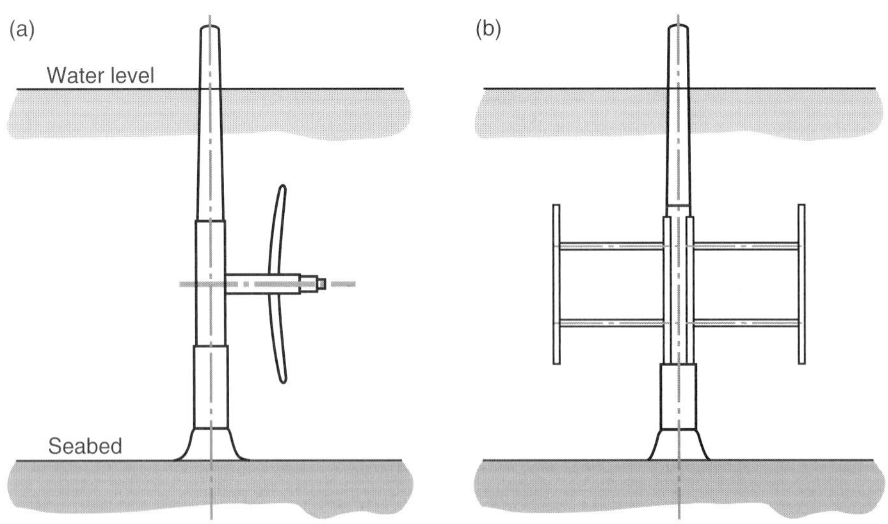

Figure 7.1-4 Horizontal and vertical axis ocean current energy converters. (a) Horizontal axis turbine (axial flow) and (b) Vertical axis turbine (cross flow).

7.1.3.2 Vertical axis turbines

The vertical axis turbine used for ocean current applications has vertical blades which are supported on struts attached to a vertical shaft. The blades are shaped so that they will rotate in a current, whatever be its direction. This is particularly useful for tidal applications where the current direction reverses regularly.

As with the horizontal axis turbine, a vertical axis machine can either be bottom mounted or fixed to a floating platform. The latter design has been tested in a 130 kW prototype in Italy.

It is possible to deploy an array of vertical axis turbines arranged like the elements of a vertical fence. When used in tidal waters this is called a *tidal fence*. The scheme allows the maximum amount of energy to be extracted from a single site. A prototype based on this concept is being planned for installation in the Philippines.

7.1.3.3 Other tidal stream energy extractors

It is possible to employ devices other than turbines to extract energy from a moving stream of water. One such device, called the *stingray*, is based on concept of the hydroplane. The stingray has a large hydroplane-like wing mounted at the end of a long, hinged arm. The angle of the hydroplane can be adjusted to control its lift and drag. If the device is placed in an ocean stream, adjusting the angle of attack of the hydroplane cyclically will force the arm to oscillate up and down, generating a hydraulic force, which can be converted into electricity. A 150 kW prototype is under development.

7.1.3.4 Ocean current environmental considerations

Ocean current energy conversion devices should normally be constructed on shore and then transported to the chosen site for installation. Seabed and marine disruption should be short and impact should be small.

More significant is the fact that an ocean current energy converter will remove energy from the ocean current, leaving it weaker. This could have a significant effect on downstream marine ecologies. A major tidal stream plant such as a tidal fence-style array of turbines would probably have a similar effect to a large tidal barrage. The effect of smaller units would be less but an environmental impact study would certainly be necessary to establish the probable extent.

The moving blades of an underwater turbine could injure or kill marine mammals and fish. Further study is required to establish how dangerous this will be. Measures similar to those needed with conventional hydropower plants are likely to be necessary in order to minimise this danger.

The other main impact of an ocean current installation will be on shipping and fisheries. Large underwater structures will form a hazard to shipping, so major shipping lanes must be avoided. Other sites may interfere with local fisheries and these too must be taken into consideration.

7.1.3.5 Cost of ocean current technology

As with the other technologies discussed in this chapter, ocean current technology is still at an early stage of development and realistic costs are difficult to establish. Some early European studies have suggested that electricity could be generated for between €0.05/kWh and €0.15/kWh. Meanwhile, a Canadian study published in 2002 concluded that Canadian technology could produce electricity for between 11 Canadian cents/kWh (for an 800 MW development) and 25 Canadian cents/kWh (for a 43 MW installation). Generation costs within these ranges would make the technology competitive with diesel generation.

End notes

1. In thermodynamics these are usually referred to as a heat source and a heat sink.
2. Ocean Energy Conversion, M.T. Pontes and A. Falcao, Lisboa, Portugal.
3. Refer *supra* note 1.
4. World Energy Council, Survey of Energy Resources, 2001.
5. Refer *supra* note 8.
6. The company which has developed this turbine, IT Power, has suggested that up to 20% of the UK's power could be derived from tidal streams.

Chapter 7.2

7.2

Tidal energy

Ian G. Bryden

Tidal power is not a new idea. Mills, which used tidal flows in bays and estuaries to drive machinery to grind cereal, were widely used in medieval times. Tides have been seriously reexamined as a potential source of energy for industry and commerce. In the United Kingdom, there were numerous proposals throughout the 20th century to exploit the tidal energy potential of the Severn Estuary. None have yet been developed, primarily as a result of the anticipated costs and concern over the environmental changes that would arise from such a large-scale development.

7.2.1 Overview

The world's first serious scheme to exploit tidal energy was constructed in France, at La Rance in Brittany between 1961 and 1967, and consists of a barrage across a tidal estuary to utilize the rise and fall in sea level induced by the tides. This scheme has proven to be highly successful despite some early teething problems. Many engineers and developers now favor, however, the use of alternative technology, which will utilize the kinetic energy in flowing tidal currents.

Like any energy resource, the prospect of tidal power exploitation has stimulated opposition. Tidal barrage proposals, like La Rance, have produced the most vocal opposition. Environmental groups, although generally in favor of the exploitation of alternative energy sources, are suspicious of the likely environmental changes large estuary based schemes would produce. Political opposition has also been considerable in some cases. One politician in the United Kingdom likened the proposed creation of a barrage across the Severn Estuary to the formation of a "large stinking lake." Similar political opposition has been voiced against any development of the tidal resource in the Solway Firth between Scotland and England. It is anticipated that public and political opposition will limit the development of tidal barrage schemes in the short term.

Opposition to schemes designed to exploit the kinetic energy in tidal currents using open turbines, analogous to wind turbines, has been less voluble, although there has as yet been no commercial demonstration of such technology. If opposition remains muted, then tidal current turbines could be installed in many parts of the world in the early 21st century.

7.2.2 The tidal phenomenon

7.2.2.1 Introduction

The tides are cyclic variations in the levels of the seas and oceans. Water currents accompany these variations in sea level, which, in some locations such as the Pentland Firth to the North of the Scottish mainland, can be extreme.

The explanation of the existence of tides represented one of the greatest challenges to early oceanographers, mathematicians, and physicists. It was not until Newton that a satisfying theory emerged to explain at least some of the properties of the tides. He formulated a theory, which has become known as the equilibrium theory of tides.

7.2.2.2 Newton's approach: the equilibrium theory of tides

The equilibrium theory of tides gives a partial description of tidal behavior for an abstract planet Earth, which is entirely and uniformly covered by water. According to the Newtonian model, the earth-moon system rotates around a common center of mass (CoMs) and the radius

Renewable Energy Focus Handbook 2009; ISBN: 9780123747051

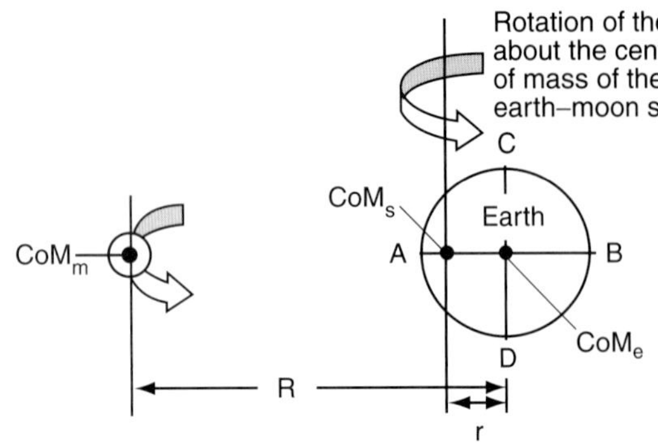

Figure 7.2-1 Schematic of the earth-moon system.

of this circulation is given by r, as shown in Fig. 7.2-1. The separation of the center of mass of the earth (CoM_e) from the center of mass of the moon (CoM_m) is given by R, which is also shown.

If the earth were not itself rotating, each point on, or in, the Earth would rotate about its own center of rotation, the radius of the rotation would also be given by r and the period of rotation would be equal to the rotational period of the earth–moon system. This results in acceleration toward the local center of rotation.

At the center of the earth, the apparent centrifugal acceleration resulting from the rotation exactly matches the gravitational acceleration. At all other points, there is an imbalance between gravitational and centrifugal effects. At the point B, as shown in Fig. 7.2-1, the centrifugal effects exceed the lunar gravitational attraction. In effect, at the surface of the earth, there will be a net flow of water from C & D to A & B. This effect results in the lunar tidal cycle. The equilibrium theory suggests, therefore, the establishment of tidal bulges in the fluid surrounding the earth, as shown in Fig. 7.2-2.

The earth rotates, and the two tidal bulges must maintain their positions with respect to the moon. They, therefore, have to travel around the earth at the same rate as the earth's rotation. The moon moves around the CoM_s every 27.3 days in the same direction that the earth rotates every 24 hours. Because the movements are in the same direction, the net effect is that the period of the earth's rotation, with respect to the earth–moon system, is 24 hours and 50 minutes. This explains why the tides are approximately an hour later each day.

During the lunar month, which is the orbital period of the moon around the earth, there will be variations in the lunar tide influence. The lunar orbit is not circular but is

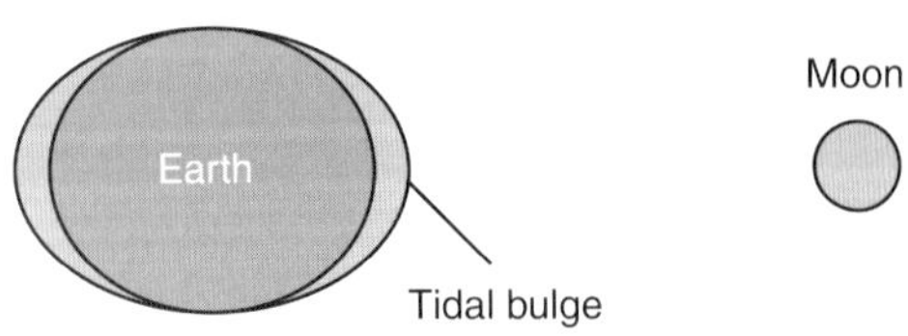

Figure 7.2-2 The tidal bulge.

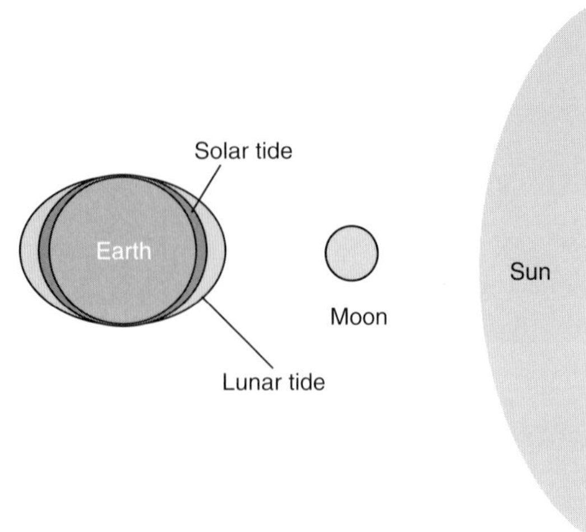

Figure 7.2-3 Earth, sun, and moon during spring tides.

elliptical in form, and the tide-producing forces vary by approximately 40% over the month. Similarly, the moon does not orbit around the earth's equator. Instead there is an angle of approximately 28° between the equator and the plane of the lunar orbit. This also results in monthly variations.

7.2.2.3 Influence of the sun on the tides

The earth–sun system is also elliptical but with only a 4% difference between the maximum and minimum distance from the earth to the sun. The relative positions of the earth, moon, and sun produce the most noticeable variations in the sizes of the tides.

In the configuration shown in Fig. 7.2-3, the influence of the moon and sun reinforce each other to produce the large tides known as spring tides, or long tides. A similar superposition also exists at the time of full moon.

When the sun and moon are at 90° with respect to each other, the effect is one of cancellation as shown in Fig. 7.2-4. This configuration results in neap tides, also known as short tides.

7.2.2.4 The presence of land and the resulting tidal dynamics

If the earth were covered entirely by water of a constant depth, the equilibrium theory of tides would give a perfectly reasonable description of water behavior. Fortunately, the oceans are not all of a constant depth and the presence of continents and islands severely influences the behavior of the oceans under tidal influences. The Coriolis force is a particularly important effect which, in the Northern Hemisphere, diverts moving objects to the right and, in the Southern Hemisphere, diverts moving objects to the left. The influence of this force, in the

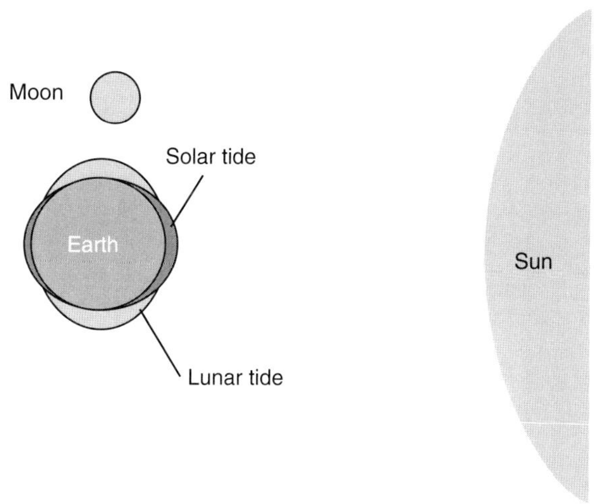

Figure 7.2-4 Earth, sun, and moon during neap tides.

presence of land, can be considered, as shown in Fig. 7.2-5, by visualizing water flowing into and out of a semi-enclosed basin in the Northern Hemisphere under the influence of tidal effects. On the way into the channel, the water is diverted to the right toward the lower boundary. When the tidal forcing is reversed, the water is diverted toward the upper boundary. This results in a substantially higher tidal range at the basin boundaries than at the center.

The net result of this effect is to generate a so-called tidal wave, which processes anti-clockwise around a point in the center of the "basin," as shown in Fig. 7.2-6.

In effect, the tides represent the terrestrial manifestation of the potential and kinetic energy fluxes present in the earth–moon–sun system. These fluxes are complicated by the presence of continents and other landmasses, which modify the form and phase of the tidal wave. As a result, substantially higher local fluxes occur in some regions of the world than in others. The Bay of Fundy in Canada and the Bristol Channel between England and Wales are two particularly noteworthy examples of high-flux regions.

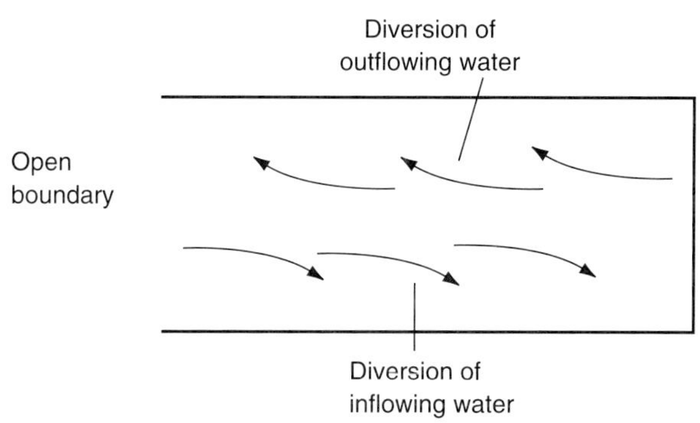

Figure 7.2-5 Flow of water into and out of a semi-enclosed basin in the Northern Hemisphere.

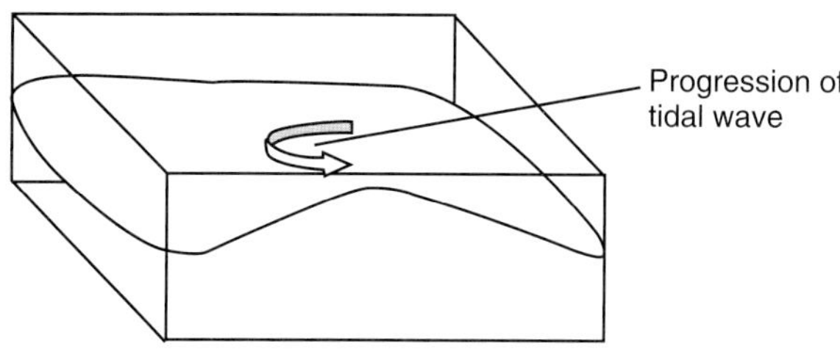

Figure 7.2-6 Progression of a tidal wave in a basin.

7.2.2.5 Global flux

It has been estimated that the total energy from the tides, which is currently dissipated through friction and drag, is equivalent to 3000 GW of thermal energy worldwide. Much of this power is in inaccessible places, but up to 1000 GW is available in relatively shallow coastal regions. Estimates of the achievable worldwide electrical power capability range from about 120 GW of rated capacity to approaching 400 GW. This is obviously a substantial energy resource, the significance of which has yet to be fully appreciated. Many enthusiasts believe these to be hopelessly pessimistic estimates. It is probably reasonable, however, to consider these estimates as representing, with an element of uncertainty, what could be exploited now using available technology.

7.2.3 Tidal barrage methods

7.2.3.1 History

There are many places in the world in which local geography results in particularly large tidal ranges. Sites of particular interest include the Bay of Fundy in Canada, which has a mean tidal range of 10 m, the Severn Estuary between England and Wales, with a mean tidal range of 8 m and Northern France with a mean range of 7 m. A tidal barrage power plant has been operating at La Rance in Brittany since 1966. This plant, which is capable of generating 240 MW, incorporates a road crossing of the estuary. Other operational barrage sites are at Annapolis Royal in Nova Scotia (18 MW), at the Bay of Kislaya near Murmansk (400 kW), and at Jangxia Creek in the East China Sea (500 kW). Schemes for energy recovery, not surprisingly, have been proposed for the Bay of Fundy and for the Severn Estuary but have never been built.

7.2.3.2 Principles of operation

The role of the barrage

The approach is essentially always the same. An estuary or bay with a large natural tidal range is identified and then

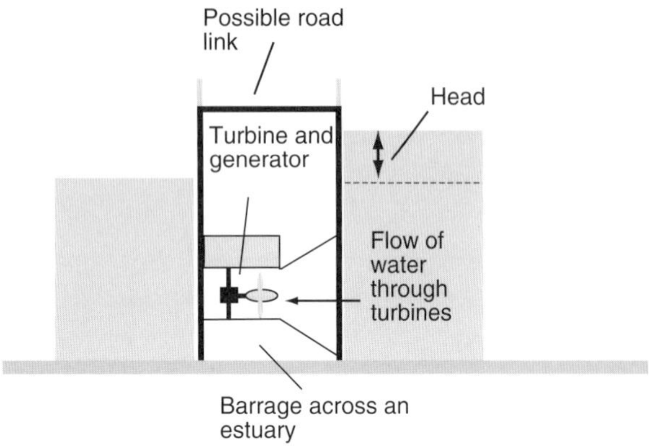

Figure 7.2-7 Schematic of a tidal power barrage.

artificially enclosed with a barrage. This would, typically, also provide a road or rail crossing of the gap in order to maximize the economic benefit. Electrical energy is produced by allowing water to flow from one side of the barrage through low head turbines, as shown in Fig. 7.2-7.

There are a variety of suggested modes of operation. These can be broken down initially into single basin schemes and multiple basin schemes. The simplest of these are the single basin schemes.

Single basin tidal barrage schemes

These schemes, as the name implies, require a single barrage across the estuary, as shown in Fig. 7.2-8. There are, however, three different methods of generating electricity with a single basin. All of the options involve a combination of sluices which, when open, can allow water to flow relatively freely through the barrage and gated turbines, the gates of which can be opened to allow water to flow through the turbines to generate electricity.

Ebb generation

During the flood tide, incoming water is allowed to flow freely through sluices into the barrage. At high tide, the sluices are closed and water is retained behind the barrage. When the water outside the barrage has fallen sufficiently to establish a substantial head between the basin and the open water, the basin water is allowed to flow out though low-head turbines and to generate electricity.

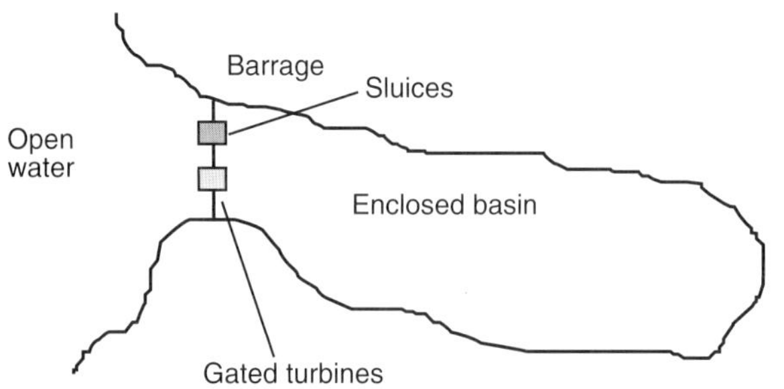

Figure 7.2-8 Schematic of the single basin generation scheme.

The system may be considered to operate in a series of steps as shown in Fig. 7.2-9.

These steps can be represented, as in Fig. 7.2-10, to show the periods of generation associated with stages in the tidal cycle.

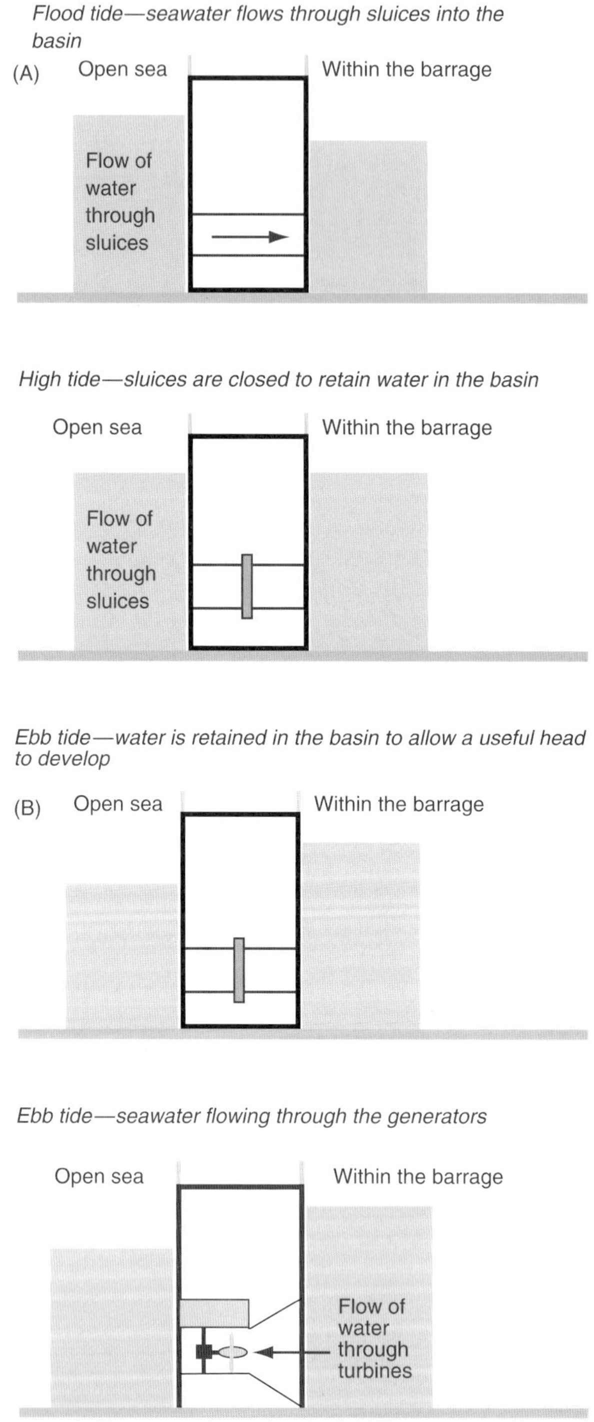

Figure 7.2-9 Operational steps in an ebb generation barrage scheme. (A) Ebb tide—water is retained in the basin to allow a useful head to develop. (B) Ebb tide—seawater flowing through the generators.

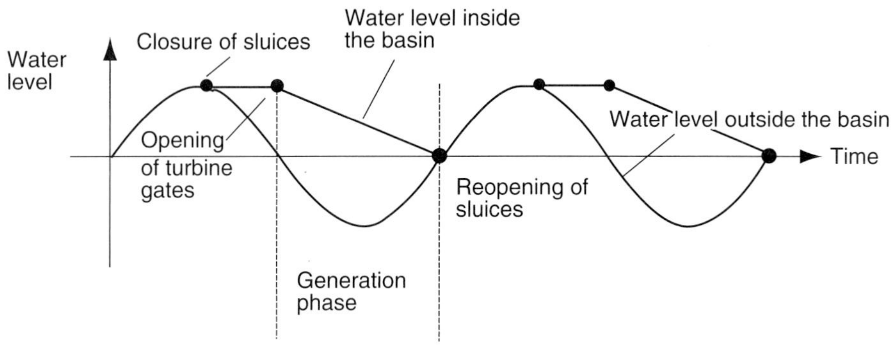

Figure 7.2-10 Water level inside and outside of an ebb generation barrage.

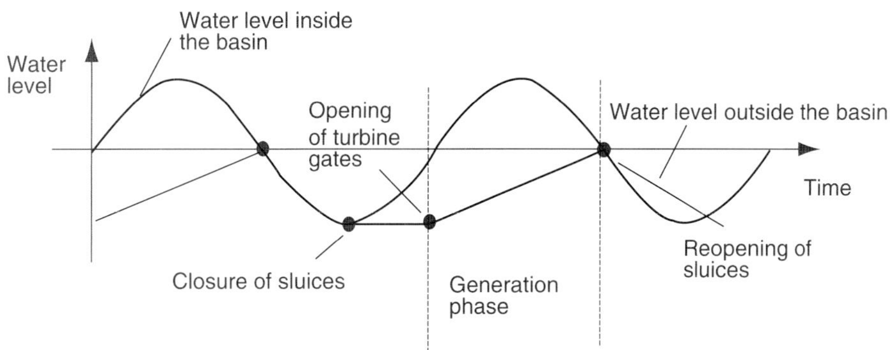

Figure 7.2-11 Water level inside and outside of a flood generation barrage.

Typically the water will only be allowed to flow through the turbines once the head is approximately half the tidal range. It is likely that an ebb generation system will be able to generate electricity for, at most, 40% of the tidal cycle.

Flood generation

The sluices and turbine gates are kept closed during the flood tide to allow the water level to build up outside of the barrage. As with ebb generation, once a sufficient head has been established, the turbine gates are opened and water can, in this case, flow into the basin and generate electricity, as Fig. 7.2-11 shows.

This approach is generally viewed as being less favorable than the ebb method because keeping a tidal basin at low tide for extended periods could have detrimental effects on the environment and shipping. In addition, the energy produced would be reduced as the surface area of a basin would be larger at high tide than at low tide, which would result in a rapid reductions in the head during the early stages of the generating cycle.

Two-way generation

In this mode of operation, use is made of both the flood and ebb phases of the tide. Near the end of the flood-generation period, the sluices would be opened to allow water to get behind the barrage and would then be closed. When the level on the open water side of the barrage had dropped sufficiently, water would be released through the turbines in the ebb generation mode, as shown in Fig. 7.2-12.

Unfortunately, computer models do not indicate that there would be a major increase in energy production. In addition, additional expenses would be associated with a requirement for either two-way turbines or a double set to handle the two-way flow. Advantages include, however, a reduced period with no generation and the peak power would be lower, allowing a reduction in the cost of the generators.

Double-basin systems

All single-basin systems suffer from the disadvantage that they only deliver energy during part of the tidal cycle and cannot adjust their delivery period to match the

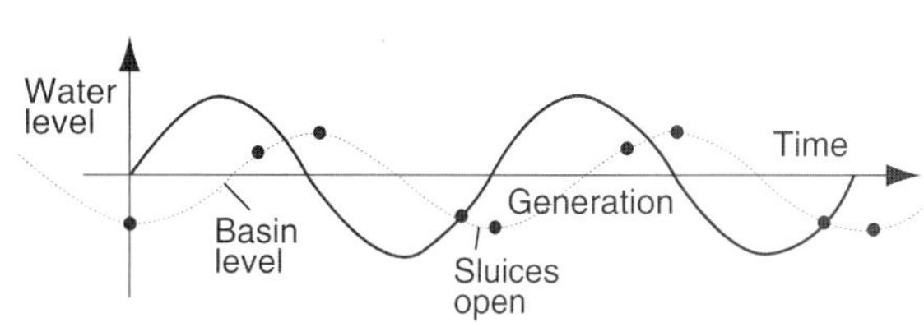

Figure 7.2-12 Water level inside and outside of a two-way generation barrage.

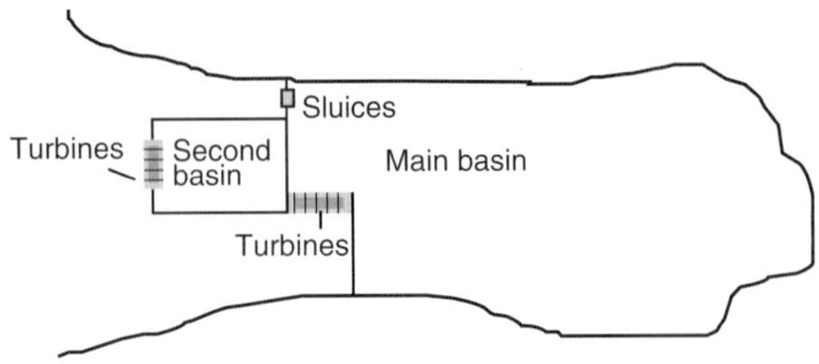

Figure 7.2-13 Schematic of a double-basin generation system.

requirements of consumers. Double basin systems, such as that shown in Fig. 7.2-13, have been proposed to allow an element of storage and to give time control over power output levels.

The main basin would behave essentially like an ebb generation single basin system. A proportion of the electricity generated during the ebb phase would be used to pump water to and from the second basin to ensure that there would always by a generation capability. The cycle can be seen in Fig. 7.2-14.

It is anticipated that multiple basin systems are unlikely to become popular, as the efficiency of low-head turbines is likely to be too low to enable effective economic storage of energy. The overall efficiency of such low-head storage, in terms of energy out and energy in, is unlikely to exceed 30%. It is more likely that conventional pump-storage systems will be utilized. The overall efficiencies of these systems can exceed 70%, which is, especially considering that this is a proven technology, likely to prove more financially attractive.

7.2.3.3 Optimal design of barrage systems

This is not a simple procedure, as the construction of a barrage will inevitably change the nature of the tidal environment. It is theoretically possible, for example, for the tidal range to be very much less, once a barrage has been constructed, than it was prior to construction. To avoid an expensive mistake, it is necessary that reliable computer models be used to predict future behavior as much as possible. Such models are available and can be used with a great degree of confidence.

Table 7.2-1 Resource estimates for selected sites

Site	Mean tidal range (m)	Barrage length (m)	Estimated annual energy production (GW h)
Severn Estuary (UK)	7.0	17,000	12,900
Solway Firth (UK)	5.5	30,000	10,050
Bay of Fundy (Canada)	11.7	8,000	11,700
Gulf of Khambat (India)	6.1	25,000	16,400

7.2.3.4 Locations for tidal barrage developments

A considerable number of sites worldwide are technically suitable for development, although whether the resource can be developed economically is yet to be conclusively determined. Some of these sites are listed in Table 7.2-1.

7.2.3.5 The environmental impact of basin systems

The construction of a barrage across a tidal basin will change the environment considerably. An ebb system will

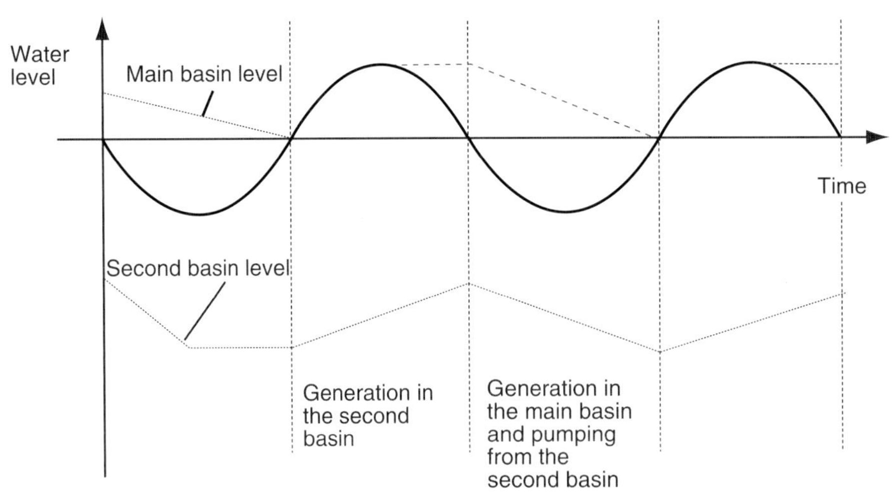

Figure 7.2-14 Water levels associated with a two-basin generation system.

reduce the time tidal sands are uncovered. This change will have considerable influences on the lives of wading birds and other creatures. The presence of a barrage will also influence maritime traffic, and it will always be necessary to include locks to allow vessels to pass through the barrage. This will be much easier for an ebb system, where the basin is potentially kept at a higher level, than it would be with a flood-generation system, in which the basin would be kept at a lower than natural level. The environmental impact from increasing the area of exposed sediment in an estuarial environment is difficult to quantify but likely to be detrimental.

7.2.3.6 The financial implications of tidal barrage development

Barrage schemes can produce a vast amount of energy. It has been suggested that the Severn Estuary could provide in excess of 8% of the United Kingdom's requirement for electrical energy. However, all barrage schemes are constrained by the massive engineering operations and associated financial investment that would be required. Construction would take several years. La Rance, for example, took 6 years, and no electricity was generated before the total project was completed. This is a major disincentive for commercial investment.

7.2.4 Tidal current systems

7.2.4.1 The resource

The public perception of tidal power has traditionally been of large barrage schemes such as that at La Rance or proposed for the Severn Estuary. In addition to changes in the sea-surface level, however, the tides also generate water currents. In the open ocean, these currents are typically very small and are measured in cm/s at most. Local geographical effects can, however, result in quite massive local current speeds. In the Pentland Firth, for example, there is evidence of tidal currents exceeding 7 m/s. The kinetic energy in such a flow is considerable. Indeed, many experts describe the Pentland Firth as the Saudi Arabia of tidal current power, which explains the considerable interest in this site which has been expressed by the UK and Scottish governments. At present, however, there is no available technology to exploit this massively energetic but hostile channel.

Other sites, in Europe alone, with large currents, include the Channel Islands and the Straits of Messina. In addition to major sites such as the Pentland Firth, numerous local sites experience rapid currents capable of generating electricity with suitable technology.

Since the late 1980s, interest in the possibility of using tidal currents to generate electricity has developed enormously. A study prepared for the Scottish government in 2001 suggested that the UK "base case" potential could exceed 33.5 TWhr per annum at reasonable cost and using available technology. An earlier study for the European Union reported in 1996 that the European Union resource should exceed 48 TW h. This figure appears low in comparison with the later UK figure, and it is suggested that, rather than the European resource outside of the United Kingdom representing only 14.5 TW h per annum potential, the state of knowledge and understanding had increased dramatically in the 5 years separating these studies. There has not yet been an authoritative study of the achievable world resource but, drawing parallels with UK and European studies, it is anticipated that the achievable world resource could exceed 270 TW h per annum using technology that is already under consideration. If problems associated with harnessing energy from currents in deep water can be overcome, then this figure might prove to be highly pessimistic, especially if technological developments make the development of sites such as the Pentland Firth feasible.

Additional studies have suggested that tidal currents could offer particular opportunities for peripheral rural regions. No one doubts the size of the worldwide potential of the tidal current resource but, unlike tidal barrage power, there has not yet been a large-scale prototype study. It is anticipated that this will soon be remedied.

7.2.4.2 Extracting energy from tidal currents

The physics of the conversion of energy from tidal currents is similar, in principle, to the conversion of kinetic energy in the wind. Many of the proposed devices have, therefore, an inevitable, though superficial, resemblance to wind turbines.

There is no total agreement on the form and geometry of the conversion technology itself. Wind power systems are almost entirely horizontal axis rotating turbines, as shown schematically in Fig. 7.2-15. In these systems, the axis of rotation is parallel to the direction of the current flow.

In horizontal axis designs, the rotational axis is parallel to the direction of the water flow. Many developers favor

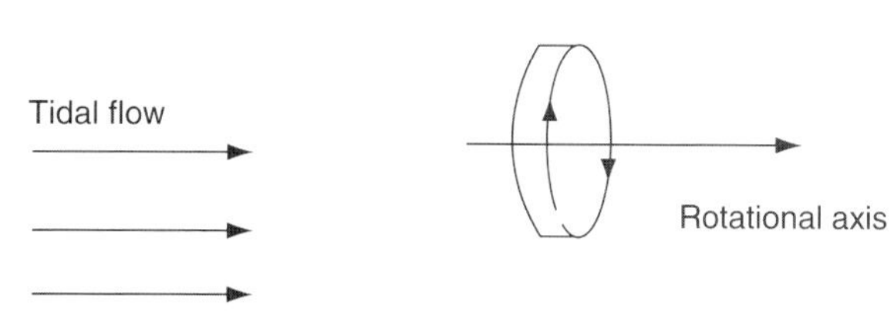

Figure 7.2-15 Schematic of a horizontal axis tidal current turbine.

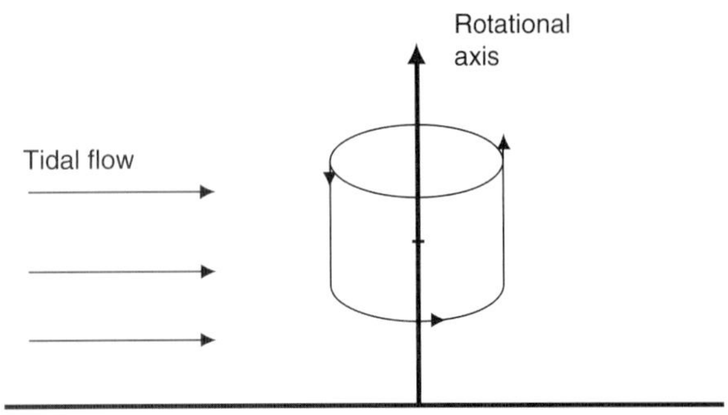

Figure 7.2-16 Schematic of a vertical axis tidal current turbine.

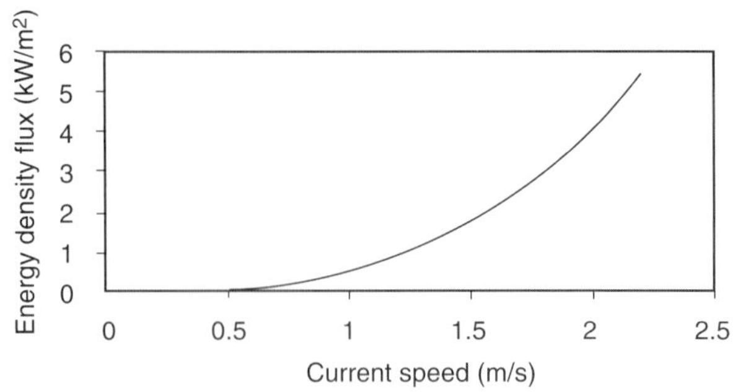

Figure 7.2-18 Influence of current speed on available energy flux density.

this geometry for tidal conversion. Vertical axis systems, such as that shown schematically in Fig. 7.2-16, in which the axis of rotation is perpendicular to the direction of current flow, have not been rejected, however.

The environmental drag forces on any tidal current energy conversion system are large when compared with wind turbines of the same capacity. This poses additional challenges to the designer. Designs exist for devices that are rigidly attached to the seabed or are suspended from floating barges, as shown in Fig. 7.2-17. It is generally accepted that fixed systems will be most applicable to shallow water sites and moored systems for deep water. There may be exceptions to this, however.

The density of seawater, at approximately 1023 to 1025 kg/m^3, is substantially greater than that of air. This, combined with the rapid currents experienced in many

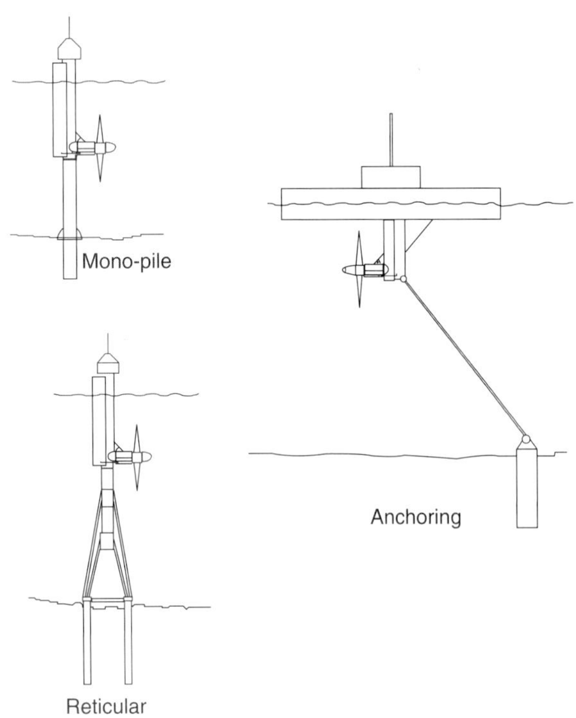

Figure 7.2-17 Alternative location systems for tidal current technology.

sites across the world, suggests that very-high-energy density fluxes will be encountered. Figure 7.2-18 shows the anticipated relationship between current speed and incoming energy flux density. Spring current speeds exceeding 3 m/s are encountered in many locations, which suggests that the exploitable resource will be considerable. The challenge for the system designer is to ensure that the system can convert energy efficiently throughout as much of the tidal cycle as possible while being robust enough to survive the inevitable environmental loading that will be experienced. The difficulty of this task should not be overestimated. In practice, the efficiency of conversion of a practical system is likely to be limited to approximately 35%, but the potential for energy conversion in tidal current speeds, which might exceed 3 m/s in spring tides, is considerable.

The environment tidal devices will operate in is very different from that experienced by wind turbines, and there are some rather difficult problems associated with installation, survivability and maintenance that need to be solved before true commercial exploitation can be achieved. Many industrial, commercial, and public bodies have suggested that there is a high degree of synergy between the development of a tidal current generation industry and the offshore oil and gas industry. This offers the intriguing prospect of a new renewables industry developing in partnership with the petroleum industry and could, perhaps, result in accelerated development, due to the availability of expertise and technology, which would otherwise have to be developed from scratch.

Unlike the wind, tides are essentially predictable as they derive from astronomic processes discussed earlier in this article. Wind power systems are dependent on random atmospheric processes, which result in it being difficult to integrate large wind power developments into strategic electricity distribution networks. The predictability of the tides will make this integration much easier. Figure 7.2-19 shows the predicted power output for one design concept over a spring/neap tidal cycle, in which the spring peak current speed is 3 m/s and the

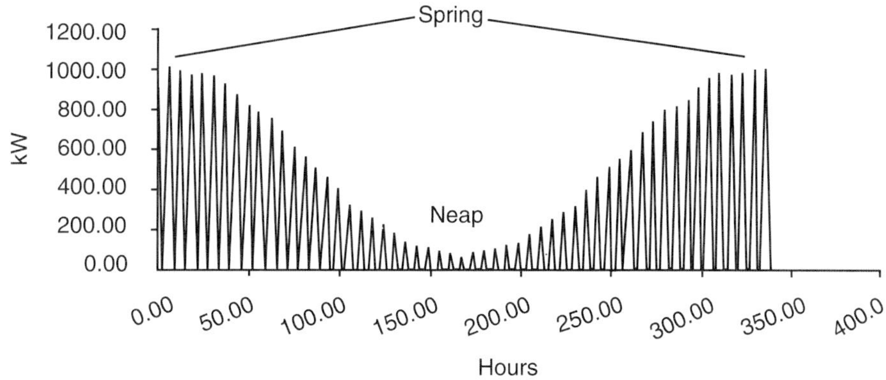

Figure 7.2-19 Predicted power output for a constant speed, 20- m diameter horizontal axis turbine.

neap peak current speed is 1.5 m/s. Although there would be some deviation from this prediction, as a result of weather related influences, the predictable nature can be clearly seen.

7.2.4.3 Testing the technology for tidal current energy extraction

At the time of writing, there is no commercial generation of electricity from tidal currents anywhere in the world. It is anticipated, however, that two large prototype systems will be installed in European waters in 2002. These devices—one of which will be located close to Lynmouth in the Bristol Channel, which lies between England and Wales, as shown in Fig. 7.2-20, and the other in Yell Sound in Shetland, as shown in Fig. 7.2-21—will be capable of producing approximately 300 kW each.

The two proposed systems are radically different in engineering concept, although both rely on the lift induced by fluid flow on an appropriately shaped lifting surface. The Lynmouth device, which is shown in Fig. 7.2-22, is a horizontal axis turbine, which resembles a traditional wind turbine.

The Yell Sound device, which is shown in Fig. 7.2-23, is a linear lift-based device, in which a reciprocating hydrofoil, which has a variable pitch angle, is used to pump oil, which then drives a hydraulic motor.

It is likely that, if tidal currents are to be commercially exploited, the generators will have to be mounted in clusters (tide farms?). If this is done, then, as with wind turbines, the devices will have to be sufficiently spread to ensure that the turbulence from individual devices does not interfere with others in the cluster. The nature of tidal currents, in which the most energetic flows are bidirectional in nature may make it possible to use relatively close spacing of devices perpendicular to the flow, with much larger spacing in the direction of the flow as seen in Fig. 7.2-24. This also shows one of the devices undergoing maintenance by a surface vessel.

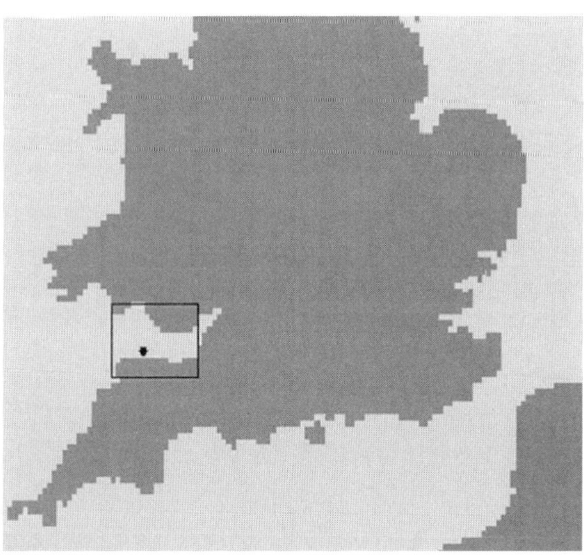

Figure 7.2-20 Location of the Lynmouth device in the Bristol Channel.

7.2.4.4 Barriers to the development of tidal current generation

At present, there is little worldwide experience in harnessing tidal current energy. Experimental turbines

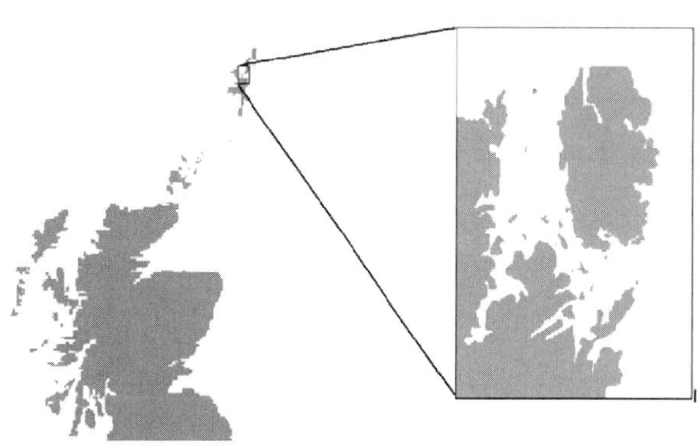

Figure 7.2-21 Location of Yell Sound in Shetland.

Figure 7.2-22 Artist's impression of the Lynmouth turbine being installed in the Bristol Channel. Illustration reproduced with the kind permission of Marine Current Turbines Ltd.

Figure 7.2-24 Artist's impression of an array of twin rotor tidal current devices. Illustration reproduced with the kind permission of Marine Current Turbines Ltd.

have been used in Japan and the United Kingdom, but as yet they have never been used to generate electricity for commercial consumption. Scottish trials in Loch Linnhe, using a floating horizontal axis system, demonstrated that

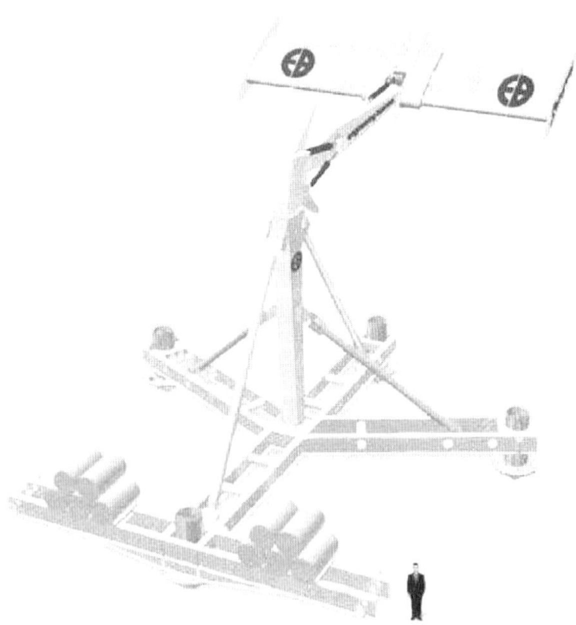

Figure 7.2-23 Artist's impression of the Stingray device being installed in Yell Sound. Illustration reproduced with the kind permission of The Engineering Business Ltd.

the technology is feasible. Experience in Japan has been largely with vertical axis systems. Until a large-scale device is commissioned and has demonstrated its performance for an extended period, there will be uncertainties about the long-term reliability and maintenance. Access to machines located in high energy tidal streams will always be problematic and could prove to require expensive maintenance procedures. Similarly, the rapid tidal currents themselves could give rise to difficulties in transferring energy via seabed cables to the coastline. Indeed some studies have suggested that the cable and its installation could be responsible for more than 20% of the total cost of a tidal current system. It is still too early to make reliable predictions of the cost of tidal current generation. There are still many uncertainties about the long-term reliability of systems and the difficulties associated with routine maintenance. The UK government's Department of Trade and Industry has indicated, however, that it anticipates that devices based on the Seaflow prototype should be able to generate electricity with a cost similar to that of onshore wind systems, assuming that problems associated with installation of multiple devices can be solved. Completion of the present program of prototype studies will allow more confidence to be given to predictions.

Environmental impacts from tidal current systems should be minimal, if projects are planned to take environmental issues into account. Energetic tidal channels do not tend to be home to many aquatic species and, although some species of marine mammals use tidal channels in their migration, the slow motion of the turbines and careful design should ensure minimal environmental impact. The most serious environmental impacts are likely to result from the extraction of kinetic energy itself, resulting in changes in the hydraulic nature of the flow environment. This could result in modification to the sediment transport systems and, of course, changes in the tidal currents which are providing the energy. Many experts assume that up to 10% of the kinetic energy in a tidal flow can be extracted without serious flow degradation.

7.2.5 The future of tidal power

The high capital costs associated with tidal barrage systems are likely to restrict development of this resource in the near future. What developments do proceed in the early 21st century will most likely be associated with road and rail crossings to maximize the economic benefit. In a future in which energy costs are likely to rise, assuming that low-cost nuclear fusion or other long term alternatives do not make an unexpectedly early arrival, then tidal barrage schemes could prove to be a major provider of strategic energy in the late 21st century and beyond. Under some local conditions, small-scale barrages might also prove attractive. The technology for tidal barrage systems is already available, and there is no doubt, given the experience at La Rance, that the resource is substantial and available.

It is likely that tidal current systems will appear in experimental form in many places around the world. Initially these will be rated at around 300 kW, such as those planned for the Lynmouth and Yell Sound demonstrations. If these schemes prove successful, then the first truly commercial developments may appear soon. Systems may not presently have the strategic potential of barrage systems but, in the short term at least, they do offer opportunities for supplying energy in rural coastal and island communities. In the longer term, massive sites such as the Pentland Firth, which has been estimated to have the potential to support a 16-GW-capacity development, could become strategically important.

Glossary

barrage An artificial barrier across an estuary or river to control the flow of water.

gigawatt (GW) The unit of power defined as 1 billion (1,000,000,000) watts.

gigawatt hour (GW h) The unit of energy defined as that delivered by a source of one GW in 1 hour.

terawatt (TW) The unit of power defined as 1000 billion (1,000,000,000,000) watts.

terawatt hour (TWhr) The unit of energy defined as that delivered by a source of 1 TW in 1 hour.

tide The motion of the world's seas and oceans induced by the gravitational and centrifugal influence of the earth, moon, and sun.

turbine A rotating mechanical device, which converts the kinetic energy of a flowing fluid into mechanical motion.

Further reading

Baker, C. (1991). "Tidal Power." Peter Peregrinus, Stevenage, UK.

Banal, M., and Bichon, A. (1991). "Tidal Energy in France. The Rance Tidal Power Station-Some Results After 15 Years in Operation." Proceedings of the Second International Symposium on Wave and Tidal Energy, Cambridge.

CENEX. (1993). "Tidal and Marine Currents Energy Exploitation." European Commission, DGXII, JOU2-CT-93–0355.

Charlier, R. H. (1983). "Alternative Energy Sources for the Centralised Generation of Electricity." Adam Hilger, Bristol, UK.

Department of Energy (UK) (1989). "The Severn Barrage Project Summary: General Report." Energy Paper 57, HMSO.

Energy Technology Support Unit (ETSU) (1993). "Tidal Stream Energy Review" (ETSU T/05/00155/REP), DTI (UK).

Sodin, H. (2001). "Scotland's Renewable Resources 2001." Scottish Executive Publications, 285/GR/04/2001, Edinburgh.

Watson, W. (1992). Tidal power politics. *Int. J. Ambient Energy* **13**(3), 1–10.

Wilson, E. (1988). A tide in the affairs of men. *Int. J. Ambient Energy* 9(3), 115–117.

Section Eight

Geothermal power

Chapter 8.1

8.1

Geothermal power

8.1.1 Introduction

A geothermal resource can be simply defined as a reservoir inside the Earth from which heat can be extracted economically (cost-wise less expensive than or comparable with other conventional sources of energy such as hydroelectric power or fossil fuels) and utilized for generating electric power or any other suitable industrial, agricultural or domestic application in the near future. A geothermal reservoir can contain heat both in the solid rock as well as in the fluids that fill the fractures and pore spaces within the rock. Estimates of geothermal resources are made on the basis of geological and geophysical data such as (i) depth, thickness and extent of geothermal aquifers, (ii) properties of rock formations, (iii) salinity and geochemistry of fluids likely present in the aquifers, and (iv) temperature, porosity and permeability of rock formations. A geothermal resource is distinct from a geothermal reserve, which refers to the part of a resource that can be extracted economically at the present price level. Reserves are confirmed on the basis of detailed reservoir datasets obtained invariably by deep drilling into potential resource areas. Therefore, the main factors in estimating reserves are the cost of drilling and the quality of available data on subsurface rock formations.

Radioactive decay of long-lived isotopes, particularly those of potassium, uranium and thorium, continuously generates heat within the Earth. The heat content of the Earth is estimated to be 1.3×10^{31} J. Heat is lost from the Earth's surface at an average rate of $\sim$80 mW m^{-2}. In most areas, this heat reaches the Earth's surface in a diffuse state, making it uneconomical to exploit this vast heat resource. It is believed that heat transfer below the lithosphere is mostly by convection and in the lithosphere by conduction. Rocks are relatively poor conductors of heat. Below nine-tenths of the Earth's surface, the thermal gradients vary from about 10°C km^{-1} to about 60°C km^{-1}. Consequently, the temperatures encountered at depths of a few kilometers are in excess of the Earth's surface temperatures by only an order of 100°C and power generators working at such small temperature differences have very low efficiency. Recent advances in binary-cycle geothermal plant technologies have made it possible to utilize such moderate temperature resources at shallow depths for low-to-moderate scale power generation. With the improvement of drilling technology, it may become economical to drill deeper, say in excess of 5 km, and reach the required high temperatures of the order of 200–300°C. Nevertheless, it is not sufficient only to reach the hot rock. Additionally, heat needs to be extracted by circulating fluids and for this process to be effective, there should exist an abundance of pore space and fissures for the fluids to permeate and circulate. At greater depths, the weight of the overlying rocks tends to close the pores and fissures, reducing the permeability considerably. Therefore, although hot rocks at depth exist almost everywhere, the current technological barriers, as well as unsuitable geological conditions, make it uneconomical to extract geothermal energy at most places.

Under some geological situations, however, such as the plate boundaries and sometimes well within the plates (as e.g., in active or geologically young volcanoes associated with mantle hotspots such as in the Hawaii island), heat may be locally transferred within a few kilometers of the Earth's surface through the process of convection by magma or molten rocks. The magma has temperatures in the neighborhood of 1000°C and interacts with the near-surface rocks, causing surficial

Renewable Energy Focus Handbook 2009; ISBN: 9780123747051

manifestations of geothermal activity such as the hot springs, geysers and fumaroles. Under certain suitable geological conditions, the heat becomes trapped, forming heat reservoirs. In such areas, after drilling a few hundred meters, temperatures of the order of 200–300°C are found and the regions could be suitable for harnessing the geothermal energy. The major producing geothermal fields of the world exploit such situations.

8.1.2 Types of geothermal systems

The essential requirements for a geothermal system to exist are (1) a large source of heat, (2) a reservoir to accumulate heat, and (3) a barrier to hold the accumulated heat. There is a suite of geological conditions that could result in a variety of geothermal systems. Consequently, all geothermal fields differ from one another. However, depending upon certain common characteristics, these can broadly be classified into the following categories: (1) vapor-dominated, (2) hot water, (3) geopressured, (4) hot dry rock (HDR), and (5) magma. The above-mentioned categories are briefly discussed here.

8.1.2.1 Vapor-dominated geothermal fields

Most of the presently exploited geothermal fields contain water at high pressures, and temperatures in excess of 100°C. When this water is brought to the Earth's surface, the pressure is considerably reduced, generating large quantities of steam, and a mixture of saturated steam and water is produced. The ratio of steam to water varies from one site to another. Some of the best-known geothermal fields, such as Cerro Prieto (Mexico), Wairakei (New Zealand), Reykjavik (Iceland), Salton Sea (U.S.A.) and Otake (Japan), belong to this category. Since steam associated with water is produced in these fields, they are known as wet steam fields. There are a few other important geothermal fields such as Larderello (Italy) and The Geysers (U.S.A.) which produce superheated steam with no associated fluids. Fields of the latter type are known as dry steam fields.

The basic requirements of the vapor-dominated geothermal fields, whether of dry steam or of wet steam type, include adequate supplies of water in addition to the three prerequisites mentioned earlier. Fig. 8.1-1 is a schematic model representation of a vapor-dominated field.

Heat source

The fact that vapor-dominated geothermal fields are situated in regions of recent (Miocene-Quaternary) volcanism, some of them being located on or close to volcanoes, has verified that magma is their source. Young, high-temperature (500–1000°C) magma intrusions to within depths of a few to several kilometers from the Earth's surface allow the necessary heat to be accumulated in economical quantities. In hard compact rocks, faulting may provide a channel for the magma to reach the surface. Soft or plastic rocks, when present, can flow and block the fault space, causing the magma to spread at the contact between the soft and the hard rocks. Active volcanoes, fumaroles, hot springs and geysers are obvious surface manifestations of recent volcanic activity. In addition, certain geological environments, such as regions of Quaternary uplift and regions of Late Tertiary and Quaternary subsidence, are indicative of shallow magmatic intrusions.

Reservoir and water supply

In order to form a heat reservoir, the anomalous magmatic intrusion should encounter porous and permeable, water-filled rock strata. Within the reservoir, convection currents of hot water and/or steam are set up, providing a good heat exchange, and the temperature difference between the top and bottom of the reservoir is not very significant. A variety of rocks have been found to constitute good reservoirs. At Larderello (Italy), it is fractured limestone and dolomite; at The Geysers (U.S.A.), it is fissured graywacke; at Wairakei (New Zealand), it is pumiceous breccia and tuff; and at Cerro Prieto (Mexico), it is deltaic sands. Good reservoirs could also be formed at geological unconformities and formation boundaries, provided that they are permeable and have good hydraulic continuity and water supply.

The origin of geothermal fluids has been debated in the past. In addition to a meteoric origin, magmatic and juvenile origins for geothermal fluids have been suggested. However, recently conducted isotopic studies in geothermal fields have shown that at least 90% of the geothermal water has a meteoric origin. The permeable aquifers forming the reservoir must therefore have hydraulic continuity with large recharge areas for the rainwater to be available in continuous supply. The freshly supplied water is heated conductively at the impermeable base of the reservoir. Withdrawing of the heated reservoir fluid through boreholes, or its upward movement through vents and fissures, disturbs the hydrological balance. This is restored, fully or partially, by the inflow of new water. An idea of the amount of the inflow can be had from the fact that a natural steam field operating a 100 MW power plant lets out between 1000 and 2000 tons of water every hour. Some of the geothermal fields, such as the Larderello in Italy, have easily identifiable recharge areas. At Larderello, the permeable reservoir terrain, consisting of Mesozoic limestones and dolomites, outcrops thereby providing an easy access to surficial water.

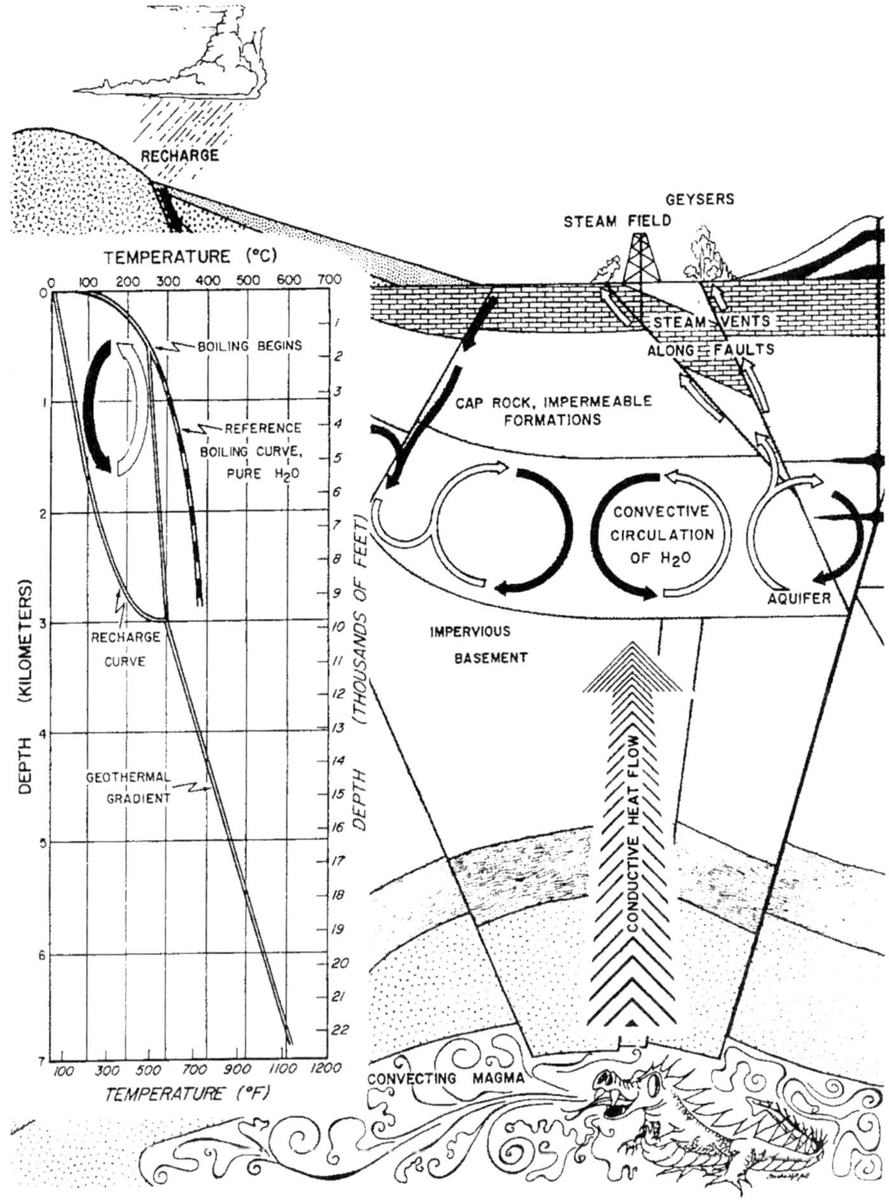

Fig. 8.1-1 Conceptual model of a steam geothermal system.

Cap rock—the barrier

An impermeable cap rock, or a cap rock with low permeability, overlying the reservoir, is necessary to prevent the escape of hot reservoir fluids through convection. The heat loss through conduction is not prevented by the cap rock. However, the amount of heat conducted is much smaller than that which could be lost through possible convection. Since volcanism is associated with tectonic movements causing fissures, ideal unfissured impermeable cap rock is nowhere to be found. The geochemical processes associated with geothermal fields, i.e., hydrothermal alteration of rocks and mineral deposition, are helpful in sealing off the fissures. Typical examples of cap rocks rendered impermeable through chemical action and deposition are seen at The Geysers and Otake geothermal fields. At The Geysers, calcite- and silica-filled fractures, up to 1 in. wide, are commonly seen. Evidence of hydrothermal alteration is presented by the bleaching of graywacke as well as by the absence of vegetation in patches. The geochemical and hydrothermal processes are complicated and vary from place to place.

At many other steam-producing fields, original impervious rocks constitute the cap rock. Examples are the lacustrine Huka Formation at Wairakei (New Zealand), the deltaic clay at Cerro Prieto (Mexico) and Salton Sea (California) and the Flysch Formation at Larderello (Italy).

8.1.2.2 Hot water geothermal systems

In hot water geothermal fields, water-convection currents carry the heat from the deep source to the shallow reservoir. The bottom of the convective cell may be heated through conduction from hot rocks. The geology of hot water geothermal fields is quite similar to that of an ordinary groundwater system. They differ from the earlier-discussed vapor-dominated geothermal fields in the fact that the hot water geothermal fields are characterized by liquid water being the continuous pressure-controlling fluid phase. Typically, the temperature of hot-water reservoirs varies from 60 to 100°C and they occur at depths ranging from 1500 to 3000 m.

Fig. 8.1-2 is a schematic representation of a hot-water field and the reference curve represents the variation of the boiling point of the pure water with depth. As shown in Fig. 8.1-2, a hot water geothermal field could develop in the absence of a cap rock, if the thermal gradients and the depth of the aquifer are adequate to maintain a convective circulation. When the cap rock is absent, the temperatures in the upper part of the reservoir cannot exceed the boiling point at atmospheric pressure, since with the convective rise, the water loses pressure and also becomes mixed with the cool groundwater.

Depending upon the temperature, chemistry and the structure of the reservoir, hot-water systems have been

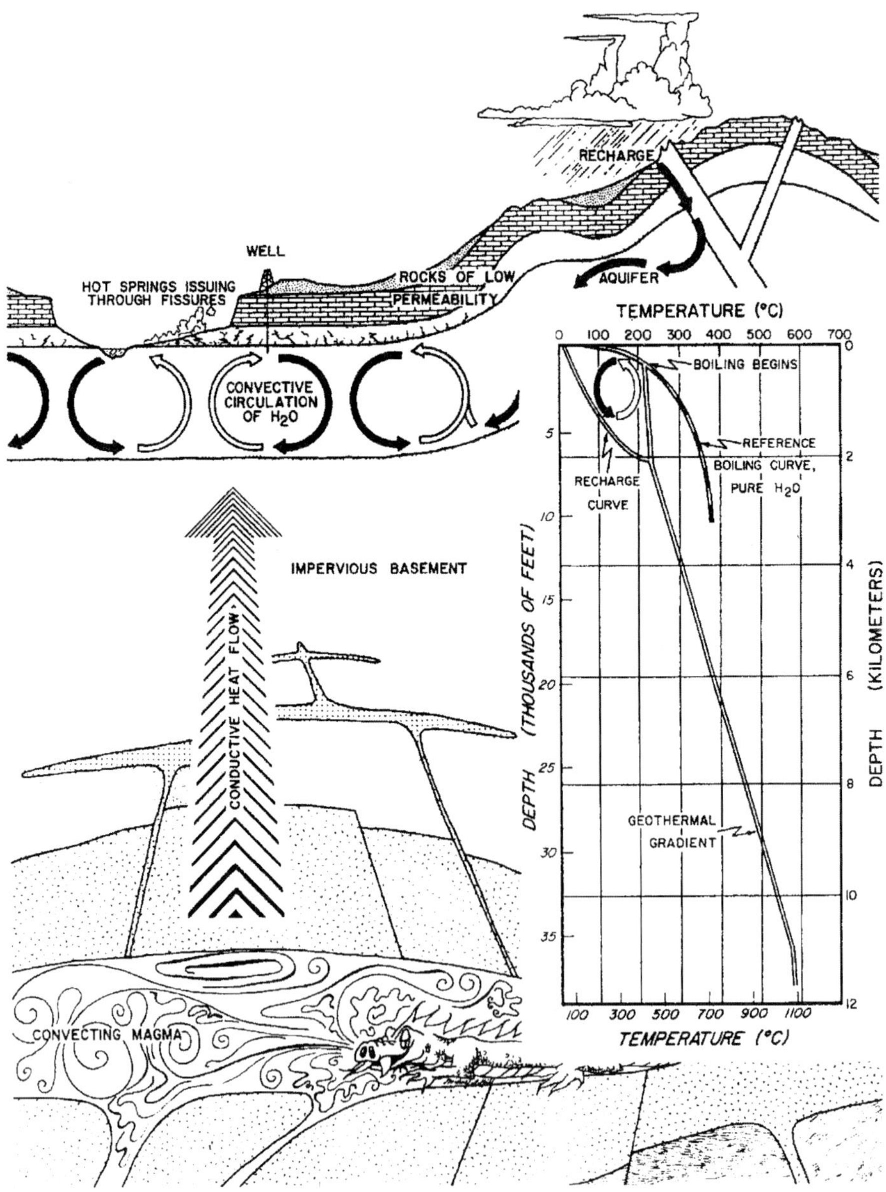

Fig. 8.1-2 Conceptual model of a hot water geothermal system.

classified into several subtypes. The following is a brief description of various subtypes:

1. Systems characterized by low-to-moderate temperatures (say, 50–150°C) and producing water with a chemical composition similar to the regional surface and shallow groundwaters.
2. Systems characterized by the presence of partly non-meteoric water. Such systems usually occur in deep sedimentary basins.
3. Systems characterized by the presence of brine of very high salinity. The chemistry can vary considerably from one field to another. The Salton Sea (California) and the Red Sea brine pools belong to this subtype and have very differing bulk chemistry of the sediments and the associated rocks, probably attributable to the difference in brine composition.
4. Systems characterized by the presence of natural cap rocks. Geothermal fields at Cerro Prieto (Mexico) and Salton Sea (California) have cap rocks constituted by fine-grained, low-permeability sediments.
5. Systems characterized by the creation of their own self-sealing cap rocks. As explained earlier in discussing vapor-dominated geothermal systems, these cap rocks are formed through chemical alteration and deposition of sediments near the surface where the temperature decreases suddenly. Wairakei (New Zealand) and Yellowstone Park (Wyoming) are typical examples.

8.1.2.3 Geopressured geothermal resources

A type of hydrothermal environment whose hot water is almost completely sealed from exchange with surrounding rocks is called a geopressured system. Such systems typically form in a basin in which very rapid filling with sediments takes place, resulting in higher than normal pressure of the hydrothermal water. Geopressured geothermal systems were first identified in the deep sedimentary layers underneath the Gulf of Mexico at a depth between 6 and 8 km with pore pressures of up to 130 MPa and temperatures in the range 150–180°C.

Systematic studies on geopressured geothermal resources were carried out during the 1970s and early 1980s in the Gulf of Mexico geosyncline. This region served as a natural laboratory for studies aimed at understanding the mechanics and geology of the formation of the geopressured geothermal systems because of the wealth of geological information available from field surveys conducted for petroleum exploration. The region has witnessed large-scale subsidence in geosynclines involving downward transformation of enormous amounts of sediments to depths of 5–10 km in relatively short geological time intervals. As evidenced in the Gulf of Mexico, normal faulting with throws of up to 1 km is the most important structural feature. This process of subsidence subjects the poorly consolidated sediments to intense heat and enormous pressure. As has been pointed out, the process of geosynclinal subsidence and rapid sedimentary deposition is analogous to igneous intrusion in an opposite sense: the motion is downward, the intruding material is cold and the rocks intruded are hot. If the sediments undergoing subsidence are constituted of clay of a swelling variety, endothermic diagenesis of clay minerals can consume much heat and thereby reduce the heat flow to the Earth's surface. This phenomenon is very noticeable in the Gulf of Mexico Basin where sediments are mostly of the swelling variety of clay and consequently the geothermal gradient map of the area shows it as a negative geothermal area (Fig. 8.1-3). Based upon measurements in holes less than 2 km deep, this map shows a decrease in geothermal gradient along the axis of the geosyncline. Obviously, being dependent on shallow heat flow measurements, the map does not indicate (or conceals) the true magnitude of the existing geothermal resource potential in the northern Gulf of Mexico Basin. It has been observed that the geothermal gradient in the hydropressured zones at depths ranging from 1 to 2 km varies from 20 to 40°C km^{-1}, and it undergoes a very sharp increase in the geopressured zone; being two to three times greater than that existing in the overlying hydropressured zone.

The deeply buried clays in the geosynclines undergo thermal metamorphism, releasing petroleum hydrocarbons and geothermal waters, the former being solute and the latter solvent, and a solution is formed. The strength of the load-bearing clay bed is drastically reduced by the process of metamorphism, resulting in the expulsion of the pore fluids to share the overburden load. Since the confining pressure decreases most rapidly in the upward direction, and the pressure applied to a confined fluid is transmitted equally in all directions, the expelled pore fluid moves out of the clay bed and moves upwards through sand beds and along the fault planes. The expelled solution of hydrocarbons and water, as it moves upward, encounters much lower pressures and temperatures. Finally, a stage is reached when the temperatures and the pressures are so low that the solute hydrocarbons can no longer remain in the solvent water. Fluid hydrocarbons and geothermal waters then appear as separate phases, and the hydrocarbons being lighter than the water seek higher points in the regional flow system and structural traps. Detailed geological mapping of structures and knowledge of sedimentary history, when combined with the isothermal, isosaline and isopressure

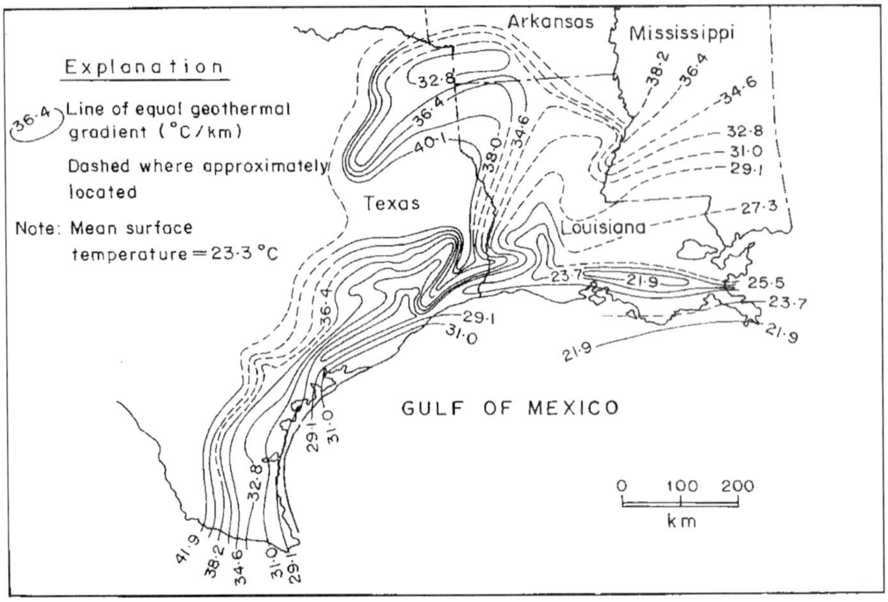

Fig. 8.1-3 Geothermal gradients in the Gulf Coast of the United States (from Jones, 1970). For details, see text.

maps are helpful tools in prospecting fluid hydrocarbon and geothermal regimes.

To date, potential geopressured geothermal fields have been discovered mainly in the Texas–Louisiana Gulf Coast region. Similar systems may exist in other hydrocarbon-bearing deep sedimentary basins elsewhere. Although the huge potential of geothermal-geopressured aquifers has been recognized, the commercial development has been considered marginally economic in only special circumstances. Availability of cheap fossil fuels has been a deterrent to further research and analysis of their energy potential. In the future, new technologies including use of binary-cycle plants may allow more efficient extraction of thermal energy from geopressured brines.

8.1.2.4 Hot dry rock geothermal systems

So far we have discussed geothermal resources associated with one or another type of fluid convection system, the fluid (mostly steam and water) being responsible for carrying the heat to the Earth's surface or to the shallower depths from which it could be exploited. Occurrences of such hydrothermal resources are restricted to countries with favourable geological conditions like in the plate boundary zones. However, there exists another category of geothermal resource where geothermal heat is stored in the hot and poorly permeable rocks at shallow depths within the Earth's crust, without any fluid availability to store or transport the heat. These resources are designated HDR. Large volumes of such rocks at high temperatures are known to exist below all major geothermal areas. Geologically young igneous intrusive bodies at shallow levels of the Earth's crust, which form potential targets for this energy resource, occur in several continental areas. After having described the *hot dry rock* geothermal resources in the above manner, three questions arise immediately: "how deep?", "how hot?" and "how dry?" Different depth and temperature ranges as well as definitions of "dryness" can be found in the literature. Here we adopt the definition given by the U.S. Energy Research and Development Administration (ERDA) in their Report No. 1 of the HDR Assessment Panel to answer these questions concerning the HDR geothermal resource. Accordingly, it is "heat stored in rocks within 10 km of the Earth's surface from which the energy cannot be economically produced by natural hot water or steam". Additionally, it cannot produce an economical volume of hot water or steam and the temperature is less than about 650°C (so as to exclude molten lava or magma). Quite understandably, HDR may be associated at some locations and/or at some depths with hydrothermal resources and/or magma.

Depending upon the cause of the HDR geothermal resources, they can broadly be classified into three categories. These are: (1) igneous related: heat being transferred from magma or stored in dry rocks surrounding the magma bodies; (2) upper mantle related: the heat being conducted to shallow crust overlying an unusually hot upper mantle, causing the anomaly; and (3) local: heat being stored locally either due to the presence of a high concentration of radioactive minerals or due to large-scale faulting and/or fracturing.

Considering the above classification of HDR based upon the different possible sources, the obvious places to look for them are the regions characterized by: (1) recent volcanism, (2) high-heat flow, and (3) localized radiometric heat sources. All magma chambers have a surrounding hot rock environment of varying size. Similarly, geothermal resource areas characterized by hydrothermal convective systems may, at some depth, be underlain by HDR. Also, high-heat flow areas, where the basement geology is favorable for low water contents, are possible HDR resource areas.

HDR technology envisages exploitation of Earth's heat stored in the high-temperature and impermeable rocks by artificially creating a fracture system at depth (that acts as a heat exchanger) and circulating water from an injection borehole towards a production borehole. Hydraulic fracturing, which involves injection of water at very high pressure into a reservoir to create new fractures or enlarge preexisting cracks, has been one of the successful methods in creating permeability in the rocks at depth. It has been estimated that cooling of 1 km^3 of hot rock by 100°C will enable operation of a 30 MWe geothermal power plant for 30 years. However, generation of large heat exchangers at depth and controlling the loss of circulation fluids present the biggest technological challenges in exploitation of HDR geothermal energy.

Detailed research and experimentation in HDR technology was first carried out at Fenton Hill, near the edge of the volcanically young Valles Caldera, near Los Alamos in New Mexico, where temperatures of the order of 195°C at a depth of 3 km were observed in the granitic basement. Scientists and technologists of the Los Alamos National Laboratory, New Mexico were able to obtain a thermal power output of 5 MW by drilling a pair of injection and production wells and creating a permeable rock volume by hydrofracturing the rocks surrounding the bottom parts of the wells. Although they were able to demonstrate the feasibility of the technology, further tests were discouraged by the high costs involved for commercial exploitation. Over the past decade or so, renewed interest in HDR geothermal energy led to several pilot projects in central Europe, Britain, Russia, Japan and Australia. The biggest success story in the last 30 years has been the one at the Soultz-sous-Forêts area located in the Upper Rhine graben near the boundary between France and Germany, where setting up a pilot HDR project to produce 6 MW electricity is nearing completion. Drilling up to 5 km depth in the area confirmed a temperature of ~200°C. This project could be the forerunner for several small HDR projects in the Upper Rhine graben region. Another pilot HDR project is underway in Cooper Basin, Australia, an area of very high heat flow. The basin is underlain by high heat producing granites (7–10 $\mu W\,m^{-3}$), and temperatures of ~240°C have been measured at depths up to 3.7 km in several oil and gas wells. The first borehole (Habanero-1), drilled under the HDR project, met with temperatures exceeding 250°C at 4.4 km.

The project holds promise for another successful demonstration of HDR technology in the next few years.

8.1.2.5 Magma

Magma, the naturally occurring molten rock material, is a hot viscous liquid, which retains fluidity till solidification. It may contain gases and particles of solid materials such as crystals or fragments of solid rocks. However, the mobility of magma is not much affected until the content of solid material is too large. Typically, magma crystallizes to form igneous rocks at temperatures varying, depending upon the composition and pressure, from 600 to 1400°C. At its site of generation, magma is lighter than the surrounding material, and consequently it rises as long as the density contrast between magma and surrounding cooler rocks continues. Eventually, magma either solidifies or forms reservoirs at some depth from the Earth's surface, or it erupts. Magma is the ultimate source of all high-temperature geothermal resources. Plate boundaries are the most common sites of volcanic eruptions. At several volcanic locales, magma is present within the top 5 km of the crust. It has been estimated that the average rate of production of magma at Kilauea Volcano, Hawaii, during 1952–1971 was about $10^8\,m^3$ $year^{-1}$. Similar estimates have been made for the Columbia Plateau and elsewhere. The heat energy available from such sources, if harvested, would constitute very large additions to the global energy inventory.

Extraction of thermal energy from magma was tested during the 1980s by drilling into the still-molten core of a lava lake in Hawaii. However, up to the present, the necessary technology has not been developed to recover heat energy from magma. Economical mining of heat energy from magma presents several practical difficulties such as locating such bodies accurately before drilling into them, the prohibitive costs of drilling and longevity of deployed plant materials in a hot corrosive environment.

Section Nine

Wind power

Chapter 9.1

9.1

Wind power

Wind is the movement of air in response to pressure differences within the atmosphere. Pressure differences exert a force which cause air masses to move from a region of high pressure to one of low pressure. That movement is wind. Such pressure differences are caused primarily by differential heating effects of the sun on the surface of the earth. Thus wind energy can be considered to be a form of solar energy.

Annually, over the earth's land masses, around 1.7 million TWh of energy is generated in the form of wind. Over the globe as a whole the figure is much higher. Even so, only a small fraction of the wind energy can be harnessed to generate useful energy.

One of the main limiting factors in the exploitation of wind power onshore is competing land use. Taking this into account, a 1991 estimate[1] put the realisable global wind power potential at 53,000 TWh/year. This figure is broken down by regions in Table 9.1-1. As the table shows, wind resources are widely dispersed and available in most parts of the world.

The figures in Table 9.1-1 are probably conservative because modern wind turbines are more efficient than those available when the survey was compiled. Even on this conservative estimate the resource is much larger than world demand for electricity. This is expected to reach 26,000 TWh, roughly half the global wind resource quoted above, by 2020.[2]

Table 9.1-2 shows estimates for the wind energy resources in the countries of Western Europe. A glance at these estimates will show that in many cases the national wind resources are again enormous. These figures, too, may represent an underestimate. For example, the potential UK generating capacity has been estimated by the UK Energy Technology Support Unit (ETSU) to be 223,000 MW, nearly four times the figure quoted in Table 9.1-2 and equivalent to an annual production of 660,000 GWh. The ETSU study used relatively conservative criteria to arrive at its estimate but did not take into account utilisation restrictions. Constraints on building close to population centres or in areas of natural beauty would severely limit available sites. Even with such constraints, the potential would remain vast.

Looking beyond Europe, a US wind potential survey was carried out in 1992.[4] It concluded that even with exclusions for environmental and land-use reasons, around 6% of the land area of the USA could be used for wind power generation. This area was judged capable, with some advance in wind turbine technology, of providing a generating capacity of 500,000 MW. The report also concluded that 12 states in the middle of the USA had sufficient potential to generate nearly four times the electricity consumed in the USA 1990.

In Asia, potential Chinese generating capacity has been put at 253,000 MW. Indian potential has been estimated at 20,000 MW but this is certainly a severe underestimate. Both countries are beginning to exploit their potential.

Wind power, though exploiting a renewable resource, is not considered beneficial by all. In the UK, and increasingly in Germany there are lobbies trying to prevent further development of onshore wind farms. This is proving a considerable handicap to wind development in the UK, at least. Under these conditions, offshore wind farming becomes increasingly attractive.

It is more expensive to build a wind farm offshore but this can be offset by higher average wind speeds. The global offshore resource has been estimated to be around 37,000 TWh.[5] Offshore sites are available in many parts of the world but the most promising for immediate development are around the coasts of northern and western Europe and of the eastern seaboard of the USA.

Renewable Energy Focus Handbook 2009; ISBN: 9780123747051

Table 9.1-1 Regional wind resources

	Available resource (TWh/year)
Western Europe	4,800
North America	14,000
Australia	3,000
Africa	10,600
Latin America	5,400
Eastern Europe and Former Soviet Union	10,600
Asia	4,600
Total	53,000

At the beginning of 2004 the global wind generating capacity was 40,000 MW[6]. It is expected to reach 150,000 by 2012. Offshore capacity at the end of 2003 was just over 500 MW.

Table 9.1-2 European wind energy resources

	Annual resource (TWh)	Potential capacity (MW)
Austria	3	1,500
Belgium	5	2,500
Denmark	10	4,500
Finland	7	3,500
France	85	42,500
Germany	24	12,000
Great Britain	114	57,000
Greece	44	22,000
Ireland	44	22,000
Italy	69	34,500
Luxembourg	–	–
Holland	7	3,500
Norway	76	38,000
Portugal	15	7,500
Spain	86	43,000
Sweden	41	20,500

Source: The figures in this table are taken from Windforce 12.[3]

9.1.1 Wind sites

The economics of wind power depend strongly on wind speed. The actual energy contained in the wind varies with the third power of the wind speed. Double the wind speed, and the energy it carries increases eightfold.

A 1.5 MW wind turbine at a site with a wind speed of 5.5 m/s will generate around 1000 MWh/year. At a wind speed of 8.5 m/s the output rises to 4500 MWh and at 10.5 m/s the annual output will be 8000 MWh. This is close the theoretical limit. Other factors will come into play at very high speeds, limiting turbine output. However, these figures indicate quite clearly that the selection of a good wind farm site is vitally important for the economics of a project.

The starting point for any wind development project, then, must be a windy site. But other factors come into play too. Wind speed varies with height; the higher a turbine is raised above the ground, the better the wind regime it will find. This will benefit larger wind turbines which are placed on higher towers, but larger turbines tend to be more efficient anyway, so additional advantages accrue.

Depending on the efficiency of a wind turbine, there is a cut-off wind speed below which wind power generation is not considered economical. This figure depends on the efficiency of wind turbine design as well as on the turbine cost. With the turbines available at the beginning of the twenty-first century, a wind speed as low as 5–5.5 m/s is considered economically exploitable at an onshore site. Since offshore costs are higher, an offshore wind speed of 6.5 m/s is needed to make a site economically attractive.

9.1.1.1 Locating a site

Prospective developers of wind energy projects will normally be able to refer to wind surveys in most of the developed countries in order to make a preliminary identification of sites suitable for wind farms. Wind energy associations exist in the UK, Europe and the USA, and the European Union (EU) also holds Europe-wide figures. In other parts of the world the wind data that is available may be less precise, though many countries are now taking greater interest in wind resources.

Once a potential site has been identified it must be studied in more detail to confirm that it is suitable. Long- and short-term wind speed measurements will normally be needed to ascertain the wind regime. Only when these figures are available can the economics of the project be determined with any accuracy. Figures for at least one full year will normally be required, longer if possible.

Offshore projects require the same attention as onshore schemes but offshore wind data is less likely to be available. European offshore surveys exist and there have been some limited surveys of offshore North American sites. It is possible to gain an estimate of the wind regime in an offshore area from satellite images. These can provide an indication of sea roughness from which wind speed can be calculated. As with an onshore site, accurate measurements over at least a year will then be needed to confirm the local wind regime.

9.1.1.2 Turbulence

When wind passes over land the unevenness of the ground and interference to wind flow from trees or undergrowth will cause a significant amount of turbulence. Turbulent air creates an additional strain on a wind turbine blade, accelerating the onset of fatigue damage. In order to limit this damage as much as possible wind turbine designers will normally place the turbine on a tower which is tall enough to raise the blades above this turbulent layer of air.

The wind offshore is generally less disturbed because the surface of the sea is smoother, resulting in a thinner turbulent layer and less overall turbulence. Waves in rough seas will increase turbulence and wave height itself needs to be taken into account offshore. Turbine blades should be lifted high enough to avoid the highest waves likely at a particular site. Generally, however, tower heights offshore can be lower than onshore. In both cases site measurements will be needed to ascertain what the optimum turbine height should be.

9.1.2 Wind turbines

Like hydropower, wind power has a long and successful history. The earliest known record is from Hero of Alexandria, who described a wind machine in the first century AD. The next recorded appearance is in Arabic texts of the ninth century which refer to a seventh century design.

Windmills soon spread from the Middle East into Europe. Post mills, in which the whole mill apparatus was mounted on a post so that it could be rotated into the wind, were known in France and England by the twelfth century. Tower mills, where only the top part of the windmill carrying the sails rotated, were introduced around the fourteenth century in France.

The industrial revolution brought refinements to windmill design but the number of mills began to decline with the advent of steam power. Even so there were still heavy concentrations of traditional mills in the Netherlands, where the Zaan district still boasted 900 windmills in the nineteenth century.

A new use for wind power developed with the invention of the wind pump, first used extensively among farmers in the USA. From the middle of the nineteenth century onwards, the classic lattice metal tower carrying a rotor with petal-shaped iron vanes crossed America and then travelled the world.

During the early part of the twentieth century there were several important experiments with the use of the wind to generate electricity, particularly in the USA and in Denmark. These failed to attract widespread attention. Finally, it was during the oil crisis of the early 1970s that modern interest in wind turbines took form. The main centres of development were the USA, particularly California, and Denmark.

From this period the basic wind energy conversion system for power generation has gradually taken shape. Today the basic system starts with a large rotor comprising two, three or four blades mounted on a horizontal shaft at the top of a tall tower. The blades intersect the wind and capture the energy it contains, energy which causes them to rotate in a vertical plane about the shaft axis. The slow rotation of the shaft is normally increased by use of a gearbox, from which the rotational motion is delivered to a generator. The electrical output from the generator is then taken through cables down the turbine tower to a substation where the power is eventually fed into the electricity grid. The mechanical components at the top of the turbine tower – the rotor, gearbox and generator – are all mounted on a platform that can pivot, or yaw, about a vertical axis so that the rotor shaft is always aligned with the wind direction. Gearbox and generator and housed within a weather-tight compartment called a *nacelle*.

9.1.2.1 Turbine size

The early wind turbines which were developed for power generation in the late 1970s and the early 1980s had generating capacities of around 30–60 kW. Hundreds of machines of this size were installed in wind farms in California.

Through the 1980s and into the 1990s, wind turbine capacities increased steadily. During the 1980s there were several pilot projects involving single wind turbines with capacities of over 1 MW but during the decade the standard turbine size tended to be between 300 and 500 kW.

During the 1990s, turbine unit size continued to increase steadily. By 1998, most new wind farms employed turbines with a capacity of between 600 and 750 kW. These modern, higher-output machines tended to provide greater efficiency than the smaller machines of the previous decade and the trend towards even larger machines continued. At the end of the 1990s, the typical wind turbine size had reached 1 MW.

As the new century dawned, manufacturers began to introduce a range of multi-megawatt machines. A unit of around 2 MW was the most common at the beginning of 2004 but larger machines are being installed and 5 MW machines are already being developed. These largest machines have blades up to 60 m in length, leading to rotor diameters of 120 m.

The largest machines are particularly popular for offshore developments where the high cost of a turbine foundation favours a large generating capacity. The end of the first decade of this century will almost certainly see wind turbines for offshore applications in the 6–10 MW range. It is not clear yet whether there is an ultimate limit to wind turbine size.

9.1.2.2 Horizontal or vertical?

As outlined above, the standard wind turbine has a vertical rotor attached to a horizontal shaft. This arrangement imposes certain restrictions on the wind turbine design. With a horizontal shaft, the rotor turns in a vertical plane and must be raised on a tower so that the blades are clear of the ground and of the turbulent layer of air next to it.

Gearbox and generator are attached directly to the turbine shaft so these, too, must be placed on the tower, high above ground. This raises the cost of both installation and maintenance. And a horizontal axis machine must include a yawing system so that the rotor and nacelle can be rotated as the wind direction changes.

There is an alternative, a vertical axis wind turbine. A vertical axis machine has all its weight supported by a ground-level bearing. Both gearbox and generator can also be placed on the ground, easing maintenance costs. And most designs for vertical axis wind turbines will operate with the wind blowing from any direction. A yawing system is unnecessary.

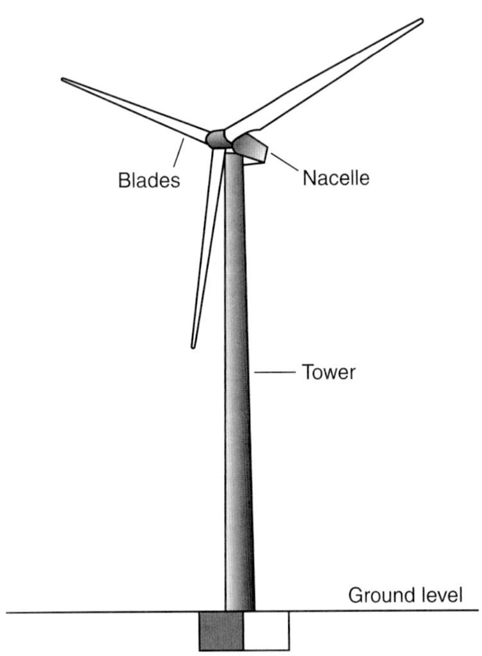

Figure 9.1-1 A horizontal axis wind turbine.

Several vertical axis designs have been tested in the past 30 years. The most exhaustively explored was the Darrieus wind turbine, which is based on a design patented by a French engineer, G.J.M. Darrieus, in 1931. Often described as the eggbeater, this wind turbine design comprises a pair of thin, curved blades with an aerofoil cross section attached to a vertical shaft and looking very much like an eggbeater.

A number Darrieus wind turbines have been built and tested and the design achieved short-lived popularity during the 1980s. Developments have included various pilot projects, a commercial machine and a prototype with an output of over 1 MW, developed in Canada. Other types of vertical axis turbine have also been built, including one with an H-shaped rotor which was operated at a test site in Wales. However, the vertical configuration has not yet achieved significant commercial success.

9.1.2.3 Rotor design

Rotors for modern wind turbines come in a variety of guises. The chief variables are the number of blades and the means used to control the rotor speed in high winds.

Rotor blades can be made from a variety of materials including metal, wood, composite materials and carbon fibre. Metal blades were frequently used in early designs but are rarely seen in modern machines. Under modern design practice, low weight is considered a desirable property and the most common blade materials in use are glass-reinforced plastic and wood epoxy. Carbon fibre-reinforced epoxy resin is being introduced by some manufacturers. Its combination of strength and lightness are extremely attractive.

The rotor shape is aerodynamically determined. It also depends on the control method used to prevent the rotor turning too fast when wind speeds are high. One option is stall control. A stall-controlled rotor uses fixed blades of a shape and orientation that causes the aerofoil to stall at high wind velocity, limiting its rotational speed. Stall-control machines tend to be more ruggedly built, and hence heavier than alternatives.

The second common option for speed control is pitch control. This involves a rotor with blades that can be twisted at different wind speeds to increase or decrease their aerodynamic effectiveness. Pitch control has an additional advantage of allowing some power-output control which can lead to better efficiency of operation. This system has become the most popular method of speed control.

Another design parameter is the number of blades on the rotor. Both weight and cost increase with the number

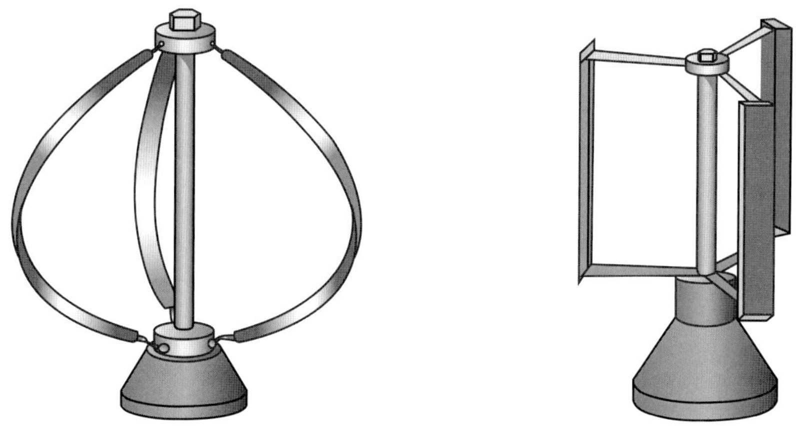

Figure 9.1-2 A vertical axis wind turbine.

of blades. Most current designs use three, though one, two and four blades have been tried.

A single-bladed machine offers the lowest-weight solution, but the single blade must be offset by a counterweight. As a result, two-bladed designs are often as light as the equivalent one-bladed machine when both rotor and nacelle are taken into account. The drawback with both one and two blades is an extra level of noise and although this can be reduced with design modification, a three-bladed rotor offers smoother rotational operation. Four blades offer good rotor balance but are heavier and less cost effective.

Rotor size is determined by the power output that the machine is designed to develop. The larger the diameter of the rotor, the more energy it can capture. A 600 kW wind turbine will have a rotor diameter of 40–50 m while a 5 MW machine needs a rotor diameter of around 120 m, rising to 140 m for 6 MW. Rotor sizes for offshore machines tend to be slightly larger than those of onshore machines for a similar output.

Rotational speed is also a factor in rotor design. It is normally necessary to keep the speed at which the tip of the rotor blades move through the air to below 70 m/s in order to minimise aerodynamic noise which can become an environmental problem. This means that as the rotor diameter increases, rotational speed must decrease. Current 2–3 MW designs normally rotate at between 5 and 20 rpm.

9.1.2.4 Tower design

Towers for wind turbines have been built in a variety of styles including steel lattice and cylindrical steel or concrete designs. The lattice style used to be common but has generally been replaced by cylindrical designs, predominantly of concrete but with steel used for some machines.

A tall tower will place the rotor in a region of high wind speed but will cost more. Thus tower height will normally be determined by the rotor diameter and by the need to avoid the layer of turbulent air close to the ground.

The bending frequency of the tower must also be taken into consideration. Excitation of this frequency could lead to structural damage in the same way as excitation of the natural frequency of a bridge can become dangerous.

9.1.2.5 Drive train and generator

As already noted, a modern wind turbine rotates at between 5 and 20 rpm while conventional generators operate at between 800 and 3600 rpm, so some form of step-up gearbox is usually necessary. This gearbox has to be extremely rugged because it must be able to withstand more than simply the rotational force from the wind turbine rotor.

Since wind speed varies with height, the force on a turbine blade at the lowest point of its orbit will be less than at the highest point. This will create a constant bending force on the rotor, which is transmitted through the shaft to the gearbox. The speed of the generator must also be controlled so that it remains synchronised with the grid. If the shaft feeding energy to the generator starts to rotate too fast there will be a countermanding force from the generator resisting this increase in speed. This will set up an additional rotational stress within the shaft. This force will also be felt within the gearbox.

To cope with this, a wind turbine gearbox needs to be a heavy-duty design. Even so the gearbox is the most likely component in a wind turbine to fail.

One solution to this problem is to eliminate the gearbox altogether and use a system where the rotor is

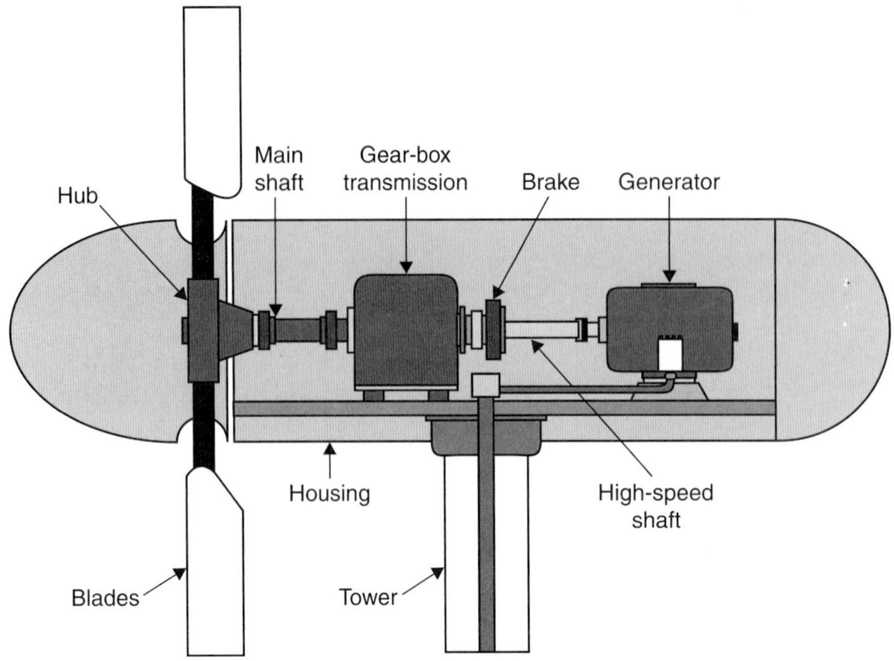

Figure 9.1-3 A wind turbine drive train.

connected directly to the generator. Direct drive generators capable of operating at the low speeds encountered in wind turbines are being developed but current designs are much heavier than more conventional generators.

Another approach which eases the load on the drive train is to use a variable-speed generator. This will generate power at whatever speed the turbine rotor turns. However, the variable frequency output must then be converted electronically to the grid frequency, adding to the cost of the wind turbine.

9.1.2.6 Wind farms and grid connection

To take full advantage of the wind, wind turbines are usually deployed in groups of from two or three to several hundred. These groupings are commonly known as *wind farms*.

When grouped together, wind turbines are usually spaced between five and ten rotor diameters apart in order to reduce interaction between adjacent machines. Even so, when machines are operating downwind of one another there will usually be some loss of output from the downwind turbines.

When this spacing is taken into account, a wind farm of twenty 500 kW turbines will occupy an area of 3–4 km^2. Of this, only around 1% is actually taken up by the turbines. The remainder can still be used as farmland.

The power from a wind farm must be delivered to the local grid. This will normally require a substation. For a small wind farm, under 100 MW, connection may be made to the local distribution system. Larger facilities, such as offshore farms, can have capacities of several hundred megawatts, possibly larger. These must be connected to the high-voltage transmission system.

Small wind farms have often relied on the grid to provide them with frequency and voltage stabilisation. However, this puts a significant strain on the grid system, which cannot support a large wind generation capacity in this way. Future wind farms will need to provide their own frequency and voltage stabilisation. This can be achieved today with standard techniques. However, for the future, electronic power-conditioning systems which convert the wind farm output to DC and then back to AC at the grid frequency and voltage will probably offer the most stable connection. This will only be cost effective for large wind farms.

9.1.3 Offshore wind technology

Pressures for land use and concerted campaigns to prevent the construction of wind farms is forcing wind farm developers in western Europe to consider building wind farms offshore. Offshore wind farming has some significant advantages. The wind regime is both more predictable and more reliable. Turbulence is lower, so wind turbines should last longer and wind farms can be sited far enough offshore to make them virtually invisible. Offshore sites also offer the possibility of building wind farms with capacities of 1000 MW or more.

Against this, the primary barrier is cost. Building a wind farm offshore costs between 40% and 100% more than building a similar farm onshore. Maintenance costs are higher too. However, the higher wind speeds available offshore mean that output will generally be higher offshore.

The main additional cost is for construction of the wind turbine foundation. This can cost up to 25% of the total installation cost offshore. Onshore it is likely to be 16% or less. Grid connection is also more expensive. As a result, while the turbine itself may account for 64% of the cost of an onshore installation, it can be 45% less of the total offshore cost.

This high foundation cost favours large wind turbines for offshore projects. In 2004, the typical offshore wind turbine was between 2 and 4 MW in capacity and larger offshore turbines were in development.

A variety of methods have been used to build offshore foundations. The most popular is a monopile, a single pillar support which is constructed using either pile driving or drilling. Tripods have also been used and experimental work is examining the plausibility of using a floating wind turbine support.

Conditions offshore are generally more severe than onshore and offshore turbines must be 'marinised' to protect them from the environment. This adds around 10% to the cost of the offshore wind turbine compared to onshore. Marinisation techniques from the offshore oil and gas industry have been exploited to protect offshore turbines. The experience from this industry is also proving valuable when installing offshore wind turbines.

9.1.4 Constraints on wind capacity

The amount of wind power that can actually be exploited is likely to be limited, eventually, by the amount of wind generated power national transmission systems can accept. The actual proportion remains a matter for debate but the Danish power industry has shown that it is possible to accept 20% wind energy without detrimentally affecting grid operation.

The problem lies in the fact that wind is not a reliable resource. The output from a wind turbine cannot be guaranteed from day to day. Over a wide area, some variability will be averaged out, but nevertheless a significant level of uncertainty will remain. Better wind forecasting will help reduce the short-term uncertainty but long-term fluctuations in wind output are unavoidable.

Faced with this uncertainty, power dispatchers cannot depend on wind for base-load generation. When wind power is available they can use it to displace other types of generation. When it is not, they must bring those other plants into service.

The situation can be alleviated somewhat by building additional storage capacity on the grid to absorb the fluctuations. Even so, there is a limit to the amount of uncertain power of this type that a system can support with ease. It is possible to operate wind farms with less uncertainty if they are rated more conservatively but that means operating at below full output for most of the time, a strategy which will increase overall cost.

Offshore wind farms present an additional problem because the power from the generating facility must be delivered to the national grid at a point on the coast. In general coastal grids are not particularly good places to make such a connection. The grid can be strengthened in order to make it capable of accepting a large input of power but this will raise the cost of offshore wind power.

9.1.5 Environmental considerations

The principal environmental advantage of wind power is that it is a renewable resource. This means that its exploitation does not lead to a depletion of a global natural resource in the way that the burning of coal or gas results in reduced reserves. As a consequence, wind power can contribute to a sustainable global energy future.

Wind power is also a clean source of energy. Its use does not lead to significant environmental or atmospheric emissions.

Table 9.1-3 contains estimates for the lifetime carbon dioxide emissions from a wind plant and from a coal- and a gas-fired power station. The lifetime assessment looks at emissions that take place during the manufacture of the components of a power plant as well as the emissions that take place during the whole of its operational service. As a result of the former, a wind plant releases 7 tonnes of carbon dioxide for each gigawatt-hour of power it generates. As the table shows, a coal plant releases well over 100 times more and a gas plant close to 70 times the amount from the wind plant.

Aside from carbon dioxide, wind power produces less sulphur dioxide, less nitrogen oxides and less of the other

Table 9.1-3 Lifetime missions of carbon dioxide for various power generation technologies

	Carbon dioxide emissions (tonnes/GWh)
Coal	964
Gas	484
Wind	7

Source: Concerted action for offshore wind energy in Europe, Delft University, 2001.

atmospheric pollutants that are emitted by coal-fired power plants, and to a lesser extent by gas-fired plants. However, wind power plants are not entirely benign. The use of wind power does have negative consequences for the environment. Key among these are visual impact and noise.

Visual impact usually attracts the most serious criticism. Wind farms cover a large area and they are impossible to hide. While actual land utilisation is low and the area occupied by a wind farm can be used for other purposes too, the sight of an array of wind turbines, often in otherwise undeveloped rural areas, is considered by many to be visually offensive.

The weight placed on the visual impact of a wind power development will vary from site to site and from community to community; this is a matter of taste and it is virtually impossible to quantify. Nevertheless it will restrict the available sites for wind power development.

The other major effect of a wind turbine is to generate noise. The noise, a low-frequency whirring, has been compared to the sound of wind in the branches of a tree but the constant frequency is likely to make it more intrusive than the sound of the wind. To this rotor noise must be added the mechanical noise emanating from the gearbox and generator and occassionally some electrical noise.

The blade noise is the most serious of these. Turbine noise is generally more intrusive when wind speeds are low but it will be masked by background noise provided the machine is far enough away from human habitation. This again will limit the possible sites for wind development.

Under certain circumstances wind turbines can also cause electromagnetic interference, affecting television reception or microwave transmission. This can normally be mitigated by simple remedial measures and by careful site selection.

The ecological impact of wind power normally centres on its effect on bird populations. The most obvious danger, of birds being injured or killed when flying through the rotor blades, appears on current evidence to offer only a small threat. This could be considered more serious if a colony of an endangered species lived in the vicinity of a proposed wind farm, but birds do seem to learn to take account of wind farms.

Site development could also damage fragile environments such as peat bogs. This should be avoidable if good construction practice is followed.

9.1.5.1 Offshore wind

The major objections to onshore wind farms are eliminated with well chosen offshore sites. These can be 15 km or more offshore, making them all but invisible. Beyond 45 km, the curvature of the earth should render a wind farm completely invisible from ground level. Noise is not usually a serious problem offshore either.

Offshore wind farms do create their own problems. They can interfere with fisheries and with shipping. Their construction will disrupt the seabed, though the effect of this appears to be temporary. There will be effects on marine life but these too appear to be small.

Offshore wind farms can also interfere with radar. This has led to a number of sites being vetoed by defence ministries on the basis of security. Ground-based radar is affected but airborne radar should not be, so more modern radar systems are generally less subject to interference.

9.1.6 Financial risks

Two primary sources of risk can be attached to wind power, a risk associated with the reliability of the wind power resource available at a particular site and the risk attached to the use of wind power equipment.

The wind resource risk, the risk that the wind will not blow as it was expected to, should be minimal provided an adequate feasibility study has been performed. While the strength of the wind on a particular day at a particular site cannot be predicted in advance, wind is normally reliable over longer periods. A windy site will not turn into a windless site, at least not over the lifetime of a wind power project; global climate change could lead to long-term changes in the wind regime.

At an operational level, improved wind forecasting will enable a wind farm operator to predict future output and this is likely to make the project more valuable to a transmission system operator. It is worth stressing again that the only way to control the wind resource risk is by carrying out a careful site study.

The other major source of risk is the wind turbine technology. Onshore wind farms have now reached a sufficient level of maturity that these risks are well understood. Wind turbines can be expected to operate for 20–30 years with availabilities of 95–98%. New technology is being introduced and this will be liable to an additional level of risk, but overall the risks must be considered manageable. Given the experience now available, onshore wind farms should be able to attract financial support without the imposition of onerous levels of interest.

Offshore wind is at an earlier stage in its development and the risks here are consequently higher. Long-term availability and the cost of maintaining offshore facilities still have to be established. However, the technology is not significantly different to that employed onshore, so performance should be broadly predictable. Offshore wind will also benefit from a major construction programme in the UK, promoted by government renewable energy targets, which is expected to lead to the construction of several thousand megawatts of installed capacity. This should establish, by the end of the first decade of this century, the viability of offshore wind farming.

9.1.7 The cost of wind power

Ever since the modern wind power industry began to develop, the main question it has had to answer is the question of cost. Can wind power compete with conventional forms of power generation?

Early development in California during the 1980s was stimulated by government financial incentives; when these were dropped, the development of projects declined too. Californian wind development was also affected by the fall in the cost of oil that started in the late 1980s. Real oil prices fell by 75% between 1980 and 1992, according to the World Bank.

In Europe the development of wind power took off more slowly but during the middle of the 1990s it became well established with Denmark and Germany its most enthusiastic early supporters, followed by Spain. Here, again, however, government incentives have helped promote wind generation. The wind power market began to grow again in the USA at the end of the 1990s, encouraged both by incentives and by regulations which required utilities in some states to provide a proportion of their electricity from renewable sources.

Continuous development since the early 1980s led to the cost of wind turbine installations falling rapidly during the 1980s and early 1990s. The World Bank estimated that wind technology costs fell by between 60% and 70% between 1985 and 1994. While prices are still falling, the rates are not so dramatic as they were. Current installation costs for an onshore wind farm are between €700/kW and €1000/kW. Offshore wind farms still cost around €1500/kW but this could drop to €1000/kW by 2010.

The installation cost is the main up-front cost of a wind farm. However, energy costs also depend on the amount of wind available at a particular site. To this must be added the cost of financing the project. Operating costs must also be included before a final figure for the cost of each kilowatt-hour of electricity can be established.

Taking these factors into account, favourable estimates suggest that at the beginning of the twenty-first century modern onshore wind farms could generate electricity for €0.03/kWh at a wind speed of 10 m/s and €0.08/kWh at a wind speed of 5m/s. Early commercial offshore wind farms generate power for between €0.05/kWh and €0.08/kWh. Generating costs have been predicted to fall by 36% between 2002 and 2010 and a further 24% between 2010 and 2020,[7] predictions which if borne out will make wind power more competitive still.

These figures imply that onshore wind is currently broadly competitive with coal-fired power generation but not with gas-fired generation. Less favourable reviews of wind power claim that it generates power for two to three times the cost of coal plants. Such reviews take account of changes needed to grid operation and the cost of strengthening transmission grids to support the input of power from regions that have previously been at the end of the supply chain.

In both cases, external costs of fossil fuel power generation, the costs attached to the effects of the atmospheric pollution these plants cause, are ignored. Such costs are difficult to estimate but a 1998 EU report put the external cost of coal-fired power generation between €0.018/kWh and €0.15/kWh while the external cost of gas-fired power generation was between €0.005/kWh and €0.035/kWh. The equivalent estimate for wind energy was from €0.001/kWh to €0.003/kWh.[8]

Adding these amounts to generation costs would make wind generated electricity relatively more competitive. Even without taking external costs into account, wind power will almost certainly become cheaper over the next 10–20 years while the cost of coal- and gas-fired generation is likely to rise.

While arguments about its cost effectiveness continue, in practice the future of wind power is likely to be determined by political decisions. Environmental concerns are increasingly leading to legislation which requires the introduction of renewable electricity generation. Aside from hydropower, wind power is the best placed renewable source to meet that need. If renewable energy is required, in many cases that renewable energy will be wind energy.

End notes

1. Global Potential for Wind Energy, A.J.M. van Wijk, J.P. Coelingh and W.C. Turkenburg, Proceedings of Amsterdam EWEC'91. The figures in Table 9.1-1 are taken from Windforce 12, a study published by the European Wind Energy Association and Greenpeace in 2002.
2. IEA World Energy Outlook 2000, quoted in Windforce 12, a study by the European Wind Energy Association and Greenpeace published in 2002.
3. Windforce 12, a study by the European Wind Energy Association and Greenpeace published in 2002. The figures appear to be taken from University of Utrecht

Study by Wijk and Coelingh, published in 1993.

4. Gridded State Maps of Wind Electric Potential, M.N. Schwartz, D.L. Elliott and G.L. Gower, presented at the 1992 American Wind Energy Association's Conference in Seattle.
5. Technical Offshore Wind Energy Potential Around the Globe, R. Leutz, T. Ackermann, A. Suzuki, A. Akisawa, T. Kashiwagi, Proceedings of European Wind Energy Conference, Copenhagen (July 2002).
6. European Wind Energy Association.
7. European Wind Energy Association, Windforce 12 report for Greenpeace, 2002.
8. Externalities of Fuel Cycles Extern E Report. Report No 10: National Implementation. DGXII, Joule, published by European Commission in 1998.

Section Ten

Hydropower

Chapter 10.1

10.1

Hydropower resources

Garold L. Sommers

This article provides a brief overview of the hydro-power resource—how this resource is assessed, current resources worldwide, and future potential development.

10.1.1 The resource

Hydroelectric power comes from water at work. To generate electricity, water must be in motion. This kinetic energy turns the blades of a water turbine, which changes the kinetic energy to mechanical (machine) energy. The turbine shaft turns a generator, which then converts this mechanical energy into electricity. Because water is the initial source of electrical energy, this technology is called hydroelectric power or "hydropower" for short.

Water constantly moves through the hydrologic cycle (Fig. 10.1-1), which is driven by solar energy. In the hydrologic cycle, atmospheric water reaches the earth's surface as precipitation. Some of this water evaporates, but much of it either percolates into the soil or becomes surface runoff. Water from rain and melting snow eventually reaches ponds, lakes, reservoirs, or oceans where evaporation is constantly occurring. The hydrologic cycle ensures that water is a renewable resource.

Hydropower plants can be located on rivers, streams, and canals, but dams are needed for a reliable water supply. Dams store water for later release for purposes such as irrigation, domestic and industrial use, and power generation. The reservoir acts much like a battery, storing water to be released as needed to generate power. The dam also creates a head, that is, a height from which water flows. A pipe (penstock) carries the water from the reservoir to the turbine.

Typical conventional hydropower projects are of three types: impoundment, diversion, and run-of-river. Impoundment projects (Fig. 10.1-2) use a dam to store water. Water is released to meet water use demands, including generation of electricity. Diversion projects (Fig. 10.1-3) channel a portion of the river through a canal or penstock to produce electrical energy. A small diversion dam may be required to channel a portion of the water from the river. Run-of-river projects (Fig. 10.1-4) use the flow of water within the natural range of the river. They may require a small impoundment to develop head. Run-of-river projects generally blend in with the natural river system.

Unconventional hydropower projects include pumped storage. These projects pump water from a lower reservoir to an upper reservoir during times when demand for electricity is low. During periods of high electrical demand, the water is released back to the lower reservoir to generate electricity.

10.1.2 How hydropower resources are assessed

The first step in assessing a potential water resource for hydropower is to calculate how much power can be produced. The estimated power is determined by the volume of water available (flow) and the vertical distance the water falls (head). A given amount of water falling a given distance will produce a certain amount of energy. In the English system of measurement, flow is commonly measured in cubic feet per second (cfs). Head is measured in feet.

Renewable Energy Focus Handbook 2009; ISBN: 9780123747051

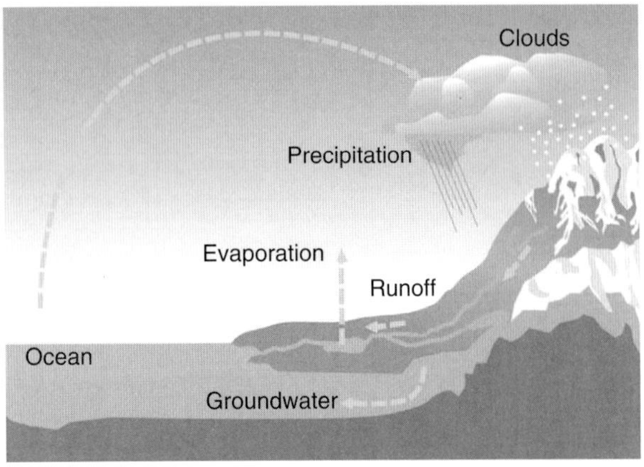

Figure 10.1-1 The hydrologic cycle.

The capacity of hydropower plants is measured in kilowatts or megawatts. Potential capacity can be calculated as follows:

$$\text{kW} = Q \times H/11.8,$$

where

- kW = electrical power;
- Q = flow rate in cubic feet per second;
- H = head in feet; and
- 11.8 = a constant (conversion of foot pounds to horsepower and then horsepower to kilowatts).

Note: the equation does not include efficiency losses in the turbine, generator, penstock, and so on.

Power potential can be converted to potential annual energy production, which is expressed in kilowatt-hours as follows:

$$\text{kWh} = \text{kW} \times 8760 \times \text{PF},$$

where

- kWh = annual energy;
- kW = project capacity;
- 8760 = 24 h per day × 365 days per year; and
- PF = plant factor.

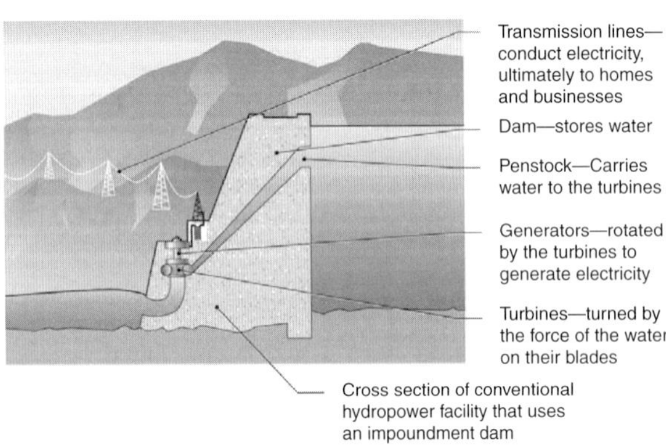

Figure 10.1-2 Impoundment hydropower project.

Figure 10.1-3 Diversion project.

Annual energy production is based on installed capacity and available water for power production. In most cases, water flow varies throughout the year. During low-flow periods, energy production is reduced. Therefore, when calculating annual energy production, a plant factor must be applied. If the actual plant factor cannot be determined based on flow conditions, an average can be used to estimate energy production. The average plant factor for hydropower projects in the United States is approximately 50% (0.5).

The potential installed capacity for a specific project is used to evaluate development potential. In general, a river with a large flow (Q) or potential drop in elevation (H) will be desirable for producing hydropower. However, specific project evaluations include issues such as project economics and accommodating the necessary legal, institutional, and environmental constraints. The sum of the installed capacity of all the projects in a region or country represents that region's or country's potential resource.

Specific issues can be applied to each project, further reducing its potential resource. Some projects cannot be developed because they are located in excluded areas.

Figure 10.1-4 Run-of-river project.

Excluded areas include wild and scenic rivers, national wilderness areas, and national parks, monuments, preserves, refuges, and historic areas. Other projects may not be developed, or developed to full potential, due to environmental, legal, and/or institutional issues. Environmental concerns may include releasing water for nonpower production to improve water quality or fish passage. Other issues could be cultural, historical, geological, recreational, scenic, loss of productive land, and displacement of people.

Undeveloped hydropower resources are generally reported by potential capacity. This capacity can be evaluated based on the various conditions and constraints that require specific site information and characteristics.

10.1.3 Global distribution

In 1998, hydroelectricity provided approximately 21.6% of the worldwide electricity capacity and 18.8% of the worldwide generation of electricity. Figure 10.1-5 illustrates the top hydroelectric generating countries in the world. Table 10.1-1 shows the installed hydropower capacity and energy production by major geographical regions worldwide.

In 1999, the developed hydropower capacity in the United States was approximately 79,700 MW. In 2000, the electrical energy production was approximately 269,000 GWh/year. This is approximately 7.1% of the total electricity in the United States.

10.1.4 Current resources

Water is one of the most valuable resources, and hydropower makes further use of this renewable treasure. The United States has invested more than $150 billion (in 1993 dollars) in hydropower facilities. This investment does not include the cost to build most of the large federal dams in the western part of the country, but it does include the hydro-power portion of these projects. The U.S. hydropower capacity is more than 94,000 MW

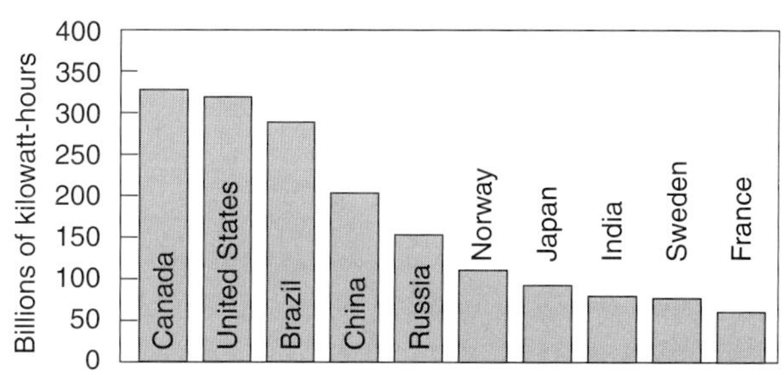

Figure 10.1-5 Top hydroelectric-generating countries in 1998. From the Energy Information Administration. (2000). "Annual Energy Review 1999," Table 11.15. EIA, Washington, DC.

Table 10.1-1 Worldwide hydropower capacity and energy production

Geographic region	Installed capacity (~MW)	Energy production (~GWh/year)
Africa	20,651	80,575
Asia (including Russia and Turkey)	241,624	793,045
Australia/Oceania	13,271	42,000
Europe	175,625	593,377
North and Central America	158,000	700,000
South America	111,459	531,168
World totals	720,630	2,740,165

Source: Aqua ~ Media International Westmead House. (2002). "The International Journal on Hydropower and Dams: World Atlas and Industry Guide" (pp. 13–15). Aqua ~ Media International Westmead House, Sutton, UK.

(includes pumped storage), enough to supply the electrical power needs of approximately 28 million households.

Hydroelectric facilities have many characteristics that favor their use and encourage upgrading existing plants and developing new projects. Hydroelectric facilities offer several benefits, including the following:

- use a renewable resource to generate power;
- are highly reliable and have low operating costs; and
- can start up quickly and have the capability to adjust their output (load following) capability and peak capacity.

As an added benefit, reservoirs have scenic and recreational value for campers, fishermen, and water sports enthusiasts. Water is the home for fish and wildlife as well. Dams add to domestic water supplies, control water quality, provide irrigation for agriculture, and avert flooding. Dams can actually improve the quality of downstream conditions by allowing mud and other debris to settle out.

Just as there is a wide variety of natural settings for hydropower sites, there is a variety of developmental schemes that have various benefits and impacts. The most common distinction is between large and small dams. Dams are often referred to as high-head and low-head dams, reflecting the concept that hydraulic head (the height from which water drops before reaching the turbine) is a key aspect of electrical generation. As a general rule of thumb, the electrical generating capacity of large hydropower projects ranges from approximately 25 to more than 10,000 MW. Small hydropower projects range from 1 to 25 MW. Smaller yet are the minihydropower

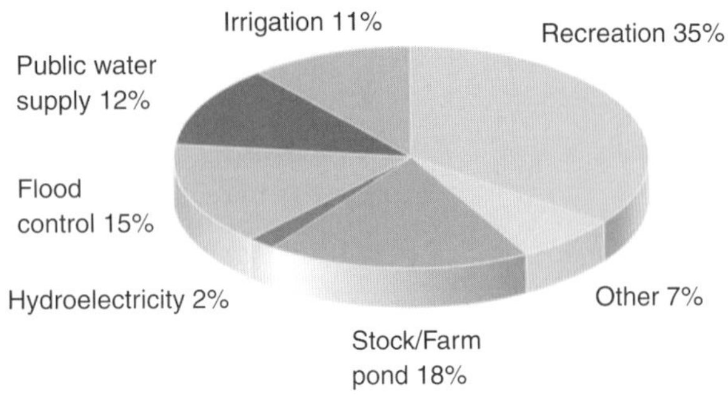

Figure 10.1-6 Primary benefits of U.S. dams. From the U.S. Army Corps of Engineers, National Inventory of Dams.

projects (less than 1 MW) and microhydropower projects (less than 100 kW). Mini-and microhydropower projects are often located on very small streams to provide decentralized electrical power in remote locations. Many of the most controversial aspects of hydroelectric development, such as the social impacts of human resettlement or inundation of terrestrial habitat, arise from the creation of large reservoirs associated with large dams. However, large dams are far less numerous than small dams. For example, the International Commission on Large Dams estimates that there are approximately 40,000 large dams (dams with a height of 15 m or more or shorter dams with high discharges) and 800,000 small dams worldwide.

Existing power plants can be upgraded, or new power plants can be added at current dam sites, without significant additional effect on the environment. Only approximately 2400 of the nation's 80,000 dams are currently used to generate power (Fig. 10.1-6). In addition, new facilities can be constructed with consideration of the environment. For instance, dams can be built at remote locations; power plants can be placed underground; and selective withdrawal systems can be used to control the water temperature released from dams. Facilities can incorporate features that aid fish and wildlife such as salmon runs and resting places for migratory birds. For a history of hydroelectricity's progression in the United States, see Table 10.1-2.

Table 10.1-2 History of hydroelectricity's progression in the United States

- July 1880: Michigan's Grand Rapids Electric Light and Power Company generates electricity by a dynamo belted to a water turbine at the Wolverine Chair Factory; it lights 16 brush-arc lamps.
- 1881: City of Niagara Falls street lamps powered by hydropower.
- 1886: Approximately 45 water-powered electric plants exist in the United States and Canada.
- 1887: San Bernardino, California, is first hydroelectric plant in the western states.
- 1889: Approximately 200 electric plants in the United States use waterpower for some or all generation of electricity.
- 1901: First federal Water Power Act is passed.
- 1907: Approximately 15% of electric generating capacity in the United States is provided by hydropower.
- 1920: Approximately 25% of U.S. electrical generation is hydropower.
- 1920: Federal Power Act establishes Federal Power Commission authority to issue licenses for hydro development on public lands.
- 1935: Federal Power Commission's authority is extended to all hydroelectric projects built by utilities engaged in interstate commerce.
- 1938: Bonneville Dam is first federal dam on the Columbia River.
- 1940: Approximately 40% of electrical generation is hydropower.
- Conventional capacity in the United States triples between 1921 and 1940 and nearly triples again between 1940 and 1980.
- Currently, approximately 7% of U.S. electricity comes from hydropower; approximately 80,000 MW is conventional capacity and 14,000 MW is pumped storage.

10.1.5 Potential of hydropower resources

The potential of hydropower resources is generally evaluated in two steps. The first looks at the resources that are technically feasible. The site physical characteristics are evaluated and generally include site access, water flow data, potential development schemes, potential head, potential installed capacity and energy production, and special site issues. Technically feasible projects require a water resource that can be developed using known engineering and construction techniques. The second step looks at other conditions that apply to developing the resource, including economic factors, environmental issues, legal issues, and institutional concerns. In most cases, economic screening is the second step in the evaluation. The economic factors include investment costs (total cost of the project), interest rate (cost of borrowing money), debt service life (duration to repay the investment with interest), energy value (value or income from the energy produced), operations and maintenance cost (annual cost to operate and maintain the plant), and escalation rates (future value of energy and future operation/maintenance costs). These engineering economic factors can be used to determine the cash flow and benefit/cost ratio. If the resource appears feasible after this step, additional screening for other conditions can be applied to further evaluate the resource. Conditions are unique to each project, and evaluation requires site-specific information. The additional conditions may also affect economics, so a new economic evaluation might be appropriate after the other conditions have been determined.

In some cases, the other screening issues can be applied after the technically feasible resources have been identified. For example, if the environmental issues prevent a resource from being developed, no further evaluation is needed.

Table 10.1-3 shows the additional hydropower potential worldwide by major geographical area, based on technically and economically feasible resources.

Table 10.1-3 Worldwide hydropower potential

Geographical region	Technically feasible potential (~GWh/year)	Economically feasible potential (~GWh/year)
Africa	1,750,000	1,100,000
Countries with major resources:		
Angola	90,000	65,000
Cameroon	115,000	103,000
Congo, Democratic Republic of	774,000	419,000
Ethiopia		260,000
Madagascar	180,000	49,000
Asia (including Russia and Turkey)	6,800,000	3,600,000
Countries with major resources:		
China, People's Republic of	2,200,000	1,270,000
India	660,000	
Japan	135,000	114,000
Russian Federation	1,670,000	852,000
Turkey	215,000	123,000
Vietnam	100,000	80,000
Australia/Oceania	270,000	107,000
Countries with major resources:		
Australia		30,000
New Zealand	77,000	40,000
Europe	1,035,000	791,000
Countries with major resources:		
Austria	54,000	50,000
France	72,000	70,000
Iceland	64,000	44,000
Italy	69,000	54,000
Norway	200,000	187,000
Spain	70,000	41,000
Sweden	200,000	130,000
North and Central America	1,663,000	1,000,000
Countries with major resources:		
Canada	981,000	536,000
Mexico	49,000	32,000
United States	528,000	376,000
South America	2,700,000	1,600,000
Countries with major resources:		
Argentina	130,000	
Bolivia	126,000	50,000
Brazil	1,300,000	764,000
Chile	162,000	
Colombia	200,000	140,000
Ecuador	133,000	106,000
Venezuela	261,000	100,000
World Totals	14,218,000	8,198,000

Source: Aqua~Media International Westmead House. (2002). "The International Journal on Hydropower and Dams: World Atlas and Industry Guide" (pp. 13–15). Aqua~Media International Westmead House, Sutton, UK.

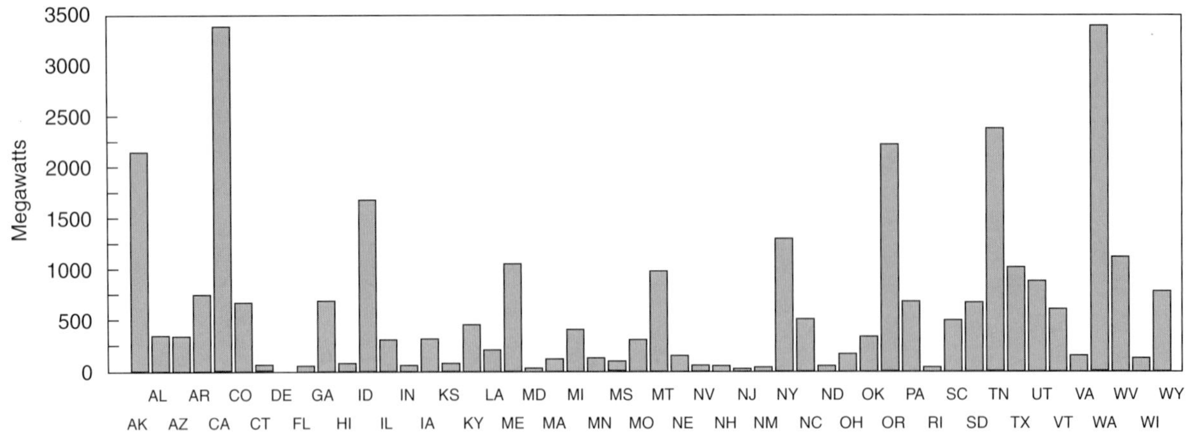

Figure 10.1-7 Undeveloped hydropower potential by state. From the Hydroelectric Power Resources Assessment database (FERC) and Hydropower Evaluation Software (INEEL). DOE has modeled the undeveloped conventional hydropower potential in the United States. This does not include developed capacity. Various state agencies have reviewed the modeled results and provided input. The 50-state undeveloped conventional hydropower potential is approximately 30,000 MW. The model includes environmental, legal, and institutional constraints to development.

The Federal Energy Regulatory Commission (FERC) has provided estimates of the undeveloped hydropower resources in the United States in its Hydropower Resource Assessment (HPRA) database. Other agencies, such as the Army Corps of Engineers, the Bureau of Reclamation, and the Power Marketing Administration (Alaska Power Administration, Bon-neville Power Administration, West Area Power Administration, Southwestern Power Administration, and Southeastern Power Administration), have conducted regional HPRA studies.

In June 1989, the U.S. Department of Energy (DOE) initiated development of a National Energy Strategy to identify the energy resources available to support the expanding demand for energy in the United States. Public hearings conducted as part of the strategy development process indicate that the undeveloped hydropower resources were not well defined. One of the issues was that some assessments did not estimate the undeveloped hydropower capacity based on site characteristics, stream flow data, and available hydraulic heads.

Therefore, the DOE developed uniform criteria to perform HPRA studies. They used the FERC HPRA database as the basis for reevaluations. Information from other databases was factored into the assessments. The information was also reviewed by various state agencies. After the review process, the specific site information was computer modeled to further screen development potential base on environmental, legal, and institutional constraints.

The DOE HPRA studies involved 5677 sites and initially determined a capacity of approximately 70,000 MW. This computer modeling of these undeveloped hydropower resources based on environmental, legal, and institutional constraints resulted in an estimated total undeveloped capacity of approximately 30,000 MW. Fig. 10.1-7 illustrates the undeveloped hydropower potential by state.

The DOE continues to conduct assessments of hydropower resources in the United States.

See also the following articles

Biomass Resource Assessment • *Hydropower Economics* • *Hydropower, Environmental Impact of* • *Hydropower, History and Technology of* • *Hydro-power Resettlement Projects, Socioeconomic Impacts of* • *Hydropower Technology* • *Oil and Natural Gas Liquids: Global Magnitude and Distribution* • *Wind Resource Base*

Glossary

benefit/cost ratio (B/C) Ratio of the present value of the economical benefit to the present value of the project cost, computed for comparable price level assumptions.

capability The maximum load that a generating unit, generating station, or other electrical apparatus can carry under specified conditions for a given period of time without exceeding approved limits of temperature and stress.

capacity The maximum power output or load for which a turbine generator, station, or system is rated.

dam A wall or structure built across a valley or river for storing water.

dependable capacity The load-carrying ability of a hydro-power plant under adverse hydrologic conditions for a specified time interval and period.

distribution system The portion of an electric system dedicated to delivering electric energy to an end user; the distribution system "steps down" power from high-voltage transmission lines to a level that can be used in homes and businesses.

diversion dam A dam used in conjunction with long tunnels, canals, or pipelines to divert water to a powerhouse located a distance from the dam.

drawdown The distance that the water surface of a reservoir drops from a given elevation as the result of withdrawal of water.

energy The capacity for performing work; the electrical energy term generally used is kilowatt-hours and represents power (kilowatts) operating for some time period (hours).

forebay The water intake area for a canal, penstock, or turbine designed to reduce water velocity and turbulence so as to settle suspended material and keep it from entering the system.

generator A machine that converts mechanical energy into electric energy.

gigawatt (GW) Unit of electric power equal to 1 million kW.

head The difference in elevation between the headwater surface above and the tailwater surface below a hydroelectric power plant under specified conditions.

horsepower A unit of the rate of doing work equal to 33,000 foot pounds per minute or 745.8 W (Britain), 746 W (United States), or 736 W (Europe).

hydroelectric plant or hydropower plant An electric power plant in which the turbine generators are driven by falling water.

hydroelectric power Electric current produced from water-power.

hydrology The scientific study of the properties, distribution, and effects of water on the earth's surface, in the soil and underlying rocks, and in the atmosphere.

hydrologic cycle Water constantly moving through a vast global cycle in which it evaporates from lakes and oceans, forms clouds, precipitates as rain or snow, and then flows back to the ocean; the energy of this water cycle, which is driven by the sun, is tapped most efficiently with hydropower.

kilowatt (kW) Unit of electric power equal to 1000 W or approximately 1.34 horsepower; for example, it is the amount of electric energy required to light 10 100-W light bulbs.

kilowatt-hour (kWh) The unit of electrical energy commonly used in marketing electric power; it is the energy produced by 1 kW acting for 1 h.

kinetic energy Energy that a moving body has due to its motion, dependent on its mass and the rate at which it is moving.

megawatt (MW) A unit of power equal to 1 million W; for example, it is the amount of electric energy required to light 10,000 100-W bulbs.

megawatt-hour (MWh) Unit of electric power equal to 1000 kWh.

microhydropower A hydroelectric plant with a rated capacity of 100 kW or less.

peaking capacity That part of a system's capacity that is operated during the hours of highest power demand.

penstock A closed conduit or pipe for conducting water to a powerhouse.

plant factor The ratio of the average output to the installed capacity of the plant, expressed as an annual percentage.

power (electric) The rate of generation or use of electric energy, usually measured in kilowatts.

pumped-storage hydroelectric plant A plant that usually generates electric energy during peak-load periods by using water previously pumped into an elevated storage reservoir during off-peak periods when excess generating capacity is available to do so; when additional generating capacity is needed, the water can be released from the reservoir through a conduit to turbine generators located in a power plant at a lower level.

reservoir An artificial lake into which water flows and is stored for future use.

turbine A machine for generating rotary mechanical power from the energy of a stream of fluid (e.g., water, steam, hot gas); turbines convert the kinetic energy of fluids to mechanical energy through the principles of impulse and reaction or a mixture of the two.

watt (W) The unit used to measure production/usage rate of all types of energy; the unit for power.

watt-hour (Wh) The unit of energy equal to the work done by 1 W in 1 h.

Further reading

Aqua ~ Media International Westmead House. (2002). "The International Journal on Hydropower and Dams: 2002 World Atlas and Industry Guide." Aqua ~ Media International Westmead House, Sutton, UK.

Conner, A.M., *et al.* (1998). "U.S. Hydropower Resource Assessment Final Report." (DOE/ID-10430.2). U.S. Department of Energy, Idaho Operations Office, Idaho Falls, ID.

Gulliver, J.S., and Arndt, R.E.A. (eds.). (1991). "Hydropower Engineering Handbook." McGraw-Hill, New York.

Hall, D.G., *et al.* (2002). "Low Head/Low Power Hydropower Resource Assessment of the Arkansas White Red Hydrologic Region." U.S. Department of Energy, Idaho Operations Office, Idaho Falls, ID.

Hall, D.G., *et al.* (2002). "Low Head/Low Power Hydropower Resource Assessment of the Pacific Northwest Hydrologic Region." U.S. Department of Energy, Idaho Operations Office, Idaho Falls, ID.

Hydrologic Engineering Center and Institute for Water Resources. (1979). "Feasibility Studies for Small Scale Hydropower Additions: A Guide Manual." U.S. Army Corps of Engineers, Hydrologic Engineering Center, Davis, CA.

U.S. Bureau of Reclamation, Power Resources Office. (2001). Hydroelectric power. www.usbr.gov/power/index.html.

U.S. Department of Energy. (1997). "Hydropower: America's Leading Renewable Energy Resource" [brochure]. U.S. Department of Energy, Idaho Operations Office, Idaho Falls, ID.

U.S. Department of Energy. (2001). "Hydropower: Partnership with the Environment" [brochure]. U.S. Department of Energy, Idaho Operations Office, Idaho Falls, ID.

Section Eleven

Power from waste

Chapter 11.1

11.1

Power from waste

The generation of power from waste is a very specialised industry and its principal aim is not to produce electricity. Power-from-waste plants are combustion plants designed to destroy or reduce in volume municipal and in some cases industrial waste.[1] As an incidental, but nevertheless valuable by-product, the processes adopted to manage these wastes may also be capable of generating electricity.

The level of exploitation of waste-to-energy plants varies from country to country. They have been used widely in parts of Europe, where waste has been burned since the end of the nineteenth century, and form a major part of Japan's waste disposal strategy. In contrast the USA has only adopted the technology patchily. In addition environmental concerns about the emissions from the plants has caused recent resistance to their construction both in the USA and elsewhere.

Where they are employed, these plants generally burn domestic and urban refuse – called in this context *municipal solid waste* (MSW) – using the resulting heat to generate steam to drive a conventional steam turbine. MSW can also be sorted and treated to produce a compacted fuel called *refuse-derived fuel* (RDF) which can be burned in a power station.

Some industrial waste may be treated in the same way. However, industrial wastes are likely to contain toxic materials which have to be handled using special procedures. Where such care is not required, they can be dealt with in the same way as urban waste.

There are a number of other categories of waste, primarily resulting from the agricultural and forestry industries, that can be used to generate electricity. These are dealt with under biomass in Chapter 12.1, which also deals with the collection and use of methane produced in landfill refuse disposal sites. However, we need to consider landfill briefly here since it offers the main alternative to waste combustion.

11.1.1 Landfill waste disposal

The landfill site – essentially an enormous hole in the ground where waste is dumped – is the main alternative to the technologies discussed in this chapter as a means of waste disposal. Though crude, its simplicity has led to it becoming the favoured method of urban waste disposal across the globe.

While landfill use remains popular in many countries, it is coming under pressure in others. This is partly a result of the demand for land which increasingly restricts that available for waste burial. More potent still are environmental concerns about the long-term effects of landfill disposal, effects resulting from the methane emissions from such sites (discussed in Chapter 12.1) and from the seepage of toxic residues into water supplies.

Such concerns have already led the European Union (EU) to legislate[2] to restrict the use of landfill waste disposal. Similar legislation is bound to follow in other parts of the world. But waste will still be produced. This is where technological solutions, such as the power-from-waste plant, enter the equation.

Power-from-waste technology is not cheap. The specialised handling that waste requires, coupled with the need for extensive emission-control systems to prevent atmospheric pollution, makes such plants much more expensive to build than any other type of combustion power plant. They are also expensive to operate.

If these plants had to survive on the revenue from power generation alone, they would never be built. Fortunately they have another source of income. Since waste

Renewable Energy Focus Handbook 2009; ISBN: 9780123747051

has to be disposed of in a regulated manner, waste disposal plant operators can charge a fee – normally called the *tipping fee* – to take the waste. The tipping fee represents the main source of income for a power-from-waste plant. Any additional income derived from power generation will benefit the economics but the plant may well be able to survive without it.

11.1.2 Waste sources

There are two principal types of waste suitable for disposal in a power-from-waste plant: urban (primarily domestic) refuse, normally referred to as MSW, and industrial waste. Some industrial waste is broadly similar in content to MSW and this can be treated in the same way as the latter. Other industrial waste must be dealt with differently because of the hazardous or valuable materials it contains. This chapter is only concerned with MSW and it will not deal with industrial waste except where it can be burned with MSW.

The main source of MSW is an urban community.[3] The quantity and size of such communities is growing rapidly. In the last two generations the number of people living in cities has increased by between 250% and 500%.[4] This has been particularly notable in the developing world where the number of urban dwellers is expected to reach 2.71 billion by 2010. A further 1 billion live in the cities of the developed world. Thus, close to half the population of the world will be living in cities by 2010.

Urban dwelling has grown particularly rapidly in South America and the Caribbean where, by 2025, 80% of the population will be living in towns. But these regions are not unique. Urban communities are growing virtually everywhere. These towns and cities constitute the source of MSW.

The amount of waste these populations produce varies from country to country and from continent to continent. In general, the city dweller in an industrialised country produces far more waste than one in a developing country. Thus a typical Californian might produce 1.3–1.4 kg each day while a city dweller in Mexico City produces only half that. A Nigerian town dweller probably produces less than 200 g of waste each day.

In the mid-1990s the International Energy Agency estimated that developed countries alone produced an estimated 426 million tonnes of waste each year. If all this was used to generate electricity, potential output would be 191 TWh/year. Annual energy demand in 2001 was 13,290 TWh.[5]

11.1.3 Waste composition

The composition of the waste varies from place to place. In general the waste from the urban household in an industrialised country will contain 30–40% paper and cardboard and up to 10% plastic. The proportions of these in the waste from a household in the Dominican Republic will be much lower but the Dominican household's waste will probably contain 80% food waste whereas the proportion in a US household waste may only be 26%.[6]

There are other important differences. The waste from households in developing countries contains a high proportion of moisture, often as high as 50%, making it difficult to burn without first reducing the moisture content by drying. In contrast, the high proportions of paper and plastic in the waste from a household in the industrial world make it much easier to burn.

All these factors affect the energy content of waste, and energy content is a crucial factor in determining the viability of a power-from-waste plant. Unless the plant can produce enough excess heat from waste combustion to raise steam, it cannot expect to generate any electricity.

Table 11.1-1 provides some figures for MSW energy content from different parts of the world. US waste has the highest energy content, 10,500 kJ/kg, approaching that of sub-bituminous coal (see Table 11.1-1). European cities and prosperous Asian cities such as Taipei generate waste with around 7500 kJ/kW. The waste from typical mid-sized Indian cities contains roughly half this amount of energy.

In the latter case the low energy content may not be entirely due to the quality of waste. In cities in India – but not them alone – much of the urban waste is collected by city sweepers. Such waste is contaminated with considerable quantities of stone, earth and sand. In Bombay, for example, the amount of non-combustible material of this type in waste may reach 30%. Not only does this reduce the energy content of the waste; it could also damage a combustion system so the design of a waste disposal plant has to take its presence into account.

Given such local variations in waste content it is vitally important, before a power-from-waste plant is built, that

Table 11.1-1 Energy content of urban wastes from different regions

	Energy content (kJ/kg)
USA	10,500
Western Europe	7,500
Taiwan (Taipei)	7,500
Mid-sized Indian cities	3,300–4,600
Sub-bituminous coal	10,700–14,900

Source: United States Agency for International Development.[7]

the waste available be carefully assessed. For that, local waste-collection procedures and organisations have to be examined.

11.1.4 Waste collection

Urban refuse collection is organised in different ways in different parts of the world. In some countries it is run by municipalities; in others it is provided by private operators. Where a municipality run waste collection as a service, the same city might build and operate its own power-from-waste plant. Under these circumstances the composition of the waste can be readily assessed and controlled if necessary.

More often waste collection is carried out by private companies. The waste that these companies provide will vary in quality. In some cases it will contain the whole range of waste, but in others it will have been sorted to remove the more valuable material. Some countries now require that glass, metal, plastic and paper be recycled. This too will affect the quality of the MSW available.

Inevitably the quality of waste will vary by season. Economic factors are also important. Waste will be poorer in a recession than in a boom. Local variations can also be significant. Richer neighbourhoods tend to produce better quality waste than poorer neighbourhoods. This has led to the suggestion that the quality of waste for a power-from-waste plant might be maintained by collecting only from prosperous areas of a city.

Whatever the strategy, knowledge of the waste, its source and its variations will form a necessary part of the management of a waste-to-energy plant. That information can only be gained with practical experience, by analysis of waste collected by the contractor that will provide waste for the plant. Even with this knowledge, it may be impossible to maintain an adequate energy content in the waste throughout the year. Then the only solution may be to add some higher-energy content fuel to the waste. Biomass waste from local sources will often be the most economical solution in this situation.

11.1.5 Waste power generation technologies

A power-from-waste plant is a power station fuelled with urban waste. As already indicated, such a facility may have, as its primary function, waste disposal. Nevertheless the technologies employed will be traditional power generation technologies as used in combustion plants. Combustion systems include grate burners, some fluidised-bed burners, and more recently gasification and pyrolysis. Heat generated in these combustion systems is used to raise steam and drive a steam generator.

Within the broad outline above, power-from-waste plants vary enormously. Much depends on the waste to be burnt, its energy content, the amount of recyclable material or metal it contains and its moisture content. Waste may be sorted before combustion or it may be burnt as received. Emission-control systems will vary too, with toxic metals and dioxins a particular target, but nitrogen oxides, sulphur dioxide and carbon monoxide emissions must all fall below local limits. Carbon dioxide emissions may need monitoring to comply with greenhouse gas emission regulations.

Once the waste has been burnt, residues remain. Power-from-waste plants will generally reduce the volume of waste to around 10% of its original. A way must then be found to dispose of this residual ash. If it is sufficiently benign, it may be used as aggregate for road construction. Otherwise it will probably be buried in a landfill. Other residues from emission-control systems will have to be buried in controlled landfill sites too.

Northern Europe has been the traditional home of waste incineration plants for power generation. Altogether there were around 250 municipal waste combustion plants in the EU in the late 1990s, most in the northern countries of the Union. Between them they had a generating capacity of around 1500 MW, almost half the global total of 3200 MW in 1997.[8] Japan has also made extensive use of waste combustion, though not always for power generation. In 1999 there were about 600 waste-to-energy plants in operation worldwide.

Europe has also developed the most widely used waste combustion technology. Two companies, Martin GmbH based in Munich and the Zurich company Von Roll, accounted for close to 70% of the market for the dominant technology, called *mass burn*, at the turn of the century.[9] The rest of the market is divided among a number of smaller companies, most based in Europe, the USA or Japan.

The dominant European technology has been widely licensed. It was the source of the technology used in most US power-from-waste plants built in the late 1970s and early 1980s. More recently several developing countries of Asia have taken interest in power from waste and European technology has been modified for use in China.

Newer technologies based on gasification and pyrolysis are being developed by a variety of companies. These are based on technologies from other industries such as power generation and petrochemicals.

11.1.5.1 Traditional combustion plants

The traditional method of converting waste to energy is by burning it directly in a special combustion chamber and grate, a process which is often called mass burning. The dominant European technologies use this system.

These involve specially developed moving grates, often inclined to control the transfer of the waste, and long combustion times to ensure that the waste is completely destroyed. Designs have evolved over 20–30 years and are generally conservative.

More recently, fluidised-bed combustion systems have sometimes been used in place of traditional grates. Such systems are good at burning heterogeneous fuel but they require the waste to be reduced to small particles first. These systems remain relatively rare.

The actual grate forms only a part of a waste treatment plant. A typical solid-waste combustion facility is integrated into a waste-collection infrastructure. Waste is delivered by the collecting trucks to a handling (and possibly a sorting) facility where it must be stored in a controlled environment to prevent environmental pollution. Recyclable materials may be removed at this stage, though metallic material is often recovered after combustion. Grabs and conveyors will then be used to transfer the waste from the store to the combustor.

Plant components, and particularly the grates, must be made of special corrosion-resistant materials. The grate must also include a sophisticated combustion-control system to ensure steady and reliable combustion while the quality and energy content of the refuse fuel varies. In some more modern systems oxygen is fed into the grate to help control combustion. The temperature at which the combustion takes place must usually be above 1000°C to destroy chemicals such as dioxins but must not exceed 1300°C as this can affect the way ash is formed and its content.

Hot combustion gases from the grate flow vertically into a boiler where the heat is captured to generate steam. The combustion process in the grate and the temperature profiles within the boiler have to be maintained carefully in order to control the destruction of toxic chemicals. Most of the residual material after combustion is removed from the bottom of the combustion chamber as slag. However, there may be further solid particles in the flue gases, some of which can be recycled into the furnace.

Upon exiting the combustion and boiler system, the exhaust gases have to be treated extensively. While the combustion chamber may utilise techniques to minimise nitrogen oxide emissions – though further reduction may prove necessary – a system to capture sulphur will be required. This will probably be designed to capture other acidic gases such as hydrogen chloride too. There may be a further capture system based on active carbon which will absorb a variety of metallic and organic residues in the flue gases. Then some sort of particle filter will be needed to remove solids. By this stage the exhaust gases should be sufficiently clean to be released into the atmosphere but continuous monitoring systems are required to make sure that emission standards are maintained.

Dust from the flue gas filters is normally toxic and must be disposed of in a landfill. Other flue gas treatment residues will probably need to be buried too. The slag from the combustor may, however, be clean enough to exploit for road construction. Modern mass-burn plants aim to generate slag that can be utilised in this way.

Mass-burn plants may burn up to 2000 tonnes/day of MSW. Where a smaller capacity is required, a different type of combustion system, called a *rotary kiln*, can be employed. As its name suggests, this system uses a rotating combustion chamber which ensures that all the waste is burned. The chamber is inclined so that the material rolls from one end to the other as it burns. Such

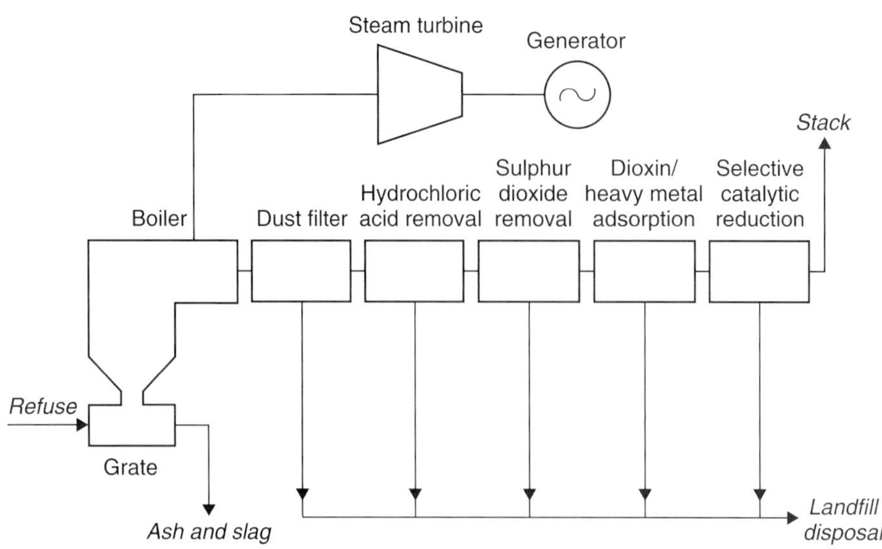

Figure 11.1-1 Block diagram of a mass-burn power-from-waste plant.

combustors are capable of burning waste with a high moisture content, perhaps up to 65%. Capacities of rotary kilns are up to 200 tonnes/day of refuse, suitable to meet the needs of small urban communities.

11.1.5.2 Gasification and pyrolysis

In recent years a number of companies have attempted to develop new waste-to-energy technologies based on both gasification and pyrolysis. These technologies are derived from the power and the petrochemicals industries.

Pyrolysis is a partial combustion process carried out at moderate temperatures in the absence of air, which usually produces a combustible gas and a combustible solid residue. Gasification uses higher temperatures and converts most of the solid material into a combustible gas. In both cases the gas will normally be burnt to generate heat and thence steam.

Typical of this type of plant is a system developed in the 1990s in Japan[10] which employs an initial pyrolysis process followed by combustion to generate heat. Waste delivered to the plant is first shredded and then fed into a rotating pyrolysis drum where it is heated to around 450°C. The heat, provided by hot air generated at a later stage in the process, pyrolyses the waste, converting it into a combustible gas and a solid residue.

The solid residue contains any metal which entered with the waste. This can be removed at this stage for recycling. Both iron and aluminium can be segregated in this way. The remaining solid slag is crushed. The gas and the crushed residue are then fed into a high-temperature combustion chamber operating at 1300°C where it is completely burnt. Combustion is controlled to limit nitrogen oxides formation. Incombustible material adheres to the walls of the combustion chamber where it flows, in liquid form, to the bottom. From here it is led out of the bottom of the furnace and immediately quenched, creating an inert granular material suitable for road building.

Hot flue gases from the combustion chamber are used to generate steam to drive a turbine. Dust is then removed from the exhaust gases and returned to the combustion system. Following this, a flue gas treatment system removes any remaining acid gases. Only this material, around 1% of the original volume of the MSW, needs to be disposed of in a landfill. The system also claims to keep residual levels of dioxins extremely low.

Waste gasification is similar to pyrolysis but conversion of waste takes place at a higher temperature in the presence of a controlled amount of air or oxygen. Depending on the process used a low- or medium-energy content synthetic gas will be produced. In a power-to-waste plant this will be burnt but it can also be used as a feed for some chemical processes or as a means of generating hydrogen.

11.1.5.3 Refuse-derived fuel

RDF is the product of the treatment of MSW to create a fuel that can be burnt easily in a combustion boiler. In order to produce RDF, waste must first be shredded and then carefully sorted to remove all non-combustible material such as glass, metal and stone. Shredding and separating is carried out using a series of mechanical processes which are energy intensive. The World Bank has estimated that it requires 80–100 kWh to process 1 tonne of MSW and a further 110–130 kWh to dry the waste.[11]

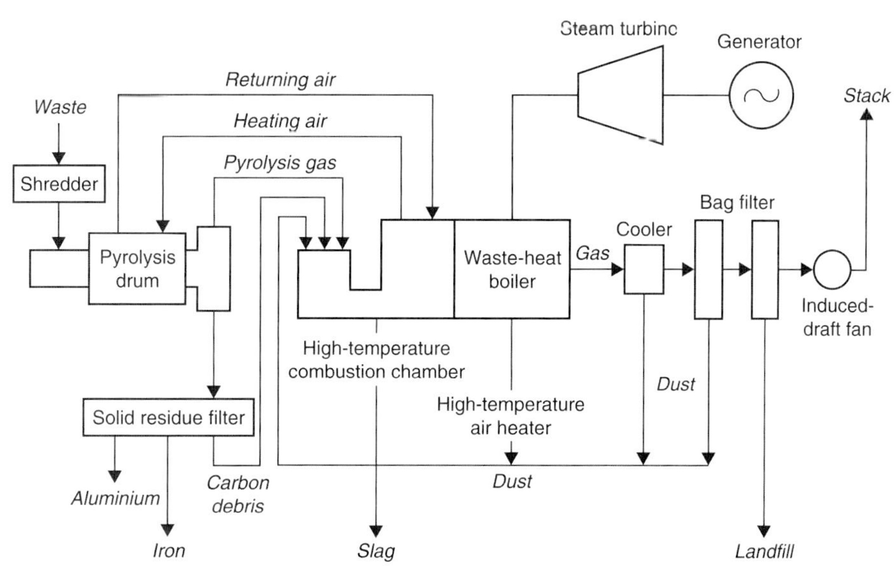

Figure 11.1-2 Block diagram of a waste pyrolysis plant.

After the waste has been shredded and separated, the combustible portion is formed into pellets which can be sold as fuel. The original intention of this process was to generate a fuel suitable for mixing with coal in coal-fired power plants. This, however, led to system problems and the modern strategy is to burn the fuel in specially designed power plants. An alternative is to mix the RDF with biomass waste and then burn the mixture in a power plant. Since RDF production must be preceded by careful sorting, this type of procedure is best suited to situations where extensive recycling is planned.

11.1.6 Environmental considerations

Urban waste, its production and its fate are major environmental issues. Modern urban living produces enormous quantities of waste in the form of paper, plastic, metals and glass as well as organic materials. How these wastes are processed is a matter of increasing global concern.

Wastes such as paper, glass and metal can be recycled, as can plastics in theory. From an environmental perspective it makes sense to reuse as much waste as possible, so environmentalists generally favour maximum recycling. Many European governments promote recycling. However, the economics of recycling are not clear cut and there are critics who consider it ineffective. Since such debates pitch sustainability against economy, the issue is not easily resolved.

While recycling offers the ideal solution, in practice there are often neither the facilities nor the infrastructure to recycle effectively. Even where recycling is employed there is still a residue of waste that cannot be reused. Thus there remains a considerable volume of waste for which an alternative means of disposal is required. The only options currently available are burial in a landfill site or combustion.

The combustion of waste would seem initially the ideal solution. Combustion reduces the quantity of waste to 10% or less of its original volume. At the same time it produces energy as a by-product and this energy can be used to generate electricity or for heating, or both. Unfortunately waste often contains traces of undesirable substances which may emerge into the atmosphere as a result of combustion. Other hazardous products may result from the combustion itself, with the waste providing the chemical precursors. So, while solving one environmental problem, waste combustion can generate others.

In the face of this, the combustion of waste is becoming increasingly subject to strict legislation. This sets limits on amounts of different hazardous materials which can be released as a result of the process. Chief among these are heavy metals such as mercury and potent organic compounds such as dioxins. Modern waste-to-energy plants appear able to meet these requirements. However, they have acquired a bad reputation in the past 20 years in some parts of the world. This has proved difficult to overcome and there are countries where power-from-waste plants are considered too unpopular to gain approval. New waste conversion technologies such as gasification and pyrolysis may be able to breach this barrier.

11.1.6.1 Waste plant emissions

A plant burning waste produces four major types of product. First there is a solid residue from the grate itself, normally termed *slag* or *ash*. Secondly there is a chemical product resulting from flue gas treatment systems. Thirdly there is a quantity of dust in the flue gases emerging from the plant boiler; this is normally captured with filters or an electrostatic precipitator. Finally there is the flue gas itself.

11.1.6.2 Ash

The nature of the ash or slag emerging from the grate of a power-from-waste plant will depend on both the type of waste being burnt and the combustion conditions. While its primary constituents will be solid, incombustible mineral material from the wastes, this residue will be contaminated with traces of a variety of metals. These traces may be in a toxic or a harmless form.

By careful control of the temperature in the furnace, it is possible to incorporate the metals into the mineral content of the ash and render them effectively harmless. This is a process called *sintering*. The effectiveness of the sintering process in rendering toxic metals harmless will be determined by measuring the amounts capable of being leached out by water. The ash may also contain some toxic organic compounds such as dioxins. Furnace conditions can minimise these too, since a sufficiently high temperature will normally destroy such compounds. The effectiveness of this will again be determined by a leaching test.

If the ash or slag is too toxic it will have to be buried in a landfill. Modern facilities aim to render it sufficiently stable and benign that it can be used for road building or for similar purposes. When they succeed, only a residual 1% of the original waste needs to be buried.

11.1.6.3 Fly ash and flue gas treatment residues

Fine solid particles called *fly ash* escape with the flue gas from a furnace. This fly ash will often contain high levels of toxic metals and must be captured. Capture is achieved either by using a fabric filter called a *bag filter*, or by employing a device called an *electrostatic precipitator*. Both should be capable of removing close to 100% of the dust from the flue gas. Once captured this dust must be safely buried in a landfill.

The same applies to the chemical residues which result from the various flue gas treatment systems used to remove harmful material from the exhaust gases of the plant. Depending on the treatment process, the residue may be a solid or a wet slurry. In the latter case, the slurry will normally be dried using the hot exhaust gases before disposal.

11.1.6.4 Flue gas

Once treated, the flue gas from a waste combustion plant should be sufficiently clean to be released into the atmosphere. The gas will usually need to be monitored continuously to ensure that emission limits are being met.

11.1.6.5 Dioxins

One of the most potent environmental concerns during the last 20–30 years has related to the release of dioxins into the atmosphere. Dioxins are undesirable by-products of the manufacture of a variety of chemicals such as pesticides and disinfectants, but one particular compound called *2,3,7,8-tetrachlorodibenzo-p-dioxin* has come to be identified as *dioxin*. This material was thought to be extremely toxic to humans, though more recent studies suggest earlier results were exaggerated.

Dioxins can be found in urban waste and there is also a danger that the compounds can be formed during waste combustion if the process is not carefully controlled. Some early waste incineration plants did not control the emissions sufficiently carefully and this led to instances of widespread contamination. Such instances have coloured the perception of waste-to-energy plants ever since.

Dioxin emission levels are now closely regulated and emissions have fallen. In the USA, the emissions of dioxins from large waste-to-energy facilities fell from 4260 g (toxic equivalent) in 1990 to 12 g (toxic equivalent) in 2000.[12] The European emission limit for dioxins is 0.1 ng/Nm^3. Power-from-waste plants built in the middle of the first decade of the twenty-first century should be capable of reducing the emission level to one-tenth of this.

11.1.6.6 Heavy metals

Heavy metals, particularly mercury, have proved another source of concern. Less mercury is used today than in the past. This combined with better filtration systems has reduced mercury emissions from power-from-waste plants in the USA to around 2 tonnes/year. Coal-fired power plants release over 40 tonnes/year.

There are other metals such as cadmium and lead which must be monitored. However, in general the emissions of metals from waste incineration plants should fall well below legal emission limits. Today, proponents of these plants would argue that they are significantly less polluting than landfills. New technologies may well be able to provide even higher-emission performance. Whether this will be sufficient to overcome the reputation which has already attached itself to such plants remains to be seen.

11.1.7 Financial risks

The traditional technology used for waste combustion is robust and extensively tested. Any risk associated with its use is small and well documented. New technologies under development such as gasification and pyrolysis have not yet been proved and the risks associated with their use are higher.

There is also a risk associated with the waste which is to provide the fuel for a power-from-waste plant. It is important to ascertain exactly what type of waste will be available to a particular project and its typical content. This can only be discovered by careful analysis of actual samples. Long-term analysis is necessary since waste content varies seasonally. However, waste quality can also vary over a longer time scale, particularly if the supplier changes or where there are demographic changes within the catchment area.

Regulations and legislation also pose a threat. Any planned project will be required to meet current regulations but these may change once the plant has been built, necessitating modifications to meet new requirements. While the legislative situation is stable in areas such as Europe, it may not be everywhere. It would seem prudent when planning a project to choose the best technology available since this is likely to meet both current and future regulations anywhere in the world.

Perhaps the greatest risk with a power-from-waste project relates to its economics. If a plant is to be operated as a public service, then the economic viability will normally be guaranteed by the public sector. If it is to be a wholly private sector project, then the viability will depend on the value to waste collectors of the service offered. The price collectors are prepared to pay will depend on the competition. Under these circumstances, long-term contracts may offer the best security.

11.1.8 The cost of energy from waste

The capital cost of equipment to generate electricity from waste is generally much higher than for conventional power generation equipment to burn fossil fuel. Plant design is specialised and must include refinements for emission control that are not necessary in the fossil fuel plant. Grate design is unique too.

Against this must be offset the revenue of the plant, not only from the electricity generated but also from the fuel itself, the waste. Industry and municipalities expect to pay to dispose of their waste. Consequently, the economics of a project should be designed so that the revenue from the waste disposal contracts is adequate to enable the power from the plant to be sold competitively.

The cost of a typical municipal waste combustion (MWC) plant is \$5000–10,000/kW, at least three times the cost of a coal-fired power plant of the same generating capacity. Smaller plants will be relatively more expensive. The cost of operating an MWC plant is probably three times that of a coal-fired power plant too. According to US government estimates, such plants generate electricity at between \$0.02 and \$0.14/kWh.

End notes

1. Modern plants often recycle any reusable material, burning only the remainder.
2. European Union, Council Directive 1999/31/EC of 26 April 1999 on the landfill of waste, Official Journal of the European Communities, pp. L182/1–19 (July 1999).
3. In the developed world the waste from rural communities is often handled in a similar way.
4. Mining the Urban Waste Stream for Energy: Options, Technological Limitations, and Lessons from the Field, United States Agency for International Development, 1996 (Biomass Energy Systems and Technology Project DHR-5737-A-00-9058-00).
5. US Energy Information Administration, International Energy Outlook, 2004.
6. Refer *supra* note 4.
7. Refer *supra* note 4.
8. This figure is from the EU DG of Energy.
9. An overview of the global waste to energy industry, Nickolas J. Themelis, Waste Management World (July–August 2003).
10. The process, called R21, was developed by Mitsui Engineering and Shipbuilding. The first plant was completed in 2000.
11. Refer *supra* note 4.
12. US Environmental Protection Agency. These figures are quoted in 'An overview of the global waste-to-energy industry' Waste Management World (July–August 2003).

Section Twelve

Bioenergy

Chapter 12.1

12.1

Bioenergy

Shang-Tian Yang

12.1.1 Introduction

Recent advances in biotechnology and public concern about environmental pollution and the sustainability of natural resources have rapidly transformed the nation's many manufacturing industries, from chemical to pharmaceutical, to become more environmentally benign and bio-based. For example, almost all major pharmaceutical companies now dedicate more than 50% of their new drug development to biotech R&D, a trend away from traditional chemical synthesis. Likewise, large chemical companies, such as DuPont and Dow Chemicals, are aggressively developing new bio-based products to replace petrochemical ones. Meanwhile, rising energy demands and oil prices have prompted large petroleum companies such as Shell to explore biofuels as alternative energy sources. Together with the agricultural industry expanding its product portfolios beyond traditional food and feed is the birth of an emerging biorefinery industry that promises reduced dependence on fossil energy and a truly sustainable economy [1–3].

This chapter will introduce some important applications of biotechnology and recent developments in bioprocessing technologies for biomass utilization with a focus on the industrial bioconversion of renewable resources to fuels and chemicals. The concept and principles of integrated biorefineries to attain the sustainable production of food, energy, and industrial products are also presented.

12.1.2 Industrial biotechnology – history and applications

Biotechnology has been described as the last great technological innovation of the twentieth century and has touched upon almost every aspect of human life, from healthcare to agriculture to the production of industrial products (Figure 12.1-1). Biotechnology, broadly defined, includes any technique that uses living organisms or parts of organisms to make or modify products, improves plants or animals, or develops microorganisms for specific uses. Based on this definition, mankind has a long history of using biotechnology; in 6000 B.C. our ancestors already knew how to make fermented foods and alcoholic beverages, although the process was not elucidated until 1857, when Pasteur proved fermentation was caused by microorganisms. In the 1910s, the fermentation industry was born and soon became the main force in the production of ethanol and solvents (mainly acetone and butanol from ABE fermentation by

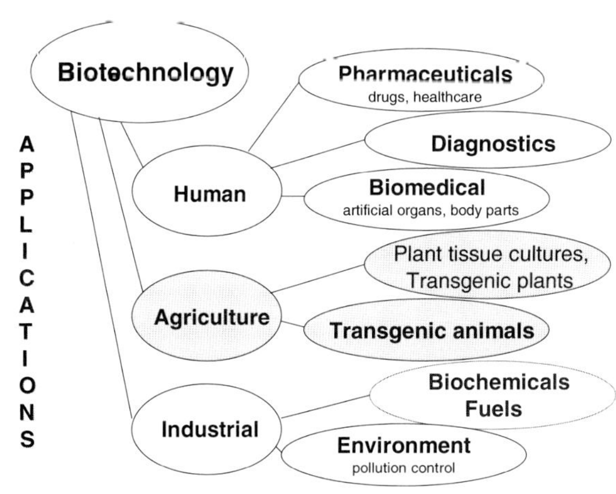

Fig. 12.1-1 Applications of biotechnology in various market sectors.

Renewable Energy Focus Handbook 2009; ISBN: 9780123747051

Table 12.1-1 Historical milestones in the development of biotechnology

Year	Historical events
6000 B.C.	Alcoholic beverages, bread, and cheese made by fermentation
1857	Pasteur proves fermentation is caused by microorganisms
1910	Fermentation industry developed (fuel & solvent production)
1923	Citric acid produced by industrial fermentation
1944	Penicillin mass-produced for Normandy landings in WWII
1953	DNA structure elucidated
1973	Recombinant DNA makes genetic engineering possible
1982	First commercial recombinant protein product (human insulin)

Clostridium acetobutyricum). The fermentation industry continued to develop, with citric acid being the first organic chemical and penicillin the first drug produced by fermentation in 1923 and 1944, respectively. However, with the discovery of oil and the rapid development of the petroleum industry in the 1950s, many of bulk chemicals and solvents, including ethanol, butanol, acetic acid, and lactic acid that previously had been predominately produced from sugars by fermentation were replaced with petroleum-based products produced by chemical synthesis [3]. Not until the first oil crisis in the 1970s did people start to realize that fossil fuels are exhaustible and that the oil-based economic development is not sustainable. Although there have been extensive efforts to develop renewable energy technologies, bio-based industrial products, with a few exceptions, have not been very successful because of relatively cheap oil. However, with recent crude oil prices exceeding $70 per barrel, bio-based products become increasingly attractive.

Until now and with only a few exceptions, most of fermentation products are drugs, foods, or animal feeds. In terms of quantity, ethanol is the leading industrial product from fermentation. Table 12.1-2 lists some of the current industrial fermentation products and their estimated global annual production.

12.1.2.1 Pharmaceutical industry

Following the elucidation of the structure of DNA by Watson and Crick in 1953, advances in molecular biology and the development of recombinant DNA technology (with the first demonstration of the transfer of heterologous genes via plasmids into *Escherichia coli* cells by Boyer in 1973) gave birth to the modern biotechnology industry. Genentech, the first biotechnology company, developed and launched, with the help of their licensing partner Eli Lilly, the first recombinant human protein (human insulin) for therapeutic applications in 1982. In the following two decades, the biotechnology industry continued to grow in the therapeutics sector to over $32 billion in annual sales in 2003 [4]. Today, the global market for biopharmaceuticals already exceeds $40 billion, which is about 10% of the total pharmaceutical market of over $400 billion. A large portion (more than 50%) of new drug development now comes from biotech R&D, a trend away from traditional chemical synthesis. In addition to the recombinant therapeutic proteins produced by fermentation, many small-molecule drugs

Table 12.1-2 Some major industrial fermentation products

	Production* (metric tons)	Microorganism	Separation method	Applications
Citric acid	1,200,000	*A. niger*	Extraction	Food
Ethanol	26,000,000	*S. cerevisiae*	Distillation	Fuel
Glutamate (MSG)	1,000,000	*C. glutamicum*	Crystallization	Flavoring
Lactic acid	400,000	*Lactobacillus sp.*	Extraction	Food, Plastics
Lysine	800,000	*C. glutamicum*	Crystallization	Feed
Penicillin	60,000	*P. chrysogenum*	Extraction	Drug
Xanthan gum	100,000	*X. campestris*	Precipitation	Food, Oil drilling

* 2006 data from personal communication with industry sources.

and drug intermediates, especially chiral compounds, are produced by biocatalysis or biotransformation using enzymes or whole cells as the catalysts.

12.1.2.2 Agriculture and food

Biotechnology also has had a major impact on the U.S. agricultural and food industries. Transgenic plants and crops have contributed to increased farm productivities and are made into food and animal feed. For example, biotech varieties of corn increased to 52% of U.S. corn acres planted in 2005, and corn production yield increased from 129.3 bushels/acre in 2002 to 142.2 bushels/acre in 2003 and 160.4 bushels/acre in 2004 [5]. Major transgenic crops in the U.S. also include soybeans (81%), cotton (73%), and canola (70%) [6]. Transgenic crops were already valued at over $20 billion in 2002 and are expected to rapidly increase in value as transgenic plants are or can be used to produce pharmaceuticals, chemicals, and fuels [7]. In the U.S. dairy industry, milk and cheese production have also increased to 177 billion lbs and 9.1 billion lbs, respectively, in 2005 [8], largely due to the improved milk production per cow resulting from the use of recombinant bovine somatotropin (BSA). Increased agriculture productivity has not only increased total production but also kept prices low, providing an opportunity and need for their increased utilization in non-food areas. Large agricultural companies, including ADM and Cargill, have aggressively expanded their business into the value-added product market beyond the traditional food and animal feed markets.

12.1.2.3 Chemical industry

The chemical industry is huge, consisting of four major subsectors: commodity chemicals, specialty chemicals, consumer care products, and pharmaceuticals, with over $2 trillion on sales worldwide [3]. Biotechnology can offer both economic and environmental benefits to the chemical industry and thus has great potential to achieve the sustainable production of existing and new products from renewable feedstocks. With few exceptions, the chemical industry has been largely built on hydrocarbon feedstocks, with nearly $24 billion worth of them being used annually. As the production of non-renewable fossil energy has reached its limit and oil and natural gas prices have skyrocketed in the recent years, interest in biobased industrial products and bioenergy has grown. The potential for biotechnology applications in the chemical and energy industries is huge, although so far the impact of biotechnology on these market sectors is very small in terms of both sales and market share. This is expected to change in the next few years, as several major chemical and agricultural companies have developed novel technologies to economically produce biobased chemicals and industrial products to replace petrochemicals and products derived from fossil fuels.

The biotechnological production of chemicals is greener than chemical methods because biocatalysts (enzymes or cells) are highly selective, resulting in higher product yield with less or no byproducts, which are usually difficult to separate. Chemical synthesis often requires toxic solvents and generates large amounts of wastes, causing disposal and pollution problems. One example is the production of cephalexin, a semisynthetic antibiotic derived from cephalosporin C. Enzymatic and direct fermentation methods developed by DSM can reduce the process steps from 10 to 4 and wastes to less than one third of those from the chemical process [9]. An enzymatic process for acrylamide developed by Mitsubishi Rayon uses ~20% as much energy as the conventional process. It also requires milder conditions and achieves greater conversion and a higher final product concentration. DuPont and Genencor have co-developed a recombinant *E. coli* fermentation for the production of 1,3-propanediol (PDO) from corn [10]. The biobased PDO, which is a key ingredient in Sorona polymer, consumes 30% to 40% less energy. A joint venture with Tate & Lyle is building a $100 million, 100,000 tons/yr Bio-PDO plant expected to be in operation in 2006 [11]. Nature Works, renamed from a previous Cargill-Dow joint venture and now solely owned by Cargill, produces polylactic acid (PLA) from lactic acid derived from corn at its 300-million-pound (140,000-metric-ton) capacity manufacturing plant and the world's largest lactic acid plant (400-million-pound or 182,000-metric-ton capacity) in Blair, Nebraska. The use of biodegradable PLA is expected to grow rapidly in the packaging material and textile fiber markets. Dow Chemical is developing vegetable-based polyol products, and ADM has announced a plan to build a polyol facility that will use carbohydrate and glycerol-based feedstocks. In addition, DSM is commercializing products derived from succinic acid produced from corn. Metabolix is developing a new generation of high-performance plastics based on polyhydroxyalkanoid (PHA) produced from renewable resources. These biopolymers can replace petroleum-based polymers and fibers, including polyesters, polyacrylics, polyamides, and polyurethanes, which have a worldwide production of 150 million tons per year [3].

12.1.2.4 Fuel and energy

Biomass has potential energy value both as a fuel for heat and power generation and as a feedstock for the production of chemicals and materials. Fuel and energy production from biomass thus represents another major market sector for biotechnology. Biofuels, including ethanol, biodiesel, butanol, methane, and hydrogen can be produced from renewable resources, replacing some fossil fuels. In the U.S., about 4.5 billion gallons of bioethanol are currently produced from corn and used along side with gasoline, which currently has an annual consumption rate of 140 billion gallons [12]. Brazil has increased its sugar cane-based ethanol output to an annual production of 15 billion liters (4 billion gallons) that can satisfy over 33% of the country's gasoline needs [13]. Domestic biofuels could eventually reduce U.S. dependence on foreign oil. In his 2006 state of the Union speech, President George W. Bush called for a 75% reduction in oil imports from the Middle East by 2025. The Energy Policy Act of 2005 mandates that 7.5 billion gallons of fuel ethanol be produced annually to replace 5% of gasoline by 2012. In fact, more than 2.2 billion gallons of annual capacity will be added in the U.S. by 2007 because demand for bioethanol is increasing rapidly as petroleum refiners phase out methyl-*tert*-butyl ether (MTBE) as an oxygenate in gasoline. The legislation also requires bioethanol use to increase to 30 billion gallons to replace 20% of gasoline by 2025. The production of biodiesel, mainly from soybean oil, also has increased rapidly from 500,000 gallons in 1999 to 75 million gallons in 2005. Biodiesel can replace petroleum-based diesel, and new U.S. legislation requires its use to increase to 250 million gallons in 2008 and 2 billion gallons in 2015.

Current bioethanol production consumes more than 12% of the corn produced in the U.S. It is clear that corn and soybeans alone will not be able to produce enough renewable fuels to displace a significant fraction of imported petroleum. Lignocellulosic materials are the most abundant renewable resources on earth, and new technologies are being developed to use them more economically as feedstocks for fuel and chemical production in the future [14]. In fact, Iogen in Canada has already built a cellulose-ethanol demonstration plant in 2004 that produces 800 liters of ethanol per day from wheat straw. Also, a Swedish ethanol plant started in summer 2005 uses sawdust as feedstock [13]. The commercial viability of cellulose-ethanol plants is promising as oil prices continue to rise and enzyme costs for cellulose hydrolysis continue to decrease as the result of new developments in this field. Biotechnology also can improve the growth yield and composition of energy crops, increasing their oil content for biodiesel production or carbohydrate content for bioethanol production, or decreasing lignin or changing cellulose crystallinity structure to facilitate faster hydrolysis. The biomass resources available for use represent 6–10 quadrillion Btu (quads) of energy.

Butanol is an important industrial solvent but also a potential liquid fuel that can be used directly to replace gasoline in current automobile engines [15]. The production of butanol totals about 350 million gal per year worldwide and 220 million gallons or 0.8 million tons in the U.S., all from petroleum. However, during World War I and until the 1950s, the "Weizmann" fermentation process was the main method for acetone and butanol production; this ceased in the U.S. and Europe during the early 1960s because of competition from petroleum-based solvents and the high price of sugar substrates [16]. The fermentation process is limited by the relatively low productivity and yield from sugars in the ABE fermentation process. The last commercial ABE fermentation plant in South Africa closed in 1981. However, recent advances in fermentation and separation technologies and rising energy prices will soon make biobutanol economically attractive again [17, 18]. Industrial ABE fermentation is now being considered in Austria [19, 20]. The extraction of butanol from the fermentation broth with biodiesel may generate a product with 18% ABE in biodiesel that can be used as a fuel without further cleaning [21].

Current technology for hydrogen production from biomass is far from economical, although hydrogen is a clean fuel and can be used in fuel cells. Biohydrogen can be produced by several routes: biophotolysis of water by algae and cyanobacteria, photodecomposition of organic compounds by photosynthetic bacteria, fermentative hydrogen production from organic compounds by anaerobic bacteria, and hybrid systems with both fermentative and photosynthetic bacteria [22]. However, none of these has the needed productivity and yield to be economically competitive at this time. Up to 12 moles of hydrogen can be obtained from glucose via ethanol fermentation followed with steam reforming of ethanol [23]:

$$\text{ethanol fermentation: } C_6H_{12}O_6 \rightarrow 2\ C_2H_5OH + 2\ CO_2$$
$$\text{steam reforming: } C_2H_5OH + 3\ H_2O \rightarrow 6\ H_2 + 2\ CO_2$$

The steam reforming process is highly endothermic. Thus, oxidative steam reforming with a reduced hydrogen yield (~5 mol) but minimal heat input is preferred. However, this fermentation-reforming process will be economical only when ethanol can be more economically produced from lignocellulosic biomass [24].

12.1.3 Bioprocessing – current status and development

A bioprocess usually consists of feedstock pretreatment, fermentation or biocatalysis, and downstream processing or separation for product recovery and purification (Figure 12.1-2). The actual bioprocess and required unit operation steps are largely dependent on the substrate and organisms used and the nature and applications of the final product. This section will briefly discuss organism choice, fermentation bioreactor design, and separation methods.

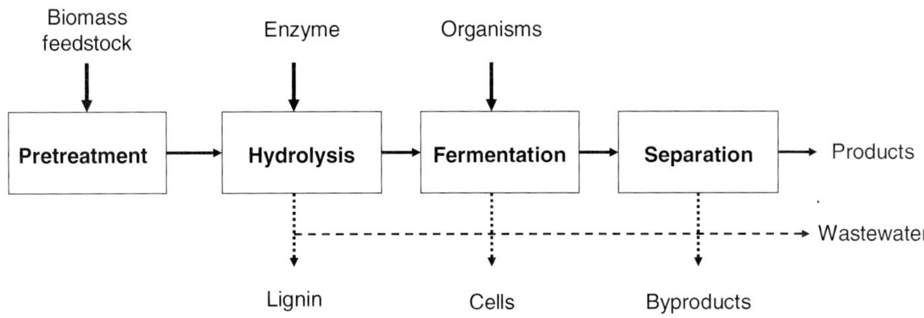

Fig. 12.1-2 A general bioprocess flowsheet.

More detailed discussions of various biomass feedstocks along with pretreatment methods and the hydrolysis of polysaccharides are given in the next section.

12.1.3.1 Organisms

As can be seen in Table 12.1-2, current industrial fermentation processes use all types of microorganisms: bacteria, yeasts, and filamentous fungi. The choice of microorganism for a fermentation process is usually based on the product, available substrate, and growth requirements, which affect the fermentation design and downstream processing. The final decision is dictated by economics. In addition to microorganisms, plant cell and hairy-root cultures also have gained increasing importance in industrial biotechnology [25]. Furthermore, marine biotechnology offers immense potential for finding new biologically active compounds that have not been thoroughly explored. Although still in their embryonic stage, marine sponge cultures offer a new type of bioprocess that one day could produce a wide spectrum of compounds for use as drugs [26,27]. It is noted that animal cells play a very important role in the biopharmaceutical and biomedical industries, but are yet to find applications in industrial biotechnology with mostly small molecules as the products.

Strain development is an important part of industrial fermentation. In recent years, conventional random mutagenesis and screening methods have largely been replaced by more rational approaches that use modern genetic engineering tools, including directed evolution and metabolic engineering [28,29]. Systems biotechnology or functional genomics using data from genomics, transcriptomics, proteomics, and metabolomics has also emerged as important tools for strain development and bioprocess analysis and optimization [30,31]. Today, genome breeding, genome engineering, and genome shuffling allow effective evolutionary whole-cell engineering of industrial strains [32]. Also, high-throughput screening techniques can speed up the discovery of new biocatalysts, organisms, and biologically active compounds.

Biocatalysts, including both enzymes and whole cells, are important in the production of specialty chemicals, especially chiral compounds that are difficult to make by chemical synthesis [33]. Chapter 12 provides some important examples of biocatalysis in the industrial production of chemicals.

12.1.3.2 Fermentation and bioreactor engineering

The majority of today's fermentation processes can be classified as submerged or solid state fermentations, with the former dominating in the Western fermentation industry. Over the last few decades, extensive research has been focused on bioreactor engineering to improve mixing and heat and mass transfer inside large-scale reactor vessels. However, new bioreactor development efforts focus on microbioreactors that allow high-throughput screening for strain and media optimization. For submerged fermentation, One key issue in industrial fermentation is improving productivity. Cell immobilization provides a viable solution to the productivity issue and has been extensively studied in the past three decades. Large-scale immobilized cell fermentation processes have been limited to industrial wastewater treatment, but will find important applications in fermentation that requires improved productivity and product titer in order to be economically competitive. Oxygen transfer is often the rate limiting step in high cell density and viscous fermentations.

12.1.3.3 Downstream processing

Except for recombinant protein therapeutics, most industrial fermentation products are separated and purified by one of several separation techniques based on differences in their size, density, volatility, solubility, and partition coefficient in two different phases [34]. Downstream processing usually consists of (1) cell (particle) separation by filtration, centrifugation, or sedimentation, (2) primary separation or enrichment by extraction, adsorption, precipitation, or evaporation, (3) secondary separation or purification by crystallization, liquid chromatography, or distillation, and (4) polishing and product packaging. Drying is used for powder products. Membrane processing, including microfiltration for particle separation and liquid sterilization, ultrafiltration for concentrating macromolecules such as proteins and polysaccharides, reverse osmosis and diafiltration to remove inorganic salts from the liquid product stream, and electrodialysis to separate ions based on their electrical charge difference, is becoming increasingly important in bioprocessing.

Table 12.1-3 shows some fermentation products and their downstream processing steps. Bioseparations using membranes with more detailed discussion on membrane fouling, which has been a major issue in biotechnological applications of membranes, are discussed earlier. Integrated fermentation and separation processes have additional advantages such as alleviating product inhibition and increasing reactor productivity [35].

12.1.4 Biomass feedstocks

Biomass currently provides over 3% of the total energy consumed in the United States and is the largest domestic renewable energy source [36]. Biomass includes any organic matter that is available on a renewable or recurring basis. Because it is renewable and abundant, biomass has the potential to offer diverse supplies of reliable, affordable, and environmentally sound energy and chemicals to replace fossil fuels and petrochemicals. The U.S. Department of Energy and the Department of Agriculture envision that biomass will provide 5% of power (heat and electricity), 20% of liquid transportation fuels (ethanol and biodiesels), and 25% of industrial products (chemicals and materials) by 2030, representing thirty percent of the current U.S. petroleum consumption, which would require over 1 billion dry tons of biomass feedstock annually [36]. Table 12.1-4 shows the potential production of various kinds of biomass, including dedicated energy crops and trees, agricultural crop residues, logging and wood processing residues, animal manures, and other waste materials.

Table 12.1-3 Some fermentation products and downstream-processing steps used in their recovery and purification

Product	Concentration (g/l)	Major steps used in downstream processing
Ethanol	70–120	Stripping, distillation
Organic acids	50–100	Precipitation/Solvent extraction, crystallization
Antibiotics (Penicillin)	10–30	Filtration, solvent extraction, crystallization, drying
Amino acids	1–100	Filtration, precipitation, crystallization, drying
Xanthan gum	25–50	Alcohol precipitation, centrifugation, drying
Enzymes	2–25	Precipitation, adsorption, chromatography
Vitamin B_{12}	0.02–0.06	Flocculation, filtration, adsorption, crystallization

Adapted from [34].

12.1.4.1 Starch and sugar crops

Currently, starch and sugar from agricultural crops are the main fermentation feedstocks used in industry. This practice will not change until the cost of using lignocellulosic biomass as feedstock has been substantially lowered. Compared to cellulose, starch is much easier to hydrolyze to glucose by either chemical or enzymatic methods. The enzymatic method involves starch liquefying and saccharification enzymes (see Table 12.1-5) and is used in the corn refinery industry, which processes more than 22% of the 11.8 billion bushels (~300 million metric tons) of corn produced annually in the U.S. into high-fructose-corn-syrup, dextrose, starch, and fuel alcohol. Dextrose or glucose derived from starch is the main substrate for industrial fermentation in the U.S. About 4 billion gallons of fuel ethanol were produced from 1.4 billion bushels of corn in 2005 in the U.S.A. On the other hand, sucrose from sugar canes and sugar beets is the main fermentation substrate in other regions. The global production and compositions of some major starch and sugar crops are listed in Table 12.1-6. Grains and processed products from these crops are mainly consumed as foods and animal feeds. Only a small fraction, the wasted crops, may be collected and used to produce fuels, chemicals and other industrial products [37]. However, the global quantity of crop residues available as potential fermentation feedstock is huge, estimated at 204 million tons for corn stover, 731 million tons for rice straw, 354 million tons for wheat straw, and 180 million tons for sugar cane bagasse [37].

12.1.4.2 Lignocellulosic biomass

Lignocelluloses are the most abundant biomass found in almost all plant-derived materials, from wood and grass to agricultural residues and municipal solid wastes. The major components of lignocelluloses are cellulose,

Table 12.1-4 Annual biomass potential from agricultural and forest resources in the United States*

Agricultural resources	(10^6 dry tons)	Forest resources	(10^6 dry tons)
Crop residues	446	Logging residues	64
Grass and woody crops	377	Excess biomass thinning	60
Municipal solid waste	111	Fuel wood	51
Grains to fuels	87	Mill processing residues	145
Animal manures	44	Urban wood residues	48
Food processing residues	44		
Total	998	Total	368

* Including both currently available and potential growth in agricultural and forest lands [36].

hemicellulose, and lignin; however, their compositions vary greatly, depending on the type of plant, cultivation conditions, and the age of the plant. Table 12.1-7 shows the compositions of some lignocellulosic materials.

Cellulose, which usually makes up the major organic components (up to 50%) of a plant, is a linear polymer of D-glucose linked by β-glucosidic bonds. The linear cellulose polymers, called elemental fibrils, are linked together by hydrogen bonds and van der Waals forces to form microfibrils, which group together to constitute cellulose fiber and are usually covered by hemicellulose and lignin. Plant cellulose appears in either crystalline or amorphous form. The former is a highly ordered form that is difficult to degrade biologically. Hemicelluloses are a group of complex heteropolysaccharides made up of various sugars (D-xylose, D-glucose, D-mannose, D-galactose, and L-arabinose) and sugar acids (D-glucuronic and D-galacturonic acids), depending on the plant species. Unlike cellulose, hemicelluloses have branches with short lateral chains of different sugars, do not form aggregates, and are easily hydrolysable. Lignin, present in the cellular cell wall, is an amorphous heteropolymer consisting of phenylpropane units (coniferyl alcohol, sinayl alcohol, and

Table 12.1-5 Enzymes (amylases) for breaking down starch

Enzyme	Reaction
α-Amylase	Randomly cuts α-1,4-glycosidic bonds in starch molecules
β-Amylase	Cleaves maltose disaccharide from the non-reducing end
Glucoamylase	Cleaves glucose from the non-reducing end
Pullulanase	Cuts α-1,6-glycosidic bonds at the branching point in amylopectin

coumaryl alcohol) joined together by different types of linkages [38]. These polymers are biodegradable but difficult to use directly as substrates in industrial fermentation.

The hydrolysis of lignocelluloses to fermentable sugars remains the greatest challenge in the development of economical plant biomass feedstock for the biorefinery industry [39]. The enzymatic hydrolysis of cellulose, which is expensive but nevertheless preferable to other methods, requires several cellulases – endoglucanases (endo-l,4-β-glucanases, EG), cellobiohydrolases (exo-l, 4-β-glucanases, CBH), and cellobiase (β-glucosidase) (see Table 12.1-8). EG hydrolyze internal bonds, while CBH work from the existing ends of cellulose, releasing cellobiose molecules, which are further broken down to two molecules of glucose by β-glucosidase. To facilitate the enzymatic hydrolysis of cellulose, it is usually necessary to pretreat the lignocellulosic materials to partially remove or degrade hemicellulose and lignin and to break up or loosen crystalline cellulose and increase its surface areas accessible for enzyme adhesion.

Table 12.1-9 lists and compares commonly used pretreatment methods [40–42]. All the pretreatment methods increase the accessibility of crystalline cellulose for enzymatic hydrolysis. They also hydrolyze hemicellulose and lignin to different extents, depending on the treatment conditions. Dilute acid pretreatment with H_2SO_4 is the most often used method in industry, but it usually generates some toxic byproducts that need to be removed before yeast fermentation.

12.1.4.3 Industrial waste

There is abundant biomass present as processing wastes requiring proper disposal to avoid pollution. One example is the corn refinery (wet milling) industry, which

Table 12.1-6 Major starch and sugar crops – global annual production and compositions (dry basis, wt%)

Crop	Production* (10^6 metric tons)	Starch	Sugar	Protein	Oil	Fiber
Corn	695	72	–	10	5	13
Wheat	628	80	–	14	–	5
Rice	619	89	–	8	1	2
Soybeans	210	16	16+	40	21	5
Sugar cane	1,290	–	55	–	–	45
Sugar beet	243	–	68	6	–	7
Sweet sorghum	59	–	50	–	–	50

* 2005 data from FAOSTAT Database available at http://faostat.fao.org.
+ including sucrose and soluble oligosaccharides.

processes more than 20% of the 11.8 billion bushels (~300 million metric tons) of com annually produced in the U.S. and generates more than 17 million metric tons of com byproducts (com fiber, etc.) that are currently of limited use and pose significant environmental problems. In the U.S. dairy industry, about 50% of the milk produced is used to produce cheese, generating ~80 billion lbs of cheese whey in 2005. Cheese whey contains about 7% total solids, of which ~70% is lactose and 13% is protein. Currently, less than 50% of the total amount of cheese whey produced in the U.S. is used to produce dry whey powder (469,000 metric tons), whey proteins, and lactose (300,000 metric tons), leaving more than 50% of the lactose in whey unused, which requires costly disposal because of its high biological oxygen demand (BOD). These abundant and inexpensive renewable resources can be readily used to produce chemicals and fuels by fermentation. The bioconversion of whey lactose to exopolysaccharides is discussed in Chapter 22. Chapter 23 describes the production of microbial polyhydroxyalkanoid (PHA) from renewable resources, including whey.

Table 12.1-7 Organic components of some lignocellulosic biomass (dry basis, wt%)

Feedstock	Cellulose	Hemicellulose	Lignin	Other
Bagasse	40	24	25	11
Corn stover	40	25	17	18
Corn cob	39	35	15	11
Corn fiber	15	35	8	42*
Rice straw	35	25	12	28
Wheat straw	38.2	21.2	23.4	17.2
Wheat chaff	38	36	16	11
Switch grass	45	31	12	12
Hard wood (hybrid poplar)	44.7	18.6	26.4	10.3
Soft wood (pine)	44.6	21.9	27.7	5.8
Waste paper	76	13	11	0

* including 23.7% starch. Sources of data: [24, 39, 40].

Table 12.1-8 Enzymes for breaking down cellulose and lignin

Enzyme	Reaction
Cellulases	
endo-l,4-β-glucanases (EG)	Hydrolyze internal β-l,4-glycosidic bonds in the amorphous region
cellobiohydrolases (CBH)	Release cellobiose disaccharides from the reducing end (CBH I) and the non-reducing end (CBH II) of the cellulose chain
β-glucosidase	Cleaves the cellobiose disaccharide to glucose
Lignin-degrading enzymes	
Peroxidases, Laccases	Depolymerize lignin by oxidizing phenolic compounds, but also non-phenolic compounds in the presence of mediators

Table 12.1-9 Some pretreatment methods for lignocellulosic biomass

Pretreatment methods	Conditions and performance
Steam explosion	Uses steam at 210–290°C, 20-50 bar for 2 min., followed with sudden pressure release; low xylose yield of 45–65%
Liquid hot water (LHW)	Uses compressed hot water 200–230°C for up to 15 min; high xylose yield (88–98%); requires recycling of water
Dilute acid pretreatment	Uses 0.5–1.5% H_2SO_4 or HCl at 160–220°C; good xylose yield: 75–90%; requires neutralization before cellulose hydrolysis; generates some toxic byproducts (acetic acid, furfural, phenolic compounds, etc.); current industrial method
Ammonia fiber explosion (AFEX)	Uses liquid ammonia (5–15%) and steam explosion (160–180°C); enhances hydrolysis of (hemi)cellulose from grass, but not as effective for soft and hard woods that contain more lignin
Alkali pretreatment	Uses lime or NaOH at lower temperatures and pressures for a longer time (hours); removes all lignin but only some hemicellulose

In addition, there are also large amounts of municipal solid wastes, waste sulfite liquor from the paper and pulp industry, and animal wastes available as inexpensive feedstocks. Chapter 25 discusses the use of animal manure to produce value-added products.

12.1.4.4 Lipids

Oils and fats are important raw materials for the production of oleochemicals, including free fatty acids, methyl esters, fatty alcohols and amines, and glycerol as a byproduct. Vegetable oils account for about 80% of the global oil and fat production, which was 105 million tons in 2000 [43]. Soybeans are the most important oil crop, followed by palm, rapeseed, and sunflower oils. About 15–17 million tons of vegetable oils are used by industry for the production of surfactants, lubricants, coatings, cosmetics, and other products. Fatty acid methyl esters have an important new use as liquid fuels. The production of biodiesel in the EU has steadily increased to 1.4 million tons in 2003, and the trend is continuing. As a result of increased biodiesel production, more glycerol, a byproduct of the production of fatty acids and esters from triglycerides, is expected to be available at a lower price in the future. Chapter 24 provides detailed discussion of vegetable oils and their industrial applications.

12.1.4.5 Proteins and nucleic acids

Proteins and nucleotides are nitrogenous compounds that are present in biomass at a relatively small percentage weight. Proteins are important nutrients in human and animal diets. Many fermentation processes require organic nitrogen sources for cell growth. Inexpensive nitrogen sources can be obtained from organic wastes containing proteins or amino acids. However, there is much less protein than carbohydrates available for non-food applications. Nucleic acids have never been considered as potential biomass feedstock. However, deoxyribonucleic acid (DNA) has unique applications as biophotonic materials [44, 45]. Although not considered as potential feedstock for chemical and fuel production, protein and DNA can be valuable products that may improve the economics of a biorefinery.

12.1.5 Biorefineries

A biorefinery is a manufacturing facility that uses biomass as feedstock to produce fuels, power, and chemicals. It is analogous to today's petroleum refineries, which use petroleum-based feedstocks, mainly oil and natural gas, to produce multiple fuels, commodity chemicals, industrial products, and commercial goods. Figure 12.1-3 illustrates the biorefinery concept. There are two different routes or platforms for biomass conversion [46]. The syngas or thermochemical platform involves the gasification of biomass at 650–900°C by reacting with air, oxygen and steam to gaseous products (CO, CO_2, H_2, CH_4). In addition, liquefaction or pyrolysis of biomass at 450–500°C in the absence of any reactive compounds or oxidants can produce pyrolysis oil. All components of biomass, including lignin, which is resistant to biological conversion, can be converted to chemical building blocks. However, the thermochemical platform cannot compete economically with fossil fuels, especially coal. On the other

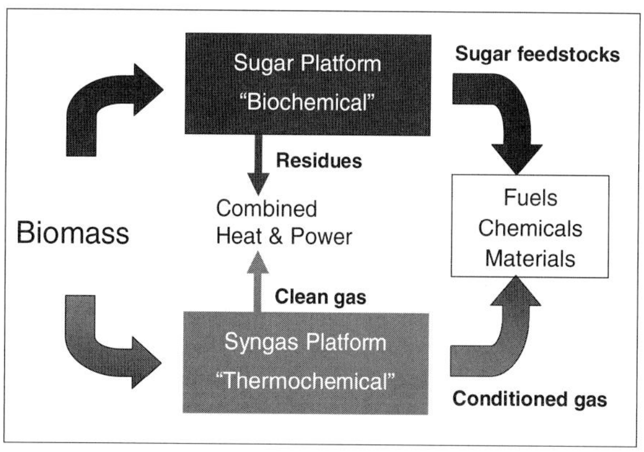

Fig. 12.1-3 Biorefinery concept. Adapted from [46].

hand, the sugar platform requires thermochemical pretreatment and enzymatic hydrolysis of lignocellulosic biomass, which at present is still too expensive because of high enzyme costs. Other economic barriers include relatively low sugar yields, low solid concentration (<20%), and impurities in the sugar solution. Advances in biotechnology one day will lead to new technologies that can produce better cellulases at lower costs [47], allowing the sugar platform to become economically attractive.

Hydrocarbon feedstocks represent the largest share of raw materials purchased in the U.S. chemical industry, nearly $24 billion in 2001. The fossil-based feedstock energy used in the production of chemicals was ~3.2 quads, equivalent to over 700 million barrels of oil [48]. In addition, over 3.6 quads were used at petroleum refineries as feedstocks for non-fuel products, of which over 55% were sold as intermediates for plastics and other chemical products. By comparison, agricultural commodities purchased by the chemical industry for conversion into products totaled about $4.5 billion in 2001. Biorefineries will have a significant impact on both the chemical and petroleum industries if biomass can be economically used to supplant fossil-based feedstocks.

12.1.5.1 Integrated biorefineries

The DOE has envisioned that by 2030, at least one third of the present transportation fuels in the U.S. will be produced from biomass [48]. In addition, the DOE's biorefinery vision includes producing high-value chemicals and materials from biomass. High-value products may represent only a small fraction of total products, but may account for most of the profits. Integrating the production of higher value bioproducts into biorefinery fuel and power output improves the overall profitability and productivity of all products. Figure 12.1-4 illustrates the concept of an integrated biorefinery that utilizes all components of the biomass feedstock to produce energy, fuels, and chemicals [46].

The DOE has also identified top chemical building blocks (see Figure 12.1-5) from biomass that have great potential to be used in current chemical synthesis processes to produce a cascade of intermediate and final industrial and consumer products [49]. These chemical building blocks are C3–C6 monomers exhibiting multiple functionalities suitable for further conversion as derivatives or molecular families. They are not already supercommodity chemicals, but can be produced from either starch or lignocellulosic biomass. These chemical building blocks include 1,4-dicarboxylic acids (succinic, fumaric, and malic acids), 2,5-furan dicaboxylic acid, 3-hydroxy propionic acid, aspartic acid, glucaric acid, glutamic acid, itaconic acid, levulinic acid, 3-hydroxybutyrolactone, glycerol, sorbitol, xylitol, and arabinitol. These chemical building blocks can be chemically converted to acrylic acid, adipic acid, acrylamide, 1,4-butanediol, and tetrahydrofuran, which are widely used in the chemical synthesis of polymers [50]. However, in order to use these chemical building blocks, their production costs from biomass need to be reduced to less than US$0.25/lb in order to compete with petroleum-based polymers. To meet this target, the cost of sugar needs to be reduced from $0.15/lb to $0.10/lb or less.

Among the chemical building blocks identified by the DOE, glutamic acid (monosodium glutamate) and

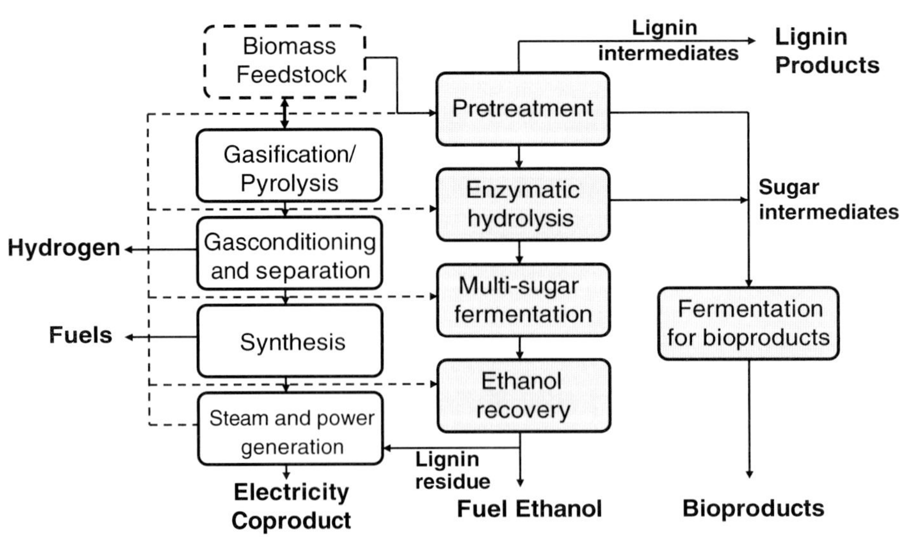

Fig. 12.1-4 An integrated biorefinery. Adapted from [46].

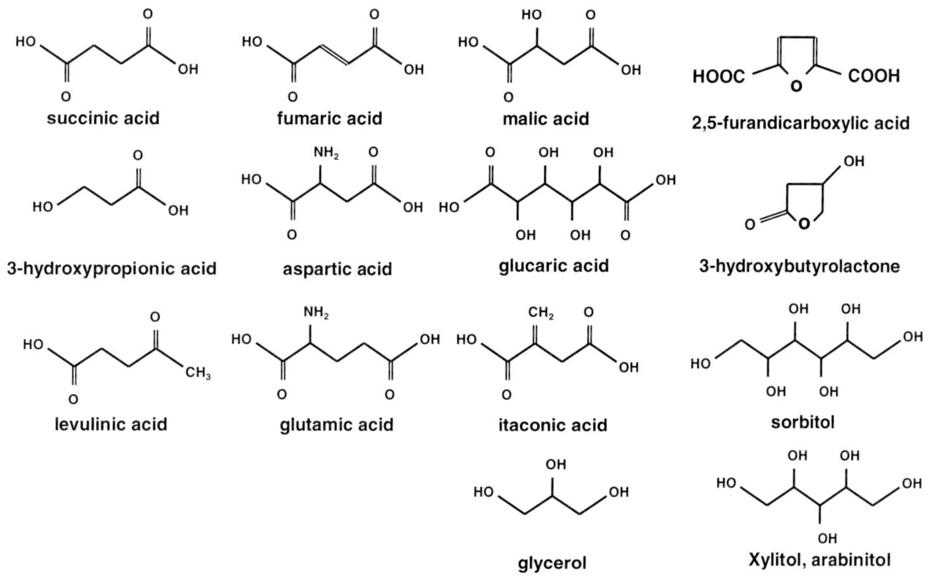

Fig. 12.1-5 Potential chemical building blocks from biomass identified by the DOE.

itaconic acid are commercially produced by glucose fermentation [51, 52], sorbitol is produced by hydrogenizing glucose, and glycerol is produced as a byproduct of the transesterification of vegetable oils with alcohol. Succinic acid and fumaric acid also can be produced from glucose by fermentation [53, 54]. There is no known microorganism capable of producing 3-hydroxypropionic acid, which is an isomer of lactic acid (2-hydroxypropionic acid) but has very distinct and unique physiochemical properties. Through metabolic engineering, Cargill and Codexis are making good progress toward developing mutants that can produce this chemical at a significant titer. Xylitol and arabinitol can be produced by hydrogenizing or fermenting pentoses (xylose and arabinose, respectively) obtained from the hydrolysis of hemicellulose [55]. Dehydrating C6 sugars leads to the production of 2,5-furandicarboxylic acid and levulinic acid, among others. Glucaric acid and 3-hydroxybutyrolactone can be produced from the oxidation of starch. However, these chemical methods need to be further developed and optimized in their selectivity in order to be economical.

Hydrocarbons, isoprenoids (natural rubber), and novel non-biodegradable polymers, such as polythioesters, also can be produced from renewable resources by fermentation [56, 57]. However, direct microbial synthesis of these chemicals and polymers is far from practical application and requires much more development work.

12.1.5.2 Corn refineries (wet milling)

In the United States, corn and soybeans are the largest biomass resources for industrial products. About 1.5 billion bushels of corn and 3.4 billion pounds of soybeans are used annually for this purpose. In addition to corn oil, starch, and feed products, various bioproducts, including ethanol, lactic, citric and itaconic acids, amino acids, and xanthan gum are currently produced by microbial fermentation in large corn wet-milling plants (see Figure 12.1-6). The expanded corn refinery plant also may include chemical conversion of glucose to sorbitol via hydrogenation, production of industrial enzymes for the conversion of starch to maltodextrins and high fructose corn syrup (HFCS), and an on-site cogeneration system providing electricity and steam for various processes. The chemical products are used in foods, detergents, and plastics. The ethanol is used as a solvent or for transportation fuels. Lactic acid can be converted to polylactic acid and used as bioplastics for packaging and textile fibers. Lactic acid and ethanol can react to form ethyl lactate ester, which can be used as an industrial "green" solvent, replacing the petroleum-based solvents currently used in the semi-conductor industry.

In addition, 1,3-propanediol and succinic acid are chemical building blocks that can be produced from corn dextrose. An emerging corn biorefinery may also produce high-value biopolymers, such as PHA and poly-γ-glutamate [58, 59]. In future corn biorefineries, corn stover and other crop residues will also be used as feedstock.

12.1.5.3 Whey processing

The U.S. dairy industry processes more than 50% of the milk produced to cheese, generating more than 80 billion pounds of whey a year. Large dairy plants produce whey

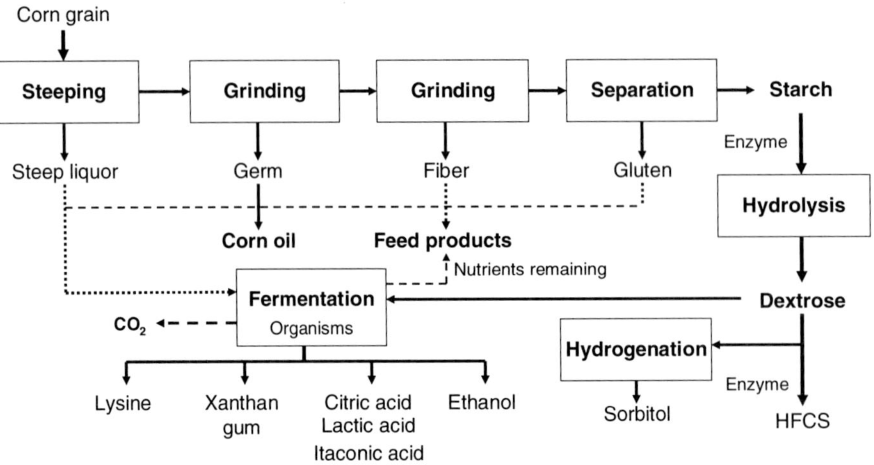

Fig. 12.1-6 Corn wet milling to various bioproducts.

powder for human and animal food uses. Whey proteins and lactose are also marketable products of whey processing (Figure 12.1-7). However, not all whey or whey permeate can be economically used for foods, and they are available as low-cost feedstock for fermentation to produce fuels and chemicals. For example, the lactose in whey (permeate) can be converted to lactic acid, propionic acid, and acetic acid via anaerobic fermentation [60–62]. The mother liquor produced in why processing is high in lactose and salt, and difficult to use as animal feed. One possible economic use of this waste stream is to produce calcium magnesium acetate as a road deicer [63]. The ultimate goal of a whey biorefinery is to achieve total utilization of all whey components and zero emission.

In an emerging whey processing plant, an immobilized enzyme reactor can be used to economically produce galacto-oligosaccharides (GOS) from lactose [64]. GOS contain 2 to 5 galactose units and one glucose. GOS are prebiotic, with wide applications in human and animal foods. As non-digestible dietary fibers, GOS are not susceptible to decomposition by human digestive enzymes, so they pass to the intestines, where they stimulate the growth of *Bifidobacteria* known to provide many health benefits, including increasing calcium absorption and the reduction of toxic metabolites and serum cholesterol. Current worldwide production of GOS is estimated at 25,000 tons per year, mainly in Japan and Europe.

However, in the GOS production process using whey lactose as the substrate, large amounts of glucose and galactose are also produced, which must be removed from GOS because these monosaccharides do not have and can reduce the prebiotic effect. It is thus desirable to convert these unwanted sugars to a high-value product, such as xanthan gum, to reduce waste while improving the process economics. Figure 12.1-8 shows an integrated bioprocess for the production of xanthan gum from glucose and galactose present in the waste sugar stream from the GOS production process. In the fermentation, a novel, rotating fibrous bed bioreactor can

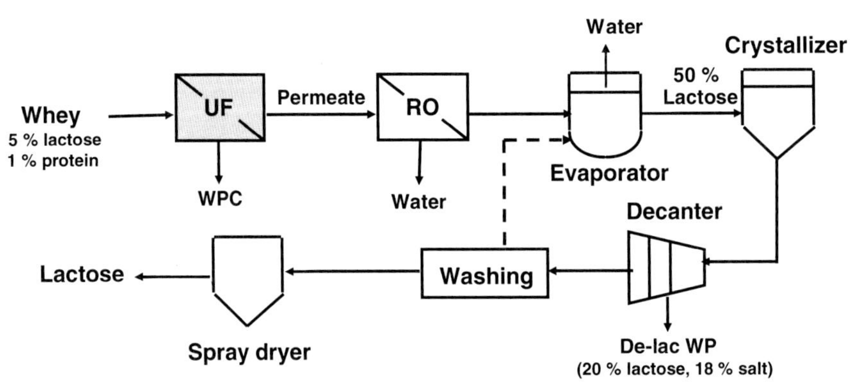

Fig. 12.1-7 A general flowsheet for whey processing to produce whey protein concentrate (WPC) and lactose from whey, which is a byproduct from cheese manufacturing. UF: ultrafiltration; RO: reverse osmosis; De-lac WP: de-lactosed whey permeate or mother liquor from crystallization.

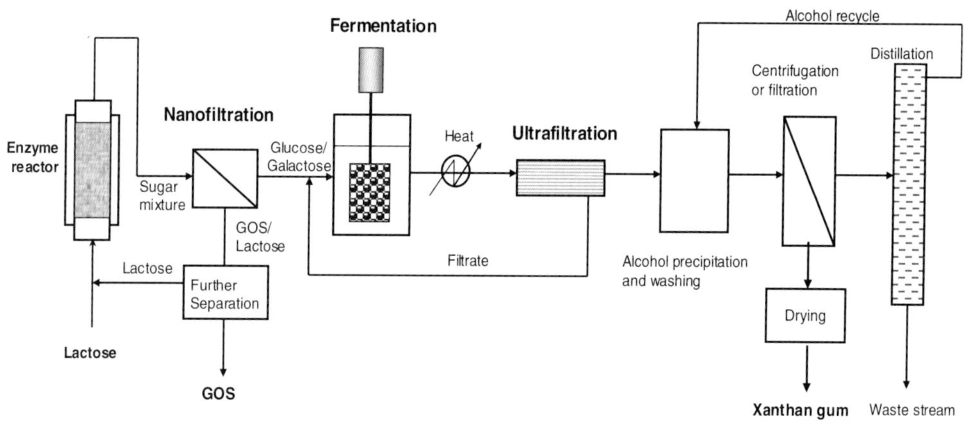

Fig. 12.1-8 An integrated process for GOS and xanthan gum production from whey lactose.

be used for cell immobilization to produce cell-free xanthan broth [65]. The centrifugal force generated from rotating the fibrous matrix separates the xanthan polymer from the immobilized cells, thus producing a cell-free xanthan broth that is then concentrated via ultrafiltration before further purification with alcohol precipitation. The permeate from ultrafiltration can be recycled and reused in the fermentation, reducing the raw material costs and the amount of spent medium. The integrated bioprocess for GOS and xanthan gum production from whey lactose not only can greatly increase the product value but also can reduce the amount of waste generated by the plant.

12.1.5.4 Lignocellulose biorefineries

Future biorefineries will use lignocellulosic materials from either crop residues or energy crops, such as switch grass, as feedstock. It has been estimated that there will be more than one billion tons of plant biomass available annually in the U.S. These lignocelluloses need to be converted to fermentable sugars in order to be used to produce fuels and chemicals. The cost of the hydrolytic enzymes is currently too high and must be reduced significantly for lignocellulose biorefineries to be economical. The production of ethanol from xylose and other pentoses remains a challenge, although several mutants have been created for this purpose [66, 67]. It has been suggested that consolidating the bioprocessing of lignocellulosic biomass by integrating cellulase production, cellulose hydrolysis, and the fermentation of both glucose and xylose into one step, would significantly reduce the production cost of ethanol [14]. However, the anaerobic bacteria capable of fermenting cellulose do not produce ethanol or organic acids at sufficiently high rates or concentrations. On the other hand, it is difficult to metabolically engineer an industrial fermentation strain to use cellulose directly.

One way to reduce the cost of enzymes is to produce and use them on site. It is also important to produce the highest-value product possible from all components of the feedstock. As an example, Figure 12.1-9 shows an integrated biorefinery producing both xylanases and cellulases for the hydrolysis of hemicellulose and cellulose, respectively. The xylose obtained from the hydrolysis of xylan can be converted to xylitol, a high-value sweetener currently used in chewing gum, whereas the glucose obtained from the hydrolysis of cellulose can be fermented to ethanol and other chemical products. The lignin and other solid residues can be burned to generate heat or steam for processing use.

12.1.6 Summary and outlook

Bioprocessing and bioproducts have gained commercial interest because of the perceived "green" advantages of using biomass rather than fossil energy for the production of chemicals and industrial products [68]. Other key benefits in the move towards bioproducts include the sustainability of renewable biomass, replacing depleted fossil energy, and reducing greenhouse gas emissions from the present petroleum-based chemical and energy industries. Currently, only 10% of the one hundred million metric tons of fine, specially, intermediate, and commodity chemicals produced annually in the U.S. are biobased. Today, hydrocarbons still dominate the U.S. economy. Currently, plant biomass provides only about 7% (by weight) of organic chemical products and 3% of transportation fuels and power. However, with rising oil prices and advances in industrial biotechnology, the potential of biomass to replace petroleum-based chemicals and fuels is huge. The potential for new bioproducts with improved performance to move into new and non-conventional markets is also substantial. The Department of Energy has predicted that the biotech-derived chemicals will exceed U.S. $100 billion in 2010

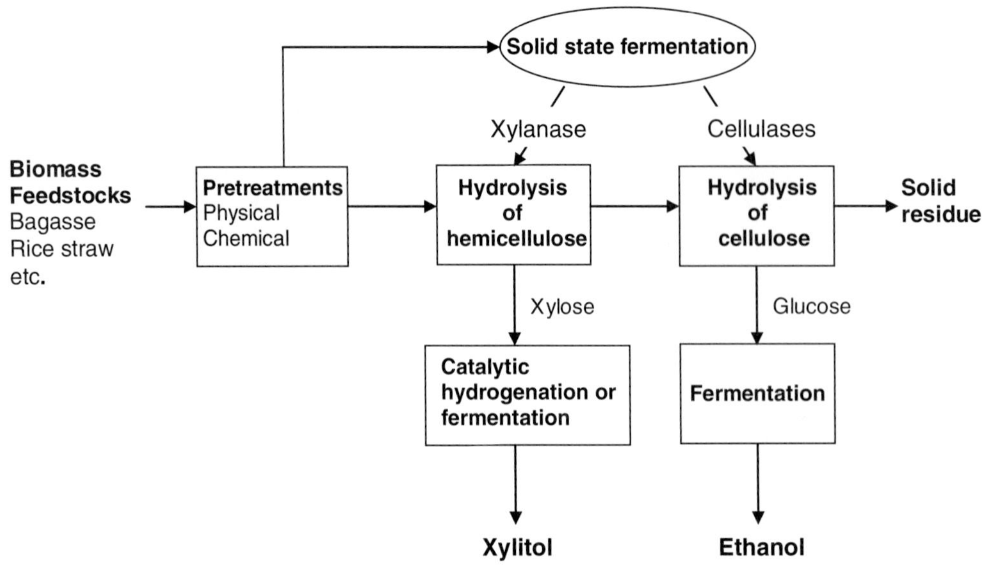

Fig. 12.1-9 An integrated biorefinery with solid state fermentation to produce xylanase and cellulase for the hydrolysis of hemicellulose and cellulose to xylose and glucose, which are then used to produce xylitol and ethanol, respectively.

and 400 billion, representing about 50% of the market for organic chemicals, in 2030. However, only with competitive prices can bioproducts make significant inroads into markets historically dominated by petrochemical products.

There are still many challenges facing the emerging biorefinery industry. In the short term, starch and sugar present in crops and processing wastes will continue to be the main substrates for fermentation. In the long run, lignocellulosic biomass offers the greatest potential as feedstock, but the cost of fermentable sugars derived from lignocelluloses must be significantly reduced to less than $0.10/lb. In order to convert carbohydrates to chemical building blocks, continued efforts to discover and engineer new and more efficient biological routes and metabolic pathways are also needed. Biorefineries must also take an integrated approach to make use of all biomass components to produce energy, fuels, chemicals, and high-value products in order to be environmentally sound and cost-competitive with traditional products. With new bioprocessing technologies under development, the biofuels and biobased chemicals and polymers are promising and will lead to a more sustainable society [69].

Acknowledgements

The author's bioprocessing research program is supported by the National Science Foundation, the Department of Energy, the US Department of Agriculture, Environmental Protection Agency, the Consortium for Plant Biotechnology Research, Inc., and Dairy Management, Inc.

References

[1] H. Ohara, Biorefinery. Appl Microbiol Biotechnol 62 (2003) 474–477.

[2] B. Kamm and M. Kamm, Principles of biorefineries. Appl Microbiol Biotechnol 64 (2004) 137–145.

[3] Th. Willke and K.-D. Vorlop, Industrial bioconversion of renewable resources as an alternative to conventional chemistry. Appl Microbiol Biotechnol 66 (2004) 131–142.

[4] A.K. Pavlou and J.M. Reichert, Recombinant protein therapeutics – success rates, market trends and values to 2010. Nat Biotechnol 22 (2004) 1513–1519.

[5] Corn Refiners Association, Corn, part of a healthy diet, Corn Refiners Association 2004 Annual Report.

[6] C.F Runge and B. Ryan, The economic status and performance of plant biotechnology in 2003: adoption, research and development in the United States. A report prepared for Council for Biotechnology Information (CBI), Washington, DC, December, 2003.

[7] M. Gavrilescu and Y. Chisti, Biotechnology – a sustainable alternative for chemical industry. Biotechnol Adv 23 (2005) 471–499.

[8] USDA National Agricultural Statistics Service, Milk Production, Disposition and Income, 2005 Summary, April 2006.

[9] E. Vandamme and C.G. Bienfait, Industrial biotechnology and

sustainable chemistry. Brussels: Royal Belgian Academy Council of Applied Science; 2004. p. 32.

[10] J.V. Kurian, A new polymer platform for the future – Sorona® from corn derived 1,3- propanediol. J Polymers and the Environment, 13 (2005) 159–167.

[11] Anonymous, Feedstock change is hot and poppin, Chemical Processing January 2006, 22–26.

[12] G. Hess, Push for biofuels seen in farm bill, Chem Eng News, May 22, 2006, 29–31.

[13] Y. Lin and S. Tanaka, Ethanol fermentation from biomass resources: current state and prospects. Appl Microbiol Biotechnol 69 (2006) 627–642.

[14] L.R. Lynd, W.H. van Zyl, J.E. McBride, and M. Laser, Consolidated bioprocessing of cellulosic biomass: an update. Curr Opin Biotechnol 16 (2005) 577–583.

[15] Environmental Energy, Inc., Butanol is an oxygenate which can replace gasoline. 2006. http://butanol.com/

[16] D.T. Jones and D.R. Woods, Acetone-butanol fermentation revisited. Microbiol. REV. 50 (1986) 484–524.

[17] T.C. Ezeji, N. Qureshi, and H.P. Blaschek, Acetone butanol ethanol (ABE) production from concentrated substrate: reduction in substrate inhibition by fed-batch technique and product inhibition by gas stripping. J Appl Microbiol 63 (2004) 653–658.

[18] N. Qureshi and H.P. Blaschek, ABE production from corn: a recent economic evaluation. J Ind Microbiol Biotechnol 27 (2001) 292–297.

[19] J.R. Gapes, The economics of acetone-butanol fermentation: theoretical and market considerations. J Mol Microbiol Biotechnol 2 (2000) 27 32.

[20] J.R. Gapes, The history of the acetone-butanol project in Austria. J Mol Microbiol Biotechnol 2 (2000) 5–8.

[21] E. Crabbe, C. Nolasco-Hipolito, G. Kobayashi, K. Sonomoto, A. Ishizaki, Biodiesel production from crude palm oil and evaluation of butanol extraction and fuel properties. Process Biochem 37 (2001) 65–71.

[22] D. Das and T.N. Veziroglu, Hydrogen production by biological processes: a survey of literature. Int J Hydrogen Energy 26 (2001) 13–28.

[23] H. Agus, F. Sun, M. Naveen, and A. Sushil, Current status of hydrogen production techniques by steam reforming of ethanol: A review. Energy & Fuels 19 (2005) 2098–2106.

[24] C.N. Hamelinck, G. van Hooijdonk, and A.P.C. Faaij, Ethanol from lignocellulosic biomass: techno-economic performance in short-, middle- and long-term. Biomass and Bioenergy 28 (2005) 384–410.

[25] C.F. Runge and B. Ryan, The economic status and performance of plant biotechnology in 2003: adoption, research and development in the United States. A report prepared for Council for Biotechnology Information (CBI), Washington, DC, December, 2003.

[26] E.H. Belarbi, A.C. Gomez, Y. Chisti, F.G. Camacho, and E.M. Grima, Producing drugs from marine sponges. Biotechnol Adv 21 (2003) 585–693.

[27] N.L. Thakur and W.E.G. Müller, Biotechnological potential of marine sponges. Curr Sci 86 (2004) 1506–1512.

[28] M. Chartrain, P.M. Salmon, D.K. Robinson, and B.C. Buckland, Metabolic engineering and directed evolution for the production of Pharmaceuticals. Curr. Opin. Biotechnol. 11 (2000) 209–214.

[29] H. Shimizu, Metabolic engineering – Integrating methodologies of molecular breeding and bioprocess systems engineering. J. Biosci. Bioeng 94 (2002) 563–573.

[30] T. Hermann, Using functional genomics to improve productivity in the manufacture of industrial biochemicals. Curr. Opin. Biotechnol. 15 (2004) 444–448.

[31] S.Y. Lee, D.-Y. Lee, and T.Y. Kim, Systems biotechnology for strain improvement. TRENDS in Biotechnol. 23 (2005) 349–358.

[32] R. Petri and C. Schmidt-Dannert, Dealing with complexity: evolutionary engineering and genome shuffling. Curr. Opin. Biotechnol. 15 (2004) 298–304.

[33] A.S. Bommarius, Biocatalysis: fundamentals and applications. New York: Wiley, 2004.

[34] S. Basu, S.T. Yang, and Bioseparations, in S. Lee (ed.), Encyclopedia of Chemical Processing, Marcel Dekker (2005), pp. 221–236.

[35] D. Stark and U. von Stockar, In situ product removal (ISPR) in whole cell biotechnology during the last twenty years. Adv. Biochem Eng/Biotechnol. 80 (2003) 150–175.

[36] Oak Ridge National Laboratory, Biomass as Feedstock for a Bioenergy and Bioproducts Industry: The Technical Feasibility for a Billion-Ton Annual Supply, U.S. Department of Energy, DOE/GO-201995-2135, 2005.

[37] S. Kim, and B.E. Dale, Global potential bioethanol production from wasted crops and crop residues. Biomass and Bioenergy 26 (2004) 361–375.

[38] J. Perez, J. Munoz-Dorado, T. de la Rubia, and J. Martinez, Biodegradation and biological treatments of cellulose, hemicellulose and lignin: an overview, Int Microbiol 5 (2002) 53–63.

[39] R.C. Brown, Biorenewable Resources: Engineering New Products from Agriculture, Iowa State Press, Ames, Iowa, 2003, Ch. 3.

[40] Badal C. Saha, Hemicellulose bioconversion, Ind Microbiol Biotechnol 30 (2003) 279–291.

[41] N. Mosier, C. Wyman, B. Dale, R. Elander, Y.Y. Lee, M. Holtzapple, and M. Ladisch, Features of promising technologies for pretreatment of lignocellulosic biomass, Bioresource Technology 96 (2005) 673–686.

[42] Y. Sun and J. Cheng, Hydrolysis of lignocellulosic materials for ethanol production: a review, Bioresource Technology 83 (2002) 1–11.

[43] J.O. Metzger and U. Bornscheuer, Lipids as renewable resources: current state of chemical and biotechnological conversion and diversification. Appl Microbiol Biotechnol 71 (2006) 13–22.

[44] J.G. Grote, D.E. Diggs, R.L. Nelson, J.S. Zetts, and F.K. Hopkins, et al., DNA photonics [Deoxyribonucleic acid]. Mol. Cryst. Liq. Cryst., 426 (2005) 3–17.

[45] L. Wang, J. Yoshida, and N. Ogata, Self-assembled supramolecular films derived from marine deoxyribonucleic acid (DNA)-cationic surfactant complexes: Large-scale preparation and optical and thermal properties. Chem. Mater. 13 (2001) 1273–1281.

[46] Office of the Biomass Program Multi Year Program Plan. U.S. Department of Energy, Office of Energy Efficiency and Renewable Energy, Washington, D.C. August 31, 2005.

[47] M.E. Himmel, M.F. Ruth, and C.E. Wyman, Cellulase for commodity products from cellulosic biomass. Curr. Opin. Biotechnol. 10 (1999) 358–364.

[48] M. Paster, J.L. Pellegrino, and T.M. Carole, Industrial Bioproducts: Today and Tomorrow. Report prepared by Energetics, Inc., Columbia, Maryland for the U.S. Department of Energy, Office of Energy Efficiency and Renewable Energy, Office of the Biomass Program, Washington, D.C. July 2003.

[49] T. Werpy and G. Petersen, Top Value Added Chemicals From Biomass, Volume I: Results of Screening for Potential Candidates from Sugars and Synthesis Gas. U.S. Department of Energy, Office of Energy Efficiency and Renewable Energy, Office of the Biomass Program, Washington, D.C. August 2004.

[50] S.Y. Lee, S.H. Hong, S.H. Lee, and S.J. Park, Fermentative production of chemicals that can be used for polymer synthesis. Macromol. Biosci. 4 (2004) 157–164.

[51] E. Kimura, Metabolic engineering of glutamate production, Adv. Biochem. Eng./Biotechnol. 79 (2003) 37–57.

[52] T. Willke and K.D. Vorlop, Biotechnological production of itaconic acid. Appl. Microbiol. Biotechnol. 56 (2001) 289–295.

[53] H. Songa and S.Y. Lee, Production of succinic acid by bacterial fermentation. Enz. Microb. Technol. 39 (2006) 352–361.

[54] N. Cao, J. Du, C.S. Gong, and G.T. Tsao, Simultaneous production and recovery of fumaric acid from immobilized *Rhizopus oryzae* with a rotary biofilm contactor and adsorption column. Appl. Environ. Microbiol. 62 (1996) 2926–2931.

[55] E. Winkelhausen and S. Kuzmanova, Microbial conversion of D-xylose to xylitol. J. Ferment. Bioeng. 86 (1998) 1–14.

[56] N. Ladygina, E.G. Dedyukhina, and M.B. Vainshtein, A review on microbial synthesis of hydrocarbons. Process Biochem. 41 (2006) 1001–1014.

[57] A. Steinbuchel, Non-biodegradable biopolymers from renewable resources: perspectives and impacts. Curr. Opin. Biotechnol 16 (2005) 607–613.

[58] C.S.K. Reddy, R. Ghai, Rashmi, and V.C. Kalia. Polyhydroxyalkanoates: an overview. Bioresource Technol. 87 (2003) 137–146.

[59] M. Ashiuchi and H. Misono, Biochemistry and molecular genetics of poly-γ-glutamate synthesis. Appl. Microbiol. Biotechnol. 59 (2002) 9–14.

[60] E.M. Silva and S.T. Yang, Continuous production of lactic acid from acid whey by *Lactobacillus helveticus* in a fibrous-bed bioreactor. J. Biotechnol. 41 (1995) 59–70.

[61] S.T. Yang, Y. Huang, and G. Hong, A novel recycle batch immobilized cell bioreactor for propionate production from whey lactose, Biotechnol. Bioeng., 45 (1995) 379–386.

[62] Y. Huang and S.T. Yang, Acetate production from whey lactose using co-immobilized cells of homolactic and homoacetic bacteria in a fibrous-bed bioreactor. Biotechnol. Bioeng., 60 (1998) 498–507.

[63] S.T. Yang, Y.L. Huang, Z. Jin, Yan Huang, H. Zhu, and W.Qin, Calcium magnesium acetate at lower-production cost: production of CMA deicer from cheese whey, Final report to Federal Highway Administration and New York State Energy and Development Authority; Publication No. FHWA-RD-98-174; April 1999.

[64] N. Albayrak and S.T. Yang, Production of galacto-oligosaccharides from lactose by *Aspergillus oryzae* P-galactosidase immobilized on cotton cloth. Biotechnol. Bioeng., 77 (2002) 8–19.

[65] S.T. Yang, Y.M. Lo, and D.B. Min, Xanthan gum fermentation by *Xanthomonas campestris* immobilized in a novel centrifugal fibrous-bed bioreactor. Biotechnol. Progress, 12 (1996) 630–637.

[66] C. Bro, B. Regenberg, J. Forster, and J. Nielsen, In silico aided metabolic engineering of *Saccharomyces cerevisiae* for improved bioethanol production. Metabolic Eng. 8 (2006) 102–111.

[67] A. Aristidou and M. Penttila, Metabolic engineering applications to renewable resource utilization. Curr. Opin. Biotechnol. 11 (2000) 187–198.

[68] EuropaBio, White biotechnology: Gateway to a more sustainable future. EuropaBio, Brussels, Belgium. (www.mckinsey.com/clientservice/chemicals/pdf/BioVision_Booklet_final.pdf)

[69] A.J. Ragauskas, C.K. Williams, B.H. Davison, G. Britovsek, J. Cairney, C.A. Eckert, W.J. Frederick Jr., J.P. Hallett, D.J. Leak, C.L. Liotta, J.R. Mielenz, R. Murphy, R. Templer, and T. Tschaplinski, The path forward for biofuels and biomaterials. *Science* 311 (2006) 484–489.

Chapter 12.2

12.2

Biodiesel fuels

Leon G. Schumacher, Jon Van Gerpen, and Brian Adams

The United States depends heavily on imported oil to fuel its transportation infrastructure. The use of alternative fuel derived from plant oils was examined by researchers in the mid-1970s to determine if internal combustion engines could be fueled from sources other than petroleum. The initial research on pure vegetable oils as a replacement for petroleum diesel fuel was met with mostly negative results. Researchers determined that transesterification of these plant- and animal-derived oils reduced the viscosity of the oil without any other significant changes to the oil. Since the new fuel was bio-derived and was used to fuel a diesel engine, the name "biodiesel" was selected to refer to the new fuel. This article focuses more specifically on how biodiesel fuel was developed, its chemical and physical properties, advantages, disadvantages, and how biodiesel is used and stored. The article concludes by reviewing the economic issues associated with its use.

12.2.1 Biodiesel fuels and their origins

12.2.1.1 What is biodiesel?

Biodiesel is made from a number of feedstocks including vegetable oil, tallow, lard, and waste cooking oils (yellow grease). "Biodiesel is defined as mono-alkyl esters of long chain fatty acids derived from vegetable oils or animal fats which conform to American Society of Testing Materials (ASTM D5453) International specifications for use in diesel engines." Biodiesel contains no petroleum, but it can be blended at any level with petroleum diesel to create a biodiesel blend. Feedstocks can be transesterified to make biodiesel using an alcohol that has been mixed with a catalyst such as potassium hydroxide or sodium hydroxide. The most commonly used alcohol for transesterification is methanol. Methanol reacts easily and is less expensive to use than most other alcohols.

Soybean derived biodiesel is the most commonly used biodiesel in the United States. The most commonly used feedstock for biodiesel production in Europe is rapeseed. Biodiesel is biodegradable, nontoxic, and essentially free of sulfur and aromatics. Biodiesel is considered a renewable resource due to the fact that it is derived from products that can be grown and produced domestically.

12.2.1.2 What biodiesel is not

In 1898 when Rudolph Diesel's compression ignition engine was demonstrated at the World's Exhibition in Paris, it ran on virgin peanut oil. Some document this event as the first use of biodiesel. However, biodiesel is not vegetable oil or animal fats. Biodiesel refers to the alkyl esters produced from a transesterification reaction between the oil or fat and an alcohol. Others refer to mixtures of biodiesel and petroleum diesel fuel as "biodiesel." This mixture is referred to as a biodiesel blend and is commonly designated in much the same way as a blend of gasoline and alcohol (E85). For example, B100 is 100% biodiesel; B20 is a blend of 20% biodiesel and 80% diesel fuel.

12.2.1.3 Beginnings

Vegetable oils were transesterified prior to the mid-1800s. Transesterification is the process of reacting a triglyceride molecule with an excess of alcohol in the presence of a catalyst (KOH, NaOH, $NaOCH_3$, etc.) to

Renewable Energy Focus Handbook 2009; ISBN: 9780123747051

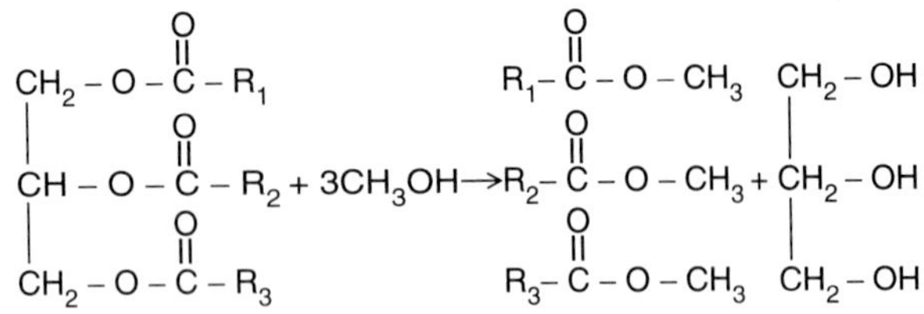

Figure 12.2-1 Chemical structure of biodiesel.

produce glycerol and fatty esters as shown in the chemical reaction in Fig. 12.2-1.

Companies such as Proctor & Gamble have used this process to make soap for years. According to Knothe, the earliest known use of alkyl esters for fuel appears in a Belgian patent granted in 1937 to G. Chavanne.

Methyl and ethyl esters are essentially by-products of this process. Ethyl esters are made using ethanol and methyl esters are made using methanol. Ethanol is made from grain such as corn. The methanol is either wood based or derived from natural gas (methane).

Peanut oil, hemp oil, corn oil, and tallow are typically transesterified in the soap-making process. Soybeans, industrial or edible rapeseed (or its cousin, canola oil), corn, recycled fryer oil, lard, and tallow are common resources for the complex fatty acids and their by-products.

It is important to find other sources of oil to enhance our ability to produce biodiesel. Research has been conducted concerning oil production from algae according to sources at the National Renewable Energy Laboratory (NREL). This oil source could have yields greater than any feedstock now known. Most biodiesel researchers worldwide agree that it would be difficult to replace more than 10% of the diesel fuel that is used for transportation purposes with biodiesel.

12.2.2 Chemical properties

Researchers have determined that the following properties characterize the chemical properties of biodiesel: fatty acid content, aromatics, olefins, paraffins, carbon, hydrogen, oxygen, and sulfur content, acid neutralization number, iodine number, and Conradson carbon residue number.

Note the chemical structure of a triglyceride (Fig. 12.2-2). R_1, R_2, and R_3 represent the hydrocarbon chain of the fatty acid elements of the triglyceride. There is a three-carbon chain called the glycerol backbone that runs along the left side of the molecule. Extending away from this backbone are the three long fatty acid chains. The properties of the triglyceride and the biodiesel fuel will be determined by the amounts of each fatty acid that are present in the molecules.

Fatty acids are designated by two numbers: the first number denotes the total number of carbon atoms in the fatty acid, and the second is the number of double bonds. For example, 18:1 designates oleic acid, which has 18 carbon atoms and one double bond. A typical sample of soybean oil based biodiesel would have the fatty acid profile shown in Fig. 12.2-3.

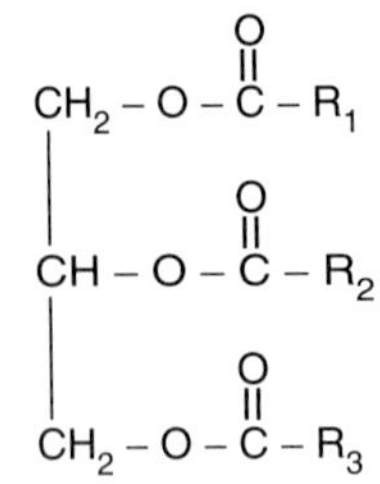

Triglyceride

Figure 12.2-2 Chemical structure of a triglyceride.

Biodiesel is essentially free of sulfur and aromatics. This is an advantage for biodiesel because sulfur poisons catalytic converter technology that is used to reduce engine exhaust emissions. The sulfur levels in biodiesel by ASTM D5453 are found to be as low as 0.00011% by mass (1ppm), where petroleum diesel is often no lower than 0.02% (200 ppm). The lack of aromatic hydrocarbons is also an advantage for biodiesel, as many of these compounds are believed to be carcinogenic. The test procedure normally used to measure aromatics in petroleum fuel (ASTM D1319) should not be used to determine the aromatics of biodiesel. This analytical procedure mistakenly identifies the double bonds commonly found in biodiesel as the resonance stabilized bonds normally associated with aromatics.

Paraffins are hydrocarbon compounds that are normally associated with petroleum diesel fuel. These compounds help increase the cetane value of the diesel fuel. However, they also typically increase cold flow problems of petroleum diesel fuel. Olefins are hydrocarbons that contain carbon-carbon double bonds. Molecules having these types of bonds are called unsaturated. Biodiesel from common feedstocks is usually 60 to 85% unsaturated. Some olefins are present in petroleum-based diesel fuel. However, the amount of olefins is usually small as they contribute to fuel oxidation.

Chain	Percentage by weight
C14:0	0.278
C16:0	10.779
C18:0	4.225
C18:1	20.253
C18:2	54.096
C18:3	9.436
C20:0	0.395

Figure 12.2-3 Fatty acid profile for soybean derived biodiesel.

The carbon content of biodiesel is nearly 15% lower than petroleum diesel fuel on a weight basis. Conversely, biodiesel has approximately 11% oxygen, on a weight basis, while petroleum diesel has almost no oxygen. Very little differences exist between biodiesel and petroleum diesel fuel concerning the weight percentage of hydrogen.

The neutralization number is used to reflect the acidity or alkalinity of an oil. This number is the weight in milligrams of the amount of acid (hydrochloric acid [HCL]) or base (potassium hydroxide [KOH]) required to neutralize one gram of the oil, in accordance with ASTM test methods. If the neutralization number indicates increased acidity (i.e., high acid number) of an oil, this may indicate that the oil or biodiesel has oxidized or become rancid. Biodiesel is allowed to have a neutralization number up to 0.8mg KOH/g.

The iodine value is a measure of the unsaturated fatty acid content of biodiesel, and reflects the ease with which biodiesel will oxidize when exposed to air. The iodine value for petroleum diesel fuel is very low, but the iodine value of biodiesel will vary from 80 to 135.

A weighed quantity of fuel is placed in a crucible and heated to a high temperature for a fixed period to determine the Conradson carbon residue of a fuel. The crucible and the carbonaceous residue is cooled in a desiccator and weighed. The residue that remains is weighed and compared to the weight of the original sample. This percentage is reported as the Conradson carbon residue value. This procedure provides an indication of relative coke forming properties of petroleum oils. No real differences are to be expected when comparing biodiesel with petroleum diesel fuel (0.02 versus 0.01).

12.2.3 Physical properties

The following properties reflect the physical properties of biodiesel: distillation curve, density/specific gravity, API gravity, cloud point, pour point, cold filter plug point, flash point, corrosion, viscosity, heat of combustion, and cetane number.

Each fuel has its own unique distillation curve. Some compare this to a fingerprint. This curve tells a chemist which components are in the fuel, their molecular weights by identifying their boiling points and their relative amounts. The temperature at which a specific volume boils off is used to establish this curve (0,10, 50, 90, and 100%). The range of boiling points for biodiesel is much narrower than petroleum diesel fuel. For example, the initial boiling point of petroleum diesel fuel is 159°C. and the end boiling point is 336°C. The initial boiling point of biodiesel is 293°C. and the end boiling point is 356°C. Thus, the distillation range for diesel fuel is 177° versus 63°C for biodiesel fuel.

The specific gravity of a product is the weight of the product compared to an equal volume of water. The specific gravity of biodiesel is slightly higher than petroleum diesel fuel. For example, the specific gravity of No. 2 petroleum diesel fuel is approximately 0.84 (7.01 pounds per gallon). The specific gravity of biodiesel is approximately 0.88 (7.3 pounds per gallon).

ASTM test procedures are used to determine cloud and pour point of biodiesel. Cloud point is the temperature at which the first wax crystals appear. Pour point is the temperature when the fuel can no longer be poured. According to ASTM specifications for diesel fuel, the cloud points of petroleum diesel fuel are determined by the season (temperature) that the fuel is used. For example, ASTM standards essentially require that the fuel distributor treat number 2 petroleum diesel fuel with cold flow improvers (CFI) for winter operation. Cold weather operation with 100% biodiesel also forces the distributor to use a CFI. However, the CFI of choice for petroleum diesel fuel is usually not as effective when used with biodiesel. The cloud point for biodiesel is higher than petroleum diesel fuel (1.6°C for biodiesel compared with −9.4° to −17.7°C for diesel fuel). Chemical companies (Lubrizol, Octell Starreon) are experimenting with CFI chemicals that work effectively with biodiesel.

An alternative method used to determine how the fuel will perform during cold weather operation is the cold filter plugging point. The cold filter plugging point is the lowest temperature at which the fuel, when cooled under specific conditions, will flow through a filter during a given period of time. The filter is a defined wire mesh. This procedure indicates the low temperature operability of a fuel with/without cold flow improver additives when cooled below the cloud point temperature. The cold filter plugging point for fuels has become an important issue due to the reduction in the size of the openings in the fuel filters (2–5 microns versus 45 microns).

The flash point of a fuel is defined as the temperature at which the air/fuel vapor mixture above the product will ignite when exposed to a spark or flame. A sample is heated and a flame is passed over the surface of the liquid. If the temperature is at or above the flash point, the vapor will ignite and a detectable flash (unsustained or sustained flame) will be observed. The flash point of biodiesel is substantially higher (159°C versus 58°C) than for petroleum diesel fuel. Biodiesel is categorized as a combustible fuel, not a flammable fuel, as is the standard classification for petroleum diesel fuel. Biodiesel is a safer fuel to transport due to its higher flash point.

As fuel deteriorates it becomes acidic. Copper is particularly susceptible to corrosion by the acids in the fuel. As a result, test procedures have been developed to detect the fuel's corrosiveness to copper. A polished copper strip is immersed in a heated sample of fuel. After a prescribed period of time, the strip is removed and

Table 12.2-1 ASTM biodiesel specification—D6751 versus ASTM LS #2 diesel fuel—D975

Property	ASTM method	Limits D6751	Units D6751	Limits D975	Units D975
Flash point	D3	130.0 min.	°C	52.0 min.	°C
Water and Sediment	D2709	0.050 max.	% vol.	0.050 max.	% vol.
Carbon residue, 100% sample	D4530	0.050 max.	% mass	N/A	N/A
Ramsbottom carbon residue	D524	N/A	N/A	0.35	% mass
Sulfated ash	D874	0.020 max.	% mass	N/A	N/A
Ash	D482	N/A	N/A	0.01	% mass
Kinematic viscosity, 400°C	D445	1.9–6.0	mm2/s	1.9–4.1	mm2/s
Sulfur	D5453	0.05 max.	% mass	N/A	N/A
Sulfur	D2622	N/A	N/A	0.05 max.	% mass
Cetane	D613	47 min.		40 min.	
Cloud point	D2500	By customer	°C	By customer	°C
Copper strip corrosion	D130	No. 3 max.		No. 3 max.	
Acid number	D664	0.80 max.	mg KOH/gm	N/A	N/A
Free glycerol	D6584	0.020 max.	% mass	N/A	N/A
Total glycerol	D6584	0.240 max.	% mass	N/A	N/A
Phosphorus content	D4951	0.001	% mass	N/A	N/A
Distillation temperature, 90% recovered	D1160	360 max.	°C	N/A	N/A
Distillation temperature, 90% recovered	D86	N/A	N/A	338 max.	°C

examined for evidence of corrosion. A standardized system is used to assign a value between 1 and 4. This number is assigned based on a comparison with the ASTM Copper Strip Corrosion Standards. No differences have been detected using this methodology between petroleum diesel fuel and biodiesel.

The viscosity of a fluid is a measure of its resistance to flow. The greater the viscosity, the less readily a liquid will flow. Test procedures are used to measure the amount of time necessary for a specific volume of fuel to flow through a glass capillary tube. The kinematic viscosity is equal to the calibration constant for the tube multiplied by the time needed for the fuel to move through the tube. The viscosity of biodiesel is approximately 1.5 times greater than petroleum diesel fuel (4.01 versus 2.6 cSt@40°C).

The heat of combustion is the amount of energy released when a substance is burned in the presence of oxygen. The heat of combustion, also known as the heating value, is reported in two forms. The higher, or gross, heating value assumes that all of the water produced by combustion is in the liquid phase. The lower, or net, heating value assumes the water is vapor. The lower heating value for biodiesel is less than for number 2 diesel fuel (37,215 kJ/kg versus 42,565 kJ/kg).

The cetane number reflects the ability of fuel to self-ignite at the conditions in the engine cylinder. In general, the higher the value, the better the performance. The cetane number of biodiesel varies depending on the feedstock. It will be 48 to 52 for soybean oil based biodiesel and more than 60 for recycled greases. This is much higher than the typical 43 to 47 observed for petroleum diesel fuel. The cetane number is typically estimated for petroleum diesel fuel using the cetane index. However, due to the fact that this index was developed for diesel fuel and not biodiesel, this ASTM test procedure provides faulty information for biodiesel. A cetane engine must be used to determine the cetane number for biodiesel. See Table 12.2-1.

12.2.4 Biodiesel production

12.2.4.1 Introduction

Biodiesel consists of the monoalkyl esters of fatty acids derived from vegetable oils or animal fats. It is most

Soybean oil (885 g) + 2X Methanol ($6 \times 32.04 = 192.24$ g)

$\xrightarrow{NaOH}$ Methyl soyate ($3 \times 296.5 = 889$ g) + Glycerol (96.12 g) + XS Methanol (92.10 g)

Figure 12.2-4 Basic transesterification reaction for soybean oil.

commonly produced through a process known as transesterification, which is a chemical reaction where an alkoxy group of an ester is exchanged with that of another alcohol to form the new ester product. The chemical reaction with methanol is shown schematically in Fig. 12.2-1.

This reaction is reversible so to force the equilibrium in the direction of the products, from 60% to 200% excess methanol is added. The reaction requires a catalyst and strong bases such as potassium hydroxide and sodium hydroxide. The actual reaction using 100% excess methanol is shown in Fig. 12.2-4.

The alkali catalysts usually result in the reaction proceeding to completion in 4 to 8 hours at ambient conditions and in 1 hour at 60°C. In general, the catalyst is dissolved in the methanol before addition to the oil to prevent direct contact between the concentrated catalyst and the oil. Since the methanol is only slightly soluble in the soybean oil, agitation is required during the early part of the reaction. The reaction proceeds through a sequence of steps involving the removal of fatty acid chains from the triglyceride to produce diglycerides, monoglycerides, and, ultimately, free glycerol. When the reaction has proceeded to the point where substantial amounts of di- and monoglycerides have been produced, agitation is less important. To further drive the equilibrium to products, the reaction is often conducted in steps. During the first step, only a portion, say 80%, of the methanol and catalyst are added. The reaction proceeds substantially to equilibrium and then the reactants are allowed to settle so the resulting glycerol can be removed. Then, the remaining methanol and catalyst are added and the reaction continued. Removal of the glycerol forces the reaction to the product side and since the alkali catalyst is selectively attracted to the glycerol, the presence of the glycerol can limit the speed of the reaction.

The most important characteristics of the triglyceride feedstocks are low water content (preferably less than 0.2%) and low free fatty acids (less than 0.5%). High free fatty acid (FFA) feedstocks can be processed but pretreatment to remove the FFAs or to convert them to biodiesel is required. This pretreatment will be described later. Water should be excluded from the reaction because it contributes to soap formation as the fatty acid chains are stripped from the triglycerides. The soap sequesters the alkali catalyst and inhibits the separation of the glycerol from the biodiesel. Excessive soap also contributes to the formation of emulsions when water washing is used at a later stage of the production process.

Another important characteristic of the feedstock is the presence of saturated and polyunsaturated triglycerides. Excessive levels of saturated fatty acid chains can produce biodiesel with a high pour point making it difficult to use at low temperatures. High levels of polyunsaturates can provide poor oxidative stability requiring the resulting fuel to be treated with an antioxidant.

While most biodiesel is made using methanol, because of its low price (and quick conversion), other alcohols, such as ethanol and isopropanol, can also be used. Higher alcohols provide superior cold flow properties but are generally more difficult to produce, requiring higher temperatures, lower levels of water contamination, and more complex alcohol recycling due to the formation of azeotropes.

As mentioned earlier, strong alkalis such as sodium hydroxide and potassium hydroxide are common catalysts. These bases form the corresponding methoxides when dissolved in methanol. Water is also formed in this reaction and is probably responsible for some soap formation although not enough to inhibit the transesterification reaction. A more desirable option is to use sodium methoxide formed using a water-free process. This catalyst is available as a concentrated solution in methanol (25% and 30%), which is easier to use because it is a liquid. Research is underway to develop heterogeneous catalysts for biodiesel production that would minimize soap formation and provide cleaner glycerol. Since most of the catalyst ends up as a contaminant in the glycerol, either as soap or free alkali, a heterogeneous catalyst would simplify glycerol refining. Research is also being conducted to find reaction conditions that do not require a catalyst. However, these conditions are at very high temperature (>250°C) and appear to produce undesirable contaminants in the biodiesel.

Figure 12.2 5 shows a schematic of a biodiesel production process. Oil enters the reactor where it is mixed with methanol and catalyst. Usually, the catalyst has been mixed with the methanol before contacting the oil to prevent direct contact of the concentrated catalyst and the oil to minimize soap formation. The reactor can be either a batch process or, as is more common with larger plants, a continuously stirred tank reactor (CSTR) or plug flow reactor. When a CSTR is used, it is common to use more than one stage to ensure complete reaction. After the reaction is complete, the glycerol is separated from the biodiesel. This separation can be accomplished with a gravity decanter or using a centrifuge. The unreacted methanol will split between the biodiesel and glycerol giving about 1 to 3% methanol in the biodiesel and 30 to 50% in the glycerol. The methanol in the biodiesel should be recovered for reuse. It may be as

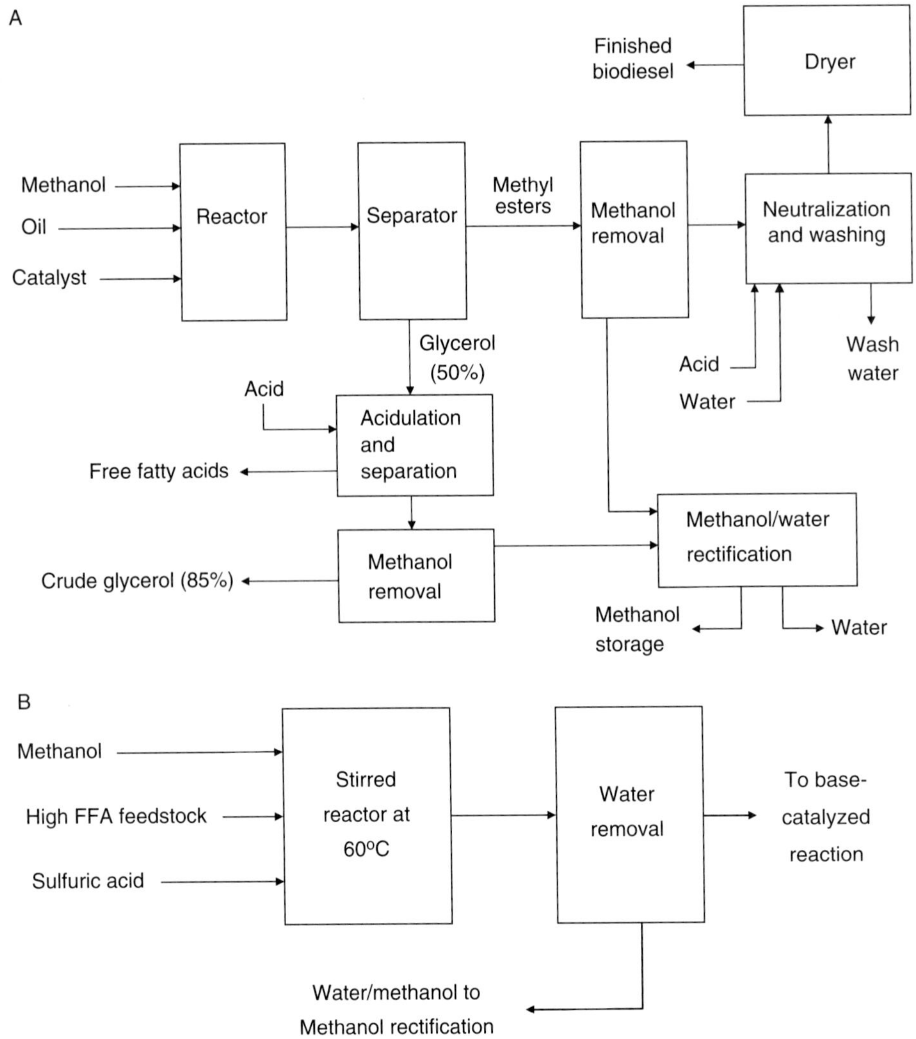

Figure 12.2-5 (A) Process flow schematic for biodiesel production. (B) Acid-catalyzed pretreatment process.

much as half of the excess methanol. This is usually accomplished by a vacuum flash process, but other devices such as falling film evaporator have also been used. The methanol-free biodiesel is then washed with water to remove residual methanol, catalyst, soap, and free glycerol. If the biodiesel contains excessive soap, this washing process can be problematic, as the soap will cause an emulsion to form between the water and the biodiesel. To minimize the formation of emulsions, a strong acid is sometimes added to the biodiesel to split the soap into free fatty acids (FFA) and salt. Without the soap, water consumption is greatly reduced and the salt, methanol, catalyst, and glycerol are removed with as little as 3 to 10% water. The water should be heated to 60°C to assist in the removal of free glycerol and should be softened to minimize the transfer of calcium and magnesium salts to the biodiesel. The final step in the process is to heat the biodiesel to remove water that may be dissolved in the biodiesel or entrained as small droplets. This is accomplished with a flash process.

Also shown in the diagram is the preliminary processing of the co-product glycerol. This glycerol contains virtually all of the catalyst and a considerable amount of soap and unreacted methanol. Usually the glycerol will be acidulated to split the soaps into FFA and salt. The FFAs are not soluble in the glycerol and rise to the top where they can be decanted and returned to the biodiesel process after pretreatment. After FFA removal, the methanol is removed by a flash process or by a thin-film evaporator leaving a crude glycerol product that is 80 to 90% pure. The balance will be salts, residual FFAs, water, and phosphotides, color-bodies, and other contaminants from the oil.

Although water is not deliberately added to the process until after the methanol has been removed, small amounts of water will enter the system as a contaminant in the oil, alcohol, and catalyst. This water will tend to accumulate in the methanol, so before it can be returned to the process, the methanol should undergo fractional distillation.

Biodiesel plants can use either batch or continuous flow processing. Batch processing is most common in small plants of less than 4 million liters/year. Batch processing provides the ability to modify the process for variations in feedstock quality. Continuous flow requires greater uniformity in the feedstock quality, generally requires 24 hour operation, 7 days per week, increases labor costs, and is most suitable for larger operations of greater than 40 million liters/year.

12.2.4.2 Pretreatment

Small amounts of FFAs can be tolerated by adding enough alkali catalyst to neutralize the FFAs while still leaving enough to catalyze the reaction. When an oil or fat has more than 3 to 5% FFA, the amount of soap formed with the use of an alkali catalyst will be large enough that separation of the glycerol from the oil may not be possible. In addition, when excess alkali is added, the FFAs are lost and not available for conversion to biodiesel.

An alternative is to convert the FFAs to methyl esters using an acid catalyst such as sulfuric acid with a large excess of methanol (>20:1 molar ratio based on the FFA content) (Fig. 12.2-5B). The acid-catalyzed reaction is relatively fast (1 hour at 60°C), converting the FFAs to methyl esters with water produced as a by-product. The water eventually stops the reaction before all of the FFAs have been converted and must be removed from the system, either by decanting with the methanol or by vaporization. After the FFAs have been converted to methyl esters, an alkali catalyst can be used to convert the triglycerides. Acid catalysis can be used to complete the transesterification reaction, but the time required is prohibitive.

12.2.4.3 Product quality

Modern diesel engines require high-quality fuels. The fuel injection system, which is often the most expensive element of the engine, can be damaged by fuel contaminants. Water and solid particles are the largest problem.

The contaminants most frequently found in biodiesel are the products of incomplete reaction and residual alcohol, catalyst, and free glycerol. Incompletely reacted biodiesel will contain monoglycerides, diglycerides, and triglycerides. These compounds are usually detected using a gas chromatograph and then the glycerol portion is summed to yield a total glycerol quantity for the fuel. ASTM standards require that the total glycerol be less than 0.24%. This means that more than 98% of the original glycerol portion of the triglycerides feedstock must be removed. Excessive amounts of monoglycerides, especially for saturated compounds, may precipitate from the fuel and plug fuel filters.

If the biodiesel is not washed with water, it may contain some unreacted alcohol. The amount will usually be small enough that it does not adversely affect the operation of the engine, but it can lower the flash point of the fuel to where it must be considered flammable and accorded the same safety requirements as gasoline. The residual catalyst can cause excessive ash formation in the engine. Free glycerol can separate from the fuel and collect in the bottom of storage tanks. This glycerol layer can extract mono- and diglycerides from the biodiesel and produce a sludge layer that may plug filters and small passages in the fuel system.

12.2.5 Advantages of biodiesel

Table 12.2-2 shows the changes that were observed in the regulated exhaust emissions of three diesel engines that were tested to produce emissions characterization data for the U.S. Environmental Protection Agency's Fuels and Fuel Additives registration program. The reductions in unburned hydrocarbons (HC) and carbon monoxide (CO) are dramatic although these specific pollutants are not generally a concern with diesel engines. The particulate matter (PM) reductions are also quite striking. Oxides of nitrogen (NO_x) were found to increase with the use of biodiesel. The increase varied depending on the engine tested, but it is clear that biodiesel may produce a NO_x increase of 5 to 13%. The reasons for this increase are still under investigation but appear to be a combination of several effects, including biodiesel's higher speed of sound and isentropic bulk modulus and the tendency of many engine fuel injection systems to advance the injection timing when greater volumes of fuel are injected. Due to biodiesel's lower energy content, a typical test protocol may demand a higher fuel flow rate when biodiesel is used, causing an inadvertent timing advance and resulting NO_x increase.

A comprehensive Life-Cycle Inventory of biodiesel conducted by the National Renewable Energy Laboratory showed that biodiesel provided 3.2 units of fuel energy for every unit of fossil energy consumed in its life cycle. Further, although some fossil-based CO_2 is released

Table 12.2-2 Changes in regulated emissions with biodiesel

Engine	HC	CO	PM	NO_x
Cummins N-14	−95.6	−45.3	−28.3	+13.1
DDC S-50	−83.3	−38.3	−49.0	+11.3
Cummins B5.9	−74.2	−38.0	−36.7	+4.3

Derived from Sharp *et al.*

during biodiesel production, mainly from the methanol consumed, the net production of CO_2 is reduced by 78%.

12.2.6 Disadvantages

12.2.6.1 Economics

One of the largest factors preventing the adoption of biodiesel is cost. The feedstock costs for biodiesel tend to be high in comparison to the cost of petroleum diesel fuel. The end result is that the cost of biodiesel is higher than that of petroleum diesel fuel. In addition, transportation costs are significantly greater for biodiesel due to the fact that the transportation infrastructure for biodiesel is in its infancy. The costs associated with the production and transportation of biodiesel fuels are discussed in more detail in the economics section of this article.

12.2.6.2 NO_x and other exhaust emissions

Biodiesel produces more NO_x emissions than diesel fuel. If B100 is used, NO_x production may be increased by 13%. If a B20 blend is used, NO_x production is only increased by 2%, and the engine will typically satisfy the EPA engine exhaust emissions requirements under the Clean Air Act. To meet the EPA emissions requirements in 2006, engine manufacturers will likely use exhaust after-treatment technology that will reduce NO_x emissions. The low sulfur levels in biodiesel fuels make them a good candidate for use with the exhaust after-treatment technologies that are available. Even though biodiesel fuels produce more NO_x emissions, they have been shown to reduce carbon monoxide, particulate matter, unburned hydrocarbons, and other pollutants.

12.2.6.3 Fuel quality

Many problems associated with biodiesel stem from poor fuel quality from the supplier. Most often this is related to the completeness of the production reaction. The ASTM has developed a quality standard for biodiesel. At this point in time, fuel manufacturer compliance with the standard is voluntary. Generally, it is a good idea to ensure that the biodiesel manufacturer sells biodiesel that meets or exceeds the ASTM specifications.

12.2.6.4 Energy content

Biodiesel fuels contain about 12.5% less energy per unit of weight than petroleum diesel fuel (37,215 kJ/kg vs. 42,565 kJ/kg). However, since biodiesel has a higher density, the energy content per unit of volume is only 8% less. As a result, the fuel economy of the diesel engine that is powered with biodiesel tends to be slightly less than when powered with petroleum diesel fuel.

12.2.6.5 Cold weather

The cloud point and cold filter plugging point are much higher for biodiesel than diesel fuel. This means that the fuel will not work in the engine as well as diesel fuel at lower temperatures. Additives can be used to reduce the cloud point and CFPP of biodiesel fuel. The cold-weather properties of biodiesel can also be improved by using a lower level blend of biodiesel fuel (i.e., B5 instead of B20). Additionally, the fuel may be blended with number 1 diesel instead of number 2 diesel to improve the cold-weather properties.

12.2.6.6 Material compatibility

Biodiesel fuel will react with some plastics and some metals in a negative manner. The plastics that seem to be compatible with biodiesel include nylon, teflon, and viton. When in contact with nonferrous metals, such as copper and zinc, biodiesel fuel can cause precipitates to form. Some of these materials can be found in fuel tank liners, fuel lines, transfer pump diaphragms, injector seals, and injection pump seals (among others).

12.2.6.7 Solvency

Biodiesel can also act as a solvent. This creates some problems when used in existing systems. The biodiesel can dissolve existing residues in fuel tanks and lines and carry them to the fuel system. Generally, after a few tanks of fuel have been used, these problems tend to be reduced.

12.2.6.8 Stability

The oxidative stability of biodiesel fuel is a major factor in determining the allowable storage time for biodiesel fuel. The iodine number can be used to estimate the oxidative stability of the fuel before any stabilizers are added. Typically biodiesel fuels can be stored for up to 6 months without problems. If biodiesel fuel needs to be stored longer, antioxidants can be added to the fuel to improve the stability. If the fuel is not stabilized, biodiesel can form gums and sediments that clog filters or form deposits on fuel system components, including fuel pumps and injectors.

Additionally, as with diesel fuel, some climatic conditions promote biological growth (such as algae), in the fuel. If this occurs, the problem can be treated with a biocide. Reduction in water contamination also reduces the amount of biological growth in the fuel since the algae grows on the water.

12.2.6.9 Warranties

Most engine manufacturers do not warranty their engines for use with a specific fuel. Consumers with engine problems that can be traced to the fuel are directed to their fuel supplier. Many engine manufacturers have developed policy statements for biodiesel that allow the use of up to 5% biodiesel but indicate that more experience is needed before fueling with higher level blends. Most require that biodiesel meet the ASTM standard. The best practice is to check with the engine and vehicle manufacturer before using biodiesel.

12.2.7 Biodiesel storage and use

12.2.7.1 Blending

Blending of biodiesel is not recommended if the temperature of either fuel is below 4.4°C. Low temperatures impact how easily the biodiesel mixes with petroleum diesel fuel. In most situations, splash blending works effectively (i.e., splashing or pouring the biodiesel into the diesel fuel), as biodiesel mixes readily with petroleum diesel fuel. Once mixed, the biodiesel tends to remain blended.

If splash blending biodiesel in a large tank, the biodiesel should be introduced after the diesel fuel has been placed in the tank or the blend should be prepared before placing the blended fuel into storage. This is due to the fact that the biodiesel is heavier than diesel fuel and will essentially rest at the bottom of the tank until some type of agitation is provided. Bulk plants or terminals may also use pumps but will most likely rely on electronic injection or in-line blending to prepare the required blend.

12.2.7.2 Transportation

Biodiesel, due to its high flash point, is *not* considered flammable. The fuel is considered combustible, just as is vegetable oil (feedstock). As such, the transportation of "neat" biodiesel may be handled in the same manner as vegetable oil (Code of Federal Regulations 49 CFR 171–173). This is not the case for a low-level blend or a B20 blend. These blends exhibit flash point tendencies that essentially mirror diesel fuel. Blends of biodiesel should be handled in the same manner as petroleum diesel fuel.

12.2.7.3 Storage tanks

Storage tanks for biodiesel can be constructed from mild steel, stainless steel, fluorinated polyethylene, fluorinated polypropylene and teflon. Biodiesel, like petroleum diesel fuel, should be stored in a clean, dry, dark environment. In the event that the container selected is made from polyethylene or polypropylene, the container should be protected from sunlight.

Some authors suggest that aluminum is suitable for use as a storage tank. However, nonferrous metals, such as aluminum, tend to react unfavorably with biodiesel by shortening the shelf life of the fuel. Much is the same for tin and zinc. Concrete lined tanks, varnish lined tanks, or tanks lined with PVC cannot be used to store biodiesel. Biodiesel reacts with each of these products, breaking down the chemical structure of each.

As with any fuel, steps must be taken to prevent water from entering the tank. Algae can grow in biodiesel just as it does with petroleum diesel fuel.

Measures should be taken to ensure that the biodiesel will flow in cold weather. This is often accomplished by mixing the fuel with either number 1 or number 2 diesel fuel. Cold flow improvers (CFI) can also be added to enhance the cold flow characteristics of biodiesel.

12.2.7.4 Material compatibility

Essentially the same materials that are used to construct a biodiesel storage tank can be used with biodiesel (stainless steel, mild steel, viton, some forms of teflon, and fluorinated polyethylene/polypropylene). Rubber elastomers cannot be used, as pure biodiesel will dissolve the rubber. The effect is lessened with lower percentage blends, but little research has been conducted to determine the long-term material compatibility of biodiesel blends.

12.2.7.5 Safety

Biodiesel is nontoxic, biodegradable, and much less irritating to the skin than petroleum diesel. However, the same safety rules that pertain to petroleum diesel fuel also apply to the use of biodiesel. The following list summarizes several of these issues:

- Store in closed, vented containers between 10°C and 50°C.
- Keep away from oxidizing agents, excessive heat, and ignition sources.
- Store, fill, and use in well-ventilated areas.
- Do not store or use near heat, sparks, or flames; store out of the sun.
- Do not puncture, drag, or slide the storage tank.

- A drum is not a pressure vessel; never use pressure to empty.
- Wear appropriate eye protection when filling the storage tank.
- Provide crash protection (i.e., large concrete-filled pipe near the storage tank).

12.2.8 Biodiesel economic considerations

The cost of biodiesel consists of five major components: the cost of the feedstock, the cost of the biodiesel, the price of glycerol by-product, and availability of biodiesel. We will examine each of these components.

12.2.8.1 Cost of the feedstock

The cost of the feedstock (oil/tallow/lard, etc.) will vary from one season to the next. The cost will also vary from one crop to the next (soybeans versus rapeseed) as well as from one year to the next, depending on supply and demand. For example, soybean oil has fluctuated from 14.15 cents/pound to 25.8 cents/pound during 1995–2002. Peanut oil has varied from 27 to 49 cents/pound during the same time period. At the same time, lard has sold for 12.5 to 23.02 cents/pound. During 2001–2002, the price of cottonseed oil varied from 14.4 to 22.3 cents per pound and corn oil varied from 11.38 to 23.02 cents/pound. During 2001–2002, the high for corn was in December and the low was in May. The high for cottonseed oil occurred in September and the low occurred in October. Some biodiesel producers have been able to purchase used restaurant grease for as little as 3 to 4 cents/pound (McDonalds, etc.), while some have been paid to haul the waste oil away (transportation was the only feedstock cost).

12.2.8.2 Cost of biodiesel

Economies of scale are a factor when buying biodiesel. Since biodiesel is not yet available on the pipeline network, most deliveries are made by truck or rail. Transportation costs can be a significant portion of the product cost.

The most significant contributor to the cost of the fuel is the cost of the oil itself. As noted in the feedstock section, the price of the oil has varied from as little as 3 to 4 cents/pound to a high of 25.8 cents/pound. If soybeans were used as the feedstock, the cost could range from \$1.73 to \$3.10/gallon. If a less expensive feedstock were used, the price range would even be greater anywhere from \$0.50 to \$63.10/gallon. (See Fig. 12.2-6.)

The next highest cost when producing biodiesel is the cost to convert the feedstock from a pure oil to

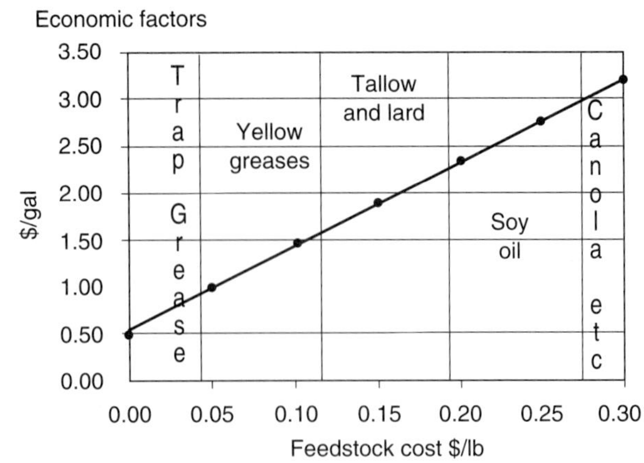

Figure 12.2-6 Production cost per gallon. Reprinted the from National Renewable Energy Laboratory, Kansas Cooperative Development Center/KSU.

a transesterified fuel. Some researchers report that this amounts to approximately 30% of the total cost (as nearly 70% of the cost is tied up in the raw materials [soybean oil feedstock, methanol, catalyst]).

The conversion costs for biodiesel can range from \$0.30 to \$0.60/gallon. One study reported that it cost \$0.58/gallon for transesterification and \$0.33/gallon for overhead. A credit of \$0.39/gallon was applied for the resale of the glycerol, bringing the total for transesterification to \$0.52/gallon.

12.2.8.3 Price of glycerol

As noted in the previous paragraph, the glycerol is a valuable by-product that must be disposed of after transesterification (credit = \$0.39/gallon). Glycerol is used in pharmaceuticals, cosmetics, toothpaste, paints, and other products. Most researchers believe that a flood of glycerol on the market (which would result from increased biodiesel production) would reduce the net credit. Alternative uses for the glycerol would develop and these uses should help maintain a solid price for the glycerol. Presently, refined glycerol prices stand in the United States at approximately \$0.65/pound.

12.2.8.4 Availability

Most economists are of the opinion that biodiesel has the capability to replace about 2 to 5% of the diesel fuel used for transportation in the United States. Although this may seem small in comparison to the volume of diesel fuel used, this would indicate that we would need to produce roughly 6 billion gallons of biodiesel each year. However, Brazil has become a major exporter in the soybean market, and large carryover stocks of soybeans have resulted. The estimated soybean oil carryover for the year 2002 was 1.2 million metric tons. The United

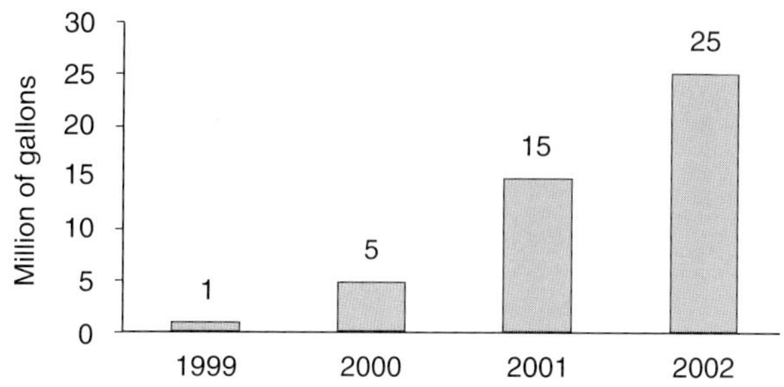

Figure 12.2-7 U.S. biodiesel production. Reprinted from the National Biodiesel Board, Kansas Cooperative Development Center/KSU.

States produces 25 million gallons of biodiesel a year. (See Fig. 12.2-7.)

See also the following articles

Alternative Transportation Fuels: Contemporary Case Studies • Biomass, Chemicals from • Biomass for Renewable Energy and Fuels • Ethanol Fuel • Fuel Cycle Analysis of Conventional and Alternative Fuel Vehicles • Internal Combustion (Gasoline and Diesel) Engines • Life Cycle Analysis of Power Generation Systems • Renewable Energy, Taxonomic Overview

Glossary

aromatics Any unsaturated hydrocarbon containing resonance stabilized carbon-to-carbon bonds characterized by benzene-type ring structure.

biodiesel A renewable energy fuel from vegetable or animal derived oil (triglyceride) that has been chemically modified to reduce its viscosity. It can be used in any concentration with petroleum-based diesel fuel in existing diesel engines with little or no modification. Biodiesel is not the same thing as raw vegetable oil. It is produced by a chemical process, which removes the glycerol from the oil.

distillation curve A measure of volatility of a fluid. More specifically, it is the temperature of the fluid (fuel) when a specific volume of the fluid has evaporated.

feedstock Raw or processed material that is chemically reacted to produce biodiesel. Note that the raw material must first be processed before the oil can be used to produce biodiesel.

gas chromatograph A chemical is heated to a gaseous state, and then the gas is passed through a cylinder and the different chemicals that make up the gas "stick" to the wall of the cylinder at different intervals. This information is then used to determine the chemical makeup of the gas.

olefin Any unsaturated hydrocarbon containing one or more pairs of carbon atoms linked by a double bond.

triglyceride A naturally occurring ester formed from glycerol and one to three fatty acids. Triglycerides are the main constituent of vegetable and animal derived fats and oils.

Further reading

Ash, M., and Dohlman, E. (2003). Bumper South American soybean crop buoys world consumption growth. *In* "Oil Crops Outlook," a USDA report in February of 2003. Found at www.ers.usda.gov.

"Biodiesel: Fueling Illinois." (2003). Illinois Soybean Association. Bloomington, IL. Found at www.ilsoy.org.

Coltrain, J. (2002). Biodiesel: Is it worth considering? Kansas State University. Found at www.agecon.ksu.edu.

National Renewable Energy Laboratory (2003). Kansas Cooperative Development Center. Department of Agricultural Economics. Kansas State University. Manhattan, KS 66506.

Sharp, C. A., Howell, S. A. and Jobe, J. (2000). The effect of biodiesel fuels on transient emissions from modern diesel engines, part I: Regulated emissions and performance. Society of Automotive Engineers Paper 2000-01-1967.

Sheehan, J., Carnobreco, V., Duffield, J., Graboski, M., and Shapouri, H. (1998). Urban bus operation. *In* "Life Cycle Inventory of Biodiesel and Petroleum Diesel for Use in an Urban Bus," pp. 171–188. Found at www.nrel.gov.

Sigmon, M. (1997). Demonstrations and economic analysis of biodiesel in fleet vehicle and marine applications. For the City of Chicago and Department of Energy. Found at www.biodiesel.org.

"Standard Specification for Diesel Fuel Oils." (2002). ASTM International. West Conshohocken, PA. Found at www.astm.org.

Section Thirteen

Storage technologies

Chapter 13.1

13.1

Storage technologies

The storage of electricity offers significant benefits for the generation, distribution and use of electric power. At the utility level, for example, a large energy storage facility can be used to store electricity generated during off-peak periods – typically overnight – and this energy can be delivered during peak periods of demand when the marginal cost of generating additional power can be several times the off-peak cost.

Energy storage plants can supply emergency back-up in case of power plant failure, helping to maintain grid stability. On a smaller scale, they can also be employed in factories or offices to take over in case of a power failure. Indeed in a critical facility where an instantaneous response to loss of power is needed, a storage technology may be the only way to ensure complete stability.

Energy storage also has an important role to play in the generation of electricity from renewable energy. Many renewable sources such as solar, wind and tidal energy are intermittent and their output often cannot be predicted with accuracy. Combining some form of energy storage with a renewable energy source helps remove this uncertainty and increases the value of the electricity generated.

Given these arguments in favour of energy storage, it may come as a surprise to learn that the use of storage plants is not widespread. One reason for the relatively small number of such plants is the availability of the technology. Another is cost. Until the late 1970s there was really only one large-scale energy storage technology and that was pumped storage hydropower. This is effective, but expensive. Since the 1980s other technologies have been developed for both utility and consumer applications but cost is still perceived as a handicap.

Yet since the 1980s there have been powerful arguments in favour of expanding storage capacity everywhere. A grid with a storage capacity of 10% to 15% of its generating capacity is much more stable and much cheaper to operate than one with virtually no storage capacity. Peaking capacity can be virtually eliminated and capacity additions can be planned more easily. But in a competitive, deregulated energy market the economics of energy storage may not appear obviously advantageous. It is probably this that has prevented greater investment.

13.1.1 Types of energy storage

Electricity normally has to be used as soon as it has been generated. This is why grid control and electricity dispatching systems are important; they have to balance the demand for electricity with the supply. Once one fails to match the other, problems occur. It would seem obvious, given this situation, that some reservoir of saved electricity would be a major boon to grid operation. Yet storing electricity has proved difficult to master.

Storing electricity in its dynamic form, amperes and volts, is almost impossible. The nearest one can get is a superconducting magnetic energy storage ring which will store a circulating DC current indefinitely provided it is kept cold. A capacitor storage system stores electricity in the form of electric charge. All other types of energy storage convert the electricity into some other form of energy. This means that the energy must then be converted back into electricity when it is needed.

A rechargeable battery may appear to store electricity but in fact it stores the energy in chemical form. A pumped storage hydropower plant stores potential energy; a flywheel stores kinetic energy while a compressed air energy storage (CAES) plant stores energy in the form of compressed air, another type of potential energy. Alternatively one might use electrolysis to turn

Renewable Energy Focus Handbook 2009; ISBN: 9780123747051

electricity into hydrogen, yet another chemical form of energy.

All these, and one or two others, represent viable ways of storing electricity. Several are commercially available, others in the development stage. And each has its advantages and disadvantages.

For large-scale utility energy storage there are three possible technologies to choose between, pumped storage hydropower, CAES and, at the low end of the capacity range, large batteries. Batteries can also be used for small- to medium-sized distributed energy storage facilities,[1] along with flywheels and capacitor storage systems. Superconducting magnetic energy storage is being used for small storage facilities and would be suitable for large facilities but is prodigiously expensive.

Some of these systems can deliver power extremely rapidly. A capacitor can provide power almost instantaneously, as can a superconducting energy storage system. Flywheels are very fast too, and batteries should respond in tens of milliseconds. A CAES plant probably takes 2–3 min to provide full power. Response times of pumped storage hydropower plants can vary between around 10 s and 15 min.

The length of time the energy must be stored will also affect the technology choice. For very long term storage of days or weeks, a mechanical storage system is the best and pumped storage hydropower is the most effective provided water loss is managed carefully. For daily cycling of energy, both pumped storage and CAES are suitable while batteries can be used to store energy for periods of hours. Capacitors, flywheels and superconducting magnetic energy storage are generally suited to short-term energy storage, though flywheels can be used for more extended energy storage too.

Another important consideration is the efficiency of the energy conversion process. An energy storage system utilises two complementary processes, storing the electricity and then retrieving it. Each will involve some loss. The round trip efficiency is the percentage of the electricity sent for storage which actually reappears as electricity again. Typical figures for different types of system are shown in Table 13.1-1.

Table 13.1-1 Round trip energy efficiencies for storage technologies

	Efficiency (%)
Capacitors	90
Superconducting energy storage	90
Flow batteries	90
CAES	80
Flywheels	80
Pumped storage hydropower	75–80
Batteries	75–90

Electronic storage systems such as capacitors can be very efficient, as can batteries. However, the efficiencies of both will fall with time due to energy leakage. Flow batteries, where the chemical reactants are separated, perform better in this respect and will maintain their round trip efficiency better over time. Mechanical storage systems such as flywheels, CAES and pumped storage hydropower are relatively less efficient. However, the latter two, in particular, can store their energy for long periods if necessary without significant loss.

13.1.2 Pumped storage hydropower

The most widespread large-scale electricity storage technology is pumped storage hydropower. This is also the oldest storage technology in use, with the first plant built at the beginning of the twentieth century. By the beginning of the twenty-first century there was probably 140,000 MW of pumped storage capacity in operation.

A pumped storage plant is like a conventional hydropower plant with a dam and reservoir but in this case there are two reservoirs rather than one (Fig. 13.1-1). These two reservoirs must be separated vertically; one must be higher than the other. The difference in height provides the head of water to drive the station's turbines.

In order to generate power, water runs from the top reservoir through a high-pressure channel to turbines at the bottom of the drop. The turbines extract the potential energy from the water and then discharge it into the bottom reservoir where it is saved. When energy is to be stored, the turbines are reversed and act as pumps, pumping water from the lower reservoir into the upper. The turbine pumps are driven using off-peak electricity so storage will normally take place at night. Once water has been pumped into the upper reservoir it is available again for power generation.

This type of plant is extremely robust and though round trip efficiency is lower than for some other technologies, long-term energy losses are low. Leakage and evaporation are the main sources of loss and if these are managed well, water loss can be kept small.

13.1.2.1 Plant design

The energy that can be extracted from a hydropower plant depends on both the volume of water available and the head of water that can be exploited. A pumped storage project will provide the most efficient and

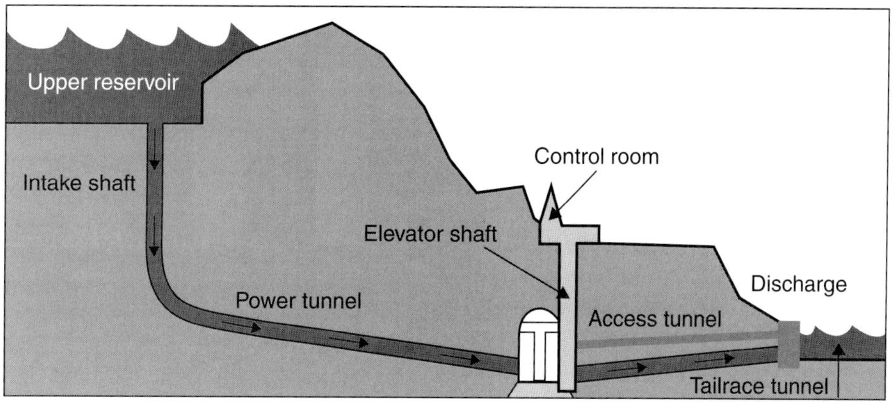

Figure 13.1-1 Cross section of a pumped storage hydropower plant.

cheapest operation when it can provide a high head between its two reservoirs. This will allow the greatest amount of energy to be stored in the smallest volume of water. That, in turn, means that pumps and turbines can be smaller, reducing the capital cost of the plant.

13.1.2.2 Turbines

The earliest pumped storage power plants used separate turbines and pumps but this made the projects costly to build. The development of reversible pump turbines made the economics of the pumped storage plant look much more attractive.

Most reversible pump turbines used in storage plants are Francis turbines. The Francis design is well suited to both generation and pumping and can pump water to a considerable height.

There is one drawback with the Francis design for this application: the turbine blade angle is fixed. A fixed blade does not provide the best efficiency for both pumping and generating power. An alternative design called the Deriaz turbine, similar to the Francis turbine in design but with movable blades, has been used in several pumped storage projects to try to achieve greater efficiency.

Propeller-like Kaplan turbines can also be used as pumps, though not to transfer water to a reservoir of any great height. The La Rance tidal power plant in France, for example, uses such turbines to pump water across its tidal barrage in order to increase efficiency of operation of this station.

The best efficiency that a hydraulic turbine can provide for generating power is around 95%. Pumps are less efficient, operating at best at around 90%. This means that the best efficiency that can be expected from a pumped storage power plant through a storage and regeneration cycle is around 86%. In practice the efficiency is normally between 75% and 80% as shown in Table 13.1-1.

Francis and Deriaz turbines can be built today to operate at heads of up to 700 m in a single stage. Beyond that it will usually be necessary to use a combination of a pump and a Pelton turbine. Several plants in Switzerland employ this configuration.

Pumped storage hydropower plants can be brought on-line extremely quickly. The 1800-MW power station at Dinorwig in Wales, for example, can be run up from zero output to 1800-MW output in around 10 s. This ability makes pumped storage extremely attractive as a system reserve to be brought into service if a major base-load unit breaks down.

13.1.2.3 Global exploitation

Of all the energy storage technologies in use, pumped storage hydropower is by far the most widely adopted. Since its introduction in Switzerland in around 1904, plants have been built in other parts of Europe, in the USA, in China, in Japan and in many other countries.

Pumped storage facilities have often been built in conjunction with nuclear power plants. This combination allows the nuclear plant to run continuously at full power, its most effective mode of operation. Electricity from the generating plant not required immediately by the grid is stored for dispatching during peak demand periods. Nuclear power plants have generating capacities up to 1300 MW; only a pumped storage plant can provide the large storage capacity needed in this situation.

The two countries with the largest pumped storage hydropower capacities are Japan and the USA. Each has around 20,000 MW. In both cases many of these plants are associated with nuclear development. In the USA, for example, the bulk of the capacity was built between 1970 and 1990 when nuclear growth was greatest. There is a further 100,000 MW distributed across the globe, providing a global capacity of close to 140,000 MW.[2]

13.1.2.4 Financial risks

Pumped storage employs the same technology and construction techniques used in conventional hydropower projects and the risks are similar. These fall into three groups: geological, hydrological and technical risks.

Geological risk will depend on the site for the project. This will have to be capable of providing two reservoirs and room for a power station. In some cases the sea can be used as the lower reservoir, simplifying the design. As with all hydropower schemes, a thorough feasibility study is vital to assess the geological conditions. Faults within the underlying rock structure could cause construction problems, leading to major cost overruns if not identified early. Risk of seismic shock must also be considered.

The hydrological risk associated with a pumped storage hydropower plant should be slight since the station will not normally depend on a supply of water from a river which may be unpredictable. However, any problems with water loss from evaporation or through leakage from the reservoirs will affect plant economics. Technical risk is minimal too. Hydro turbine technology is well established and should not lead to any problems.

13.1.2.5 Costs

Capital costs are likely to be broadly in line with those for a conventional hydropower project – a unit cost of between \$800/kW and \$3,500/kW – although the specialised nature of the pump turbines, or the need for separate pumps and turbines could push the cost of the plant up. Some pumped storage plants place the lower reservoir underground. This is likely to increase construction costs.

13.1.3 Compressed air energy storage

CAES is exactly as its name suggests: air is compressed and stored under pressure. Release of the pressurised air is subsequently exploited to generate electricity. Although the storage of compressed air is clearly a means of storing energy, it is only when it is considered in conjunction with the gas turbine that it makes complete sense from a power generation perspective.

A gas turbine consists of two major components. These are a compressor and a turbine. Conventional gas turbines used in aero applications or for power generation have the two components mounted on a single drive shaft.

During conventional operation, air is drawn into the compressor and compressed. This compressed air is then directed into a combustion chamber where it is mixed with fuel and ignited. Heating the compressed air increases its energy content significantly. The hot compressed gas is then released through the machine's turbine blades, causing them to rotate and generate electricity or motive power.

Although a gas turbine normally has the compressor and turbine closely integrated, there is no reason, in principle, why compression should not be carried out separately, and at a different time to power generation. This is the crux of the CAES plant.

In a CAES plant the compressor and the turbine are separated. By use of a system of clutches, each can be linked, separately, to a motor generator. In storage mode the compressor stage of the gas turbine is driven by the reversible motor generator using off-peak power from the grid system. The product, compressed air, is stored in a special cavern (Fig. 13.1-2).

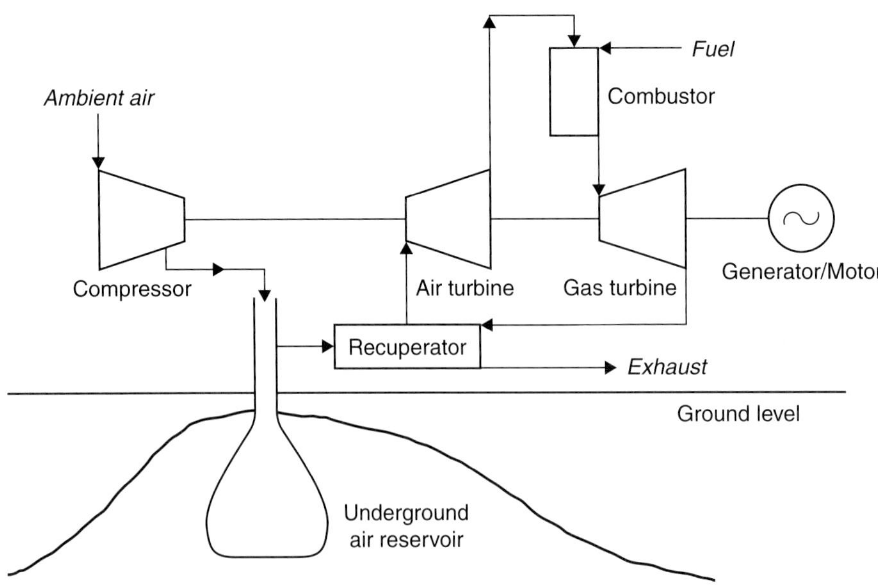

Figure 13.1-2 Diagram of a CAES plant.

When the power is required, air is released from the cavern into a combustion chamber and mixed with fuel; this is ignited and allowed to expand through the turbine section of the system. Under these conditions the motor generator is used in generation mode to produce electricity.

13.1.3.1 Storage caverns

The most important part of a CAES plant is somewhere to store the compressed air. Small-scale CAES plants – with storage capacities of up to 20 MWh – can use overground storage tanks but large, utility-scale plants need underground caverns in which to store the air. The natural gas industry has used underground storage caverns for years to store gas; these same caverns can provide ideal storage facilities for a CAES plant. However, the demand for such a cavern limits the development of CAES to places where such storage caverns are available.

A number of different types of underground cavern can be exploited. The simplest is a man-made rock cavern. This must be sited in an impervious-rock formation if it is to retain the compressed air without loss.

Salt caverns have been commonly used for gas storage. These are created by dissolving or dry mining salt to create a suitable enclosure. Salt deposits suitable for such caverns occur in many parts of the world.

A third type of underground storage is found in porous rock bounded by an impervious barrier. Examples can be found in water-bearing aquifers, or as a result of oil and gas extraction. Aquifers can be particularly attractive as storage media because the compressed air will displace water, setting up a constant pressure storage system. With rock and salt caverns, in contrast, the pressure of the air will vary as more is added or released.

All three types of storage structure require sound-rock formations to prevent the air from escaping. They also need to be sufficiently deep and strong to withstand the pressures imposed on them. It is important, particularly in porous-rock storage systems, that there are no minerals present that can deplete the oxygen in the air by reacting with it. Otherwise the ability of the air to react with the fuel during combustion will be affected, reducing the power available during the generation phase of the storage generation cycle.

13.1.3.2 Turbine technology

A CAES plant uses standard gas turbine compressor and turbine technology but because the two units operate independently, they can be sized differently in order to match the requirements of the plant. The larger the compressor compared to the turbine, the lesser the time it requires to charge the cavern with a given amount of energy. Thus a plant built in Germany required 4 h of compression to provide an hour of power generation whereas a plant in Alabama needs only 1.7 h of compression for an hour of generation.

As a result of compression and generation being separated, a CAES plant turbine can operate well at part load as well as full load. More complex operation is also possible. The Alabama plant, for example, uses two turbine stages with the exhaust from the last turbine used to heat air from the cavern before it enters the first turbine.[3] Fuel is not actually burnt in the compressed air until it enters a combustion chamber between the first and second turbine stages.

A key feature of a CAES plant is that it generates more electricity than was actually consumed when the air was stored. This is a result of the fuel burnt in the compressed air during the generation part of the cycle. A typical plant will deliver 30–35% more electricity to the grid that was originally consumed during storage.

13.1.3.3 Global exploitation

CAES has had a short history of limited development. The largest project yet built was a 290-MW power plant constructed at Huntorf in Germany in 1978. This plant operated for 10 years with 90% availability and 99% reliability, providing storage for a nuclear plant. Even so the German utility decommissioned the project. Interest in CAES then shifted to the USA where the Electric Power Research Institute (EPRI) began to promote the technology in the latter half of the 1980s.

EPRI saw CAES as a useful technology to enable small US utilities to limit their need for expensive peaking power stations. It estimated that rock formations capable of providing reliable underground storage caverns exist across 75% of the USA.

As a result of EPRI's work, a 110-MW commercial project was built by the Alabama Electric Cooperative. The plant entered service in May 1991 and has operated ever since. It cycles once or twice each day, and can store 2600 MWh of energy.

At around the time the Alabama plant was built, Italy tested the technology in a 25-MW installation. No plant was built. More recently a 2700-MW plant has been proposed in the USA but not yet constructed.

13.1.3.4 Financial risk

There should be little technical risk associated with a CAES project. Gas turbine technology is well understood and relatively cheap. Gas storage techniques are also well tested. The combination of the two remains novel but practical experience suggests that the technique is both robust and reliable.

13.1.3.5 Costs

There is little experience with CAES so any cost estimate must be considered tentative. However, it would appear to be an economically attractive option for energy storage. Installation costs of around $400/kW have been mooted for the USA.

13.1.4 Large-scale batteries

The traditional way of providing electricity storage has been the battery. This is an electrochemical device which stores energy in a chemical form so that it can be released as required.

A battery comprises a series of individual cells, each of which is capable of providing a defined current at a fixed voltage. Cells are joined both in series and in parallel to provide the required voltage and current rating for a particular application.

Each cell contains two electrodes, an anode and a cathode. These are immersed in an electrolyte.[4] At its simplest, the electrodes are made of materials which will react together spontaneously but the electrolyte in which they are immersed will allow the passage of only one of the components required to complete the reaction.[5] An electrical connection must be made between the two electrodes to allow the passage of electrons from one electrode to the other in order to complete the reaction. This is the source of electrical power.

There are two different types of traditional batteries: the primary cell and the secondary cell. A primary cell can only be discharged once, after which it must be discarded. A secondary cell can be discharged and recharged many times. Only the second type is of any use for energy storage systems.

Secondary cells can further be divided into shallow discharge and deep discharge cells. A shallow discharge cell is only partially discharged before being recharged again; an automotive battery would typify this type of cell. A deep discharge cell is normally completely discharged before recharge. This is the type which is most attractive for large-scale electricity storage.

Traditional electrochemical storage systems boast a best case conversion efficiency of 90% but a more typical figure would be 70%. Most batteries also suffer from leakage of power. Left for too long, the cell discharges itself. This means that battery systems can only be used for relatively short term storage.

An additional problem with batteries is their tendency to age. After a certain number of cycles, the cell stops holding its charge effectively, or the amount of charge it can hold declines. Much development work has been aimed at extending the lifetime of electrochemical cells but this remains a problem.

To their advantage, batteries can respond to a demand for power almost instantaneously. This property can be used to good effect to improve the stability of an electricity network. It is also valuable in both distributed generation and for back-up power applications.

Traditional batteries are completely self-contained. However, there is another type called a *flow battery* in which chemical reagents involved in the generation of electricity are held is tanks separated from the actual electrochemical cell. In this type of device the reagent is pumped through the cell as needed. Such cells suffer less from energy leakage. Several types are being developed for utility electricity storage.

13.1.4.1 Lead acid batteries

Lead acid batteries are the best known of all rechargeable batteries. These are the cells used in automobiles worldwide as well as for small-scale energy storage in homes and offices. Advanced lead acid cells have been developed for utility storage applications, the largest being for a 10-MW plant in California.

Lead acid batteries operate at ambient temperature and use a liquid electrolyte. They are extremely heavy and have a poor energy density but neither of these is a major handicap for stationary applications. They are also cheap and can be recycled many times, though they should not be completely discharged as this can cause problems.

13.1.4.2 Nickel–cadmium batteries

Nickel–cadmium batteries have higher energy densities and are lighter than lead acid batteries. They also operate better at low temperatures. However, they tend to be more expensive. This type of battery was used widely in portable computers and phones but has now been superseded by lithium ion batteries. The largest nickel–cadmium battery ever built is a 40-MW unit in Alaska which was completed in 2003. It occupies a building the size of a football field and comprises 13,760 individual cells.

13.1.4.3 Sodium–sulphur batteries

The sodium–sulphur battery is a high-temperature battery. It operates at 300°C and contains liquid sodium which will explode if allowed contact with water. Safety is a major issue with these batteries. However, the battery has a very-high-energy density which makes it attractive, particularly for automotive applications.

The battery is being developed for utility applications in Japan. Demonstration and early commercial projects have ranged in size from 500 kW to 6 MW. Most are in Japan but a small unit was commissioned in the USA in 2002.

13.1.4.4 Flow batteries

A flowing-electrolyte battery, or flow battery, is a cross between a conventional battery and a fuel cell (Fig. 13.1-3). It has electrodes like a conventional battery where the electrochemical reaction responsible for electricity generation or storage takes place in an electrolyte. However, the chemical reactants responsible for the electrochemical reaction and the product of that reaction are stored in tanks separate from the cell and pumped to and from the electrodes as required, much like a fuel cell.[6]

Two types of flow battery have been developed for utility applications, the polysulphide–bromide battery and the vanadium redox battery. Both designs have been developed to the demonstration stage. Capacities of up to 15 MW have been proposed. Response time from zero to full power is expected to be around 100 ms.

13.1.4.5 Financial risks

While battery technology is over a century old, the types of cell proposed for utility storage are novel and experience with them is limited. Most of the promising designs are in the demonstration or early commercialisation stage. This uncertainty about the technology means that there is a significant technological risk. Some operating lead acid storage plants are now over a decade old, providing early feedback about cell lifetime as well as operating experience. Much more is needed to establish a good measure of their potential.

13.1.4.6 Costs

Initial estimates suggest that lead acid batteries will cost around $500/kW to install. Sodium–sulphur batteries are expected to cost around $1000/kW while flow batteries should cost between $800/kW and $900/kW. Costs for both can be expected to fall if demonstrations prove successful and lead to commercial uptake.

13.1.5 Superconducting magnetic energy storage

Superconductivity offers, in principle, the ideal way of storing electric power. The storage system comprises an electromagnetic coil of superconducting material which is kept extremely cold. Off-peak electricity is converted to DC and fed into the storage ring, and there it stays, ready to be retrieved as required. Provided the system is kept below a certain temperature, electricity stored in the ring will remain there indefinitely without loss.

The key to the superconducting magnetic energy storage device is a class of materials called *superconductors*. Superconductors undergo a fundamental change in their physical properties below a certain temperature called the *transition temperature*, which is a characteristic of each material. When a material is cooled below its transition temperature it becomes superconducting. In this state it has zero electrical resistance. This means that it will conduct a current with zero energy loss.

Unfortunately the best superconducting materials only undergo this transition at below 20°K (−253°C).

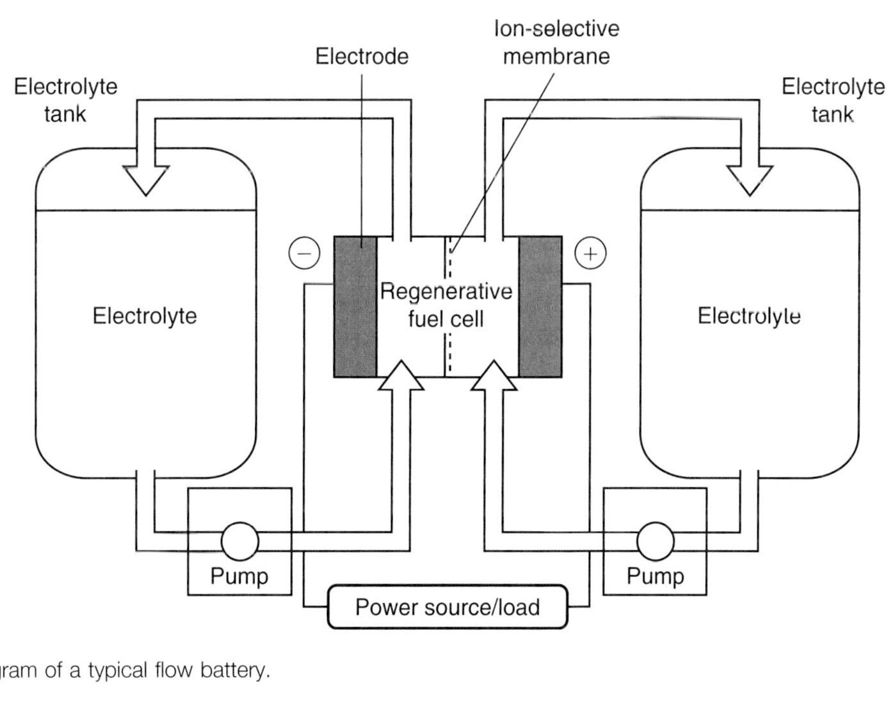

Figure 13.1-3 Diagram of a typical flow battery.

Temperatures this low can only be maintained by cooling the superconducting coil with liquid hydrogen or liquid helium, in either case an expensive process.

In recent years scientists have discovered materials that become superconducting at relatively high temperatures, temperatures accessible by cooling with liquid nitrogen. (Liquid nitrogen boils at 98°K, −175°C.) Most of these materials have proved to be rather brittle ceramics which are difficult to work but techniques are being found to exploit them. This is helping make superconductivity more economically attractive for a range of utility applications including storage.

Superconductors store DC current without loss but losses occur in converting the off-peak AC current to DC and then back to AC when required. The round trip efficiency is around 90%. A superconducting magnetic storage device can respond extremely quickly, delivering its rated power in about 20 ms.

A number of small superconducting storage rings have been built for use as power-conditioning systems. One of 10 MW capacity has been tested on a utility system in the USA where its primary role was to improve transmission system stability. Such systems are extremely expensive.

The unit cost of storing power in a superconducting ring decreases as the size of the plant increases so large storage devices would be preferred for utility applications. The superconducting ring for a 5000 MW device would be roughly 1600 m in diameter. The magnetic fields associated with such a device would be enormous and it would have to be built into rock to ensure it did not collapse under the force generated.

13.1.5.1 Financial risks

Superconducting magnetic energy storage involves a range of advanced technologies, most of which have not been proved beyond the experimental or small-scale demonstration stage. New, high-temperature superconducting materials are being developed but these remain experimental. The technological risk associated with the use of this technology is high. It is unlikely to be appropriate for wide scale use for several years.

13.1.5.2 Costs

Superconducting rings are costly to make and costly to operate. Although figures of around $2000/kW have been mooted for 1000 MW installations, smaller units cost in excess of $3000/kW.

13.1.6 Flywheels

A flywheel is a simple mechanical energy storage device comprising a large wheel on an axle fitted with frictionless bearings. A flywheel stores kinetic energy as a result of its rotation. The faster it rotates, the more energy it stores. In order for a flywheel to be effective as an energy storage device there must be a way of feeding energy into the flywheel and a means of extracting it again.

Simple flywheel energy storage devices are fitted to all piston engines to maintain smooth engine motion. The engine flywheel is attached physically to the engine camshaft and as the pistons cause the camshaft to rotate they feed energy into the flywheel. For electricity storage applications, energy will normally be fed into the flywheel using a reversible motor generator.

The faster a flywheel rotates, the more energy it will store. Conventional flywheels are fabricated from heavy metal discs made of iron or steel. However, these discs are only capable of rotating at low speeds. For power applications, new lighter composite materials are being developed, capable of rotating at 10,000–100,000 rpm without fracturing under the immense centrifugal force they experience. Such devices must be housed in exceedingly strong containers which will prevent the pieces of the flywheel scattering like shrapnel in the event of a catastrophic failure.

Energy storage systems must operate with low energy loss. This is accomplished in flywheel systems by using magnetic bearings to eliminate bearing friction and by operating the flywheel in either a vacuum or in a container filled with a low-friction gas such as helium.

One of the problems with flywheel energy systems is that the flywheel will rotate at varying speeds depending on how much energy it contains. If a conventional motor generator is used to extract electrical energy from the flywheel, this will translate into a variable frequency output. Grid electricity, however, must be generated at a constant AC frequency. Various electromechanical and electronic means of overcoming this difficulty have been found.

Flywheels have the attraction of virtually zero maintenance and infinite recyclability. They have proved to be one of the best and cheapest ways of maintaining power quality during power failure or network voltage or frequency dip. Response time is fast and in the case of power failure a flywheel system can bridge the period between the power outage and a long-term back-up system such as a generator set coming on line.

The largest flywheel system so far built is a 1-MW unit comprising 10 100-kW flywheels used to maintain system voltage on the New York transit system. Storage capacity is 250 kWh, sufficient to provide 1 MW for 15 min.

13.1.6.1 Financial risks

Flywheels represent a conventional technology extended by the use of unconventional materials. They have been under development for many years. Units are now available commercially and their characteristics appear to be predictable. The technological risks associated with this type of storage technology would appear to be fairly low.

13.1.6.2 Costs

The capital cost of a flywheel storage system may be as high as $2000/kW, though costs should fall below this for standard modular units. However, against this high cost must be balanced the fact that they are virtually maintenance-free, can be cycled indefinitely and are extremely predictable. The energy contained in the system can always be determined. In applications where medium capacity, short-term storage is required, flywheels offer one of the best solutions.

13.1.7 Capacitors

Capacitors are used extensively in electrical and electronic circuitry. In power networks they have been used to enhance system stability. More advanced capacitors are now being developed specifically for energy storage.

The classic capacitor comprises two parallel metal plates with an air gap between them. When a voltage is applied to the plate a positive charge collects on one plate and a negative charge on the other.

A number of different capacitor types exist. Those being considered for energy storage are called *electrochemical capacitors*. These utilise a solid electrode and an electrolyte. Charge collects at the interface between the two. These devices, sometimes called *super capacitors* or *ultra capacitors*, can store a very large energy density, probably the highest of any storage device. They can respond in tens to hundreds of milliseconds and are most suited to short-term energy storage applications.

The technology is relatively new and there is little cost data available. Nor is lifetime or operational experience available, but static electrical devices of this type should show good long-term stability and should be relatively maintenance-free.

13.1.8 Hydrogen

When a sufficiently high voltage is applied to water using two electrodes, the effect is to cause the water to decompose into its two elemental constituents, hydrogen and oxygen. One gas appears at one electrode, the other at the second, so it is relatively simple to separate the two. This can form the basis of an energy storage system.

When water is electrolysed in this way the oxygen is normally discarded while the hydrogen is retained. Hydrogen is an excellent and versatile fuel which can be burnt cleanly in a power plant to regenerate electricity or used in a variety of other ways such as fuel for motor vehicles. Ideally the hydrogen would be burned in a fuel cell, a device capable of up to 60% energy conversion efficiency – perhaps rising to 75% in a combined cycle configuration. When burnt, the product of combustion is water.

A major problem with hydrogen as a storage medium is round trip efficiency. Hydrolysis of water is generally only around 70% efficient, though some companies have claimed up to 80%. Assuming a hydrogen to electricity conversion efficiency of 75%, and ignoring other losses, the round trip efficiency would be 60%. This is optimistic; when storage and other losses are taken into account, it would probably result in a round trip efficiency of closer to 50%.

While this is obviously a major handicap, the advantages of hydrogen may eventually make such losses acceptable. Fuel technology based on hydrogen is being developed but is not currently commercially viable. However, the concept of a hydrogen economy[7] could take off later this century.

13.1.9 Environmental considerations

Each of the energy storage technologies considered in this chapter has an environmental impact related to the technology and techniques it employs. Pumped storage hydropower, for example, will entail many of the same considerations that apply to conventional hydropower while CAES will involve similar emission considerations as those relevant when considering a gas turbine.

Large battery energy storage systems may involve the use of toxic materials such as cadmium or lead which need to be handled and eventually recycled with care. The sodium in a sodium–sulphur battery is particularly dangerous if not handled carefully. Flow battery systems contain reagents which should not be allowed to escape into the environment either.

High-technology storage systems such as superconducting magnetic energy storage and super capacitors will also involve novel, possibly toxic materials. However, these will usually be costly to produce and there will be a strong incentive to recycle them. Flywheels are probably the most benign of the technologies with little environmental impact unless treated extremely carelessly.

There are, however, two aspects of storage technologies that have wider ranging impacts. The first is their ability to improve overall system efficiency and the second is the advantages that accrue to their use in conjunction with renewable technologies.

Adding energy storage capacity to a transmission or distribution network makes it easier to manage. As already indicated, storage capacity can be used to store off-peak electricity generated in cheap base-load generating plants, electricity that can then be used when demand rises beyond the capacity of the base-load units.

This mode of operation is economical because it replaces peak-load generation with base-load generation and the latter is normally the cheapest source of electricity. It is also more efficient since it will allow a utility to base the majority of its generation on its most energy efficient units. This is of environmental benefit since the most efficient generation results in the lowest atmospheric pollution – or should if regulation is operating correctly.

13.1.9.1 Renewable energy

Improved energy efficiency is one consequence of the use of energy storage. However, electricity storage can also have a profound effect on the economics and utility of renewable energy sources. Wind power, solar power, tidal power, wave power – these are all either intermittent or unpredictable sources of electricity, or both. Both features are a handicap which makes the energy less valuable to a power network operator and less easy to accommodate in large quantities. There is a limit to the amount of unpredictable power a network can accept while still providing a good service.

If energy storage is added alongside these renewable sources, the situation becomes completely different. Now the energy from the wind or solar plant can either be used if required or stored. The output from these plants is averaged out. Both peaks and troughs are accommodated by the storage unit. As a consequence the energy source becomes predictable. This makes the energy much easier to dispatch and it also allows larger quantities to be accepted without affecting the quality of the network supply.

There is a price to pay and that is the cost of the storage system. At the beginning of the twenty-first century that makes the combination an uneconomical prospect. But as the cost of renewable energy drops and that of fossil fuel rises, and as the general benefits of larger energy storage capacities are accepted, the economics are likely to look less of a disadvantage.

Table 13.1-2 Capital costs for energy storage systems

	Cost ($/kW)
Flywheel	2000
Superconducting magnetic energy storage	2000–3000
Pumped storage hydropower	800–3500
CAES	400
Battery storage	500–1000

13.1.10 Costs

The costs of energy storage systems vary widely. Some, like pumped storage hydropower are inherently expensive to build, while others, like superconducting magnetic energy storage are expensive because they are new. One or two, like CAES, are relatively cheap.

Table 13.1-2 presents some tentative costs for the different technologies discussed here. They suggest that, as already noted, CAES is the cheapest to install though battery storage could also be inexpensive. The figures should be treated with caution, however, particularly because many of the technologies are under development and costs are likely to fall significantly once they become widely available commercially.

When considering the economics of a storage system the round trip efficiency will also be a consideration. This will determine how much of the electricity used to charge the storage plant can actually be returned to the system.

With the exception of CAES, a storage plant does not use any fuel. Thus there are usually no fuel costs to consider. Many of the technologies are relatively easy to operate and maintain too.

Overall, however, it is the conversion of off-peak electricity into peak-period electricity that dominates the economics of a storage plant. It is this equation, therefore, that will determine whether the plant is economical or not.

End notes

1. Distributed storage facilities may be used by utilities to improve local grid stability or they may be used by consumers to make their own supplies more secure.
2. The Commercial World of Energy Storage: A Review of Operating Facilities (under construction or planned), Septimus van der Linden, 1st Annual Conference of the Energy Storage Council, Houston, 2003.
3. This process is called recuperation and it reduces fuel consumption by around 25%.
4. This is a simplification. Some advanced cells utilise pastes or solid electrolytes.
5. Normally the electrolyte will permit a charged molecule, an ion, to pass but will not conduct electrons.
6. The process occurring here is somewhat different to the simple electrochemical process described above but the principle is similar.
7. A hydrogen economy is one in which hydrogen replaces oil, gas and coal as the main fuel in automotive, heating and many power generation applications. It is seen as one option for a sustainable energy future.

Index

Renewable Energy Focus Handbook 2009; ISBN: 9780123747051

CONVERSION OF UNITS—STRESS AND PRESSURE*

	MN m^{-2}	dyn cm^{-2}	lb in^{-2}	kgf mm^{-2}	bar	long ton in^{-2}
MN m^{-2}	1	10^7	1.45×10^2	0.102	10	6.48×10^{-2}
dyn cm^{-2}	10^{-7}	1	1.45×10^{-5}	1.02×10^{-8}	10^{-6}	6.48×10^{-9}
lb in^{-2}	6.89×10^{-3}	6.89×10^4	1	703×10^{-4}	6.89×10^{-2}	4.46×10^{-4}
kgf mm^{-2}	9.81	9.81×10^7	1.42×10^3	1	98.1	63.5×10^{-2}
bar	0.10	10^6	14.48	1.02×10^{-2}	1	6.48×10^{-3}
long ton in^{-2}	15.44	1.54×10^8	2.24×10^3	1.54	1.54×10^2	1

CONVERSION OF UNITS—ENERGY*

	J	erg	cal	eV	Btu	ft lbf
J	1	10^7	0.239	6.24×10^{18}	9.48×10^{-4}	0.738
erg	10^{-7}	1	2.39×10^{-8}	6.24×10^{11}	9.48×10^{-11}	7.38×10^{-8}
cal	4.19	4.19×10^7	1	2.61×10^{19}	3.97×10^{-3}	3.09
eV	1.60×10^{-19}	1.60×10^{-12}	3.38×10^{-20}	1	1.52×10^{-22}	1.18×10^{-19}
Btu	1.06×10^3	1.06×10^{10}	2.52×10^2	6.59×10^{21}	1	7.78×10^2
ft lbf	1.36	1.36×10^7	0.324	8.46×10^{18}	1.29×10^{-3}	1

CONVERSION OF UNITS—POWER*

	kW (kJ s^{-1})	erg s^{-1}	hp	ft lbf s^{-1}
kW (kJ s^{-1})	1	10^{-10}	1.34	7.38×10^2
erg s^{-1}	10^{-10}	1	1.34×10^{-10}	7.38×10^{-8}
hp	7.46×10^{-1}	7.46×10^9	1	15.50×10^2
ft lbf s^{-1}	1.36×10^{-3}	1.36×10^7	1.82×10^{-3}	1

*To convert row unit to column unit, multiply by the number at the column-row intersection, thus 1 MN m^{-2} = 10 bar